P9-DTG-963

DROPS AND BUBBLES

THIRD INTERNATIONAL COLLOQUIUM

THIRD INTERNATIONAL
COLLOQUIUM ON
DROPS & BUBBLES - 1988

AIP CONFERENCE PROCEEDINGS 197

RITA G. LERNER
SERIES EDITOR

DROPS AND BUBBLES

THIRD INTERNATIONAL COLLOQUIUM

MONTEREY, CA 1988

EDITOR:
TAYLOR G. WANG
VANDERBILT UNIVERSITY

American Institute of Physics New York

L.C. Catalog Card No. 89-46360
ISBN 0-88318-392-7
DOE CONF 8809192

Printed in the United States of America.

Contents

FOREWORD

In the early seventies, as novices to the field of drops and bubbles, my co-investigators, M. Saffren and D. D. Elleman, and I asked ourselves if anyone other than a few die-hard physicists (and ourselves, of course) were interested in this area of study. To our surprise, a limited survey showed that many of our colleagues in the fields of combustion, astrophysics, meteorology, materials science, and chemical engineering all had a similar curiosity about drops and bubbles, and, as a matter of fact, many of them were working on similar equations without knowing of each other's efforts.

Not knowing any better, we thought it would be nice to have everyone talk to each other. This is how the International Colloquium on Drops and Bubbles was born.

The first Colloquium held at Caltech, in Pasadena, CA in 1974, was an enormous success. We should have quit while we were ahead, but we held a second Colloquium in Monterey, CA in 1981, with equal success.

After 14 years, still a novice, but now General Chairman of the *third* Drops and Bubbles Colloquium, I understand why no one wanted to organize a conference like this one. It is a *lot* of work: mailing lists are obsolete, the discipline has changed, earlier participants' interests may have shifted, and a hundred other details which would make such a conference unattainable, had we not been able to draw on the experience of the organizing committee and the support of the local team:

Daniel D. Elleman, a member of the original Drops and Bubbles team, provided the continuity at JPL. Dr. Elleman, a physicist, served as chairman of the Poster Session.

Norman Lebovitz, a member of the first two Drops and Bubbles Colloquium Steering Committees, provided experience and stability to this organizing committee, and is one of its most conscientious and hard-working members. Professor Lebovitz, a physicist, served as Chairman of the Astrophysics Session.

Clive Saunders, also a member of the two previous steering committes, provided continuity and international support to this committee. Professor Saunders, a meterologist, served as chairman of the Meteorology Session.

Jack Salzman, a new participant in the Drops and Bubbles Colloquium, provided new insight and enthusiasm to the conference. Dr. Salzman, a combustion expert and another of the hardest-working members of the Colloquium, chaired the Combustion Session.

I. Dee Chang was a new participant in the Drops and Bubbles Colloquium. His mild mannerisms and scholarly approach contributed to the congenial atmosphere of the committee. Professor Chang, a fluid dynamicist, chaired the Hydrodynamics and Physics of Droplets Session.

Mark C. Lee and **Bradley Carpenter** jointly provided NASA's input to this committee, maintaining a coherent theme to the conference. Dr. Lee participated in the Steering Committee, and Dr. Carpenter chaired the Microgravity Science and Space Experiments Session.

Robert A. Brown, a member of the second Colloquium, did an excellent job of reviewing and maintaining the quality of the proceedings, even though he was often limited to participation-by-mail in the steering committee. Professor Brown chaired the Computational Fluid Dynamics Session.

Robert Bayuzick, a first-timer, provided input from the materials science community to the Colloquium. His steadfastness contributed to the tenor of the conference. Professor Bayuzick chaired the Undercooling and Solidification Session.

Robert Apfel and **Xavier Avula** jointly chaired the Bubbles, Shells, and Encapsulations Session, and provided the central theme for the Colloquium. Bob's good-natured sense of humor made the conference even more interesting.

This Third Colloquium has turned out to be a very great success. The conference was well-attended, and top quality papers were presented. More important, a rigorous peer-review proceedings volume is being published by the American Institute of Physics to preserve the knowledge and optimism for the future.

The conference, held in Monterey, California, was sponsored by the National Aeronautics and Space Administration (NASA), Microgravity Science and Applications Division (MSAD), and hosted by the Jet Propulsion Laboratory (JPL). Thanks for the success of this conference, and seemingly effortless manner in which it was conducted, is due largely to the efforts of Ms. Gail Yepez and her dedicated staff: Margaret A. York, James A. Fox-Davis, Angela Belcastro, Pamela Distaso, and Bobbi Grable. A special thanks to Gail's Editorial Assistant, Ms. A. Marina Fournier, for her patience and understanding in the final editing of this proceeding.

Taylor G. Wang
General Chairman
Third International Colloquium on
Drops and Bubbles

INTRODUCTION TO THE STEERING COMMITTEE AND SESSION CHAIRS

Taylor G. Wang (*Colloquium Chair*)

Professor Taylor G. Wang is Centennial Professor in the Dept. of Materials Science and Engineering at Vanderbilt University. In addition, he is Director of the Center for Microgravity Research and Applications. The research projects conducted at the Center were transferred in Sept. 1988 from the Caltech/Jet Propulsion Laboratory, where he had been employed since 1972. Prof. Wang was manager of JPL's Microgravity Science & Applications program, of which the Material Processing in Space program is a subset.

Prof. Wang was selected by NASA in June, 1983 to train as a space shuttle astronaut-scientist for Spacelab, a research facility flown in the cargo bay of the space shuttle. Dr. Wang flew aboard the Challenger as one of a seven-member crew on the successful SpaceLab-3 (STS-51) mission, April 29-May 5, 1985.

During the flight, Dr. Wang studied the dynamic behavior of rotating spheroids in zero-gravity in his experiment called the Drop Dynamics Module (DDM).

Prof. Wang is President of the *Assoc. of Space Explorers-USA*, a non-profit organization of space flight crewmembers, whose prime objective is to promote public awareness and support of space flight as important to our nation's future.

Prof. Wang is the inventor of the acoustic levitation and manipulation chamber for the DDM, is the author of approximately 130 articles in open literature, as well as the holder of nearly twenty U. S. patents.

Prof. Wang was born on June 16, 1940, in Shanghai, China. He later moved to Taipei, Taiwan, and attended college in the United States, at the Univ. of California, Los Angeles, where he received his Ph.D. in physics in 1971. He became a citizen of the United States in 1975. Professor Wang, his wife (the former Beverly Fung), and their younger son live in Nashville, TN. Their eldest son is currently studying at the Univ. of California, Los Angeles.

Robert E. Apfel (Co-chair, *Bubbles Shells, & Encapsulations* session)

Robert E. Apfel was born in New York City on March 16, 1943. He received his B. A. in Physics in 1964 from Tufts Univ., and his M. A. and Ph. D. degrees in applied physics in 1967 and 1970, respectively, from Harvard.

He is Professor of Mechanical Engineering, and Chairman of the Council of Engineering, at Yale University. He has published approximately 70 referreed publications, primarily in ultrasonics and radiation detection. He consults in architectural acoustics and is president of Apfel Enterprises, Inc., New Haven CT, which is producing a neutron detector based on his patents in this area.

Dr. Apfel is a member of the *Am. Physical Socy.*, the *Acoustical Socy. of America*, the *Am. Socy. of Mechanical Engineers*, and the *Am. Assoc. of Physics Teachers.*

Xavier J. R. Avula (Co-chair, *Bubbles Shells, & Encapsulations* session)

Xavier J. R. Avula, born in Kavali, Andrha Pradesh, India, on January 8, 1936, and received his B. S. from the Indian Institute of Technology (Kharagpur, India) in 1960. He then took an M. S. in 1964 at Michigan State Univ. in East Lansing, before going on to receive his Ph. D. in 1968 from the Iowa State Univ. of Science & Technology at Ames, IA. In 1967, he joined the faculty of the University of Missouri at Rolla, and became a Professor in the Dept. of Mechanical and Aerospace Engineering and Engineering Mechanics there in 1977.

Prof. Avula's research interests include solid-fluid interaction problems, non-linear problems in liquid drop formation, and control problems in human body dynamics. He has been an IBM Visiting Scientist (1983-4, Charlotte, NC), has twice served as a Visiting Scientist at the Aerospace Medical Research Lab, Wright-Patterson AFB (1974-76 and 1987-88), and is the founder and Editor-in-Chief of the journal *Mathematical and Computer Modelling*, published by Pergamon Press.

Prof. Avula is himself the author of over fifty publications, including journal articles and contributions to conferences and books. His research has been supported by the National Science Foundation, Air Force Office of Scientific Research, National Institutes of Health, National Academy of Sciences, and IBM Corp.

Robert J. Bayuzick (Chair, *Undercooling & Solidification* session)

Robert J. Bayuzick received his B. S. in Metallurgical Engineering from the Univ. of Pittsburgh in 1961, his M. S. in Physical Metallurgy at the Univ. of Denver in 1963, and his Ph. D. in Materials Science from Vanderbilt University in 1969, where he studied with Robert S. Goodrich. Dr. Bayuzick began his professional career in 1961 as a Metallurgical Engineer at Bell Aerosystems in Buffalo NY, where he did x-ray diffraction studies and mechanical property studies on aerospace materials. After obtaining his M. S., he worked for the Battelle Memorial Institute, where he conducted research on high-temperature nuclear fuels, primarily ceramics.

Dr. Bayuzick joined Vanderbilt's faculty in 1968, directing his research efforts toward applications of Field Ion Microscopy in Physical Metallurgy. Areas of concentration were the structure and properties of grain boundaries, phase transformations and thin films. In 1977-78, he was a Visiting Research Fellow at Cambridge Univ. (England) studying the segregation of residual elements to transformer steels.

Dr. Bayuzick is presently Professor of Materials Science at Vanderbilt, and for several years, his research interests have been in the area of materials processing under microgravity conditions. There has been a particular emphasis on the structure and properties of alloys resulting from deep undercooling through containerless solidification. Dr. Bayuzick is also the Director of the Center for the Space

Processing of Engineering Materials at Vanderbilt, one of the NASA-supported Centers for the Commercial Development of Space. The Vanderbilt Center involves major U. S. corporations in projects for the development of new materials by the application of the microgravity environment.

Robert A. Brown, (Chair, *Computational Fluid Dynamics* session)

Robert A. Brown is a native of San Antonio, Texas. He did his initial studies in chemical engineering at the Univ. of Texas at Austin, where he received a B.S. in 1973, and an M.S. in 1975. He completed his Ph.D. in 1979 at the Univ. of Minnesota, working with L. E. Scriven on the shape and stability of drops and bubbles. He joined the MIT faculty in 1979, and presently holds the Arthur D. Little Professorship there. He became Department Head of MIT Chemical Engineering in 1989.

Professor Brown's research interests are in fluid mechanics, numerical analysis, and materials processing. He has worked extensively on problems in drop dynamics, and in modelling transport processes in material processing systems, especially solidification phenomena and non-Newtonian fluid mechanics, and on the development of numerical methods for analysis of these systems.

He has received the Allan P. Colburn Award from the Am. Inst. of Chemical Engineers, the Young Author Award from the Am. Assoc. of Crystal Growth, and the Camille and Henry Dreyfus Teacher-Scholar Award in recognition for his research contributions.

Brad Carpenter (Chair, *Microgravity Science and Space Experiments* session)

Dr. Bradley M. Carpenter, a native of Redwood City, CA, received his B. S. in chemistry in 1975 from Univ. of California, Berkeley, an M.S. in Chemical Engineering at the Univ. of Virginia in 1981. He was awarded a Ph. D. in Chemical Engineering in 1987 from Stanford Univ., where he studied with G. M. Homsy. He then went on to a post-doc with Robert Sami, at the Ctr. for Low-Gravity Fluid Mechanics & Transport Phenomena at the Univ. of Colorado, Boulder.

Currently, Dr. Carpenter is a Senior Staff Scientist with the Bionetics Corp., a contractor for the Microgravity Science & Applications Div., at NASA HQ. He is the MSAD discipline scientist for fluid dynamics, combustion science, and biotechnology.

I. D. Chang (Chair, *Hydrodynamics and Physics of Droplets* session)

Dr. I-Dee Chang is a Professor of Aeronautics and Astronautics at Stanford University. He received a Ph.D in Aeronautics and Mathematics from the California Institute of Technology in 1959. After staying as a research fellow at CalTech for two years, he joined the Aeronautics & Astronautics faculty at Stanford in 1961.

His main research interest is in fluid mechanics: he has published many papers on the stability of liquid layers, deformation and breakup of accelerating liquid drops, and motion of suspended particles and droplets. For many years, he has taught courses on low Reynolds number flows and the dynamics of suspensions. He has published over 100 papers and reports, and is a member of the *Am. Physical Socy.* and the *Am. Inst. of Aeronautics and Astronautics.*

Daniel D. Elleman, (Chair, *Poster Session*)

Daniel Elleman presently serves as the Assistant Section Manager for Microgravity Experiments, in the Applied Science and Microgravity Experiments Section at the Jet Propulsion Laboratory, and has been active in the area of the drop physics for over ten years. He was a co-investigator on a number of drop dynamics projects that were flown on the space shuttle, such as the DPM experiment, the ACES experiment, and the 3AAL experiment.

Dr. Elleman has also been active in the area of the fluid dynamic properties of superfluid helium in the low gravity environment provided by the space shuttle and was a co-investigator for the Superfluid Helium experiment that flew on SpaceLab III. He now serves as the project scientist for the Lambda Point Experiment, which will study the heat capacity of superfluid helium in a low gravity environment.

Norman Lebovitz (Chair, *Astrophysics* session)

Norman R. Lebovitz was born in New York City, but attended public school in California, receiving his undergraduate training in physics at UCLA, graduating in 1956. His graduate education, also in physics, was at the University of Chicago, where he received his Ph.D., under the directorship of S. Chandrasekhar, in 1961.

He was C. L. E. Moore Instructor of Mathematics at MIT from 1961 to 1963, when he returned to Chicago to join the applied-mathematics program in the mathematics department. He continues at the University of Chicago, where he is currently chairman of the Committee on Applied Mathematics. His research interests include astrophysics, fluid dyna-mics, and singular-perturbation methods for ordinary differential equations.

Mark C. Lee *(Steering committee)*

Dr. Lee received his B.S. degree in Physics from the National Taiwan Univ. in 1965, and his M. S. and Ph. D. in experimental physics from UCLA in 1968 and 1971, respectively. After post-doctoral work under an Air Force fellowship at the Univ. of Wisconsin, he joined Wyle Laboratories as a Senior Specialist and Program Manager, and later joined Aerojet ElectroSystems. He joined Jet Propulsion Laboratory as a Senior Scientist in 1978, responsible for containerless materials investigation and processing in the microgravity environment.

He joined NASA headquarters, Washington, D. C. in January, 1989, after a two-year assignment as a JPL detailee to the Microgravity Science & Applications Divn. His primary responsibility at NASA Headquarters is to manage U.S. flight and ground-based programs; and to serve as the Program Scientist for the United States Microgravity Laboratory (USML) and United States Microgravity Payload (USMP) Mission Series.

He has published over 70 technical papers in various professional journals and delivered numerous invited speeches. He holds eight U. S. Patents in the area of materials processing in space. He is the NASA Principal Investigator on *Metallic Glass Research in Space*. He was also the DOE Principal Investigator on *Plastic Microshell Technology*. He served as Chairman for numerous symposiums, conferences and technical sessions. Since September, 1985, he has served as the Director of the *Am. Assoc. for Promotion of Science in China.*

Jack Salzman , NASA Lewis Research Center (Chair, *Combustion* session)

Mr. Salzman is currently Chief of the Microgravity Science and Technology Branch at the NASA Lewis Research Center in Cleveland, OH. The charter of the Branch is to conduct and guide research in the area of microgravity fluid physics/dynamics and combustion science, with the goal of defining and utilizing space-based experiments. This work is sponsored by the Microgravity Science and Applications Program, and includes in-house as well as sponsored research programs in the industrial and academic communities.

Mr. Salzman joined NASA in 1964 to pursue research in low-gravity fluid behavior, to support the development of liquid propellant management systems for upper-stage launch vehicles. After leaving this technical area in 1976 to accept research and management positions in other space applications programs, he returned to low-gravity research via his current position in 1985. He holds a B.S. in Physics from Baldwin-Wallace College; and an M.S. in Engineering Science, from Toledo University.

Clive Saunders, (Chair, *Meteorology* session)

Dr. C. P. R. Saunders is in the Atmospheric Physics Research Group in UMIST, Manchester, England. The research interests of this large group cover a wide range of important atmospheric problems such as the greenhouse effect, climate change,

the acidification of rain by the ozone/SO_2 reaction, dual polarization radar studies of the initiation of convective clouds, atmospheric electricity and the ocean/atmosphere interface. An interest in "drops and bubbles" is apparent in all of the above areas.

Recently, in collaboration with Dr. Hallett of the Desert Research Institute, there have been flights on the NASA KC135, which provides periods of 25 seconds of low gravity, each of which is followed by a period of 2g. The effect of convection on the growth of ice crystals from the vapor has shown that, in the absence of gravity, the crystal growth rate slows, while in 2g the increased rate of latent heat removal leads to faster growth. In a second experiment, the freezing of supercooled water drops was studied in order to determine whether ice splinters are ejected, which could help account for the high concentration of ice particles in some clouds. Further work on this is under way.

Third International Colloquium on Drops and Bubbles
❦ *Monterey, California*

1. Hydrodynamics & Physics of Droplets

I-Dee Chang
Session Chair

DEFORMATION AND BURST OF SINGLE DROPS IN SHEAR FLOWS

Andreas Acrivos
Department of Chemical Engineering, Stanford University, Stanford CA 94305

ABSTRACT

The deformation and break-up of single drops which are freely suspended in shear fields is examined both theoretically and experimentally. For creeping flows, a comprehensive theory is summarized which combines small deformation analysis, a slender-body asymptotic expansion, as well as exact numerical computations using an integral equation formulation of the solution to the relevant mathematical system. Theoretical predictions are shown to be in excellent agreement with experimental results.

INTRODUCTION

Imagine a liquid drop of viscosity $\lambda\mu$ and volume $V = 4\pi a/3$, surrounded by another fluid of viscosity μ, with which it is immiscible. When the ambient fluid is sheared, the initially spherical drop will deform and, for large enough shear rates, it will often break into several fragments. This phenomenon is of considerable fundamental importance in fluid mechanics as an example of a free boundary value problem and as a prototype for a flow induced deformation of a variety of flexible bodies such as red blood cells. It is also of relevance to the design of commercial blenders and emulsifiers where it can be used to estimate the power needed to insure that all the drops of the dispersed phase remain below a given maximum size.

When the drop is sufficiently small for inertia and gravity effects to be negligible, the extent of the deformation is governed by two dimensionless groups in addition to the type of flow: i) the viscosity ratio λ, and ii) the capillary number $Ca=G\mu a/\gamma$, where G denotes the strength of the impressed shear field, and γ is the interfacial tension.

In what follows, I shall summarize both the various theories which have been developed over the years to predict the drop's deformation and the conditions for its break-up, as well as the relevant experimental findings. Further details are provided in some recent reviews[1-3], and in the references mentioned below.

THEORY (CREEPING FLOW)

The appropriate mathematical system consists of the creeping flow equations which are assumed to apply within both fluids, subject to the boundary conditions of continuity of velocity and shear stress at the boundary of the two phases. Continuity of shear stress implies, of course, the absence of surface active agents, i.e.; we suppose that the interface is clean. In addition, the normal stress suffers a jump across the interface, from the inside to the outside, equal to the product of γ with the principal curvature of the surface. Also, on the assumption that the equivalent radius of the drop is much smaller than a characteristic length scale over which the impressed macroscopic velocity field varies, we can express the undisturbed velocity $u_i^\infty(\vec{x})$ relative to a moving origin at 0, the center of the drop, simply as

$$u_i^\infty(\vec{x}) = E_{ij}\, x_j + \tfrac{1}{2}\varepsilon_{ijk}\, \Omega_j\, x_k \qquad (1)$$

where $\vec{x}$ is the position coordinate measured from 0, and E_{ij} and Ω_j are, respectively, the rate of strain tensor and the vorticity of the applied shear flow both

[a]*Current address:* Levich Institute, City College of CUNY, 138th Street at Convent Avenue, New York, NY 10031

evaluated at 0. Equation (1) serves as the boundary condition for the velocity u_i, as $r = |\vec{x}| \to \infty$. Finally, we require that, at steady state,

$$u_j n_j = 0 \quad \text{on the interface } S \tag{2}$$

where n_j denotes the unit outer normal to S.

It is important to appreciate that, although the creeping flow equations are linear, the mathematical problem as posed is in fact highly non-linear owing to the *a priori* unknown shape of S which must be determined as part of the solution. Nevertheless, it is possible to proceed theoretically by taking advantage of the linearity of the basic equations and the fact that often, prior to break-up, the drop is either almost spherical or long and slender. The relevant theoretical results will now be summarized.

SMALL DEFORMATION ANALYSIS

When the capillary number Ca is small, the drop shape is almost spherical since, in the absence of flow, the drop is exactly a sphere. Hence, as first pointed out Taylor[4], one can solve the creeping flow equations for flow past a sphere subject to all the boundary conditions except for the normal stress balance which is then used to compute the deformation to first order in Ca. In fact, it is easy to show without constructing a detailed solution that, to this order, the sphere deforms into an ellipsoid with its principal axes coincident with those of E_{ij}.

Taylor's theory, which has subsequently been refined[5,6], leads to an expression for the deformation which remains surprisingly accurate considering the limitations of the analysis. Unfortunately, it cannot lead to a criterion for break-up which can only be obtained by extending the solution to higher order in Ca. The resulting analysis[7] leads to the prediction that, under certain conditions, a steady state solution to the boundary value problem stated above cannot exist if the capillary number exceeds a critical value Ca^*, dependent only on λ and the type of the impressed shear flow, which is thereby identified as the sufficient condition for drop break-up. More specifically, it is found that the theoretically computed deformation with increasing Ca either approaches a steady finite limit as $Ca \to \infty$, or that it reaches a steady limit point at Ca^* beyond which is increases indefinitely with time. Although this analysis[7] is only approximate, it predicts values for Ca^* which are in surprisingly close agreement with more exact numerical results and the corresponding experimental findings, to be discussed below, as long as λ is not too small, i.e.; typically in excess of 10^{-1}, for under these conditions the drop shape at the point of break-up does not differ all that much from that of a sphere.

SLENDER-BODY THEORY

Taylor[8] appears to have been the first to realize that drops of very low viscosity relative to that of the ambient fluid ($\lambda<<1$) could become long and slender prior to break-up and that the critical capillary number Ca^* could be calculated in these cases by adapting to creeping flows the well-known slender body analysis of aerodynamics. Specifically, since the length of a slender-drop is much greater than its breath, spatial derivatives along the axis of maximum extension of the drop are then much smaller than those normal to it and hence the flow within the drop is quasi-parallel[9,10]. Also, the external velocity can be represented by either a line distribution of Stokeslets and sources along the axis[11], or more simply via a matched asymptotic analysis[10]. It is found[9,10] that, when the impressed shear field consists of an axixymmetric pure straining flow,

$$Ca^* = 0.148\,\lambda^{1/6} \quad \text{as } \lambda \to 0 \tag{3}$$

and that, for the analogous case of a two-dimensional hyperbolic flow, the corresponding expression for Ca^* is identical except for the proportionality constant[12], which now equals 0.145. In fact, since two-dimensional flows can be represented by

$$u_x = G_x \text{ and } u_y = G\alpha y\,,$$

where α=1 corresponds to the case of pure strain, α = 0 to simple shear, and α = -1 to pure rotation, it is easy to show[13] that, for α> 0 and $\lambda \to 0$,

$$Ca^* = 0.145\,\alpha^{1/2}\,\lambda^{-1/6}\ . \tag{4}$$

The analogous expression for a simple shear flow[14] is

$$Ca^* = 0.055\,\lambda^{-2/3}\ .$$

A surprising and aesthetically pleasing feature of slender drops is that their ends are sharp and very pointed[8].

NUMERICAL SOLUTION

In principle, it is always possible of course to solve the equations of motion numerically and thereby obtain "exact" theoretical predictions for the drop deformation and the conditions for its break-up by the application of standard finite-difference or finite-element techniques. Nevertheless, it appears more logical to take advantage of the linearity of the creeping flow equations and base the numerical procedure on the fact that these equations admit a general solution in closed form. The problem thereby reduces to the solution of the integral equation[15,16]

$$\frac{1}{2}(1+\lambda)\,u_i\,(\vec{x}) + (1-\lambda)\,\int_S K_{ijk}\,(\vec{s})\,u_j(\vec{y})\;n_k(\vec{y})\,dS_y = u_i^\infty(\vec{x}) - \frac{\gamma}{8\pi\mu}\int_S J_{ij}\,(\vec{s})n_j\,(\vec{y})\,\kappa(\vec{y})dS_y \tag{5}$$

subject to $u_j n_j$ = 0 on S, where $\vec{x}$ and $\vec{y}$ are points on S, the interface of the drop; κ is its local principal curvature, and

$$K_{ijk} = -\frac{3}{4\pi}\frac{s_i s_j s_j}{|\vec{s}|^5}, \qquad J_{ij} = \frac{\partial_{ij}}{|\vec{s}|} + \frac{s_i s_j}{|\vec{s}|^3}, \qquad \vec{s} = \vec{x} - \vec{y}\ . \tag{6}$$

The advantage of using Equation (6) is that the domain of integration is finite since it extends only over the interface and that the dimension of the integral equation problem is one lower than that of the original system. For example, a three-dimensional flow problem is thereby reduced to a two-dimensional integral equation, while an axisymmetric problem reduces to a one-dimensional integral equation.

Computations using Equation (6) have been performed[15] for the case of an axisymmetric pure straining flow at infinity and for $0.3 \leq \lambda \leq 100$, while, for λ = 1, results have been presented[16] for a range of two-dimensional linear flows including a simple shear flow.

As expected, the computations are in close agreement with the findings of the small deformation analysis when $\lambda = O(1)$ or larger, and with the predictions of the slender body theory when λ is small.

EXPERIMENTAL RESULTS

The first relevant experiments were those of Taylor[17]. These were performed in a "four-roll mill", a versatile instrument which, in principle, is capable of generating a variety of planar shear flows corresponding to Equation (4), although, in Taylor's case, it was used only to produce a hyperbolic flow (α = 1). Taylor[17] also studied deformation and break-up in simple shear flows (α=0) which he produced in a parallel band apparatus. Additional experiments in hyperbolic and simple shear flows were subsequently performed for a wide variety of physical system[18-20].

The most extensive and reliable experiment to date are those reported recently by Bentley and Leal[21] and by Stone, Bentley and Leal[22], who were able to employ the "four-roll mill" to its fullest potential for plane shear flows in the range $0.2 \leq \alpha \leq 1.0$.

These authors showed conclusively that the small deformational analysis accurately predicts the conditions for drop break-up as long as λ exceeds approximately 0.1, and that slender-body theory describes this phenomenon quantitatively if λ is below about 0.01.

Thus on the whole, there is excellent agreement between theory and experiment.

REFERENCES

1. A. Acrivos, *Ann. N.Y. Acad. Sci.* **404**, 1 (1983).
2. J. M. Rallison, *Ann. Rev. Fluid Mech.* **16**, 45 (1984).
3. A. Acrivos, **Physicochemical Hydrodynamics** (M. G. Velarde ed.), Plenum Publishing Corp., 1988.
4. G. I. Taylor, *Proc. Roy. Soc. London* **A138**, 41 (1932).
5. R. G. Cox, R. G., *J. Fluid Mech.* **37**, 601 (1969).
6. J. M. Rallison, *J. Fluid Mech.* **98**, 625 (1980).
7. D. Barthes-Biesel, and A. Acrivos, *J. Fluid Mech.* **61**, 1 (1973).
8. G. I. Taylor, **Proceedings of the 11th International Congress of Applied Mechanics**, Munich, p. 790 (1964).
9. J. D. Buckmaster, *J. Appl. Mech.* **E40**, 18 (1973).
10. A. Acrivos, and T. S. Lo, *J. Fluid Mech.* **86**, 641 (1978).
11. J. D. Buckmaster, *J. Fluid Mech.* **55**, 385 (1972).
12. E. J. Hinch and A. Acrivos, *J. Fluid Mech.* **91**, 401 (1979).
13. D. V. Kahkhar and J. M. Ottino, *J. Fluid Mech.* **166**, 265 (1986).
14. E. J. Hinch and A. Acrivos, *J. Fluid Mech.* **98**, 305 (1980).
15. J. M. Rallison and A. Acrivos, *J. Fluid Mech.* **89**, 191 (1978).
16. J. M. Rallison, *J. Fluid Mech.* **109**, 465 (1981).
17. G. I. Taylor, *Proc. Roy. Soc. London* **A146**, 501 (1934).
18. F. D. Rumscheidt, and S. G. Mason, *J. Coll. Sci.* **16**, 238 (1961).
19. S. Torza, R. G. Cox, and S. G. Mason, *J. Coll. Int. Sci.* **38**, 395 (1972).
20. H. P. Grace, **Engng. Found. 3rd Res. Conf. Mixing**, Andover NH (1971).
21. B. J. Bentley and L. G. Leal, *J. Fluid Mech.* **167**, 241 (1986).
22. H. A. Stone, B. J. Bentley, and L. G. Leal, *J. Fluid Mech.* **173**, 131 (1986).

THE CIRCULATION PRODUCED IN A DROP BY AN ELECTRIC FIELD: A HIGH FIELD STRENGTH ELECTROKINETIC MODEL

J. C. Baygents[a] and D. A. Saville
Department of Chemical Engineering, Princeton University, Princeton, NJ 08544-5263

ABSTRACT

The circulation produced in a drop by an electric field is examined. Here the leaky dielectric, used in the prototypal solution of Sir Geoffrey Taylor, is replaced with an electrokinetic model of the charge transport. Singular perturbation methods are employed to obtain a description of the macroscopic response of the drop to the field. The remarkable result is that complete agreement is found between the predictions of the two models. Previously, it was thought that effects due to space-charge accumulation at the drop surface, which had presumably been omitted from the leaky dielectric, explained extant discrepancies between theory and experiment. The electrokinetic model shows that, in an aggregate sense, the leaky dielectric accounts for the space charge, and, so, the differences with the data remain unexplained.

1. INTRODUCTION

A prototypal problem of electrohydrodynamics concerns the circulation produced in a drop by an electric field (Melcher & Taylor 1969). Experiments had been performed on the deformation of liquid drops suspended in liquid dielectrics in electric fields; for dielectric drops, steady shapes in the form of prolate and oblate spheroids were found (Allan & Mason 1962). Extant electrostatic models, which treated the low conductivity liquids as perfect dielectrics, were unable to explain the oblate deformations. Using a model called the "leaky" dielectric, Taylor (1966) predicted drop shapes qualitatively consistent with experimental observation. The fundamental postulate of the Taylor analysis was that the liquids, though dielectrics and relatively insulating, still passed small amounts of current by Ohmic conduction. Taylor showed that, by ignoring the conduction processes, the previous theories had completely overlooked electrical shear stresses that would result from the imposed field acting at the drop surface. The shear stresses produced circulation in and about the drop which, in certain cases, led to oblate deformations.

Subsequent investigations have revealed, though, that calculations of the magnitude of drop deformation based on the Taylor model do not agree with laboratory measurements (Torza *et al.* 1971). A strong rationale for the quantitative discrepancy between theory and experiment has been that, except for the limiting case of unipolar conduction, the leaky dielectric ostensibly gives an inconsistent result for an important factor influencing electrohydrodynamic flows: the distribution of charge near the liquid-fluid interface. More specifically, the criticism is that the leaky dielectric inherently rules out space charge distributions.

It is well-established that the electric field induces at the drop surface an interfacial charge to compensate for the change in the normal component of the electric displacement across the phase boundary. Since liquids described by the leaky dielectric are ionic as well as Ohmic conductors—*i.e.*, conduction involves at least two types of charge-carriers (Melcher 1973)—the charged species in solution interact electrostatically with the induced surface charge. In the steady state, a diffuse space charge forms about the interface. The ion cloud can be thought of as a non-equilibrium analog to the Gouy-Chapman layer that accompanies colloidal particles in

[a]Currently affiliated with Universities Space Research Association, NASA Space Science Laboratory, Marshall Space Flight Center, Huntsville, AL 35812.

aqueous suspensions (Loeb *et al.* 1961).

As the situation is depicted by the leaky dielectric, there is an induced surface charge, but no space charge is manifested. The balance laws are predicated on a lumped parameter formulation that does not consider the charge-carrying solutes separately and all charge is transferred by Ohmic conduction. Consequently, phases which are uniformly neutral prior to imposition of the field, remain so; any spatial accumulation of charge must be short-lived. Say, for instance, that ϵ is the dielectric constant, ϵ_0 the permittivity of a vacuum, and σ_∞ the electrical conductivity. Then the charge relaxation transient is $O(\epsilon\epsilon_0/\sigma_\infty)$, which, in nonaqueous media, is about 10^{-2} seconds (Taylor & Melcher 1969).

In the study synopsized here, we have examined the role of space charge in the response of drops to applied electric fields. To do so, we have resolved the problem for the circulation produced in a drop by an electric field, replacing the leaky dielectric with a high-field electrokinetic model. The electrokinetic equations, which incorporate effects due to ion diffusion and convection, provide a more representative picture of the charge transport. We have thus been able to describe in a consistent fashion the interplay between hydrodynamic and electrostatic stresses in the neighborhood of the drop surface; our results include the details of the space charge distribution. In the sections to follow, we outline our analysis, emphasizing some of the essential features. A complete exposition of the work will be published later[1].

The presentation is organized as follows. We begin with the balance laws and boundary conditions in dimensionless form; the scaling is for high field strengths and the interfacial boundary conditions are chosen to permit comparisons with Taylor (1966). The diffusion terms drop out of the ion balances, so singular perturbation methods are required to construct uniformly-valid solutions. We are interested in the case where drop deformation due to the imposed field is slight, and where: $a \gg k_B T/eE_\infty \gg \kappa^{-1}$ and $\bar{\kappa}^{-1}$. In the notation of the text: a is the drop size; E_∞ the applied field strength; e the fundamental charge; k_B the Boltzmann constant; and T the absolute temperature; κ^{-1} and $\bar{\kappa}^{-1}$ are the Debye screening lengths for the exterior and interior phases, respectively. We first solve the outer problem, which involves processes that predominate away from the interface on a length scale typified by a. Over the outer domain, the liquid phases behave as conceived in the leaky dielectric; *i.e.*, as isotropic dielectrics that posses a finite Ohmic conductivity. Next we set up an inner problem that accounts for the space charge. Diffuse charge accumulates on each side of the interface and the thickness of the charge cloud is characterized by κ^{-1} and $\bar{\kappa}^{-1}$. Efforts to match the inner and outer solutions, however, fail because, as the analysis reveals, there are concentration boundary layers located between the inner and outer domains. At distances $O(k_B T/eE_\infty)$ from the interface, sharp concentration gradients develop as ion diffusion becomes comparable to convection and electromigration. The dramatic concentration changes are related to those elucidated by Acrivos & Goddard (1965) for laminar forced-convection heat and mass transfer. We solve the intermediate problem, then match inner, intermediate and outer solutions. Finally we work out the drop shape and discuss the results. The solutions reported are for a simple z-z electrolyte. When a is *circa* 1 mm, the asymptotic constraints limit E_∞ roughly to values from 1 to 10^4 V/cm.

Quite remarkably, the electrokinetic and leaky dielectric models yield equivalent expressions for the drop deformation. Contrary to previous thinking, the results show that, in an aggregate sense, the leaky dielectric does indeed account for the space charge that sandwiches the induced surface charge. Since $a\kappa$ and $a\,\bar{\kappa} \gg 1$, the space charge occupies a region that is very narrow compared to the local surface

[1] Interested parties are invited to contact the authors.

curvature; from a macroscopic perspective, this means *all* of the charge appears to be confined to the drop surface. The "apparent" surface charge density—the actual surface charge density that is not neutralized by the diffuse charge—equals the induced surface charge density given by the the leaky dielectric. Effects due to the finite thickness of the accumulated charge are obscured and, because of the disparity in length scales, cannot be resolved on the macroscopic level.

The results point toward several conclusions. The most obvious is that, with respect to drop behavior in electric fields, accounting explicitly for the space charge adjacent to the drop surface does not alone reconcile the theory to the experiments. Secondly, the results indicate the need for further experimentation. The data of Torza *et al.* (1971) contain no information regarding the individual charge-carrying species; a deformation study that reported bulk ion concentrations and mobilities would, for example, be quite useful. Lastly, even though in this instance the models yield the same macroscopic behavior, the work demonstrates the strength of the electrokinetic formulation vis-à-vis the leaky dielectric. Electrokinetic studies at low fields strengths show unequivocally that events which occur on the micro-scale of the Debye length influence macroscopic behavior. The details of this region cannot be delineated or interpreted with a lumped parameter model like the leaky dielectric. The work here is the first wherein an electrokinetic model has been applied to a high-field problem involving ion transport normal to the charged interface; moreover, the results provide a substantive basis for future investigation of such factors as: a native surface charge with an attending Gouy-Chapman layer; dissociation-association processes involving the charge-carrying species; and interfacial tension gradients caused by composition variations.

Applications for the work exist in a number of areas. The manipulation of fluid interfaces with strong electrostatic fields is relevant to: heat and mass transfer enhancement (Morrison 1977, Griffiths & Morrison 1979); bio-separations (Brooks & Bamberger 1982, Brooks *et al.* 1984); and containerless processing in microgravity (Barmatz 1982, Rhim *et al.* 1982). Melcher (1973) lists technologies that kindled early developments in electrohydrodynamic theory. High-field electrokinetic phenomena are used as analytical tools in electrophoresis (Kuo & Osterle 1967, Simonova *et al.* 1972, Stotz 1972, Hair & Landheer 1982, Fixman & Jagannathan 1983) and molecular electro-optics (O'Konski 1976, Fixman 1980, Rau & Charney 1983).

2. THE ELECTROKINETIC MODEL: BALANCE LAWS AND INTERFACIAL BOUNDARY CONDITIONS

Consider an immiscible drop suspended freely in an unbounded liquid. The viscosity, dielectric constant, and ion mobilities for each phase are independent of position. Outside of the interfacial region, the number density of the kth solute is c_∞^k and $\bar{c}_\infty^k$ for the drop exterior and interior, respectively. The (signed) valence of the kth species is z^k. There are N distinct charge-carrying species; all are formed from ionogenic solutes that dissociate completely. When there is no external electric field: the drop radius is a; the interfacial tension γ_0, is uniform over the drop surface; and there are no spatial nor interfacial charge accumulations. If the capillary number $Ca \equiv a\epsilon\epsilon_0 E_\infty^2/\gamma_0$, is small compared to unity, the applied field will not deform the drop extensively and a streamline pattern like that in Figure 1 develops (Taylor 1966). As shown, the surface flow is from pole to equator. But, depending on the physical properties of the two phases, the circulation could also be from equator to pole. The drop will either remain a sphere or be deformed into an oblate or prolate spheroid.

For the sake of brevity, we will simply list the governing equations here. Explanations of the electrokinetic balance laws (Saville 1977) and the interfacial boundary conditions (Baygents & Saville 1988) appear elsewhere. The scaled balance laws read:

(Ion Conservation)

$$\beta Pe\, \mathbf{v} \cdot \nabla n^k = \frac{\omega^k}{\beta} \nabla^2 n^k + z^k \omega^k \nabla \cdot (n^k \nabla \Phi), \qquad k = 1, 2, \ldots, N; \qquad (2.1)$$

(Gauss' Law)

$$\beta \nabla^2 \Phi = -(a\kappa)^2 \sum_{k=1}^{N} z^k n^k; \qquad (2.2)$$

(Momentum Balance)

$$\mathbf{0} = -\nabla p + (\nabla^2 \Phi) \nabla \Phi - \nabla \times \nabla \times \mathbf{v}; \qquad (2.3)$$

(Continuity)

$$\nabla \cdot \mathbf{v} = 0. \qquad (2.4)$$

The dependent variables are: n^k, the number density for the kth species; Φ, the electrostatic potential; $\mathbf{v}$, the fluid velocity; and p the pressure. The dimensionless groups are: $Pe \equiv \epsilon\epsilon_0 k_B T/\mu e^2 \omega^0$, a Péclet number characterizing the relative importance of convection and diffusion near the interface; $a\kappa \equiv (a^2 e^2 c^0/\epsilon\epsilon_0 k_B T)^{1/2}$ the drop size divided by the Debye screening length in the exterior phase; ω^k, the mobility of the kth species normalized to ω^0, a typical ion mobility; and $\beta \equiv aeE_\infty/k_B T$, the scaled magnitude of the applied field. The reference scales are: length, a; potential, aE_∞; stress, $\epsilon\epsilon_0 E_\infty^2$; velocity, $a\epsilon\epsilon_0 E_\infty^2/\mu$; and concentration, $c^0 \equiv \sum_{k=1}^{N} (z^k)^2 c_\infty^k$. The inertial terms are omitted from the momentum balance because the fluid motion is slow. Note carefully that, in contrast to electrokinetic problems where the external field is small, the scale factors now are for high field strengths ($\beta \gg 1$). Expressions similar to Equations (2.1)-(2.4) apply to the drop interior.

After Taylor (1966), we are interested in the circumstance where $Ca \ll 1$, meaning there is little deformation of the drop. For this case, the *actual* drop shape can be determined through $O(Ca)$ from the solution for the circulation produced in a drop that remains spherical upon imposition of the electric field (Torza *et al.* 1971).

The interfacial boundary conditions are greatly simplified when the drop does not deform. To begin with, the velocity can be separated into an r-component v_r, and a θ-component v_θ, so that at $r = 1$,

$$v_r(1,\theta) = \overline{v}_r(1,\theta) = 0 \qquad (2.5)$$

and

$$v_\theta(1,\theta) = \overline{v}_\theta(1,\theta)\,, \qquad (2.6)$$

where overbars are placed on quantities that pertain to the drop interior. The origin of the coordinate system is at the drop center; θ is the angle measured from the direction of the applied field to the position vector $\mathbf{r}$. The constraints on the potential are:

$$\frac{\partial \Phi}{\partial \theta} = \frac{\partial \overline{\Phi}}{\partial \theta} \qquad (2.7)$$

and

$$-\frac{\partial \Phi}{\partial r} + \frac{\overline{\epsilon}}{\epsilon} \frac{\partial \overline{\Phi}}{\partial r} = \frac{(a\kappa)}{\beta} \mathcal{Q}, \qquad (2.8)$$

with

$$\mathcal{Q} = \sum_{k=1}^{N} z^k K^k n^k(1,\theta) = \sum_{k=1}^{N} z^k \overline{K}^k \overline{n}^k(1,\theta)\,. \qquad (2.9)$$

In (2.9), $\mathcal{Q}$ is the surface charge density; K^k and $\overline{K}^k$ are the linear adsorption coefficients for the kth species normalized to $1/\kappa$. The tangential stress balance is

$$\frac{\partial v_\theta}{\partial r} - \frac{\overline{\mu}}{\mu} \frac{\partial \overline{v}_\theta}{\partial r} - \frac{1}{r^2}\left(1 - \frac{\overline{\mu}}{\mu}\right) v_\theta + \frac{1}{r} \frac{\partial \Phi}{\partial \theta}\left(\frac{\partial \Phi}{\partial r} - \frac{\overline{\epsilon}}{\epsilon} \frac{\partial \overline{\Phi}}{\partial r}\right) + \frac{1}{rCa} \frac{\partial(\gamma/\gamma_0)}{\partial \theta} = 0. \qquad (2.10)$$

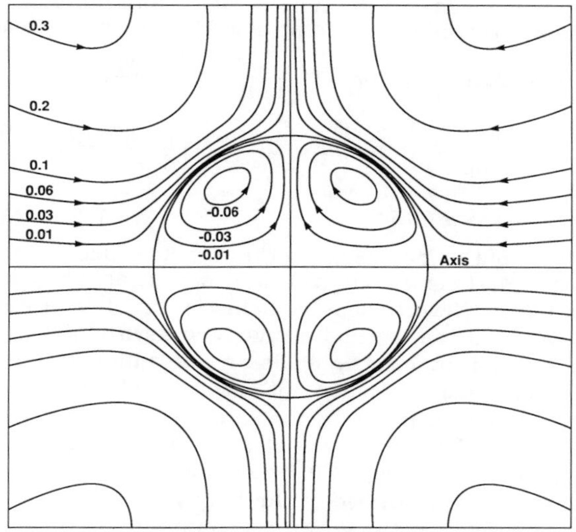

Figure 1

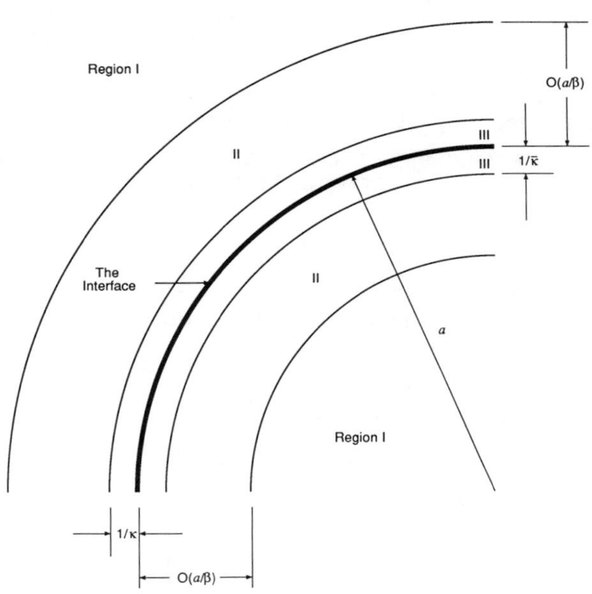

Figure 2

The normal stress balance is, for the time being, inconsequential. Finally, conservation of kth species at the interface is represented by a surface analog to (2.1), *viz.*

$$0 = (a\kappa)\omega^k\left(\frac{1}{\beta}\frac{\partial n^k}{\partial r}+z^k n^k\frac{\partial \Phi}{\partial r}\right) - (a\kappa)\,\overline{\omega}^k\left(\frac{1}{\beta}\frac{\partial \overline{n}^k}{\partial r}+z^k\overline{n}^k\frac{\partial \overline{\Phi}}{\partial r}\right) - \frac{1}{r\sin\theta}\frac{\partial}{\partial\theta}\left\{\sin\theta\left(\beta Pe\,v_\theta\Gamma^k - \frac{\omega_s^k}{r}\left(\frac{1}{\beta}\frac{\partial n^k}{\partial\theta}+z^k n^k\frac{\partial\Phi}{\partial\theta}\right)\right)\right\}. \qquad (2.11)$$

Here

$$\Gamma^k = K^k n^k = \overline{K}^k\overline{n}^k, \qquad k = 1, 2, ..., N; \qquad (2.12)$$

and ω_s^k is the mobility of the kth species within the interface.

The governing equations are coupled and nonlinear, so exact analytical solutions are not available. Nevertheless, satisfactory analytical approximations can be derived; we have used the method of matched asymptotic expansions. Extant data on drop deformation by electric fields are for millimeter-sized drops in fields of 1 kV/cm and higher (Allan & Mason 1962, Torza *et al.* 1971). Under these experimental conditions, $\beta \gg 1$ and the diffusion terms effectively vanish from Equation (2.1); *i.e.*, the highest order derivatives of the ion balances are lost at high field strengths. Clearly uniformly-valid solutions cannot be extracted from the terms of the electrokinetic model that survive in the limit $\beta \to \infty$. In order to satisfy the interfacial boundary conditions, the governing equations need to be rescaled in the vicinity of the interface. In particular the ion balances should reflect that, there, diffusion is prevalent.

We set β to infinity in Equations (2.1)-(2.4) and work out a solution without regard for Equations (2.5)-(2.12); we call this the outer problem. Next, with physical arguments, we devise a new scaling that is appropriate adjacent to the interface; this is the inner problem which accounts for the space charge and satisfies the interfacial boundary conditions. The inner and outer solutions do not match and we deduce the existence of an intermediate region that is situated between the inner and outer domains. With the intermediate solution, we then match contiguous regions and establish the macroscopic behavior of the drop.

3. THE OUTER PROBLEM: ISOTROPIC DIELECTRICS WITH UNIFORM OHMIC CONDUCTIVITIES

As $\beta \to \infty$, Equation (2.1) becomes

$$\beta Pe\,\mathbf{v}\cdot\nabla n^k = z^k\omega^k\nabla\cdot(n^k\nabla\Phi) + O(1/\beta) \qquad k = 1, 2, ..., N. \qquad (3.1)$$

Both the electromigrational and convective terms are retained in Equation (3.1) because, far into the supporting liquid, disturbances due to the drop disappear, *i.e.*,

$$\left.\begin{array}{l}\mathbf{v}\to\mathbf{0}\,,\\ \Phi\to -\,r\cos\theta\,,\end{array}\right\} \quad \text{as } r\to\infty. \qquad (3.2)$$

For the ion distributions, it follows that

$$\lim_{r\to 0} n^k = n_\infty^k \equiv c_\infty^k/c^0. \qquad (3.3)$$

According to Equation (2.2), electroneutrality prevails locally, *i.e.*,

$$\sum_{k=1}^{N} z^k n^k = 0 + O\left(\frac{\beta}{(a\kappa)^2}\right), \qquad (3.4)$$

so that conservation of the net charge requires:

$$0 = \sum_{k=1}^{N}(z^k)^2\omega^k\nabla\cdot(n^k\nabla\Phi) + O\left(\frac{1}{\beta},\frac{\beta}{(a\kappa)^2}\right). \qquad (3.5)$$

A solution for the n^k that is implied by Equations (3.1), (3.4), and (3.5), and which does satisfy Equation (3.3), is merely

$$n^k(r,\theta) = n^k_\infty, \qquad k = 1, 2, ..., N; \tag{3.6}$$

with

$$\nabla^2 \Phi = 0 \tag{3.7}$$

or, equivalently,

$$\Phi(r,\theta) = \left(\frac{D^0}{r^2} - r\right)\cos\theta. \tag{3.8}$$

The constant D^0 is an unknown that corresponds to the dipole strength of the drop.

Based on Equation (3.7), the electric potential is harmonic and the electrical body force vanishes from Equation (2.3). The fluid motion, therefore, is a simple Stokes flow where the correct stream function Ξ, is

$$\Xi(r,\theta) = \frac{A}{2}\left(\frac{1}{r^2} - 1\right)\sin^2\theta\cos\theta\ , \tag{3.9}$$

and

$$v_r(r,\theta) = -\frac{1}{r^2\sin\theta}\frac{\partial\Xi}{\partial\theta}; \qquad v_\theta(r,\theta) = \frac{1}{r\ \sin\theta}\frac{\partial\Xi}{\partial r}. \tag{3.10}$$

The "apparent" peak interfacial velocity $U \equiv A/2$, must come from matching with the solutions for the velocity that hold closer to the drop surface; a second constant of integration has already been eliminated from Ξ because v_r must tend to zero as $r \to 1$. The sign of A determines the gross sense of the circulation; viz., when $A > 0$, the flow is from equator to pole; when $A < 0$, the flow is from pole to equator.

4. THE INNER PROBLEM: DIFFUSE CHARGE LAYERS ADJACENT TO THE INTERFACE

The solution to the outer problem fits the concept of the leaky dielectric. That is: the concentration field is uniform and electrically neutral; the electric potential is harmonic; the momentum balance contains no electrical body force; and the conduction processes are Ohmic. Closer to the interface, though, the diffusion terms omitted from Equation (2.1) are significant and the outer solution does not apply.

The surface charge density induced by the applied field is $O(\epsilon\epsilon_0 E_\infty)$, the magnitude of the electric displacement at the interface. Ions in solution interact with the surface charge. Concentration gradients develop as coulomb forces attract counter-ions to, and repel co-ions from, the interface. Eventually, Brownian forces grow to the point where the evolving concentration profiles are arrested and a steady space charge is established.

Let $a\delta$ be the thickness of the charge configuration in the exterior phase. The concentration variations are then $O(\epsilon\epsilon_0 E_\infty / ea\delta)$ or, in terms of the characteristic bulk concentration, $O(\beta c^0 / \delta a^2 \kappa^2)$. The electrical field, which is $O(E_\infty)$, acts on the space charge, producing an electro-osmotic flow that is $O(a\delta\epsilon\epsilon_0 E_\infty^2/\mu)$ down. For the space charge to persist, Brownian forces must counteract the currents driven by Coulomb interactions. Thus, $\delta \equiv 1/(a\kappa)$.

The scale analysis and the behavior of the outer solution as $r \to 1$ form the basis for the inner expansions. When Equations (2.1)-(2.12) are properly rescaled, a collection of linear differential equations results, where the derivatives are taken with respect to w, a radial coordinate defined implicitly by

$$r \equiv 1 + \frac{w}{a\kappa}. \tag{4.1}$$

The solution to the inner problem is straightforward and involves only elementary techniques. To a large degree, the inner formulation resembles a well-known electrokinetic effect: electro-osmotic streaming past a flat surface (Ehrlich & Melcher 1982). There is a charge cloud that decays exponentially as $w \to \pm\infty$ and, accordingly, the potential is non-harmonic. The fluid motion consists of the Stokes flow, with a superimposed eletro-osmosis that moves wholly tangent to the drop surface.

After application of the interfacial boundary conditions, matching the inner and outer solutions requires that

$$D^0 = \frac{\overline{\omega}^k \overline{n}_\infty^k - \omega^k n_\infty^k}{\overline{\omega}^k \overline{n}_\infty^k + 2\omega^k n_\infty^k}, \qquad k = 1, 2, ..., N. \tag{4.2}$$

Obviously, Equation (4.2) overspecifies D^0 since the ion mobilities and bulk concentrations are fixed quantities. In physical terms, this means that the dipole strength D^0 cannot be prescribed in a manner that will insure conservation of all N ionic species at the interface.

5. THE INTERMEDIATE PROBLEM: MASS TRANSFER BOUNDARY LAYERS

The inner and outer solutions do not match because there is a region between them that, *a priori*, is difficult to see. In the outer problem, ion transport is controlled by electromigration and convection; in the inner problem, electromigration and diffusion prevail. When, in the outer region, r tends asymptotically toward unity, or when, in the inner region, w goes to plus or minus infinity, a transition, or intermediate, region is encountered; in the intermediate region all three transport modes (convection, diffusion, and electromigration) are important. A match is possible with the outer and intermediate solutions, and with the intermediate and inner solutions, but not with the inner and outer solutions.

If y is defined as

$$r = 1 + y/\beta, \tag{5.1}$$

then at distances which are $O(a/\beta)$ from the interface, the ion concentrations of a simple z-z electrolyte are governed by

$$\frac{1}{2} Pe \frac{\omega^+ + \omega^-}{\omega^+ \omega^-} \left(v_2(y,\theta) \frac{\partial n^\pm}{\partial y} + u_1(\theta) \frac{\partial n^\pm}{\partial \theta} \right) = \frac{\partial^2 n^\pm}{\partial y^2}, \tag{5.2}$$

where

$$u_1(\theta) \equiv -A \sin\theta \cos\theta \tag{5.3}$$

and

$$v_2(y,\theta) \equiv 2Ay\left(\frac{3}{2}\cos^2\theta - \frac{1}{2}\right). \tag{5.4}$$

The diffusion processes, which were subordinate in Equation (3.1), are now comparable to convection and electromigration; the effect of electromigration on the ion distribution is buried in the coefficient of the convective terms on the left-hand side of Equation (5.2). Concentration variations, therefore, develop from mass-transfer effects, as well as from electrostatic interactions with the induced surface charge.

Equation (5.2) has the form of the boundary layer equations for laminar forced-convection heat and mass transfer. Acrivos & Goddard (1965) have discussed at length the treatment of equations of this type and we have drawn on their work to obtain the intermediate solution.

6. SUMMARY OF RESULTS

We have found that once the mass-transfer boundary layers of the intermediate region are resolved, the inner, intermediate and outer solutions can be matched. The principal results are the expressions for the dipole coefficient D^0 and the "apparent" peak interfacial velocity U, viz.

$$D^0 = \frac{R - 1}{R + 2}, \tag{6.1}$$

and

$$U = \frac{9/10}{1 + \overline{\mu}/\mu} \frac{R - \overline{\epsilon}/\epsilon}{(R+2)^2}, \tag{6.2}$$

with

$$R = \frac{(\overline{\omega}^+ + \overline{\omega}^-)\overline{n}_\infty}{(\omega^+ + \omega^-)n_\infty} = \frac{\overline{\sigma}_\infty}{\sigma_\infty}. \tag{6.3}$$

The dipole strength D^0 and the peak surface velocity U are not easy to measure in the laboratory; the drop shape and the extent of deformation under the imposed field are more accessible experimentally. As we noted earlier, the drop shape can be calculated from the solution for the circulation produced in a drop that remains spherical. Let the total deformation Δ_0 be defined by

$$\Delta_0 = \frac{l_1 - l_2}{l_1 + l_2} < 1 \ , \tag{6.4}$$

where l_1 and l_2 are the drop axes parallel and perpendicular to the applied field, respectively. Then the normal stress balance implies

$$\Delta_0 = \frac{9}{16}\, Ca\, \frac{f_d\,(R\,,\overline{\epsilon}/\epsilon,\overline{\mu}/\mu)}{(R+2)^2}, \tag{6.5}$$

with

$$f_d\,(R\,,\overline{\epsilon}/\epsilon,\overline{\mu}/\mu) = 1 - 2\frac{\overline{\epsilon}}{\epsilon} + R^2 + \frac{3}{5}(R - \frac{\overline{\epsilon}}{\epsilon})\,\frac{2+3\overline{\mu}/\mu}{1+\overline{\mu}/\mu}\,. \tag{6.6}$$

The function f_d discriminates between drop shapes; viz.:

$$f_d \begin{cases} > 0, & \text{prolate spheroid;} \\ = 0, & \text{sphere;} \\ < 0, & \text{oblate spheroid.} \end{cases} \tag{6.7}$$

7. DISCUSSION AND CONCLUDING REMARKS

The singular perturbation solution of the electrokinetic model is represented schematically in Figure 2; the analysis hinges on the changes in the modes of ion transport and in the electrostatics. Equation (6.1) for the dipole strength of the drop, Equation (6.2) for the peak interfacial velocity, and Equation (6.5) for the drop deformation are equivalent to the results reported by Taylor (1966) and Torza *et al.* (1971).

The observable drop response predicted with the electrokinetic model indicates that the differences between theory and experiment do not stem simply from a failure to allow for space-charge accumulations in the governing equations. Indeed, the details provided by the electrokinetic model make clear that the leaky dielectric *does* incorporate at least some of the effects due to the diffuse charge. With the leaky dielectric, a space charge distribution is not evident; nevertheless, insofar as the macroscopic behavior is concerned, the net influence of the space and surface charge is properly evaluated.

As was stated in the introduction, one way to understand why the electrokinetic and leaky dielectric models agree is by recognizing that the space charge distribution is very narrow relative to the surface curvature. Thus, effects associated with the finite thickness of the intermediate and inner regions are not noticeable to the would-be observer. For example, the magnitude of the tangential electrical stress imposed on the drop surface differs from that given by the leaky dielectric. However, because the diffuse charge layers are thin, electro-osmosis resulting from the space charge contributes viscous stresses that compensate *exactly* for the difference. The aggregate mechanical response, then, is the same irrespective of the models.

The advantage of the electrokinetic model over the leaky dielectric is that the individual charge-carrying species are accounted for separately. While this does complicate the analysis, it also enables physicochemical processes that cannot be included in the leaky dielectric to be considered. This is likely to be an important attribute of the model since the deformation calculations and the experimental findings remain at odds. The solution outlined here provides a basis for additional theoretical developments.

ACKNOWLEDGEMENT

Support for this work was provided by NASA contract NAG3-259, as part of the NASA PACE program, administered through the Lewis Research Center, Cleveland, OH.

REFERENCES

A. Acrivos & J. D. Goddard, 1965. "Asymptotic expansions for laminar forced-convection heat and mass transfer. Part 1. Low speed flows", *J. Fluid Mech.* **23**, 273-291.

R. S. Allan & S. G. Mason, 1962. "Particle behaviour in shear and electric fields. I. Deformation and burst of fluid drops", *Proc. R. Soc. London* **A267**, 45-61.

M. Barmatz, 1982. "Overview of containerless processing technologies", *Mater. Res. Soc. Symp. Proc.* **9** , 25-37.

J. C. Baygents & D. A. Saville, 1988. "The migration of charged drops and bubbles in electrolyte gradients: diffusiophoresis", *PhysicoChem. Hydrodynam.* **10**, 543-560.

D. E. Brooks & S. Bamberger, 1982. "Studies on aqueous two phase polymer systems useful for partitioning of biological materials", *Mater. Res. Soc. Sym. Proc.* **9**, 233-40.

D. E. Brooks, K. A. Sharp, S. Bamberger, C. H. Tamblyn, G. V. F. Seaman, & H. Walter, 1984. "Electrostatic and electrokinetic potentials in two polymer aqueous phase systems", *J. Colloid Interface Sci.* **102**, 1-13.

R. M. Ehrlich & J. R. Melcher, 1982. "Bipolar model for traveling-wave induced non-equilibrium double-layer streaming in insulating liquids", *Phys. Fluids* **25**, 1785-93.

M. Fixman, 1980. "Charged macromolecules in external fields. 2. Preliminary remarks on the cylinder", *Macromolecules* **13**, 711-716.

M. Fixman & S. Jagannathan, 1983. "Spherical macroions in strong fields", *Macromolecules* **16**, 785-99.

S. A. Griffiths & F. A. Morrison Jr., 1979. "Low Peclet number heat or mass transfer to a drop in an electric field", *Trans. Am. Soc. Mech. Engrs, C, J. Heat Transfer* **101**, 484-88.

M. L. Hair & D. Landheer, 1982. "Particle charge in nonaqueous dispersions", *Am. Chem. Soc. Symp. Ser.* **199** , 313-25.

S. Kuo & F. Osterle, 1967. "High field electrophoresis in low conductivity liquids", *J. Colloid Interface Sci.* **25**, 421-8.

A. L. Loeb, P. H. Wiersema & J. Th. G. Overbeek, 1961. **The Electrical Double Layer Around a Spherical Colloidal Particle**. Cambridge, MIT Press. 375 pp.

J. R. Melcher, 1973. "Electrohydrodynamics", *Proc. IUTAM 13th Int'l. Cong. Theo. Appl. Mech.*, Moscow Univ., 1972, eds. E. Becker & G. K. Mikhailov, pp. 240-63.

J. R. Melcher & G. I. Taylor, 1969. "Electrohydrodynamics: a review of the role of interfacial shear stresses", *Ann. Rev. Fluid Mech.* **1**, 111-46.

F. A. Morrison Jr., 1977. "Transient heat and mass transfer to a drop in an electric field", *Trans. Am. Soc. Mech. Engrs, C, J. Heat Transfer* **99**, 269-73.

D. C. Rau & E. Charney, 1983. "High-field saturation properties of the ion atmosphere polarization surrounding a rigid, immobile rod", *Macromolecules* **16**, 1653-1661.

C. T. O'Konski, 1976. "Electric birefringence and relaxation in solutions of rigid macromolecules", in **Molecular Electro-Optics. Part 1: Theory and Methods**, ed. C. T. O'Konski, pp. 63-120. Dekker, New York. 528 pp.

W. K. Rhim, M. M. Saffren, & D. D. Elleman, 1982. "Development of electrostatic levitator at JPL", *Mater. Res. Soc. Symp. Proc.* **9**, 115-20.

D. A. Saville, 1977. "Electrokinetic effects with small particles", *Ann. Rev. Fluid Mech.* **9**, 321-37.

T. S. Simonova, I. T. Gorbachuk, & S. S. Dukhin, 1972. "Nonlinear electrophoresis of solid spherical particles", *Fiz. Tverd. Tela, Kiev Gos. Pegag. Inst.* , pp. 113-23.

S. Stotz, 1978. "Field dependence of the electrophoretic mobility of particles suspended in low-conductivity liquids", *J. Colloid Interface Sci.* **65**, 118-30.

G. I. Taylor, 1966. "Studies in electrohydrodynamics. I. The circulation produced in a drop by an electric field", *Proc. R. Soc. London* **A291**, 159.

S. Torza, R. G. Cox & S. G. Mason, 1971. "Electrohydrodynamic deformation and burst of liquid drops", *Philos. Trans. R. Soc. London* **269**, 259-319.

FIGURE CAPTIONS

Figure 1. The circulation produced in a drop by an electric field. The applied field is directed parallel to the axis, and the streamlines show the sense of the flow is from pole to equator (Taylor 1966). The numbers denote values of the stream function normalized to $A/2$.

Figure 2. Schematic results for the singular perturbation analysis of the electrokinetic model when $\beta \gg 1$, with κ and $\bar{\kappa} \gg \beta/a$. Regions are delineated on the basis of ion transport. In region I: the outer problem, the dominant modes are convection and electromigration. In region II: the intermediate problem, convection, diffusion and electromigration are all significant. Diffusion and electromigration balance in region III: the inner problem. Regions I and II are electrically neutral, region III contains the diffuse charge layer.

LABORATORY MEASUREMENTS OF THE EFFECT OF CHARGE ON COALESCENCE AND TEMPORARY COALESCENCE OF SMALL RAINDROPS

Robert R. Czys and Harry T. Ochs III
Climate and Meteorology Section, Illinois State Water Survey, 2204 Griffith Drive, Champaign, IL 61820

ABSTRACT

The first measurements of the effects of charge on the coalescence of small precipitation drops initially at terminal velocity in widely separated pairs have been obtain in a laboratory experiment. Observations were taken for a small drop (190 μm radius) with a constant positive charge interacting with a larger drop (340 μm radius) with both positive and negative charge. Data were also obtained for both drops carrying minimal charge and for a case with the small drop having a larger positive charge. Coalescence, bounce, temporary coalescence and temporary coalescence with satellites were observed. The results indicate that $|Q_R - Q_r|$ may be adequate to characterize the outcomes of drop collisions. New features of charged drop interactions were also observed: a unique impact angle dividing the coalescence from the non-coalescence region that is independent of charge; a two-order of magnitude range of relative charge between drop bounce and complete charge-induced coalescence; a wide range of charge where a satellite drop occurs with temporary coalescence; and at higher charge levels, the beginning of charge-induced permanent coalescence at the most grazing collisions.

1. INTRODUCTION

It is well established that charge and/or electric field can aid in the coalescence of water drops. Rayleigh (1879) observed that drops bouncing in the breakup of a water jet coalesced in the presence of a weakly electrified metal rod. A similar influence has been found in more contemporary studies involving drop-streams impacting flat water surfaces (Jayaratne & Mason 1964), drop-streams impacting other drop-streams (Sartor & Abbott 1972; Brazier-Smith *et al.* 1972) and drops impacting a suspended drop (List & Whelpdale 1969; Whelpdale & List 1971). All have found that sufficient levels of charge will cause bouncing drops to coalesce. However, because of the sensitivity of drop interaction results to experiment conditions, a wide range ($3x10^{-14}$ to $4000x10^{-14}$ Coulombs) of charges required to cause coalescence has been measured. The difficulty of applying previous results to rain producing processes has been demonstrated by discrepancies between results using uncharged drop streams or supported drops and those using drops in free fall at terminal velocity (Ochs *et al.* 1986; Ochs & Beard 1984; Beard & Ochs 1983). The results of the present investigation further serve to emphasize that experiments using drops in free fall at terminal velocity are necessary since the charge required to induce coalescence differed from previous results at similar drop sizes and size ratios by an order of magnitude and new features of charged drop interactions were observed.

This paper presents the first observations of the influence of charge on coalescence for collisions between isolated pairs of dissimilarly sized drops in free fall at terminal velocity. More than 5500 interactions between isolated 340 and 190 μm radius drops were photographed which provided 915 recorded collisions during 13 experiment runs. No data exist in this size area for drops falling freely at terminal velocity. The range of charges used in this experiment coincide with charges observed in nature: about $1x10^{-16}$ C for drops in developing cumulus to about $1x10^{-12}$ C for drops in thunderstorms (Takahashi 1972). Observations were taken using both positive and negative charges on the large drop and fixed positive charge on the small drop. Collisions between drops with like sign were emphasized since drizzle drops in clouds are likely to have the same polarity (Takahashi 1972). Observations were also taken with minimal charge on both drops to establish the uncharged collision results of the drop pair.

2. EXPERIMENT APPARATUS AND METHODS

A computer controlled, dual drop generator system, advanced from a system described by Ochs *et al.* (1986), was used to photograph collisions between isolated, dissimilarly sized drops falling freely at terminal velocity using incandescent and stroboscopic light. Figure 1 is an orthographic view of the dual-generator/dual-camera system and shows the position of the drop generators relative to the high voltage electrodes and arrangement of the photographic equipment. Collisions occurred at laboratory temperature and pressure within a humidified, plexiglas chamber 1 m tall and 10 cm on each of four sides (not shown in Figure 1). The impact angle of each collision was calculated from the distance separating drops measured from the photographic record of two cameras which viewed the collisions orthogonally. Collision outcomes were stratified according to ascending order of impact angle and the critical impact angle for coalescence was determined from the stratified list for each relative drop charge studied.

The computer-controlled drop interaction system used in this study has several improvements over the system described by Ochs *et al.* (1986). For example, the computer facilitated the use of two drop generators which permitted study of drop collisions at size ratios smaller than could be produced with the single drop generator system described by Ochs *et al.* (1986). Their investigation was limited to drop size ratios greater than 0.65, since they required that the two dissimilar drops, produced at equal velocities from a single jet, interact at near terminal velocity. Drop interactions for this investigation have been extended to smaller size ratios by producing small and large drops from separate water jets with different diameters and flow rates. Thus by choosing an appropriate flow rate for each generator, both drops were produced with velocities very near to their terminal speed. Hence, only short distances, a few tens of centimeters, were required for either drop to reach within 5% of its terminal value.

Another improvement over the system used by Ochs *et al.* (1986) was the multiplexing of up to three cameras with three strobe lights, which was made possible by computer control of the camera shutter synchronized to drop production. It was possible to compensate for different shutter opening lags in software and assure simultaneously open shutter and strobe flash for every interaction event that was recorded.

To adequately model the forces acting during drop collisions in clouds, laboratory drop collisions must occur with nearly vertical initial trajectories. To minimize the angle between vertical fall trajectories the drops must be generated at nearly the same location in space. This is not a problem when producing drops of different size from a single water jet. However, it does become a problem if the horizontal spacing between two drop generators becomes too large. Thus, small drop generators were designed that permit the water jets to be spaced approximately 1 cm apart. The generators were positioned approximately 120 cm above the location of the drop collisions, and there was no more than 0.5° between the vertical fall trajectories.

The principle of drop generation is similar to that described by Adam *et al.* (1971). The design was based on a hollow cylindrical piezoelectric transducer that vibrates radially. Figure 2 shows a vertical cross-section of the generator. Water first passes through a flow constriction, then through the transducer chamber and

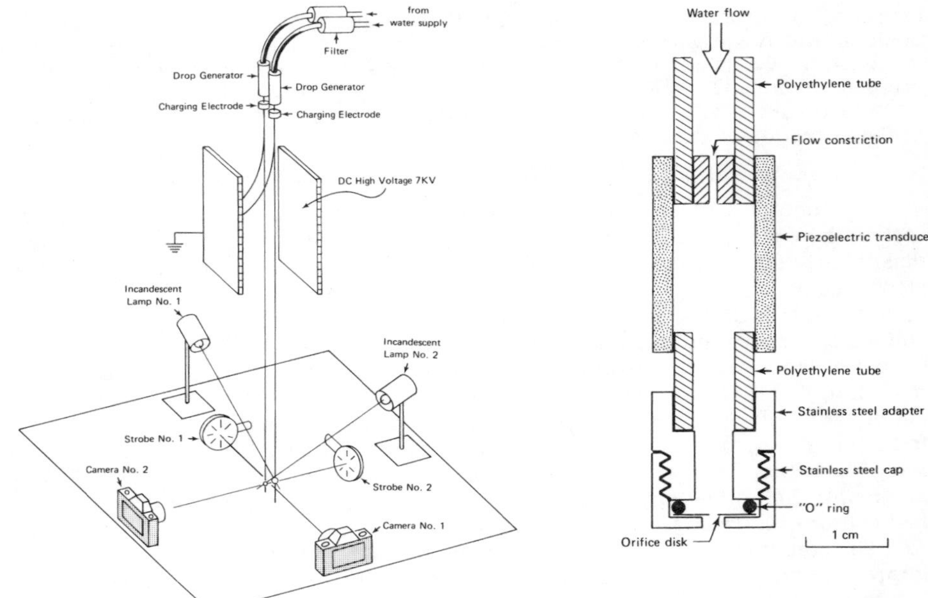

Figure 1

Figure 2

finally through an orifice disk seated in an electrically grounded stainless steel cap. This design results in strong coupling between transducer vibration and capillary waves on the liquid jet. Thus, stable streams of uniformly sized drops could be produced for long periods of time.

The spherical drop radius was calculated from the jet flow rate and the drop generation frequency. The jet flow rate was measured by directing one of the drop streams into a clean, dry polyethylene cup for 2 minutes. The cup was then weighed and the flow rate calculated. The mass of a single drop was obtained to within 1% from division of the flow rate by drop generation frequency. The same procedure was used to determine the drop size in the other drop stream.

Drops for collision were selected from streams of drops highly charged by a cylindrical electrode around the point of jet breakup (see Figure 1). Deionized water was used with sodium nitrate in a concentration of 8.75 mg l^{-1} to increase the electrical conductivity with negligible effect on surface tension. The highly charged drops were deflected by a strong horizontal electric field between parallel electrodes and collected in a gutter. A switching circuit controlled by the computer periodically imposed a pulse of prescribed duration and height on the drop charging voltage. Coincidental with this pulse, individual drops (i.e., pulsed-out drops) were produced that had lower charge and thus fell between the high voltage electrodes and into the interaction section of the experiment chamber. The computer controlled the time between the periodic selection of a small drop from one generator and a large drop from the other to produce isolated repetitive pairs of drops that interact in the field of view of the cameras. Micromanipulators were used to independently adjust the angle of the jets to optimize the number of collisions.

The charge on the drops was measured often during the course of observation using a laboratory-built electrometer and computer interfaced digital storage oscilloscope. From laboratory tests of the electrometer it was determined that drop charges as small as 5×10^{-15} C could be detected. However, a practical lower limit of 10^{-15} C exists for measured charge. As drops pulsed-out from only one of the drop streams would strike the electrometer sensor time-voltage traces would appear on the oscilloscope screen that could be digitally stored on command. The area beneath each trace was proportional to the charge carried by the drop. The digital oscilloscope traces were retrieved by computer for immediate computation of drop charge. Thus charge levels were monitored closely during the course of experiment runs and adjustments made as warranted. The drop charge record was later used to exclude data from analysis when charge on either drop varied by more than 10% between charge measurements.

Interactions were recorded by two cameras which viewed the drops from orthogonal directions (see Figure 1). Incandescent lamps angled 30° above each camera's optical axis illuminated the falling drops from behind. The light focuses to a spot on the front surface of each drop and was recorded on the film as the drops fell past the camera. Thus, two streak-lines were created on the film that corresponded to the path followed by each drop.

The distance between the streak-lines represents a projection of the true horizontal distance separating the drops since the vertical plane defined by the fall trajectories was usually randomly oriented. The projected horizontal drop separations recorded by each camera were measured (±2 μm) on the film just above the point of impact using a microscope with a micrometer eyepiece.

Digital recording databacks imprinted the time of interaction to the nearest second on the film and thus allowed unambiguous matching of the frames of film from each camera. The true horizontal distance separating the drop centers just before impact is given by the relationship, $x = (x_1{}^2 + x_2{}^2)^{1/2}$, where x_1 and x_2 are the orthogonally projected distances between drop centers before impact. The impact angle Φ defines the acute angle between the vertical and a line joining the centers of the drops at initial impact. The impact angle was calculated from, $\Phi = \sin^{-1}(x/(R+r))$, where R and r are the spherical radii of the large and small drops, respectively. The ordered pairs of computed impact angles and collision results were sorted according to ascending order of impact angle. A critical impact angle, Φ_c, for each relative charge was determined as the angle which distinguished between collisions that resulted in coalescence and those that resulted in noncoalescence.

The computer was also used to control the operation of strobe lamps positioned behind the interactions and facing into the camera lens along the optical axis (see Figure 1). The intensity of the strobe and the incandescent light were balanced so that the light from the strobe flash would not obscure the sharp definition of the streak lines. The electronic controls allowed very fine adjustment to cause the strobe to flash a few milliseconds after drop collision. Hence, a back-lit image was obtained and used to determine the number and size of any satellite drops that may have been produced from the collision.

III. RESULTS

This experiment is composed of 13 runs, each conducted at a different combination of drop charges. Table 1 gives a summary of the parameters for each experiment. Listed by experiment run are the number of observed collisions (N), the large drop size and its charge (R and Q_R), small drop size and charge (r and Q_r), mean relative drop charge (Q_R-Q_r) and critical impact angle Φ_c. Critical impact angle is not listed for runs 8 and 9 since charge levels were sufficient to cause almost all collisions to result in permanent coalescence.

A minimal amount of charge was used on both drops in run 1 as a control observation. The positive charge on the small drop was held nearly constant for 11 of the runs (2 through 12) and averaged 2.5×10^{-14} C with standard deviation of 5×10^{-15} C. For the 11 runs with fixed small drop charge, a positive charge was used on the large drop in 8 runs and a negative charge in 3 runs. The last run listed in Table I was conducted with an increased small drop charge and an intermediate large drop charge to test the hypothesis that collision results can be characterized by absolute mean relative charge and impact angle.

Table I.
Summary of Experiment Parameters.
Charge values have been rounded to three significant figures or less.

Experiment	N	R μm	r μm	Q_R $\times 10^{-14}$C	Q_r $\times 10^{-14}$C	Q_R-Q_r $\times 10^{-14}$C	Φ_c deg.
1	66	329	193	1.0	-0.1	1.1	44
2	37	340	191	3.9	3.0	0.9	42
3	45	341	193	0.2	2.1	-2.0	44
4	81	342	192	11.0	2.3	8.7	42
5	42	331	186	35.2	1.8	33.4	44
6	77	336	192	41.0	1.7	39.4	44
7	93	340	191	84.4	2.8	81.5	43
8	50	341	197	129.7	3.8	125.4	—
9	160	339	192	208.0	2.9	205.0	—
10	47	339	198	-3.5	2.3	-5.7	40
11	45	338	197	-19.6	2.3	-21.9	42
12	42	351	195	-99.2	3.0	-102.1	44
13	130	341	194	40.1	23.4	16.7	42

Three basic types of drop collision results were observed: coalescence, bounce and temporary coalescence. Examples of the resulting streak photographs for each are shown in Figure 3. The corresponding orthogonal frames are not shown and the imprinted time has been cropped. Examples without the low intensity strobe flash have been chosen to clearly illustrate the streak lines.

A miss (Figure 3a) occurs when the large drop overtakes and does not hit the small drop. A coalescence (Figure 3b) occurs when two drops collide and permanently unite to form a single drop. Waviness of the streak line below the collision point indicates oscillations of the coalesced drop.

Figure 3c shows a bounce. (Brazier-Smith *et al.* 1973) have suggested that bounce occurs when the drops impact and the air between them is not sufficiently expelled to allow their surfaces to make contact. The rate of air film drainage decreases as the distance separating the drops decreases, because the viscous force opposing drainage increases. The drainage rate is further reduced because the deformation of the near drop surfaces increases the area of the air film. If the interaction time is short compared to the drainage time, the air film will not be sufficiently thinned for the drops to make contact and the drops will bounce apart.

Temporary coalescence (Figure 3d and 3e) occurs when the drops make contact but only temporarily form a single drop because the resulting rotational energy is sufficient to pull them apart (Brazier-Smith *et al.* 1972). In the case of temporary coalescence, the interaction time is comparable to or larger than the air film drainage time and the drops contact. Upon coalescing a drop of dumbbell shape rotates around its center of mass. If the centrifugal force opposing merger is sufficient, the drop pulls apart and stretches a filament of water. When the filament breaks, two new drops are formed that have a horizontal component of velocity proportional to the speed of rotation. An analysis of these horizontal velocities (Czys 1987) shows that the streak lines for temporary coalescence should intersect below the point of impact (see Figure 3d and 3e).

A feature sometimes seen with a temporary coalescence signature is the appearance of a third streak-line as shown in Figure 3e. This streak-line was created by a satellite drop. If the water filament, produced during the final stage of temporary coalescence, breaks in two places a satellite drop is formed. An average satellite radius of 80 μm was measured from the frames with a strobe exposure. No more than one satellite drop was ever observed with temporary coalescence.

Collision results for the 340 and 190 μm radius drops used in this experiment are displayed as a function of impact angle (Φ) and magnitude of mean relative charge in Figure 4. This figure was constructed from the data for positively charged drops (runs 2 through 9). Two general areas of collision results can be seen in Figure 4, one for coalescence (unhatched) and the other for noncoalescence (hatched). Within each run that showed a charge influence there was no clear organization in the occurrence of temporary coalescences and bounces with impact angle. The division between coalescence and noncoalescence collision results was the only clearly discernible boundary in the data (see Czys 1987). Thus, the regions in Figure 4 are schematic and specific efficiencies for various collision outcomes as a function of charge level could not be determined.

Figure 4 clearly shows an abrupt transition from collisions that result in coalescence to collision results depending on charge at a critical impact angle of 43° $\pm 1 \Phi$ for the drop sizes used in this experiment. The error bracket for the critical angle encompasses the entire transition region from coalescence to noncoalescence for the eight runs that were used to construct Figure 4. The critical impact angle corresponds to a coalescence efficiency of 47%. Collisions occurring for impact angles less than critical always resulted in coalescence regardless of the amount of charge on the drops.

For impact angles greater than 43°, the magnitude of the mean relative charge must exceed a value of about 2×10^{-14} C before charge will influence the result of collision. Below this value of charge all collisions with impact angle ≤43° resulted in bounce (denoted by hatching sloping down from left to right in Figure 4). Above

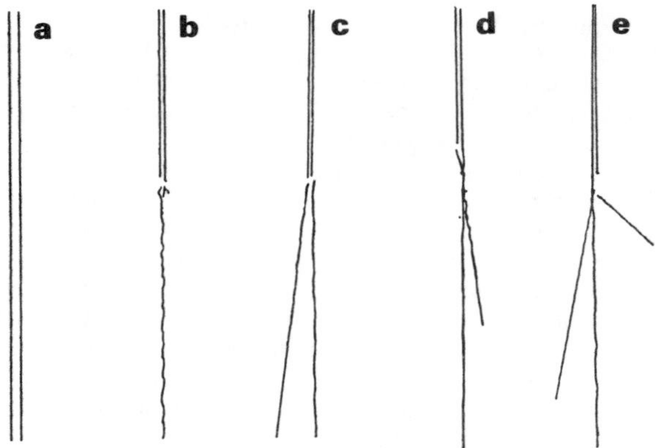

Figure 3

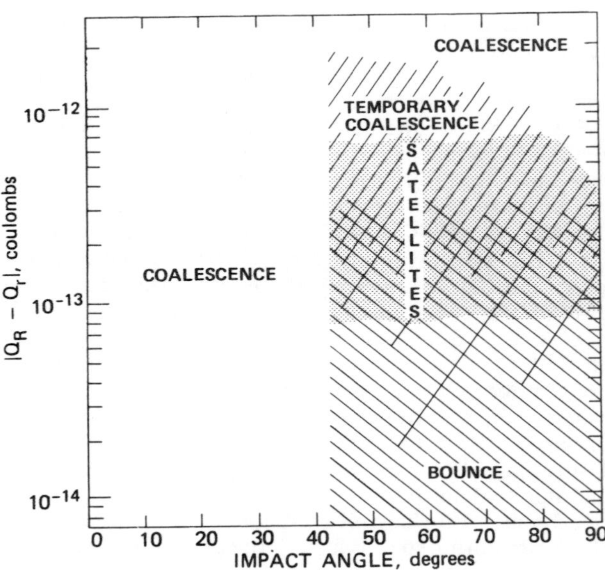

Figure 4

about $2x10^{-14}$ C, temporary coalescence (denoted by hatching sloping up from left to right) occurred with increasing frequency. In runs 3 and 4, temporary coalescence was infrequent with only one temporary coalescence in run 3 and two in run 4. However in run 5, with a mean relative charge of about $3x10^{-13}$C, bounce was almost completely suppressed in favor of temporary coalescence (see Czys 1987). These trends are reflected in the way the hatching overlap has been drawn in Figure 4.

The shaded area in Figure 4 covers the impact angles and charges where a satellite drop could be produced with temporary coalescence. The likelihood that a satellite drop would be produced with temporary coalescence decreased with increasing mean relative charge. As shown in the top right portion of Figure 4, when charge reached a level just sufficient to begin to cause permanent coalescence the data show a clear trend for coalescence to initiate at glancing impact angles and progress toward the critical impact angle with increasing charge.

Three runs were conducted using about the same mean charge on the small drop as in runs 2-9, but with negative charge on the large drop. Large drop charges were selected so that the results of each run could be compared to the results for positively charged drops to determine if $|Q_R - Q_r|$ can be used to characterize the results. The data were ranked according to $|Q_R - Q_r|$ for runs 2 through 11 and the details of the results of runs 10 through 12 compared to those for runs 2 through 9.

The collision results using oppositely charged drops were found to be generally consistent with the results using positively charged drops. Runs 10 through 12 have about the same critical impact angle, fit the overall trend for bounce to be replaced initially by temporary coalescence and do not contradict the tendency for charge induced coalescence to begin for the most glancing collisions. Two inconsistencies were found in this comparison. First, run 12 had satellite production when the data for positively charged drops suggested none and, second, run 10 was the only case in the series of observations reported with Φ_c slightly outside the range 43° ≥1°.

Two additional runs were conducted with a charge on the small drop different from that used in runs 2 through 12. Run 1 was conducted with minimum charge on each drop. The results of this run clearly indicated that the collision results for "uncharged" drops are no different from the results for weakly charged drops where collision results are divided between coalescence and bounce by the critical impact angle.

Finally, run 13 was conducted with an increased charge on the small drop and intermediate charge on the large drop to provide further information on using $|Q_R - Q_r|$ to generalize the results. As can be determined from Table 1, the absolute value of the difference in mean relative charge for run 13, falls in a region where the data suggest a transition from mostly bounce to mostly temporary coalescence with a satellite. Data from run 13 had a critical impact angle of 43° and had a mixture of temporary coalescence and bounce that could be expected from the trends in the data from runs 4, 5 and 11. In contrast to the other runs in which charge influenced collision outcome, the results of run 13 (see Table II) have a high degree of organization of collision outcome with impact angle. Collision results showed temporary coalescence with a satellite drop at impact angles between critical and about 6° followed by temporary coalescence without a satellite drop. Bounce events occurred between about 56° and 74°. The degree of this organization with greater charge on the small drop may have resulted from less drop-to-drop charge variability that led to more repeatable collision results.

To summarize, the data from runs 1 through 13 was found to organize reasonably well with the absolute value of the mean relative charge and impact angle as displayed in Figure 4. However, more data are needed before these results can be generalized to other combinations of drop sizes and charges.

IV. DISCUSSION

Previous data on the influence of charge on drop interaction have not come from experiments using freely falling drops at terminal velocity. Observed threshold charges have ranged over three orders of magnitude. The lowest value of charge for coalescence, $3\text{x}10^{-14}$ C, was reported by Jayaratne and Mason (1964) for a stream of drops impacting a flat water surface. The reflected drop stream was observed to flicker as the charge neared threshold, thus, indicating temporary coalescence. Whelpdale and List (1971) have reported the highest value of charge required for coalescence, $3\text{x}10^{-11}$ C, for droplets striking a supported drop from beneath. They observed a progression from bounce, to temporary coalescence and finally to coalescence with successive increases in relative charge. Their data also showed no significant difference between results for negatively charged drops and for oppositely charged drops.

Table II
Collision Results for Run 13
Ranges of impact angles where chosen to correspond to equal concentric projected collision cross-section areas. Collision results are given for coalescence (C), bounce (B), temporary coalescence without a satellite (T) and temporary coalescence with a satellite (TS). The number preceding each type of collision result indicates occurrences in each range of impact angle.

Impact Angle (Degrees)	Collision Result
0.0 - 11.5	9C
11.5 - 16.4	8C
16.4 - 20.3	8C
20.3 - 23.6	2C
23.6 - 26.6	9C
26.6 - 29.3	5C
29.3 - 31.9	12C
31.9 - 34.4	7C
34.4 - 36.9	6C
36.9 - 39.2	6C
39.2 - 41.6	2C
41.6 - 43.9	4TS
43.9 - 46.1	1T,4TS
46.1 - 48.4	7TS
48.4 - 50.8	2TS
50.8 - 53.1	3TS
53.1 - 55.6	3TS
55.6 - 58.1	3B,3TS
58.1 - 60.7	2B,2TS
60.7 - 63.4	4B
63.4 - 66.4	2B,1T
66.4 - 69.7	1B,2T
69.7 - 73.6	3B,2T
73.6 - 78.5	5T
78.5 - 90.0	2T

Intermediate charge thresholds have been reported from studies of collisions between two streams of drops. Park (1970) has discussed the unpublished results of Howarth and Crosby obtained for two similarly sized drop streams angled at each other. These data showed a dependence of critical charge on impact velocity and drop size. A critical charge of $3\text{x}10^{-13}$ C was reported for 350 μm radius drops. A similar threshold can be inferred from the drop stream experiments of Sartor and Abbott (1972).

Our data for freely falling drops at terminal velocity demonstrate that effects of charge on drop coalescence are significantly more complicated than indicated by previous laboratory observation. For example, charge-induced temporary coalescence has been observed (Jayaratne and Mason 1964; Park 1970, and Whelpdale and

List 1971), but only Brazier-Smith *et al.* (1973) observed satellite production with temporary coalescence. An aspect of the results which is not yet understood is that the critical impact angle dividing permanent coalescence from bounce or temporary coalescence is independent of charge. Since drops colliding with impact angles just less than 43° always coalesce, it is reasonable to expect that drops colliding with impact angles just greater than 43° almost coalesce. Thus, as charge increases, coalescence should first be induced for collisions near the critical impact angle, and the critical impact angle should steadily increase with charge until permanent coalescence occurs for the largest impact angle. The new data clearly prove this reasoning to be incorrect.

Our data also show that the lowest charge to induce permanent coalescence occurs at the largest impact angles (i.e., when the drops follow grazing trajectories). This, too, is unexpected since collisions with grazing trajectories generally have the shortest interaction time and the largest angular momentum; conditions conducive to temporary coalescence.

The results of this investigation also show that coalescence was promoted by charge even though both drops carry positive charge. This result is consistent with the findings of Sartor and Abbott (1972) who found that the charges on the drops caused increased coalescence regardless of the relative sign. Electric field theory (Davis 1964) indicates that when two conducting spheres carrying the same polarity of net charge are placed in close proximity, the sphere with the largest charge can induce the opposite charge on the near surface of the other. For a size ratio of 0.5, the charge on the larger sphere must exceed the net charge on the smaller by a factor of about three to induce an area of opposite charge. Therefore, in the area of the trapped air film the oppositely charged drop surfaces may enhance air drainage by electrostatic attraction (Foote 1974).

An electrical instability caused by a strong electric field between the near drop surfaces may also aid coalescence (Brazier-Smith *et al.* 1972). We have applied the theoretical treatment of an electrically-induced instability of a flat liquid surface given by Melcher (1963) using conditions for deformable drops. The dispersion relation describing the relationship between frequency and wave length for disturbances on a liquid surface is:

$$\omega^2 = k^2(V_g^2 + V_c^2 - V_b^2) \tag{1}$$

where, V_g, V_c and V_b are the phase velocities of gravity, capillary and electrohydrodynamic waves. This equation shows that a surface instability will develop when $V_b^2 > (V_g^2 + V_c^2)$. Hence it can be used to estimate the charge required to aid coalescence.

$$V_g^2 = g/k \tag{2}$$

$$V_c^2 = \sigma k/\rho_w \tag{3}$$

$$V_b^2 = \epsilon_o E^2 \coth(kb)/\rho_w \tag{4}$$

where, g is gravity, k the wavenumber, σ the surface tension of water, ρ_w the density of water, ϵ_o the permittivity of air, E the electric field strength, and b the minimum air film thickness just before the drop surfaces rupture and coalescence begins.

Scale analysis shows that V_c^2 and V_b^2 are comparable in magnitude and much larger than V_g^2. Thus, the critical condition for the development of an electrically enhanced surface instability is met when $V_c^2 = V_b^2$.

The relationship, $\epsilon_o E = Q_R/A$, was used in Equation 4 to expresses the phase velocity in terms of charge on the large drop, Q_R, where the surface charge density is, Q_R/A and $A = \pi a^2$ is the area of drop deformation with radius, a, over which the charge is distributed. Hence, the critical charge on the large drop for surface instability is given by:

$$Q_R = [(\sigma \varepsilon k \pi^2 a^4)/\coth(kb)]^{1/2} \quad (5)$$

In order to evaluate Equation 5 and obtain an order of magnitude estimate of the charge sufficient to aid coalescence it is necessary to make assumptions regarding how the charge on the large drop distributes, about the wavelength of the disturbance to obtain a value of k and about the thickness of the air film when drop contact initiates. For our calculation, all of the charge on the large drop was assumed to concentrate within the area of deformation and the opposite charge was assumed to be imaged on the facing surface of the small drop. A radius of deformation of half the radius of the smaller drop (*i.e.*, $a = r/2$) was deduced from high speed motion pictures of the drop collisions. A value of $b = 0.1$ μm (Whelpdale and List 1971) was chosen as the air film thickness when contact is initiated.

If the instability wave number, k, is restricted by the diameter of deformation, the calculated charge required for instability is $2.7\text{x}10^{-13}$ C. This value is consistent with the charge in our data for transition from bounce to temporary coalescence. The extent of agreement between this calculated value and our results suggests that an electrically induced instability may aid coalescence initiation for charged drop collisions with deformation. Clearly, this calculation of threshold charge is sensitive to the fraction of total charge that is assumed to concentrate in the area of deformation, becomes smaller or larger with selection of the instability wave number and can be about an order of magnitude smaller or larger depending on whether the air film ruptures at a thickness of 0.01 μm or at 1.0 μm. Hence, even with this uncertainty in the calculation, the agreement with observation suggests that a charge induced instability of the drop surface may be responsible for initiating coalescence of colliding drops.

V. CONCLUSION

A computer-controlled experiment has allowed the first study of the influence of charge on coalescence for collisions between dissimilarly sized water drops in free fall at terminal velocity. Previous experiments on the influence of charge on drop coalescence have not adequately modeled the conditions of drop collisions in clouds, and previously reported values of drop charge required to cause coalescence have ranged over three orders of magnitude. This wide range of charges suggests that experiment conditions may have contributed to the differences. Our data for freely falling drops showed a constant critical impact angle separates collisions that always result in coalescence from collisions with charge dependent outcomes. Collisions with impact angles greater than critical had results that were charge dependent. Our data also indicate that the outcome of drop collisions varies from bounce, to temporary coalescence with a satellite, to temporary coalescence, to coalescence over the natural range of drop charges found in clouds. Such variability suggests important consequences for precipitation evolution, particularly since moderate amounts of charge can induce temporary coalescence with the production of a satellite drop which, in turn, can serve as an embryo for a new precipitation drop. Clearly, additional experiments at different combinations of drop sizes and charges and with imposed electric fields are needed before the effect of a cloud's electrical state on the evolution of precipitation can be adequately evaluated.

ACKNOWLEDGEMENTS

The authors wish to thank Kenneth Beard for his design of the cylindrical drop generator shown in Figure 2.

This research was supported by the National Science Foundation under grant NSFATM 8601549.

REFERENCES

(1) J. R. Adam, R. Cataneo & R. G. Semonin, 1971, "The Production Of Equal And Unequal Size Droplet Pairs," *Rev. Sci Instrum.* **42**, 1847-1849.

(2) K. V. Beard & H. T. Ochs, "Measured Collection Efficiencies For Cloud Drops," *J. Atmos. Sci.* **40**, 146-153, 1983.

(3) P. R. Brazier-Smith, S. G. Jennings, & J. Latham "Raindrop, Interactions And Rainfall Rates Within Clouds," *Qrtly. J. R. Met. Soc.* **99**, 260-272, 1973.

(4) P. R. Brazier-Smith, S. G. Jennings, & J. Latham, "The Interaction Of Falling Water Drops: Coalescence," *Proc. R. Soc. London* A**326**, 393-408, 1972 .

(5) R. R. Czys, "A Laboratory Study Of Interactions Between Small Precipitation-Size Drops In Free Fall," Ph. D. thesis, University of Illinois, 131 pp., 1987.

(6) M. H. Davis, "Two Charged Spherical Conductors In A Uniform Electric Field: Forces and Field Strength," *Qrtly. J. Mech. Appl. Math.*, **27**, 499-511, 1964.

(7) B. G. Foote, "A Theoretical Investigation Of The Dynamics Of Liquid Drops". Ph. D. thesis, University of Arizona, 205 pp., 1974.

(8) O. W. Jayaratne & B. J. Mason, "The Coalescence And Bouncing of Water Drops At An Air/Water Interface," *Proc. R. Soc. London* A**280**, 545-565, 1964.

(9) J. R. Melcher, "A Preliminary Investigation Of Factors Affecting The Coalescence Of Colliding Water Drops," *J. Atmos. Sci.* **26**, 305-308, 1963.

(10) J. R.Melcher, **Field-Coupled Surface Waves: A Comparative Study Of Surface-Coupled Electro-Hydrodynamic And Magnetohydrodynamic Systems**. Cambridge MA, M.I.T. Press, 190 pp., 1963.

(11) H. T. Ochs, R. R. Czys, & K. V. Beard, "Laboratory Measurements Of Coalescence Efficiencies For Small Precipitation Drops," *J. Atmos. Sci.*, **43**, 225-232, 1986 .

(12) H. T. Ochs & K. V. Beard, "Laboratory Measurements Of Collection Efficiencies For Accretion," *J. Atmos. Sci.*, **40**, 863-867.

(13) R. W. Park, 1970 "Behavior of Water Drops Colliding in Humid Nitrogen," Ph. D. thesis, University of Wisconsin, 577 pp., 1984.

(14) Lord Rayleigh, "The Influence Electricity On Colliding Water Drops," *Proc. R. Soc. London* **28**, 406-409, 1879.

(15) J. D. Sartor & C. E. Abbott , "Some Details Of Coalescence And Charge Transfer Between Freely Falling Drops In Different Electrical Environment," *J. Rech. Atmos.*, **6**, 479-493, 1972.

(16) T. Takahashi, "Electric Charge Of Cloud Droplets and Drizzle Drops In Warm Clouds Along The Mauna Loa-Mauna Kea Saddle Road Of Hawaii Island," *J. Geophys. Res.*, **77**, 3869-3878,1972.

(17) D. M. Whelpdale, & R. R. List, "The Coalescence Process In Raindrop Growth," *J. Geophys. Res.*, **76**, 2836-2856, 1971.

FIGURE CAPTIONS

Figure 1 Orthographic view of experiment equipment.

Figure 2 Cross section of cylindrical water drop generator.

Figure 3 Streak strobe signatures for (a) miss, (b) coalescence, (c) bounce, (d) temporary coalescence without satellite and (c) temporary coalescence with satellite.

Figure 4 Collision outcome as a function of impact angle and absolute mean relative charge. The charge on the small drop was fixed at 2.5×10^{-14} C and the large drop charge varied from about 1.5×10^{-15} to 2.1×10^{-12} C as indicated in Table 1. The clear area delineates the region of coalescence. The hatched area sloping down from left to right indicates where bouncing occurs. The hatched area sloping up from left to right delineates the region of temporary coalescence. The stippled area covers the drop charges and impact angles for which satellite production is likely if temporary coalescence occurs. The concentration of hatched area overlap suggests the relative occurrence of bounce and temporary coalescence in the transition region.

NEAR-CONTACT HYDRODYNAMICS OF TWO VISCOUS DROPS

Robert H. Davis
Department of Chemical Engineering and
The Center for Low Gravity Fluid Mechanics
and Transport Phenomena,
University of Colorado Boulder, CO 80309-0424

ABSTRACT

The hydrodynamic force resisting the relative motion of two drops moving along their line-of-centers is determined for Stokes flow conditions. The drops are assumed to be in near-contact and to have sufficiently high interfacial tension that they remain spherical. The squeeze flow in the narrow gap between the drops is analyzed using lubrication theory, and the flow within the drops near the axis of symmetry is analyzed using a boundary integral technique. The two flows are coupled through the nonzero tangential stress and velocity at the interface.

Depending on the ratio of drop viscosity to that of the continuous phase, and also on the ratio of the distance between the drops to their reduced radius, three possible flow situations arise, corresponding to nearly rigid drops, drops with partially mobile interfaces, and drops with fully mobile interfaces. The results for the resistance functions are in good agreement with an earlier series solution using bispherical coordinates. The new results for near-contact motion have important implications for droplet collisions and coalescence.

The theory is also extended to consider the normal motion of a drop toward a solid boundary. As expected, the resistance to this motion is less than that experienced by a solid particle moving toward a solid boundary. In particular, the force on a drop as it becomes very close to the boundary approaches one-fourth of that on a solid sphere with the same size and relative velocity.

I. INTRODUCTION

The relative motion of droplets of one fluid dispersed in a second, immiscible fluid plays an important role in a variety of natural and industrial processes, including rain drop formation, liquid-liquid extraction, and liquid phase miscibility gap materials processing. In addition, the relative motion of a drop toward a solid boundary is important in droplet pushing and capture during directional solidification of liquid phase miscibility gap materials.

In order to determine the nature of the singularity in the hydrodynamic resistance to relative motion as two drops become close, we consider here a novel method for analyzing the near-contact interaction. Lubrication theory is used to solve for the flow in the narrow gap separating the drop interfaces, and boundary integral theory is used for the flow within the drops near the axis of symmetry. As the drops move together, the resulting radially-outward squeeze flow in the gap exerts a tangential shear stress on the drop interfaces. This causes a tangential motion of the drop surfaces and drives a flow in the drops. This in turn has a significant effect on the magnitude of the force resisting the approach of the drops. Knowledge of the dependence of this force on the drop sizes, the drop and continuum phase viscosities, the relative velocity and the distance between the drops is needed for predicting criteria for coalescence and capture.

II. THEORETICAL DEVELOPMENT

Two spherical drops of radii a_1 and a_2, respectively, and equal viscosity $\lambda\mu$, move toward one another along their line-of-centers with relative velocity $W = V_1 - V_2$ through an immiscible fluid of viscosity μ, as depicted in Figure 1a. It is assumed that the closest separation h_o between the drop surfaces is small compared with both a_1 and a_2. The force resisting the relative motion is therefore dominated by a small lubrication region (sketched in Figure 1b) near the axis of symmetry. Our objective is to determine the velocity and pressure profiles in this region, and then to use them to calculate the hydrodynamic force on the drops. Since inertia is assumed small, this force must balance any external or interparticle forces. In a later section, the development is extended to the case of a drop approaching a solid plane boundary.

II.A. LUBRICATION FLOW IN THE GAP

The key difference between the present problem and that for solid spheres is that the tangential velocity may be nonzero at the drop surfaces. It proves convenient to account for this by decomposing the radial velocity profile into two parts: $u(r,z) = u_t(r) + u_p(r,z)$, where $u_t(r)$ is a uniform flow, and $u_p(r,z)$ is a parabolic flow driven by the local pressure gradient and satisfying $u_p = 0$ at the interfaces ($z = z_1, z = z_2$). Therefore, u_p is given by

$$u_p = \frac{1}{2m}\frac{\partial p}{\partial r}(z - z_1)(z - z_2) \ . \tag{1}$$

The tangential stress exerted by the fluid in the gap on each of the drop interfaces is thus:

$$f_t(r) \equiv \mu\frac{\partial u}{\partial z}\bigg|_{z=z_1^+} = -\mu\frac{\partial u}{\partial z}\bigg|_{z=z_2^-} = -\frac{h}{2}\frac{\partial p}{\partial r} \ . \tag{2}$$

where p is the dynamic pressure. The dynamic pressure drives the flow of fluid out of the gap and has its maximum value at $r = 0$ and a value of zero at $r \to \infty$.

The pressure distribution in the fluid between the drops may be obtained by integrating Equation (2):

$$p(r) = 2\int_r^\infty (f_t(r)/h(r))dr \tag{3}$$

Once the dynamic pressure is known, the magnitude of the hydrodynamic force resisting the relative motion of the drops may be obtained by integrating the pressure distribution:

$$F = 2\pi\int_0^\infty p(r)rdr \ , \tag{4}$$

because, for $h_o/a \ll 1$, the pressure is large compared to the z-component of the shear stress on the drop interfaces in the close contact region. The close contact region has a radial length scale of $\sqrt{ah_o}$, since this is the radial distance for which $h(r)$ changes by $O(h_o)$. The upper limit of Equations (3) and (4) should be in the range $\sqrt{ah_o} \ll r \ll a$, but since f_t, p for $r \gg \sqrt{ah_o}$, it is valid to replace this by $r = \infty$. Near the axis of symmetry, the spherical surfaces of the drops may be approximated as paraboloids, and so the film thickness profile for undeformed drops is:

$$h(r) \equiv z_2 - z_1 = h_o + r^2/2a \ , \tag{5}$$

where $a = a_1 a_2/(a_1 + a_2)$ is the reduced radius.

Finally, a mass balance on the fluid being squeezed out of the gap between the two approaching drops yields:

$$\pi r^2 W = 2\pi r\int_{z_1}^{z_2} u\,dz = 2\pi r\left\{hu_t + \frac{h^2}{6\mu}f_t\right\} \ . \tag{6}$$

Equations (1) and (2) were used in performing this integration.

II.B. BOUNDARY INTEGRAL THEORY FOR FLOW IN DROPS

Equation (6) represents one relationship between the unknown tangential velocity, u_t, and tangential shear stress, f_t, at the droplet interfaces. A second such

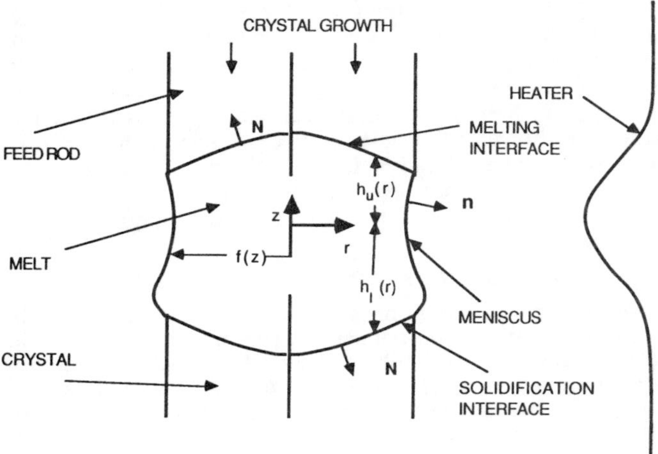

Figure 1

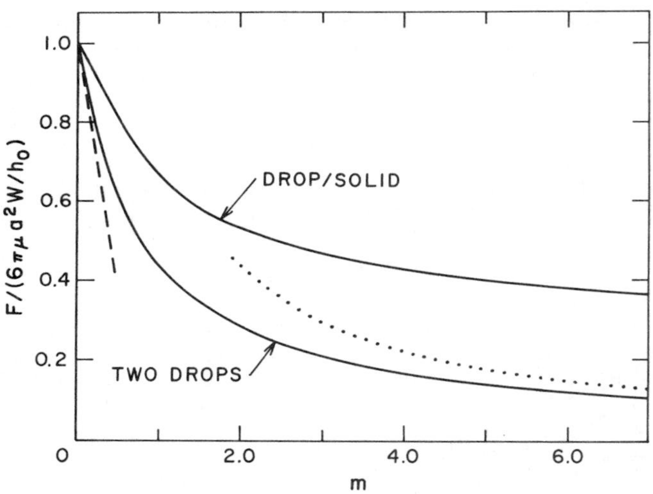

Figure 2

relationship must be found by considering the flow in the droplet phase. Because information is needed only at the drop surfaces, the solution is expedited by using the boundary integral form of the Stokes flow equations (Rallison & Acrivos (1978), Ladyzhenskaya (1969)):

$$\boldsymbol{u}(\boldsymbol{r}) = \frac{1}{8\pi\lambda\mu}\iint_S \left\{\frac{\boldsymbol{I}}{d} + \frac{\boldsymbol{dd}}{d^3}\right\} \cdot \boldsymbol{f}(\boldsymbol{r}')dA' + \frac{3}{4\pi}\iint_S \boldsymbol{u}(\boldsymbol{r}') \cdot \frac{\boldsymbol{ddd}}{d^5} \cdot \boldsymbol{n}\, dA' \ , \qquad (7)$$

where S represents the surface bounding the flow field, $\boldsymbol{n}$ is the unit normal to S, $\boldsymbol{r}$ is a position vector specifying a point within the flow field, $\boldsymbol{r}'$ is the variable of integration and specifies a point on S, $\boldsymbol{d} \equiv \boldsymbol{r} - \boldsymbol{r}'$, $d \equiv |\boldsymbol{d}|$, $\boldsymbol{f}$ is the stress vector, and $\boldsymbol{I}$ is the second-order unit tensor. If $\boldsymbol{r}$ is on S, then the left-hand-side of Equation (7) should be multiplied by 1/2, as a result of the well-known jump condition.

On the drop interfaces, the tangential components of $\boldsymbol{u}$ and $\boldsymbol{f}$ are non-negligible only within an $O(ah_o)^{1/2}$ radius of the axis of symmetry. Since, when $h_o/a \ll 1$, this is small compared to the radius of curvature of either drop, the interfaces may be treated as flat for the purpose of solving for the drop flow field in the near-contact region. By restricting our attention to $\boldsymbol{r}$ on the interface and taking advantage of the flow being axisymmetric, the tangential component of Equation (7) becomes

$$u_t(r) = \frac{1}{\lambda\mu}\int_0^\infty \phi(r',r)\, f_t(r')dr' \ , \qquad (8)$$

where

$$\phi(r',r) = \frac{1}{2\pi}\frac{r'}{\left(r^2 + r'^2\right)^{1/2}}\int_0^\pi \frac{\cos\theta d\theta}{\left(1 - k^2\cos\theta\right)^{1/2}} \qquad (9)$$

is an elliptic-type Green's function kernel having an integrable (logarithmic) singularity at $r' = r$ ($k = 1$, where $k^2 = 2rr'/(r^2 + r'^2)$).

II.C. SCALING CONSIDERATIONS

In the narrow gap between the approaching drops, the axial length scale is h_o the radial length scale is $\sqrt{ah_o}$ the axial velocity scale is W, and the radial velocity scale is $W\sqrt{ah_o}$ (as required by mass conservation). Using these to nondimensionalize the governing equations, the lubrication force on the drops is found to depend on the single dimensionless parameter,

$$m \equiv \lambda^{-1}\sqrt{a/h_o} \ . \qquad (10)$$

The physical significance of m may be deduced from the continuity of tangential stress across each interface

$$\mu\ddot{c} + \alpha\dot{c} + \beta c = 0\mu\ddot{c} + \alpha\dot{c} + \mu\left.\frac{\partial u}{\partial z}\right|_{z=z_2^- \text{ or } z_1^+} = \mu\left.\frac{\partial u_p}{\partial z}\right|_{z=z_2^- \text{ or } z_1^+} = \lambda\mu\left.\frac{\partial u}{\partial z}\right|_{z=z_2^+ \text{ or } z_1^-} c = 0 \qquad (11)$$

For flow within the drops, the relevant length scale for *both* the radial and axial directions is $\sqrt{ah_o}$. From Equation (11), it follows that

$$u_t/u_p = O(\lambda^{-1}\sqrt{a/h_o}) = O(m) \ , \qquad (12)$$

Therefore, when $m \ll 1$, $u_t \ll u_p$ and the interfaces are nearly immobile; whereas when $m \gg 1$, $u_t \gg u_p$ and the interfaces are essentially fully mobile. Thus, m has the character of an *interface mobility*.

We note that the mobility considered here is purely a hydrodynamic effect involving the ratio of viscosity between the continuous and drop phases. It is assumed that there are no temperature gradients or adsorbed species on the interfaces which give rise to a surface tension gradient.

III. RESULTS AND DISCUSSION

The technique which we have used is to solve Equations (6) and (8) simultaneously for the tangential velocity distribution, $u_t(r)$, and the tangential stress distribution, $f_t(r)$. Once this was accomplished, Equations (3) and (4) were used to find the dynamic pressure distribution, $p(r)$, and the hydrodynamic lubrication force, F.

III.A. NEARLY RIGID DROPS ($m \ll 1$)

When the drops are very viscous relative to the continuous phase ($\lambda \gg \sqrt{a/h_o}$), then they behave as nearly rigid particles. The radial squeeze flow in the gap is dominated by the parabolic portion of the flow ($u_p = O(W\sqrt{a/h_o})$); the uniform portion of the flow is small ($u_t = O(mu_p) = O(\lambda^{-1}Wa/h_o)$). For $m \to 0$, the rigid sphere solution (Jeffrey, 1982) is recovered:

$$u_t = 0, \qquad f_t = 3\,\mu rW/h^2, \qquad p = 3\,\mu aW/h^2, \qquad F = 6\pi\mu a^2 W/h_o \tag{13}$$

When m is small but nonzero, Equations (2)-(9) admit a regular asymptotic expansion in powers of m. The first few terms of the expansion for the lubrication force are:

$$\frac{F}{6\pi\mu a^2 W/h_o} = 1 - 1.31m + 1.78m^2 - 2.46m^3 + 3.44m^4 - 4.83m^5 + O(m^6). \tag{14}$$

III.B. DROPS WITH FULLY MOBILE INTERFACES ($m \gg 1$)

When the viscosity of the drops is comparable to or smaller than that of the continuous phase ($\lambda \ll \sqrt{a/h_o}$), then the drops offer relatively little resistance to the radial flow in the gap. The uniform portion of the radial velocity profile in the gap ($u_t = O(W\sqrt{a/h_o})$) dominates over the parabolic portion ($u_p = O(m^{-1}u_t) = O(\lambda W)$).

As $m \to \infty$, the magnitude of the nearly flat velocity profile in the gap is given directly by the mass balance (Equation (6)):

$$u_t = Wr/h \tag{15}$$

Using this expression in Equation (8) yields a Fredholm integral equation of the first kind that can be inverted to solve for $f_t(r)$. However, it is possible instead to develop an alternate expression that directly gives the stress field as an integral involving the velocity distribution on the boundary (Jansons & Lister (1988), Davis *et al.* (1989)). Using either approach, it is found that the hydrodynamic force resisting the relative motion of the drops is

$$F/(6\pi\mu a^2 W/h_o) = 0.876/m, \qquad \text{or} \quad F = 16.5\,\lambda\,\mu aW\sqrt{a/h_o}, \tag{16}$$

There are two notable features of this result. First, the hydrodynamic resistance on the drops is proportional to the drop viscosity rather than the surrounding fluid viscosity. Second, the relative velocity under the action of a constant applied force decreases only in proportion to $h_o^{1/2}$, rather than in proportion to h_o as in the rigid sphere case, and so coalescence can occur in a finite time without the requirement of an attractive force (such as the London-van der Waals force) that increases in magnitude as the gap-size decreases.

III.C. DROPS WITH PARTIALLY MOBILE INTERFACES ($m = O(1)$)

When λ is of order $\sqrt{a/h_o}$, then the velocity scales for both the uniform and parabolic portion of the radial flow in the gap are $O(W\sqrt{a/h_o})$, and the drop flow and the gap flow are fully coupled. Our solution method was to substitute Equation (6) into Equation (8) and then to numerically solve the resulting Fredholm integral equation of the second kind for $f_t(r)$. The lubrication force resisting the relative motion of the drops was then calculated and is plotted in dimensionless form in Figure 2 as a function of the interface mobility, m. Also shown are the limiting forms for $m \ll 1$ and $m \gg 1$. A Padé-type approximate expression, which is accurate to within one or two percent for all values of m, is

$$\frac{F}{6\pi\mu a^2 W/h_o} = \frac{1 + 0.38m}{1 + 1.69m + 0.43m^2} \tag{17}$$

In Figure 3, the results are recast as a plot of the hydrodynamic resistance vs. the distance between the two drops for different values of the viscosity ratio, λ. For comparison, the exact completed results using bispherical coordinates are also shown. The latter were obtained computationally by using the analysis of Haber *et al.* (1973)

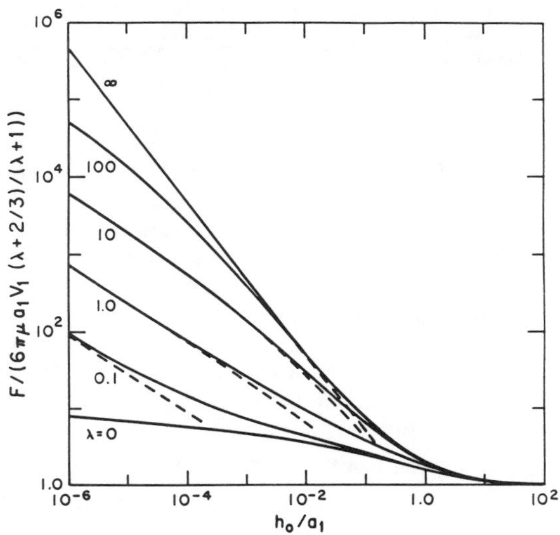

Figure 3

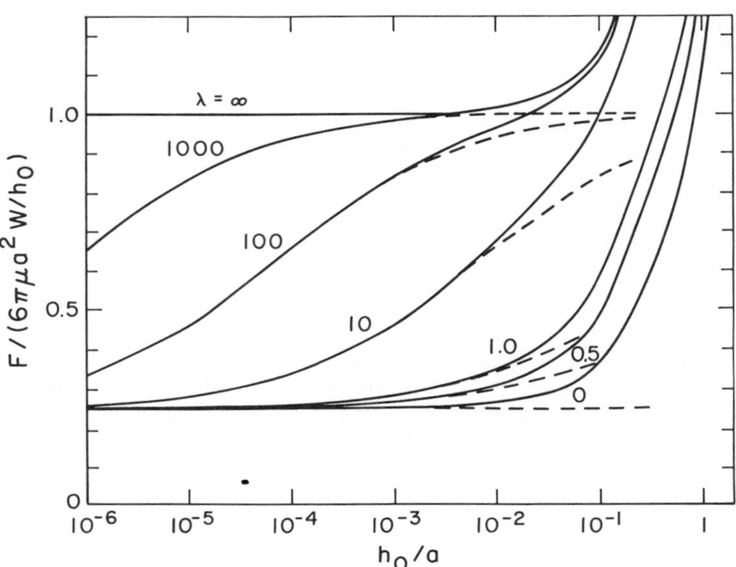

Figure 4

and keeping a large number of terms in order to achieve convergence when the gap was made small. The symmetric case of $a_1 = a_2 = 2a$ and $V_1 = -V_2$ was chosen for these computations. The hydrodynamic resistance force is made dimensionless using the Hadamard-Rybczynski formula for an isolated drop.

For very viscous drops ($\lambda \gg 1$), an interesting feature of the curves shown in Figure 3 is that the slope of these log-log plots changes from -1 to -1/2 as h_o/a decreases. This represents the transition from the drops exhibiting rigid sphere behavior for moderate separations to their interfaces becoming mobile at very small separations. The new lubrication results agree with the exact results when h_o/a is sufficiently small. For very viscous drops with immobile or partially mobile interfaces ($\lambda \geq O\,(ah_o)^{1/2}$), this is expected to be the case when $h_o/a \ll 1$. For drops of moderate relative viscosity and fully mobile interfaces ($\sqrt{h_o/a} \ll \lambda \ll \sqrt{a/h_o}$), the more stringent condition $\sqrt{h_o/a} \ll \lambda$ is required. Finally, when $\lambda \leq O\,(\,h_o/a)^{1/2}$, such as may be the case for two interacting gas bubbles, then the lubrication force does not dominate over the contribution to the force in the outer region, and the near-contact results described in this paper do not apply.

IV. INTERACTION OF A DROP AND A SOLID BOUNDARY

We now consider the case where drop 2 is replaced by a solid phase. In the limit of $a_2 \to \infty$, our problem then becomes that of a drop moving toward a solid plane boundary. The development presented earlier still holds, except that the no-slip condition is applied on the solid surface. The pressure-driven velocity profile in the gap is then given by

$$u = \frac{1}{2\,\mu}\frac{\partial p}{\partial r}\,(z - z_1)\,(z - z_2) + u_t\,\frac{z_2 - z}{z_2 - z_1}\,, \tag{18}$$

here u_t is the tangential velocity at the drop interface.

The pressure force pushing the fluid out of the gap between the drop and the solid must balance the sum of the tangential shear forces on the drop and solid surfaces (the two tangential stresses are equal only for two drops of equal viscosity). Expressions for these stresses are obtained by differentiating Equation (18) with respect to z. The pressure gradient may be eliminated using the mass balance as before by substituting (18) into the integral in Equation (6). Finally, the required relationship between the tangential stress and velocity at the drop interface is given by Equation (8), the boundary integral equation applied on the drop boundary. The details of this solution are given by Barnocky and Davis (1989). The final result for the force on the drop approaching the solid is

$$\frac{F}{6\pi\mu a^2 W / h_o} = \frac{1}{4} + \frac{3}{4}\,F'(m) \tag{19}$$

where $F'(m)$ is obtained numerically by inverting the boundary integral equation for the tangential stress on the drop surface, integrating Equation (3) with $2f_t$ replaced by the sum of the tangential stresses on the drop and solid surfaces in order to obtain the pressure in the gap, and then using Equation (4) to obtain the hydrodynamic force resisting the relative motion. This result is also shown in Figure 2. A Padé-type approximate expression is given by Barnocky and Davis (1989):

$$F'(m) = \frac{1 + 0.25m}{1 + 1.13m + 0.19m^2} \tag{20}$$

A key result to note is that, when the drop is sufficiently close to the solid so that $\sqrt{a/h_o} \gg \lambda$, the drop interface is fully-mobile ($m \ll 1$) and the force on the drop to leading-order is one-fourth that on a rigid sphere approaching a solid—independent of the viscosity of the drop. This is shown in Figure 4 where the dimensionless lubrication force given by Equation (19) is plotted vs. the dimensionless gap between a spherical drop of radius a and a solid plane boundary for various values of the ratio of drop viscosity to continuous-phase viscosity. The solid lines are the series solution of Wacholder and Weihs (1972), and the dashed lines are the present lubrication results. In the limit as $h/a \to 0$, the force asymptotes to one-fourth of that for a solid sphere ($\lambda = \infty$) regardless of whether the drop is a bubble ($\lambda = 0$) or has a finite viscosity. This conclusion was also drawn by Beshov *et al.* (1978) by examining the limiting form of the series solution as $h/a \to 0$. Finally, it is seen in Figure 4 that highly viscous drops ($\lambda \gg 1$) behave as rigid spheres for sufficiently large separations ($h_o/a \gg \lambda^{-2}$) but that they behave as bubbles with freely mobile interfaces for very small separations ($h_o/a \ll \lambda^{-2}$).

V. CONCLUDING REMARKS

The hydrodynamic force resisting the axisymmetric relative motion of two spherical drops, or of a drop and a solid, in close contact has been determined using a combination of lubrication theory and boundary integral theory. Very viscous drops at moderate separations ($\sqrt{a/h_o} \ll \lambda$) behave as rigid spheres and experience a force resisting their motion that is inversely proportional to the minimum distance between the drop surfaces. In contrast, significant tangential motion of the drop interfaces occurs due to the radial squeeze flow in the gap for drops with moderate relative viscosity at small separations ($\lambda \leq O\,(a/h_o)^{1/2}$). In the limit $\sqrt{h_o/a} \ll \lambda \ll \sqrt{a/h_o}$, the drop interfaces are fully mobile, and the force resisting the relative motion of drops is inversely proportional to the square root of the minimum distance between the drop surfaces. This permits contact between two drops to occur in a finite time when they are subject to a constant force pushing them together, and thus has important implications in droplet coalescence. In contrast, the hydrodynamic resistance force on a drop approaching a solid is always at least one-fourth of that on a solid sphere under otherwise identical conditions, even in the limit of $\lambda \to 0$. A constant applied force is therefore insufficient to push the drop into contact with a solid boundary in a finite time, and an additional mechanism—such as the action of attractive van der Waals forces—is required.

In the analysis, inertia effects have been neglected relative to viscous effects. For the flow inside the drops, this requires that $\rho_d Wa/\lambda\mu \ll 1$, where ρ_d is the density of the fluid comprising the drops. For flow in the narrow gap between the drops, inertia effects are negligible when $Re \equiv (h_o/a) \ll 1$ for the nearly rigid and partially mobile cases, and when $Re\sqrt{h_o/a}/\lambda \ll 1$ for the fully mobile case, where $Re \equiv \rho Wa/\mu$ is the Reynolds number and ρ is the fluid density of the continuous phase.

In addition, it is assumed that the drops remain spherical. In order for any flattening of the drop interfaces to be small relative to the gap size, a normal stress balance on the interfaces requires that $(a/h_o)^2\,Ca \ll 1$ for the nearly rigid and partially mobile cases, and that $\lambda(a/h_o)^{3/2}\,Ca \ll 1$ for the fully mobile case, where $Ca \equiv \mu W/\gamma$ is the capillary number and γ is the interfacial tension. Each of these conditions is met when the drops are sufficiently small ($a \leq 10\ \mu m$, typically). However, when larger drops become close together, it is expected that the pressure which builds up to squeeze the fluid out from between the drop surfaces will also cause deformation of the interfaces. The combination of lubrication theory and boundary integral theory developed in this work will still apply when deformation is important, but the normal stress balance must also be used to infer the gap thickness profile, $h(r,t)$.

ACKNOWLEDGMENT

A large portion of this work was undertaken in the Department of Applied Mathematics and Theoretical Physics at the University of Cambridge while the author was supported by NSF grant CBT-8451014 and NASA grant NAGW-951 from the Microgravity Sciences and Applications Division of the Office of Space Sciences and Applications. Several discussions with E. J. Hinch, J. M. Rallison, and J. A. Schonberg were instrumental in the development of this work. Many of the numerical computations were performed by G. Barnocky.

REFERENCES

G. Barnocky & R. H. Davis (1989) *J. Multiphase Flow* (in press).
V. N. Beshkov, B. P. Radoev & I. B. Ivanov (1978) *J. Multiphase Flow*, **4**, 563.
R. H. Davis, J. A. Schonberg, & J. M. Rallison (1989) *Phys. Fluids A*, **1**, 77.
S. Haber, G. Hetsroni & A. Solan (1973) *J. Multiphase Flow*, **1**, 57.
K. M. Jansons & J. R. Lister (1988) *Phys. Fluids*, **31**, 1321.
D. J. Jeffrey (1982) *Mathematika*, **29**, 58.
A. Ladyzhenskaya (1969) **The Mathematical Theory of Viscous Incompressible Flow**, 2nd ed. (Gordon and Breach), p. 57.
J. M. Rallison & A. Acrivos (1978) *J. Fluid Mech.*, **89**, 191.
E. Wacholder & D. Weihs (1972) *Chem. Eng. Sci.*, **27**, 1817.

FIGURE CAPTIONS

Figure 1 Schematic of two spherical drops in relative motion, including (a) side view, and (b) close-up of near-contact region.

Figure 2 The dimensionless hydrodynamic resistance force as a function of the interface mobility parameter; ____ numerical results; --- asymptotic results for $m \ll 1$ given by the first two terms of Equation (14); ... asymptotic results for $m \gg 1$ given by Equation (16).

Figure 3 The dimensionless hydrodynamic resistance force on two equal drops as a function of the dimensionless gap-size for different viscosity ratios; ____ exact results using bispherical coordinates; --- near-contact results.

Figure 4 The dimensionless hydrodynamic resistance force on a drop approaching a solid plane as a function of the dimensionless gap-size for different viscosity ratios; ____ exact numeral results using bispherical coordinates; --- near-contact results.

EXPERIMENTAL STUDY ON THE INSTABILITY OF CONDUCTIVE DROPLETS

H.A. Elghazaly
Cairo University, Dept. of Electrical Engineering, Giza, Egypt

G.S.P. Castle
The University of Western Ontario, Dept. of Electrical Engineering, Faculty of Engineering Science, London, Ontario N6A 5B9, Canada

ABSTRACT

In previous studies, the authors introduced an analytical model to predict the final state for both single sibling and multi-sibling breakups. In order to verify the validity of this model, charged droplets of water were formed at the tip of capillary tube raised to high potential and subjected to external electric fields. The nozzle characteristics were tested to identify the different ejection modes. Under some conditions of electric field at the nozzle tip, breakup of the ejected droplets resulted. These droplets were collected on water sensitive paper and then examined to estimate the droplets size. The results showed good agreement with the analytical predictions and also supported the validity of the concept of treelike secondary breakups introduced in the model.

INTRODUCTION

It is well known that an electrically charged liquid drop becomes unstable when the electrical force of repulsion exerted by the charges exceeds the surface tension force[1]. The drop then emits one or more highly charged droplets[2-4] and thereby loses both mass and charge. In order to understand this phenomenon, many investigators have carried out both theoretical and experimental studies[5-9].

A mathematical model has been introduced by the authors to predict the final state for both single[10] and multi-sibling breakup[11]. In these studies, a numerical technique, based on scanning all the possible radii of the sibling droplets, was introduced for a drop charged to its Rayleigh limit with known surface tension and initial size. For a single sibling disintegration, the solutions satisfied the conservation of energy as well as the Rayleigh limit criteria for both the residual drop and the sibling. A further condition for minimum final potential energy was used to identify the most probable solution.

This model was extended to model the multi-sibling instability on the basis of treelike secondary breakups which leads to a residual drop and n siblings of different sizes and charges. This extended model clarified the role of the external forces on the breakup process and on the forced disintegration of the drop below its Rayleigh limit.

The demarcation between the modes of single and multi-sibling breakup has been also clarified in terms of the sibling mass ratios (the ratio of the total mass of all the siblings to the initial drop mass). It was predicted that for the case of sibling mass ratios >11.1%, single sibling breakup occurred. For sibling mass ratios <11.1%, multi-sibling breakup was predicted so that for certain ranges of the sibling mass ratio, a number of siblings was estimated. In the present study, the validity of these analytical models has been examined experimentally.

GENERAL DESCRIPTION OF THE APPARATUS

The experimental set-up used is shown schematically in Figure 1. It consists of a

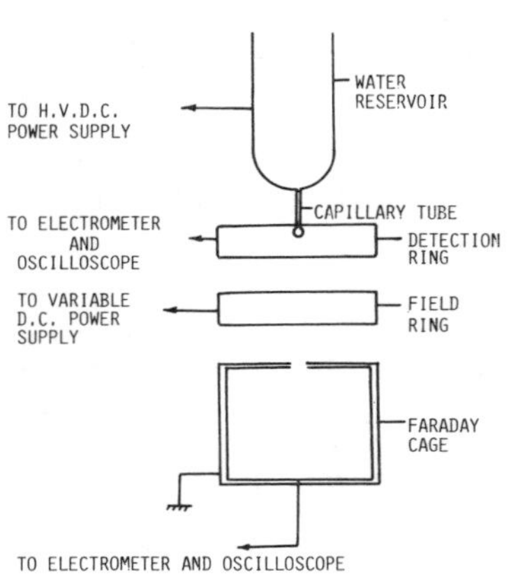

FIGURE (1) SCHEMATIC DIAGRAM OF THE EXPERIMENTAL SET-UP

stainless steel hypodermic capillary tube of 150 um inside diameter, and 450 um outside diameter. Tap water, coloured with dye, was fed from the reservoir to the capillary tube. The height of the reservoir was varied in order to control the hydrostatic pressure, which in this experiment was established to be very close to zero, i.e., no dripping for uncharged liquid. A high voltage power supply was connected to the liquid reservoir to charge the water by conduction. Two identical copper rings, 3.8 cm inner diameter and 1.3 cm height, were mounted separated by a vertical distance of 1.5 cm. The upper edge of the first ring (detection ring) was aligned with the capillary tube tip. This ring was grounded through an electrometer set to its current mode. The electrometer measurement was amplified and traced on an oscilloscope to sense any dripping and breakup of the water drops. Another D.C. power supply was connected to the second ring (field ring). This ring allowed the fine adjustment of the electric field around the end of the capillary tube. It also served the purpose of centering the droplets to be captured by a double shielded Faraday cage located 6.2 cm below the capillary tip. The Faraday cage was also grounded through a similar electrometer and its signals were traced on the second channel of the oscilloscope allowing the number of droplets entering the cage to be counted.

To reduce the external electrical noise, mainly from the 60 Hz pickup, the capillary tube, the two rings and the Faraday cage were mounted in an electrically shielded chamber. A band reject filter (40 - 100 Hz) was connected in the detection circuit between the electrometer and the oscilloscope to minimize the 60 Hz noise.

DETERMINATION OF NOZZLE CHARACTERISTICS

Some preliminary experiments were carried out to determine the characteristics of the drops ejected from the nozzle under different charging voltages and ring voltages (i.e., different fields around the nozzle tip). In each of these experiments either the charging voltage or the ring voltage was kept constant and the other was changed in steps to cover a broad range of drop sizes.

During the experiments, drops which formed at the nozzle tip became charged then fell through the ring system into the Faraday cage. The second channel of the oscilloscope was used to sense the arrival of a drop in the cage and to estimate the time between two consecutive drips. Droplet size was measured by collecting 60 samples of the ejected droplets at each combination of charging and ring voltages. Fifteen samples were collected on water sensitive paper, placed on the top of the Faraday cage, and 45 samples were collected in oil (nondrying immersion oil for microscopy type B). The oscilloscope signal detected by the upper ring was used simultaneously with the collecting process to check that the collected number of droplets for either single sibling or multi-sibling breakups was equal for each sample. If any difference in droplet number occurred, the sample was considered to be faulty and rejected. A total of 60 acceptable samples was obtained for each of the 30 separate experiments carried out. Since the oscilloscope used had the capability of expanding the trace, all the measurements were stored originally at 2 sec/division and then the required part was enlarged to determine the number of siblings.

The collected droplets were examined under a microscope (6.4X to 80X magnifications) to measure the drop size and the spreading factor of the water sensitive paper (w.s.p.). These experiments also served to check the reproducibility of the results and the measurement of the time between two consecutive drips. In the case of a drop disintegration, the initial drop size was determined by measuring the diameter of each individual drop and calculating the total mass of these droplets. This allowed the calculation of the initial drop diameter.

Plate 1 shows examples of typical oscilloscope traces measuring the detection ring current for the three different modes of drop formation, *i.e.*, single drop, breakup and spray.

The interpretation of these traces can be understood by recalling that the current (I) can be represented as $I = dQ/dt = (dQ/ds) \cdot (ds/dt)$, where Q represents charge, t represents time, and s represents the distance between the drop and the detecting ring.

For the single drop case, the current increases to a maximum as the drop separates from the nozzle and the velocity increases (labeled as segment A in the photograph). The current then decreases as the flux component originating from the drop (dQ/ds) decreases as the drop leaves the vicinity of the detection ring (labeled as segment B in the photograph). This trace is characteristic of a single drop with no disruption.

For some combinations of charging and ring voltages, it was found that the droplet would undergo disruption after its formation. This condition can occur for a droplet with charge close to its Rayleigh limit which is perturbed by an external force such as aerodynamic, gravitational or electric (see discussion in earlier paper[11]).

The droplet disruption could be observed visually through the stereo microscope. Attempts to photograph the process were unsuccessful due to the lack of adequate high speed flash equipment. However, the disruption was also observed indirectly on the oscilloscope traces for the detection ring current. Two typical examples are shown in Plate 1. The disintegrations were interpreted as producing a sudden change in the droplet velocity caused by the repulsion between the droplets. This then produces a sudden change in the current superimposed upon the normal decay. By comparison between the oscilloscope traces and the droplets collected on the w.s.p. it was established that the number of siblings could be determined by counting the number of sudden increases in the current trace in segment B of the oscillograph. The numbered sections shown in Plate 1 illustrate examples for five and three siblings respectively.

At higher values of charging voltage individual droplets were no longer formed and a spraying mode developed. This produced a characteristic repetitive signal of markedly higher frequency. A typical example of this condition is illustrated in the last photograph shown in Plate 1.

These three regimes are also illustrated in the results of a typical nozzle characteristic experiment given in Figure 2. This shows that for zero ring voltage the drop size reduced with increasing charging voltage. As the charging voltage was further increased, the drop started to disintegrate after its formation. In this breakup region, the number of collected droplets increased rapidly with increasing charging voltage. As the charging voltage approached a certain value, slightly higher than that required for drop disintegration, the breakup mode changed to the spraying mode. In the spraying mode, the residual drop which remained attached to the needle suddenly became conical shaped ,and very fine droplets were emitted from its tip

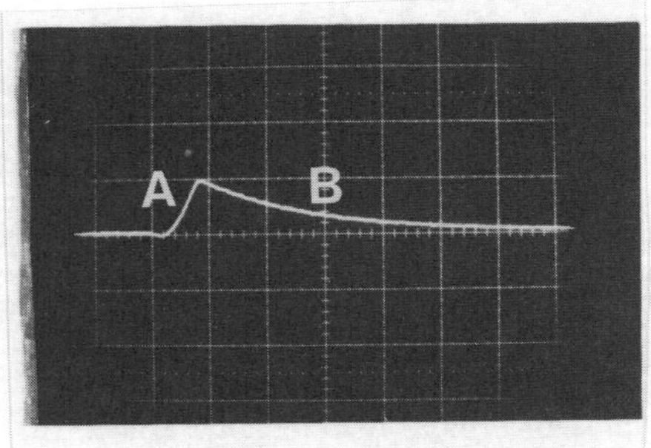

Single Drop (t = 50 ms/div)

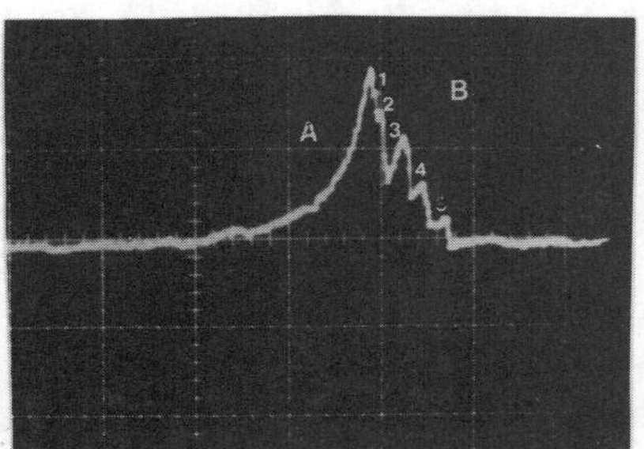

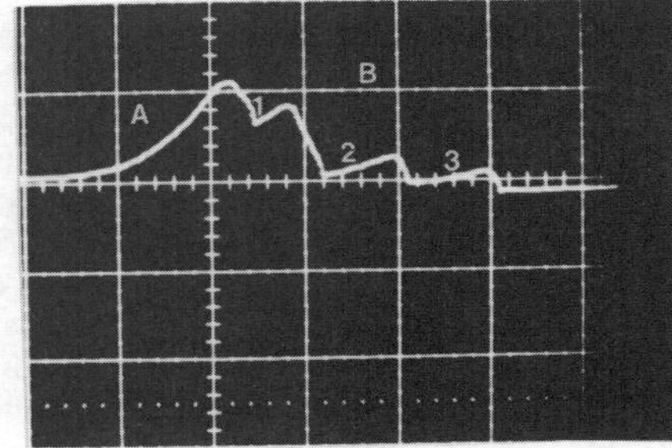

Break-up (t = 50 ms/div)

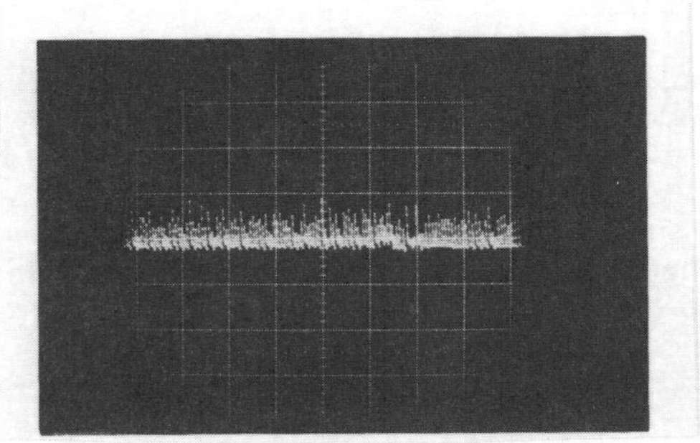

Spray (t = 5 s/div)

PLATE (1) OSCILLOSCOPE TRACES FROM THE DETECTION RING

(in both the single drop and breakup mode, the original drops were ejected directly from the needle tip). Figure 2 also shows that the time between two consecutive drips decreased with increasing charging voltage.

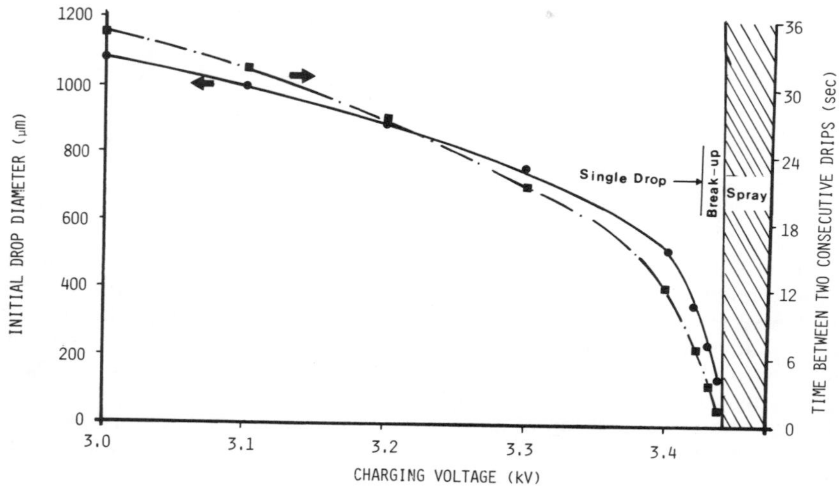

FIGURE (2) NOZZLE CHARACTERISTICS AT ZERO RING VOLTAGE

In order to obtain better control in the breakup mode, the electric field at the capillary tip was varied in precise increments. This was accomplished by setting the charging voltage close to that required for breakup. The ring voltage was then varied in steps, with either positive or negative polarity, until breakup started. This fine adjustment of the ring voltage allowed a fairly wide range of control. The results show that the more positive the ring voltage (the less the electric field concentration at the tip), the larger the drop size. The contours of equal size droplets for different charging and ring potentials are given in Figure 3 which also shows the voltage combinations at which the breakup occurred. These contour lines represent the average results of approximately 1000 data points generated from 18 different conditions of voltage combinations.

The spreading factor of the water sensistive paper was also measured at each step, by comparing the average drop diameter of those collected on w.s.p., with those collected in oil. For the drop sizes used in performing the breakup tests, the spreading factor was found to vary between 2.19 and 2.31 with an average of 2.25. Although the spreading factor showed a trend to increase with the increase of the drop size, for the narrow range of drop sizes used in the breakup tests it was decided to use the average spreading factor in measuring the drop sizes. The w.s.p. was much more convenient than the use of the oil in capturing the droplets and it also offered an extra magnification of about 2.25 which made the measurement of the very fine droplets easier and more precise.

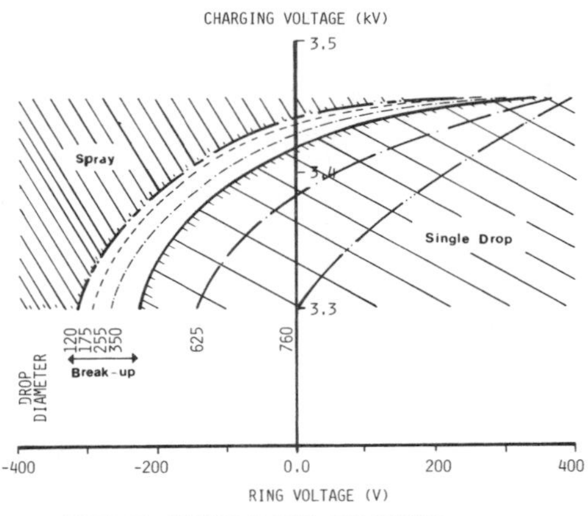

FIGURE (3) CONTOURS OF EQUAL SIZE DROPLETS

The objective of these experiments was to determine the number of siblings for different sibling mass ratio intervals, and compare the results with the theoretical predictions. Both the charging voltage and the ring voltage were changed within the range of values required to produce breakup, as shown in Figure 3, to allow different disintegration conditions. Eleven different combinations of the charging voltage and the ring voltage were selected to cover conditions producing different numbers of siblings and a range of initial drop sizes between 200 μm to 350 μm diameter. In choosing these conditions, the time between two consecutive drips was restricted to be longer than 3 sec. This restriction was the practical limit required to ensure individual collection of drops.

The samples at each voltage combination were collected on w.s.p. The oscilloscope trace was simultaneously stored and examined to count the number of siblings ejected from the initial drop and to verify the collected sample. If any difference in droplet number showed up, the faulty sample was rejected. After the collection process, the individual sibling diameters and the residual diameter were measured for each sample using a microscope of 80X magnification and the sibling mass ratio was calculated.

Of a total of 144 experimental samples collected, 34 samples were rejected because there was either an error in collecting the droplets or because the initial droplet did not disintegrate.

Although many attempts were made to eliminate the experimental errors, some sources of error could not be removed, and are described below:

A) The presence of unpredictable space charge and external acoustic noise near the nozzle may introduce random fluctuations to the external force.

B) The error in measuring normal droplet diameters of about 50 μm using the graticule of the microscope with 80X magnification (±5 μm resolution based on half a division) was in the range of ±10%. The error in measuring the large drops, 200 m or larger, was less than ±2.5%. The error translates to an error of about ±30% in calculating the mass of the normal droplets and about ±7.5% for the large drops. For small siblings less than 50 μm the error in measuring their mass may be considerably higher (±100% for 16 μm droplets). This error will dramatically affect the accuracy of the sibling mass ratios especially at the transition between the different ranges of sibling mass ratios.

Although using the average spreading factor of the w.s.p. was another source of error in calculating the drop mass (± 8%), it offered an extra magnification of

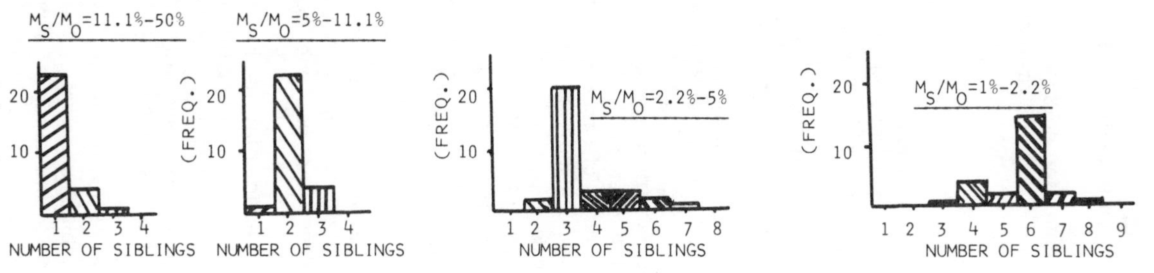

FIGURE (4) THE EXPERIMENTAL RESULTS OF THE DISTRIBUTION OF THE NUMBER OF SIBLINGS FOR DIFFERENT SIBLING MASS RATIOS.

TABLE I

Analysis of experimental results' distribution

Sibling Mass Ratio	Number of Samples	Summary of the Experimental Results: Theoretical Prediction for Number of Siblings	Distribution of the Number of Siblings								Percentage Agreement with Theory
			1	2	3	4	5	6	7	8	
11.1 - 50.0%	28	1	23	4	1	--	--	--	--	--	82%
5.0 - 11.1%	27	2	1	22	4	--	--	--	--	--	81%
2.0 - 5.0%	31	3	--	2	20	3	3	2	1	--	65%
1.0 - 2.0%	24	6	--	--	1	4	2	14	2	1	58%

2.25. This magnification in turn reduced the total measuring error of the drop mass for a normal drop (50 μm) from ±30% to about ±21%, while for 16 μm drop, the total measuring error reduced from ±100% to ±53%.

C) Capturing the drop before reaching its final stable condition could be another source of error. For the few samples in which an oval shape of a drop appeared, the average diameter was calculated and the drop was counted as a single drop.

RESULTS

Plate 2 shows some typical magnified collected samples. In each of these photographs a relatively large drop, the residual drop, can be observed. One or more smaller droplets (siblings) are in most cases distributed to one side of the residual drop. These photographs also show that all the breakups were asymmetric as none of the samples showed a disruption of the initial drop to *n* identical drops. This phenomenon supports the assumption of a treelike disintegration used in the theoretical study[11].

Although in a few photographs the distance between the residual drop and some of the siblings was very small in comparison with the drop size, this does not contradict the catenary assumption used in calculating the breakup distance[10]. Since the distance on the w.s.p. represents only the horizontal component of the separation there could in fact have been considerably more vertical displacement.

Figure 4 presents four histograms showing the distributions of experimental results of the number of siblings for different sibling mass ratios. The results of the analysis of these distributions are summarized in Table I. It is clear that the mode of each distribution always satisfies the theoretical prediction of number of siblings. Table 1 also shows that more than 80% of the collected samples verify the theoretical estimate of the number of siblings for the sibling mass ratios covering the range between 5% to 50%. For the sibling mass ratios between 1% and 5%, this percentage agreement with the theory drops to approximately 60%. This lower agreement with the theory is not unexpected. In the analysis the assumption was made that the external forces affect the minimum energy condition only for the primary and not the secondary breakup[11]. It can also be related to the error in measuring the droplet mass due to the small sibling sizes in the lower ranges of sibling mass ratios.

Figure 5 presents a graphical representation of all the experimental results of the number of siblings collected in the breakup test as a function of the sibling mass ratio as well as the results of the theoretical prediction[11]. It is clear that the experimental results show good agreement with the theoretical prediction especially in the high ranges of the sibling mass ratios. This figure also presents the calculated maximum error in measuring the sibling mass ratios for some typical cases of the collected samples. It shows that as the sibling mass ratio decreases, the maximum error increases. As stated earlier, this is because, for low sibling mass ratios the error in measuring the masses of the small siblings increases considerably. For some of the samples which did not satisfy the theoretical predictions, the calculated maximum error was higher than the error for those samples that match the predictions. Although the theoretical analysis predicts no breakup which yields 4, 5, 7, or 8 siblings, in some samples these numbers were observed. This may be due to capturing the drops before their final stability, or due to the effect of the external forces on the secondary breakup, which may lead to more disintegrations.

Due to the extremely small sibling sizes involved there are no experimental observations for the sibling mass ratios less than 1%. However, it is expected that the error in matching the theory would be much higher. This limits the practical use of the theoretical prediction of the number of siblings to cover the range of sibling mass ratios between 1% and 50%. This range is quite wide for most of the electrostatic applications where the charge of the drop can be considered the main force driving the breakup process.

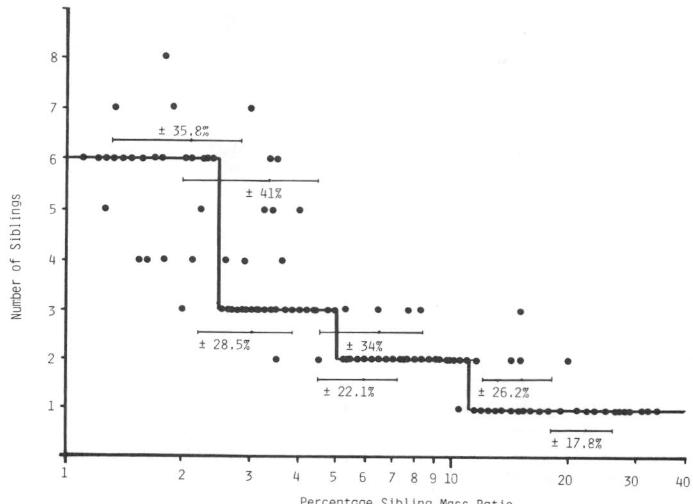

FIGURE (5) THE EXPERIMENTAL RESULTS OF THE NUMBER OF SIBLINGS AT DIFFERENT SIBLING MASS RATIOS

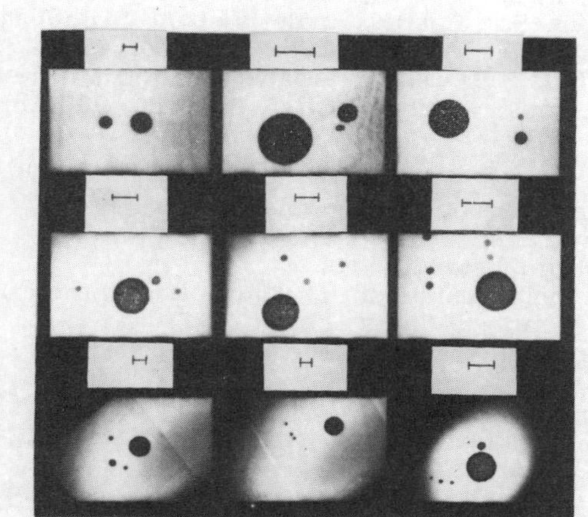

PLATE (2) SAMPLES OF MAGNIFIED COLLECTED DROPLETS (CONTACT PRINTS)

*DISTANCE BETWEEN TICK MARKS REPRESENTS 200 μm

CONCLUSIONS

The experimental results of the collected number of siblings at each range of the sibling mass ratio show good agreement with the analytical model, and support the validity of the concept of the tree-like secondary breakups. For the sibling mass ratios covering the range between 5% to 50%, the percentage agreement with the theory is more than 80%. This agreement drops to about 60% for the sibling mass ratios between 1% and 5%. The experimental results also clarify the demarcation between the modes of single sibling and the multi-sibling breakup at the sibling mass ratio of 11.1

ACKNOWLEDGEMENT

The authors acknowledge with thanks the financial support of the Natural Science and Engineering Research Council of Canada.

REFERENCES

1 Lord Rayleigh, "On the Equilibrium of Liquid Conducting Masses Charged with Electricity", *Philo. Mag.*, **14**, 184-186, (1882).

2 J.W. Schweizer and D.N. Hanson, "Stability Limit of Charged Drops", *J. Colloid Interface Sci.*, **35**, 417-423, (1971).

3 M.A. Abbas & J. Latham, "The Instability of Evaporating Charged Drops", *J. Fluid Mech.*, **30**, 663-670, (1967).

4 A. Doyle, D.R. Moffett & B. Vonnegut, "Behaviour of Evaporating Electrically Charged Droplets", *J. Colloid Sci.*, **19**, 136-143, (1964).

5 M.A. Abbas, A.K. Azad & J. Latham, "The Disintegration and Electrification of Liquid Drops Subjected to Electrical Forces", **Proc. of the Second Conference on Static Electrification**, 69-77, (1967).

6 S.A. Ryce & R.R. Wyman, "Asymmetry in the Electrostatic Dispersion of Liquids", *Can. J. Phys.*, **42**, 2185-2194, (1964).

7 D.G. Roth & A.J. Kelly, "Analysis of the Disruption of Evaporating Charged Droplets", *IEEE Trans. IAS*, **IA19**, 771-775, (1983).

8 B. Vonnegut & R.L. Neubauer, "Production of Monodisperse Aerosol Particles by Electrical Atomization", *J. Colloid Sci.*, **7**, 515-522, (1952).

9 S.A. Ryce & D.A. Patriarche, "Energy Consideration in the Electrostatic Dispersion of Liquids", *Can. J. Phys.*, **43**, 2192-2199, (1965).

10 H.A. Elghazaly & G.S.P. Castle, "Analysis of the Instability of Charged Liquid Drops", *Trans. Ind. Appl.*, **IA22**, 892-895, (Sept./Oct. 1986).

11 H.A. Elghazaly & G.S.P. Castle, "Analysis of the Multi-sibling Instability of Charged Liquid Drops", **IEEE-IAS Conf. Record**, 1985 Annual Meeting, 1337-1342, (1985).

FIGURE CAPTIONS

Figure 1 Schematic diagram of the experimental set-up
Figure 2 Nozzle characteristics at zero ring voltage
Figure 3 Contours of equal size droplets
Figure 4 The experimental results of the distribution of the number of siblings for different sibling mass ratios.
Figure 5 The experimental results of the number of siblings at different sibling mass ratios.
Plate 1 Oscilloscope traces from the detection ring
Plate 2 Samples of magnified collected droplets (contact prints). Distance between tick marks represents 200 μm.

NOZZLELESS DROPLET FORMATION WITH FOCUSED ACOUSTIC BEAMS

S. A. Elrod, B. Hadimioglu, B. T. Khuri-Yakub[a], E. G. Rawson and C. F. Quate
Xerox Palo Alto Research Center, 3333 Coyote Hill Rd. Palo Alto, CA 94304

N. N. Mansour
Computational Physics, Inc. Hillsborough, CA 94010

T. S. Lundgren
Department of Aerospace Engineering and Mechanics, University of Minnesota, Minneapolis, MN 55455

ABSTRACT

We report the use of focused acoustic beams to eject discrete droplets of controlled diameter and velocity from a free liquid surface. No nozzles are involved. Droplet formation has been experimentally demonstrated over the frequency range of 5 to 300 MHz, with corresponding droplet diameters from 300 to 5 microns. The physics of droplet formation is essentially unchanged over this frequency range. For acoustic focusing elements having similar geometries, droplet diameter has been found to scale inversely with the acoustic frequency. We summarize the results of a simple model which is used to obtain analytical expressions for the key parameters of droplet formation and their scaling with acoustic frequency. We also describe a more detailed theory which includes the linear propagation of the focused acoustic wave and the subsequent non-linear hydrodynamics of droplet formation. This latter phase is modeled numerically as an incompressible, irrotational process using a boundary integral vortex method. For simulations at 5MHz, this numerical model is very successful in predicting the key features of droplet formation.

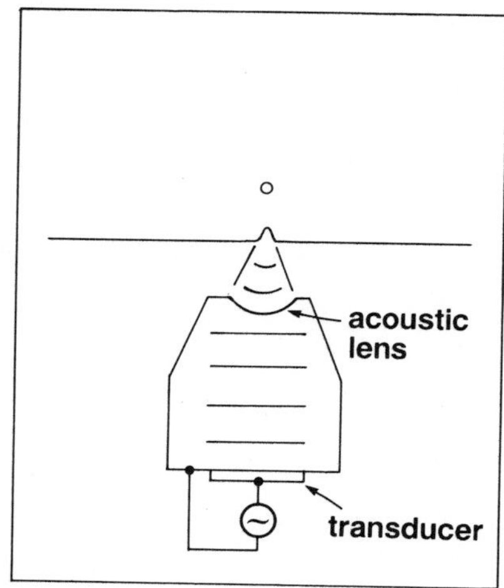

Figure 1

INTRODUCTION

Nozzleless droplet generation using focused acoustic beams has been previously described[1,2]. Wood and Lumis[1] describe the use of a high intensity sound beam to overcome the restraining force of surface tension. In their case, the acoustic energy was continuously applied, and the resultant droplet formation was described as "a volcano", consisting of droplets of random sizes. Lovelady and Toye[2] suggest the use of a single tone burst of acoustic energy to generate a single droplet of controlled diameter and velocity. However, their account is fragmentary and does not appear in the literature.

---

[a]Also of Edward L. Ginzton Laboratory, Stanford University, Stanford, CA 94305

using tone bursts of focused acoustic energy. The geometry of interest is shown in Figure 1. Some suitable transducer and lens system is used to generate a spherically converging acoustic beam. The figure is specialized to the case of a high frequency acoustic microscope lens[3]; other types of focusing elements could be employed. The liquid surface is adjusted to be at the focal plane of the acoustic beam, where the beam is confined to a focal spot whose diameter is typically on the order of one acoustic wavelength λ. In cases of practical interest, the time duration of the acoustic energy burst T is much longer than the period of the acoustic radiation $1/f$, yet much shorter than the time scale for bulk motion of the liquid. Following the burst of acoustic energy, a mound rises up from the liquid surface, and a single droplet is expelled at a velocity of several meters per second. After droplet ejection, the surface relaxes and a capillary wave[4] propagates away from the focal spot. The ejected droplets have been found to be very stable in size and velocity. For lenses with low spherical aberration and an F-number of one, the diameter of the ejected droplet is found to scale inversely with acoustic frequency f.

It is the radiation pressure[5,6] associated with the acoustic beam which acts to overcome the restraining force of surface tension and expel the droplet from the liquid surface. As noted in the review article by Chu and Apfel[5], the theory of radiation pressure has been in a state of confusion, even for the case of plane waves. For the purposes of the simple model presented below, we will use the expression for the radiation pressure (Ω) given in Ref. 6 for a plane wave reflected from a free surface:

$$\Omega = \frac{2I_i}{c} \tag{1}$$

where I_i is the average intensity of the incident acoustic wave and c is the acoustic velocity in the liquid. In the vicinity of the acoustic focus, the phase fronts of the incident beam are approximately parallel to the liquid surface[7]. We are therefore justified in generalizing the plane wave result (Equation (1)) to the case where the incident intensity varies with radial displacement from the center of the focal spot.

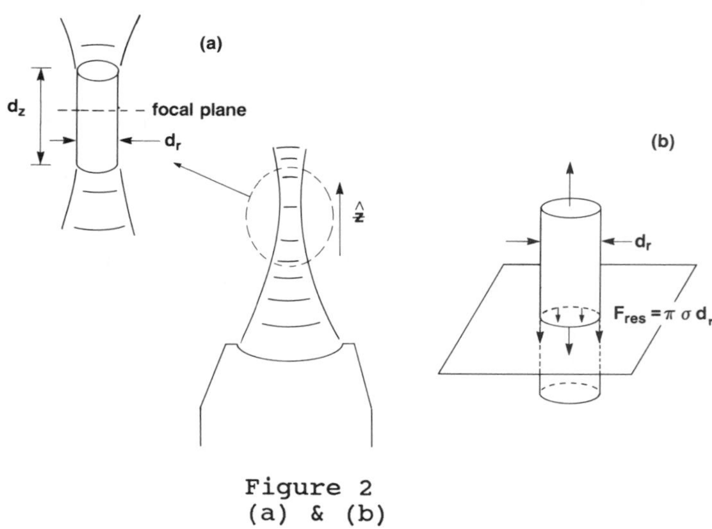

Figure 2
(a) & (b)

I. SIMPLE THEORY

The goal of our simple model is to use elementary equations of motion to deduce frequency scaling laws for parameters of interest. These include the droplet diameter d, the usable range of liquid heights Z, the droplet separation time t_{sep}, the initial surface velocity v_{init}, and the maximum height of the liquid mound h_{rise}. The last three of these will initially be expressed as functions of both energy and frequency of the acoustic pulses. In Section III, the empirical scaling of threshold energy with frequency will be used to express these three, as functions of frequency alone. Figure 2 illustrates the basic features of our model. We assume that the radiation pressure Ω acts during the time of the acoustic pulse T to impart an initial momentum per unit area given by the impulse:

$$M_{init} = \Omega T \tag{2}$$

For ease of calculation, we make the simplifying assumption that the initial momentum is uniformly distributed in a cylinder whose surface intersects the full-width at half maximum (FWHM) of the acoustic beam, as shown in Figure 2(a). Taking d_r to be the FWHM diameter of the focal spot at the focal plane and d_z to be the FWHM depth of focus, we calculate the volume of fluid in the portion of this cylinder that lies below the focal plane:

$$V = \frac{d_z}{2}\pi\left(\frac{d_r}{2}\right)^2 \tag{3}$$

For a diffraction limited beam, expressions for d_r and d_z are given in the literature[7]:

$$\text{(a)}\quad d_r = 1.02\lambda F \qquad\qquad \text{(b)}\quad d_z = 7.1\lambda F^2 \tag{4}$$

where F is the F-number of the acoustic lens. Unless otherwise stated, the data presented in this paper is for lenses having an F-number equal to one.

Making the plausible assumption that the droplet diameter is determined by the size of the focal spot, we anticipate from Equation (4a) that the droplet diameter will scale as f^{-1}.

For a fixed power level, there is only a certain range of liquid levels over which the intensity in the focal spot is sufficient to expel a droplet. From Equation 4b, we expect the usable range of liquid heights Z to scale as f^{-1}.

Following the initial impulse, the cylinder of fluid will rise up out of the surface as shown in Figure 2b. It will experience a constant decelerating force due to surface tension σ given by[8]:

$$F_{res} = \pi d_r \sigma \tag{5}$$

We use the static value of surface tension for water (σ=74 X 10^{-3} Nt/m), the fluid used for all of the experiments presented in this paper.

Given the initial impulse and the restoring force of surface tension, elementary equations of motion can be used to calculate the key parameters of interest[9]:

$$v_{init} = 0.7\,\frac{Ef^3}{\rho c^4} \propto Ef^3 \tag{6}$$

$$t_{sep} = \frac{2Ef}{c^2\pi\sigma} \propto Ef \tag{7}$$

$$h_{rise} = 0.21\,\frac{E^2 f^4}{c^6\pi\sigma} \propto E^2 f^4 \tag{8}$$

Here, E is the energy in the acoustic pulse at the focal plane, and ρ is the density of the liquid.

II. CHARACTERISTIC ENERGIES AND PRESSURES

It is instructive to compute the characteristic energies and pressures for a representative case of droplet formation at 5 MHz. For a 20 μsec burst of acoustic energy, the threshold energy E_{th} to create a free droplet having zero velocity is measured to be 50 μJ. This implies a power during the burst of 2.5 W. At focus, this power is concentrated in an area approximately one wavelength in diameter (λ = 300 μm), with a corresponding average intensity of 3.5 kW-cm^{-2}. The peak acoustic pressure at focus has a large value of 70 atm. The radiation pressure, which is responsible for droplet formation, has the intuitively reasonable value of 0.4 atm. The kinetic and surface energies associated with droplet generation are small compared to the energy in the incident pulse. The energy required to break the surface and create a free droplet at rest is on the order of σS, where S is the surface area of the drop[8]. At 5 MHz, this energy is 20 nJ. The kinetic energy of the droplet is even lower (7 nJ) for the observed ejection velocity of 1 m/sec. These results are consistent with the fact that most of the acoustic energy is reflected from the liquid surface.

III. FREQUENCY SCALING RELATIONSHIPS

In general, there is a fairly narrow range of incident pulse energies for which the droplet beam is stable. We have found that there is an upper limit on the allowable energy which is typically 20-40% higher than the threshold energy. Above this value, secondary droplets are generated, rendering the primary beam significantly less stable. The experimental results presented below all correspond to an intermediate incident energy which is approximately 13% above the threshold energy.

Our simple model does not predict the threshold energy, so we must use the empirical result. A least-squares fit to the data yields a frequency dependence of $f^{-2.3\pm0.2}$ for short acoustic pulses.

Given the empirical relationship between energy and frequency, scaling laws for v_{init}, t_{rise} and h_{rise} can be computed using Equations (6), (7), and (8). Column I of Table I summarizes the predictions of our simple model for all of the parameters of interest. The uncertainties quoted follow directly from the error in the threshold energy measurement.

TABLE I

Comparison between experimentally observed frequency scaling relationships and the predictions of two theoretical models

Parameter	I. Simple Theory	II. Experiment	III. Numerical Model
Drop Diameter (d) Initial Surface	f^{-1}	$f^{-0.9\pm0.1}$	f^{-1}
Velocity (v_{init})	$f^{0.7\pm0.2}$	$f^{0.2\pm0.1}$	$f^{0.5}$
Drop Separation Time (t_{sep})	$t_{sep} \sim f^{-1.3\pm0.2}$	$f^{-1.3\pm0.1}$	$f^{-1.5}$
Maximum Height of Liquid Mound (h_{rise})	$f^{-0.6\pm0.4}$	$f^{-0.9\pm0.1}$	f^{-1}
Ejection Threshold Energy (E_{th})	–	$f^{-2.3\pm0.2}$	$f^{-2.5}$
Usable Depth of Focus (Z)	f^{-1}	$f^{-1.5\pm0.3}$	–

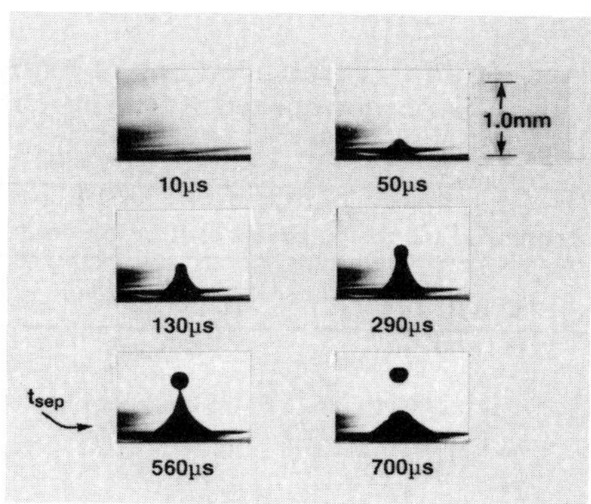

Figure 3

IV. COMPARISON BETWEEN EXPERIMENT AND SIMPLE THEORY

The time evolution of droplet formation is shown in Figure 3 for an acoustic frequency of 5 MHz, a pulse width of 20 μsec and a pulse energy of 55 μJ. Observations were made by stroboscopically illuminating the process from behind and viewing it through a microscope objective. Each photograph represents a superposition of thirty successive droplets, thus the image sharpness attests to the stability of the drop formation process. We find that the process is qualitatively similar over the entire frequency range from 5 to 300 MHz. At 300 MHz, we have successfully generated stable droplets having a diameter of only 5 μm.

In order to make comparisons to our simple theory, the parameters of interest (v_{init}, h_{rise}, t_{sep}, Z, and d) must be defined for the experimental case. While some of the definitions are arbitrary, they are chosen to be as reasonable and as consistent as possible. The height of the liquid mound h_{rise} is taken to be the height of the liquid mound (including the droplet) just prior to droplet separation. Experimentally, we find that the surface velocity increases rapidly following the acoustic pulse, and then slowly decreases as the mound rises against the restoring force of surface tension. The initial velocity v_{init} is taken to be the maximum value of the surface velocity. Evaluated at an incident energy of approximately 1.13 E_{th}, the usable depth of focus Z is taken to be the range of heights over which free droplets are generated. Droplet diameter and separation time have the obvious meanings.

With these definitions, we measure the frequency dependencies summarized in Column II of Table I. Agreement between experiment and the simple theory is favorable, particularly for drop diameter and separation time. For completeness, a quantitative comparison of experiment and our simple theory is included in Table II for a frequency of 5 MHz.

V. NUMERICAL SIMULATIONS

Below, we briefly summarize a more detailed numerical model which includes three parts: (1) the propagation of the acoustic wave to focus; (2) the crossover from acoustic fields to an initial surface velocity potential; and (3) the subsequent non-linear hydrodynamics of droplet formation.

The focusing properties of spherical transducers have been treated extensively[7]. Using a thin lens approximation, we have computed the acoustic intensity in the focal plane for an F=1 lens operating at 5 MHz. Plotted in Figure 4 is the intensity in the focal plane normalized to the intensity at the transducer surface.

TABLE II

Comparison between experimental results at 5MHz and predictions of two theoretical models

Parameter	I. Simple Theory	II. Experiment	III. Numerical Model
Drop Diameter (d)	300 μm	220 μm	190 μm
Initial Surface Velocity (v_{init})	1.0 m/sec	5.3 m/sec	5.0 m/sec
Drop Separation Time (t_{sep})	t_{sep}= 1050 μsec	560 μsec	360 μsec
Maximum Height of Liquid Mound (h_{rise})	470 μm	810 μm	790 μm
Ejection Threshold Energy (E_{th})	–	50 μj	30 μj

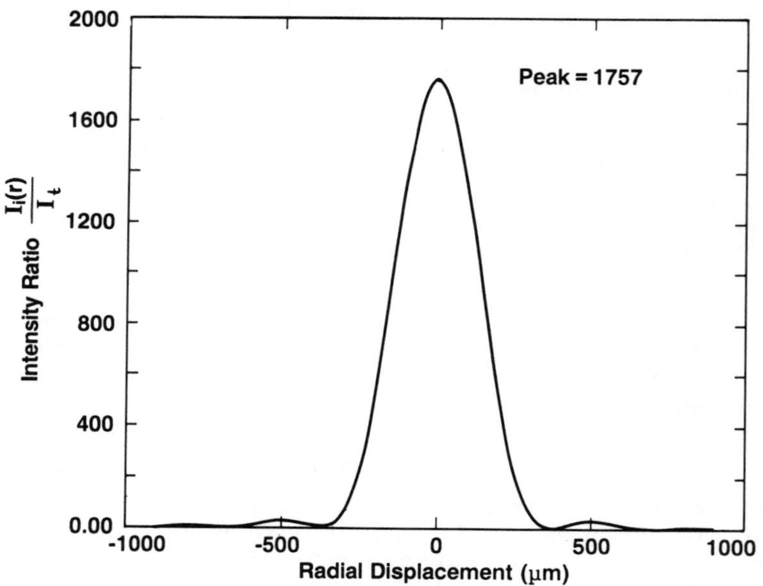

Figure 4

A detailed consideration[9] of the full equations for nonviscous hydrodynamics leads to an expression for the initial velocity potential at the liquid surface following the acoustic pulse:

$$\phi(0) = \frac{2I_i\,(r)T}{\rho c} \tag{9}$$

In deriving this expression, we assume that the acoustic pulse is much shorter than the time required for the surface to move an appreciable distance.

The motion which results from this initial velocity potential is assumed to be an irrotational, incompressible, axially symmetric flow constrained by surface tension. The flow may be represented by singular dipole solutions of Laplace's equation distributed over the deforming interface. The numerical method described by Lundgren and Mansour[10] for the motion of free drops has been adapted to this problem. Only a brief description of this boundary integral method will be outlined here.

A numerical grid of Lagrangian tracer points is introduced on the moving surface. The dipole integral representation allows the entire computation to take place on this surface. Given the axial symmetry of the problem, the computation is made along a meridional line. With knowledge of the dipole density at all grid points, one can compute the velocity $\mathbf{v}$ at each point $\mathbf{r}$ by evaluating surface integrals. To obtain the changing shape of the interface, each point is evolved in time as:

$$\frac{d\mathbf{r}}{dt} = \mathbf{v} \tag{10}$$

As the shape changes, the dipole density must change in such a way that the pressure difference across the interface is balanced by surface tension. Using the Bernoulli equation[11] for the pressure and the appropriate expression for the surface tension force[8], we find an equation for the evolution of the velocity potential ϕ on the surface:

$$\frac{d\phi}{dt} = \frac{1}{2}\mathbf{v}\cdot\mathbf{v} - \frac{\sigma}{\rho} = \left(\frac{1}{R_1} + \frac{1}{R_2}\right) \tag{11}$$

R_1 and R_2 are the principal radii of curvature computed from the local shape of the surface. The dipole density μ may be determined from ϕ by solving an integral equation:

$$\phi(\mathbf{r}) = \frac{1}{2}\mu(\mathbf{r}) = P.V.\int_S \mu' \frac{\delta g\,(\mathbf{r},\mathbf{r}')}{\delta n'} dS' \tag{12}$$

where $g(\mathbf{r},\mathbf{r}')$ is the Green function solution of Laplace's equation in an infinite domain. The above equations are to be solved with the initial condition on the velocity potential given in Equation (9).

Sample computations for a spherically focused transducer at 5 MHz yield the time evolution shown in Figure 5. The shape of the potential function was taken from the intensity profile of Figure 4, with the initial potential adjusted to be approximately 13% above the threshold value. A comparison between the experimental results (Figure 3) and the numerical model (Figure 5) is very favorable. A quantitative comparison for the 5 MHz case is included in Table II. All quantities of interest, including the value for the threshold energy, are within a factor of two of experiment quantities.

The numerical computations are carried out in dimensionless form. For intensity profiles which are of the same shape, the results of the numerical computation can simply be scaled by appropriate normalization constants. A detailed analysis[9] of these normalization factors allows us to predict scaling laws for v_{init}, E_{th}, d, h_{rise} and t_{sep}. These are summarized in Column III of Table I. The agreement with experiment is very good.

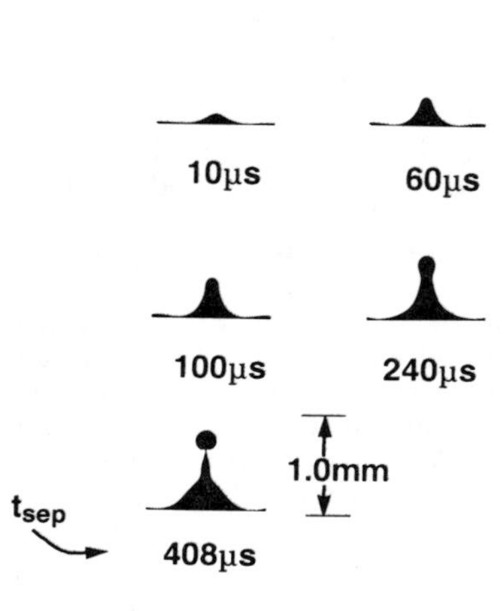

Figure 5

CONCLUSION

Detailed experimental results have been presented for nozzleless droplet formation using focused acoustic beams. Two theoretical models have been described, both of which yield reasonable agreement with experiment. The simpler model captures the key physics of the droplet formation process, and allows us to derive analytical expressions for the scaling of key parameters with frequency. The more elaborate numerical model predicts the actual time evolution of the liquid surface with reasonable accuracy. In addition, the dimensionless nature of the computations allows simple scaling arguments to be made. These are also in agreement with the experimental case.

ACKNOWLEDGEMENTS

We gratefully acknowledge R. G. Sweet for his contributions to our early experiments in the physics of acoustic droplet formation. We are also grateful to J. C. Zesch, J. A. Gasbarro, and T. R. VanZandt for significant contributions to this work.

REFERENCES

(1) R. W. Wood and A.L. Loomis, *Phil. Mag.* S. 7. 4, 417 (1927).

(2) K. T. Lovelady and L. F. Toye, U.S. Patent #4,308,547 (1981).

(3) C. F. Quate, A. Atalar and H. K. Wickramasinghe, **Proc. IEEE.** 67, 1092 (1979).

(4) W. Eisenmenger, *Acustica.* 9, 327 (1959).

(5) B. Chu and R. E. Apfel, *J. Acoust. Soc. Am.*, 72, 1673 (1982).

(6) B. P. Hildebrand and B. B. Brenden, **An Introduction to Acoustical Holography**, Plenum Press, New York (1972).

(7) G. S. Kino, **Acoustic Waves: Devices, Imaging and Analog Signal Processing**, Ch. 3, Prentice Hall, New Jersey (1987).

(8) H. N. V. Temperley and D. H. Trevena, **Liquids and their Properties**, Ch. 9, John Wiley and Sons, New York (1978).

(9) To be submitted for publication in *J. Appl. Phys.*

(10) T. S. Lundgren and N. N. Mansour, *J. Fluid Mech.* 194, 479 (1988).

(11) A. L. Fetter and J. D. Walecka, **Theoretical Mechanics of Particles and Continua**, Ch. 12, McGraw Hill, New York (1980).

FIGURE CAPTIONS

Figure 1 Focused acoustic beam used to expel droplets from a free liquid surface.

Figure 2 (a) Acoustic beam in the vicinity of the focal plane. The lateral and axial full-width at half maximum points are specified.
(b) Rising cylinder of liquid, decelerated by the restoring force of surface tension.

Figure 3 Time evolution of droplet formation at an acoustic frequency of 5 MHz. Each image represents a superposition of thirty successive droplets, thus the image quality attests to the stability of the droplet formation process.

Figure 4 Acoustic intensity versus radial displacement in the focal plane normalized to the intensity at the surface of the spherical acoustic transducer. The data corresponds to an acoustic frequency of 5 MHz.

Figure 5 Numerical simulation of droplet formation at an acoustic frequency of 5 MHz. See Figure 3 for comparison with experiment.

TABLE I Comparison between experimentally observed frequency scaling relationships and the predictions of two theoretical models

TABLE II Comparison between experimental results at 5 MHz and predictions of two theoretical models

THERMAL ACOUSTICAL INTERACTION AND FLOW PHENOMENON

E. W. Leung, E. Baroth, C. K. Chan[a], T. G. Wang[b]
Jet Propulsion Laboratory, California Institute of Technology, Pasadena CA 91109

ABSTRACT

In containerless science for material processing, the acoustic field is used to levitate and to control the position of a heated or cooled sample. The interaction between the temperature and the acoustic fields leads to complicated fluid flow phenomena, resulting in the perturbation of the sample position and the heat transfer process. The physical mechanisms in this thermal-acoustic field were investigated using the technique of holographic interferometry and thermometry. Of particular interest was the heat transfer rate from the sample associated with the sound intensities, normal frequencies of the acoustic standing wave field, and gravitational effects. For metallic spheres with high thermal conductivity, the surface temperature was found to be uniform. The thermal flow phenomenon, which is associated with the circulating flow inside the resonant chamber, was recorded. The heat transfer coefficient at the sample surface was correlated with the acoustic and the gravitational parameters, based on the classical theory of convective heat transfer. These correlations can be used to predict the heat transfer from a spherical object in a zero-gravity environment.

INTRODUCTION

Heat transfer rate and temperature distribution on a levitated sample are of interest in the study of sample heating and cooling for the processing of materials. Currently, furnaces used for materials processing in space reach temperatures of about 1000°C. The heating and cooling processes in these furnaces are usually a combination of convection and thermal radiation. In an acoustically levitated high temperature furnace that is operated on the ground, the sample temperature is affected by gravity, acoustic streaming, nonsymmetrical wall geometry, sample motion, nonisotropic radiative heating, and phase-change inside the sample. A basic understanding of the interaction of these factors with the heat transfer process would be essential in the design of future experimental facilities such as the drop physics module (DPM) and the modular containerless processing facility (MCPF), which will provide uniform heating and cooling. This basic understanding is especially critical for flight facility applications, because the facility is usually tested on the ground where the heat transfer process is influenced by gravity.

As a first step to resolve this complicated problem, the heat transfer process from a spherical object on the ground was experimentally investigated. A sphere of high conductance was chosen deliberately so the surface temperature was maintained uniformly. The heat transfer rate due to free convection was measured. The coupling effect between gravity and acoustic streaming on the heat transfer was then investigated by comparing the flow profile with the heat transfer data and by studying the limits of the heat transfer correlation. In addition, correlation for zero-gravity application is also suggested.

EXPERIMENTAL SET-UP

The experiment was performed in a rectangular acoustic chamber, as shown in

[a] *current address:* TRW, M/S 01-2050, 1 Space Park, Redondo Beach, CA 90278
[b] *current address:* Center for Microgravity Research and Applications, Vanderbilt University, Nashville, TN 37235

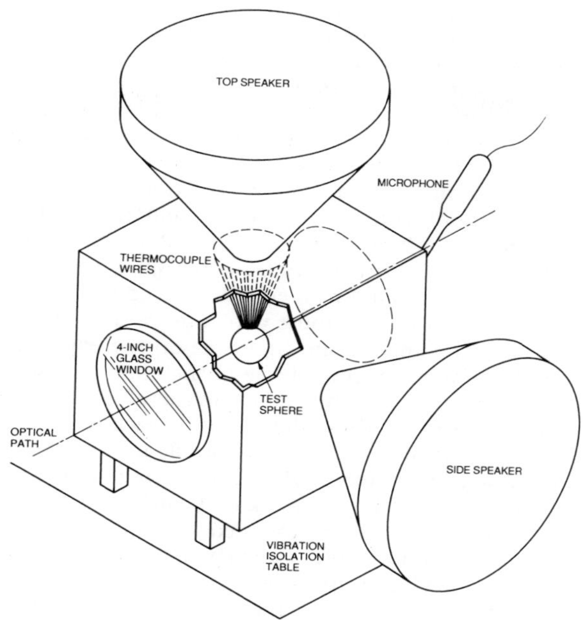

Figure 1

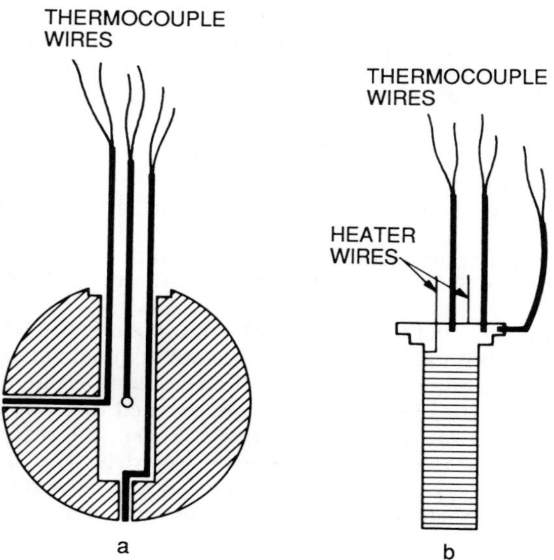

Figure 2

Figure 1. The chamber has a dimension of 15.3 x 15.3 x 12.7 cm. Two JBL loudspeakers (model no. 2445J) were mounted at right angles to each other. The sound wave was made to propagate parallel to or perpendicular to the gravitational force by activating either speaker. To measure the acoustic pressure, a Bruel and Kjaer 1/4-inch microphone with a type 2618 pre-amp was mounted at the end corner, along the diagonal of the chamber. The test spheres used were made of copper and measured 2.54 cm and 1.27 cm in diameter; a heater was inserted at the center. A total of seven chromel-alumel thermocouple junctions were used to monitor temperatures; six for the sphere temperature and one for the wall temperature. The locations of these thermocouples are shown in Figure 2 (a and b). The heater assembly that was inserted into the central portion of the sphere was made out of macor (Figure 2b). It had a cylindrical core with equally spaced small groves cut on the surface and a slightly larger cap. Nichrome wire was wound along the grooves. The three thermocouples on the sphere and the three on the cap were directed through the cap to suspend the sphere in the middle of the chamber, so the majority of the heat input onto the sphere would have to be conducted to the surface of the sphere and removed from there, and not through the support wires.

When the sphere was heated, the thermal profile in the gaseous medium around the sphere was obtained by a holographic interferometer (Leung and Baroth[1]) as shown in Figure 3. A 20 mw helium-neon laser beam was split into equal lengths of reference and object path, which were expanded, collimated and made to intersect at the holographic camera. An erasable, thermoplastic plate was used to record the hologram. The camera was aligned to bisect the 30° angle of the beams to obtain the maximum performance of the plate. A lens and two mirrors were positioned behind the holography camera to image the sphere at the viewing screen. The laser and the optical components were mounted on an optical table, which was covered with an attached top to remove random temperature fluctuations along the optical paths due to vents in the laboratory. The table top also served to shield the camera from exterior light and dampen the sound from the speakers. A phase-lock feedback system, used successfully before (Leung and Wang[2]), consisting of the HP 85 computer, the HP 3326 synthesizer, the Crown D60 power amplifier, the EG&G 5206 lock-in amplifier, and the B&K microphone with the 2618 preamplifier, was used for acoustic resonance tracking. An HP 3497 control unit was used for data acquisition and reduction.

PROCEDURE

A typical experimental procedure consists of:

(1) establishing the acoustic resonant condition of the highest sound pressure level (SPL),

(2) turning the heater on a fixed power setting, and

(3) maintaining the SPL within one-percent of the microphone voltage as heat is applied.

As heat is applied, the phase-lock feedback system keeps track of the change in resonant frequency. Without the feedback system, a quasi-steady state could not be maintained, and the temperature of the sphere would fluctuate rapidly, attenuating the sound energy. As the sphere temperature reached the steady state (*i.e.*, when it varied by less than 0.1%), the thermocouple voltages, the resonant frequency, the microphone voltage, and the heater power were recorded. The SPL was then reduced and the procedure was repeated. Twenty values of SPL were set to complete the run for a fixed heater power setting. A total of 180 measurements (nine heater power settings, 20 SPLs) for the 2.54 cm diameter sphere and 60 measurements (three heater power settings, 120 SPLs) for the 1.27 cm diameter sphere were made for each direction (parallel or perpendicular) of the acoustic field. The free convection case was achieved when the speaker was turned off.

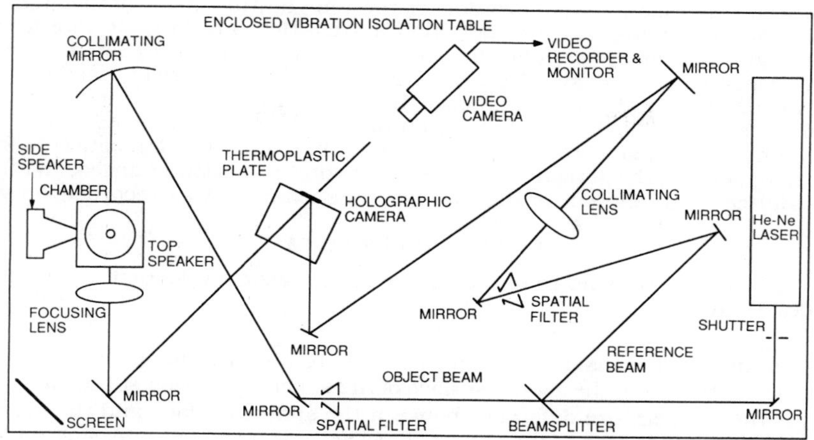

Figure 3

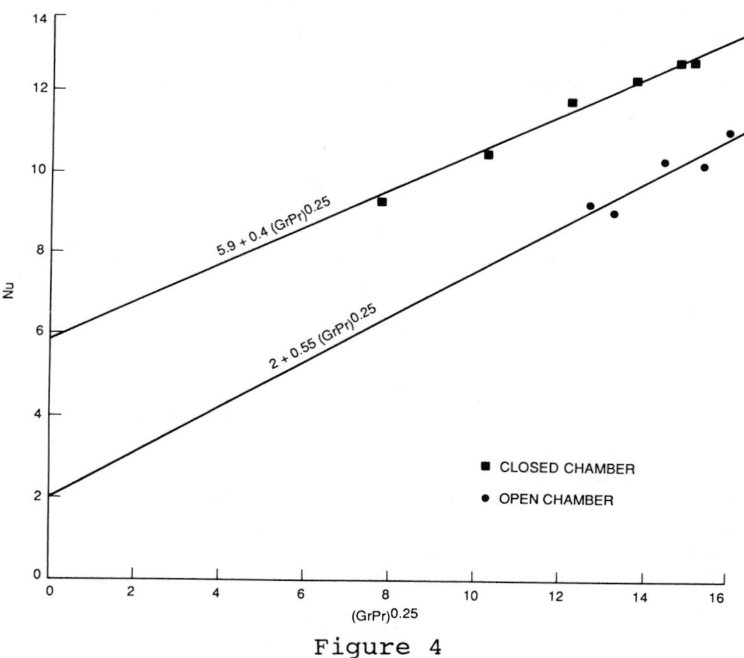

Figure 4

In these experiments, visualization of the temperature field, associated with thermal acoustic interaction, was obtained by real-time holographic interferometry. A hologram of the sphere and the acoustic chamber in an unheated state was recorded. Upon reconstruction, the image was directed onto a viewing screen. As the sphere was heated, the density and, hence, the index of refraction of air varied. This resulted in changes in path length between the heated state occurring in real time and the unheated state recorded earlier. Changes in the path length caused a fringe pattern to appear on the screen. As the experimental condition changed, the fringe pattern shifted while the entire thermal flow phenomenon was recorded on video tape.

HEAT TRANSFER DATA

The rate of heat transfer from the sphere in the resonant acoustic environment can be described as

$$Q = h(\Delta T_s)A_s \tag{1}$$

where Q is net heat transfer rate from the sphere surface when the power loss, through areas other than the copper surface of the sphere, has been subtracted.

ΔT_s is the temperature difference between the sphere surface and the chamber wall. A_s is the surface area of the sphere, h is the heat transfer coefficient, which is the primary objective of the present experiment.

The basic assumption in equation (1) is uniform temperature on the sphere surface. To ensure that this assumption was fulfilled in the present experiments, temperatures at three locations shown in Figure 2a were measured. Measurements indicated that the maximum temperature variation on the copper sphere was less than 1°C for the various orientations of the sound direction relative to the gravitational force.

The following non-dimensional parameters were introduced:

Nusselt No.

$$Nu = hD/k_f \tag{2}$$

Grashof No.

$$Gr = g\,\beta_f(\Delta T_s)D^3/v_f^2 \tag{3}$$

Reynolds No.

$$Re = DU_{ta}/u_f \tag{4}$$

where U_{ta} is the thermal acoustic streaming velocity (Lee and Wang[3]), which is related to U_{ta}, the isothermal-acoustic streaming velocity and the vibrational particle velocity U as

$$U_{ta} = U_{ia}\,(T_f/T_o) \tag{5}$$

where

$$U_{ia} = 2U^2/\omega D \tag{6}$$

D is the sphere diameter. k_f is the thermal conductivity; υ_f the kinematic viscosity; and β_f the volumetric expansion coefficient of the air.

The Nusselt numbers for the free convection case (i.e., SPL = 0) were computed and are shown in Figure 4 as a function of Grashof number. In the case of the closed chamber, the wall was found to have major effects on the heat transfer rate and the correlation did not agree with the classical free convection data. However, when the chamber wall was opened, the data agreed with the classical data. Hence,

for closed chambers:

$$Nu - 2 = 0.37(Gr)^{0.5} \tag{7}$$

for open chambers:

$$Nu - 2 = 0.5(Gr)^{0.25} \tag{8}$$

A typical flow pattern of natural convection around a heated sphere is shown in Figure 5.

In the presence of the acoustic field, the flow pattern changed to a four-quadrants pattern as shown in Figure 6 for a vertical sound field, and in Figure 7 for a horizontal sound field.

NATURAL CONVECTION

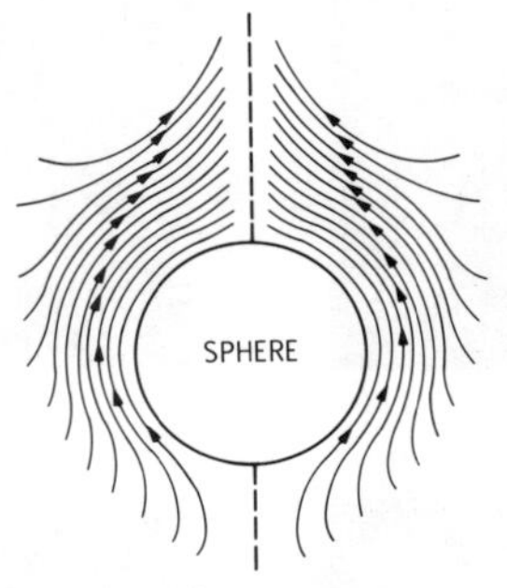

Figure 5

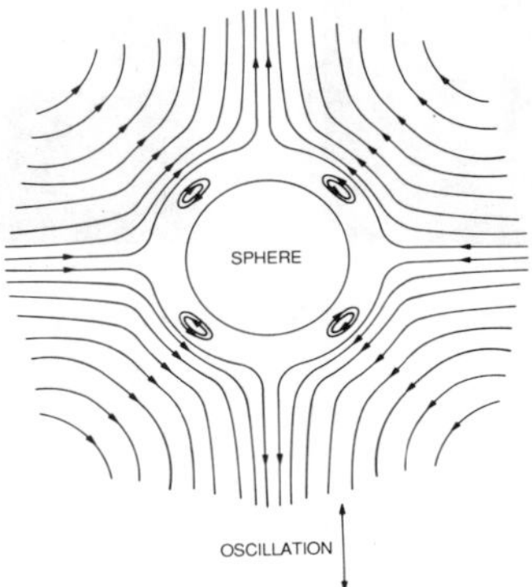

Figure 6

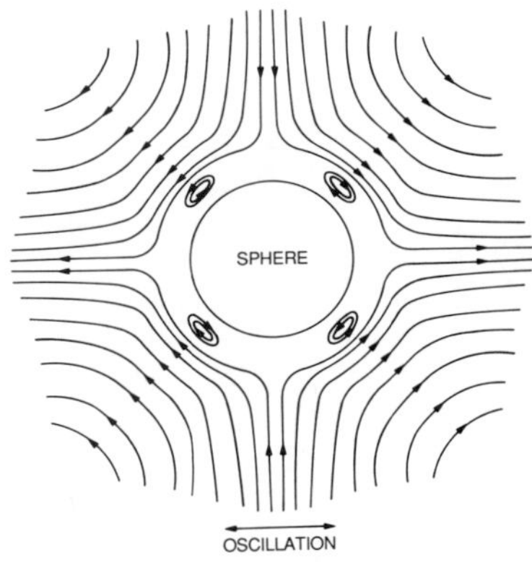

Figure 7

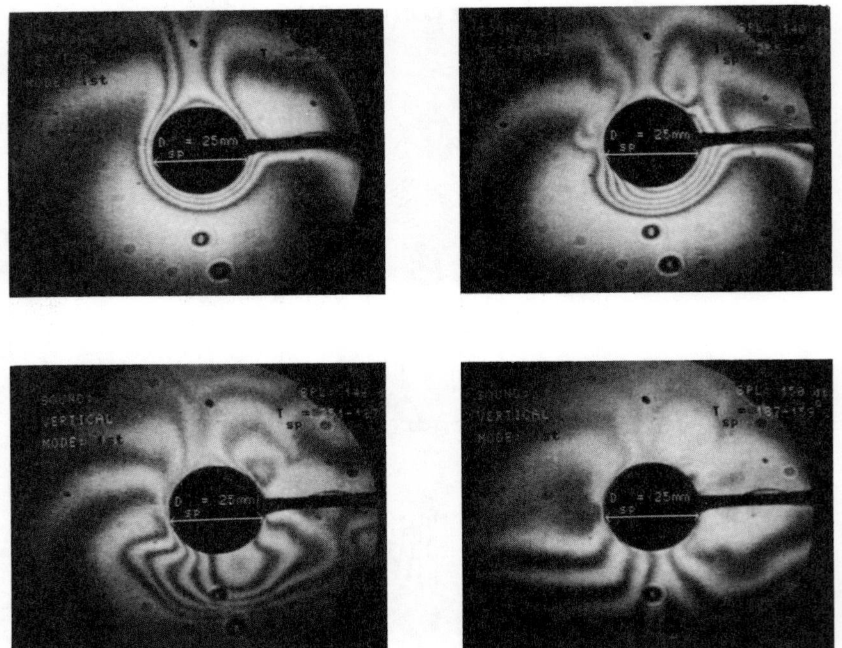
MODE: 1st
D
sp
= 25mm
VERTICAL
MODE: 1st
D
sp
= 25mm
VERTICAL
D
sp
= 25mm

Figure 8

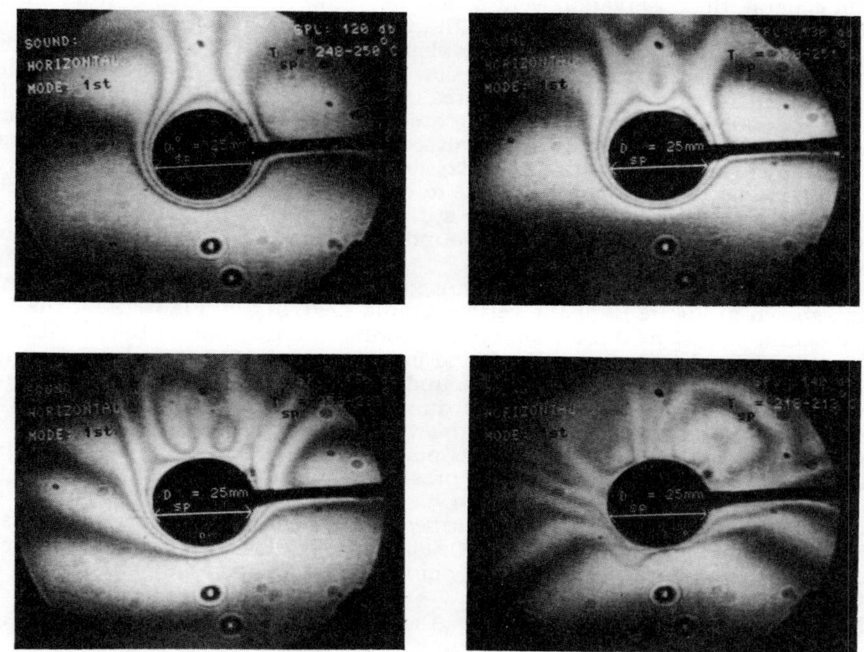
SOUND:
HORIZONTAL
MODE: 1st
SPL: 120 db
T = 248-250 °C
sp
HORIZONTAL
MODE: 1st
D = 25mm
sp
HORIZONTAL
MODE: 1st
D = 25mm
sp
HORIZONTAL
MODE: 1st
sp
D = 25mm
sp

Figure 9

In general, the observation went as follows. Flow regimes change from laminar to vortex shedding to turbulent state with increasing SPL. In the laminar state, the temperature gradient is confined to the vicinity of the sphere to form a thick thermal boundary layer. This results in a slow heat transfer rate. In the vortex shedding state, heated air pockets start breaking away from the sides of the sphere surface, extending into the surrounding area, and the heat transfer rate increases. However, the breaking of this heated air pocket is not continuous, but occurs at certain frequencies (normally a few Hertz), depending on the SPL. Consequently, the thermal boundary layer was not stable. In the turbulent state, heated air, which is continuously released from the sphere surface, extends to the entire region of the chamber. For this case, the thermal boundary layer is thin, and the heat transfer rate is fast.

Experimentally determined, the change of thermal flow pattern with increasing SPL is shown in Figure 8 for the vertical sound field, and in Figure 9 for the horizontal sound field. The discussion on how sound intensity affects the instability of the thermal boundary layer can be found in Reference 1. The only important relevance is the following: In the vertical sound field, the thermal flow pattern changes from laminar to vortex shedding to turbulent state. The three flow regimes are bounded by well defined sound pressure levels (below 136 dB, between 136 dB and 146 dB and above 146 dB). In the horizontal sound field, flow pattern changes from laminar to turbulent state without the presence of vortex shedding state. The flow regime changes at 146 dB.

The flow observation was also confirmed by the heat transfer data. The Nusselt number is plotted as a function of the Grashof number for constant Reynolds number, as shown in Figures 10 and 11. By curve fitting, the correlation of *Nu*, *Gr*, and *Re* is found to be of the following form

$$ln(Nu - 2) = A + B ln(1 + Gr) \qquad (9)$$

where

$$A = a_0 + a_1 \, ln(1 + Re) \qquad (10)$$

and

$$B = b_0 + b_1 l_n(1 + Re) \qquad (11).$$

The constants a_0, a_1, b_0, b_1 are found as in Table 1.

Table I:
Correlation of heat transfer parameters

Direction of Sound Field	a_0	a_1	b_0	b_1	Re	Gr
Vertical	-0.72	0.46	0.23	-0.01	7.2 - 3.17	3600- 10450
Vertical	-3.45	1.78	0.44	-0.11	2.0 - 7.2	49270 - 112500
Vertical	-1.73	0.55	0.33	-0.05	0.6 - 2.0	7100 - 112700
Horizontal	-0.25	0.31	0.22	-0.004	17.0 - 31.9	3500 - 91580
Horizontal	-2.22	0.94	0.36	-0.05	0.6 - 17.0	4200 - 11130

Figures 12 and 13 show the results of correlation ((10) and (11)).

The limit of *Re* and *Gr*, which separates the values of *a*'s and *b*'s, agrees with that in the flow visualization data which separates the flow regimes. Equation (9) should assume the classical results when the following limits are applied:

I. As *Re* approaches 0, Equation (9) becomes

$$ln(Nu\text{-}2) = a_0 + b_0 \, ln(1+Gr).$$

In all of our experiments $Gr >> 1$,

$$ln(Nu\text{-}2) = a_0 + b_0 \, ln \, Gr \qquad (12)$$

or

$$Nu = 2 + e^{a_0} Gr^{b_0}$$

which is the same as Equation (8) for the case of free convection.

II. As *Gr* approaches 0 and $Re >> 1$, Equation (9) becomes

$$ln(Nu\text{-}2) = a_0 + a_1 \, lnRe,$$

or

$$Nu = 2 + e^{a_0} Re \qquad (13)$$

which is the form of force convection.

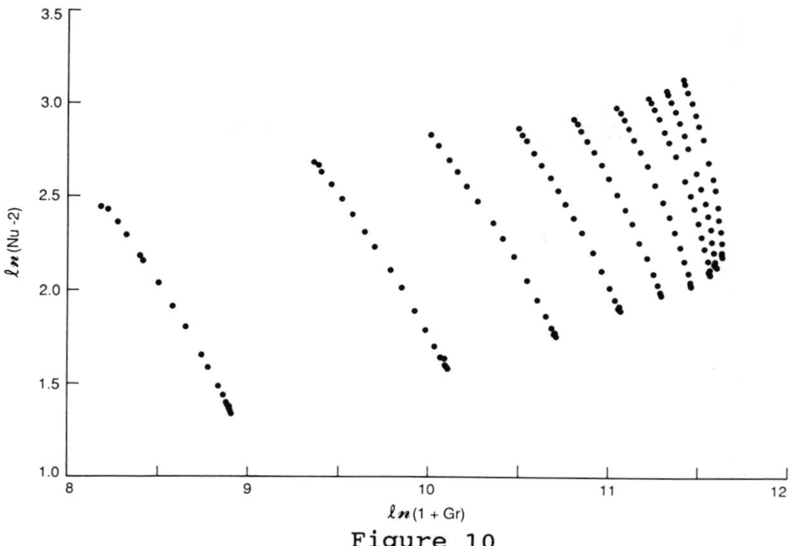

Figure 10

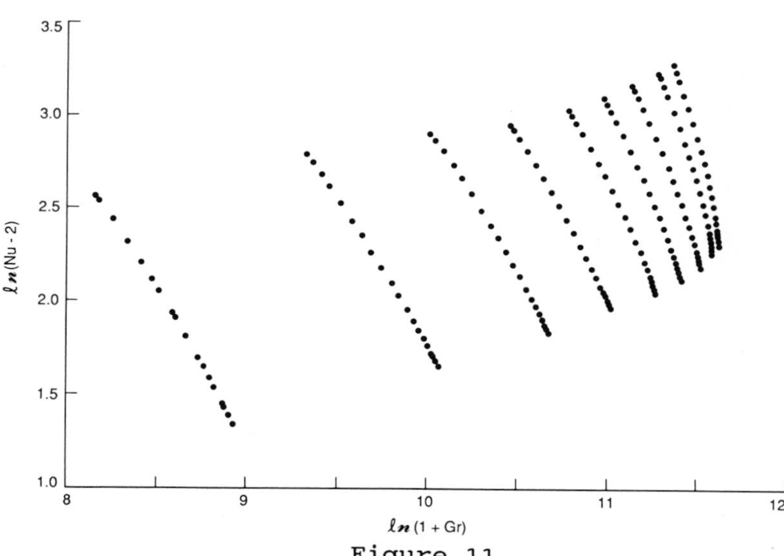

Figure 11

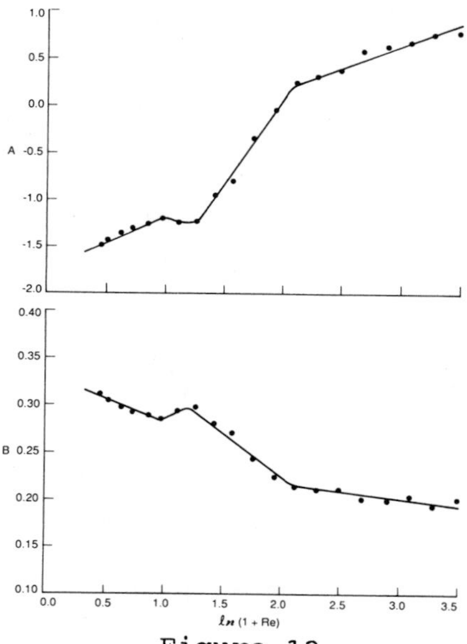

Figure 12

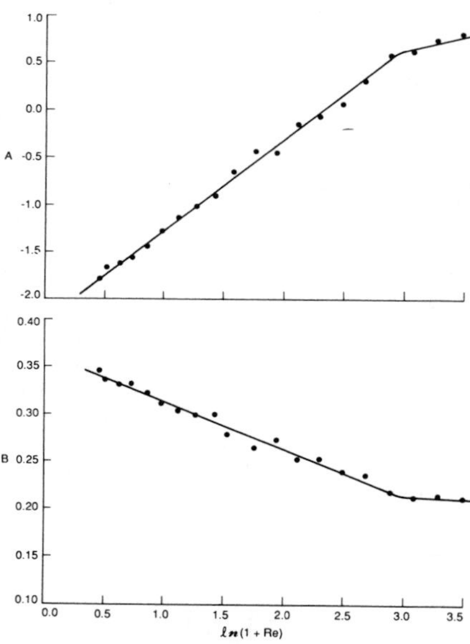

Figure 13

III. As Gr approaches 0, and Re approaches 0, Equation (9) becomes

$$\ln(Nu-2) = a_0$$

or

$$Nu = 2 + e^{a_0} \quad (14).$$

The theoretical result shows $Nu = 2$ for pure conduction. In order for the second term to be small compared to the first term, a_0 has to be a large negative number. Our measurement of a_0 is -1.73 for the vertical mode and -2.22 for the horizontal mode, with an equivalent Nu of 2.2 and 2.1, respectively. It is conceivable that if Gr is reduced by many orders of magnitude, a_0 will be smaller. Therefore, it is safe to say as Re approaches 0, Gr approaches 0, as in the case of pure conduction,

$$Nu = 2. \quad (15)$$

CONCLUSION

Correlation of heat transfer coefficient from the surface of the sphere with gravitational and acoustical parameters were found to be in the form of

$$\ln(Nu-2) = A + B\ \ln(1 + Gr)$$

where A and B are functions of Re with constants a_0, a_1, b_0, and b_1

$$A = a_0 + a_1\ \ln(1 + Re)$$

$$B = b_0 + b_1\ \ln(1 + Re).$$

Flow visualization indicates there should be two flow regimes for horizontal sound field, and three flow regimes for the vertical sound field. Analysis of the heat transfer rate verifies these facts.

In a zero gravity environment, the heat transfer rate without the radiation effect can be characterized as

$$\ln(Nu-2) = a_0 + a_1\ \ln(1 + Re) \quad (16)$$

with a_0 and a_1 determined for $Gr = 0$. Before a space experiment becomes available, in order to predict the heat transfer rate with acoustic effects alone, we must make use of what has been learned on Earth. Since acoustic streaming is the only flow field in a space environment, we suggest to use the value of a_0 and a_1 obtained for a horizontal sound field at large Reynolds numbers, for minimum gravitational effect on the heat transfer process. In principle, the Reynolds number can be computed for all temperature ranges as shown in Equations (4) and (5). However, as the temperature of the sphere increases, the radiation effect becomes more important. The next logical step is to study the interactions between acoustic and radiation effects on heat transfer.

ACKNOWLEDGMENTS

The authors are grateful to Mr. George Tanent for his ingenious skill in making the sphere-heater assembly. Our thanks to Mr. Martin Funches for the dedication in taking the experimental data.

This paper presents the results of one phase of research carried out at the Jet Propulsion Laboratory, California Institute of Technology under contract with the National Aeronautics and Space Administration.

REFERENCES

1. E. W. Leung & E. C. Baroth, "An Experimental Study Using Flow Visualization on the Effect of an Acoustic Field on Heat Transfer from Spheres," Symposium on Microgravity, Fluid Mechanics, FED, Vol. 42, D. J. Norton, ed.. book No. HOO357.
2. E. W. Leung & T. G. Wang, "Force on a Heated Sphere in a Horizontal Plane Acoustic Standing Wave Field," *J. Acoust. Soc. Am.* **77** (5), 1985.
3. C. P. Lee & T. G. Wang, "Acoustic Radiation Force on a Heated Sphere Including Effects of Heat Transfer and Acoustic Streaming," *J. Acoust. Soc. Am.* **83** (4), 1988.

FIGURE CAPTIONS

1. Details of the test chamber.

2. Assembly of the test sphere
3. Holographic Interferometry Flow Visualization Set-Up
4. Nusselt Numbers as a Function of Grashof Number and Prandtl Number for Free Convection
5. Free Convection Flow Pattern
6. Acoustics Streaming in a Vertical Sound Field
7. Acoustics Streaming in a Horizontal Sound Field
8. Interferogram of Thermal Acoustic Streaming in a Vertical Sound Field
9. Interferogram of Thermal Acoustic Streaming in a Horizontal Sound Field
10. Nusselt Number as a Function of Grashof Number for Fixed Reynolds Number in the Vertical Sound Field
11. Nusselt Number as a Function of Grashof Number for Fixed Reynolds Number in the Horizontal Sound Field
12. Results of Equation (9) for a Vertical Sound Field
13. Results of Equation (9) for a Horizontal Sound Field

TERRESTRIAL LEVITATION, DEFORMATION AND DISINTEGRATION (ATOMIZATION) OF LIQUIDS AND MELTS IN A ONE-AXIAL ACOUSTIC STANDING WAVE

E.G. Lierke and D. Lühmann
Battelle-Institute, Frankfurt, Germany

E.W. Leung
Jet Propulsion Laboratory, Pasadena, California, USA

ABSTRACT

When a liquid drop is levitated on earth in a one-axial acoustic levitator, it deforms under the action of radiation and Bernoulli pressure. With increasing product of Bond number $Bo = \rho_s/\sigma_s a_o^2 g_o$ (σ_s —surface tension, ρ_s —density, a_o —radius, $g_o = 9.81\ m/s^2$) and levitation safety factor ϕ_s , its shape varies from spheroid to "doughnut", until, at a critical value of $Bo{\cdot}\phi_s$ the drop will either disintegrate or (at large viscosity) self-inflate to a shell or a multibubble foam structure. Measurements of the drop aspect ratio are presented and compared with the theoretical prediction for an optimized levitator. Drop disintegration and self-inflation is demonstrated with video recordings.

Practical applications such as acoustical measurement of the Bond number and atomization of drops or liquid jets, fed continuously into the tuned standing wave, are presented. The "contactless standing wave atomization" is a new, valuable technique for containerless supercooling of melts, providing a wide spectrum of droplet size and respective rates with small quantities of any desirable material of technical or scientific interest.

INTRODUCTION

The objective of our activities at JPL, Pasadena and Battelle Frankfurt in a common effort on terrestrial drop levitation has been concentrated on the determination of the dynamic range and optimum conditions for drop levitation in a one-axial acoustic levitator considering drop deformations up to disintegration.

In a normalized approach the shape of the drop (aspect ratio) can be described by three factors:

- the Bond-number: $Bo = \rho_s/\sigma_s{\cdot}a_o^2 g_o$ as ratio of hydrostatic versus capillary pressure with ρ_s the density, σ_s surface tension, a_o initial radius of the sample and $g_o = 9.81\ ms^{-2}$ gravity constant.
- the levitation safety factor $\phi_s = f_{ac,max}/m_s\, g_o$ with $f_{ac \cdot max}$ axial acoustic levitation force and $m_s\, g_o$ sample weight
- and a function of the product ka_o of wave number $k = 2\,\pi/\lambda$ and initial sample radius a_o.

THEORETICAL CONSIDERATIONS

The axial acoustic levitation force fac on a spherical sample (radius a_o) in a one-axial standing wave levitator with the pressure and velocity amplitude variation

$$P_z = P \sin kz\,, \qquad \upsilon_z = \upsilon \cos kz \tag{1}$$

follows from the integration of the radial projection of the stress tensor around the sample which consists of radiation pressure and Bernoulli components

$$p_{rad} = \frac{p^2}{2\rho c^2}\cdot \sin^2 k_z z, \qquad p_{Bern} = -\rho/2 \cdot \upsilon^2 \cos^2 k_z z \tag{2}$$

The integration[1,2] results in

$$f_{ac} = f_{King} \cdot f_1\ (ka_o) \quad (3)$$

with

$$f_{King} = 5\pi/6\ \frac{p^2}{2\rho c^2}\ ka_o{}^3 \sin 2kz_o \quad (4)$$

for small samples and a geometrical force factor

$$f_1\ (ka_o) = \frac{3}{(2ka_o)^2}\ \left(\frac{\sin\ 2ka_o}{2ka_o} - \cos 2ka_o\right) \quad (5)$$

The balance of levitation force f_{ac} and sample weight $m_s g_o$ results in a downward displacement z_o of the sample center from the pressure node ($z = o$), which would be the stable position under zero gravity conditions. This defines the levitation safety factor

$$\phi_s = \sin^{-1} 2kz_o \approx \frac{p^2}{\rho c^2}\ \frac{k}{3 \cdot 2\rho_s g_o}\ f_1(ka_o) \quad (6)$$

When replacing ρc^2 by either the compressibility β_o or the product $p_o\ \gamma$ of static pressure p_o and specific heat ratio of the gas which surrounds the sample we find from Equation (6)

$$\frac{p^2}{p_o \gamma} = p^2\ \beta_o = \frac{3 \cdot 2\ \rho_s g_o}{k}\ \frac{\phi_s}{f_1(ka_o)} \quad (7)$$

This equation allows to optimize the levitator for a spherical sample of radius a_o and leads to the smallest required sound pressure level[1] at $ka_o = \pi/3$.

A liquid sample, when levitated in a one-axial acoustic levitator, is deformed against the stabilizing capillary pressure

$$p_{\sigma,o} = \frac{2\ \sigma_s}{a_o} \quad (8)$$

by the axial radiation pressure and the radial Bernoulli underpressure (Figure 1) and assumes at small aspect ratio a nearly oblate spheroidal shape, with the half axes and aspect ratio

$$a = a_o y \qquad b = a_o y^{-2} \text{ and } \qquad a/b = y^3 \quad (9)$$

respectively.

We refer to the paper by Trinh and Hsu[3], who calculated and measured drop deformations for aspect ratios $a/b<2$ and $ka_o<0.4$.

Their results may be summarized in the equation

$$y = \frac{a}{a_o} \approx 1 + \frac{6}{64}\ \frac{p^2 \beta_o}{p_{\sigma,o}}\ (1 + 0.35 \sin^2 2ka_o) \quad (10)$$

We replaced $7/5\ (ka_o)^2$ in Trinh's equation by $0.35 \sin^2\ (2ka_o)$ to get better results for $ka_o > 0.5$. Inserting Equation (7) into Equation (10) leads to

$$y \approx 1 + Bo^* \cdot \phi_s f_2\ (ka_o) \quad (11)$$

with

$$f_2\ (ka_o) = \frac{(2k\ a_o)^3 (1 + 0.35 \sin\ 2\ k\ a_o)}{40\ (\sin 2\ k\ a_o / 2\ k\ a_o - \cos 2\ k\ a_o)} \quad (12)$$

and

$$Bo = \rho_s / \sigma_s\ a_o{}^2 g_o = Bo^* (ka_o)^2 \quad (13)$$

The Bond number Bo is the ratio of hydrostatic and capillary pressure of the drop. A modified Bond number $Bo^* = Bo(ka_o)^{-2}$ which is independent of the drop radius a_o for given wavenumber k is used in equation (11) and shown in Table 1 for several liquids when levitated in ambient air at 20 kHz.

Table I

$Bo^{*}(a_{o,o})$

Aniline	1.72
Acetone	2.5
Benzol	2.22
Butanol	2.56
Ethanol	2.56
Methanol	3.12
Mercury	2.12
Octane	3.56
Water	1.0
Xylol	2.12

The Bond number Bo^* in Equation (11) correlates drop deformation with the gravitation level g_o = 9.81 cms^{-2} and allows—after experimental calibration—the prediction of the maximum diameter of drops which can be levitated on earth in a one-axial acoustic levitator. It also shows the chances for microgravity levitation of large drops when deviations from spherical shape are intolerable.

Since aspect ratio a/b and axial displacement z_o relative to the pressure node are easy to measure, the Bond number can be determined with sufficient accuracy and used for surface tension measurements as demonstrated by Trinh and Hsu[3] and by Rhim, Chung, and Elleman[4].

When the levitation safety factor for a given bond number and ka_o is increased from $\phi_s = 1$ the aspect ratio will increase according to Equation (11) until, at a critical value depending on ka_o the drop will disintegrate.

Equation (11) describes the shape variation of drops with $ka_o \leq 1$ and $B_o{}^* \phi_s \leq 1$ with sufficient accuracy. For larger aspect ratios and $\phi_s > 1$, we experimentally verified and empirical equation

$$y = \frac{a}{a_o} \approx 1 + Bo^* f_2\,(ka_o) + \frac{2}{3}\,\frac{Bo^*(\phi_s - 1)}{ka_o} \tag{14}$$

This allows the prediction of a dynamic range $1 \leq \phi_s \leq \phi_{s,max}$ for levitation if the tolerable drop deformation y_{max} at the limit of disintegration is measured as function of ka_o.

EXPERIMENTAL PROCEDURE AND PRELIMINARY RESULTS

For the experimental determination of the dependence of the aspect ratio vs. Bond number Bo^*, levitation safety factor ϕ_s and ka_o, we used a 20 kHz standing wave levitator with 35 mm frontface diameter, and a 40 mm diameter concave reflector separated by 34 mm ($1 = 4 \cdot \lambda/2$) in ambient air (Figure 2).

As sample liquids we used distilled water and glycerin-water mixtures of different concentrations and some hydrocarbons of Table 1. A calibrated piezoelectric pick-up transducer at the center of the reflector (pressure antinode) enabled the recording of the SPL.

The aspect ratio and the safety factor were determined from sample shape and displacement by video recordings orthogonally to the levitator axis.

The effective sample diameter a_o was varied between 1 and 7 mm which resulted in a ka_o range between 0.2 and 1.3.

Figure 3 shows typical sample deformations at the smallest and largest tolerable levitation safety factor $\phi_s = 1$ and $\phi_{s,max}$ The respective ka_o and Bond number decrease from upper left to lower right of the figure. At the upper limit, the sample assumes a doughnut shape, and the aspect ratio a/b is not described by Equation (14). We therefore measured $y = a/a_o$.

Figure 4 shows the results of the dynamic range measurements $y_{min} < y < y_{max}$ vs. ka_o with the lower curves calculated from Equation (11) for $\phi_s = 1$ and $Bo^* = 1$

and Bo^* = 2, respectively. The measured upper curve indicates the limit for drop disintegration or drop self-inflation as will be shown in a short video recording.

The measurements approximately allow for the prediction of the largest diameter (ka_o) of a drop which can be levitated on earth for a given Bond number Bo^* at $\phi_s \geq 1$.

The effect of drop disintegration depends on surface tension σ_s , density ρ_s and radius a_o of the sample and can be qualitatively described as follows:

- Small samples (small $ka_o<0.5$ and small Bo^*) will reach a relatively large aspect ratio ($a/b>10$) and disintegrate by capillary waves from the center diaphragm of a doughnut shaped sample (Figure 5).
- Low viscous samples of large $ka_o>0.5$ and Bo^* will disintegrate by a coaxial shock wave which penetrates the doughnut shaped sample through a relatively thick central section before a capillary wave can be established (Figure 6).
- Large samples with $ka_o> 0.5$ and large Bond number will not be disintegrated by the central shock wave but rather self inflated, forming a reproducible and reversible multi-segment foam structure with increasing number of segments at increasing SPL (Figure 7).

ATOMIZATION OF LIQUIDS AND METAL MELTS IN AN ACOUSTIC STANDING WAVE

A typical spin off application of our standing wave levitator reported elsewhere[5,7], is the contactless atomization of metal melts, which however requires SPL above 180 dB and operation at elevated inert gas pressure ($p_o>$ 3 bar). The technique can be used for atomization of small quantities (single batch) up to several 100 kg/h in a continuous process, and produces droplets with diameters d between 5 and 150 μm with the maximum of the distribution shifting to smaller values with increasing SPL. A typical apparatus is shown schematically in Figure 8. The melt is poured from a melting furnace through a 2 mm heated feeding tube into the pressure node of the horizontally aligned atomizer. The atomized drops are blown out radially with a velocity of 1-3 m/s by acoustic convection and solidify within 10-100 ms. At melting temperatures up to 1200°C and a droplet size below 100 μm the cooling rate is mainly determined by forced convection rather than by radiation and is proportional to the square of the drop diameter d. The resulting cooling rates for droplets between 5 and 150 μm range from 10^4°K/s to 10^7°K/s. Typical values for 50 μm drops at a Nusselt number[8] Nu = 3 are shown in Table 2.

Table II Cooling rates for d = 50 μm drops at Nu = 3 near melting temperature

	Pb	Al	Ag	Cu	Fe
dT/dt (°K/s)	1.7x104	1.2x105	2x105	1.7x105	3x105

The large range of cooling rates make this technique ideally suited for containerless undercooling experiments with representative fractions of the particle distribution easily separated by sieving.

ACKNOWLEDGEMENT

This work represents part of research carried out at the Jet Propulsion Laboratory, Pasadena, sponsored by the National Research Council.

REFERENCES

1. E. G. Lierke *et al.*: "Acoustic Positioning for Space Processing of Materials Science Samples in Mirror Furnaces". **Proc. of the IEEE Ultrasonic Symposium** Atlanta, Nov. 1983, pp. 1129-1139.
2. E. Leung, N. Jacoby, & T. Wang: "Acoustic Radiation Force on a Rigid Sphere in a Resonance Chamber", *J. Acoust. Soc. Am.* **70**(6), 1981, pp. 1762-1767.
3. E. .H. Trinh & C. J. Hsu: "Equilibrium Shapes of Acoustically Levitated Drops", *J. Acoust. Soc. Am.* **79**(5), 1986, pp. 1335.

4. D. D. Ellemann: W. K. Rhim, S. K. Chung, & D. D. Elleman: "Experiments on rotating charged liquid drops", **Proc. 3rd Intl. Colloq. on Drops and Bubbles** [*this vol. Chapter 1.*]
5. E. G. Lierke *et al.*: "Standing Wave Atomizer", German Patent 2.656.330,19.
6. E. G. Lierke: "Ultrasonic Atomization of Liquids under Special Consideration of Metal Melts", JPL-Presentation May 1988.
7. M. Hohmann, S. Jönsson & E. G. Lierke: "Metal powder production by capillary and standing wave technique", Proc. of the Intl. Powder Metallurgy Conf., *Mod. Devel. Powder Metall.*, **18-21**, 1988..
8. K. Bauckhage & H. M. Liu: "Models for the Transport Phenomena in a Spray Compacting Process," 4th Int. Conf. on Liquid Atomization and Spray Systems, Sendai, Japan. August 22-24, 1988

LIST OF FIGURES

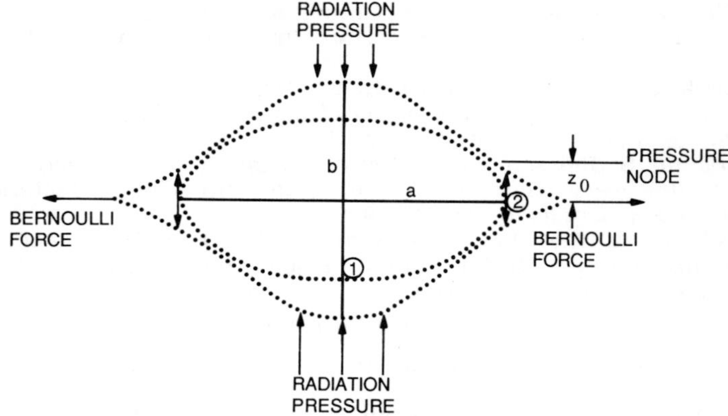

Figure 1. Drop deformation by axial radiation pressure and radial Bernoulli underpressure

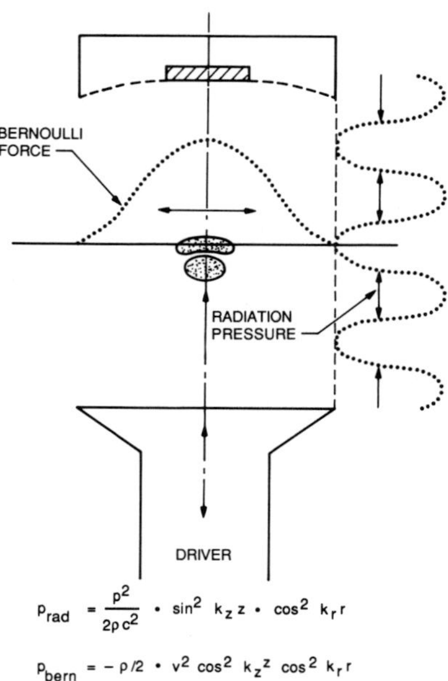

Figure 2. Drop levitation in an one-axial acoustic levitator (schematical)

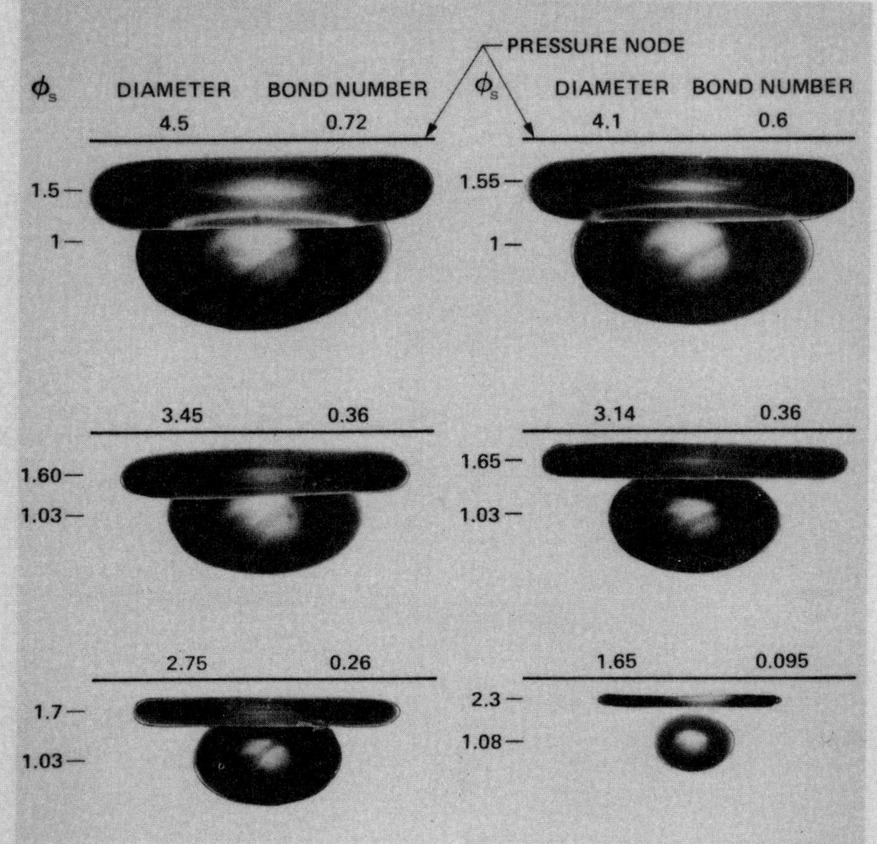

FIGURE 3

TYPICAL AXIAL DISPLACEMENT AND DEFORMATION OF WATER DROPS OF DIFFERENT INITIAL RADIUS a_o AT SMALL AND LARGE LEVITATION SAFETY FACTOR Φ_s IN A ONE-AXIAL 20 kHz ACOUSTIC LEVITATOR IN AIR (λ = 17mm)

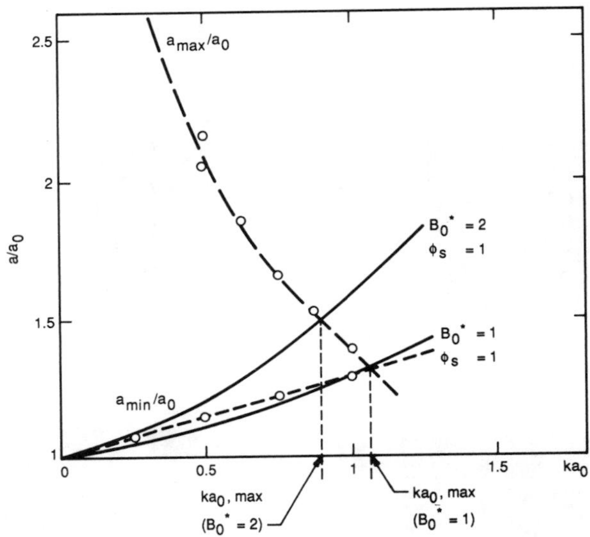

Figure 4. Measured dynamic range for terrestrial drop levitation as function of ka_0 with theoretical low limit according to equation (11)

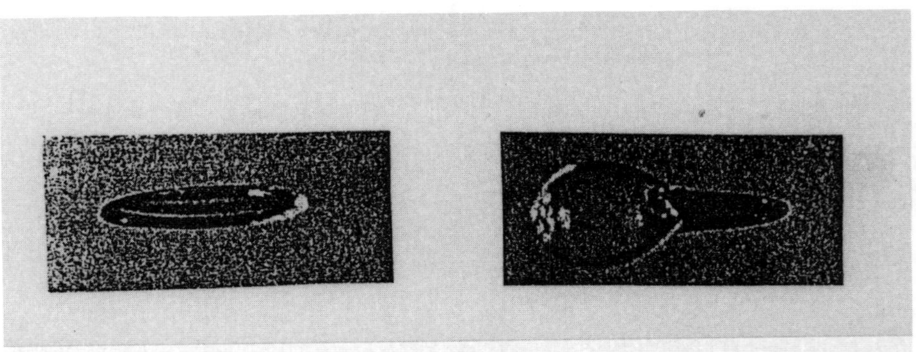

Fig. 5: Glycerine drop with $ka_o = 0.5$ prior and after selfinflation at $\Phi_{s.max}$

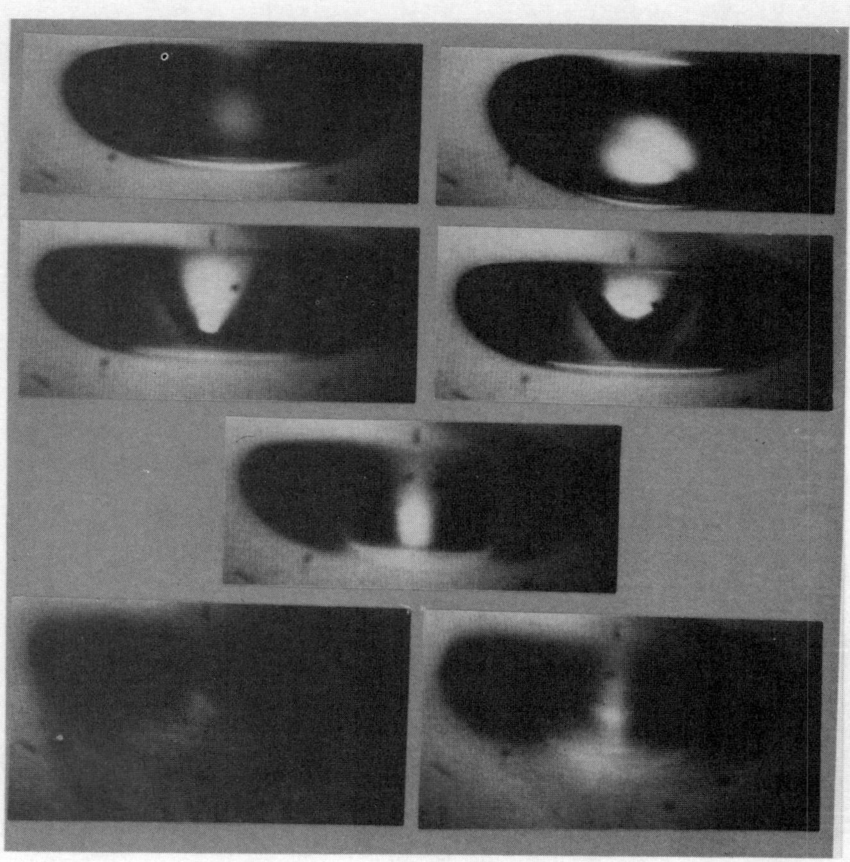

Fig. 6: "Shockwave" generation prior to water drop disintegration at large ka_0

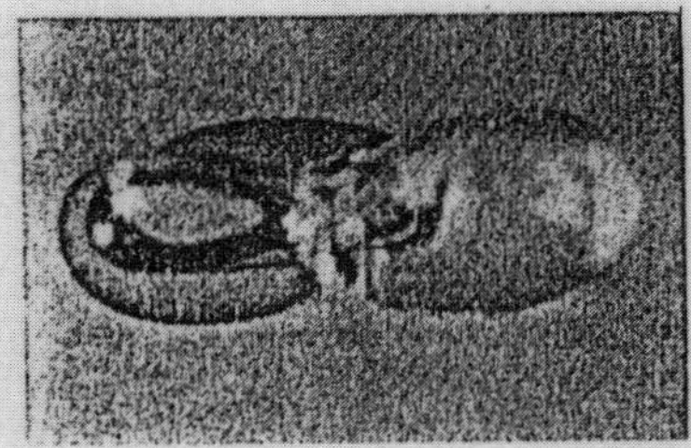

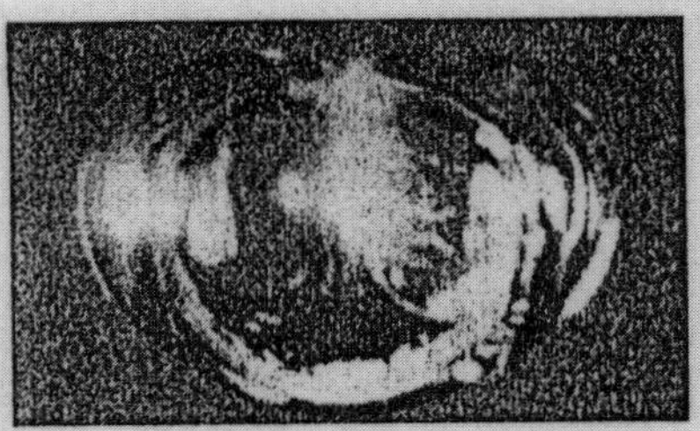

Fig. 7: Trigonal shapes of water drops in an acoustic levitator after self-inflation

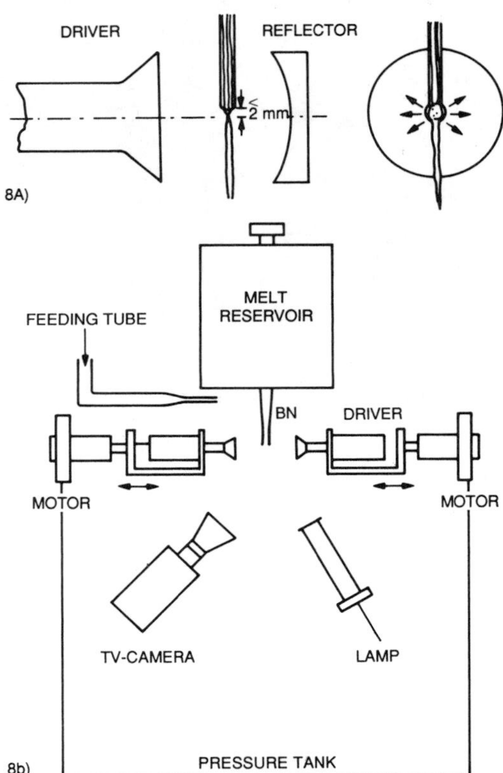

Figure 8. Ultrasonic standing-wave-atomizer
a) principle b) experimental apparatus

SHAPE OSCILLATIONS OF DROPS FOR STUDYING INTERFACIAL PROPERTIES

Hui-Lan Lu and Robert E. Apfel
Department of Mechanical Engineering, Yale University
New Haven, CT 06520

ABSTRACT

An acoustic method for suspending a drop in another liquid and deforming the drop has been combined with an optical detection scheme to measure the free quadrupole oscillation frequency and damping constant of the drop. The data for the pure water system support the theoretical predictions. For hexane drops in SDS (surfactant) aqueous solutions, it is found that the most important interfacial property is the Gibbs elasticity. Furthermore, the measured damping constants for SDS concentrations lower than 20% of the critical micelle concentration (CMC) are well described by employing an ideal equation of state for surfactants at the interface. The acoustic method, which is nonperturbative and requires very little amount of sample, may supplement other methods to measure the dynamic properties at a liquid-liquid interface.

1. INTRODUCTION

The characteristics of surface waves are strongly influenced by the presence of surfactants or contaminants. There are, therefore, numerous attempts to measure the dynamic interfacial properties by studying the propagation characteristics of surface waves[1-7]. In the work described below, we describe a non-perturbative acoustic method for inferring dynamic interfacial properties.

2. EXPERIMENTAL SETUP

A schematic drawing of the experimental setup adapted from previous work by Marston & Apfel[8], Trinh *et al.*[9], and Hsu[10], is shown in Figure 1, and is described extensively in Reference 14. This particular configuration of the levitation cell has several advantages over its predecessors: the completely fused glass cell reduces the number of cavitation sites, avoids problems with cleaning out chemicals, surfactants, etc.; the fixed external location of the transducer further reduces the cavitation sites, and the experimental conditions are easily reproducible.

An immiscible drop is levitated at a position where the acoustic radiation force balances the buoyancy force. Usually the drop is levitated at the pressure maximum if it is more compressible than the host liquid.

The levitated drop can be further triggered into free quadrupole oscillations by giving another acoustic forcing (due to modulated acoustic radiation pressure) in a form of a tone burst of a signal frequency near the quadrupole resonance frequency of the drop. The input voltage to the transducer thus has the form[11],

$$V_T = V_L + V_C V_M \ .$$

Here, V_T is the total input voltage, V_L the levitation voltage, V_C the deformation (carrier) voltage, V_M the modulation voltage. V_L and V_C are sinusoidal signals, but of different frequencies corresponding to different resonances of the system. The frequency of V_L is about 45.4 kHz and that of V_C is about 405 kHz. V_M is in the form of a tone burst. The signal frequency is between 50 and 100 Hz, depending on the quadrupole resonance frequency of the drop, pulse repetition rate 0.2 Hz, and duty cycle about 6%. Under these conditions, $V_C V_M$ does little to effect the levitation force on the drop.

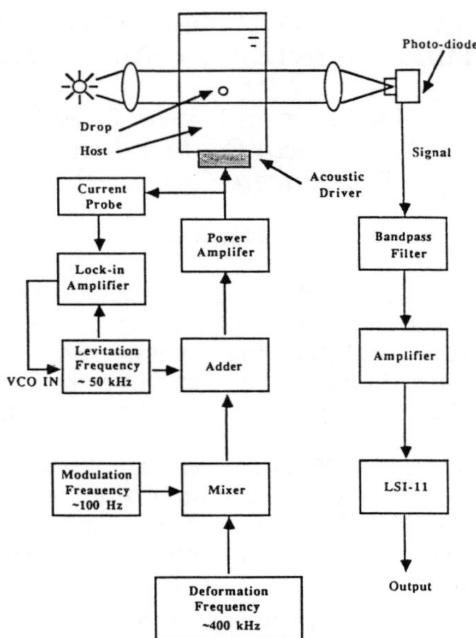

Figure 1

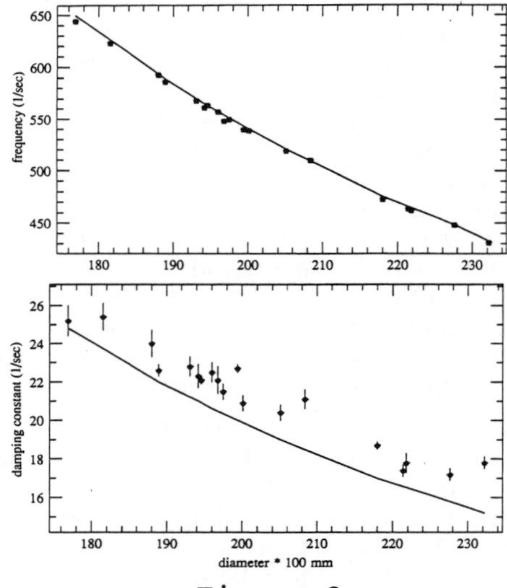

Figure 2

The quadrupole oscillations of the drop are detected by impinging a parallel light beam on the oscillating drop, and measuring the total intensity of the unscattered light[12]. Since the wavelength of the light is much smaller than the size of the drop, the extinction cross section is proportional to the geometrical cross section of the drop[13]. Therefore, when the drop is oscillating, the total intensity of the unscattered light oscillates accordingly. A major advantage of this detection scheme is that the signal from the photodiode is insensitive to the position of the drop as long as it is within the light beam. The light beam has a width of about three to four times of the size of the drop. After the light passes through the drop, the unscattered light is collected by a photodiode situated at the focal point of a concave lens, and then the resulting signal is sent to a computer for processing.

The data can be fit numerically to an exponentially damped function,

$$A(t) = A_0 + A\, e^{-bt} \cos(\Omega t),$$

to deduce the free oscillation frequency Ω and the damping constant b.

The sample systems studied are n-hexane drops in water and in SDS aqueous solutions. The n-hexane (from Aldrich, 99% purity) and sodium dodecyl sulfate (SDS, from Kodak) were used as received. The water was distilled, twice deionized, and filtered down to 0.2 mm by a 4-module NANOpure system (from Barnstead). To reduce cavitation during the experiment, the water is also degassed. The SDS was added afterwards to make the aqueous solution.

3. RESULTS AND DISCUSSIONS

The measured free quadrupole oscillation frequency, Ω, and damping constant, b, can be compared with the theoretical relation[14],

$$b + i\Omega = i\,\omega^* \left(1 + \varepsilon^{(1)} + \varepsilon^{(2)} + \ldots\right). \tag{1}$$

Here

$$\varepsilon^{(1)} = \frac{\dfrac{-i_s(2L+1)^2\sqrt{\hat{\rho}\hat{\eta}}}{2\Upsilon x_i^2} + A_d\varepsilon_d + A_s\varepsilon_s + \dfrac{4ii_s A_d A_s}{2\Upsilon}}{\dfrac{1+\sqrt{\hat{\rho}\hat{\eta}}}{x_i} + ii_s\left(A_d + A_s\right)},$$

$$\varepsilon^{(2)} = \frac{ii_s\left[f + f_s + f_d - 4i_s A_d A_s\left(L\sqrt{\hat{\rho}/\hat{\eta}} + (L+1)\right)\right]}{2\Upsilon\left[\dfrac{1+\sqrt{\hat{\rho}\hat{\eta}}}{x_i} + ii_s\left(A_d + A_s\right)\right]},$$

$$\varepsilon_d = \frac{(L+2)^2\sqrt{\hat{\rho}\hat{\eta}} + (L-1)^2}{2\Upsilon x_i},$$

$$\varepsilon_s = \frac{L^2\sqrt{\hat{\rho}\hat{\eta}} + (L+1)^2}{2\Upsilon x_i},$$

$$A_d = \frac{\Upsilon B}{(L-1)(L+2)(1+i_s\lambda)} - \frac{iL(L+1)\beta_d}{x_i^2},$$

$$A_s = \frac{-i(L-1)(L+2)\beta_s}{x_i^2},$$

$$f(x_i,\hat{\rho},\hat{\eta}) = \frac{4\Upsilon}{x_i}\left(1+\sqrt{\hat{\rho}\hat{\eta}}\right)\left(\varepsilon^{(1)}\right)^2 - \frac{i\,2\Upsilon}{x_i^2}\left[(L+1)\sqrt{\hat{\rho}\hat{\eta}} + 2(1-\hat{\eta}) + L\sqrt{\hat{\eta}/\hat{\rho}}\right]\varepsilon^{(1)}$$

$$- \frac{i\,(2L+1)^2}{x_i^2}\sqrt{\hat{\rho}\hat{\eta}}\,\varepsilon^{(1)} + \frac{2i_s(2L+1)}{x_i^3}\left(\hat{\eta}-1\right)\left[\sqrt{\hat{\rho}\hat{\eta}}\,L\,(L+2)+1-L^2\right],$$

$$f_d = A_d \left\{ 3\Upsilon\varepsilon^{(1)2} + \frac{2i_s\Upsilon\varepsilon^{(1)}}{x_i}\left[L\sqrt{\hat{\eta}/\hat{\rho}} + (L+1)\right] + \frac{i_s\varepsilon^{(1)}}{2x_i}\left[(L+2)^2\sqrt{\hat{\rho}\hat{\eta}} + (L-1)^2\right] \right.$$

$$\left. - \frac{i}{x_i^2}\left[4(L-1)(L+2)(L-\hat{\eta}) - (L+1)(L+2)^2\sqrt{\hat{\rho}\hat{\eta}} - L(L-1)^2\sqrt{\hat{\eta}/\hat{\rho}}\right] \right\}$$

$$+ \left[\frac{2\Upsilon\beta}{(L-1)(L+2)(1+i_s\lambda)} - \frac{iL(L+1)\beta_d}{x_i^2}\right]\left[2\Upsilon\varepsilon^{(1)} + \frac{i_s}{x_i}\left[(L+2)^2\sqrt{\hat{\rho}\hat{\eta}} + (L-1)^2\right]\right]\varepsilon^{(1)},$$

$$f_s = A_s \left\{ 5\Upsilon\varepsilon^{(1)2} + \frac{2i_s\Upsilon\varepsilon^{(1)}}{x_i}\left[L\sqrt{\hat{\eta}/\hat{\rho}} + (L+1)\right] + \frac{3i_s\varepsilon^{(1)}}{2x_i}\left[(L+1)^2\sqrt{\hat{\rho}\hat{\eta}} + L^2\right] \right.$$

$$\left. + \frac{i}{x_i^2}\left[4L(L+1)(1-\hat{\eta}) + (L+1)L^2\sqrt{\hat{\rho}\hat{\eta}} + L(L+1)^2\sqrt{\hat{\eta}/\hat{\rho}}\right] \right\}$$

$$- 4\varepsilon^{(1)}A_s\left[\frac{\Upsilon\beta}{(L-1)(L+2)(1+i_s\lambda)} + A_d\right].$$

x_i^2	$\equiv$	$R^2 r_i w^*/h_i$, Reynolds number,
w^*	=	$\sqrt{L(L^2-1)(L+2)\gamma_0/\Upsilon\rho_i R^3}$, Lamb frequency,
L	=	the mode of the oscillation, $L = 2$ for quadrupole oscillations,
β	=	E/γ_0, surface dilatational elasticity number,
E	=	$(\partial_\gamma/\partial lnC)_0$, Gibbs elasticity,
b_d	=	x_s/Rh_i, surface dilatational viscosity number,
b_s	=	h_s/Rh_i, surface shear viscosity number,
Υ	=	$L\hat{\rho} + (L+1)$, reduced density ratio,
$\eta_i \eta_0$	=	bulk shear viscosity of the inner and outer fluid respectively,
ζ	=	bulk dilatational viscosity,
η_s	=	surface shear viscosity,
ξ_s	=	surface dilatational viscosity,
$\hat{\eta}$	=	h_o/h_i, bulk shear viscosity ratio,
γ_0	=	equilibrium interfacial tension,
R	=	radius of the drop,
ρ_i and ρ_o	=	the density of the inner and outer fluids respectively
i_s	=	$(1+i)/\sqrt{2}$,
λ	$\equiv$	$\sqrt{\frac{D}{\omega^*}}\left[\frac{\partial C}{\partial \Gamma}\right]_0$, diffusion-adsorption number.

4. N-HEXANE DROPS IN WATER

We have found that both the free oscillation frequency and damping constant are time dependent. The damping constant almost doubles within several hours while the frequency increases slightly in this period, but after twenty hours decreases considerably, about ten percent. This time dependent phenomenon indicates that there are contaminants present in the system. The contaminants are probably airborne and difficult to avoid unless in a dust free room. The amount of contaminants, however, is little, so their effects do not show instantaneously. We therefore compare only data taken within five minutes after the drops are introduced with the theoretical predictions for a pure system. As shown in Figure 2, these data are in good agreement with the theoretical predictions. This supports the validity of the hydrodynamic analysis.

5. N-HEXANE DROPS IN SDS AQUEOUS SOLUTIONS

We have measured the free oscillation frequency and damping constant for hexane drops of different radii in SDS aqueous solution of (apparent) concentrations[15] ranging from 0.164 to 5.9 mM—cf. CMC 7.6 mM at 25°C. Usually the solution stands for at least one hour before measurements are taken, so that the surfactants can dissolve completely in water.

Again, the free oscillation frequency and damping constant in the presence of surfactants are time dependent; they decrease with time. The time scale involved, hours, is so long that it cannot be described by a simple diffusion process of surfactants[16-18]. Because of this time dependence, it is very difficult to perform experiments under the full equilibrium condition without worrying about the effects of contaminants. The best thing we can do is to record also the time elapsed after a hexane drop is introduced, when a measurement is conducted. For each drop we take at least two measurements: one within five minutes after it is introduced, the other about one hour after.

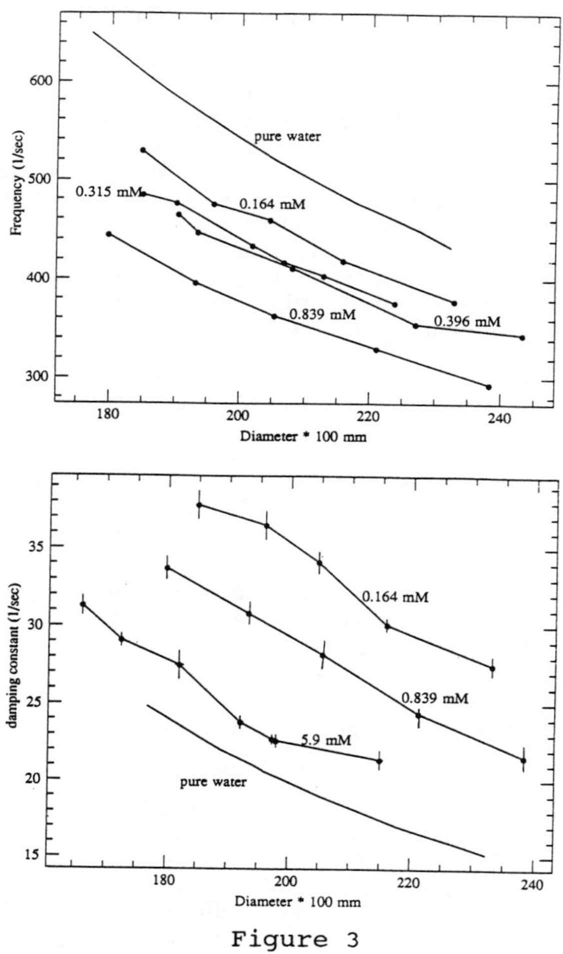

Figure 3

Figures 3a and 3b display data taken one hour after hexane drops were introduced in SDS solutions of different concentrations. The fluctuation of the data is due to the different local surfactant concentration near the interface for each measurement. We find that the free oscillation frequency decreases from the value for the pure system as the surfactant concentration increases. This agrees with the fact that the interfacial tension decreases with the increasing concentration of surfactants. The damping constant also decreases as the concentration increases. Nonetheless, it does not decrease from the value for the pure system; *it decreases toward it.* If the interface does not manifest any other interfacial properties besides the equilibrium interfacial tension, we would expect a completely different trend. The damping constant should be always lower in a system with surfactants than that in a system without surfactants, since the free oscillation frequency is lower owing to the reduction of the interfacial tension. Furthermore, the higher the surfactant concentration is, the lower the damping constant becomes. The seeming anomaly of the damping constant data therefore indicates that the hexane-water/SDS interface exhibits additional interfacial properties besides the interfacial tension.

To describe fully the effects of the SDS surfactant, we need to include other surface parameters such as Gibbs elasticity, surface dilatational viscosity, surface adsorption, etc. In principle, we should be able to fit Equation (1) with the measured frequency, and damping constant to infer the interfacial properties. Yet because of the limited range of the radius and the fluctuation of the data, we cannot, with confidence, fit Equation (1) with measured data to infer the surface parameters. Therefore, we take a different approach. We make certain assumptions first and find out if the data support these assumptions.

Since the interfacial viscosities are less important than the Gibbs elasticity in the range of concentrations studied here[14], they are assumed to be zero. Moreover, the complexities of the problem can be further reduced if the two-dimensional equation of state or the adsorption isotherm is known for the SDS surfactants at concentrations far below the CMC.

For dilute surfactant solutions, interactions among surfactant molecules may not be important. The variation of interfacial tension with the surfactant concentration is linear[19]. Thus

$$\lambda = \lambda_{pure} - \text{const}\ C_s$$

In our case, C_s is time dependent and different from the equilibrium bulk concentration, C_O, unless complete equilibrium of the system is established. Applying Gibbs' adsorption isotherm and the assumption of an ideal bulk solution[20],

$$d\gamma = -2\Gamma RT\, d\ln C_s$$

where Γ is the surface concentration of surfactants, and the factor two is specific to SDS used here because of its ionic nature. We then obtain the two-dimensional equation of state for the surfactants,

$$\pi = 2\Gamma RT, \tag{2}$$

where π is the surface pressure, $\pi \equiv \gamma_{pure} - \gamma$.

Based on the equation of state, the Gibbs elasticity becomes

$$E = -\left[\frac{\partial\gamma}{\partial \ln\Gamma}\right]_Q = \pi.$$

Therefore, the Gibbs elasticity and the interfacial tension are determined simultaneously. The adsorption equation can also be obtained, which relates the surface concentration to the bulk concentration of surfactants,

$$\Gamma = C_s l, \tag{3}$$

where l is a constant which together with the diffusion constant will determine the order of magnitude of the diffusion-adsorption number, $\lambda = (dC/d\Gamma)\sqrt{D\omega^*}$. We first assume that λ is negligible compared with 1. The film thus behaves like an

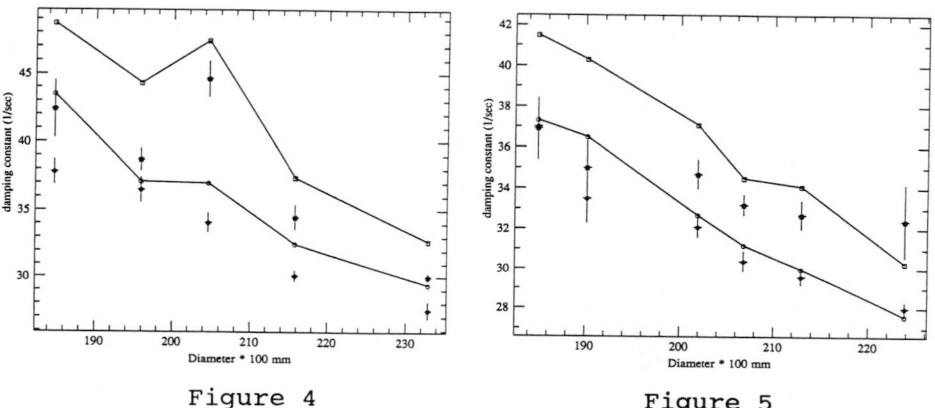

Figure 4

Figure 5

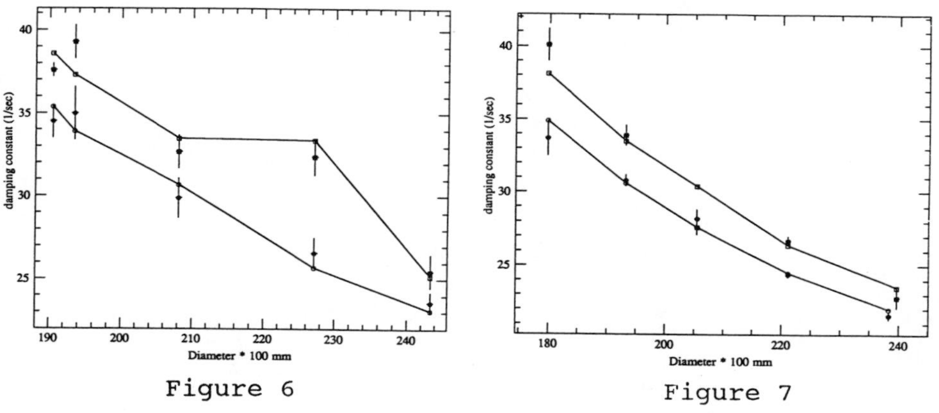

Figure 6

Figure 7

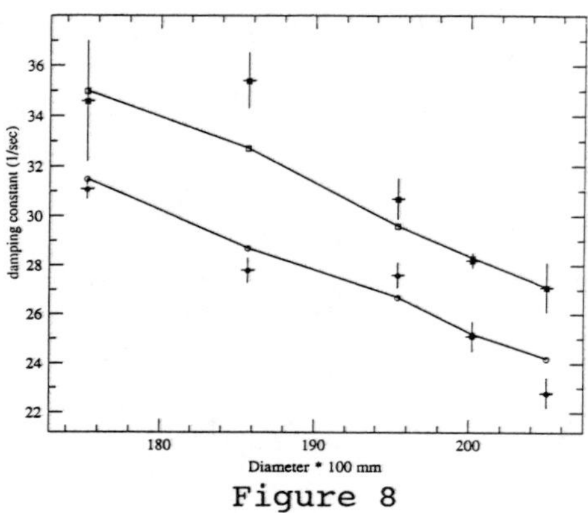

Figure 8

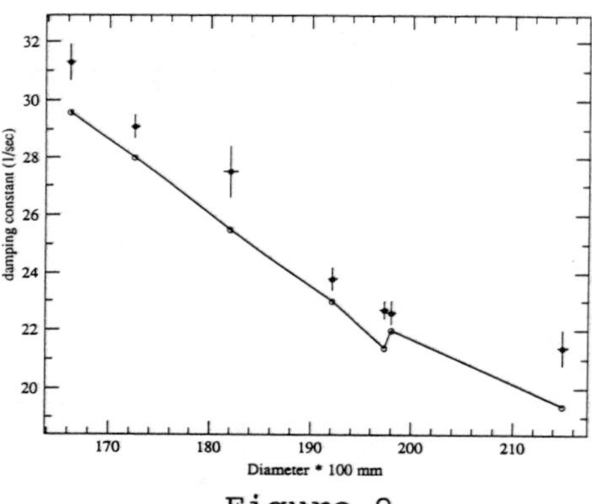

Figure 9

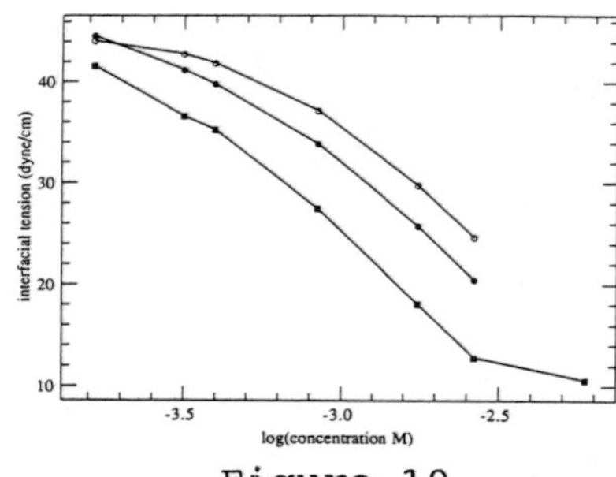

Figure 10

insoluble one. We check the consistency of this assumption after the equation of state is tested.

With the aid of the equation of state, Equation (2), we use Equation (1) and the measured frequency to infer the equilibrium interfacial tension. The interfacial tension is obtained on a MicroVax work station according to a procedure described in reference 14 (pp. 103-105). Once the interfacial tension is known, the damping constant can be calculated. In Figures 4 to 9, we compare the measured and calculated damping constant at different SDS concentrations.

For moderate (apparent) concentrations, 0.396 to 1.75 mM, the data agree well with the calculated values. But for the lower concentrations, 0.164 to 0.315 mM, the calculated values tend to be higher than the measured ones. The discrepancy, in fact, is an indication of the existence of the surface dilatational viscosity[14]. If we assign a surface dilatational viscosity ranging from 0.005 to 0.05 surface poise to the different hexane-water interfaces, the calculated damping constant becomes in good agreement with the experimental one. The introduction of the surface dilatational viscosity to the interface, however, has little effect on the calculated damping constant for the moderate concentrations, within 3 percent, since the effect of surface elasticity dominates in this region. The deviation of the damping constant from the calculated one for higher concentrations may be attributable to the fact that the "perfect gas" equation of state, Equation (2), is no longer satisfied because of interactions among surfactant molecules.

The interfacial tension obtained based on the equation of state is shown in Figure 10. Compared with the data obtained by Rehfeld[21] using drop's weight method, our results are lower. This may be due to impurity of the surfactants we used, or acoustic streaming associated with our method which helps the surfactants to aggregate onto the interface faster.

The results of interfacial tension are then substituted into Equation (2) to determine the surface concentration Γ. The adsorption length constant, ι, and the diffusion-adsorption number, λ, then can be estimated by assuming that the local bulk concentration near the interface, C_s, is the same as the apparent concentration, C, for $C = 0.164$ mM. It is found that $\iota \approx 7.9$ mm and $\lambda \approx 0.1$. Using the new value for λ and recalculating the interfacial tension and damping constant by the same logic outlined in Figure 7, we find that the change in interfacial tension is negligible, and the change is damping constant is about one percent. Therefore, the adsorbed film on the interface behaves like an insoluble one.

6. CONCLUDING REMARKS

Although we have been able to explain the response of an oil drop oscillating in an aqueous solution of surfactants or contaminants by imposing certain surface properties, it is impractical to infer the surface properties by the measured free oscillation frequency and damping constant alone. This is due to the fact that many surface parameters are involved, and their effects are intertwined. To reduce the complexities of the problem and to infer more meaningful results, additional information on some of the surface parameters are required. For example, in conjunction with a static method to measure the interfacial tension and its dependence on surfactant concentration, the technique can be a diagnostic tool for low to moderate interfacial viscosities.

ACKNOWLEDGMENT

This work was supported by the Office of Naval Research.

REFERENCES

1. R. S .Hansen & J. A. Mann, *J. Appl. Phys.* **35**, 152, 1964.
2. M.Tempel & R. P. Riet, *J. Chem. Phys.* **42**, 2679, 1965.
3. J. Lucassen & R. S. Hansen, *J. Colloid Interface Sci.* **22**, 32, 1966.
4. J. Lucassen & R. S. Hansen, *J. Colloid Interface Sci.* **23**, 319, 1967.
5. H. Löfgren, R. D. Neuman, L. E. Scriven, & H. T. Davis, *J. Colloid Interface Sci.* **98**, 175, 1983.
6. L Ting, D. T Wasan, and K. Miyano, *J. Colloid Interface Sci.* **107**, 345, 1985.
7. C. Hsu & R. E. Apfel, *J. Colloid Interface Sci.* **107**, 467,1985.
8. P. L. Marston, *J. Acoust. Soc. Am.* **67**, 15, 1980.
9. E. Trinh, A. Zwern, & T. G. Wang, *J. Fluid Mech.* **115**, 474,1982.
10. C. Hsu, Ph. D. Diss., "Interfacial Tension between Water and Selected Superheated Liquids by Quadrupole Oscillations of Drops", Yale University, 1983.
11. P. L.Marston, *J. Colloid Interface Sci.* **68**, 280, 1979.
12. P. L. Marston, & Trinh, E. H., *J. Acoust. Soc. Am. Supp. 1*, **77**, S20, 1985.
13. H. C. van de Hulst, **Light Scattering by Small Particles** (Dover), 263-5, 1981.
14. H. Lu, "Study of Interfacial Dynamics of Drops in the Presence of Surfactants or Contaminants", Ph. D. Diss., Dept. of Mechanical Engin., Yale Univ., 1988.
15. S. J. Rehfeld, *J. Phys. Chem.* **71**, 738,1966.
16. K. Durham, **Surface Activity and Detergency** (Macmillan & Co, London), 1961.
17. H. Kimizuka, L. G. Abood, T. Tahara, & K. Kaibara, *J. Colloid Interface Sci.* **40**, 27, 1972.
18. A. Yousef & B. J. Mccoy, *J. Colloid Interface Sci.* **94**, 497.
19. A. W. Adamson, 1982, **Physical Chemistry of Surfaces** (John Wiley & Sons), 4th edn., pp. 10-39, 56-8, 84, 112 -4, 117, 120-1, 484-485, 1983.
20. L. E. Reichl, **A Modern Course in Statistical Physics** (Univ. Texas Pr., Austin), 42-4, 66-8, 1980.
21. S. J. Rehfeld, *J. Phys. Chem.* **71**, 738, 1966.

FIGURE CAPTIONS

Figure 1 Schematic diagram of the experimental setup.

Figure 2 Frequency and damping constant versus diameter for hexane drops in pure water at 24 ± 2°C. The solid curves are the theoretical predictions based on Equation (1).

Figure 3 Frequency (3a) and damping constant (3b) versus diameter for hexane drops in SDS aqueous solution of different concentrations at 2.5 ± 1ΥC.

Figures 4-9 Damping constant versus diameter for hexane drops in the indicated SDS aqueous solution at 25 C. The open symbols connected by the line represent calculated results. The filled symbols represent measured results. The squares and circles denote data taken within five minutes and one hour after the drop is introduced into the solution, respectively.

Figure 10 Interfacial tension at different SDS concentrations. Circles and squares denote results obtained within five minutes and one hour after the drop is introduced into the solution, respectively. The open circles represent results measured by Rehfeld using drop weight method. Rehfeld's data were also taken within five minutes after the drop was formed.

EXPERIMENTS ON ROTATING CHARGED LIQUID DROPS

Won-Kyu Rhim, Sang Kun Chung, and Daniel D. Elleman
Jet Propulsion Laboratory, California Institute of Technology
4800 Oak Grove Drive, Pasadena, California 91109

ABSTRACT

Shapes and stabilities of freely rotating charged drops are investigated experimentally. Liquid drops approximately 3 mm in diameters were suspended in air by an electrostatic levitator and rotated by a torque which was acoustically generated in the levitation chamber. As the drop angular velocity was increased from the static state, families of axisymmetric, triaxial, dumbbell shapes, and eventual fissioning have been observed. With the assumption of "effective surface tension" by which the surface charge simply modified the surface tension of neutral liquid, the results from drops carrying low level surface charges agree quantitatively well with the Brown and Scriven's prediction. The normalized angular velocity at the bifurcation point agrees with the predicted value of 0.56 to within 3%. However, as the sample charges approached the Rayleigh limit, the results indicated a marked deviation from those of low charges. Drops of high charges showed asymmetric dumbbell shapes and the drop break-up directly from the axisymmetric shapes. Also discussed in this paper are methods of measuring drop charges and surface tension of highly viscous liquids in microgravity laboratories.

I. INTRODUCTION

As Swiatecki[1] eloquently stated, the problem of a rotating drop held together by a surface tension is "a special case of a single mathematical structure...which embraces in a unified manner the equilibrium of rotating masses representing at one extreme idealized atomic nuclei, at the other idealized heavenly bodies, and covering in between engineering applications in weightless space laboratories." Since Rayleigh's[2] initial calculation of axisymmetric shapes of rotating drops, the theoretical predictions have been brought to a rare degree of accuracy by Chandrasekhar[3], and more recently by Brown and Scriven[4] in their numerical studies on the stability of various equilibrium shapes of uncharged drops. However, the efforts to experimentally verify these predictions have been rather limited. Although Plateau's[5] experiment showed the existence of various drop shapes, they cannot be correlated with the theoretical predictions because they do not satisfy the basic condition of solid body rotation. Results from the drop rotation experiment performed in the microgravity laboratory by Wang *et al.*[6] show quantitative disagreement from the predictions, adding more puzzlement to the studies of the rotating drops.

In this paper we hope to clarify some of these puzzles by discussing our results obtained from charged drop rotation experiments. Drops were quietly suspended in air by the electrostatic forces, and the drop oscillation or rotation were acoustically induced. A brief description of our apparatus will be followed by a discussion on the drop oscillation modes and their characteristic frequencies, thus setting the stage for the analysis of the rotating drop experiments. After discussing rotating drops carrying high level charges, this paper will be concluded with a proposal for measuring surface tension of highly viscous liquids.

II. EXPERIMENTAL APPARATUS

The experimental system used in the present experiments was essentially the same electrostatic-acoustic hybrid system reported earlier[7,8] except for a slight modification to improve positioning stability. The electrostatic-acoustic hybrid system, as the name indicates, is composed of two parts, one for the electrostatic levitation, and the other for the drop manipulation by acoustic forces. As shown in

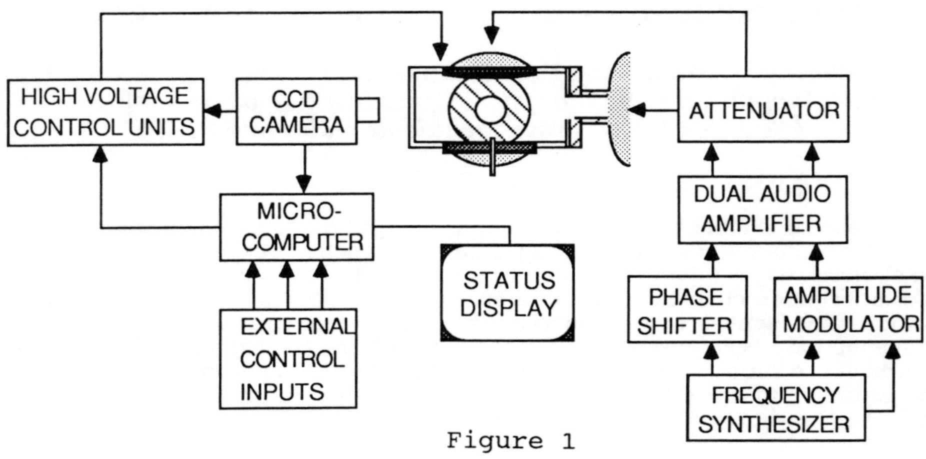

Figure 1

Figure 1, the levitation part comprises a pair of electrodes, a position sensing camera, a microprocessor, and a high voltage amplifier. The acoustic part uses an acoustic resonance chamber, a dual audio amplifier, a phase shifter, an amplitude modulator, and a frequency synthesizer. Figure 2 shows the schematic drawing of the levitation chamber used in the present experiments. The chamber was made of lucite, with inside dimensions of 15 cm x 15 cm x 4 cm. A pair of cylindrical electrodes were mounted at the centers of the top and bottom faces and two acoustic transducers were mounted on the two orthogonal side faces of the chamber. While we refer to the previous reports[8] for the detailed descriptions of design and operation of this system, we wish to mention here that drops approximately 3 to 4 mm in diameter can be routinely levitated in this system with a vertical position instability of less than 50 microns.

Assuming a pair of horizontal parallel electrodes, the balancing equation is given by

$$mg = QV/d \tag{1}$$

if fringing fields are neglected, and where m is the mass of the drop carrying charge Q, and V is the voltage difference between electrodes which are separated by d. For $Q = 4 \times 10^{-10}$ Coul., $m = 1.4 \times 10^{-5}$ kg, and $d = 3 \times 10^{-2}$ m, then V should be approximately 10 kv.

On the acoustic side, the outputs from the two-phase and frequency-locked synthesizers were amplified before driving transducers. Two transducers produced mutually orthogonal standing waves in the chamber which exerted appropriate acoustic stresses which induce drop rotation or oscillation. The acoustic torque produced on a spherical sample by the orthogonal acoustic waves is given by[9]

$$\tau = (3/2)\ (2\nu/\omega)^{1/2} (P_x P_y/2\ \rho c^2)\ A \sin \phi_o \tag{2}$$

where τ is the resulting torque, $(2\nu/\omega)^{1/2}$ is the viscous length with kinematic viscosity of air ν, P_x and P_y are the pressure amplitudes of acoustic waves, ρ is the density of the medium, and c is the velocity of sound in the medium, A is the surface area of the sample, and ϕ_o is the relative phase angle between the two waves. Following this formula in our experiments, a varying level of acoustic torque could be generated either by adjusting the sound pressure level while holding the relative

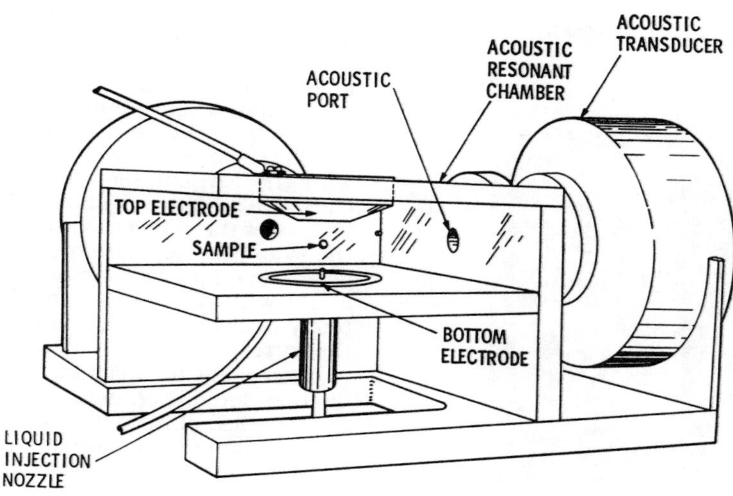

Figure 2

phase at 90° or by holding the pressure level fixed while varying the phase angle. In order to induce pure oscillation the relative phase angle was set at 0° and the wave amplitudes were sinusoidally modulated[10]. Referring to the earlier reports[7] for the system performance, we will close this section by mentioning that the system produced sufficient levels of pressure and torque required for the present work.

III. OSCILLATING DROPS

The oscillation frequency of a conducting charged liquid drop under the restoring force of its own surface tension was first derived by Lord Rayleigh[11]. Wong and Tang[12] arrived at the same results in the case of a non-conducting charged drop. They concluded that, in a non-viscous liquid drop, the hydrodynamic transport of surface charge is responsible in maintaining an equipotential surface at the surface of the drop at any instant. Rayleigh's expression for the small amplitude oscillation of a free conducting drop carrying charges Q is given by

$$\omega = [8\sigma(1-Q^2/16\pi R^3\sigma)/R^3\rho]^{1/2} \tag{3}$$

where σ and ρ are the surface tension and the density of the liquid, and R is the drop radius. This equation can be expressed in the following form;

$$(\omega/\omega_o)^2 = 1 - (Q/Q_R)^2 \tag{4}$$

where

$$\omega_o = (8\sigma/R^3\rho)^{1/2} \tag{5}$$

is the characteristic angular frequency of uncharged drop and

$$Q_R = (16\pi R^3\sigma)^{1/2} \tag{6}$$

is the Rayleigh critical charge. Equation 3 (or Equation 5) represents the characteristic frequencies of a charged (or uncharged) spherical drop which are degenerate with respect to all second harmonic modes. However, this degeneracy may be lifted if the drop shape deviates from a perfect spherical shape. Trinh *et al.*[13] reported the observation of three modes appearing at the three distinct frequencies in an oblate ellipsoidal liquid. They observed a well-defined pulsation mode. However, the other two modes showed complex shapes.

Our electrostatically levitated drops have slightly prolate spheroidal shapes with aspect ratios (defined by the vertical diameter/horizontal diameter) less than

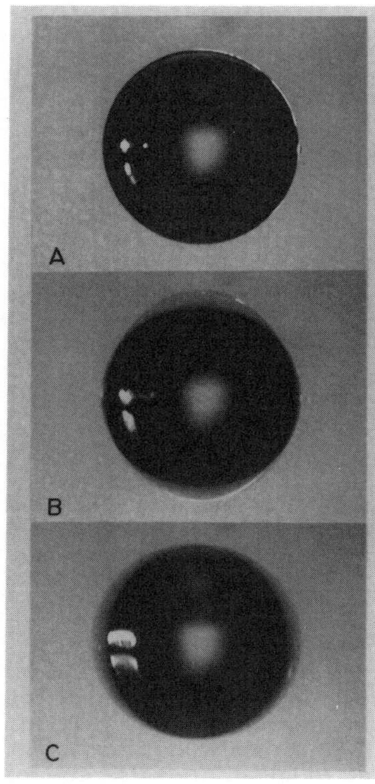

Figure 3

1.03. In each drop, we could identify both the pulsation and the toroidal modes which were appearing at the two distinct frequencies. The pulsation mode shows axisymmetric spheroidal shape along the vertical axis, and the toroidal mode shows triaxial shape with the vertical semi-axis remaining constant. Figure 3 shows the actual shapes of these two modes viewed from the side. The separation of these characteristic frequencies amounts to several Hz with the toroidal mode consistently showing higher frequencies than the corresponding pulsation mode.

According to Rosenkilde[14], a conducting liquid drop deformed by a uniform applied electric field should reveal three resonant modes with three distinct characteristic frequencies. These three modes are degenerate when the drop is spherical and they separate widely as the aspect ratio increases. In our experiments we have not observed the third mode, the so-called transverse shear mode. However, the other two modes are generally in agreement with the predictions in their modal shapes as well as in their characteristic frequencies (Fig. 3 of Ref. 14). One can observe in the same referenced figure that, in the small aspect ratio region, the toroidal frequency increases at a much slower rate than the rate at which the pulsation frequency decreases. This means that the drop oscillation frequencies measured from toroidal modes will give closer approximations to the resonance frequencies of spherical drops.

Figure 4 shows our experimental values for an evaporating water drop along with the Rayleigh's formula for the spherical charged drops (Equation 3). In this figure, the observed frequencies of charged drops were normalized by the fundamental frequency of uncharged free drops (Equation 5) of same sizes having surface tension of 67 dynes/cm. The observed charges were normalized by the Rayleigh critical value shown in Equation 6. As expected, the characteristic frequencies of the toroidal mode closely follows the theoretical line while the frequencies of the pulsation mode falls 2 to 5 Hz below the line in terms of the real frequency.

According to Adornato and Brown[15], a levitated liquid drop which is evaporating, while maintaining its surface charge, starts with a large aspect ratio which decreases as the drop loses its mass until it reaches the vicinity of its Rayleigh limit. Then the aspect ratio increases rapidly and the drop breakup takes place. This is shown in Figure 5 along with the theoretical curves from the reference. In this figure, Q and E are the dimensionless charge and electric field which are defined by $E = \widetilde{E}\,(\sigma/4\pi\varepsilon R)^{-1/2}$ and $Q = \widetilde{Q}\,(4\pi\varepsilon\sigma R^3)^{-1/2}$, where $\widetilde{E}$ and $\widetilde{Q}$ are dimensional quantities. Curves (a) and (b) are traces through which two evaporating charged drops will follow, each having different sizes and surface charges. Data points on curve (b) are from an evaporating water drop with its initial radius of 1.63 mm. Dashed

Dashed lines shown in the figure are equi-aspect ratio curves which coarsely divide the levitatable region into regions having different aspect ratios. From this figure we can see how the aspect ratio changes as the evaporation proceeds, therefore, its effect in the mode frequencies separation. Although the existing theories justify our observation qualitatively, the quantitative analysis requires accurate measurements of aspect ratios. We will postpone the further analysis to a future report. In the following section we will use the oscillation frequency so obtained to normalize the rotation frequency of the same drop in order to compare with theoretical predictions.

In closing, a comment can be made for the method of determining the drop charges of a free liquid drop in microgravity laboratories. An electrically neutral drop of known size can be positioned in an acoustic positioner and its resonant oscillation frequency can be measured. Then the surface tension is obtained from Equation 5. After injecting an unknown amount of charge to the same drop the frequency measurement is repeated, and with the help of Equation 3 the drop charge can be determined.

IV. ROTATING DROPS

A. Drops with low surface charges

For the rotating drop it was observed that as the drop gained angular momentum, the initial prolate spheroid passed through the spherical shape and then turned into an oblate spheroid with a decreasing aspect ratio until the bifurcation point was reached. When an actual drop shape is compared with an ideal spheroid having the same aspect ratio, only a minimal shape distortion can be noticed as shown in Figure 6. A slight flattening near the top is observed, while the bottom is slightly narrowed. Such deviation from an ideal spheroid must have been caused by the fact that the distribution of electrical stress was not symmetric over the surface of a charged drop in a uniform electric field and gravity.

The top and side views of a rotating water drop shown in Figure 7 were obtained using a high speed video recorder (Spin Physics 2000), equipped with two synchronized cameras at the frame rate of 1000 frames/sec. Starting from the static drop, point A in the figure, the acoustic torque was gradually increased. The general behavior of the shape evolution of a rotating drop is shown in Figure 8, in which the normalized drop sizes R(MAX)/R(0) are plotted against the normalized rotational frequencies Ω/ω, where R(MAX) and R(0) are the drop radius of the rotating drop measured in the equitorial plane and the corresponding dimension when the drop is static. The angular velocity Ω is normalized to the toroidal oscillation frequencies of the static drop for reasons described above. As the drop gains angular velocity, the initial prolate spheroid becomes spherical. Then, it turns into an oblate ellipsoid, with its eccentricity increasing along with further increase in velocity. When the drop reaches point C, it becomes unstable, and the drop shape evolves into a triaxial ellipsoid. From the bifurcation point on, the rotational rate gradually decreases for increasing torque. As the torque is further increased a patch of concavity grows around the axis of rotation, and the drop acquires a dumbbell shape. Then the lobes become more spherical as the connecting neck becomes thinner. Finally, it fissions into two identical drops, and in the case of a water drop, an approximately 200 micron-size satellite is produced at the center.

The top views of a rotating anhydrous glycerol drop are shown in Figure 9. The pattern of shape evolution is essentially same as water drops, except that the neck was thinned out instead of producing a satellite. Detailed analysis of the viscous liquid drops will be given in a future report.

The actual shape evolution of a rotating water drop can be seen more clearly if we plot an axis ratio a_3/a_1 against another ratio a_2/a_1, as shown in Figure 10, where a_1, a_2 and a_3 are the three orthogonal axes of the drop satisfying the relationship, $a_1 \geq a_2 \geq a_3$. The theoretical result by Chandrasekhar[16] is compared with the experimental point in this figure, for an ellipsoidal fluid mass of uniform density, rotating with a uniform angular velocity, and subjected only to the force of its

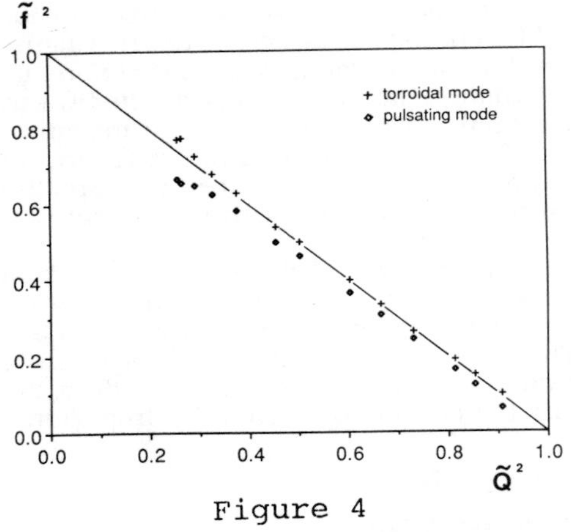

Figure 4

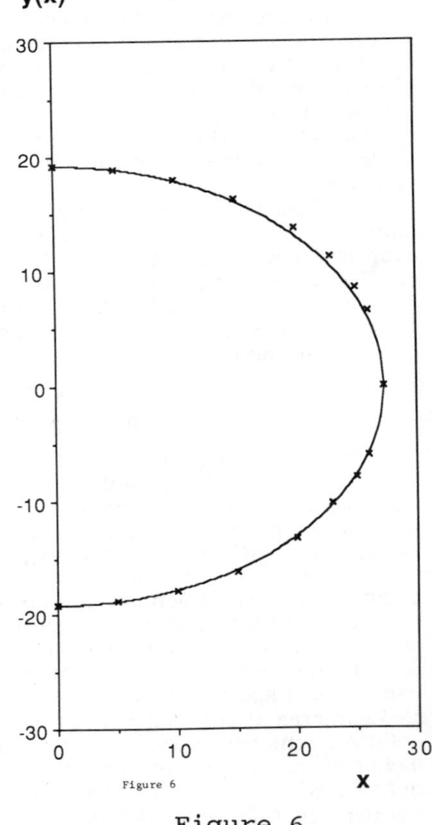

Figure 6

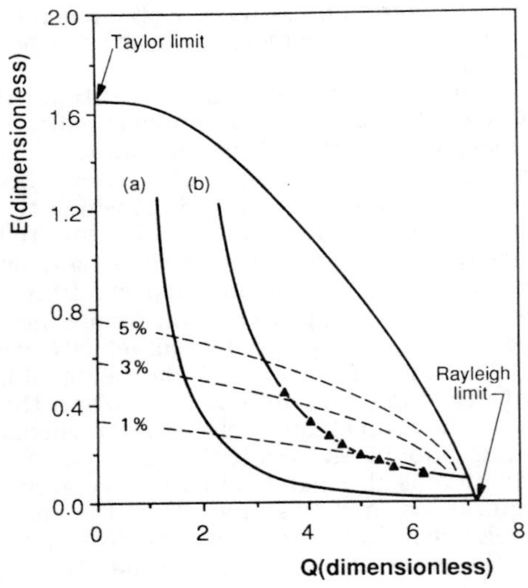

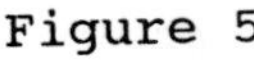

Figure 5

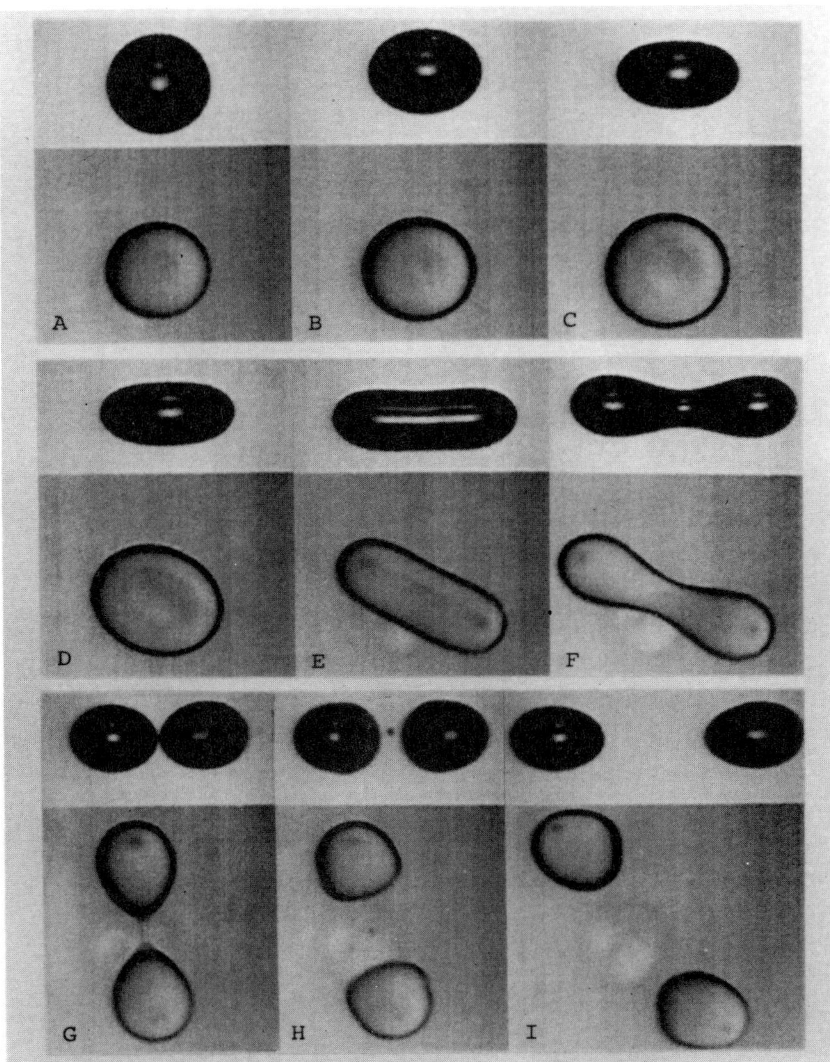

Figure 7

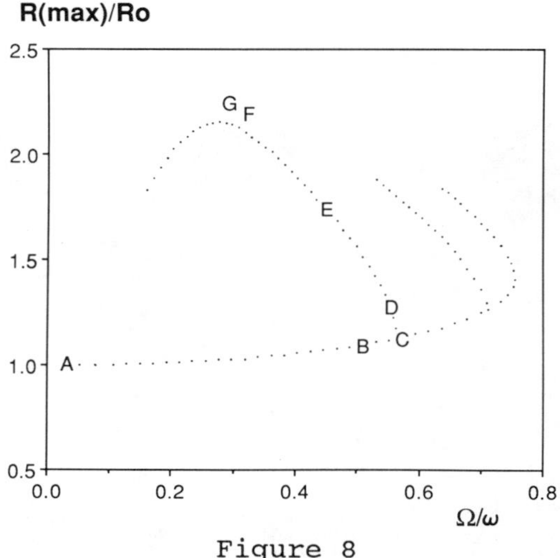

Figure 8

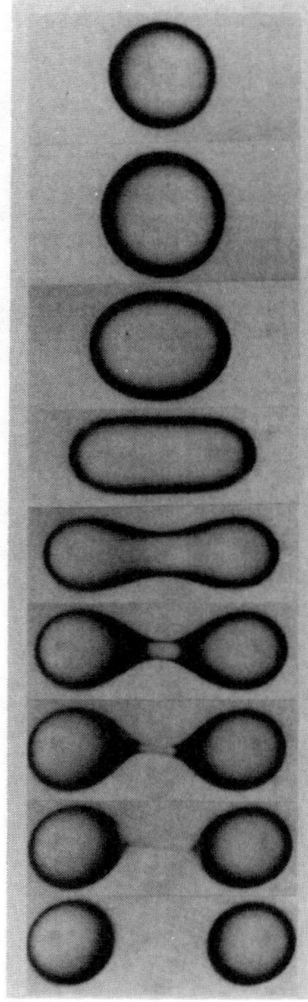

Figure 9

own gravitation. Also shown in alphabetical symbols are the corresponding points of shape evolution indicated in Figs. 7 and 8: points A, B, and C represent a sequence of the axisymmetric spheroids; point C (a_2/a_1 = 1 and a_3/a_1 = 0.5827) is the point of bifurcation; and points C, D, and E represent the triaxial ellipsoids. Starting from point E, the drop acquires a dumbbell shape with $a_2 \approx a_3$. Point F is the experimentally determined fissioning point which lands at a substantially lower point than $a_2/a_1 \approx 0.44$ for the self-gravitating ellipsoid. Two naturally occurred lunar globules (Swiatecki[1], Ross et. al.[17]) are also indicated in the figure. In the case of water the experimental points stay below the theoretical curve in the earlier portion of the triaxial branch, however, it is interesting to observe that these two different systems essentially follow a same curve.

The whole rotation process depicted in Figure 7 took approximately twelve seconds. Therefore, one might question the validity of a solid body rotation in this particular result. In a separate experiment, the acoustic torque was incremented or decremented, if necessary, in small steps and the drop was allowed to come to equilibrium for several seconds between steps to ensure the attainment of solid body rotation. With this approach it was possible to maintain the steady rotational rate close to the fissioning point. As shown in Figure 11, the agreement of this result with theory is excellent. In fact the small deviation of our experimental points may be reduced further if oscillation frequency from a spherical drop could be measured and used for frequency normalization. According to Rosenkilde[14], the oscillation frequency of a spherical drop should be about 1% less than the toroidal mode of prolate spheroid having an aspect ratio of 1.03, as was the case in the present experiment. If we make this correction to the frequency normalization, the agreement with the theory becomes nearly perfect.

B. Water drops with high surface charges

The above description on the shape evolution for drops with a small total charge holds true even for the drops carrying surface charges as high as 83% of Rayleigh charge, i.e. $Q/Q_R = 0.83$. However, drops with higher charge levels began showing marked deviation from the above description. In our experiments the levitated drops were allowed to evaporate until they acquired preset levels of surface charge. Measurements of charge were made by intermittently monitoring the drop sizes and the resonant frequencies. When the desired charge level was reached, the high speed video recorder was initiated, and the whole shape evolution was recorded as the acoustic torque was gradually increased. Figure 12 shows the results of five experiments when the charge ratio Q/Q_R was progressively increased. As the charge ratio approaches 1, the general trends are:

(i) the bifurcation point occurs progressively at the lower values of Ω/ω. In the case of $Q/Q_R = 0.93$, which was the highest charge level we could obtain prior to the break-up (the sequence E in Figure 11), the bifurcation point was at $\Omega/\omega \approx 0.485$ which was substantially less than the predicted value of 0.556;

(ii) In sequences B and C, the drops passed through triaxial ellipsoidal shapes. However, they evolved into asymmetric dumbbell (or peanut) shapes with unequal lobes. When the neck broke off, a satellite was observed in sequence B while it was absent in sequence C;

(iii) In sequences D and E, the drops became unstable upon bifurcation, and lost charge as one end became sharply pointed, which resembled the phenomena we have observed in an evaporating static drop when the drop approached the Rayleigh limit. From these observations the critical charge Q_c, beyond which no stable triaxial drop shapes can exist, is estimated to be approximately $Q/Q_R = 0.9$.

At the present time there exists no theoretical work for the quantitative comparison to our results. Our observations of various drop shapes carrying low surface charges and the drop break-ups which occurred near the high charge limit are at least qualitatively in agreement with the predictions emerged from the existing theories, which were well summarized in a diagram by Natarajan and Brown[18] as shown in

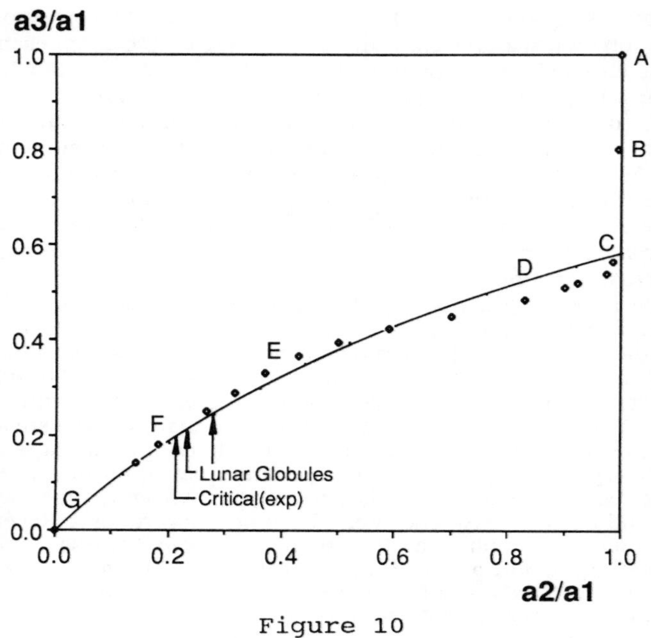

Figure 10

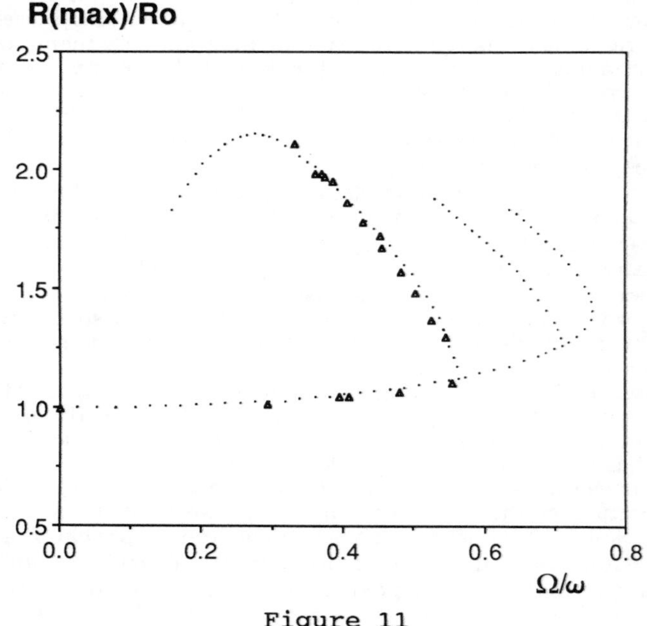

Figure 11

Figure 13. We close this section by quoting a statement by these authors: "The locus of limit points for drops in an electric fields and the locus of bifurcation points to non-axisymmetric shapes for rigidly rotating drops bound the region in parameter space where axisymmetric drop shapes are stable. Stable non-axisymmetric drop shapes do not exist near the Rayleigh limit. For rigidly rotating drops, the bifurcation that changes the stability of the oblate shapes is subcritical with respect to charge and angular momentum, and the resulting triaxial shapes are unstable. Stable triaxial shapes exist only for $Q < Q_c$, where the bifurcation is subcritical."

V. SURFACE TENSION MEASUREMENTS OF HIGHLY VISCOUS LIQUIDS

In the previous section, we have discussed the resonant drop oscillation of inviscid drop using Rayleigh's equation, Equation 3. This equation allows us to measure surface tension of inviscid liquids from the observed characteristic frequencies if the drop sizes and the surface charges are known. However, in the case of a highly viscous drop, the resonance is over damped, and the use of this equation for the surface tension measurement will not be possible.

Since the theory of rotating drops assumes solid body rotation and the viscosities of liquids do not come into play except for their practical role in achieving solid body rotation, the result shown in Figure 10 should hold regardless the liquid viscosity. One of the authors, D. D. Elleman[19], proposed earlier that the surface tension of a highly viscous liquid may be determined from the measurements of a drop rotation frequency and the corresponding shape of freely rotating drops. Since the drop angular frequency of an axisymmetric drop is difficult to measure in the homogeneous liquid, they proposed its measurement in the vicinity of bifurcation point on the triaxial branch. Having seen close agreements of the theory with the experimental observation, one can now rely on the dimensionless angular velocity, 0.559, at the bifurcation point and extract the surface tension using $\Omega_b/\omega_o = 0.559$, where Ω_b is the observed rotational angular velocity at the bifurcation point and ω_o is the oscillational angular velocity given by Equation 5 in the case of uncharged drops.

In the case of charged drops the characteristic frequencies for uncharged drops should be replaced by Equation 3. However, to do so requires the knowledge of the drop charges. For drops levitated in Earth laboratories the charge information can be readily obtained from Equation 1, the force balance equation, regardless of the viscosity of liquids. In microgravity laboratories, the drop charges and surface tension of low viscosity liquids can be measured according to the prescription given in the previous section. However, it will require a new method if one tries to obtain the drop charges of high viscosity liquids.

In order to confirm this idea we chose two liquids, water and anhydrous glycerol, and followed the prescription described above. The drop sizes and rotational frequencies were measured using a high speed video recorder. The drop charges were obtained by Equation 1 using the observed levitation voltage. The surface tensions so obtained were 67 dynes/cm in water and 62 dynes/cm in anhydrous glycerol showing fair agreements in both cases with the Handbook values of 72 dynes/cm and 63.4 dynes/cm respectively. No special care has been given to reduce measurement errors in the parameters used in the present experiments.

VI. CONCLUSION

Using liquid drops carrying various charge levels, the oscillating and rotating liquid drops have been experimentally investigated. Liquid drops approximately 3 mm in diameter were suspended in air by an electrostatic levitator and the drop oscillations and rotations were induced either by modulating acoustic pressure or torque produced in the levitation chamber. The pulsation and the toroidal oscillation modes have been identified and discussed on the bases of theories by Rosenkilde, and Adornato and Brown. Families of axisymmetric shapes, triaxial shapes, dumbbell shapes, and eventual fissioning have been observed. With the assumption of "effective surface tension", in which the surface charge simply modified the surface

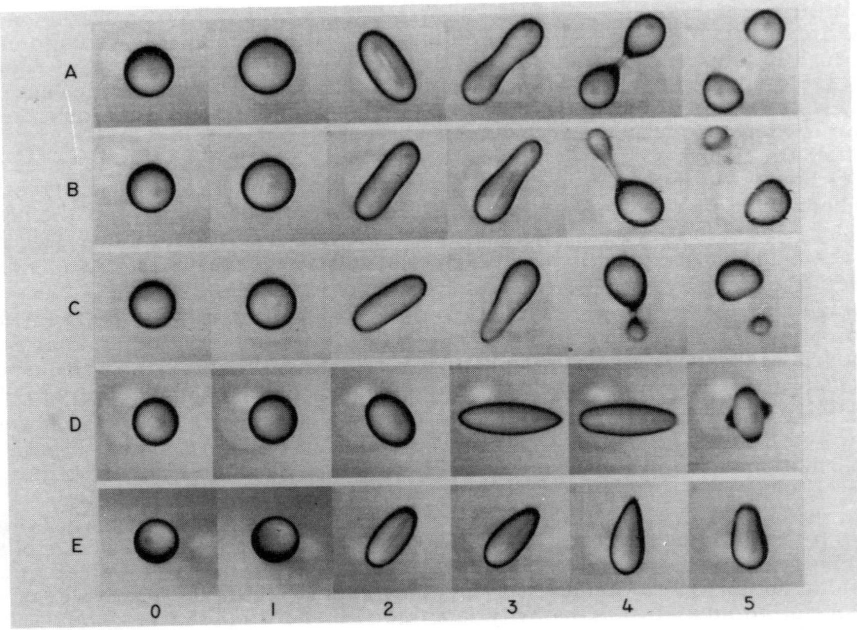

Figure 12

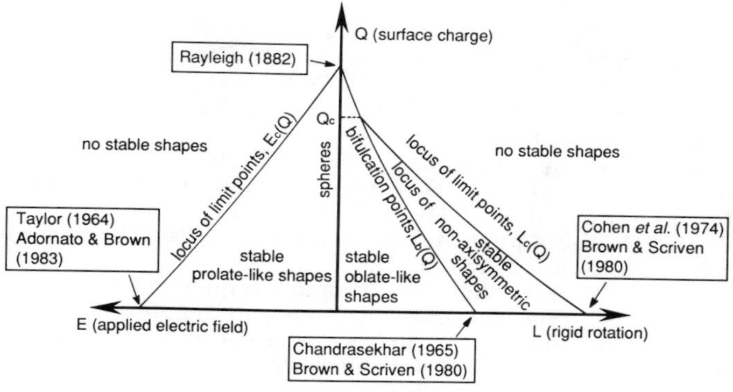

Figure 13

tension of the neutral liquid, the results from drops carrying low level surface charges agreed quantitatively well with Brown and Scriven's prediction. Especially the normalized rotational velocity at the bifurcation point agrees within 3% of the predicted value 0.56, when toroidal mode frequency was used for the normalization. However, as the sample charges approached the Rayleigh limit, we observed asymmetric dumbbell (or peanut) shapes, and also the direct drop break-up from the axisymmetric shapes. Various drop break-ups that occurred near the high charge limit are in qualitative agreement with the prediction emerging from the existing theories, although no exact theory exists at this time which can account for the present experimental results. We believe, with today's computational facilities, numerical analysis of our experiments is quite possible, and we hope that our present work may stimulate of such efforts.

Also discussed in this paper are possibilities of measuring drop charges of low viscous liquids by drop oscillation method, and of measuring the surface tension of highly viscous liquids by drop rotation method, which are applicable in the microgravity laboratories.

ACKNOWLEDGEMENTS

The authors would like to acknowledge the extraordinary assistance of Dr. E. Trinh, Mr. D. Barber , and Ms. A. Belcastro.

This work was carried out at the Jet Propulsion Laboratory, California Institute of Technology, under contract with NASA.

REFERENCES

1. W. J. Swiatecki, "The Rotating, Charged or Gravitating Liquid Drop, and Problems in Nuclear Physics and Astronomy," **Proc. Int. Coll. on Drops and Bubbles, 1**, 52, 1974.
2. Lord Rayleigh, *Philos. Mag.* 28, 161, 1914.
3. S. Chandrasekhar, "The Stability of a Rotating Liquid Drop," *Proc. R. Soc. London*, **A 286**, 1, 1965.
4. R. A. Brown & L. E. Scriven, "The Shape and Stability of Rotating Liquid Drops," *Proc. R. Soc. London*,**A 371**, 331, 1980.
5. A. F. Plateau, **Annual Report of the Board of Regents of Smithsonian Institution**, 270, 1863.
6. T. G. Wang, E. H. Trinh, A. P. Croonquist, & D. D. Elleman, "Shapes of Rotating Free Drops: Spacelab Experimental Results," *Phys. Rev. Lett.* **56**, 452, 1986.
7. W. K. Rhim, M. Collender, M. T. Hyson, W. T. Simms, & D. D. Elleman, "Development of an Electrostatic Positioner for Space Material Processing," *Rev. Sci. Instrum.* **56**, 307, 1985. *Also* W. K. Rhim, S. K. Chung, M. T. Hyson, & D. D. Elleman, "Charged Drop Levitators and Their Applications," *Proc. Mat. Res.*

Soc. Symp. **87**, 103, 1987.
8. W. K. Rhim, S. K. Chung, M. H. Hyson, E. H. Trinh, & D. D. Elleman, "Large Charged Drop Levitation against Gravity," *IEEE Trans. Ind. Appl.* **IA-23**, 975, 1987. *Also* W. K. Rhim, S. K. Chung, E. H. Trinh, & D. D. Elleman, "Charged Drop Dynamics Experiment using an Electrostatic-Acoustic Hybrid System," *Proc. Mat. Res. Soc.* **87**, 329, 1987.
9. F. H. Busse & T. G. Wang, *J. Fluid Mech.* **158**, 317, 1985.
10. P. Annamalai, E. T. Trinh, & T. G. Wang, "Experimental Studies of the Oscillation of a Rotating Drop," *J. Fluid Mech.*, **158**, 317, 1985.
11. Lord Rayleigh, "On the Equilibrium of Liquid Conducting Masses Charged with Electricity," *Philos. Mag.* **14**, 184, 1882.
12. C. Y. Wong & H. H. K. Tang, "Vibrational Frequency of a Non-Conducting Charged Liquid Drop", **Proc. Int. Coll. on Drops and Bubbles**, **1**, 79, 1974.
13. E. H. Trinh, P. L. Marston, & J. L. Robey, "Acoustic Measurement of the Surface Tension of Levitated Drops," *J. Colloid Interface Sci.*, **124**, 95, 1988.
14. C. E. Rosenkilde, "A Dielectric Fluid Drop in an Electric Field," *Proc. R. Soc. London*, **A 312**, 473, 1969.
15. P. M. Adornato & R. A. Brown, "Shape and Stability of Electrostatically Levitated Drops," *Proc. R. Soc. London*, **A 389**, 101, 1983.
16. S. Chandrasekhar, **Ellipsoidal Figures of Equilibrium**, Yale Univ. Press, New Haven, 1969, *and also* N. R. Lebovitz, "The Fission Theory of Binary Stars," **Proc. Int. Coll. on Drops and Bubbles**, **1**, 1, 1974.
17. J. Ross, John Bastin, & K. Stewart, "The Numerical Analysis of the Rotational Theory for the Formation of Lunar Globules," **Proc. Int. Coll. on Drops and Bubbles**, **2**, p 350, 1981.
18. R. Natarajan & R. A. Brown, "The Role of Three Dimensional Shapes in the Break-up of Charged Drops," *Proc. R. Soc. London*, **A 410**, 209, 1987.
19. D. D. Elleman, T. G. Wang, & M. Barmatz, "Acoustic Containerless Experiment System: A Non-Contact Surface Tension Measurement," NASA Tech. Memo. 4069, **Vol. 2**, 557, 1988.

FIGURE CAPTIONS

Figure 1 Block diagram of the Apparatus.

Figure 2 Schematic drawing of the electrostatic-acoustic chamber.

Figure 3 Side views of a drop approximately 3 mm in diameter. A is the static shape, B shows the drop shape when the drop undergoes pulsation mode of oscillation, and C is when the drop is in the toroidal mode of oscillation (C).

Figure 4 Normalized oscillation frequencies of pulsation and toroidal modes vs. normalized charges in an evaporating charged water drop. The solid line is from Equation 3, the Rayleigh's formula for a charged drop, where $\tilde{f} = \omega/\omega_o$, $\tilde{Q} = Q/Q_R$, $Q = 6.68 \times 10^{-10}$ Coul., and 67 dynes/cm was used for the surface tension.

Figure 5 Locus of stability limit points of electrostatically levitated charged drops expressed in dimensionless Q vs. E space from ref. 14. Curves (a) and (b) are traces through which two evaporating drops will follow each of which carry surface charges 2×10^{-10} Coul. and 6.6×10^{-10} Coul. respectively. Data points on (b) are from our experiment, with a drop having charges 6.6 $\times 10^{-10}$ Coul. and its initial radius 1.63 mm. Also shown here are equi-aspect ratio curves expressed in percentage.

Figure 6 Comparison of a rotating axisymmetric water drop with an oblate spheroid of same aspect ratio. Here, R = 1.61 mm, $Q = 6.69 \times 10^{-10}$ Coul., E = 254 Kv/m, s = 67×10^{-3} Newton/m, and the rotation frequency was 25 Hz.

Figure 7 Top and side views of a rotating water drop. The diameter of an equivalent sphere is R = 1.61 mm and $Q = 6.69 \times 10^{-10}$ Coul., A is the static drop, C is the drop near the bifurcation point, and the rest are self explanatory.

Figure 8 Comparison of Figure 6 with Brown and Scriven's theory. A, B, C, D,... in this figure correspond to Figure 6, and R(max)/Ro and Ω/ω are as defined in the text.

Figure 9 Top views of a spinning anhydrous glycerol drop having R = 1.69 mm and $Q = 7.2 \times 10^{-10}$ Coul. The surface tension measured from the bifurcation point was 62.1 dynes/cm which agrees well with the Handbook value 63.4 dynes/cm.

Figure 10 a_3/a_1 vs. a_2/a_1 from Figure 6, where $a_1 \geq a_2 \geq a_3$ are semi-axis of the rotating drop and the solid line is by Lebovitz in the reference 13. $a_2/a_1 \approx 0.21$ is the experimentally observed critical point at which drop fission took place. Also shown here are two naturally occurred lunar globules shown in references (1) and (17).

Figure 11 Shape evolution of a spinning water drop. The normalized shape dimensions vs corresponding normalized angular frequencies are compared with the theory. The rotation frequencies were obtained ensuring solid body rotation and the experimentally obtained toroidal oscillation frequencies were used for normalization. Here, R = 1.65 mm and $Q/Q_R = 0.51$.

Figure 12 Rotating charged water drops near Rayleigh limit. The measured radius (in mm) and Q/Q_R in each sequence are A(1.2, 0.83), B(1.19, 0.87), C(1.13, 0.89), D(1.15, 0.92), and E(1.11, 0.93).

Figure 13 Schematic of transitions for charged, conducting drops caused by an applied electric field and by rigid rotation (from ref. 18).

DYNAMICS OF ROTATING AND OSCILLATING FREE DROPS

T. G. Wang
Center for Microgravity Research and Applications
Vanderbilt University, Nashville, TN 37235

E. H. Trinh, A. P. Croonquist and D. D. Elleman
Jet Propulsion Laboratory, California Institute of Technology
Pasadena, CA 91109

ABSTRACT

The experimental observation of the behavior of acoustically rotated and oscillated free drops in the microgravity environment of low earth orbit has yielded quantitative results on the gyrotational equilibrium shapes and the oscillation frequency of a liquid spheroid. Positioning techniques using the effects of acoustic radiation pressure were used during the Spacelab 3 flight to carry out this classical fluid mechanics experiment.

INTRODUCTION

The surface tension controlled shapes of liquid drops in gyrostatic equilibrium have been theoretically considered as part of the more general problem treating the dynamics of rotating masses under the dominant effects of gravitational, electrical, or even nuclear forces. The modeling of the behavior of planets, distant stars, and atomic nuclei is the obvious motivation for undertaking the solution of this problem, and an elegant summary of the subject has been presented by Swiatecki.[1] Until very recently, however, experimental verification of these theoretical predictions has not been possible to obtain even for the prosaic case of a liquid drop under the influence of surface tension and not submitted to an overwhelming gravitational field. The availability of experimental instrumentation aboard the NASA Space Shuttle and the implementation of acoustic positioning techniques[2], have allowed the investigation of the surface tension-dominated equilibrium shapes of rotating drops with gravitational acceleration reduced by a factor of at least 10^{-3}.

ACOUSTIC POSITIONING AND TORQUE APPLIED TO LIQUID DROPS IN MICROGRAVITY

Available experimental evidence appears to indicate that the steady-state acceleration levels in the Spacelab 3 module during a gravity gradient posture is on the order of 0.001 (where the gravitational level is at sea level on earth). Our observations reveal that the acoustic positioning forces generated in the experiment chamber (DDM chamber) were capable of containing a 7 cc liquid drop having a density of 1.18 g/cc when subjected to the steady-state level of residual acceleration, but were unable to restrain liquid drops during spacecraft maneuvers characterized by peak acceleration levels of 0.1 g.

The acoustic restoring force in a one dimensional standing wave may be expressed as[1]

$$F=\frac{5\pi}{6}\left(\frac{P_i^{\,2}}{\rho c^2}\right)k R^3 \sin 2kx,$$

where P_i is the pressure amplitude, ρ and c are the density and sound velocity characteristic of air, k is the wave vector, and R is the radius of a spherical sample. Theoretically, the restoring force is reduced to zero in the middle of the chamber, and increases sinusoidally as the sample deviates from that position. In practice, the force profile is not sinusoidal, and is fairly distorted due to scattering effects from the sample.

During the operation of the experiment, the acoustic pressure levels were kept between 135 and 145 dB, and little static distortion of the drop was observed during normal operating sequences. This absence of static, acoustically induced distortion is crucial to the viability of acoustical manipulation techniques as experimental tools for material science and fluid dynamics investigations.

Should a freely suspended drop receive an impulse from a transient acceleration spike, it will undergo translatory oscillations within the potential known. Figure 1 illustrates such translational oscillations. The magnitude of the oscillations along the Y axis (normalized to the drop radius) is plotted as a function of time. The frequency of the oscillations correlate with the SPL of the standing wave along the same axis: higher frequencies of oscillations are associated with relatively higher SPL.

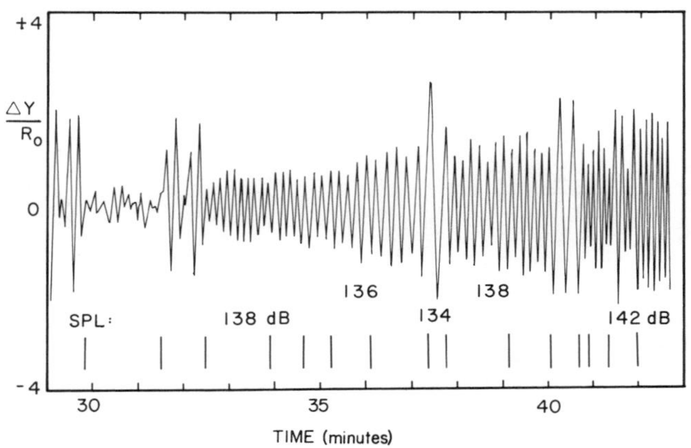

Figure 1 Transitional drop motion in the Y direction

When two sides of the acoustic resonant cavity have equal length, it is possible to generate a steady-state torque with appropriate phasing of the associated acoustic waves. When the two waves along the x and y axes are related by +90°(or -90°), a torque with direction vector in the +z (or -z) direction is induced, and drives a clockwise (counterclockwise) rotation of a sample suspended in the center of the chamber. The theoretical expression of this torque is given by[2]

$$T = \frac{3}{2} L_\eta \left(\frac{P_x P_y}{2\rho c^2} \right) A \sin \phi_o$$

where T is the torque, L_η the acoustic boundary layer (or viscous length) defined as $(2v/w)^{1/2}$, P_x and P_y the pressure amplitudes of the waves in the x and y direction respectively, A the total surface area of the sample, v the kinematic viscosity of the air, w the angular frequency of the sound wave, and ϕ_o the phase angle between the waves along the x and y axes.

By knowing the torque acting on the liquid drop, the deformation of the drop as it rotates, and the drag coefficient of the air, it is possible to plot the rotation rate of a given sample as a function of time. The results of such a calculation is reported in Figure 2 for a 3 cc drop of water/glycerin mixture with a kinematic viscosity of 100 centiStokes. The theoretical curve has been obtained by fitting the experimental

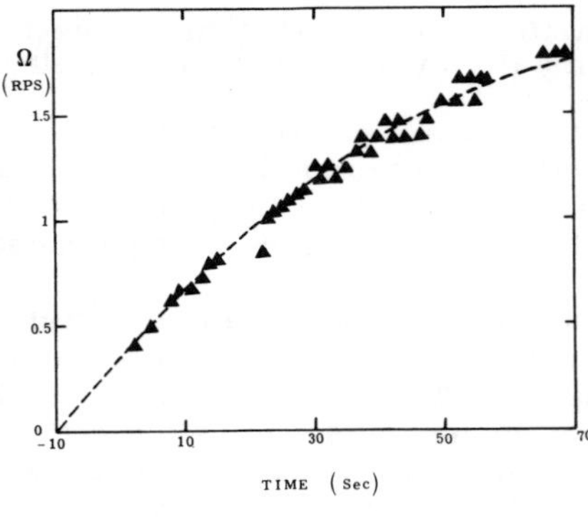

Figure 2

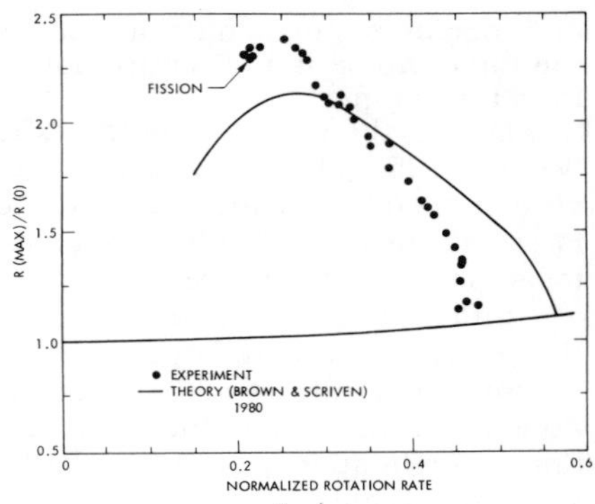

Figure 5

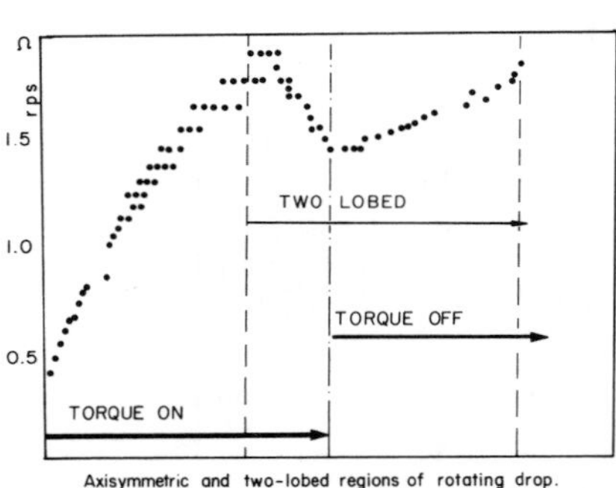

Axisymmetric and two-lobed regions of rotating drop.

Figure 3

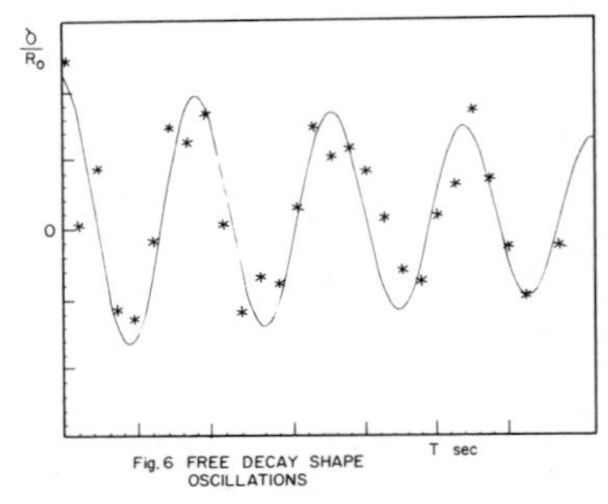

Fig. 6 FREE DECAY SHAPE OSCILLATIONS

Figure 6

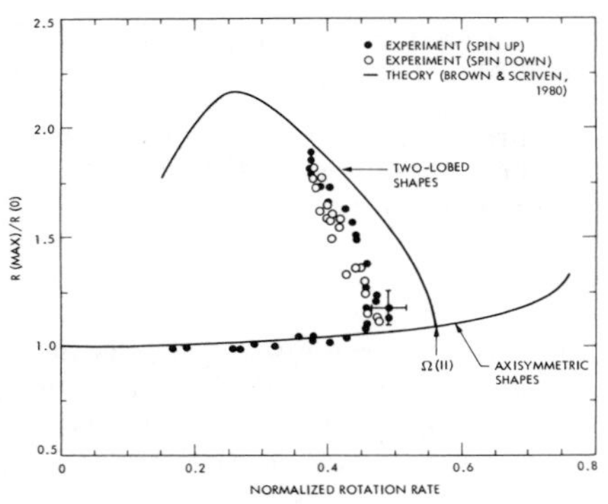

Figure 4

data with the air drag coefficient as adjustable parameter (the data included in Figure 2 are for a drop having a shape axisymmetric with respect to the axis of rotation).

EXPERIMENT DESCRIPTION

An initially non-rotating drop was subjected to an acoustically generated torque which caused it to spin up with increasing velocity. The rotational velocity of the liquid was determined through the motion of immiscible tracer particles suspended within the primary drop. The time variation of the shape of the rotating drops was determined by recording the drop profile along three orthogonal views on 16 mm motion picture film.

A typical experimental sequence involves several spin-ups and spin-downs of a 1100cSt drop.

EQUILIBRIUM SHAPES OF ACOUSTICALLY ROTATED DROPS

Figure 3 reproduces the experimentally measures rotation velocity of a 3 cc water/glycerine drop as the acoustic torque is applied and removed. The horizontal axis displays time. As the drop is spun up, it flattens at the pole and bulges out at the equator, remaining axisymmetric and gaining in rotation speed. At the bifurcation point, the two-lobed shape becomes the stable equilibrium geometry and the drop slows down because the moment of inertia increases as well as the surface area. Although the speed decreases, the largest dimension of the drop cross section continues to increase leading to an eventual fission of the liquid into two separate drops. In this particular case, however, the data of Figure 3 shows that the acoustic torque is turned off before fission is allowed to occur, and the rotation speed increases again as the drop stretch is reversed. The axisymmetric equilibrium shape is recovered when the same rotation speed than that measured at bifurcation during the spin-up phase is reached. No "hysteresis" has been detected with the present experimental uncertainty.

Figure 4 reproduces experimental data plotted together with available theoretical predictions. In this case, the largest dimension in the rotating drop equatorial cross section (measured perpendicularly to the rotation axis) divided by the radius of the nonrotating drop, is plotted as a function of the rotation rate divided by the frequency of the fundamental mode of shape oscillation. This frequency is given by

$$F = \frac{1}{2\pi}\left(\frac{8\sigma}{\rho^* c^2}\right)^{\frac{1}{2}}$$

where σ is the surface tension of the liquid and ρ^* its density. The liquids used in these experiments were water, a series of water and glycerin mixtures of increasing viscosity (from 10 to 500 cSt), and finally Silicone oil. The density was between 1.0 and 1.18 g/cc, and the surface tension had values between 20 and 70 dynes/cm.

The results in Figure 4 have been obtained with a 100 cSt liquid drop 3 cc in volume, using rotational acceleration of about 0.01 revolution/second. Under these particular circumstances, solid body rotation is easily attained, and the effects of differential rotation are minimized. The behavior of the drop should then approach that of a fluid mass rotating at constant velocity since the rotation rate changes very little during the characteristic time required for reaching solid body rotation. This assumption would not be valid should the viscosity significantly decrease.

The data reveal a very good agreement with the theoretical predictions corresponding to the axisymmetric regime. On the other hand, a quantitative confirmation of the theoretical prediction was not obtained for the specific value of the reduced rotation rate at the bifurcation point. Experimental evidence suggests that the onset of secular instability for the axisymmetric shape is located at a lower rotational velocity than theoretically predicted.

Figure 5 reproduces both experimental and theoretical results for an experimental sequence including a drop fission. Once again a qualitative agreement is obtained, but experimental data reveal a much faster increase in deformation with the decrease in rotation velocity. Fission generally produces two main drops of equal volume and a satellite droplet arising from the breakup of the liquid bridge forming the central region of the stretched rotating single drop. Assessment of the volumes of the drops resulting from fission is difficult to obtain, but their volumes were probably equal to within 10%. No evidence for two-lobed configurations with greatly different volume for the lobes has been obtained.

SHAPE OSCILLATIONS AND MEASUREMENT OF SURFACE TENSION AND VISCOSITY

The part of the experiment dealing with shape oscillation studies has been curtailed due to the subnominal performance of phase control which resulted in the inability to completely null out the acoustic torque. It was nevertheless possible to obtain free decay measurement of the frequency and damping of shape oscillations in order to carry out surface tension and viscosity measurement.

Figure 6 reproduces the experimental results obtained for a freely decaying 4.5 cc drop of water/Glycerin mixture having a viscosity of 10 cSt. A fit of the experimental data using an exponentially decaying sine wave yielded a surface tension of 60 dynes/cm and a viscosity of 12 cSt. The experimental uncertainty is primarily due to the relatively low frame rate used for the 16 mm camera.

Nonlinear effects arising during large amplitude oscillations of the drops were among the unfulfilled experimental goals due to lack of operation time as well as sub-nominal control over the nulling of the acoustic torque. In ground-based experiments, the resonance frequencies for shape oscillation decrease as the amplitude grows larger. The scant data available from this flight does not appear to confirm this finding, but hints to the existence of a hard non-linearity. The results are very inconclusive, however, and these experiments must be repeated in possible subsequent flight. Additional phenomena to be studied in the future include the oscillary dynamics of a rotating drop, the behavior of a compound drop, and detailed behavior of particles inside a rotating and oscillating drop.

SUMMARY

These experimental data obtained in the low-gravity environment of space have yielded a comforting confirmation of the various analytical and numerical predictions regarding the axisymmetric equilibrium shapes of rotating drops and a surprising development in the early onset of secular instability experienced by those shapes. This may mean that the drop shape is less stable to fluctuations than the calculations have predicted. That no mode higher than the two-lobed shape was observed also lends support to the diminished stability argument.

CONCLUDING REMARKS

It must be reiterated that the availability of greatly reduced gravitational acceleration condition is essential to the rigorous performance of this experiment. This seemingly simple problem involves boundary conditions which are theoretically elementary, but are an earthbound experimentalist's nightmare. Neutral buoyancy systems may be used to remove the effects of the gravitational field, but change the boundary conditions through viscous and inertial stresses. Levitation of free drops in a gaseous atmosphere on earth introduces non-negligible drop distortion as well as artifacts due to the levitating acoustic or electrical fields.

In addition to the results mentioned above, the operations of the Drops Dynamics Module during the flight of STS 51B has yielded yet another confirmation of the benefits of man's intervention when a performance-threatening malfunction has occurred in the hardware. In this case, it was only the intervention of the experimenter (payload specialist) under the guidance of a ground engineering team which allowed the apparatus to perform to the extent where meaningful scientific data could be gathered.

ACKNOWLEDGEMENTS

This work is part of the Spacelab III mission. The authors wish to acknowledge the support provided to them by the DDM team and the MSFC POCC team. They are indebted to D. McFarland for his continuous support. The authors also wish to thank R. White, W. Hodges, A. Villamil and J. Robey for their contributions.

The research described in this paper was carried out at the Jet Propulsion Laboratory, California Institute of Technology, under contract with the National Aeronautics and Space Administration.

REFERENCES

1. E. Leung, N. Jacobi, & T. G. Wang, *J. Acoust. Soc. Am.* **70**, 1762 (1982).
2. F. H. Busse, T. G. Wang, *J. Acoust. Soc. Am.* **69**, 1634 (1981).
3. E. Trinh, T. G. Wang, J. Fluid Mechs. 122, p. 315 (1982).
4. W. J. Swiatecki, "The rotating, charged, or gravitating liquid drop and problems in nuclear physics and astronomy," in **Proc. of the (First) Int. Colloquium on Drops and Bubbles** (D. J. Collins, M. S. Plesset, & M. M. Saffren, eds.) Jet Propulsion Laboratory, Pasadena (1974).
5. T. G. Wang, M. M. Saffren, & D. D. Elleman, "Drop dynamics in space," in **Materials Sciences in Space with Applications to Space Processing** (Ed. L. Steg) American Institute of Aeronautics and Astronautics, New York (1977).
6. S. Chandrasekhar, "The stability of a rotating drop," *Proc. R. Soc. London* *A***286**, 1-26 (1965).
7. D. K. Ross, "The stability of a rotating liquid mass held together by surface tension," *Austr. J. Phys.* **21**, 837-844 (1968).
8. R. A. Brown & L. E. Scriven, "Shape and stability of rotating liquid drops," *Proc. R. Soc. London A***371**, 331-357 (1980).
9. J. A. F. Plateau, "Experimental and theoretical researches on the figures of equilibrium of a liquid mass withdrawn from the action of gravity," **Annual Report of the Board of Regents of the Smithsonian Institution**, 270-285, Washington, D. C. (1963).
10. R. Tagg, L. Commack, A. Croonquist, & T. G. Wang, "Rotating liquid drops: Plateau's experiment revisited," Report 900-954 Jet Propulsion Laboratory, Pasadena, (1979).
11. F. H. Busse & T. G. Wang, "Torque generated by orthogonal acoustic waves theory," *J. Acoust. Soc. Am.* **69**, 1634-1639 (1981).
12. T. G. Wang, E. H. Trinh, A. P. Croonquist, & D. D. Elleman, , "Shape of rotating free drops: Spacelab experimental results," *Phys. Rev. Lett.*, **56**, 452 (1986).

FIGURE CAPTIONS

Figure 1. Translational drop motion in the *Y* direction.
Figure 2. Rotational rate of axisymmetric drop. Theoretical fit for drop spin-up. Data indicated by ▲, theory by -----.
Figure 3. Axisymmetric and two-lobed regions of rotating drop.
Figure 4. Comparison of experimental with theoretical results for a 100cSt liquid drop.
Figure 5. Experimental data for drop fission due to rotation.
Figure 6. Free decay of shape oscillation.

2. Microgravity Science & Space Experiments

Brad Carpenter
Session Chair

RESPONSE OF CONVECTIVE-DIFFUSIVE TRANSPORT TO SPATIAL AND TEMPORAL VARIATIONS IN EFFECTIVE GRAVITY

J. Iwan D. Alexander, Jalil Ouazzani and Franz Rosenberger
Center for Microgravity and Materials Research
The University of Alabama in Huntsville, Huntsville Alabama 35899

ABSTRACT

The reduced gravity environment on board a spacecraft in low earth orbit gives materials scientists and fluid physicists the opportunity to undertake experiments under conditions that reduce or eliminate buoyancy driven fluid convection in comparison to earth based conditions. As a consequence, the relative importance of heat and mass transport by diffusion is increased. In crystal growth this might be expected to lead to a more uniform crystal composition than would be obtained under terrestrial conditions. The process of crystal growth by the Bridgman technique is chosen as a case study. Two and three dimensional numerical models are used to examine the response of heat, mass and momentum transport to conditions characteristic of the microgravity environment. It is shown that the orientation of the experiment with respect to the steady component of the residual gravity is a crucial factor in determining the suitability of the spacecraft as a means to suppress or eliminate unwanted effects caused by buoyant fluid motion. The process is also extremely sensitive to transient disturbances. For example, a 3×10^{-3} g impulse of one second duration acting parallel to the interface of a growing crystal produces a response in the solute field which lasts for nearly 2000 seconds. Consequently lateral and longitudinal compositional variations occur over a length of nearly 6 mm in the grown crystal.

1. INTRODUCTION

For over two decades, particularly since the operation of Skylab in the early seventies, the prospect of undertaking experiments aboard a spacecraft in low earth orbit has appealed to scientists from several disciplines for a variety of reasons. In addition to serving as useful base for cosmologists and astronomers, an orbiting spacecraft provides an opportunity to study physical phenomena in the absence of the 9.8 ms^{-2} (1 g) acceleration experienced in an earth based laboratory. The various phenomena that have been the subject of space experiments have been described in a number of recent books and articles (Abduyevsky 1985; Feuerbacher *et al.* 1986; Hazelrigg & Reynolds 1986; Walter 1987).

Physical systems which respond to buoyancy forces will exhibit different behavior in space than in an earth based laboratory. A simple example is the behavior of bubbles or liquid drops with densities that differ from the host liquid. At the earth's surface, these move up or down in response to gravitational acceleration. In a spacecraft they will no longer exhibit such predictable behavior. More complex examples include the surface shapes of liquid films, bridges and drops, and mass transfer in crystal growth systems (which we shall examine in this work). Under earth-based conditions mass transfer conditions can be strongly affected by buoyancy driven convection in the (fluid) nutrient phase. In the reduced gravity environment of a space laboratory, mass transfer is more likely to be controlled by diffusion (Rosenberger 1979; Sekerka & Coriell 1979, Hurle *et al.* 1987).

As the number and level of sophistication of space experiments has increased, our understanding of the characteristics of the space laboratory environment has also changed. What was initially referred to as "zero-g" has become "micro-g", and more recently transient acceleration measurements revealed "milli-g" perturbations caused by crew activities, machinery vibrations, attitude changes etc. (Chassay & Schwaniger 1987). Indeed, it has become clear from measurements of the acceleration environment in the Spacelab (Chassay & Schwaniger 1987; Hamacher *et al.* 1987), that the residual gravity levels on board a spacecraft in low earth orbit can be significant and should be of concern to experimenters who wish to take advantage of the low gravity conditions on future Spacelab missions and on board the planned Space Station. While accelerations may be orders of magnitude lower than that experienced at the earth's surface, they are nonetheless finite and, thus, pose potential problems for certain types of experiments.

To date, analyses of the effects of a low gravity environment have, with a few exceptions (Kamotani *et al.* 1981; Chang & Brown 1983; Polezhaev 1984; McFadden *et al.* 1988), been restricted to either idealized assumptions or order of magnitude estimates (Spradley *et al.* 1975; Camel & Favier 1984, 1986; Langbein 1984, 1987; Boudreault 1984; Monti & Napolitano 1984; Monti *et al.* 1987). The validity of the various estimates has only been demonstrated for a few special cases (Rouzaud *et al.* 1985).

As a case study, we examine the effects of time-dependent and spatial variations in the effective gravity vector on solute redistribution in a simplified model crystal growth by the Bridgman-Stockbarger technique. In Section 2, we discuss the phenomenon of segregation which, when influenced by unsteady and spatially varying transport conditions in the nutrient, leads to undesirable compositional variations. In Section 3, we discuss the basic heat, mass and momentum transport model of the Bridgman-Stockbarger technique. The results obtained from our numerical models of the effects of residual steady and time-dependent acceleration during the melt growth of a doped semi-conductor crystal are presented in Section 4 and discussed in Section 5. Our results show that for this specific directional solidification system, residual g-levels of 10^{-6} g are adequate only if an optimal alignment of the system with respect to the residual gravity vector is maintained.

1.1 Solute Redistribution

Consider a two component solution system. The equilibrium between the solid and liquid phases in such a system can be represented by a binary phase diagram (Chalmers 1977). Except for the case of either a pure or a congruently melting material, the liquidus and solidus lines do not coincide. This is illustrated in Figure 1, which depicts the equilibrium phase diagram for a solid solution of A-B, where B is the dilute species (solute). The separation of liquidus and solidus indicates the compositional difference between the solid and the liquid with which it is in equilibrium. Such compositional changes on solidification occur in the majority of systems of interest to the materials science community, particularly those materials needed for device fabrication (Rosenberger 1979). During growth of crystals from melts containing two or more components this characteristic gives rise to a phenomenon termed *segregation* or *redistribution*.

Segregation is of great importance for many aspects of materials preparation (Wilcox 1971; Flemings 1974; Rosenberger 1979) including, for example, the purification of materials, or the predetermination of the composition of the nutrient phase in order to achieve the desired dopant distribution in the solid. When a solid phase is solidified from a melt with an initially uniform composition, the rate of advance of the solid into the liquid is rarely slow enough to allow the liquid phase to adjust its composition throughout in accordance with the equilibrium phase diagram. This is because the mass transfer rates in the liquid are typically slower than the solidification rate. As a consequence, while *local equilibrium* may persist at the phase boundary, compositional gradients will still be established in both the solid and liquid phases (see Figure 2). For cases in which local equilibrium conditions prevail at the phase boundary the interfacial transfer of a component may

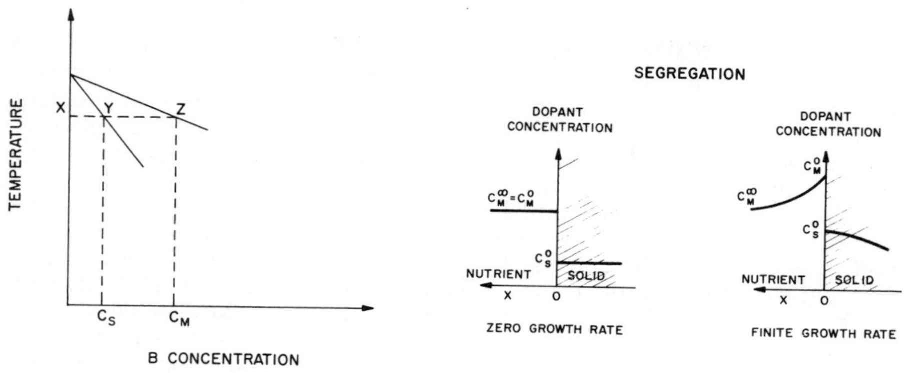

Figure 1

Figure 2

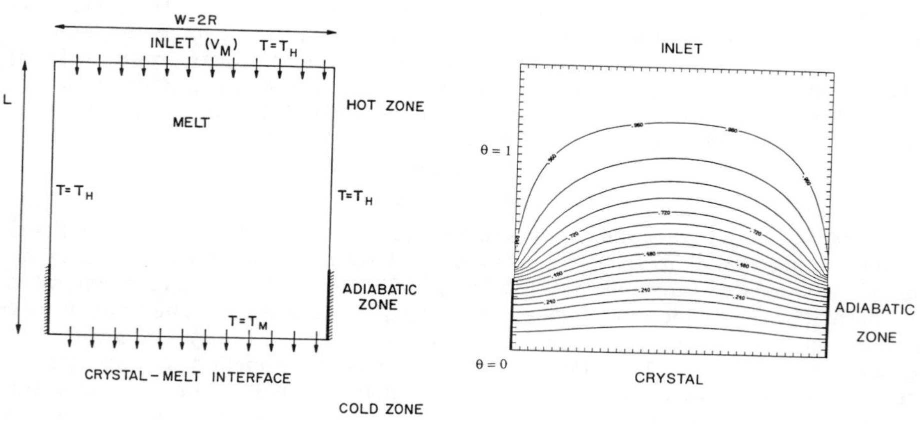

Figure 3

Figure 4

be described in terms of the equilibrium distribution coefficient (Rosenberger 1979). For the dilute binary alloy illustrated in Figure 1, this distribution coefficient is given by $k = c_S/c_M$, where c_s and c_M are respectively, the solute (B) concentrations in the solid and liquid. The value of k reflects whether the solute is preferentially rejected or incorporated upon solidification. The equilibrium distribution coefficient depends only on the thermodynamic equilibrium properties of the system, and is independent of the mass transfer kinetics.

In order to predict the operating conditions under which desirable dopant uniformities can be obtained, for example for doped semi-conductor crystals, it is necessary to understand the interplay between the various processes affecting mass transfer and their effect on the composition of the growing solid. For situations in which the equilibrium distribution coefficient reflects the local mass transfer conditions at the interface it is clear that any time dependence or non-uniformity in the growth rate and mass transfer conditions will result in compositional non-uniformities in the crystal.

It has been recognized for some time that convection in the nutrient phase can, in some cases, result in undesirable compositional variations in melt grown semi-conductor crystals (Pimpuktar & Ostrach 1981; Müller 1982; Müller *et al.* 1984). In many of these cases, for example growth by the Bridgman-Stockbarger technique, the convection is buoyancy driven (Langlois 1985). An attractive feature of the low effective gravity environment afforded by a spacecraft, is the possibility of substantially reducing or eliminating buoyancy driven convection, and thus creating diffusion controlled conditions (Hurle *et al.* 1987) which can favor compositional uniformity in the crystal. In the following model of crystal growth by the vertical Bridgman-Stockbarger technique we shall investigate the compositional inhomogeneities introduced into the crystal as a consequence of buoyancy driven convection under steady and time-dependent reduced gravity conditions.

2. FORMULATION

2.1 The Basic Model

The basic setup to be considered is depicted in Figure 3. It is based on a model used by Chang and Brown (1983). Directional solidification takes place as an ampoule is translated through fixed "hot " and "cold" zones. The zones are separated by a thermal barrier which is modelled using adiabatic sidewalls. The temperature conditions are chosen such that the upper and lower parts of the system are molten and solid, respectively. Translation of the ampoule is modelled by supplying a doped melt of dilute bulk composition c_∞ at a constant velocity V_M at the top of the computational space, and withdrawing a solid of composition c_S (which in general will be a function of both space and time) from the bottom. The crystal-melt interface is located at a distance L from the top of the computational space. The temperature at the interface is taken to be T_M, the melting temperature of the crystal, while the upper boundary is held at a higher temperature T_H. In an actual experiment, owing to the finite length of the ampoule there is a gradual decrease in length of the melt zone during growth. In this model transient effects related to this change are ignored. Thus, it is assumed that the ampoule is sufficiently long for these effects to be negligible. The only transient effects to be considered will arise directly from the time-dependent nature of the residual gravity field. We also assume that the contribution of the solute (dopant) to convection is negligible. Convection is driven only by thermal gradients. It has been shown that curvature of the solid-liquid interface can result in significant lateral compositional non-uniformity (Coriell & Sekerka 1979; Coriell *et al.* 1981). We wish to focus attention on the influence of convection on the composition of the crystal and thus, have chosen to examine the planar interface case. For the two-dimensional model the dilute binary melt is assumed to occupy a rectangular region, Ω, which is bounded by planar surfaces (see Figure 3). In the three-dimensional model the rectangular region is replaced by a circular cylinder with a diameter equal to W.

The governing equations are cast in dimensionless form using L, the effective melt length, κ/L (where κ is the melt's thermal diffusivity), $\rho_M\kappa^2/L^2$, T_H-T_M, and c_∞ to scale the lengths, velocity, pressure, temperature and solute concentration, respectively. The dimensionless equations governing momentum, heat and solute transfer in the melt are

$$\frac{\partial \mathbf{u}}{\partial t} + (\text{grad } \mathbf{u})\mathbf{u} = -\text{grad } p + Pr\Delta\mathbf{u} + RaPr\Theta\mathbf{g}(t), \tag{1}$$

$$\text{div } \mathbf{u} = 0 \tag{2}$$

$$\frac{\partial \theta}{\partial t} + \mathbf{u} \cdot \text{grad } \Theta = \Delta\Theta \tag{3}$$

$$\frac{\text{Sc}}{\text{Pr}}\left(\frac{\partial C}{\partial t} + \mathbf{u} \cdot \text{grad } C\right) = \Delta C \tag{4}$$

where, $\mathbf{u}(\mathbf{x},t)$ represents the velocity, $\theta = (T(\mathbf{x},t) - T_M))/(T_H\text{-}T_M)$ the temperature (where T_H-T_M is the temperature difference between the hot zone and the crystal interface), and C represents the solute concentration. The parameters $Pr=\nu/\kappa$, $Ra= g\,\beta(T_H\text{-}T_M)L^3/\kappa\nu$ and $Sc=\nu/D$ are respectively the Prandtl, Rayleigh and Schmidt numbers. The term $\mathbf{g}(t)$ in Equation (1) represents the (time-dependent) gravity vector. The value of g in Ra is taken to be 980 cm s^{-2}, i.e., equal to the terrestrial acceleration. Thus, the magnitude of $\mathbf{g}(t)$ represents the ratio between the actual residual acceleration and g. Table I lists the forms of $\mathbf{g}(t)$ we have examined to date.

TABLE I

Selected forms of the acceleration vector examined in this work.

Steady

$\mathbf{g}_o = g_{ox}\mathbf{i} + g_{oz}\mathbf{k}$;

$\|\mathbf{g}\| = 10^4, \sqrt{2}(10)^{-5}, 10^{-5}, 5\sqrt{2}(10)^{-6}, 10^{-6}, \sqrt{2}(10)^{-6}, 10^{-7}$

Time dependent

$\mathbf{g}(t) = \mathbf{g}_o + \mathbf{g}_n cos(2\pi\omega_n t)$; $\mathbf{g}(t) = \mathbf{g}_o + \Sigma\mathbf{g}_n cos(2\pi\omega_n t)$;

$\omega_n = 10^{-4}, 10^{-3}, 10^{-2}, 10^{-1}, 1, 10$ Hz,

$\|\mathbf{g}_n\| = \sqrt{2}(10)^{5}, 5\sqrt{2}(10)^{-6}, \sqrt{2}(10)^{-6}$

Impulse

$\mathbf{g}(t) = \mathbf{g}_B\, t < t_1$, $\mathbf{g}(t) = \mathbf{g}_B + \mathbf{g}_I$, $t_1 < t < t_2$, $\mathbf{g}(t) = \mathbf{g}_B\, t > t_2$ $\qquad$ $\mathbf{g}_B = \sqrt{2}(10)^{-6}$, $\mathbf{g}_I = 3(10)^{-3}$

The following boundary conditions apply at the crystal-melt interface:

$$\Theta = 0, \tag{5}$$

$$\mathbf{u}\cdot\mathbf{N} = Pe/\sigma, \tag{6}$$

$$\mathbf{N} \times \mathbf{u} \times \mathbf{N} = 0 \tag{7}$$

$$\frac{\partial C}{\partial z} = \frac{PeSc}{Pr}(1\text{-}k)\, C, \tag{8}$$

where $\mathbf{N}$ points into the melt and is the unit vector perpendicular to the planar crystal melt interface, $Pe = V_M L/\kappa$ is the Péclet number, $\sigma = \rho_M/\rho_s$ and k is the distribution coefficient. We define the measure of compositional non-uniformity in the crystal at the interface to be the lateral range in concentration given by

$$\xi = \frac{(c_{smax} - c_{smin}) \times 100\%}{c_{av}},$$

where c_s is the (dimensional) solute concentration in the crystal, and c_{av} is the average concentration. At the "inlet" (z=0) the following boundary conditions are applied

$$\frac{\delta C}{\delta z} = \frac{PeSc}{Pr}(C-1), \quad (9)$$

$$\Theta = 1, \quad (10)$$

$$\mathbf{u} \cdot \mathbf{N} = Pe\sigma, \quad (11)$$

$$\mathbf{u} \times \mathbf{N} = 0\,. \quad (12)$$

Equations (8) and (9) express conservation of mass at the crystal-melt interface and the "inlet", respectively. Equations (6) and (10) guarantee continuity of the melt with the crystal and with the supply of melt at the "inlet", while Equations (7) and (11) ensure no-slip tangent to the interface and the top surface. At the side walls the following conditions are applied:

$$\text{grad}\,\mathbf{C} \cdot \mathbf{e}_w = 0, \qquad \mathbf{u} \cdot \mathbf{N} = Pe\,\sigma, \qquad \mathbf{e}_w \cdot \mathbf{u} = 0 \quad (13)$$

along with

$$\Theta = 1, \quad (14)$$

in the isothermal zone and

$$\text{grad}\,\Theta\, \mathbf{e}_w = 0 \quad (15)$$

in the adiabatic zone. Here $\mathbf{e}_w$ is the normal to the ampoule wall.

While the above model does not strictly apply to a specific furnace (for example, details of the heat transfer at the ampoule walls are neglected), it nonetheless serves as a reasonable "generic" model with which to carry out a preliminary analysis of a directional solidification experiment under conditions characteristic of the low gravity environment of space.

Our calculations are limited to thermo-physical properties corresponding to dilute gallium-doped germanium. The values of the thermo-physical properties and the associated dimensionless groups and operating conditions are given in Table II. For all the calculations discussed here, the temperature difference T_H-T_M was taken to be 115°C, and the diameter of the furnace, W, and the effective melt length, L, were taken to be 1 cm.

2.2 Method of Solution

The governing equations were solved using the code PHOENICS (Spalding 1981; Rosten & Spalding 1986). PHOENICS embodies a finite volume or finite domain formulation (Patankar 1980). It represents the governing equations introduced in the previous section as a set of algebraic equations. These equations represent the consequence of integrating the differential equation over the finite volume of a computational cell (and, for transient, problems over a finite time) and approximating the resulting volume, area and time averages by interpolation. For the 2-D calculations discussed in this report we employed a 40x39 grid. The 3-D calculations were performed in a circular cylindrical domain with 20 nodes in the radial direction, 12 in the azimuthal direction and 39 in the axial direction. We found that for the 2-D calculation less than 40 points in the direction parallel to the interface resulted in poor convergence of the solute field.

The scheme employed has the same accuracy as a finite difference scheme which is of the order Δx, where Δx is the distance between the grid nodes. The time scheme is implicit and thus unconditionally stable. However, small time steps are required to obtain accurate solutions. For highly non-linear flows the use of under-relaxation is necessary to eliminate divergence and to ensure good convergence of the solutions.

TABLE II
Thermo-physical properties characteristic of gallium-doped germanium
(after Chang & Brown 1983)

Property	Value
Thermal conductivity of the melt	0.17 WK^{-1} cm^{-1}
Heat capacity of the melt	0.39 $Jg^{-1}K^{-1}$
Density of the melt	5.6 g cm^{-3}
Density of the solid	5.6 g cm^{-3}
Kinematic viscosity of the melt (n)	1.3 $(10)^{-3}$ cm^2s^{-1}
Melting temperature (T_M)	1231 K
Solute diffusivity (D)	1.3$(10)^{-4}$$cm^2s^{-1}$
Thermal diffusivity of the melt (κ)	1.3$(10)^{-1}$ cm^2s^{-1}
Segregation coefficient (k)	0.1
Thermal expansion coefficient (β)	2.5 $(10)^{-4}$ K^{-1}
Operating conditions	
Hot zone temperature (T_H)	1346 K
Distance between inlet and interface (L)	1 cm
Height of adiabatic zone	2.5 mm
Ampoule width (diameter)	1 cm
Translation (supply) rates (V_M)	6.5 μm s^{-1},0.65 μm s^{-1}
Associated dimensionless parameters	
Prandtl number $Pr = \nu/\kappa$	0.01
Peclet number $Pe = V_M L/k$	$5(10)^{-3}$ and $5(10)^{-4}$
Schmidt number $Sc = \nu/D$	10
Density ratio σ	1.0

We compared our results (to be discussed in detail in the following sections) with the results of Chang and Brown's (1983) axisymmetric calculations. Our 3-D axisymmetric computations were found to be in agreement with their work. In addition we carried out full 3-D (*non-axisymmetric*) steady calculations in order to calibrate our 2-D results. For one set of examples, we found that with $\sqrt{2}(10)^{-5}$ g parallel to the interface the percentage compositional non-uniformity predicted by the 2-D calculation (ξ = 152%) was 50% higher than predicted by the full 3-D calculation. This difference may be attributed to the increased surface to volume ratio which increases the effect of the "no-slip" boundary condition. The presence of the rigid walls somewhat retards the flow and, thus, in comparison to the 2-D case, the degree of solute redistribution is decreased. A comparison between 2- and 3-D calculations for $\sqrt{2}(10)^{-6}$ g revealed no significant difference in ξ. At this magnitude of the gravity vector, convection is localized and weak. The increase in surface to volume ratio for this 3-D case has little influence on the transport conditions.

3. RESULTS

3.1 Steady accelerations: 2-D

The temperature field (Figure 4) was found to be insensitive to the slow convective flows. This is due to the low Prandtl number of the melt. Note that there are lateral, as well as longitudinal, temperature gradients in the system. The consequent density gradient is responsible for driving convection in the melt even when residual acceleration magnitudes are as low as 10^{-6} times that experienced under terrestrial conditions.

TABLE 3

Compositional non-uniformity ξ [%] for computed 2-D (3D) steady state cases

Residual Acceleration Magnitude	Orientation N	Orientation $\mathbf{e}_g$	Ampoule Width 1 cm, Growth Rate [μm s^{-1}] 6.5	1 cm, 3.25	1 cm, 0.65	2 cm, 6.5	0.5 cm, 6.5
10^{-4} g	↑	↓	(36)				
10^{-4} g	↑	↑	(32)				
$\sqrt{2}(10^{-5})$ g	↑	↘	110				
$\sqrt{2}(10^{-5})$ g	↑	→	152				
	↑	→	(91)				
10^{-5} g	↑	→					12
10^{-5} g	↑	↓	7.5	4.6	0.7		
$5\sqrt{2}(10^{-6})$ g	↑	↘	57				
$\sqrt{2}(10^{-6})$ g	↑	↘	10				
$\sqrt{2}(10^{-6})$ g	↑	→	22				
	↑	→	(26)				
$\sqrt{2}(10^{-6})$ g	↑	↑	4.0				
$\sqrt{2}(10^{-6})$ g	↑	↓	2.0				
10^{-6} g	↑	↓	0.7	0.4	0.0	3.8	
10^{-7} g	↑	↑	1.0	0.5	0.2		

$\mathbf{e}_g \equiv$ unit vector parallel to **g**, **N** $\equiv$ normal vector to interface pointing into melt, H = 1 cm for all cases

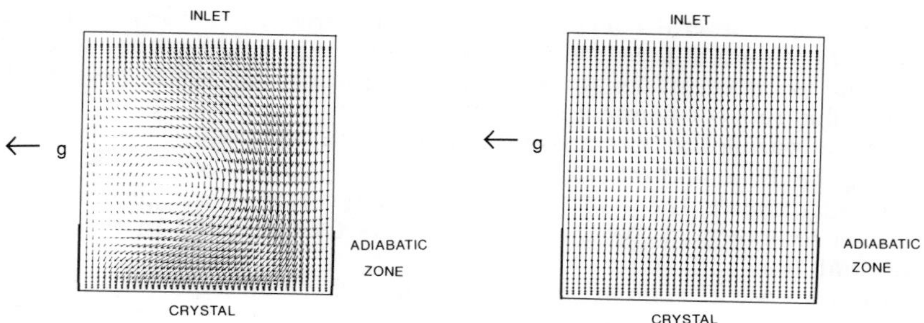

Figure 5

Figure 7

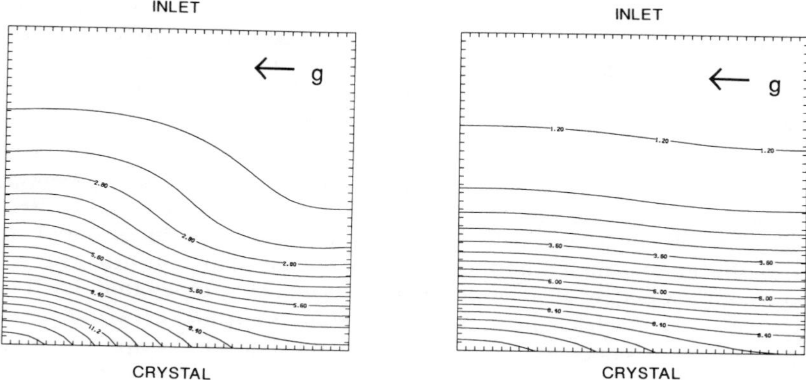

Figure 6

Figure 8

The amount of lateral concentration non-uniformity varied with both magnitude and orientation of the residual acceleration. When no residual acceleration is present the isoconcentrates are all parallel to the planar crystal-melt interface. Figures 5-8 illustrate the solute and velocity fields for two cases where the residual acceleration is parallel to the crystal-melt interface. The values of lateral non-uniformity ξ for these and other cases are listed in Table III along with the orientation and magnitude of the associated acceleration vector. For a given magnitude, accelerations oriented parallel to the crystal interface resulted in the maximum compositional non-uniformity.

For the purpose of comparison with results of calculations corresponding to terrestrial gravitational conditions we refer to the results of Chang and Brown (1983) for growth under axisymmetric conditions (i.e., gravity is parallel to the ampoule axis). Their results show that the amount of compositional nonuniformity in the crystal varies nonlinearly with increasing magnitude of the gravity-vector. The maximum nonuniformity occurs at 10^{-2}g, an intermediate value between the value of gravitational acceleration experienced at the earth's surface, and the 10^{-6}-10^{-5}g values characteristic of the quasi-steady component of the effective gravity in a spacecraft. Unless the acceleration is lowered (with respect to terrestrial conditions) by four orders of magnitude, the compositional uniformity may not be improved. Similarly, we have evaluated the conditions under which lateral nonuniformities are a maximum when the residual gravity vector is parallel to the interface. In this case, for the particular system under consideration, the maximum lateral non-uniformities will occur for accelerations of the order 10^{-4} g.

3.2 Steady accelerations: 3-D

Five three-dimensional calculations were undertaken in order to examine the influence of a more realistic geometry. Three runs were axisymmetric, with $\mathbf{g}_o$ anti-parallel to the solidification direction. Two were fully three-dimensional with $\mathbf{g}_o$ parallel to the crystal-melt interface. The compositional non-uniformity ξ was found to be approximately 10% lower for the axisymmetric cases than for their 2-D analogs. The axisymmetric calculations were carried out for $||\mathbf{g}|| = (\mathbf{g} \cdot \mathbf{g})^{1/2} = 10^{-4}$, 10^{-3} and 10^{-2}. The isoconcentrates in the crystal are shown in Figures 9 and 10 for a section cut perpendicular to the ampoule axis . The fully 3-D cases were carried out for $||\mathbf{g}|| = \sqrt{2}(10)^{-5}$ and $\sqrt{2}(10)^{-6}$. At the higher value of the residual acceleration, $\xi =$ 91%, which is approximately half the 2-D value. At $\sqrt{2}(10)^{-6}$ g, ξ = 26%. Thus, for this case, there was little difference between the 2 and 3-D predictions.

3.3 Time-dependent accelerations:

3.3.1 Single frequency; combined steady + single frequency accelerations

A number of different types of periodic disturbances were examined. Single frequency disturbances of the form $\mathbf{g}(t) = \mathbf{g}_o + \mathbf{g}_n \cos(2\pi\omega_n t)$ were examined with $\mathbf{g}_o = 0$, $\sqrt{2}(10)^{-6}$ and $\sqrt{2}(10)^{-5}$, oriented parallel to, perpendicular to, and at 45° to the crystal-melt interface. The range of frequencies examined was $\omega_n = 10^{-4}$, 10^{-3}, 10^{-2}, 10^{-1}, 1 and 10 Hz. For frequencies greater than 10^{-2} Hz, there were no discernible effects on the solute fields. The velocity field did, however, respond to the oscillatory disturbances. For the case of 10^{-3} Hz (at 5×10^{-6} g) the response of the solute field was significant. Lateral and longitudinal non-uniformity levels in excess of 15% were calculated. Figures 11 and 12 show the lateral non-uniformity as a function of time, and highlight the additive effect of oscillatory and steady components of the residual acceleration.

The effect of a multiple frequency disturbance is illustrated in Figure 13. The acceleration has three components, consisting of steady and periodic contributions

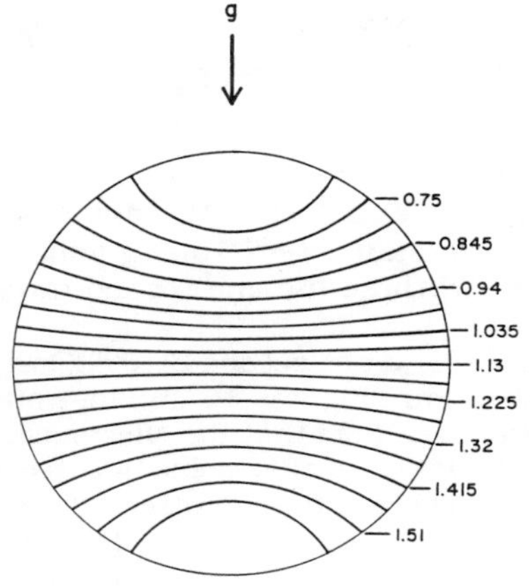

Figure 9

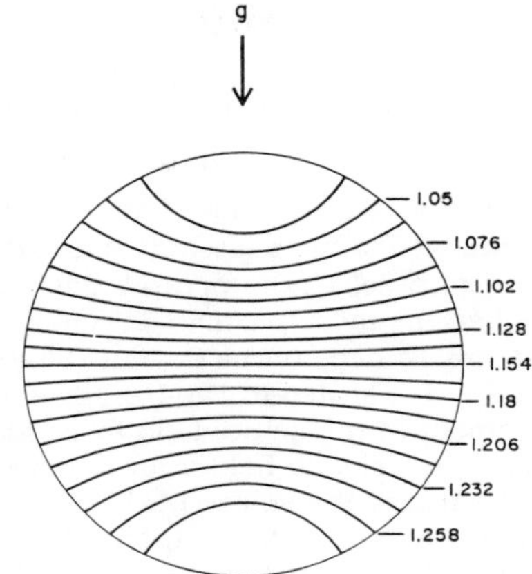

Figure 10

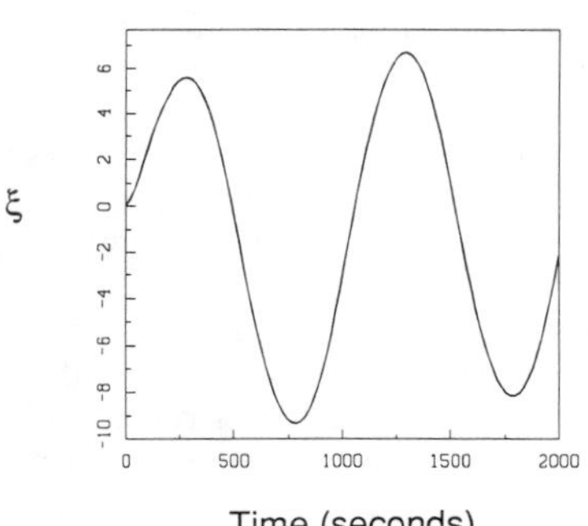

Figure 11

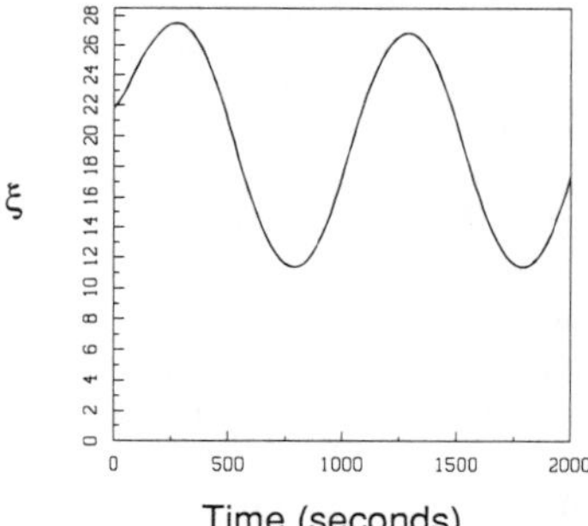

Figure 12

and has the form $\mathbf{g}(t) = \mathbf{g}_o + \mathbf{g}_1 \cos(2\pi 10^{-3}t) + \mathbf{g}_2 \cos(2\pi 10^{-2}t)$, where $||\mathbf{g}_o|| = \sqrt{2}(10)^{-6}$, $||\mathbf{g}_1|| = 3\sqrt{2}(10)^{-6}$ and $||\mathbf{g}_2|| = 3\sqrt{2}(10)^{-5}$. The magnitude of ξ is seen to vary with the frequencies of the acceleration.

3.3.2 Impulse-type accelerations

Four cases of impulse-type disturbances were examined. All impulses were superimposed onto a steady background acceleration of $\sqrt{2}(10)^{-6}$ g which was oriented parallel to the crystal-melt interface. Of these the one second duration pulses had the most dramatic effects.

Figure 14 depicts the flow field immediately after a one second $3(10^{-3})$g impulse oriented anti-parallel to the background acceleration. Figures 15-20 illustrate the development of the solute field following the impulse. Note that the effects are long lasting. The velocity field relaxes back to the initial state after some 300 seconds. The response of the solute field lags behind. The effect of the impulse is to initially re-orient the flow field (compare Figures 7 and 14). At first this has the effect of reducing the lateral compositional non-uniformity. 45 seconds after the termination of the impulse, the composition non-uniformity is reduced to zero. Subsequently, it increases in magnitude, but has the opposite sense in comparison to the initial non-uniformity. This change in sense can be seen upon comparison of Figures 16 and 17. At approximately 260 seconds after the termination of the impulse, the lateral segregation reaches a maximum value of 26%. It then decreases in value eventually reaching zero at 1017 seconds, changing sense and slowly increasing toward its initial steady level of 21.5% after more than 2000 seconds have elapsed.

A shorter duration (10^{-1} second) pulse resulted, 350 seconds after the termination of the pulse, in a maximum deviation of the lateral non-uniformity of only 5% from the initial steady level.

The effects of two one second pulses separated by one second were also calculated. The magnitude of the pulses was $3(10)^{-3}$ g, and they were oriented parallel to the crystal interface. Their main effect was to drive the lateral segregation from 22% (the initial value) to 76% after 225 seconds.

In addition double pulses were also examined. A pulse anti-parallel to the background steady acceleration followed by an equal but opposite pulse does not result in a "null" effect. While the flow generated by the first pulse is reversed by the second pulse there is a net flow following the termination of the second pulse. This flow is in the same sense as the initial steady flow and results in an increase of the lateral non-uniformity in composition to a maximum of 24% at 100 seconds, whereupon it decays slowly to its initial value.

4. SUMMARY

The salient results of our calculations for materials with properties and growth conditions similar to those listed in Table 2 can be summarized as follows:

1) For a fixed growth rate, the amount of lateral non-uniformity in composition is very sensitive to the orientation of the steady component of the residual gravity vector. The worst case appears to be when the acceleration vector is parallel to the crystal interface. At growth rates on the order of microns per second, this orientation can lead to non-uniformities of 22% when the magnitude of the acceleration is $\sqrt{2}(10)^{-6}$ g. If, however the growth rate is lowered by an order of magnitude, the non-uniformity is reduced significantly (down to 4-5% in this case).

2) A steady background level on the order of 10^{-6} - 10^{-5} g can be tolerated provided that the acceleration vector is *aligned with the axis of the growth ampoule,* and provided that no accelerations with frequencies less than 10^{-2} Hz (and amplitudes of the order of the steady component) are present.

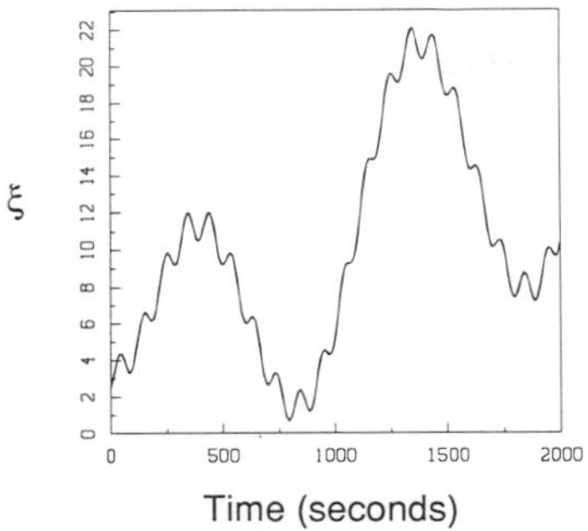

Figure 13

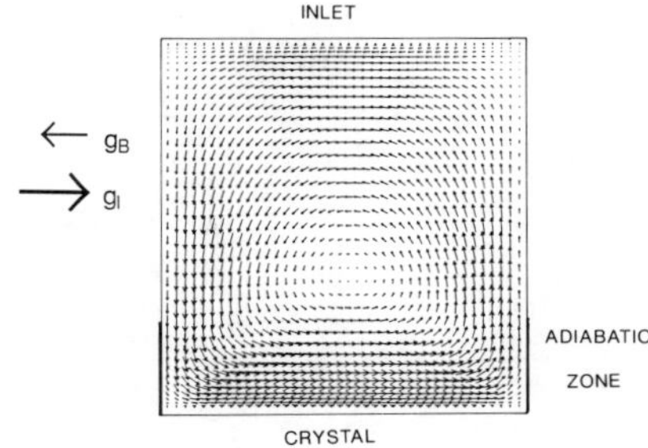

Figure 14

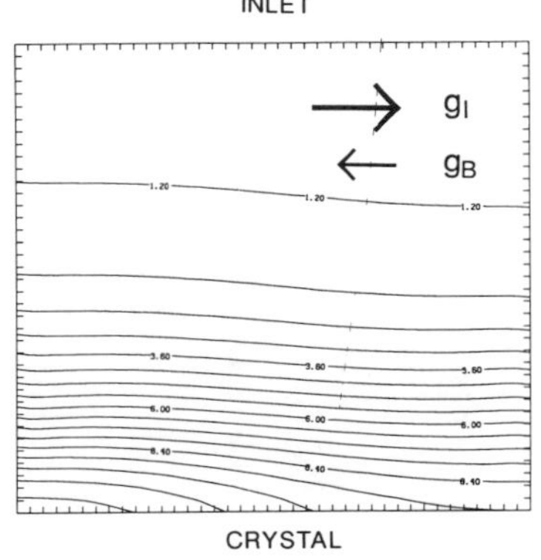

Figure 15

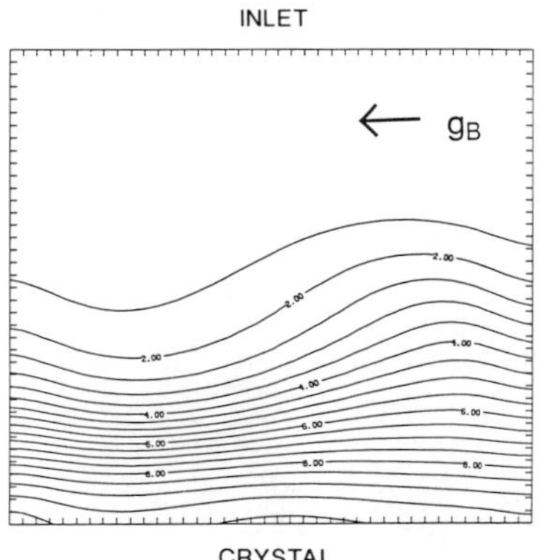

Figure 16

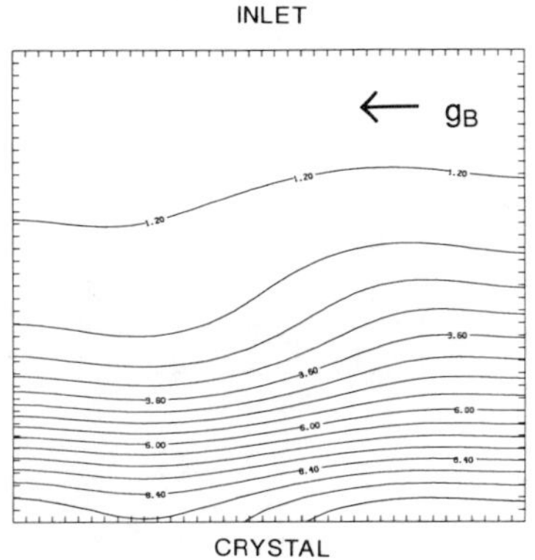

Figure 17

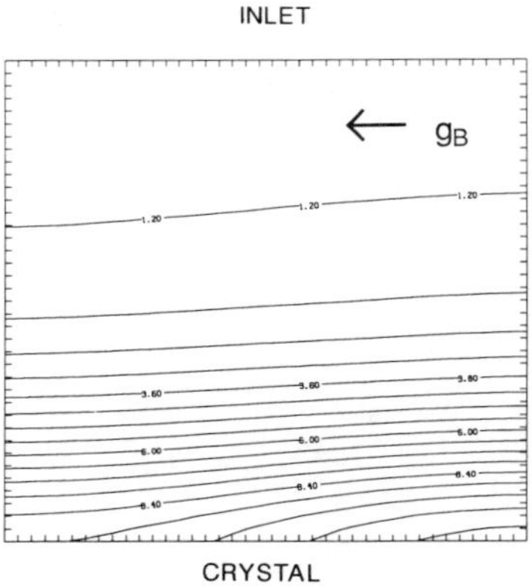

Figure 18

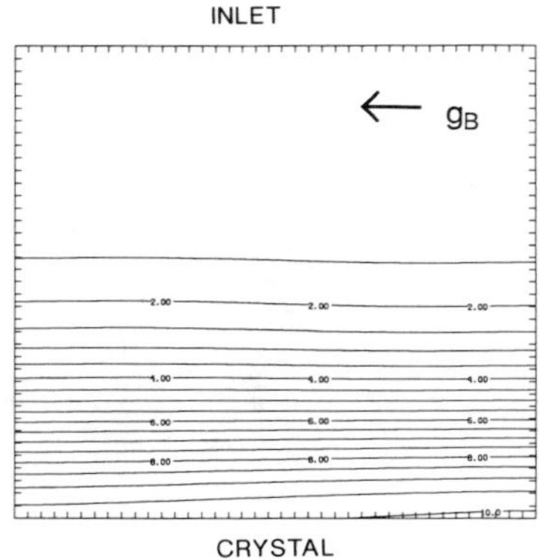

Figure 19

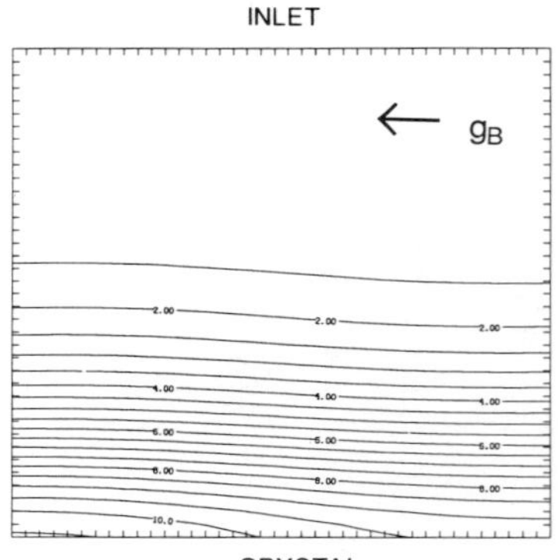

Figure 20

3) The response of the solute field, and the lateral non-uniformity, to oscillatory accelerations varies from no response at all (at frequencies above 1 Hz with amplitudes below 10^{-3} g) to a significant response at 10^{-3} Hz at amplitudes on the order of 10^{-6} g. In addition, additive effects were observed for combinations of a steady component and low frequency components. These additive effects gave rise to significant lateral and *longitudinal* non-uniformities in concentration.

4) The effects of impulse-type disturbances can be severe and can extend for a long time (on the order of 10^3 seconds) after the termination of the impulse. For example a pulse with a one second duration, or a combination of such pulses has a drastic effect on the segregation levels at pulse amplitudes of 10^{-3} g. The nature of the response depends on the magnitude, direction and duration of the impulse, and whether sequential opposing impulses are involved. A so-called "compensating" double pulse will not result in completely offsetting effects. For the case we examined, however, the resulting compositional non-uniformity was not as severe as for sequential pulses with the same orientation. Further investigation of more realistic impulses (g-jitter) is necessary since the response of the system appears to depend on the nature of the impulse and our results indicate that impulses appear to have important consequences for transient behavior in crystal growth systems.

It should be borne in mind that our calculations have only covered a small part of a large parameter space. In particular, it should be noted that for a given level of residual acceleration the amount of lateral segregation can be expected to vary according to the magnitude of the Schmidt number and distribution coefficient k. We have also examined the effect of the growth rate and have found that a reduction in growth rate by an order of magnitude will result in a significant reduction in non-uniformity (for the range of parameters we have studied). This is consistent with the results of Adornato and Brown (1987).

ACKNOWLEDGEMENT

This work was supported by the National Aeronautics and Space Administration (NAG8-684), and by the State of Alabama, through the Center for Microgravity and Materials Research.

REFERENCES

P. M. Adornato & R. A. Brown, 1987. "Convection and Segregation in directional solidification of dilute and non-dilute binary alloys: Effects of ampoule design" *J. Cryst. Growth* **80**, 155-190.

V. S. Abduyevsky, S. D. Grishin, L. V. Leskov, V. E. Polezhaev, & V. V. Savitchev, 1984. **Foundations of Space Manufacturing**, MIR Publishers Scientific, Moscow.

Boudreault. R. "Numerical simulation of convections in the μg environment", **Proc. 5th European Symposium Materials Sciences under Microgravity**, Schloss Elmau FRG November 1984, ESA SP-222, 259-264

D. Camel & J.J. Favier, 1986. "Scaling analysis of convective solute transport and segregation in Bridgman crystal growth from the doped melt." *J. Phys.* **47**, 1001-1014.

B. Chalmers, 1977. **Principles of Solidification,** Krieger Publishing Co.

C. J. Chang & R. A. Brown, 1983. "Radial segregation induced by natural convection and melt/solid interface shape in vertical Bridgman growth." *J. Cryst. Growth* **63**, 353-364.

R. P. Chassay & A. J. Schwaniger, Jr., 1986. "**Low g measurements by NASA**", NASA-TM 86585.

S. R. Coriell, R. F. Boisvert, R. G. Rehm, & R. F. Sekerka, 1981. "Lateral solute segregation during directional solidification with a curved solid-liquid interface." *J. Cryst. Growth* **54**, 167-175.

S. R. Coriell & R. F. Sekerka, 1979. "Lateral solute segregation during unidirectional solidification of a binary alloy with a curved solid liquid interface." *J. Cryst. Growth* **46**, 479-482.

B. Feuerbacher, R. J. Naumann, & H. Hamacher, 1986. **Materials sciences in space. A contribution to the scientific basis of space processing**, Springer-Verlag.

M. Flemings, 1974. **Solidification processing**, Mcgraw-Hill.

H. Jilg Hamacher & R. U. Mehrbold, 1987. "Analysis of microgravity measurements performed during D-1." **Proc. 6th European symp. on materials sciences under microgravity conditions,** Bordeaux, France Dec. 2-5 1986, ESA SP-256, 413-420.

George A. Hazelrigg & Joseph M. Reynolds (editors), 1986. **Opportunities for Academic Research in a Low Gravity Environment,** American Institute for Aeronautics and Astronautics Inc.

J. T. Hurle, G. Müller & R. Nitsche, 1987. "Crystal growth from the Melt", in **Fluid Sciences and Materials Science in Space. A European Perspective.** (ed. H. U. Walter) pp. 315-354, Springer-Verlag.

Y. Kamotani, A. Prasad, & S. Ostrach, 1981. "Thermal convection in an enclosure due to vibrations aboard a spacecraft." *AIAA Journal* **19**, 511-516.

D. Langbein, 1984. "Allowable g-levels for microgravity payloads, Final report for ESA contr. No. 5.504/83/F/FS(SC)", September 1984, pp.1-29 Batelle Frankfurt.

D. Langbein, 1987. "The sensitivity of liquid columns to residual accelerations." **Proc. 6th European symposium on materials sciences under microgravity conditions**, Bordeaux, France Dec 2-5 1986, ESA SP-256, 221-228.

W. E. Langlois, W. E. 1985 Buoyancy driven flows in crystal growth melts. *Ann. Rev. Fluid Mech.* **17**,191-215.

G. B. McFadden & S. R. Coriell (to appear), "Solutal convection during directional solidification" in *Proc. 1st National Fluid Dynamics Congress*, held in Cincinnati, Ohio, July 1988.

R. Monti, 1987. *ESA Contr. Report, R-66.525, Technosystems Rept.TS-7-87*, April 1987.

R. Monti, J. J. Favier & D. Langbein, 1987. "Influence of residual accelerations on fluid physics and materials science experiments" in **Fluid sciences and materials science in space, a European perspective.** (ed. H. U. Walter) pp. 637-680. Springer-Verlag.

R. Monti, & L. Napolitano, 1984. "g-level threshold determination, Final report for ESA, Contr. no. 5.504/83/F/FS/SC" *Technosystems Report TS-7-84*, June 1984.

G. Müller, 1982. "Convection in melts and crystal growth", in **Convective transport and instability phenomena** (eds. J. Zierep & H. Oertel) pp. 441-468, Braun Verlag.

G. Müller, G. Neumann, & W. Weber, 1984. "Natural convection in vertical Bridgman configurations." *J. Cryst. Growth* **70**, 78-93.

S. V. Patankar, 1980. **Numerical Heat Transfer and Fluid Flow**, Hemisphere Pub.

S. M. Pimpuktar & S. Ostrach, 1981. "Convective effects in crystals grown from melt". *J. Crystal Growth* **55**, 614.

V. I. Polezhaev, 1984. "Hydrodynamics, heat and mass transfer during crystal growth" in **Crystals 10** (ed. H. C. Freyhardt) pp. 87-150, Springer-Verlag

V. I. Polezhaev, A. P. Lebedev & S. A. Nikitin, 1984. "Mathematical simulation of disturbing forces and material science processes under low gravity", **Proc. 5th European symposium on materials sciences under microgravity,** Schloss Elmau FRG, ESA SP-222, 237.

A. Rouzaud, D. Camel, & J. J. Favier, 1985. "A comparative study of convective solute transport and segregation in Bridgman crystal growth from the doped melt." *J. Cryst. Growth* **73**, 149-166.

F. Rosenberger, 1979. **Fundamentals of Crystal Growth I.** *Springer Series in Solid-State Sciences*, Vol. 5, Springer-Verlag.

H. I. Rosten & D. B. Spalding, 1986. "Numerical simulation of fluid flow and heat and mass transfer processes", in **Lecture Notes in Engineering** (eds. C. A. Brennia and S. A. Orszag) Vol 18, pp. 3-29, Springer Verlag.

R. F. Sekerka & S. R. Coriell, 1979. "Influence of the space environment on some materials processing phenomena", **Proc. 3rd European symposium on materials sciences in space**, Grenoble France, April 1979, ESA SP-142, 55-65.

D. B. Spalding, 1981. "A general purpose computer program for multi-dimensional one- and two-phase flow." **Mathematics and computers in simulation Vol. XXIII,** North Holland Press 267-268.

L. W. Spradley, S. W. Bourgeois, & F. N. Lin, 1975. "Space processing convection evaluation, g-jitter of confined fluids in low gravity," *AIAA Paper No.* 75-695.

H. U. Walter, editor, 1987. **Fluid Sciences and Materials Science in Space. A European Perspective.** Springer-Verlag.

W. A. Wilcox, 1971. "The role of mass transfer in crystallization processes", in **Preparation and Properties of Solid State Materials,** (ed. R. A. Lefever) M. Dekker.

FIGURE CAPTIONS

Figure 1. The temperature-composition phase diagram for a dilute binary system *A-B*, where *B* is the dilute species. The compositions in the solid (*S*) and liquid (*L*) are respectively given by c_S and c_M. The equilibrium distribution coefficient, *k*, is given by the *XY/XZ*.

Figure 2. Concentration profiles in the liquid and solid caused by segregation:
a) The zero growth rate (equilibrium) case.
b) In this case the solute is rejected at finite growth velocities. This results in compositional gradients in both phases.

Figure 3. The prototype directional solidification model.

Figure 4. The dimensionless temperature field, θ, for all 2-D cases discussed in this paper.

Figure 5. The steady flow field produced by a residual acceleration with a magnitude $\sqrt{2}(10)^{-5}$ g acting parallel to the crystal melt interface. The maximum speeds are approximately twice the growth speed.

Figure 6. The dimensionless solute field, *C*, associated with the flow depicted in Figure 5. For this case ξ= 152%.

Figure 7. The steady flow field produced by a residual acceleration with a magnitude $\sqrt{2}(10)^{-6}$ g acting parallel to the crystal melt interface. The maximum speeds are slightly greater than the growth speed.

Figure 8. The dimensionless solute field, *C*, associated with the flow depicted in Figure 7. For this case x= 22%.

Figure 9. The steady solute distribution ($C_s=c_S/c_\infty$) in the crystal consequent to $\sqrt{2}(10)^{-5}$ g acceleration oriented parallel to the interface for the 3-D case. The cross section is taken perpendicular to the ampoule axis. ξ = 91%.

Figure 10. The steady solute distribution ($C_S=c_S/c_\infty$) in the crystal consequent to $\sqrt{2}(10)^{-6}$ g acceleration oriented parallel to the interface for the 3-D case. The cross section is taken perpendicular to the ampoule axis . ξ = 26%.

Figure 11. Lateral non-uniformity in composition, ξ, plotted as a function of time for an oscillatory residual acceleration with a maximum magnitude of $3\sqrt{2}(10)^{-6}$ g and a frequency of 10^{-3} Hz, acting parallel to the crystal-melt interface. The initial state was purely diffusive.

Figure 12. Lateral non-uniformity in composition, ξ, plotted as a function of time for a residual acceleration consisting of a steady part with a magnitude of $\sqrt{2}(10)^{-6}$ g and an oscillatory part with a maximum magnitude of $3\sqrt{2}(10)^{-6}$ g and a frequency of 10^{-3} Hz, acting parallel to the crystal-melt interface. The calculation was started from a steady flow associated with a $\sqrt{2}(10)^{-6}$ g acceleration acting parallel to the interface.

Figure 13. Lateral non-uniformity in composition, x, plotted as a function of time for a multi-component disturbance consisting of a steady low g background, plus two periodic components: $\mathbf{g}(t) = \mathbf{g}_o+\mathbf{g}_1 cos(2\pi 10^{-3}t)+\mathbf{g}_2 cos(2\pi 10^{-2}t)$, where $||\mathbf{g}_o||=\sqrt{2}(10)^{-6}$, $||\mathbf{g}_1||=3\sqrt{2}(10)^{-6}$ and $||\mathbf{g}_2||=3\sqrt{2}(10)^{-5}$. The calculation was started from a steady flow associated with a $\sqrt{2}(10)^{-6}$ g acceleration acting perpendicular to the interface.

Figure 14. The velocity field after a one second pulse of $3(10)^{-3}$ g superimposed on a steady flow caused by a $\sqrt{2}(10)^{-6}$ g acceleration, both parallel to the crystal melt interface. The impulse, $\mathbf{g}_I$, is in the opposite direction to the background acceleration, $\mathbf{g}_B$. Note that the maximum velocity magnitudes are of the order $3(10)^{-2}$ cm s^{-1}, i.e., about 500 times those in Figure 7.

Figure 15. Dimensionless solute field, *C*, immediately after the termination of the impulse. The compositional non-uniformity, ξ, is 21.5%.

Figure 16. Dimensionless solute field, *C*, 31 seconds after the termination of the impulse. The compositional non-uniformity, ξ, is 6.1%.

Figure 17. Dimensionless solute field, C, 81 seconds after the termination of the impulse. The compositional non-uniformity, ξ, is 11.4%, and has changed sense.

Figure 18. Dimensionless solute field, C, 431 seconds after the termination of the impulse. The compositional non-uniformity, ξ, is 21.7%.

Figure 19. Dimensionless solute field, C, 881 seconds after the termination of the impulse. The compositional non-uniformity, ξ, is 4.5%.

Figure 20. Dimensionless solute field, C, 1781 seconds after the termination of the impulse. The compositional non-uniformity, ξ, is 12.5%.

FINITE ELEMENT ANALYSIS OF MELT CONVECTION AND INTERFACE MORPHOLOGY IN EARTHBOUND AND MICROGRAVITY FLOATING ZONES

Jacques L. Duranceau and Robert A. Brown
Dept. of Chemical Engineering and Materials Processing Center
Massachusetts Institute of Technology Cambridge, MA 02139

ABSTRACT

Calculations of convection driven by crystal and feed rod rotation and by surface-tension-gradients are presented for small scale silicon floating zones operating under microgravity and earthbound conditions. The analyses are based on a thermal-capillary model of the floating zone process that presents a self-consistent analysis of convection in the melt, heat transport in the melt, feed rod and growing crystal, the shape of the melt/gas meniscus and the shapes of the solidification and melting interfaces. Results for small-scale silicon floating zones demonstrate show the intense convection driven by thermocapillary forces caused by the axial temperature gradients that are needed to sustain the molten zone. Sample results illustrate the difference between the shape and flows in floating zones on earth and in microgravity.

INTRODUCTION

The major advantages of melt crystal growth in outer space hinge on the benefits of the microgravity environment for reduced convection in the melt and on the increased stability of melt/ambient interfaces due to the absence of hydrostatic pressure. The suppression of buoyancy-driven convection has been thought to remove unwanted solute striations in the crystal and lead to diffusion-controlled axial segregation of dopants. The reduction of the deformation of liquid surfaces by hydrostatic pressure is particularly important because it opens the way to processing methods in microgravity with liquid/fluid interfaces too large to be maintained with earth's gravity. Both of these effects influence the floating zone method for crystal growth and hence this technique has received a great deal of attention as a possible candidate for use in microgravity; see the references in Brown[1,2]. As originally designed by Pfann[3] small-scale floating zones are formed by translating a short circumferential heater along the axis of a cylindrical polycrystalline feed rod, as shown in Figure 1. A molten zone forms just ahead of the heater and is held in place by surface tension acting against gravity. The melt is resolidified into a single crystal at the stern of the heater. The size of the zone is controlled by heat transfer between the melt and solid phases with the surrounding ambient and heater and by surface tension through the shape of the melt.

Stabilization of the molten zone and suppression of buoyancy-driven convection in microgravity is not enough to guarantee the growth of compositionally uniform crystals. Intense thermocapillary convection driven by surface-tension gradients is still present in floating zones where buoyancy-driven flows are negligible, as has been demonstrated by many researchers[4-6]. Thermocapillary motion, as well as that caused by differential rotation of the feed rod and the crystal, lead to convective mixing and both radial and axial segregation of solutes, as discussed by Harriott and Brown[7]. The thermocapillary motion can be intense enough to lead to transitions to three-dimensional oscillatory flows which again give striations in the crystal, as documented by Preisser *et al.*[6]. Also, convective heat transport in the melt strongly influences zone length and the shapes of the melt/crystal and melt/feed rod interfaces, thereby coupling the pattern and intensity of the flow to the shape of

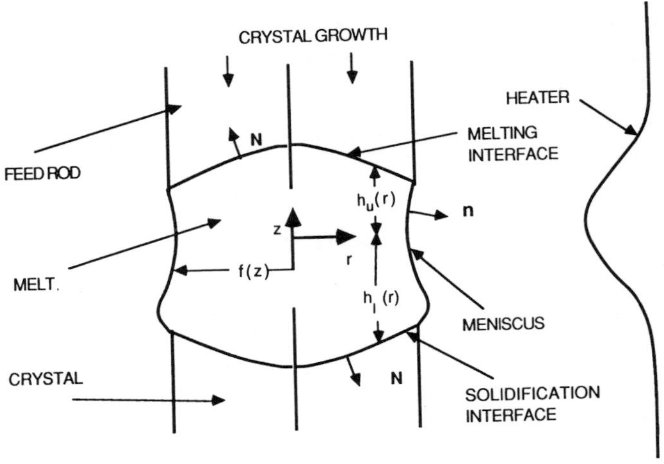

Figure 1

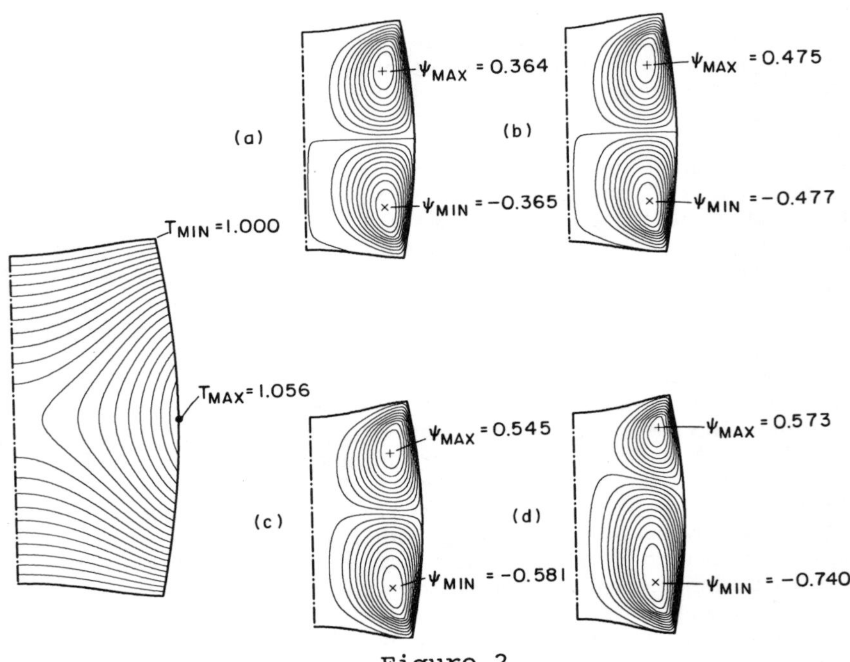

Figure 2

the molten region.

The purpose of our research has been to develop a self-consistent analysis of the transport processes and interface shapes in small-scale floating zones under both earthbound and microgravity conditions. The Hydrodynamic Thermal-Capillary Model (HTCM) described in the next section includes axisymmetric, steady-state convection in the melt, heat transport in the melt, crystal and feed rods and the determination of the shapes of the melt/solid and melt/ambient interfaces from interfacial energy and momentum balances, respectively. The complete HTCM for the floating zone process comprises a complex free-boundary problem for the field variables and interface shapes. This paper describes a finite-element/Newton method for solution of the coupled system. The methodology used here is an extension to include hydrodynamics of the analysis of the conduction-dominated model for the floating zone process developed by Duranceau and Brown[8]; we will refer to this work as DB in this manuscript. Because of its brevity, this paper can only give an overview of the numerical method and the results of the analysis for small-scale floating zones of silicon. A more complete report of the work will be presented elsewhere[9].

The calculations presented in Section 3 describe convection and zone shape in a small-scale floating zone in zero gravity and on earth. The important features of the analysis are the influence of the details of the heat transfer on the flow pattern and the coupling between the flow and the shape of the melt/crystal interface caused by convective heat transport. This coupling is missing from the previous computational studies[10-12] of rotational and thermocapillary flows in idealized floating zones with cylindrical melt/gas interfaces and flat solidification fronts. For silicon, the high surface tension and the relatively low melt viscosity makes the coupling between the flow and the shape of the melt/gas interface weak, as first pointed out by the perturbation analysis by Harriott and Brown[13]. Although the meniscus shape is computed in our analysis by a self-consistent balancing of stress along this surface, the shapes are essentially those for a hydrostatic interface.

ANALYSIS

Hydrodynamic Thermal Capillary Model

Steady-state crystal growth in a resistively heated floating zone is modelled by extending the conduction-dominated, thermal-capillary model introduced in DB to include axisymmetric melt motion. The resulting Hydrodynamic Thermal-Capillary Model (HTCM) is described in detail by Duranceau and Brown[9]. Here we focus only on the modifications to the conduction-dominated model necessary for the analysis described here. The regions of melt, crystal and feed rod represented in this analysis are shown in Figure 1 along with the notation for each phase and the shape functions used to approximate the shapes of the two melt/solid interfaces and the meniscus.

The shape functions and the equations listed below are written in a cylindrical polar coordinate system with its center located at the middle of the surrounding resistive heat source. Variables are made dimensionless by scaling lengths with the radius of the feed crystal R_f, temperature with the melting temperature of the material T_m, velocity components in the melt with α_m/R and pressure with $\nu\rho_m\alpha_m/R_f^2$, where α_m is the thermal diffusivity of the melt and ρ_m is the melt density. The calculations presented here are limited to cases where the growing crystal has the same radius as the feed rod.

The equations describing the heat balances in the melt, feed rod and crystal, and momentum and mass conservation in the melt are listed below:

<u>crystal (c) and feed rod (f).</u>

$$Pe\,\mathbf{v}\cdot\nabla\Theta + K_i\nabla\cdot k(\Theta)\,\Theta = 0 \quad , \qquad i = c, f; \tag{1}$$

melt (*m*),

$$Pr^{-1}\,\mathbf{v}\cdot\nabla\mathbf{v} - Ra\cdot\Theta\mathbf{e}_z + \nabla P - \nabla^2\mathbf{v} = 0\,, \qquad (2)$$

$$\nabla\cdot\mathbf{v} = 0, \qquad (3)$$

$$\mathbf{v}\cdot\nabla\Theta + \nabla^2\Theta = 0\,, \qquad (4)$$

where the dimensionless variables are the velocity $\mathbf{v}(r,z)$, the temperature $\Theta(r,z)$, and the pressure $P(r,z)$ fields. Convection of the solid rods caused by steady-state crystal growth is accounted for by the inclusion of the term in Equation (1) multiplied by the dimensionless growth rate $Pe \equiv V_g R_f/\alpha_m$, where V_g is the displacement rate of both rods. The thermal conductivity in the solid phases is taken to be a function of temperature $k(\Theta)$ according to the expression given in DB. The dimensionless parameters that appear in these and other equations are listed in Table I. Values for these parameters are based on a silicon floating zone with radius of 0.5 cm.

The boundary conditions that specify the temperature field are the same as discussed in the conduction-dominated model in DB. The zone is assumed to receive radiative energy from a resistive heater that is modelled by a Gaussian temperature profile with a dimensionless peak temperature Θ_m. Both the crystal and the feed rod are taken to be long enough that the temperature field far from the solidification and melting surfaces varies only in the axial direction. Radially-averaged energy equations are used to describe the temperature field in this region which is spliced to the two-dimensional temperature fields computed in the feed and crystal rods close to the interfaces.

The boundary conditions on the temperature field at the melt/crystal and melt/feed rod surfaces are that the temperature equals the equilibrium melting point for the material and that a local heat balance including latent heat release is satisfied. These conditions are written as

$$\Theta = 1 \text{ on } \partial D_{Ic} \text{ and } \partial D_{If}\,, \qquad (5)$$

$$\kappa[\mathbf{N_i}\cdot\nabla\Theta]_m - k(\Theta)[\mathbf{N_i}\cdot\nabla\Theta]_i = -\,Pe\,St\,(\mathbf{N_i}\cdot\mathbf{e}_z) \text{ on } \partial D_{Ii},\ i = c,f \qquad (6)$$

where the brackets [] signify a quantity evaluated on the interface in the indicated phase. The Stefan Number *St* measures the importance of latent heat and is defined in Table I.

Table I.

Dimensionless group and operating values used in the analysis of floating zone crystal growth for silicon.

Dimensionless Group	Definition	Value Used
Dimensionless Growth Rate	$Pe \equiv V_g\alpha_m/R_f$	0.4
Thermal Conductivity Ratio	$\kappa \equiv k_m/k_s$	2.9
Prandtl Number	$Pr \equiv \nu/\alpha_m$	0.013
Rayleigh Number	$Ra \equiv g\beta_t T_m R_f^3/\alpha_m\nu$	0.0
Stefan Number	$St \equiv \Delta H/c_p T_m$	1.07
Marangoni Number	$Ma \equiv -(d\sigma/dT)T_m R_f\ /\ \rho_m\alpha_m\nu$	0-150 000
Modified Marangoni Number	$Ma^* \equiv Ma\cdot\Delta T/T_m$	0-3000
Bond Number	$Bo \equiv \rho_m g R_f^2/\sigma$	0-0.875
Capillary Number	$Ca \equiv \nu\alpha_m\,\rho/\sigma R_f$	0.0
Rotational Reynolds Numbers for crystal ($i = c$) and feed ($i = f$) rods	$Re_i \equiv \Omega_i R_f^2/\nu$	0-400
Reference Pressure Difference	$\lambda \equiv \Delta p^o R_f/\sigma$	——

The two boundary conditions (5) and (6) are needed on the melt/solid surfaces in order to set the temperature and the interface shape there. Analogously, three conditions are required along the melt/gas meniscus to set the two components of the velocity and the shape of this surface.

The tangential component of the velocity field is set on both solid surfaces by the no-slip condition. The normal component of the velocity there is fixed by the dimensionless solidification rate V_g. The azimuthal motion caused by rotation of the crystal and feed rods is introduced through a no- slip boundary condition on the azimuthal component of the axisymmetric velocity field. These conditions introduce the rotational Reynolds Numbers Rei that scale the importance of the forced convection caused by rotation of crystal (i=c) and feed rod (i=f); see definitions in Table I.

At the melt/gas meniscus no normal velocity is allowed,

$$\mathbf{n} \cdot \mathbf{v} = 0 \,, \tag{7}$$

where n is the normal vector to the meniscus. Also, both the tangential and normal components of the interfacial momentum balance must be satisfied. These balances are written on ∂D_{Im} as

tangential;

$$\mathbf{tn}: \tau - Ma\mathbf{t} \cdot \nabla_{II}\Theta = 0 \,, \tag{8}$$

normal;

$$2H + Bo\, z - 2\lambda + \mathrm{Ca}(\mathbf{nn}: \tau - \mathrm{P}) = 0 \,, \tag{9}$$

where H is the mean curvature of the meniscus, Bo is the Bond Number which measures the effect of hydrostatic pressure to surface tension in influencing the meniscus shape, Ca is the Capillary number or the ratio of viscous to surface tension forces in the normal force balance and Ma is the Marangoni number which scales the thermocapillary force due to the tangential gradient of the temperature field. The hydrostatic pressure has been subtracted from the dimensionless pressure field $P(r,z)$. The formulation (8) of the normal force balance assumes that the surface tension varies linearly with temperature.

The constant λ in Equation (9) is a dimensionless reference pressure difference across the interface. In model floating zone systems formed by drops captive between inert solid surfaces, this constant is set by the drop volume; see Reference 13. In a floating zone with solidification the actual volume of the melt cannot be set a priori, because it depends on the interaction of heat transfer and capillarity. As described in DB, the reference pressure λ is determined by the additional constraint that the contact angle formed at the junction of the crystal, melt and ambient has a specified value; this condition is written as

$$[df/dz] = \tan \phi_o \,, \text{ at } \partial D_{Im}\, \partial D_{Ic} \quad . \tag{10}$$

For silicon growing on the <111> crystallographic plane this angle is approximately 11° (Surek and Chalmers, 1976).

When the mean curvature H is expressed in terms of the cylindrical coordinate shape function $r = f(z)$ the normal force balance is recognized as a second-order differential equation for the meniscus shape. We use this condition to determine the interface shape, as was done in DB and in the perturbation analysis of Harriott and Brown (1982). The boundary conditions on this equation specify that the edges of the contact line formed by melt, solid (either feed rod or crystal) and the ambient correspond to the junction of the melt/solid and melt/gas surfaces, *i.e.*, that this curve is a three phase junction.

In the problem description, Equations (1)-(10), the zone length may change as any of the parameters in the model are varied. Operationally, the tendency of the zone shape to change with variation in operating conditions would be counterbalanced in practice by changing the heater power and thus the ambient temperature distribution seen by the zone. We model this control action by adding the constraint that the length of the melt, measured at the meniscus surface can be specified as a dimensionless constant L_m, scaled with the crystal radius. This constraint is used as an additional equation, i.e.

$$L_m = constant, \tag{11}$$

to find the heater temperature Θ_m.

Finite-Element Analysis

The methodology used to solved the HTCM is an extension of the finite-element/Newton technique described in DB for the conduction-dominated model and is based on the method first described by Ettouney and Brown[15] for solving free-boundary problems. The free-boundary problem is first mapped to a fixed domain in the coordinates $(\zeta(r,z),(\omega(r,z))$ by a sequence of non-orthogonal transformations that are outlined in DB.

The resulting differential equations and boundary conditions are discretized by Galerkin's method using mixed finite element representations for field variables and interface shapes. The temperature in each phase and the velocity components in the melt are approximated by Lagrangian biquadratic basis functions and the pressure in the melt is written in terms of a linear discontinuous approximation in each element. The interface shape functions $(f(\zeta),h_1(\omega),h_2(\omega))$ are approximated by expansions of one-dimensional Lagrangian quadratic functions that are consistent with the isoparametric mapping for that element; these approximations are listed in DB.

Galerkin weighted residual equations for the momentum, continuity and energy equations are written for the field variables. One-dimensional residual equations are developed for the meniscus shape from the normal stress condition (9), and from the isotherm condition (5) for the melt/solid interface shapes, just as developed by Ettouney and Brown[15]. The set of nonlinear algebraic equations that results is solved simultaneously by Newton's method using frontal elimination techniques to solve the linear equation set at each iteration.

An alternative to the finite-element/Newton analysis used here has been developed by Chang and Brown[16]; also see Reference 17. There the formal non-orthogonal mapping is replaced by the isoparametric mappings inherent to the description of each deformed finite element in the physical coordinate system. The algebraic equations resulting from Galerkin approximations again are solved by Newton's method.

The sample calculations described below are for two different finite element meshes; the first has 12 elements equally spaced across the radial coordinate in the transformed coordinate system and 36 elements distributed axially; 12 of these elements each in the crystal and feed rod and twelve elements are in the melt. The second mesh has 24 elements across the radius in all three phases, 12 axial elements in the crystal and feed rods and 24 elements axially in the melt. The number of unknowns associated with the algebraic equation sets for the two discretizations are approximately 5000 and 14 000, respectively. Flow fields are displayed as stream function computed by solving the appropriate Poison equation using the velocity field from the calculation as input; see Reference 12 for details.

SURFACE-TENSION-DRIVEN FLOW

Because of the limited space in this manuscript, we concentrate the computations discussed here on the flows driven by rotation of the feed and crystal rods and surface-tension differences; buoyancy-driven flows are eliminated so $Ra = 0$. The thermophysical properties of silicon used in the simulations are the same as listed in Table 1 of DB.

Calculations are listed below for varying Marangoni Number Ma and changing rotational Reynolds Numbers Re_f and Re_c. Because the melting point is used as the temperature scale, the Marangoni Number defined in Table I does not scale the driving force for surface-tension-driven flow, which more appropriately should depend on the temperature difference along the melt surface ΔT. The additional parameter

$$Ma^* \equiv Ma \cdot (\Delta T/T_m) \tag{12}$$

is listed in the figures and is a more appropriate measure of the intensity of the surface-tension-driven flow. Because ΔT is computed as a result of the analysis, Ma^* is an output of the calculation, exactly as it is in an experimental system.

The calculations using Ma as a control parameter do not correspond to changing a controllable parameter in the process because only thermophysical properties appear in the definition. Alternatively, we think of increasing Ma as corresponding to increasing the sensitivity of the surface tension on temperature, *i.e.*, increasing $d\sigma/dT$.

Microgravity Zone With No Crystal Growth

Zone shapes and flow and temperature fields computed for a silicon floating zone without growth ($Pe = 0$) are shown in Figure 2 for increasing values of Ma. In these results the zone length L_m is held fixed at $L_m = 2.5$ and the heater temperature is adjusted to satisfy this constraint. The flows correspond to small values of Ma^* and hence correspond to flows dominated by a linear balance of viscous and pressure forces. The zone shapes are symmetric about the plane corresponding to the center of the ambient temperature distribution, $z = 0$. The isotherms are not deformed from those for a conduction-dominated zone for the range of Ma shown and so only one temperature field is displayed. The melt/solid interfaces are deflected symmetrically inward showing that the melt is losing heat for this zone length and heater temperature profile. The meniscus bulges out from the rods symmetrically about the heater in order to satisfy the condition of a 11° angle for crystal growth at the bottom crystal.

For low values of Ma^*, the flow driven along the surface by this symmetrical temperature field moves melt from the hot spot at the middle of the zone to the solid surfaces and results in two, axially-stacked toroidal cells shown in Figure 2. The intensity of the flow, as measured by the circulation rate, scales linearly with the surface-tension-driving Ma^*, for $0 \leq Ma^* \leq 40$.

Significant deviation from linearity in noticeable at high values of Ma^* as inertial boundary layers begin to form in the flow. The flow fields at $Ma^* = 47.6$ and 75.0 (Figures 2c and 2d) show the development of an asymmetry in the motion with the bottom flow cell becoming more intense. Although we have not yet explored fully the structure of these flows, there is strong evidence that the asymmetry is a result of a bifurcation from flows with symmetric cells to asymmetric flows with one cell larger than the other. A similar nonlinear transition was observed by Harriott and Brown[12] for rotationally-driven flows in a cylindrical zone.

Microgravity Zone With Crystal Growth

The introduction of latent heat at the two melt/solid interfaces and the translation of the solid rods caused by crystal growth ($Pe = 0$) destroys the symmetry of the zone shape and temperature field about the centerline of the heater and causes the molten zone to translate down relative to the heater. This asymmetry is demonstrated by the zone shapes and temperature and velocity fields shown in Figure 3 for increasing values of Ma and crystal growth at a rate of 0.5 cm/min. In these calculations the length of the molten zone L_m has been allowed to vary and the maximum temperature of the heater is held fixed at 1.5 times the melting point. The three values of Ma^* are 69.2, 122, and 244, respectively. The two cell flow structure seen without growth are distorted by the asymmetry in the temperature field; the bottom cell fills most of the melt and the top cell is confined to the upper corner. The melt/solid interface shapes are also affected. The melting (upper) interface is drawn further into the heater, thereby increasing its deflection. The solidification (bottom) interface is pushed further outside the heater and the temperature field becomes

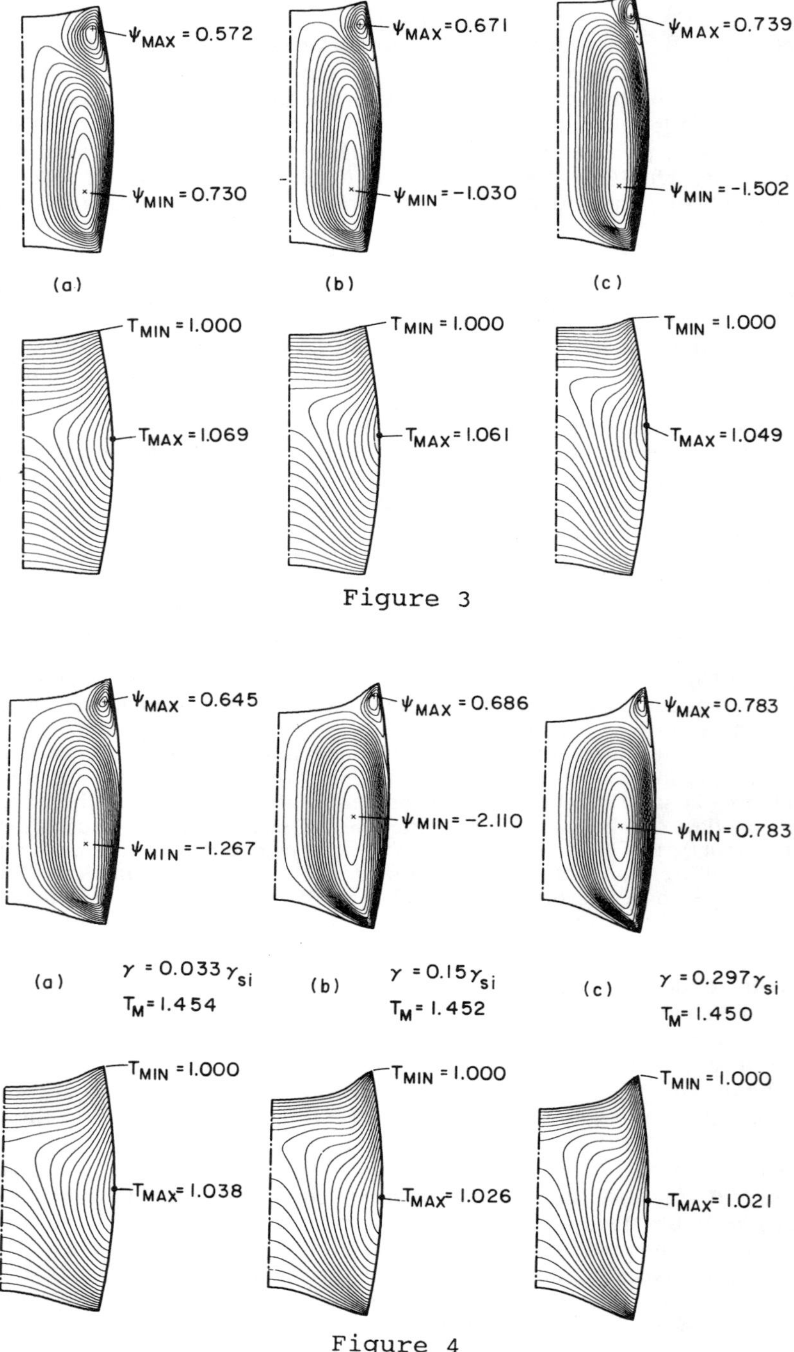

ψ_{MAX} = 0.572
ψ_{MIN} = 0.730
ψ_{MAX}=0.671
ψ_{MIN}= −1.030
ψ_{MAX}=0.739
ψ_{MIN}= −1.502
(a)
(b)
(c)
T_{MIN} = 1.000
T_{MAX} = 1.069
T_{MIN} = 1.000
T_{MAX} = 1.061
T_{MIN} = 1.000
T_{MAX} = 1.049
ψ_{MAX} = 0.645
ψ_{MIN} = −1.267
ψ_{MAX} = 0.686
ψ_{MIN}= −2.110
ψ_{MAX}= 0.783
ψ_{MIN}= 0.783
(a) γ = 0.033 γ_{si} T_M= 1.454
(b) γ = 0.15 γ_{si} T_M= 1.452
(c) γ = 0.297 γ_{si} T_M= 1.450
T_{MIN} = 1.000
T_{MAX}= 1.038
T_{MIN} = 1.000
T_{MAX}= 1.026
T_{MIN} = 1.000
T_{MAX}= 1.021

Figure 3

Figure 4

more uniform, thereby flattening the interface.

Increasing Ma intensifies the flow and leads to the formation of inertial boundary layers along the meniscus and to significant convective heat transfer in the melt. The maximum temperature difference in the melt decreases with increasing Ma from 0.0692 (116°K) at $Ma^* = 69.2$ to 0.0489 (82°K) at $Ma^* = 244$.

Calculations with varying zone length become extremely difficult with increase Ma because of the rapid deformation of the melt/solid surfaces with increasing convection intensity. This deformation is demonstrated in Figure 4 by calculations for a fixed zone length of $L_m = 2.5$ with values of Ma that correspond to $d\sigma/dT$ equal to approximately 1/6 and 1/3 of the value appropriate for silicon. The influence of the convection on the shapes of the melt/solid interfaces is obvious. The hot fluid carried by the cells to the interface "melts away" the surface at the edges. This mechanism accentuates the concavity of the melting surface and leads to an inflection in the solidification surface.

Viscous and thermal boundary layers form near the meniscus for high levels of convection. Boundary layer scaling predicted previously[18] for these flows have not yet been compared to our calculations.

Earthbound Zone with Crystal Growth

Including gravity ($Bo = 0.875$) causes deformation of the meniscus and distortion of the temperature field in the melt, as discussed in detail in DB for calculations without convection. Calculations shown in Figure 5 show the flow structure for calculations with crystal growth and a fixed zone length of $L_m = 2.5$.; $98 \leq Ma^* \leq 840$. The deformation of the two cell structure is extreme, as the surface-tension-driven flow tries to conform to the shape of the meniscus. At the highest value of Ma, which is about a factor of 3.3 too low for a clean silicon meniscus, the primary flow cells are confined close to the meniscus and secondary toroidal cells have been spawned closer to the axis of the melt.

The influence of convective heat transport is noticeable in the isotherms shown in Figure 4. The solidification interface (bottom) has developed an inflection point because of the intense recirculation cell located adjacent to only the outer one third of the surface.

EFFECT OF EQUAL COUNTER-ROTATION ON SURFACE-TENSION-DRIVEN FLOW

Introducing rotation of the feed and crystal rods causes azimuthal motions and meridional flows that modulate the flow driven by surface-tension gradients. Because the basic secondary flow driven by rotation of both solid surfaces moves melt outward along each solid surface this motion counteracts the cellular motion driven by the surface tension gradient which is in the opposite direction. A competition between the two mechanisms for flow results and the extent of the melt involved in each type of motion depends on the relative magnitudes of the two driving forces.

This is demonstrated in Figure 6 where three solutions computed for fixed zone length ($L_m = 2.5$) and $Ma = 1000$ are shown for different values of the Rotational Reynolds Number $Re = Re_c = Re_f$ with equal counter-rotation of the feed and crystal rods. The flow at the lowest value of Reynolds Number is very similar to the motion driven solely by surface tension gradients. The two flow cells are not symmetric about the center of the heater because crystal growth is included. Increasing the rotation of the surfaces decreases the intensity of the primary flow cells and, at the highest value of Re, new cells caused by the rotation appear close to the axis of the melt. The complex flow structure caused by the interaction of surface-tension-gradients and forced convection should be observable in the radial composition variation of crystals grown under these conditions; Harriott and Brown[7] discuss the link between cellular structure and radial segregation.

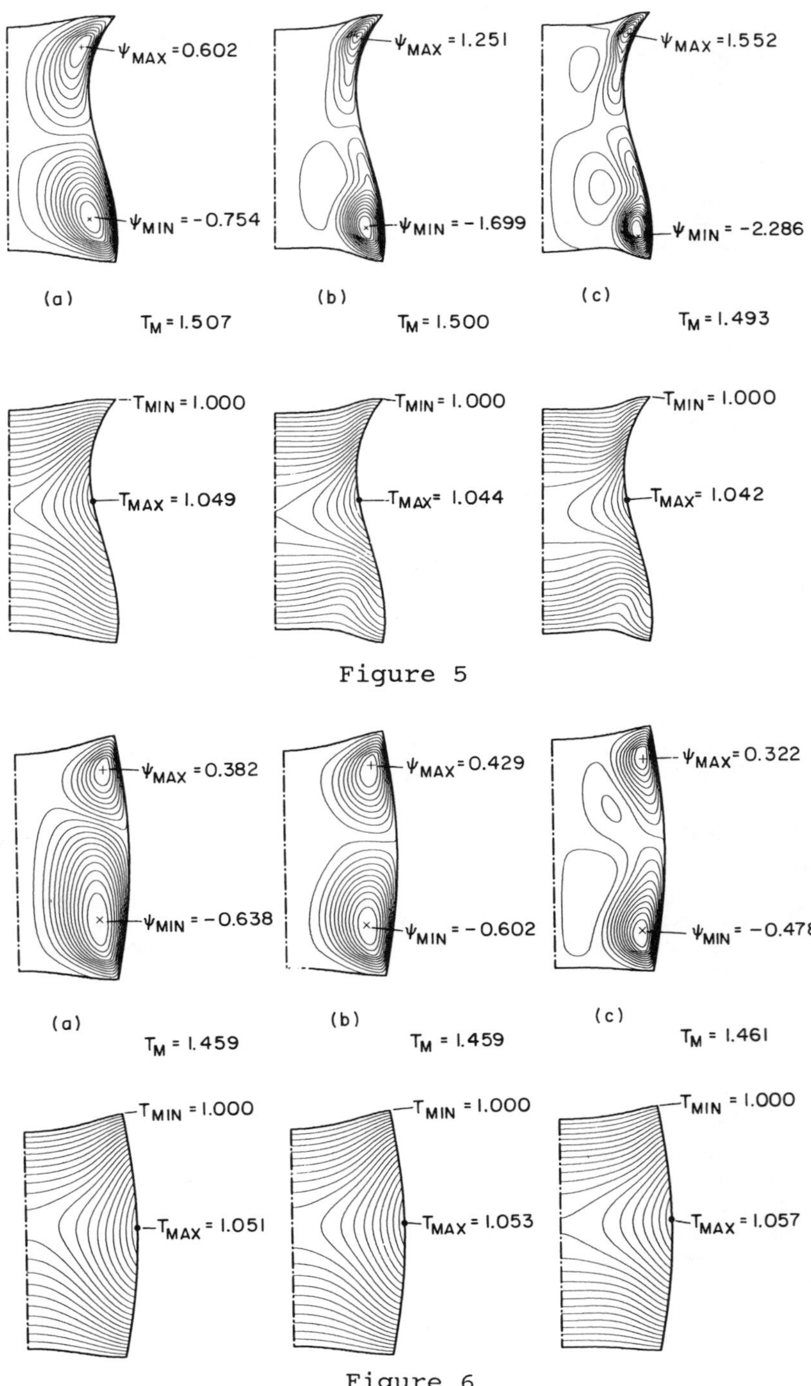

Figure 5

Figure 6

DISCUSSION

Although the calculations presented in this brief report are far from complete they do serve to make two important points. First, and most importantly, these calculations give the first results for the interaction between the flow structure and the shape of the floating zone. The calculations demonstrate the large asymmetries in the flow structure that result because of asymmetry in the temperature field caused by solidification and gravity-induced deformation of the meniscus. Moreover, the intense flows near the melt/gas surface driven by surface-tension gradients lead to deflection of the solid interfaces by uneven convective heat transport in the melt.

These effects are only predicted by analysis of a detailed Hydrodynamic Thermal-Capillary Model (HTCM) that accounts for flow in the melt and heat transfer in all the phases with a self-consistent treatment of the free surface shapes. The analysis of the HTCM with the finite-element/Newton algorithm and comparison of the calculations to asymptotic descriptions of transport processes promises to yield a great deal of insight into the operation and potential for optimization of small-scale floating zones.

ACKNOWLEDGEMENTS

This research was supported by the Microgravity Sciences and Applications Program of the US National Aeronautics and Space Administration and by a grant for the use of the ETA computer system at the John von Neumann National Supercomputer Center.

REFERENCES

1. R. A. Brown; B. Feuerbacher, H. Hamacher, R.J. Naumann, eds., Springer-Verlag, Berlin, 55-94 (1986).
2. R. A. Brown, *A.I.Ch.E. J.* **34**, 881 (1988).
3. W. G. Pfann, **Zone melting**, Krueger, Huntington (1978).
4. Ch.-H. Chun, *J. Crystal Growth* **48**, 600 (1980).
5. D. Schwabe, A. Scharmann, G. Preisser & R. Oeder *J. Crystal Growth* **43**, 305 (1978).
6. G. Preisser, D. Schwabe & A. Scharmann *J. Fluid Mech.* **126**, 545 (1982).
7. G. M. Harriott & R. A. Brown *J. Crystal Growth* **69**, 589 (1984).
8. J. L. Duranceau & R. A. Brown *J. Crystal Growth* **75**, 367 (1986).
9. J. L. Duranceau & R. A. Brown, *J. Crystal Growth* submitted (1989).
10. P. A. Clark & W. R. Wilcox *J. Crystal Growth* **50**, 461 (1980).
11. N. Kobayashi & W. R. Wilcox *J. Crystal Growth* **59**, 616 (1982).
12. G. M. Harriott & R. A. Brown *J. Fluid Mech.* **144**, 403 (1984).
13. G. M. Harriott & R. A. Brown *J. Fluid Mech.* **126**, 269 (1983).
14. T. Surek & B. Chalmers *J. Crystal Growth* **29**, 1 (1975).
15. H. M. Ettouney & R. A. Brown *J. Comput. Phys.* **49**, 118 (1983).
16. C. J. Chang & R. A. Brown,*J. Comput. Phys.* **53**, 1 (1984).
17. P. A. Sackinger J. J. Derby & R. A. Brown *Int. J. Numer. Meths. Fluids* **xx**, xxx (1988).
18. A. Zebib, G. M. Homsy & E. Meiburg *Phys. Fluids* **28**, 3467 (1985).

FIGURE CAPTIONS

Figure 1. Schematic of small-scale floating zone system used for the HTCM.

Figure 2. Microgravity zone shapes, streamlines and isotherms for surface-tension-driven flows with increasing intensity in the absence of crystal growth and with the length of the zone held fixed at $L_m = 2.5$. The figures correspond to Ma^* values of (a) 23.8, (b) 35.7, (c) 47.6 and (d) 75.0. Twenty streamlines are shown with values equally spaced between the minimum and maximum values and zero. The temperature field is unchanged by increasing Ma^* in this range and only one plot of the isotherms are shown. Twenty isotherms are shown equally spaced between the maximum value and the melting point. This representation scheme is used for all the plots in this paper.

Figure 3. Microgravity zone shapes, streamlines and isotherms for surface-tension-driven flows with increasing intensity with crystal growth and with the length of the zone allowed to vary. The figures correspond to Ma^* values of (a) 69.2, (b) 122.0 and (c) 244.0.

Figure 4. Microgravity zone shapes, streamlines and isotherms for surface-tension-driven flows with increasing intensity with crystal growth and with the length of the zone held fixed at $L_m = 2.5$. The figures correspond to Ma^* values of (a) 187.5 (b) 583 and (c) 910.

Figure 5. Earthbound zone shapes, streamlines and isotherms for surface-tension-driven flows with increasing intensity with crystal growth and with the length of the zone held fixed at $L_m = 2.5$. The figures correspond to Ma^* values of (a) 98, (b) 444, and (c) 840.

Figure 6. Microgravity zone shapes, streamlines and isotherms for surface-tension-driven flows with equal counter-rotation of the two solid surfaces. There is crystal growth and the length of the zone is held fixed at $L_m = 2.5$. The figures correspond to $Re = Re_f = Re_c$ with values of (a) 72.7, (b) 181.8 and (c) 363.8. The Marangoni number Ma is held fixed at 1000; Ma^* values are (a) 50.8, (b) 52.3 and (c)

3. Computational Fluid Dynamics

Robert A. Brown
Session Chair

COMPUTATIONAL STUDIES OF DROP AND BUBBLE DYNAMICS IN A VISCOUS FLUID

L. G. Leal[a]
*Dept. of Chemical Engineering, California Institute of Technology
Pasadena, CA 91125*

ABSTRACT

This paper reviews recently developed numerical methods for the solution of free-boundary problems in fluid dynamics. A sample of recent CFD studies from our research is then presented, with the goal of illustrating the potential of "computational experiments" in elucidating the dynamics of bubbles and drops in viscous fluids at both zero and nonzero Reynolds numbers.

INTRODUCTION

The development of new and improved numerical techniques and the recent accessibility of powerful computer systems have had a major impact on research in many areas of science and engineering. Indeed, for many fields, the ability to directly simulate physical phenomena by numerical solution of the governing equations has led to a third branch of investigative methodology that is a co-partner of the classical approaches of laboratory experiment and mathematical analysis. Nowhere is this more evident than in fluid mechanics, as is perhaps reflected by the existence of the widely-recognized acronym CFD ("Computational Fluid Dynamics") and an explosion of research publications and journals.

Although based upon a theoretical framework, the computational method of investigation is (or should be) more closely related to an experimental study. The advantages are that physical parameters (or, preferably, dimensionless groups) can be varied independently—often difficult or impossible in a laboratory experiment–and the computational experiment can probe problems that are characterized by very small length or time scales (though resolution at very small scales for a problem that is not inherently small scale can be difficult and/or expensive). Furthermore, we can consider well-characterized, fundamentally important flows (such as uniaxial extension) that are difficult to realize in an "ideal" form experimentally. Finally, we can study classes of solutions that reveal important physical mechanisms, but are subject to instabilities, and are thus not realizable in the laboratory. The primary disadvantage is that the accessible parameter range can often be restricted by numerical considerations unrelated to physical phenomena. In addition, there is the implicit assumption that the equations (and boundary or initial data) actually describe the physical system of interest, but no inherent means, apart from comparison with the "real" system, to verify that this is true.

For many fluid mechanics systems, the governing equations and boundary conditions are well known. However, for fluid flows that involve non-Newtonian fluids or fluid interfaces with surfactant, the connection between the real physics and the mathematics is often unknown. This is both a liability and an opportunity. It is a liability in the sense that we lack an *a priori* predictive capability for these important classes of fluid mechanics problems. From a computational point of view, however, it is also an important opportunity in the sense that numerical solutions may provide the only meaningful basis for comparisons between model predictions and experimental data.

Among all fluid mechanics problems, one of the most difficult for analytical

---

[a]*current address*: Dept. of Chemical and Nuclear Engineering, University of California at Santa Barbara, Santa Barbara, CA 93106

theory has been free-boundary problems in which two or more distinct fluids are separated by an interface. The source of difficulty is well known. The shape of the interface changes in response to motion of the contiguous fluids and is therefore one of the unknowns that must be determined as part of the solution. As a consequence of the unknown interface shape, free-boundary problems are always nonlinear even in the limit $R \to 0$, where the linear Stokes equations approximate the nonlinear Navier-Stokes equations. Not surprisingly, analytical solutions are inevitably restricted to limiting cases when the shape of the interface is either identical with or asymptotically close to its equilibrium form. For bubbles and drops, nearly all analytical solutions therefore assume that the shape is either spherical or nearly-spherical so that domain perturbation methods can be used (one exception is the slender-body theory for bubbles and inviscid drops pioneered by G. I. Taylor[1] and developed in recent years by Buckmaster[2,3] and Acrivos and co-workers[4]). Because of the difficulties of good experiments or analytic theories, numerical (CFD) experiments can play a critical role for many problems involving the motions of bubbles and drops.

The general attributes of the CFD approach, cited earlier, are especially pertinent to bubble and drop dynamics problems. It is often relatively easy to visualize the shape of a bubble or drop experimentally. However, it is always extremely difficult (or impossible) to get detailed data on the velocity fields, inside or outside, and even shape visualization can be difficult if changes occur very rapidly or on a small length scale (a classic example is cavitation bubble collapse, where framing rates of 10^6 may not achieve adequate resolution, Lauterborn & Timm[5]). One experimental problem is the interface which interferes with optical techniques of flow visualization or measurement. But, of course, none of these factors is an issue with "data" from a numerical experiment. Another experimental difficulty is that bubbles and drops are usually translating with respect to laboratory coordinates. This not only makes detailed measurement difficult, but, with the exception of simple shear flow in a Couette device or translation through a quiescent fluid, the bubble or drop is almost always subject to a Lagrangian time-variation of the velocity field that makes data interpretation extremely difficult. Again, the ability to study the dynamics of a stationary bubble or drop in a well-characterized homogeneous flow is not a problem from the computational point of view. Last, but not least, the computational experiment has the potential for a clear delineation of the effects of different types of interfacial properties. For example, the dynamics of a bubble or drop with a "clean" interface, characterized by a constant interfacial tension alone, can be studied easily via CFD, with none of the practical worries of contamination, nonisothermal conditions, etc., that often create ambiguities in the interpretation of laboratory experiments. At the same time, the CFD "experiment" holds the potential to understand the independent influence of such non-ideal features as surface tension gradients, and interface viscosity or elasticity, which are often entwined to such an extent in the laboratory experiment that it is unclear which is even present in a given system.

In this paper, I discuss the application of CFD methods to studies of the dynamics of bubbles and drops in a viscous liquid. My primary purpose is not a detailed discussion of methodology. Thus, my description of methods is limited to a brief outline of the boundary-integral and finite-difference, boundary-fitted coordinate techniques that have been used in my own research group. Alternative approaches such as the finite-element techniques, are not discussed at all. In part, this is because there are others far more qualified than I to discuss this material. More importantly, however, it is because the central issue is not the techniques at all, but rather the kind of research that they allow. In an attempt to illustrate the potential of CFD for bubble and drops dynamics, I conclude by discussing a number of problems that we have considered in our group.

COMPUTATIONAL METHODS FOR THE SOLUTION OF BUBBLE AND DROP PROBLEMS

A. Governing Equations and Boundary Conditions

We wish to consider the motion of a single bubble or drop in an incompressible Newtonian fluid which undergoes some undisturbed motion far from the bubble or drop, denoted symbolically $\mathbf{u}_\infty$. In some cases, $\mathbf{u}_\infty = 0$, and the motion is driven entirely by surface forces at the interface. In others, the drop may move in the vicinity of an external wall or boundary, so that the fluid domain is not unbounded in all directions. Nevertheless, to provide a frame of reference for the following discussion, we pose the governing equations and boundary conditions for a single drop that is stationary at the origin of our coordinate reference system, with the velocity $\mathbf{u}_\infty$ imposed in the far-field. If the governing equations and boundary conditions are nondimensionalized with respect to a characteristic velocity u_c, a characteristic length scale a (the undeformed radius of the drop), and a characteristic pressure p_c, the result is

$$Re\left[\frac{\partial \mathbf{u}}{\partial t}+\mathbf{u}\cdot\nabla\mathbf{u}\right]=-\left[\frac{ap_c}{\mu u_c}\right]\nabla p+\nabla^2\mathbf{u} \tag{1}$$

$$\frac{\kappa Re}{\lambda}\left[\frac{\partial \hat{\mathbf{u}}}{\partial t}+\hat{\mathbf{u}}\cdot\nabla\hat{\mathbf{u}}\right]=-\left[\frac{ap_c}{\mu u_c}\right]\frac{1}{\lambda}\nabla\hat{p}+\nabla^2\hat{\mathbf{u}} \tag{2}$$

$$\nabla\cdot\mathbf{u}=\nabla\cdot\hat{\mathbf{u}}=0 \tag{3}$$

with

$$\mathbf{u}\rightarrow\mathbf{u}_\infty \quad \text{as} \quad |\mathbf{x}|\rightarrow\infty \tag{4}$$

$$\mathbf{u}=\hat{\mathbf{u}} \quad \text{for} \quad |\mathbf{x}|\,\varepsilon\,\mathbf{x}_s \tag{5}$$

$$\frac{\partial \mathbf{x}_s}{\partial t}=\mathbf{n}(\mathbf{n}\cdot\mathbf{u}) \quad \text{for} \quad |\mathbf{x}|\,\varepsilon\,\mathbf{x}_s \tag{6}$$

and

$$-\mathbf{n}(p-\hat{p})\left[\frac{ap_c}{\mu u_c}\right]+(\mathbf{n}\cdot\boldsymbol{\tau}-\lambda\mathbf{n}\cdot\hat{\boldsymbol{\tau}})=\frac{\sigma}{\mu u_c}(\nabla_s\cdot\mathbf{n})\mathbf{n}-\mathbf{n}\,\frac{(\rho-\hat{\rho})ga^2}{\mu u_c}(\mathbf{i}_g\cdot\mathbf{x}_s) \text{ for } \mathbf{x}\,\varepsilon\,\mathbf{x}_s \tag{7}$$

For the fluid inside the drop, the variables are denoted with "^". The dimensionless parameters that appear are

$$Re=\frac{\rho u_c a}{\mu},\quad \lambda=\frac{\hat{u}}{\mu}\quad \text{and } \kappa=\frac{\hat{\rho}}{\rho}.$$

in addition to those shown explicitly in the equations and boundary conditions. In addition $\boldsymbol{\tau}$ and $\hat{\boldsymbol{\tau}}$ are the viscous contributions to the stress, i.e.

$$T=-p\mathbf{I}+\hat{\boldsymbol{\tau}},$$

while $\mathbf{x}_s$ denotes a position vector at a point on the interface, and $\mathbf{i}_g$ is a unit vector in the direction of gravitational acceleration. The pressure p (and $\hat{p}$) is the dynamic pressure, i.e. $p=P-\rho g(\mathbf{i}_g\cdot\mathbf{x})$ [and $\hat{p}=\hat{P}-\hat{\rho}g(\mathbf{i}_g\cdot\mathbf{x})$].

The equations and boundary conditions, as stated in Equations (1) - (7), apply for both high and low Reynolds number flows. For a low Reynolds number flow, the characteristic pressure is $p_c=\mu u_c/a$ and the two additional parameters that appear in Equation (7) are the capillary number $Ca\equiv\mu u_c/\sigma$ and the buoyancy parameter $Cg\equiv\frac{\mu u_c}{\Delta\rho g a^2}$. For $Re\geq O(1)$, on the other hand, $p_c=\rho u_c^2$ and the added dimensionless parameters are the Weber number $We\equiv\rho u_c^2 a/\sigma$ and the inertial buoyancy parameter $Wg\equiv CgRe\equiv\rho u_c^2/ga\Delta\rho$. When $\mathbf{u}_\infty\neq 0$, we can use the magnitude of $\mathbf{u}_\infty$ as a characteristic velocity, say $u_c=u_\infty$. When $\mathbf{u}_\infty=0$, on the other hand, the motion is capillary-pressure driven and $u_c=\sigma/\mu$.

The numerical problem is to solve Equations (1) - (7) for the velocity and pressure fields, and for the shape of the drop. There are two distinct types of numerical method for this classical free-boundary problem. The first can be used for the limiting approximations $Re = 0$ (creeping flow) or $Re = \infty$ (potential flow) when the governing differential equations reduce to a linear form. In this case, a fundamental solution can be obtained and the problem is reduced to the application of boundary conditions at an interface whose shape must also be determined. Although this latter problem is still nonlinear, as a consequence of the intrinsic nonlinearity of the boundary conditions at $\mathbf{x}_s$, it is much simpler than solving the full problem in Equations (1) - (7). The resulting numerical technique is known as the Boundary-Integral (BI) method. The second class of numerical methods applies for all finite Reynolds numbers, *Re*. For these cases, we must solve the full nonlinear Navier-Stokes equations, and this requires a finite-difference, finite-element, spectral or other numerical technique that is suitable for nonlinear partial differential equations.

B. Boundary-Integral Techniques

The first application of the boundary-integral technique to numerical solution of a low Reynolds number flow problem seems to be due to Youngren & Acrivos,[6] who built upon the pioneering work of Ladyzhenskaya[7]. The first application to bubble dynamics problems was also due to Youngren & Acrivos,[8] who examined bubble shapes in an axisymmetric extensional flow, while Rallison & Acrivos[9] were the first to consider the generalization to viscous drops. The basic description given here closely follows Stone & Leal[10] who also provide reference to a large number of the earlier studies using the BI formulation for low Reynolds number flows.

The basic idea of the low Reynolds number BI method follows from the fact that the governing Equations (1) and (2) are linear for the creeping flow limit and thus a general solution can be obtained in which the velocity at any point in the two fluids can be represented exactly in the form

$$\mathbf{u}(\mathbf{x}) = \mathbf{u}_\infty(\mathbf{x}) - \int_s \mathbf{n}\cdot\mathbf{T}\cdot\mathbf{J}dS(y) - \int_s \mathbf{n}\cdot\mathbf{K}\cdot\mathbf{u}\,dtS(y) \tag{8}$$

$$\hat{\mathbf{u}}(\mathbf{x}) = \int_s \mathbf{n}\cdot\hat{\mathbf{T}}\cdot\mathbf{J}dS(y) + \int_s \mathbf{n}\cdot\mathbf{K}\cdot\hat{\mathbf{u}}\,dS(y) \tag{9}$$

where

$$\mathbf{J} = \frac{1}{8\pi}\left[\frac{\mathbf{I}}{|\mathbf{x}-\mathbf{y}|} + \frac{(\mathbf{x}-\mathbf{y})(\mathbf{x}-\mathbf{y})}{|\mathbf{x}-\mathbf{y}|^3}\right] \tag{10a}$$

$$\mathbf{K} = -\frac{3}{4\pi}\frac{(\mathbf{x}-\mathbf{y})(\mathbf{x}-\mathbf{y})(\mathbf{x}-\mathbf{y})}{|\mathbf{x}-\mathbf{y}|^5} \tag{10b}$$

Here, for the problem (expressed in Equations (1) - (7)) of a single drop in an unbounded fluid, *S* represents the surface of the drop, and thus $\mathbf{n}\cdot\mathbf{T}$ and $\mathbf{u}$ in the integrands are the stress vector and the fluid velocity on *S*. Hence, provided that the interfacial stress, the interfacial velocity and the drop shape are known, the velocity field can be calculated anywhere in the fluid domain using Equation (8) or (9). A similar formula can also be derived for the pressure in terms of surface integrals involving the velocity and stress at the interface, and hence the pressure can again be calculated anywhere in the domain once the interface stress and velocity distributions are known.

The key to solution of drop dynamics problems via the BI method is to note that Equations (8) and (9) can be combined at the interface in conjunction with the boundary conditions in Equations (5) and (7) to produce a *single integral equation for the interfacial velocity* alone,

$$\frac{(1+\lambda)}{2}\hat{\mathbf{u}}(\mathbf{x}_s) = \mathbf{u}_\infty(\mathbf{x}_s)$$

$$-\int_s \mathbf{J}\cdot\mathbf{n}\left[\frac{1}{Ca}(\nabla_s\cdot\mathbf{n})\mathbf{n} - \frac{1}{Cg}\mathbf{n}(\mathbf{i}_g\cdot\mathbf{x}_s)\right]dS(y) - (1-\lambda)\int_s \mathbf{n}\cdot\mathbf{K}\cdot\hat{\mathbf{u}}\,dS(y) \tag{11}$$

Here, the term in square brackets in the integrand of the first of the two integral is nothing more than the right-hand side of Equation (7) for the low Reynolds number limit, where $p_c = \mu u_\infty / a$.

For a given interface shape, the only unknown in Equation (11) is the interfacial velocity, which can thus be obtained by solving (11) numerically. The solution of (11) satisfies all of the boundary conditions at the interface except for the kinematic condition in Equation (6). Hence, once the interfacial velocity is known from (11), (6) can be integrated forward to obtain an update of the interface shape. In the form described above, the BI method for low Reynolds number flows is particularly convenient for drop dynamics studies because the evolution of the drop shape with time can be followed on the basis of the interfacial velocity alone, from (11), with no need to calculate velocities, pressure or stress elsewhere in the fluids.

The details of implementing the BI scheme can be found elsewhere.[10] Here, we offer only a few general comments. First, Equation (11) is lower order than the original differential equation, and this is an important advantage of BI methods. If the problem of interest is fully three-dimensional, we must solve a 2D integral in Equation (11), but if the problem is axisymmetric, the azimuthal integration can be performed analytically and the surface integrals are reduced to line integrals. The integral equation can be reduced to a discretized set of algebraic equations using standard collocation techniques, then solved via standard iterative procedures. After solving Equation (11), the interface shape is updated by integrating the kinematic condition forward one time step. Frequently, this is done using a simple explicit Euler scheme, though implicit methods have also been used with some improvement in numerical stability. The most important aspect of the discretization for studies of drop dynamics is accurate representation of the drop shape and of the curvature of the interface. This is accomplished, in part, by judicious placement of boundary elements to achieve high density in regions of high curvature, and in part by the use of cubic splines to generate twice continuously differential representations of the interface. The only technical details of the basic discretization scheme that are worth mentioning here are: the integrand is singular for $\mathbf{y} \to \mathbf{x}$ and this requires special care, including a local analytical integration in the vicinity of the singularity to maintain accuracy; second, the majority of the computation time is used in evaluating the coefficient matrix that involves the integration referred to above for each collocation point $\mathbf{x}_s$, on S; third, the resulting coefficient matrix is dense, rather than sparse as in other numerical CFD methods, and this requires some added time to solve the system of algebraic equations.

One generalization of the basic ideas is worth mentioning here, because it is particularly pertinent to the motions of bubbles, drops or small particles near a boundary of fixed shape. One way to solve such a problem is to distribute singularities corresponding to the fundamental solution of Equation (10a) over all of the boundaries. Hence, the integral equation at the drop surface must be supplemented by a second integral equation at the bounding wall, with S now standing for both the drop surface and the bounding walls. However, rather than basing the solution on the fundamental (Stokeslet) velocity field for a point force in an *unbounded* fluid (Equations (10) and (10b)), it may be possible to derive a fundamental solution (or, more accurately, a Green's function) that already satisfies boundary conditions at the fixed boundary. One case where this has been done is the solution for a point force in a fluid near a flat, no-slip wall which was solved by Blake.[11] In this case, an integral equation can be derived for the interfacial velocity on the drop that is identical in form to Equation (11). However, in place of the kernel functions of Equations (10a) and (10b), we use the results from Blake's solution. As a consequence, the surface of integration S is now restricted only to the surface of the drop, and there is no need for an additional integral equation on the fixed boundary. This generalization of Equation (11) automatically satisfies boundary conditions on the fixed boundary for any interfacial velocity distribution and any drop shape. This approach has been applied in our group by Ascoli *et al.*[12] for the motion of a drop near a wall. Details of implementation can be found in this reference or in the PhD theses of Ascoli[13] & Dandy[14] which also include a number of other applications.

The application of boundary-integral methods to bubble and drop dynamics in the potential flow limit $R = \infty$ is simpler than in the low Reynolds number limit. Steady solutions for the motion of bubbles and drops have been reported by Miksis and coworkers,[15,16] while transient calculations have been carried out by Blake & Gibson,[17] Blake *et al.*,[18] and Dommermuth & Yue.[19] The basic idea in this limit is to formulate the problem in terms of the velocity potential, $\mathbf{u} = \Delta\phi'$, with governing equation $\Delta^2\phi' = 0$, and the kinematic condition Equation (6) and the normal component of the stress balance Equation (7) applied at the interface, with $p_c = \rho u_c^2$, so that

$$-(p-\hat{p}) = \frac{1}{We}(\nabla_s \cdot \mathbf{n}) + \frac{1}{Wg}(\mathbf{i}_g \cdot \mathbf{x}_s) \tag{12}$$

for this limiting case, with

$$p + \frac{\partial \phi'}{\partial t} + \frac{1}{2}\nabla\phi' \cdot \nabla\phi' = G(t) \tag{13}$$

where $G(t)$ is a function of time only. If we consider only the disturbance flow problem, *i.e.*

$$\phi \equiv \phi' - \phi_\infty$$

where $\Delta\phi_\infty = \mathbf{u}_\infty$, then a general solution for ϕ can be derived in the form

$$\alpha(\mathbf{x},t)\,\phi(\mathbf{x},t) = \int_s \left[\frac{\partial\phi}{\partial n} - \phi\frac{\partial}{\partial\eta}\right]\frac{1}{|\mathbf{x}-\mathbf{y}|}\,dS(y) \tag{14}$$

where $\alpha(\mathbf{x},t)$ is the included solid angle at $\mathbf{x}$.

Two basic solution schemes have been developed for the solution of potential flow problems based upon the above formulation. The first, used by Miksis and coworkers[16] for steady-state solutions, is simply to solve Equations (12), (14) and the kinematic condition simultaneously via a full Newton's scheme. The second is a mixed Euler-Lagrange method for unsteady deformation problems such as those considered by Blake and coworkers.[18] In this scheme, a single iteration step consists of three steps: first, update ϕ using the normal stress condition ($\partial\phi/\partial t = \ldots$); second, solve Equation (14) to determine ($\partial\phi/\partial n$; and, third, update the interface shape by applying the kinematic condition in Equation (6). The details of these solution schemes can be found in the original publications referred to above.

C. Finite-Difference Techniques

For problems involving bubble or drop motions in viscous fluids at *finite* Reynolds number, we must solve the full nonlinear Navier-Stokes equations. The basic difficulty in solving any free-boundary problem by techniques suitable for the full Navier-Stokes equations is that the shape of the fluid domain is generally complicated, unknown and changing during the course of the computation (either because the domain shape is actually time-dependent, or because we seek a steady but unknown domain shape via an iterative technique that starts from an initial guess). Finite element methods seem ideally suited to the discretization of domains of complicated shape, and a number of successful implementations of finite-element methods for the solution of free-boundary problems have been reported in the literature,[20] including at least two that have been applied to study the dynamics of gas bubbles in a viscous fluid at finite Reynolds number (Bousfield *et al.*[21]; Tsukada *et al.*[22]). The development of finite-difference methods for this class of problems has been more recent, but these techniques have been applied to a more extensive set of problems involving bubbles and drops.

The fundamental development that has enabled the use of finite-difference techniques for problems with complex boundary geometry (including free-boundary problems) is the existence of methods for numerical generation of boundary-fitted coordinate grids. This allows complex boundaries to coincide with coordinate lines (or surfaces), which is a necessity for accurate application of boundary conditions. Many of the developments in numerical grid generation are described in a recent book.[23] In our own research, we have developed a technique called "Orthogonal Mapping" that is designed to produce *orthogonal* grids for two-dimensional or axisymmetric domains and is particularly well suited to the solution of free-boundary problems (Ryskin & Leal[24,25]).

The basic idea of finite-difference methods for the solution of flow problems involving the motion of a bubble or drop is thus to solve the Navier-Stokes equations, along with the boundary conditions, on a boundary-fitted orthogonal coordinate system (ξ,η,ϕ). For problems involving the motion of a bubble, we require only a coordinate system exterior to the bubble surface which, for axisymmetric geometries, can be connected with common cylindrical coordinates (x,σ,ϕ) via a pair of mapping functions $x(\xi,\eta)$ and $\sigma(\xi,\eta)$. These mapping functions are obtained numerically by solving a pair of covariant Laplace equations

$$\frac{\partial}{\partial\xi}\left[f\,\frac{\partial x}{\partial\xi}\right]+\frac{\partial}{\partial\eta}\left[\frac{1}{f}\,\frac{\partial x}{\partial\eta}\right]=0$$

$$\frac{\partial}{\partial\xi}\left[f\,\frac{\partial\sigma}{\partial\xi}\right]+\frac{\partial}{\partial\eta}\left[\frac{1}{f}\,\frac{\partial\sigma}{\partial\eta}\right]=0 \qquad (15)$$

where $f(\xi,\eta)$ is the so-called distortion function that represents the ratio of scale factors h_η/h_ξ for the boundary-fitted coordinates. The fluid mechanics part of the problem, then, is to solve the Navier-Stokes Equations (1) and (2), the continuity Equation (3), and the boundary conditions in Equations (4)-(7), using a finite-difference approximation on the boundary-fitted coordinates (ξ,η,ϕ).

We have applied a solution scheme of this type to both steady and unsteady problems involving axisymmetric motions of bubbles and drops. To date, no fully 3D problems have been considered. There are two fundamental issues in the transition to fully three-dimensional flows. First is the universal difficulty of a much larger numerical problem that is common to all CFD techniques. Second, the mapping techniques are much more difficult in 3D. For problems involving two-dimensional or axisymmetric geometries, it is always possible, in principle, to obtain an *orthogonal* map via solution of the pair of Equations (12) with associated boundary conditions.† As discussed in our original paper on the "orthogonal" mapping technique,[24] the difficulty for three-dimensional geometries is that the constraint on orthogonality cannot be enforced while simultaneously requiring the coordinate system to be boundary fitted (i.e. requiring that the boundaries of the fluid domain coincide with a coordinate surface). Hence, the fluid mechanics problem must be solved on a nonorthogonal grid. For bubbles, we do not anticipate any intrinsic difficulty in the transition to 3D, though some care must be taken to maintain numerical accuracy in regions of strong nonorthogonality. For drops, on the other hand, it is necessary to generate a coordinate grid both inside and outside the drop (see below), and accurate approximation of the interfacial boundary conditions requires that the coordinate lines for the two grids match at the interface. For two-dimensional or axisymmetric problems, both grids are orthogonal at the interface and it is not difficult to generate a pair of coordinate maps that satisfy this constraint. For 3D, however, this difficulty has not been addressed and it is unclear whether coordinate lines can be forced to match in value, and in their angle of inclination relative to the interface.

† In fact, even for 2D and axisymmetric problems, a comprehensive existence proof does not exist. However, existence for a reasonably broad class of problems was demonstrated recently by Ascoli *et al.*[26]

During the past several years we have developed two quite distinct numerical algorithms for the solution of steady, axisymmetric free-boundary problems. The fundamental difficulty is that the steady-state geometry of the flow domain (which determines the coordinate map) is unknown. Hence, in general, we must begin with an initial guess of the drop shape and develop a method to iterate toward the final steady-state solution. For problems that have a stable steady-state solution this can be done either via a full time-dependent calculation (discussed below in Sect. 3) or by a scheme of successive approximation as developed originally by Ryskin & Leal[25] and described below in Section 1. The other more powerful alternative that also allows investigation of unstable steady solutions is a full global iteration technique based upon Newton's method. A scheme of this type has been developed recently in our group (Dandy & Leal[27]) and will be discussed in Section 2.

(1) Iterative Algorithms for Steady, Axisymmetric Flows via Successive Approximations

The basic premise of the successive approximation scheme, developed by Ryskin & Leal[25,28,29] for the motion of deformable gas bubbles and extended by Dandy & Leal[30] for viscous drop problems, is that the flow problem and the mapping problem are decoupled. Specifically, for an initial bubble shape, we first seek a boundary-fitted coordinate map by solution of the mapping Equations (12) together with appropriate boundary conditions; once this is done, we solve the flow problem on the resulting coordinate system. Since we seek a steady solution, the kinematic condition Equation (6) is applied in the form $\mathbf{u} \cdot \mathbf{n} = 0$ at $\mathbf{x} \varepsilon \mathbf{x}_s$. Further, for the limiting case of a bubble (or void), the tangential component of the stress balance (7) reduces to the condition of zero shear stress. These two conditions are sufficient to completely determine the flow field for a bubble of specified shape. Of the remaining two conditions at $\mathbf{x} \varepsilon \mathbf{x}_s$, the velocity continuity balance of Equation (5) is redundant. The normal component of the stress balance of Equation (7), on the other hand, must be satisfied by the final steady-state solution. However, since the initial guess for the bubble shape is not likely to be correct, the normal stress condition will generally not be satisfied, and the mismatch between the left- and right-hand sides of Equation (7) is then used to increment the bubble shape to a new (and hope-fully better) approximation and the process is continued. In the successive approximation scheme it is thus the normal stress condition that provides the connection between the fluid mechanics and mapping problems.

The main difficulty with this procedure is the specification of a new bubble shape without overspecifying the coordinate map. Two quite distinct forms of the orthogonal mapping technique were developed: one is called the "strong constraint" map and involves *a priori* specification of the distortion function f (which is usually held fixed throughout a complete time-dependent or iterative solution cycle) and is particularly advantageous for free-boundary problems—note that the mapping equations with f specified are linear; the other technique is known as the "weak constraint" map and is designed for orthogonal maps in which the positions of coordinate lines on the boundary are completely specified (note that the distribution of coordinate lines at the boundary in the strong constraint method is determined completely by f in conjunction with the boundary shape and cannot otherwise be controlled), but suffers from the major disadvantage that the mapping equations are nonlinear since f can only be specified by the map itself via its definition in terms of $x(\xi,\eta)$ and $\sigma(\xi,\eta)$. The key to using the strong constraint map in the successive approximation scheme is to increment the bubble shape without directly specifying the updated positions of points on the boundary. In the technique developed by Ryskin,[25] for example, the interface shape is changed in response to the normal stress imbalance by incrementing the scale factor h_η of the coordinate map at the interface, rather than directly specifying the incremented positions of points on the boundary.

The successive approximation scheme developed by Ryskin for gas bubbles has also been generalized to the case of viscous drops by Dandy & Leal.[30] The basic new difficulty for drops is that a coordinate map must also be developed for the interior fluid domain and the grid lines for this interior map must match exactly the grid lines from the exterior map in order to insure accurate approximation of the interfacial boundary conditions. In the scheme developed by Dandy & Leal,[30] the mapping problem is therefore split into two successive steps. First, for a given interface shape (either specified directly at the initial guess, or indirectly by increment of h_η), the strong constraint technique is used to generate an external map. Second, with the position of grid lines at the interface for the external map now known, the weak constraint technique is used to generate an internal map with a precise one-to-one correspondence of coordinate lines. This scheme worked without fundamental difficulties. However, the nonlinear internal mapping problem was quite unstable, and this required strong underrelaxation and a significant incremental increase in computation time.

(2) Iterative Solution for Steady, Axisymmetric Flows via Newton's Method

Although the successive approximation method described in the previous section has proven to be extremely robust and useful for the class of problems considered, the decoupled successive solution of the mapping and flow problems requires many iterations to coverage, and is particularly slow for the solution of problems involving viscous drops where extreme underrelaxation is required to insure stability of the "weak constraint" interior mapping problem. Further, these successive approximation methods are not optimal for solution of problems that may exhibit limit points, multiple solutions or instabilities (bifurcations to other solution branches). When the successive approximation scheme fails to converge, it is difficult to distinguish between numerical difficulties and a true approach to a limit point or a branch point to time-dependent solutions of the same class. Further, when an instability or bifurcation to a different type of solution is encountered, it cannot be detected at all with the successive approximation scheme.

For these reasons we have recently developed an alternative method, based on a global iteration using Newton's method in which we simultaneously iterate on all of the unknown variables at each step—including the flow-field, the coordinate map and the interface shape. The details of this development for a gas bubble are reported by Dandy & Leal.[27] The latter represents a modest improvement of the scheme developed by Dandy.[14] The application to viscous drops has not yet been published, but is discussed here briefly.

Due to space limitations, we restrict our present discussion to a few comments and comparisons with the successive approximation scheme. The most important is that the global iteration greatly simplifies and streamlines the determination of the drop or bubble shape, as well as the corresponding coordinate mapping problem for viscous drops. The basic reason is that a separate step to calculate the interface explicitly from the normal stress condition is no longer necessary. The coordinate mapping functions and the flow field variables are obtained simultaneously. At the interface, the boundary conditions in Equations (5), (6), and (7) plus the orthogonality condition

$$\frac{\partial x}{\partial \xi} - \frac{1}{f}\frac{\partial \sigma}{\partial \eta} = 0$$

are completely sufficient to determine both the flow field and the mapping functions. Thus, though the shape of the interface can always be determined from the mapping functions $x(\xi,\eta)$ and $\sigma(\xi,\eta)$ (corresponding in our applications to the coordinate line $\xi = 1$), it is not calculated explicitly as was necessary in the decoupled, successive approximation scheme. Since the interior and exterior mapping problems are now solved simultaneously, the strong constraint method, which is linear, can be used in both regions. Rather than specifying the boundary correspondence of the interior map by specifying the positions of the coordinate lines from the exterior map (which forced us to use the weak constraint technique in our successive approximation scheme), we now simply require that the interior and exterior maps are "continuous" at the interface. This weaker boundary condition on the interior mapping problem allows it to be solved by the strong constraint formulation.

Another comparison with the decoupled successive approximation scheme is computation time. It is well known that a properly implemented Newton's scheme will exhibit quadratic convergence. We have tested the present algorithm for several real problems. One of these is the buoyancy-driven motion of a gas-bubble through a viscous fluid, a problem that we had solved previously over a wide range of Reynolds and Weber numbers (*R* and *W*) using the decoupled, successive approximation scheme. The convergence rate for the Newton's scheme for this problem is illustrated in Table I, which is reproduced from Dandy and Leal.[27] This table shows the maximum norm of the residual at the end of each of four global iterations for a variety of Reynolds and Weber numbers on a 61x61 mesh, with the initial condition in each case obtained by a standard arc length continuation from the previous solution. After only four iterations, the maximum norm is reduced by approximately 10 orders of magnitude from the initial condition. Execution times for these cases were approximately 22.5 CPU seconds per Newton step on a Cray XMP-48, using a frontal method developed by Harwell (MA32) for solution of the large sparse Jacobian matrix system. This computation time could be decreased dramatically with access to a solid-state storage device (SSD) so that the temporary files created by the frontal method would not have to be transferred to the regular local disks. Execution times for the *O*(1000) iterations required for the same accuracy via the decoupled, successive approximation scheme are estimated to be approximately 6 sec. per iteration

(3) *Numerical Algorithm for Unsteady, Axisymmetric Problems*

From the algorithms described in the preceding sections for steady, axisymmetric flows, it is a relatively small step to a scheme for the solution of time-dependent problems. This generalization is discussed by Kang & Leal.[31,32] In brief, we discretize the time derivatives that appear in the equations of motion and in the kinematic condition and, beginning from a prescribed initial condition, we can solve the problem at each time step using one of the steady-state solution algorithms. In our work to date, we have used a slight modification of the decoupled, successive approximation (ADI) scheme of Ryskin & Leal[25] at each time step.

The main source of possible confusion in the use of boundary-fitted coordinates for time-dependent flow problems is the calculation of time derivatives. In particular, since the numerical calculations are carried out on the boundary-fitted coordinates, we must remember to transform properly between partial time derivatives at fixed points in space and partial time derivatives in the (ξ,η) system. The source of the difference between these two quantities is that the mapping is changing as a function of time as the boundary shape changes and as a consequence a fixed point in (ξ,η) is actually moving relative to the fixed coordinate system. The relationship between the two time-derivatives for an arbitrary dependent variable w is

$$\left[\frac{\partial w}{\partial t}\right]_{(x,\sigma)} = \left[\frac{\partial w}{\partial t}\right]_{(\xi,\eta)} - \left[\frac{\partial \mathbf{x}}{\partial t}\right]_{(\xi,\eta)} \cdot \nabla w$$

The transient solution algorithm that we have developed is based upon a fully implicit, backward time-differencing scheme. Although this results in a large system of highly-coupled nonlinear algebraic equations for each time step, the scheme is absolutely stable and this allows an arbitrary time step subject only to an acceptable discretization error.

A SAMPLE OF CFD STUDIES FOR BUBBLE OR DROPS

As suggested in the Introduction, numerical methods have now developed to the extent that the most important issue is the type of problems that can be investigated advantageously via the computational experiment, rather than the numerical methodology. The major limitation at this point is that almost all studies have been restricted to axisymmetric geometries. Very few fully 3D problems have been considered, and only in the creeping flow limit. As noted earlier, this is partly a consequence of the large "cost" associated with the application of current techniques on current hardware, and partly a reflection of the necessity (or at least desirability) for additional developments of numerical technique. In this section, I describe briefly several problems that have been considered in our research group where CFD has yielded new insight and understanding that could not have been achieved in any other way.

A. Rising Bubbles and Drops in a Quiescent Fluid

The buoyancy-driven translation of a single bubble or drop through a quiescent liquid has received more attention than perhaps any other problem in fluid mechanics. On the theoretical side, there is the creeping flow solution of Hadamard & Rybczynski, the boundary-layer analysis for slightly nonspherical bubbles or drops due to Moore,[33] and the famous result of Davies & Taylor for the velocity of rise of a spherical cap bubble (among many other investigations). Experimentally, there have been literally hundreds of studies with the primary goals of correlating the bubble/drop velocity and shape with the dimensionless parameters of the system. Much of this previous work is summarized in the book of Clift *et al.*[34]

Despite intensive study, however, there are many questions that have not (and possibly cannot) be answered with existing experimental and theoretical (analytical) techniques. For example, very little of a systematic nature is known about the velocity fields for either bubbles or drops, even for circumstances when the bubble or drop maintains a steady shape and a rectilinear trajectory. As a consequence, qualitative understanding of transport processes involving bubbles or drops is difficult, as is the nature of the interactions between two (or more) bubbles or drops of different size. Furthermore, though it is well known experimentally that rising bubbles and drops in most liquids exhibit a transition with increase of size from steady, rectilinear motion to time-dependent (initially periodic and later more chaotic) motions, a satisfactory theoretical explanation for this phenomena has never been given (cf. Meiron[35]). Finally, in the creeping flow limit, though it is well known that a spherical shape is a proper steady-state solution that exactly satisfies the normal stress balance, a satisfactory resolution of the stability of this solution has not been reported.

In our own studies, we have been able to resolve many of these outstanding problems and questions via CFD-based computational experiments. For example, in two separate but related investigations, we have been able to use the boundary-fitted coordinate, finite-difference technique to obtain a comprehensive set of numerical solutions for a rising bubble, and for a rising drop, in an unbounded viscous fluid. These solutions were constrained to be steady and axisymmetric, but include an exact calculation of the bubble or drop shape, in addition to the complete details of the flow fields. The complete set of bubble shapes, obtained in Ryskin & Leal,[28] is reproduced here as Figure 1. Apart from the existence of solutions at the larger Reynolds and Weber's numbers where experimental observation shows unsteady, oscillatory shapes and trajectories, there is nothing seemingly remarkable about the results shown in Figure 1. At the lower Reynolds numbers, there is a smooth transition toward spherical cap shapes, while at higher Reynolds numbers the deformation is toward a flattened pancake shape that is clearly reminiscent of the highly flattened, though unsteady shapes, that are observed experimentally in the intermediate *We* regime. If we examine the flow-fields, however, a rather surprising result is that there there is a strong recirculating wake downstream of the bubble for most Reynolds numbers at sufficiently large Weber number. The fact is that some limited experimental studies have also shown closed streamline wakes for moderately deformed bubbles. However, neither the low Reynolds number Hadamard-Rybzynski solution nor the high Reynolds boundary-layer solution for a bubble with a clean interface show any sign of such a flow structure, and it has often been assumed that such a phenomenon could not exist in the absence of surface contamination or some other mechanism to produce a departure from the zero shear-stress condition that is characteristic of a "clean" interface. The numerical solutions show that this assumption is false. The zero shear-stress boundary condition was used for all of the solutions shown in Figure 1! Not only that, but a comparison with experimentally observed bubble shapes and wake structures due to Hnat & Buckmaster[36] shows almost perfect agreement, as illustrated for one case in Figure 2. The existence of closed streamline wakes for bubbles at finite Reynolds numbers has important implications for transport processes and for the possible modes of interaction between two bubbles of different size. Most importantly, however, it provides an obvious basis to explain the onset of time-dependent oscillations of the bubble shape and the bubble trajectory. If it is assumed that closed-streamline wakes do not occur for clean bubbles, one must sup-pose that the time-dependent oscillations are due to an instability of shape, but many theories of this type over the years have failed to provide a satisfactory explanation of the experimental observations. We believe that the transitions observed experimentally are of two types: first, at a moderate Weber number, the recirculating wake becomes unstable via a vortex shedding mechanism in a manner closely analogous to the wake instability observed for a solid sphere—once the wake flow is unstable, the bubble trajectory becomes unsteady along with the bubble shape; second, at a higher Weber number, the steady solution branch will exhibit an *instability of shape* representing a limit point beyond which additional steady axisymmetric solutions cannot exist—and this may be related to the onset of more chaotic time-dependent fluctuations of shape. The critical point, however, is that no inviscid mechanism that excludes the dynamics of the recirculating wake can be expected to provide an explanation of the observed instabilities. We are currently involved in a detailed analysis of the stability of the steady axisymmetric solutions of Ryskin & Leal[28] to corroborate these speculations.

Although space limitations prevent a discussion of the solutions obtained for viscous drops, these again produce the surprising result of closed streamline wakes for a wide parameter range, but with the added observation that these wakes are physically *detached* from the drop. The interested reader may refer to Dandy's thesis[14] or the forthcoming paper by Dandy & Leal[30] for a complete description of these results.

The dynamics of a viscous drop rising through a quiescent fluid at low Reynolds number presents another interesting problem that can be studied profitably via CFD methods, this time using the boundary-integral technique. Previous analytic investigation has shown that an exact steady solution for this problem is the Hadamard-Rybzynski result for a spherical drop. Indeed, Taylor & Acrivos[37] showed some years ago that the normal stress condition was satisfied exactly by this solution for arbitrary values of the interfacial tension (or Capillary number), including zero.

Numerical solution of free-boundary problems. Part 2

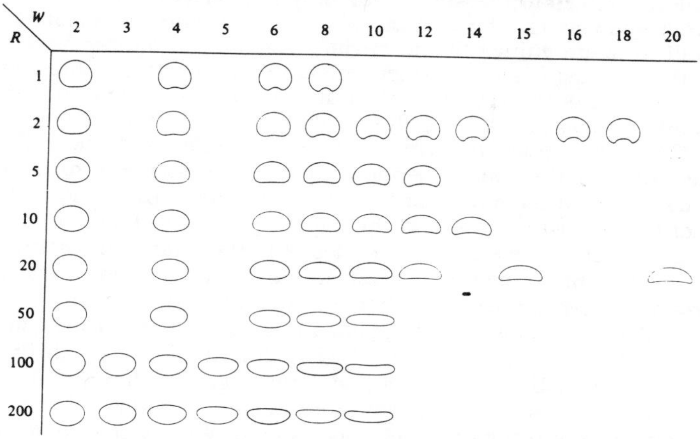

W: 2 3 4 5 6 8 10 12 14 15 16 18 20

R: 1 2 5 10 20 50 100 200

FIGURE 2. Computed steady, axisymmetric shapes of rising bubbles as a function of R and W.

Figure 1. Reprinted from Ryskin and Leal.

G. Ryskin and L. G. Leal

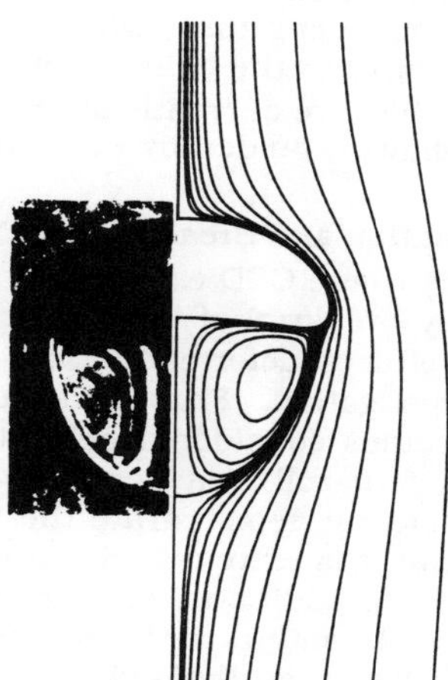

FIGURE 6. Comparison of the experimental photograph by Hnat & Buckmaster (1976) for $R = 19.4$, $W = 15.3$ ($C_D = 3.44$) with present results for $R = 20$, $W = 15$ ($C_D = 3.55$).

Figure 2. Reprinted from Ryskin and Leal.

Thus, the sphere is an equilibrium solution. The only remaining question is whether it is a stable solution. A correct analysis of the linear stability problem for infinitesimal disturbances in the shape was only reported recently by Kojima *et al.*[38] Their analysis shows that the spherical shape is *stable* to infinitesimal disturbances for all nonzero values of the surface tension *no matter how small*, but is *unstable* in the limiting case of surface tension equal to *zero*. This somewhat perplexing result suggests that the spherical drop may be unstable to disturbances of finite amplitude, at least for arbitrarily small but nonzero values of σ.

To test this hypothesis, we have carried out a series of numerical experiments in which we calculate the time-dependent motion and shape of drops that are initially deformed into either oblate or prolate spheroids. In particular, for a given fixed value of the viscosity ratio, and the capillary number based upon the velocity of the undeformed drop, we examine a series of initial shapes in which we systematically increase the initial degree of deformation. For finite capillary numbers, we find that drops sufficiently near to spherical are stable and return monotonically to a spherical final shape. This would appear to be consistent with the stability to infinitesimal disturbances that is predicted by Kojima *et al.*[37] for all finite capillary numbers. On the other hand, our numerical investigation shows that for each finite capillary number there is a critical minimum degree of initial deformation, beyond which the drop will *not* return to a spherical shape. With *increase* of *Ca*, this critical initial deformation decreases as would be expected from the linear stability theory. Detailed plots showing the critical deformation as a function of the capillary number and the viscosity ratio are presented in a forthcoming paper by Koh & Leal.[39] Here, we consider only two figures that illustrate the time-dependent deformation that ensues when we begin with a sufficiently deformed initial shape. In Figure 3, the initial shape is prolate and we see that the drop develops a tail that eventually pinches off. On the other hand, in Figure 4, the initial shape is oblate and the drop develops an indentation of increasing magnitude. In both cases, the upper surface of the drop appears to be extremely stable. The initial deformation seems to be focused at the rear of the drop with increasing time, where it increases rapidly in magnitude. Qualitatively, this is similar to the behavior predicted by the linear stability analysis for the limit $Ca = \infty$. We are currently attempting to develop an analytical prediction of the observed behavior, guided by the results of our numerical experiment.

B. Drop Deformation and Breakup in Extensional Flows

Another class of problems where CFD experiments have played a critical role is the deformation (and possibly breakup) of bubbles and drops in extensional flow fields. The general problem of drop deformation and breakup has again been the subject of extensive prior investigation. From an experimental point of view, however, the vast majority of studies considered only steady, simple-shear flows or plane extensional flow in a four-roll mill beginning with the original work by Taylor.[40] The latter results clearly demonstrate the vital importance of flow-type in the deformation process, but it is extremely difficult to devise other experiments in which the drop is subjected to additional flow types that are not transient from the Lagrangian point of view. Recently, for the *creeping flow limit*, a modified version of Taylor's four-roll mill has been devised that allows drop deformation and breakup to be studied experimentally in steady, two-dimensional linear flows with a (nearly) arbitrary ratio of strain rate to vorticity. However, no corresponding experiment exists for studies at finite Reynolds number, and our experimental knowledge of this problem is thus limited to results in simple Couette flow and in a few other rather poorly characterized flows that are time-dependent from the point of view of the drop (*e.g.* flow through a contraction). Even in the creeping flow limit, our inability to make detailed measurements of the velocity fields either inside or outside the

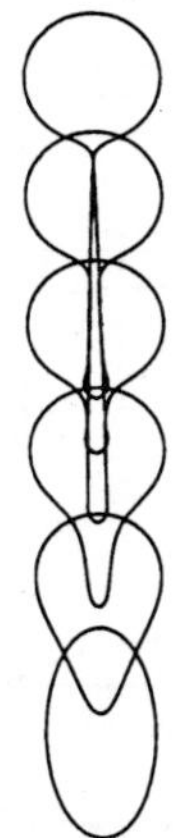

$$\mathrm{Ca} = 2.0 \qquad \lambda = 0.5 \qquad \Delta \equiv \left| \frac{L-B}{L+B} \right| = \frac{1}{3}$$

(dimensionless time between each shape = 2)

Figure 3. Evolution of an initially prolate drop that is translating through a quiescent fluid at zero Reynolds number for $\lambda = 0.5$, Ca = 2.0.

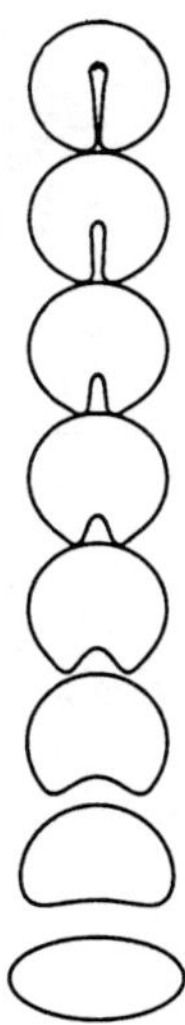

$$\mathrm{Ca} = 4.0 \qquad \lambda = 0.5 \qquad \Delta \equiv \left| \frac{L-B}{L+B} \right| = \frac{1}{3}$$

(dimensionless time between each shape = 2)

Figure 4. Evolution of an initially oblate drop that is translating through a quiescent fluid at zero Reynolds number for $\lambda = 0.5$, Ca = 4.0.

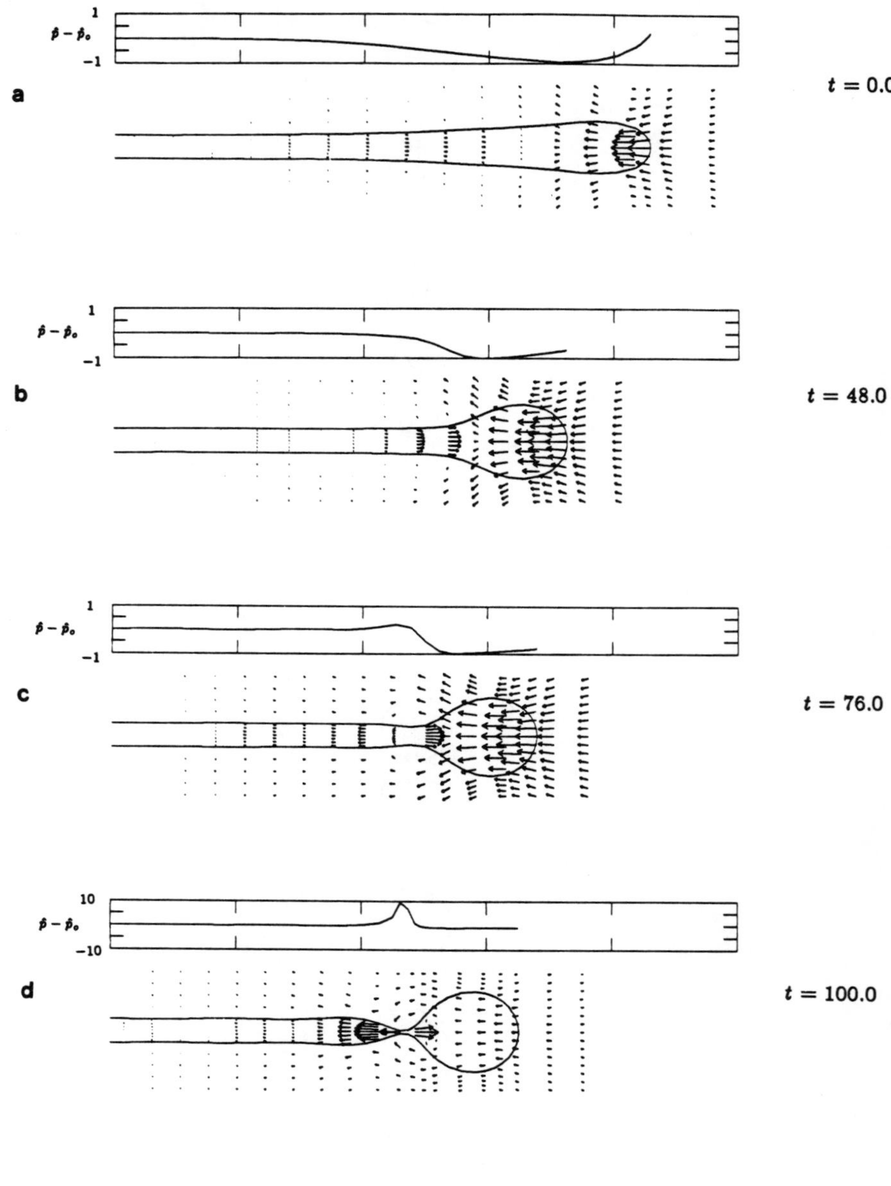

Figure 5. Evolution of an elongated drop due to surface tension-driven motion in an otherwise quiescent fluid for $\lambda = 1$.

drop means that it is difficult to understand the observed mechanisms of deformation or breakup.

The experimental observations in our lab of deformation and breakup in low Reynolds number flows show that the drop is basically elongated in the flow, and then undergoes breakup by a capillary-pressure-driven mechanism if the flow is suddenly stopped or otherwise decreased abruptly in strength, cf. Stone *et al.*[41] and Stone & Leal.[10] This may not in itself be too surprising. What is surprising, however, is that the mode of breakup that ensues is not generally a consequence of capillary wave instabilities as assumed in all existing theories, but rather results from a deterministic capillary-pressure-driven flow in which the drop breaks up piece by piece from the ends. We have termed this mode of capillary-pressure-driven breakup as "end pinching". An important property of end-pinching is that it is inhibited if the drop viscosity is large relative to that of the suspending fluid. The obvious questions that cannot be answered by experimental study of drop shapes alone, however, are: how does "end pinching" occur, and why is it inhibited for large viscosity ratios? To study these questions, we carried out a series of numerical experiments. Using the boundary-integral technique, we started with an elongated drop with a shape taken from experiments and then followed the capillary-pressure-driven motion in an unbounded quiescent fluid. The details of this investigation are reported by Stone & Leal;[10] a typical series of results is shown in Figure 5. Here, we show a time-dependent series of shapes for a drop with $\lambda = 1$ starting from an experimentally-observed shape at $t = 0$, up to a point just prior to pinch-off of the ends, corresponding to the mechanism of breakup that we have termed "end pinching". Also shown is the velocity fields and the pressure distribution along the centerline of the drop for the first three cases.

It can be seen that the variations in surface curvature lead to a strong, capillary-pressure-driven flow. At the end of the drop there is a pressure maximum which corresponds to a local maximum in surface curvature, and this drives an internal flow toward the center of the drop, as the end of the drop bulbs outward toward a spherical shape. In the central region of the drop the curvature is nearly constant and the pressure is uniform so that there is relatively little motion. In the transition zone, however, the curvature passes through a local maximum and there is a corresponding flow from the center of the drop towards the end. It is this flow which initially produces a local pinch in the drop shape. Once formed, the pinch leads to a local maximum in the capillary pressure which accelerates the decrease in drop radius and leads to pinch-off.

The key to the role of viscosity ratio in "end pinching" is that it controls the relative magnitude of the *outward* flow in the transition zone and the *inward* flow from the ends of the drop. The primary resistance to the latter is the translation of the bulbous end through the *exterior* fluid (velocity gradients inside the drop are relatively small in this region). The primary resistance to the outward flow in the middle of the drop is the viscosity of the interior fluid (velocity gradients inside the drop are relatively large and the flow locally resembles Poiseuille flow). Thus, when the interior viscosity is increased relative to the exterior (*i.e.* λ is increased), the flow field is dominated by the inward motion at the ends, no pinch develops and the drop returns to a spherical equilibrium shape. On the other hand for smaller λ, the outward flow in the central region is increased in relative importance and the tendency toward breakup by "end pinching" is enhanced. Although these explanations may seem straightforward in retrospect, they could not have been achieved in the absence of numerical solutions that provide a detailed view of the velocity fields inside and outside the drop.

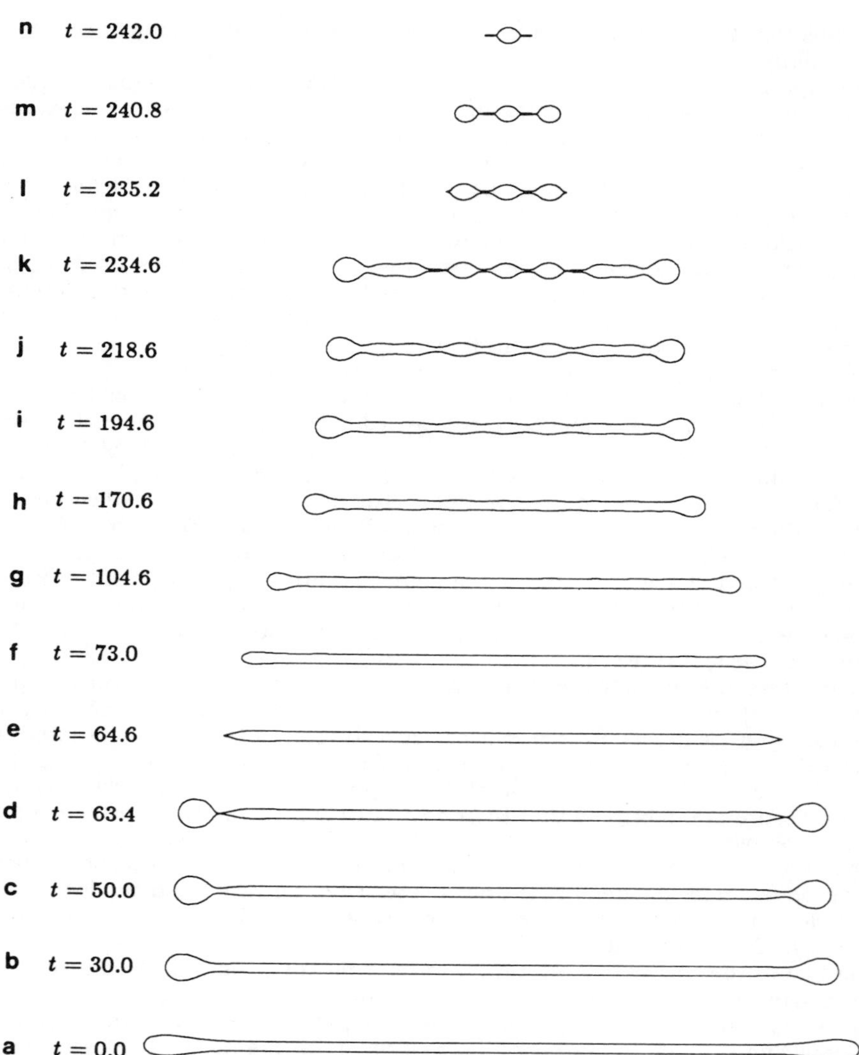

Figure 6. Evolution of an elongated drop due to surface tension-driven motion in an otherwise quiescent fluid for $\lambda = 1$. The initial shape is given a small initial disturbance comprised of capillary waves of amplitude $\sim 10^{-3}$ times the initial cylindrical radius.

Two other important insights were also achieved via our numerical experiments. First, the quantitative agreement achieved between computational and experimental results provides clear evidence, difficult to obtain from experiments alone, that surface-tension gradients play no fundamental role in the end-pinching process. Numerical experiments also provide a basis to understand the *competing* effects of "end pinching" which represents a consequence of a *deterministic* capillary-pressure-driven flow, and capillary wave growth which represents a capillary-pressure-driven instability.

For an infinite cylindrical thread without ends, capillary wave instability is the only mechanism for breakup, and even for a highly elongated drop, capillary-wave instabilities must always be present. However, in the latter case, they essentially compete with the "end pinching" process that apparently occurs on a time scale that is often too short for significant growth of capillary waves. On the other hand, breakup via "end pinching" is observed to be a *sequential* process with a total time to breakup of the whole drop that increases as the drop becomes more elongated and the number of end-pinching events increases. Thus, it may be expected that capillary waves should begin to play a role in breakup for a sufficiently elongated drop, and the numerical experiment provides a convenient way to investigate this phenomena by systematic increases in the degree of initial stretch and in the magnitude of initial capillary wave disturbances.

We have not yet carried out a complete study, but we have considered one case for $\lambda = 1$ and an initial drop length that is 73 times the diameter of the cylindrical region. In this case, the cylindrical midsection is given a small initial disturbance of amplitude, $\sim 3 \times 10^{-3}$ times the radius, with a combination of wavelengths that are both stable and unstable according to the linear capillary wave instability theory. The unstable initial modes have wave numbers $\omega = 0.44$, 0.66 and 0.88. The most unstable mode predicted by linear theory is $\omega = 0.56$ for $\lambda = 1$, and the linearly unstable range is $0 < \omega < 1$. A schematic of the numerical results for this case are sketched in Figure 6, which is reproduced from Stone & Leal[10], and shows the calculated drop shape at a number of dimensionless times (here, time is nondimensionalized with $t_c = R_o \mu/\sigma$ where R_o is the initial midsection radius). In proceeding beyond the shape at $t = 63.4$ (which is exact), an *ad hoc* procedure was used. Once the drop reaches a shape where an additional increment of time would move collocation points across the centerline, the ends of the drop are assumed to be broken off, the radius at the point of fracture is set equal to zero and the calculation is continued with only the central thread. The daughter droplets are thus neglected completely in ensuing calculations. Though this is not necessary in the boundary-integral formulation, it simplifies the calculation considerably.

The interesting features demonstrated in Figure 6 are that capillary waves do appear on the central thread (first clearly visible at $t = 170.6$), and these eventually cause breakup of the residual thread into three large and two small satellite drops at $t = 242$. Furthermore, in spite of the fact that the drop is continually shortening, there is very little flow produced in the central thread, and the capillary waves grow for a large portion of their lifetime (well into the finite amplitude regime) at a rate predicted by linear stability theory. The dominant mode leading to breakup is not the fastest growing linear mode, however, but is one of the nearby modes ($\omega = 0.66$) that was initially included as a disturbance. The growth rates in the vicinity of the fastest growing mode are relatively insensitive to ω, and in this case the disturbance with largest initial amplitude remains dominant up to the point of breakup. We have already noted that the linear theory is remarkably accurate through almost the entire capillary wave growth process. Nonlinear effects only finally appear at the last moment when they are responsible for the appearance of the small satellite drops. Other details of the calculation and results can be found in Stone & Leal.[10]

Finally, we turn to the problem of drop deformation and breakup at *finite* Reynolds numbers. As indicated earlier, it is extremely difficult to study this problem via laboratory experiments because it is almost impossible to achieve well-characterized flows with a time history (as seen by the bubble or drop) that is controllable independent of other features of the flow. Although this may seem to argue for the importance of studying drop breakup in the more complicated "real" flows, it is necessary to begin with simpler, more highly controlled flows in order to gain the understanding necessary to interpret drop breakup data in more complex flows. For example, most attempts at data correlation in "real" flows are predicated upon the existence of a critical Weber number for breakup. However, though this is an obvious implication of dimensional analysis, the likely dependence of the critical Weber number on flow type has not been studied even for steady flows. In addition, there is an obvious need to understand the role of transients in order to relate breakup in complex flows to the critical Weber number breakup criteria for steady flows. Obviously, this is a problem ideally suited to computational experiments, since we can easily study well-characterized steady and unsteady flows that are difficult to achieve in the lab.

To initiate our investigation of bubble and drop breakup at finite Reynolds number, we have used the finite-difference boundary-fitted coordinate algorithm described earlier to study the steady and time-dependent deformation of a gas bubble in two types of axisymmetric extensional flow—uniaxial and biaxial. Details of these studies are reported in a series of papers, Ryskin & Leal,[29] and Kang & Leal.[31,32] The results can be summarized as follows:

1 For uniaxial extensional flow, there exists a critical Weber number for Reynolds number ≥ 10, beyond which steady bubble shapes are not possible. The critical Weber number increases smoothly from $W_c \sim 1$ at $R = 10$ to the asymptotic, potential flow value $W_c \sim 2.7$ at $R \to \infty$, and there is also a smooth transition in bubble shapes from $R = 10$ to $R = \infty$ (Ryskin & Leal[29]).
2. The critical Weber number for *steady uniaxial*, extensional flow is an *upper* bound, at any fixed R, for transient uniaxial flow conditions. Lower critical Weber numbers were found in transient calculations, for example, if the bubble was already deformed prior to subjecting it to the uniaxial flow, or if the bubble was deformed starting from an initially spherical shape. Furthermore, if a bubble is subjected to an *inviscid* straining flow, there is a time-dependent oscillation of shapes with a frequency of the principle (P_2) deformation mode that decreases with increase of Weber number, and becomes zero at a Weber number that is exactly the limit point value for the existence of steady solutions (i.e. $W = 2.7$) (Kang & Leal[31]).
3. For steady *biaxial* extensional flow, the bubble shapes and the critical Weber number are identical to the uniaxial flow results for the potential flow limit $R \to \infty$. However, for finite Reynolds numbers in the numerical experiment, the steady shapes obtained are *fundamentally different* from the potential flow limit. Furthermore, unlike the uniaxial flow problem, steady solutions were found up to quite high Weber numbers $O(10)$, with no sign of a critical Weber number corresponding to a limit point for existence of steady solutions for all Reynolds numbers up to $R = 200$. At $R = 400$, there is evidence of an apparent transition toward the expected potential flow limit, but with a critical Weber number $W \sim 6$ (Kang & Leal[32]).

These results indicate clearly that the bubble deformation and breakup criteria is a very sensitive function of flow type, at least for finite Reynolds numbers. The difference between bubble deformation in uniaxial and biaxial extensional flow is a consequence of enhanced disturbance vorticity levels in the latter case, caused by the presence of a vortex line-stretching mechanism in the motion around the bubble.

CONCLUSIONS

Computational methods currently exist to allow the solution of steady or unsteady free-boundary flow problems, including the deformation and motion of bubbles or drops in a viscous or inviscid liquid. This allows computational experiments, in addition to laboratory and analytical studies, as a means of understanding many important phenomena for this class of problems. The examples discussed here are intended to give a flavor for the type of problems and questions that can be considered, and of the role that computational studies can play in this general area of research. It is clear, however, that we have only just scratched the surface with these examples and there are *many* important areas remaining to be studied. Among those being investigated in our own research at the present time are: (1) the role of surfactant transport processes in bubble and drop dynamics; (2) cavitation bubble dynamics, including the stability of shape for collapsing cavitation bubbles; (3) chaotic oscillations of bubble and drop shape in time-periodic flows; and (4) the growth and breakaway of vapor bubbles at a heated boundary (boiling).

ACKNOWLEDGEMENT

This work has been supported by grants from the Fluid Mechanics program at NSF, and from the Applied Hydrodynamics program at ONR. The author is grateful for this support.

REFERENCES

1. G. I. Taylor,1964 **Proc. of the 11th Congress Applied Mechanics**, Munich.
2. J. Buckmaster, 1972, *J. Fluid Mech.* **55**, 385.
3. J. Buckmaster, 1973, *J. Appl. Mech.* **E40**, 18.
4. A. Acrivos & T. S. Lo, 1978, *J. Fluid Mech.* **86**, 641.
5. W. Lauterborn & R. Timm, 1980, in **Cavitation and Inhomogeneities in Underwater Acoustics**, W. Lauterborn (ed.), Springer, Berlin, p. 42.
6. G. K. Youngren & A. Acrivos, 1975, *J. Fluid Mech.* **69**, 377.
7. O. A. Ladyzhenskaya, 1963, **The Mathematical Theory of Viscous Incompressible Flow**, Gordon & Breach, New York.
8. G. K. Youngren & A. Acrivos, 1976, *J. Fluid Mech.* **76**, 433.
9. J. M. Rallison & A. Acrivos, 1978, *J. Fluid Mech.* **89**, 191.
10. H. A. Stone & L. G. Leal, 1988, "Relaxation and Breakup of an Initially Extended Drop in an Otherwise Quiescent Fluid", *J. Fluid Mech.*, to appear.
11. J. R. Blake, 1971, *Proc. Camb. Phil. Soc.* **70**, 303.
12. E. P. Ascoli, D. S. Dandy, & L. G. Leal, 1988, "Buoyancy-Driven Motion of a Deformable Drop Toward a Planar Wall at Low Reynolds Number", to appear. Please contact author for details.
13. E. P. Ascoli, 1988, PhD Thesis, California Institute of Technology.
14. D. S. Dandy, 1987, PhD Thesis, California Institute of Technology.
15. M. Miksis, J.-M. Vanden-Broeck, & J. B. Keller, 1981, *J. Fluid Mech.* **108**, 89.
16. M. Miksis, J.-M. Vanden-Broeck, & J. B. Keller, 1982, *J. Fluid Mech.* **123**, 31.
17. J. R. Blake & D. C. Gibson, 1981, *J. Fluid Mech.* **111**, 123.
18. J. R. Blake, B. B. Taib, & G. Doherty, 1986, *J. Fluid Mech.* **170**, 479.
19. D. G. Dommermuth & D. K. P. Yue, 1987, *J. Fluid Mech.* **178**, 195.
20. R. Keunings, 1986, *J. Comp. Phys.* **62**, 199.
21. D. W. Bousfield, R. Keunings, & M. M. Denn, 1987, *J. Non-Newt. Fluid Mech.* **27**, 205.
22. T. Tsukada, M. Hozawa, N. Imaishi, & K. Fujinawa, 1984, *J. Chem. Eng. Japan* **17**, 246.
23. J. F. Thompson, Z. U. A. Warsi, & C. W. Mastin, 1985, **Numerical Grid Generation - Foundations and Applications**, North-Holland/Elsevier, New York.
24. G. Ryskin & L. G. Leal, 1983, *J. Comp. Physics* **50**, 71.
25. G. Ryskin & L. G. Leal, 1984a, *J. Fluid Mech.* **148**, 1..
26. E. P. Ascoli, D. S. Dandy, & L. G. Leal, 1987, *J. Comp. Phys.* **72**, 513.

27. D. S. Dandy & L. G. Leal, 1988a, "A Newton's Method Scheme for Solving Free-Surface Flow Problems", *J. Comp. Phys.*, submitted.
28. G. Ryskin & L. G. Leal, 1984b, *J. Fluid Mech.* **148**, 19.
29. G. Ryskin & L. G. Leal, 1984c, *J. Fluid Mech.* **148**, 37.
30. D. S. Dandy & L. G. Leal, 1988b, **Buoyancy-Driven Motion of a Deformable Drop Through a Quiescent Liquid at Intermediate Reynolds Numbers**, *J. Fluid. Mech.*, submitted.
31. I. S. Kang & L. G. Leal, 1987, *Phys. Fluids* **30**, 1929.
32. I. S. Kang & L. G. Leal, 1988, "Numerical Solution of Axisymmetric, Unsteady Free-Boundary Problems at Finite Reynolds Number. II. Deformation of a Bubble in a Biaxial Straining Flow", *Phys. Fluids*, to appear.
33. D. W. Moore, 1963, *J. Fluid Mech.* **16**, 161.
34. R. Clift, J. R. Grace, & M. G. Weber, 1978, **Bubbles, Drops and Particles**, Academic Pr., New York.
35. D. I. Meiron, 1988, **On the Stability of Gas Bubbles Rising in a Inviscid Fluid**, *J. Fluid Mech.*, submitted.
36. J. G. Hnat & J. D. Buckmaster, 1976, *Phys. Fluids* **19**, 182.
37. T. D. Taylor & A. Acrivos, 1964, *J. Fluid Mech.* **18**, 466.
38. M. Kojima, E. J. Hinch, & A. Acrivos, 1984, *Phys. Fluids* **27**, 19.
39. C. Koh & L. G. Leal, 1989, "The Stability of Shape for a Translating Drop at Zero Reynolds Number", to appear. Please contact author for details.
40. G. I. Taylor, 1934, *Proc. R.. Soc. London* **A146**, 501.
41. H. A. Stone, B. J. Bentley, & L. G. Leal, 1986, *J. Fluid Mech.* **173**, 131.
42. E. J. Hinch & A. Acrivos, 1979, *J. Fluid Mech.* **91**, 401.

LIST OF FIGURES

Figure 1. Computed steady, axisymmetric shapes of rising bubbles as a function of R and W.

Figure 2. Comparison of the experimental photograph by Hnat & Buckmaster (1976) for $R = 19.4$, $W = 15.3$ ($C_D = 3.44$) with present results for $R = 20$, $W = 15$ ($C_D = 3.55$).

Figure 3. Evolution of an initially prolate drop that is translating through a quiescent fluid at zero Reynolds number of $\lambda = 0.5$, $Ca = 2.0$.

Figure 4. Evolution of an initially oblate drop that is translating through a quiescent fluid at zero Reynolds number for $\lambda = 0.5$, $Ca = 4.0$.

Figure 5. Evolution of an elongated drop due to surface tension-driven motion in an otherwise quiescent fluid for $\lambda = 1$.

Figure 6. Evolution of an elongated drop due to surface tension-driven motion in an otherwise quiescent fluid for $\lambda = 1$. The initial shape is given a small initial disturbance comprised of capillary waves of amplitude $\sim 10^{-3}$ times the initial cylindrical radius.

NONLINEAR DYNAMICS AND BREAK-UP OF CHARGED DROPS

John A. Tsamopoulos
Department of Chemical Engineering,
State University New York, Buffalo, Buffalo, NY 14260

I. ABSTRACT

A rigorous mathematical framework for studying the nonlinear dynamics of charged drops has been developed and summarized here. A combination of domain perturbation and multiple timescale methods are systematically used to compute the evolution of axisymmetric, inviscid and charged drops that exhibit a number of nonlinear phenomena.

Nuclear physics has contributed theoretical analysis and impetus for experimental study of liquid drops, since Bohr and Wheeler began modeling atomic nuclei as uniformly charged liquid drops with surface tension. When moderate amplitude oscillations of charged drops are considered, it is shown that the increased inertia of the system slows down the motion by decreasing the frequency of the oscillation. The analysis also demonstrates the possibility of resonance between the fundamental mode of oscillation and one of its harmonics for particular values of the net charge on the drop. Both frequency and amplitude modulation of the oscillations are predicted for drop motions starting from general initial conditions. This effect cannot be anticipated from the linear analysis and proves that Rayleigh's solution for small-amplitude oscillations can actually be unstable.

The dynamics of break-up of a charged drop is a long standing issue, although the neutrally stable states have been known since the early sixties. Rayleigh calculated the maximum charge that a spherical drop can carry before it becomes unstable due to electrostatic repulsion. The present analysis shows that the first axisymmetric family that bifurcates from the spherical shape evolves transcritically, so that the drop will be either unstable for elongated prolate shapes or stable for flat oblate shapes. The evolution of drop shapes leading to break-up is also analyzed, and the dependence of the amount of charge on the amplitude of the deformations is computed. The asymptotic analysis for the static shapes is in very good agreement with the finite element calculations for even large amplitude deformations of the drop. Recently, it has been shown that oblate spheroids are unstable with respect to non-axisymmetric disturbances and, thus, are not observable.

II. FORMULATION

The mathematical formulation for the motion of a charged inviscid drop is given below. We consider the irrotational and incompressible motion of an electrically conducting inviscid drop with volume $\tilde{V} = 4\pi R^3/3$, density ρ, surface tension σ and net electrical charge Q. The motion of the drop in a tenuous surrounding medium is caused by initially introducing a small axisymmetric deformation. The surface of the drop is described by $RF(\theta,t)$ where $F(\theta,t)$ is the dimensionless shape function of the drop and θ is the meridional angle in spherical coordinates. Appropriate scales are used to define the dimensional velocity potential $(\sigma R/\rho)^{1/2}\phi(r,\theta,t)$, pressure $(2\sigma/R)P(r,\theta,t)$ and time $(\rho R^3/\sigma)^{1/2}t$ in terms of their dimensionless counterparts. The inviscid equations of motion and boundary conditions are:

$$\nabla^2\phi = 0 \qquad (0 \le r \le F(\theta,t),\ 0 \le \theta \le \pi), \qquad \text{(II.1)}$$

$$\frac{\partial\phi}{\partial r} = 0 \qquad (r = 0,\ 0 \le \theta \le \pi), \qquad \text{(II.2)}$$

$$2P+\frac{\partial \phi}{\partial t}+\frac{1}{2}\left[\left(\frac{\partial \phi}{\partial r}\right)^2+\left(\frac{1}{r}\frac{\partial \phi}{\partial \theta}\right)^2\right]=G(t) \qquad (0 \leq r \leq F(\theta,t),\ 0 \leq \theta \leq \pi), \qquad \text{(II.3)}$$

$$\frac{\partial \phi}{\partial r}=\frac{\partial F}{\partial t}+\frac{1}{r^2}\frac{\partial \phi}{\partial \theta}\frac{\partial F}{\partial \theta} \qquad (r=F(\theta,t),\ 0 \leq \theta \leq \pi), \qquad \text{(II.4)}$$

$$\Delta P_O+2P+\frac{1}{2\pi}\left(T^e_{n2}-T^e_{n1}\right)=-2H \qquad (r=F(\theta,t),\ 0 \leq \theta \leq \pi), \qquad \text{(II.5)}$$

$$\int_0^\pi F^3(\theta,t)\sin(\theta)\,d\theta=2, \qquad \text{(II.6)}$$

Equation (II.1) is the Laplace equation governing irrotational flow; (II.2) is the condition for zero radial velocity at the center of the drop; (II.3) is Bernoulli's equation for the pressure everywhere in the drop; (II.4) is the kinematic condition relating the surface velocity to the velocity field of the material there. Equation (II.5) is the balance of static, dynamic and capillary pressure, with normal electric stress across the interface, where the right-hand-side of this equation is the negative of twice the local mean curvature of the interface; (II.6) is the constraint for constant volume of the drop. The static pressure difference across the interface is ΔP_O.

The medium surrounding the drop is assumed to be electrically insulating, and the dimensionless electrostatic potential, $V(r,\theta,t)$, and the uniform potential in the conducting drop, V_O, are both scaled with $(4\pi\varepsilon_m/\sigma R)^{-1/2}$, where ε_m is the permittivity of the medium. The electric field is related to the potential as $\underline{E}=-\nabla V$ and is scaled with $(4\pi\varepsilon_m R/\sigma)^{-1/2}$; this results in scaling the net charge Q with $(4\pi\varepsilon_m R^3)^{-1/2}$. The equations and boundary conditions governing the electrostatic potential are

$$\nabla^2 V=0 \qquad (F(\theta,t) \leq r \leq \infty,\ 0 \leq \theta \leq \pi \qquad \text{(II.7)}$$

$$V \to 0 \qquad (r \to \infty,\ 0 \leq \theta \leq \pi \qquad \text{(II.8)}$$

$$\underline{n}\cdot\nabla V=-4\pi q(\theta,t) \qquad (r=F(\theta,t),\ 0 \leq \theta \leq \pi), \qquad \text{(II.9)}$$

$$\underline{t}\cdot\nabla V=0 \qquad (r=F(\theta,t),\ 0 \leq \theta \leq \pi), \qquad \text{(II.10)}$$

$$2\pi\int_0^\pi qF\,(F^2+F_\theta^2)^{1/2}\sin(\theta)\,d\theta=Q. \qquad \text{(II.11)}$$

in which $q(\theta,t)$ is the local surface charge density and $\underline{n}$ and $\underline{t}$ are the unit vectors normal and tangential to the drop surface. In formulating (II.11) we have assumed that charge is confined to the interface and equilibrates in a time much shorter than the characteristic time of the fluid motion. Equation (II.10) guarantees that the tangential component of the electric field is continuous across the interface.

The electric stress caused by the external electric field is

$$\underline{\underline{T}}^e_2\ \underline{E}\underline{E}-\frac{1}{2}|\underline{E}|^2\underline{\underline{I}} \qquad \text{(II.12)}$$

where $\underline{\underline{I}}$ is the identity tensor, and $|\underline{E}|$ is the magnitude of $\underline{E}$; see Melcher and Taylor[1]. The component of this stress normal to the surface of the drop,

$$T^e_{n2}=\underline{n}\underline{n}:\ \underline{\underline{T}}^e_2=\frac{1}{2}(\underline{n}\cdot\underline{E})^2 \qquad \text{(II.13)}$$

appears in the normal-stress balance (II.5) and couples together the fluid flow and electrostatic problems. The spatially uniform potential inside the conducting drop forces $\underline{\underline{T}}^e_1$ to be zero, hence $T^e_{n1}\equiv 0$.

The dynamical problem for the velocity and electrostatic potentials and the drop shape are solved for motions originating with an initial deformation of the drop. We describe initial deformations which satisfy conservation of mass (II.6) and which have no initial velocity, *i.e.*,

$$\frac{\partial F}{\partial t}(\theta,0)=0. \qquad \text{(II.14a)}$$

We define the amplitude of the oscillation e in terms of this initial deformation as

$$F(\theta,0)=1+\varepsilon P_n(\theta)+O(\varepsilon^2), \qquad \text{(II.14b)}$$

where $P_n(\theta)$ is the Legendre polynomial of nth order. The amplitude ε is taken to be a small parameter in the analysis that follows.

III. METHODS OF SOLUTION FOR MOVING-BOUNDARY PROBLEMS

The solution to the above stated system of equations depends on the interface shape which is not known *a priori* but is determined as part of the problem. Depending on whether the boundary in this and a whole class of similar problems is stationary or moving, such problems are usually referred to as free- or moving-boundary problems[2]. In all these cases nonlinear partial differential equations are augmented by boundary conditions that apply on the "unknown" boundary and which further complicate the problem.

III.1 Domain Perturbation and Nonlinear Dynamics

Mathematically rigorous calculations of the existence of two-dimensional progressive waves, which are analytic in the amplitude of the wave, can be found in Stoker's book[3]. These analyses rely heavily on the theory of analytic functions of a complex variable and cannot be extended to three dimensional wave problems or to problems with more complicated boundary or initial conditions, like the one that is treated here. The higher-order theory of waves uses a perturbation theory which represents solutions to problems as a power series in the amplitude of the wave. Joseph[4] gave a systematic development of the procedure which has now been applied to general free- and moving-boundary problems. He named the method "domain perturbations" - a name that will be used hereafter.

In this formulation a one parameter family of domains is used on which the field equations and boundary conditions are solved. The solution must be known in some reference (possibly static) domain onto which all higher order problems can be mapped. In the present class of problems the boundary shape is immobilized as a sphere by introducing the change of coordinates $r = \eta F(\theta,t)$. The solution in the perturbed domain is developed in a power series about the spherical shape in terms of a perturbation parameter, ε as

$$\begin{bmatrix} F(\theta,t;\varepsilon) \\ \phi(r,\theta,t;\varepsilon) \\ V(r,\theta,t;\varepsilon) \end{bmatrix} = \sum_{k=0}^{\infty} \frac{\varepsilon^k}{k!} \begin{bmatrix} F^{(k)}(\theta,t) \\ \phi^{[k]}(\eta,\theta,t) \\ V^{[k]}(\eta,\theta,t) \end{bmatrix} \tag{III.1}$$

where for example the superscripts are defined as

$$F^{(k)}(\theta,t) \equiv \frac{d^k F(\theta,t;0)}{d\varepsilon^k} \quad , \quad \phi^{[k]}(\eta,\theta,t)\, \frac{d^k \phi(\eta,\theta,t;0)}{d\varepsilon^k} \tag{III.2}$$

Using the chain rule for differentiation, each term $\phi^{[k]}(\eta,\theta,t)$ in the expansion for the potential can be written as a sum of a contribution evaluated on the spherical domain $(0 \le \eta \le 1,\ 0 \le \theta \le \pi)$, and terms that account for the deformation of the domain at each order of ε. The first two relationships evaluated at the boundary, $\eta = 1$, are

$$\phi^{[0]}(\eta=1,\theta,t;0) \equiv \phi(0)(\eta=1,\theta,t;0), \tag{III.3a}$$

$$\begin{aligned} \phi^{[1]}(\eta=1,\theta,t;0) &\equiv \frac{\partial\phi}{\partial\varepsilon}(\eta=1,\theta,t;0) + F^{(1)}(\theta,t)\,\frac{\partial\phi}{\partial\eta}(\eta=1,\theta,t;0) \\ &\equiv \phi(1)(\eta=1,r,t) + F^{(1)}(\theta,t)\,\left.\frac{\partial\phi^{(0)}}{\partial\eta}\right|_{\eta=1} \end{aligned} \tag{III.3b}$$

where terms like $\phi^{[k]}(\eta,\theta,t) \equiv \partial^k\phi/\partial\varepsilon^k$ are always defined on the spherical shape. The values of the coefficients on the boundary are computed using the properties of the mapping on the boundary only, so that its nature in the interior need not be known. As a result, the solution in the perturbed domain depends only on the boundary values of the perturbation method and can be computed by solving a hierarchy of linear problems.

The existing nonlinear stability theories are based on the ideas of Landau[5] and the later work by Stuart[6]. The essential point is that a finite amplitude perturbation close to the curve of neutral stability can be described as a wave of slowly varying amplitude. Thus, for weakly nonlinear disturbances, evolution equations for the long time behavior of the amplitude can be derived by singular perturbation methods, such as the Linstedt-Poincaré technique, the method of multiple time scales, etc., which are described in standard texts[7]. Formally, the dependent variables are assumed to be also functions of a slow timescale generally related to the actual time t as $T \equiv \varepsilon^{l/m} t$. The different timescales are introduced into the governing equations by expanding the temporal partial derivative $\partial/\partial t$ as

$$\frac{\partial}{\partial t} = \frac{\partial}{\partial t} + \varepsilon^{l/m} \frac{\partial}{\partial T} \qquad \text{(III.4)}$$

As a result, the drop shape is expanded as

$$F_n(\theta,t) = 1 + \varepsilon\,[A(T)e^{i\omega_n t} + c.c]\; P_n(\theta) + O(\varepsilon^2) \qquad \text{(III.5)}$$

Then, the amplitude of the motion evolves in the slow timescale according to the equation resulting from the solvability condition of the problem at each order in the perturbation expansion; these conditions are commonly known as Landau equations and require that the right-hand-side of the linearized differential operator is orthogonal to the adjoint eigenvector of the homogeneous problem arising at $\theta'(\varepsilon)$.

The so-called harmonic resonance is a nonlinear mechanism through which energy can be transferred from one oscillation mode to another and has been the subject of much interest in the last twenty years; see Phillips[8]. In this situation the higher harmonic in the nonlinear system resonates with the fundamental oscillating mode and so draws or continuously exchanges energy with it. Another possibility which can initiate quite dramatic effects is provided by physical systems in which two frequencies can assume almost equal values. This secular behavior has been called direct resonance, by Akylas and Benney[9]. We have followed this formalism in analyzing the dynamics close to the stability limit of a charged drop.

III.2 Numerical Methods

Finite element methods are systematic techniques for constructing approximations to functions which are themselves solutions to partial differential equations. To formulate these approximations, the domain of the function is divided into subdomains or elements and in each element the unknown functions are approximated by low-order polynomials. The coefficients of the polynomials are chosen so that the function has a specified degree of inter-element continuity and so the weighted residuals for the equation governing the behavior of the function are zero. In this way, the original problem defined for continuous variables is reduced to a system of discrete algebraic equations for the coefficients of the polynomials. The finite element methods developed for solving free- and moving-boundary problems differ in the technique used to determine the location of the free-surface, and in the numerical iteration scheme used to solve the resulting set of nonlinear algebraic equations. Ettouney and Brown[10] and Beris *et al.*[11] transformed the domain with a free-boundary to a fixed domain before discretizing the problem with a fixed finite-element mesh. This transformation has the advantage of making explicit the nonlinearities inherent in the unknown location of the free surface. This advantage was exploited by using Newton's method to solve the nonlinear equation set that resulted from the finite element formulation. Saito and Scriven[12] have solved similar problems in the original domain. Instead of the global mapping of Ettouney and Brown, they took advantage of the isoparametric mapping for each element onto a square element where the dependence of the basis functions on the moving surface is expressed explicitly. Patzek *et al.*[13] and Basaran *et al.*[14] have calculated the inviscid axisymmetric oscillations of the free and conducting drop with or without electrical charge.

Finite difference schemes have been developed for free-surface flows using either Lagrangian techniques (particle based coordinates) or Eulerian techniques (fixed coordinates). The Marker-and-Cell method (MAC) developed by Harlow and Welsch[15] solves the full Navier-Stokes equations for an incompressible, viscous fluid with free- or moving-surface. Velocity components and pressure are defined over a staggered Eulerian mesh. A Lagrangian system of marker particles is defined and these markers are moved through the grid at interpolated local fluid speeds, behaving like dye particles in actual experiments.

More recently, the Boundary Integral Equation Methodology has been employed by Miksis *et al.*[16] for steady calculations, and by our group and others for unsteady, free surface problems. In short, this method is based on integral equation representation of the governing differential equations of motion. Numerical solution of these equations written for the boundaries of the problem gives a system of coupled equations between velocities and stresses. This method is particularly useful for inviscid problems since (i) it requires solution only on the domain boundaries thus reducing the dimensionality of the problem by one and (ii) the radiation condition associated with unbounded domains is automatically accounted for, if the exact fundamental solutions are used. However, the solution procedure will still be iterative due to the nonlinear boundary conditions and the necessary implicit schemes for stable time integration.

IV. NONLINEAR OSCILLATIONS OF INVISCID DROPS AND BUBBLES

The problem of single drop or bubble oscillations was undertaken in order to understand the mechanics of coupling of spherical modes. In this section we present the asymptotic analysis for moderate-amplitude axisymmetric oscillations of inviscid and incompressible liquid globes. As pointed out by Miller & Scriven[17], the analysis of the motion of an interface separating two inviscid liquids leads to a discontinuity between the components of fluid velocity tangential to and on either side of the interface. We avoid the physically unacceptable results associated with inviscid liquid/liquid systems by limiting our study to cases where one phase is either a vacuum or a tenuous gas, so that its hydrodynamical effects can be neglected. The two limits of liquid internal and external to the closed interface are denoted as drops and bubbles respectively.

The linear equation set for an oscillating drop has an infinite number of solutions, each of the form

$$
\begin{aligned}
F^{(1)}(\theta,t) &\equiv F_n^{(1)}(\theta,t) = \cos(\omega^{(0)} t)\, P_n(\theta), \\
\phi^{(1)}(\eta,\, \theta,t) &\equiv \phi_n^{(1)}(\eta,\, \theta,t) = \frac{-(n-1)(n+2)}{\omega^{(0)}}\, \eta^n \sin(\omega^{(0)} t)\, P_n(\theta) \qquad \text{(IV.1)} \\
\omega^{(0)} &\equiv \omega_n^{(0)} = [n(n-1)(n+2)]^{1/2} \quad (n=2,3,..),
\end{aligned}
$$

which corresponds to the linear modes of the oscillation analyzed by Rayleigh[18]. The corrections to the drop shape and potential caused by mode coupling at second order in amplitude are predicted for two-, three- and four-lobed motions; see Tsamopoulos and Brown[19]. A graphic example of the mode coupling at higher orders in the perturbation expansion is given in Figure IV.1 where the most common normal modes are excited by the initial deformation.

The first nonzero correction $\omega^{(2)}$ is determined so that the solutions to the third order equations contain no secular terms.

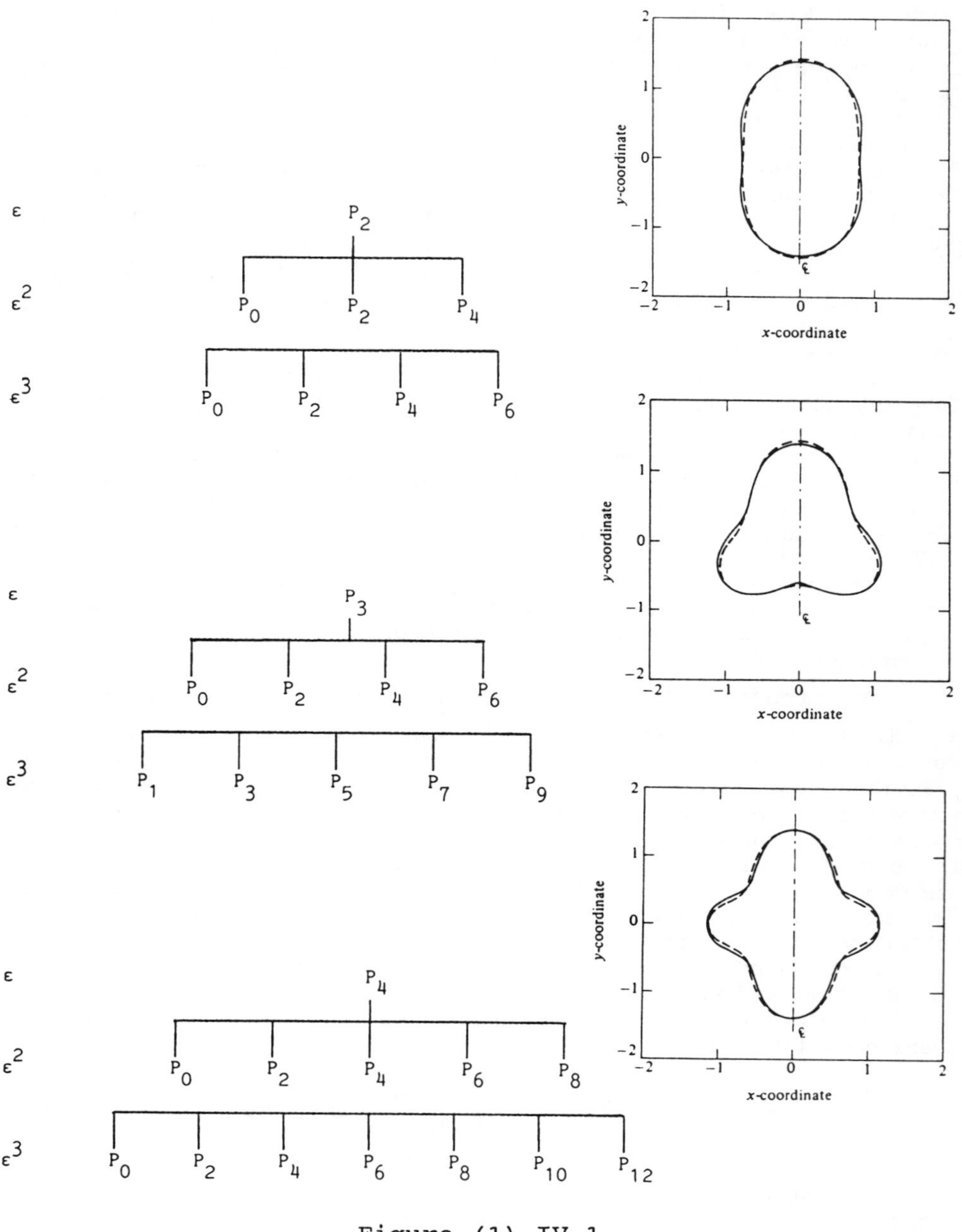
ε
P_2
ε²
P_0 P_2 P_4
ε³
P_0 P_2 P_4 P_6
ε
P_3
ε²
P_0 P_2 P_4 P_6
ε³
P_1 P_3 P_5 P_7 P_9
ε
P_4
ε²
P_0 P_2 P_4 P_6 P_8
ε³
P_0 P_2 P_4 P_6 P_8 P_{10} P_{12}
y-coordinate
x-coordinate
2
1
0
−1
−2
2
1
0
−1
−2

Figure (1) IV.1

For $n = 2$, $\omega^{(2)} = -\frac{34409}{29400}\omega^{(0)} \cong -1.17037\,\omega^{(0)}$, (IV.2)

for $n = 3$, $\omega^{(2)} = -\frac{783899}{396396}\omega^{(0)} \cong -1.97757\,\omega^{(0)}$, (IV.3)

for $n = 4$, $\omega^{(2)} = -\frac{181430960793}{64865536736}\omega^{(0)} \cong -2.79703\,\omega^{(0)}$, (IV.4)

Since the correction $\omega^{(3)}$ will be identically zero, Equations (IV.2-4) give predictions for the frequency correction that are valid up to the fourth order in the amplitude.

A number of asymptotic results for oscillating drops can be compared directly with the numerical calculations of Foote[20] and Alonso[21] for viscous drops oscillating in the fundamental mode. The effect of viscosity on the frequency of oscillation and on the shape of the drop is small when the product $(\sigma R/\rho \nu^2)^{1/2}$ is large, where ν is the kinematic viscosity; see Lamb[22]. In the calculations of Foote and Alonso this ratio was 35.4 and 3.3 respectively, thus the comparison of Foote's calculations with our inviscid theory is reasonable, while Alonso's results may deviate substantially solely because of viscosity.

A drop undergoing $n = 2$ oscillations spends a considerably longer part of each period in a prolate form than in an oblate one. The percentage of excess time is shown in Figure IV.2 as a function of the amplitude of the oscillation, as measured by $\alpha \equiv L/W$, where L and W are the major and minor axis of the drop, respectively. Also shown in this figure are the results of Foote[20]. The agreement is reasonable. Bubbles exhibited only a slight tendency to stay in prolate forms, as shown by the line on Figure IV.2.

The quadratic decrease in frequency with amplitude predicted here is compared on Figure IV.3 to the numerical results of Foote and Alonso. The asymptotic results are within 4% of Foote's viscous calculations over the entire range of amplitude $0 \leq \alpha \leq 1.8$ presented by that author. The single value calculated form Alonso's report differed more significantly from our results. Finally, the inviscid predictions are compared in Figures IV.2 and IV.3 to experimental results of Trinh & Wang[23] and Trinh *et al.*[24] for almost neutrally buoyant drops of silicone oil and carbon tetrachloride suspended in distilled water. In the limits of moderate-amplitude oscillations and large drops ($R \cong 1$ cm), the oscillation frequencies measured this way were expected to be near those of an inviscid liquid/liquid system.

Trinh & Wang's measurements for the percentage of time spent in prolate shapes by the neutrally buoyant drop are shown on Figure IV.2, and, as expected, are bracketed by the inviscid calculations for drops and bubbles. Experimental data for the dependence of frequency on amplitude for drops with radii of 0.62 cm (□) and 0.49 cm (•) are shown in Figure IV.3. the data for the larger drop are again described by an asymptotic result intermediate to the calculations for drops and bubbles for a less than 1.7. The experimental measurements for the smaller drop differ systematically from the inviscid results; this difference may represent the coupling between viscosity and the finite-amplitude motion.

V. RESONANT OSCILLATIONS OF INVISCID CHARGED DROPS

It is anticipated that the linear oscillation frequency of a charged drop will be less than that for an uncharged one. The physical reason behind it, is that the restoring capillary force, is now opposed by the electrostatic repulsion. The results presented in the previous section are readily extendable for charged drops, see Tsamopoulos and Brown[25], except for those values of electric charge where resonance occurs.

Resonant interactions between the fundamental mode and secondary and tertiary harmonics can completely change the pattern of the oscillation. The regular forms of the weakly nonlinear oscillations described in the previous section are not

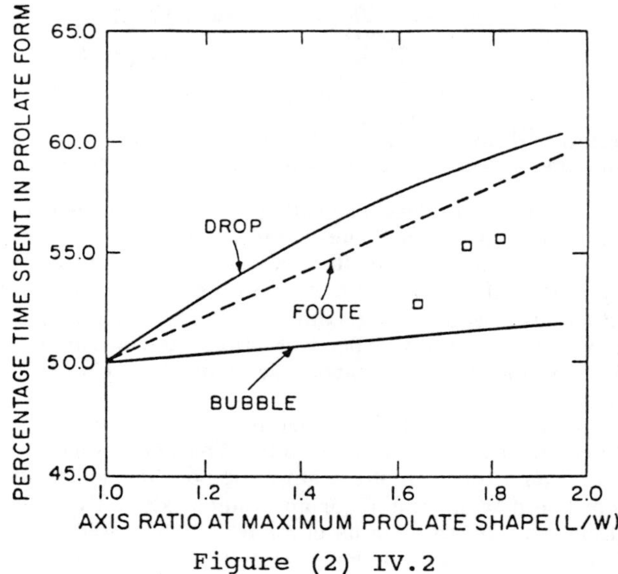

Figure (2) IV.2

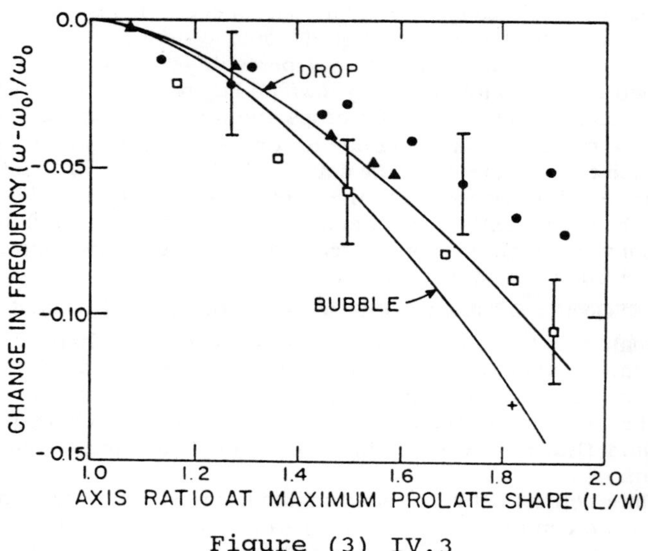

Figure (3) IV.3

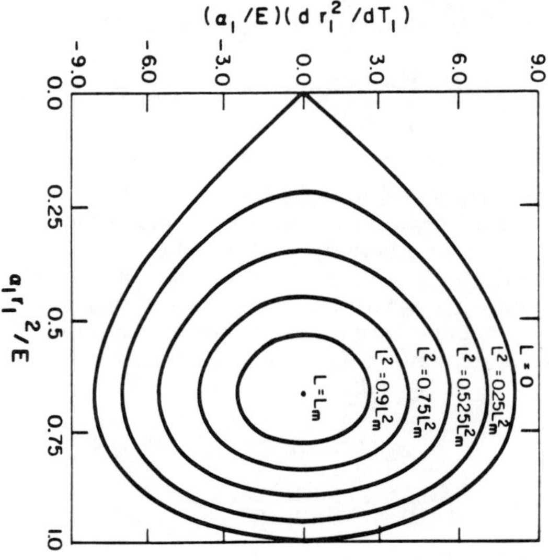

(a1/E)(dr1²/dT1)
9.0
6.0
3.0
0.0
-3.0
-6.0
-9.0
0.0
0.25
0.5
0.75
1.0
a1r1²/E
L=0
L²=0.25L²m
L²=0.525L²m
L²=0.75L²m
L²=0.9L²m
L=Lm

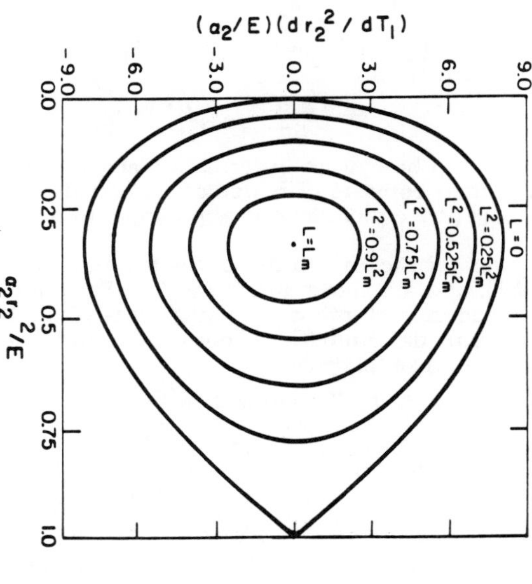

(a2/E)(dr2²/dT1)
9.0
6.0
3.0
0.0
-3.0
-6.0
-9.0
0.0
0.25
0.5
0.75
1.0
a2r2²/E
L=0
L²=0.25L²m
L²=0.525L²m
L²=0.75L²m
L²=0.9L²m
L=Lm

Figure (4) V.1

valid when the frequencies of the higher harmonics of the fundamental mode are close to integral multiples of its frequency. This does not occur for uncharged drops up to second order in amplitude, but happens for charged drops that satisfy the condition

$$\frac{m(m-1)(m+2)-Q^2/4\pi}{\omega_n^2} = (\text{integer})^2,\ m \neq n. \tag{V.1}$$

Resonance is detectable for four-lobed oscillations when the net charge is near $Q_r = (32\pi/3)^{1/2} < Q_c$, where $\omega_6^2 = 4\omega_4^2$ and Q_c is the Rayleigh stability limit. Then, the four-lobed fundamental resonates with the six-lobed harmonic form, and the two modes exchange energy in a periodic or aperiodic fashion, depending on the initial deformation of the drop. The calculation involves a multiple-timescale expansion coupled with domain perturbation. The solvability conditions that result form equating to zero the secular terms provide the evolution equations for the complex amplitudes of the fundamental ($A(T_1)$) and resonating harmonic ($B(T_1)$):

$$\alpha_1\left(\frac{dA}{dT_1} - \alpha_3 iA\right) = iA^*Be^{-iNT_1}, \tag{V.2}$$

$$\alpha_2\left(\frac{dB}{dT_1} - \alpha_4 iB\right) = iA^2e^{iNT_1}, \tag{V.3}$$

where $\alpha_1 \equiv \dfrac{143}{267\sqrt{10}},\ \alpha_2 \equiv \dfrac{88}{89\sqrt{10}},\ \alpha_3 \equiv \dfrac{\lambda}{\sqrt{10}},\ \alpha_4 \equiv \dfrac{9\lambda}{4\sqrt{10}}.$

Introducing the substitutions $A(T_1) \equiv r_1(T_1)e^{i\,\theta_1(T_1)}$ and $B(T_1) \equiv r_2(T_1)\,e^{i\,\theta_2(T_1)}$, where the $\{r_i, \theta_i\}$ are real functions of the slow timescale $T_1 = \varepsilon t$, decouples these equations, through the energy-like relation; McGoldrick[26]

$$\alpha_1 r_1^2 + \alpha_2 r_2^2 \equiv E, \tag{V.4}$$

into the form

$$\frac{\alpha_1}{2}\frac{dr_1^2}{dT_1} = -\frac{1}{\sqrt{\alpha_2}}\left(\alpha_1 r_1^6 + Er_1^4 - \alpha_2 L_2\right)^{1/2} \tag{V.5a}$$

$$\frac{\alpha_2}{2}\frac{dr_2^2}{dT_1} = \frac{1}{\alpha_1}\left(\alpha_2^2 r_2^6 + 2E\,\alpha_2 r_2^4 + E^2 r_2^2 - \alpha_1^2 L^2\right)^{1/2} \tag{V.5b}$$

$$\alpha_1 r_1^2\frac{d\theta_1}{dT_1} = \alpha_2 r_2^2\frac{d\theta_2}{dT_1} = L \tag{V.5c}$$

The general solution can be written in terms of Jacobi's elliptic functions[25]. The first-order solution then consists of periodic or aperiodic amplitude modulations together with periodic phase modulations superimposed on a slow linear frequency shift.

Phase-plane plots for the amplitudes of the two modes are readily constructed using Equation (V.5), and are shown in Figure V.1. For initial conditions with only the fundamental mode excited, the individual phases of the two interacting modes are constant and the modal amplitudes follow the outermost trajectory in Figure V.1. The general solution simplifies to

$$r_1(T_1) = R_1 \operatorname{sech}\frac{R_1T_1}{(\alpha_1\alpha_2)^{1/2}} \tag{V.6a}$$

$$r_2(T_1) = R_1\left(\frac{\alpha_1}{\alpha_2}\right)^{1/2}\tanh\left[\frac{R_1T_1}{(\alpha_1\alpha_2)^{1/2}}\right] \tag{V.6b}$$

$$\theta_2 - 2\theta_1 = \pm\frac{\pi}{2}. \tag{V.6c}$$

At exact resonance a purely four-lobed oscillation of any amplitude cannot persist, but transforms into a six-lobed oscillation within less than three periods of the initially excited mode. Drop shapes for this case are shown in Figure V.2 for the initial phase conditions $\theta_1(0) = 0$ and $\theta_2(0) = \pi/2$ with $\varepsilon = 0.2$. The inner trajectories in

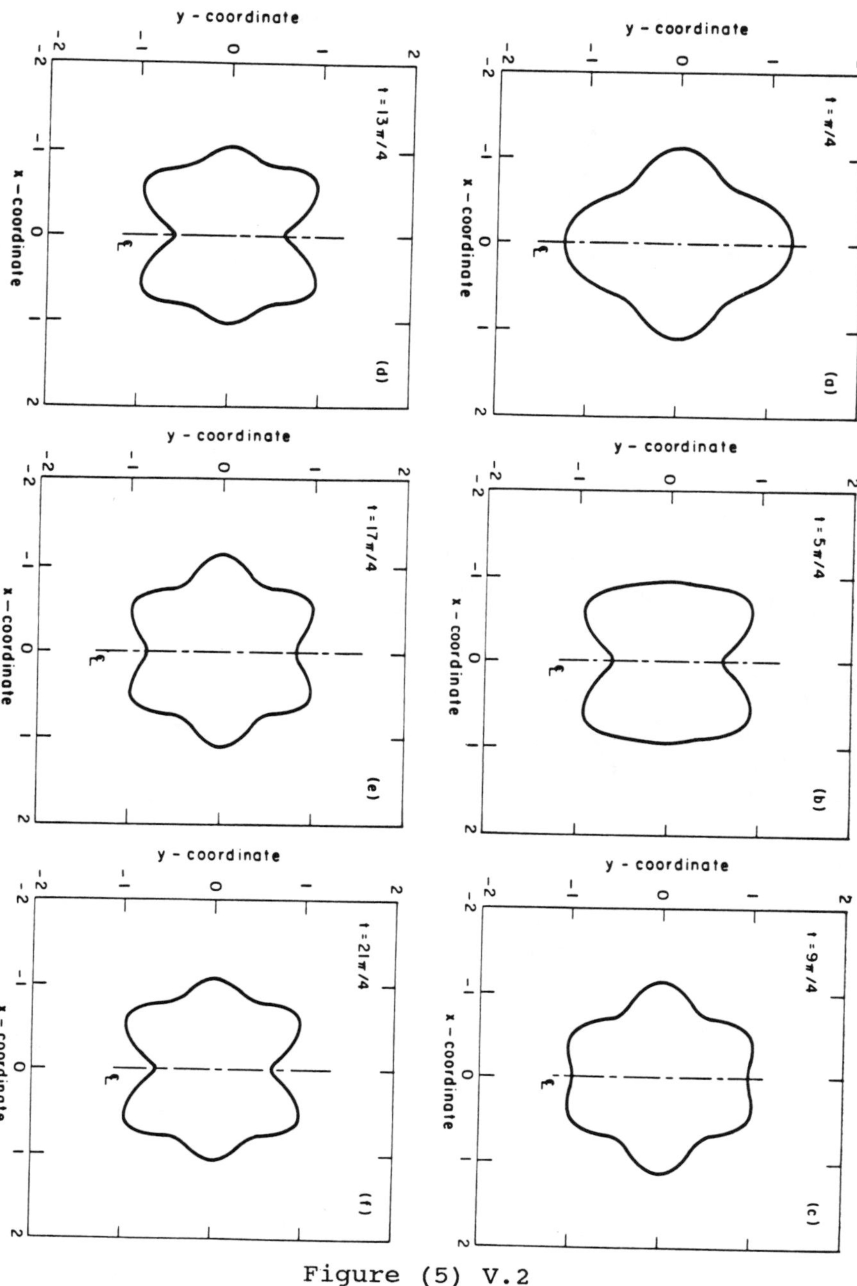

y - coordinate
x - coordinate
t = π/4
(a)
t = 5π/4
(b)
t = 9π/4
(c)
t = 13π/4
(d)
t = 17π/4
(e)
t = 21π/4
(f)

Figure (5) V.2

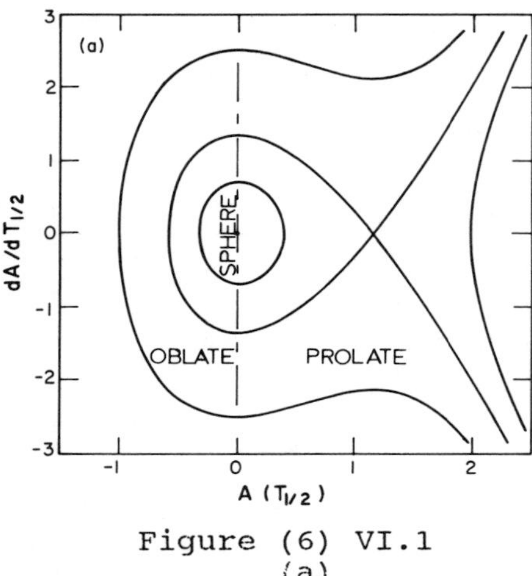

Figure (6) VI.1
(a)

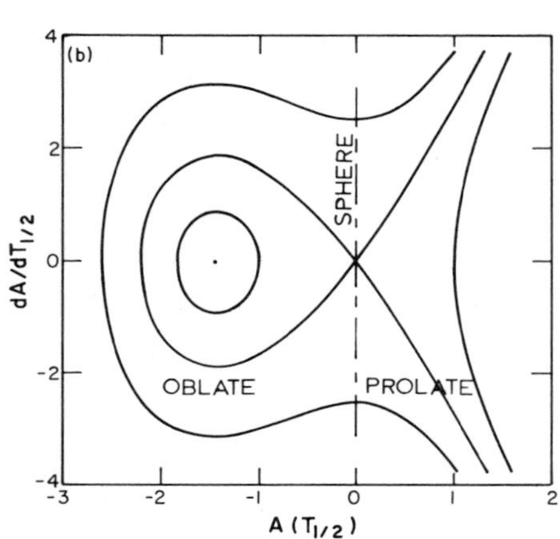

Figure (6) VI.1
(b)

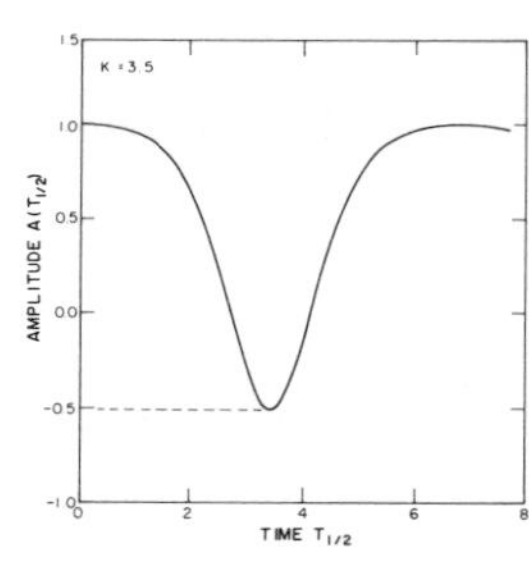

Figure (7) VI.2
(a)

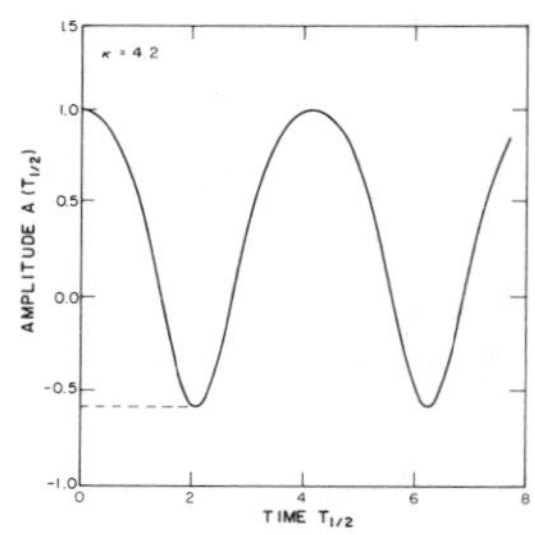

Figure (7) VI.2
(b)

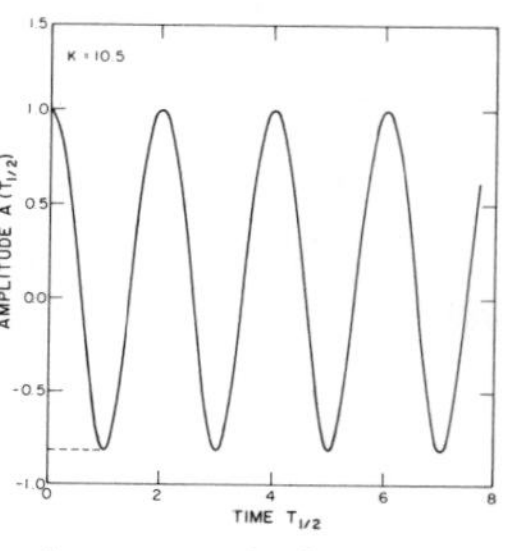

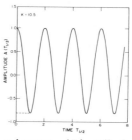

Figure (7) VI.2
(c)

Figure V.1 correspond to initial conditions that combine four- and six-lobed shapes. These modes continuously exchange energy during oscillations at a frequency which is slightly modulated about the mean value. For initial conditions represented by the single point on each of the phase-plane plots shown in Figure V.1 both the amplitude and the phase modulations vanish entirely.

VI DYNAMICS OF CHARGED DROP BREAKUP

Lord Rayleigh[27] considered the stability of charged spherical drops to infinitesimal shape disturbances described by Legendre polynomials and determined that the critical amount of charge just necessary to disrupt the drop due to the nth axisymmetric mode was

$$\widetilde{Q}_c^{(n)} = 4\pi\{\varepsilon_m \sigma R^3 (n+2)\}^{1/2}, \; n \geq 2, \qquad \text{(VI.1)}$$

The mode number n indicates the number of lobes on the deformed drop that is neutrally stable at the value of charge given by $\tilde{Q} = \tilde{Q}_c^{(n)}$. The two-lobed shape perturbation becomes unstable at the lowest value of $\tilde{Q} = \tilde{Q}_c^{(2)}$ and corresponds to the limit of stability for the spherical shape computed by Rayleigh above which the charged drops fission; Taylor[28]. Bohr and Wheeler[29] first recognized the importance of the prolate axisymmetric forms in the theory of nuclear fission, as estimates for the energy barrier separating stable spherical forms from fission.

To remove the singularity as $\omega^2 \to 0$ we rescale the variables in ε so that this parameter simultaneously determines the relationship between the difference of the net charge on the drop and the charge at the Rayleigh limit, and the deformation of the drop. The appropriate scaling is $\omega^2 = (\kappa\varepsilon)^{1/2}$ where $\kappa = O(1)$ when the drop starts from rest with the initial conditions (Equation (II.14)).

This dependence of frequency on amplitude is valid only for a range of values for Q close to the Rayleigh limit given by

$$Q = 4\sqrt{\pi}\,(1 - \frac{\kappa}{16}\varepsilon - \ldots) \equiv Q^{(0)} + \varepsilon Q^{(1)} + \varepsilon^2 Q^{(2)} + \ldots \qquad \text{(VI.2)}$$

The method of multiple timescales is applied. The solvability condition yields the evolution equation for the slowly varying amplitude of the deformation $A(T_{1/2})$ in terms of the slow timescale $T_{1/2} = \varepsilon^{1/2} t$.

After a first integration it becomes

$$\left(\frac{dA}{d\,T_{1/2}}\right)^2 = -(A \pm 1)\,[-\kappa(A \pm 1) + \frac{16}{7}(A^2 \pm A + 1)]. \qquad \text{(VI.3)}$$

Static drop shapes are given by the roots of the polynomial on the right-hand-side of the expression. The family of spherical shapes corresponds to the stationary point $A = 0$ for all values of κ (charge). Prolate static forms exist for $A = 7\kappa/24$ and $\kappa > 0$, or values of $Q < 4\sqrt{\pi}$. Oblate forms are possible for $\kappa < 0$ and are given by the root $A = 7\kappa/24$.

The stability of the static shapes follows from a phase-plane analysis of Equation (VI.3); plots of $A'(T_{1/2})$ versus $A(T_{1/2})$ are shown in Figure VI.1 for fixed values of κ greater than (Figure VI.1A) and less than (Figure VI.1B) zero. For small amplitude perturbations, the spherical form $(A,A') = (0,0)$ is a stable center for $\kappa > 0$ and a saddle point for $\kappa < 0$. The oblate static shapes $(A,A') = (7\kappa/24, 0)$, $\kappa < 0$, are stable centers and the prolate forms $(A,A') = (7\kappa/24, 0)$, $\kappa > 0$, are saddle points. These results agree with bifurcation analysis of only the static forms. The stable oscillations of the amplitude $A(T_{1/2})$ with time are shown in Figure VI.2 for $A(0) = +1$ and values of $\kappa > \kappa_c \equiv +24/7$. Away from κ_c the motion resembles the simple sinusoidal oscillation between prolate and oblate forms. The evolution of an initially prolate perturbation ($A(0) = +1$) for drops with $\kappa < \kappa_c$ is shown in Figure VI.3. The deformation of the drop grows slowly until a point where it suddenly increases almost exponentially in time. Combining Equations (VI.2) and (VI.3) shows that the correction to the Rayleigh limit for loss-of-stability of a spherical form is

$$Q^{(1)} = -6\sqrt{\pi}/7\,, \qquad \text{(VI.4)}$$

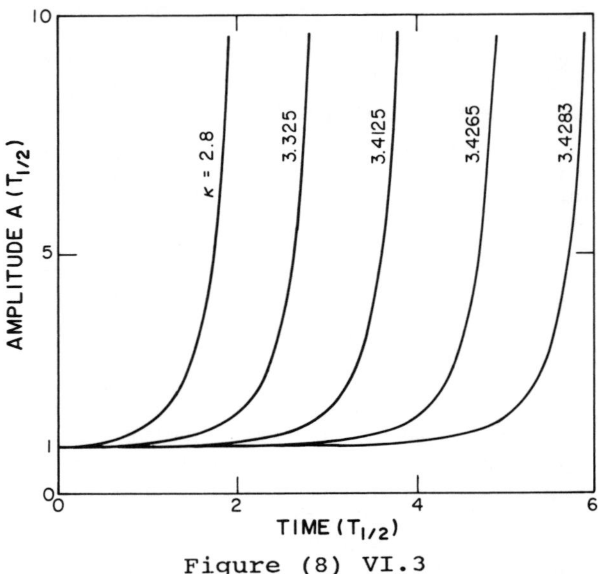

Figure (8) VI.3

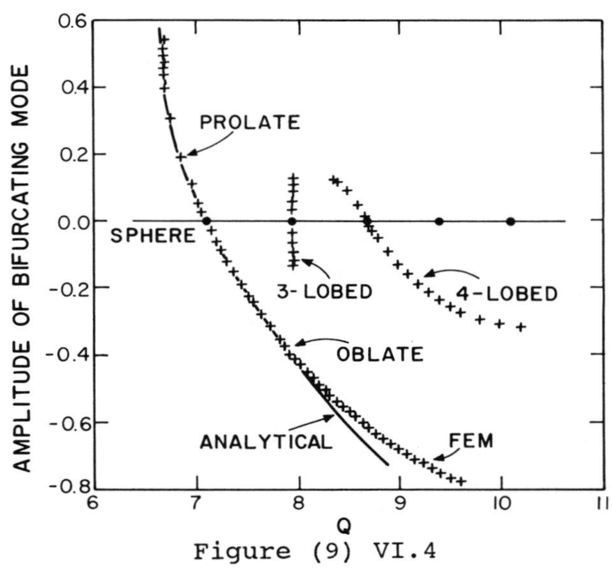

Figure (9) VI.4

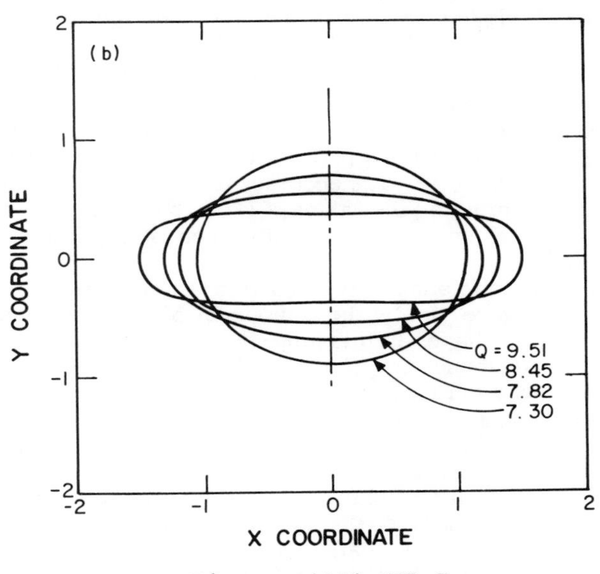

Figure (10) VI.5
(a)

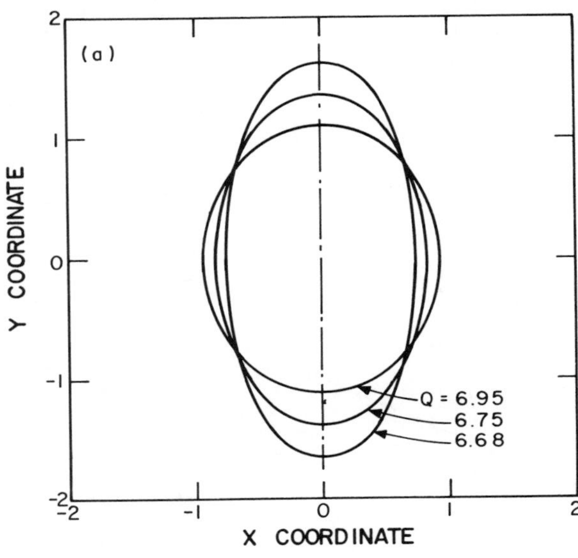

Figure (10) VI.5
(b)

indicating that finite amplitude prolate disturbances destabilize the drop, whereas the drop is more stable to oblate perturbations. The nonzero coefficient at $O(\varepsilon)$ in Equation (VI.2) also confirms the transcritical bifurcation.

VI.1 Finite Element Analysis of Static Shapes

The region of validity of the perturbation results presented in the last section was established by comparing the analytically calculated static drop shapes to results of finite element calculations performed using the same methodology presented by Adornato and Brown[30]. The accuracy of the finite element approximations was determined by computing the values of Q for the lowest four bifurcation points between the family of spheres and multi-lobed static forms; see Table 1 in Tsamopoulos *et al.*[31]

The families of static shapes evolving from the first three bifurcation points are represented in Figure VI.4 by the component of the corresponding bifurcating mode in the computed shape. In this projection, the transcritical bifurcation of the prolate and the oblate shapes is approximated very well by the asymptotic analysis. The three-lobed forms bifurcate supercritically (to higher values of Q) from the spheres. Since the spheres are stable for $Q < 4\sqrt{\pi}$, the oblate forms are stable and the prolate ones unstable as computed in the dynamical analysis above. The three- and four-lobed shapes are all unstable.

Both the finite element calculations and the asymptotic analysis of the bifurcating prolate shapes predict that this family passes through a limit point where it turns towards higher values of charge Q. The value of Q at the limit point Q_1 is predicted analytically to be 6.6504 and 6.678, from the finite element calculations. Prolate and oblate drop shapes computed by finite element analysis are shown in Figure VI.5 for several values of Q.

A quantitative comparison between these forms and the perturbation results is made by decomposing the shapes in finite element representation into a Legendre-Fourier series. The values of the first eight coefficients for the two-lobed shapes and of the net charge are compared to similar coefficients $\{c_j\}$ and charge values resulting from the perturbation solution at $\theta'(\varepsilon^2)$. The agreement is very good; see Table 2 in Tsamopoulos *et al.*[31] The analytical result is also plotted on Figure VI.4 for comparison with the numerical results.

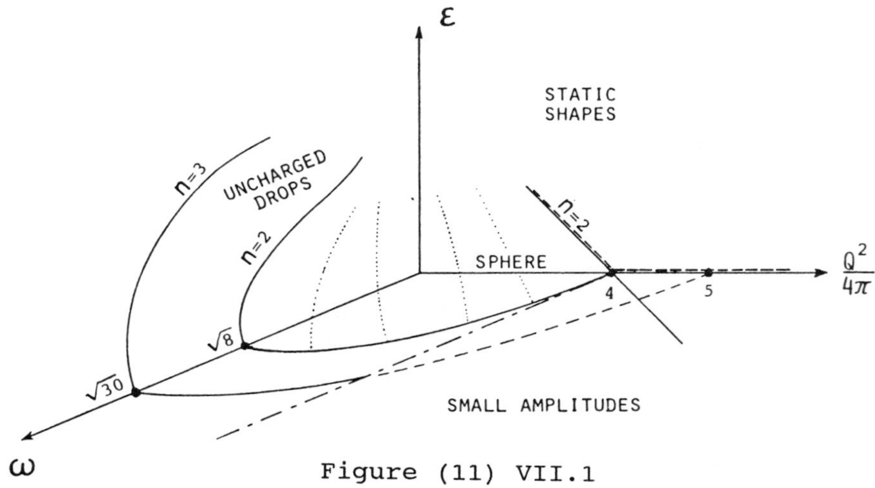

Figure (11) VII.1

VII. POSTSCRIPT

Moderate-amplitude oscillations of drops display an array of nonlinear dynamic phenomena, which are summarized in Figure VII.1. The plane of $\varepsilon = 0$ reproduces results of Rayleigh's linear theory; it shows the quadratic decrease of the frequency of the oscillation as the net charge on the surface increases. The frequency becomes zero according to this linear theory at the Rayleigh limit ($Q = Q_c = 4\sqrt{\pi}$), where the capillary force is exactly balanced by the electrostatic repulsion. The decrease of the eigenfrequencies is proportional to the square of the amplitude as shown in any plane of constant charge Q ($0 \le Q \le Q_c$). This last result is caused by the fluid inertia and has been verified experimentally for uncharged drops[23]. At specific values of the dimensionless charge, Q_r, harmonic resonance is induced by the nonlinear interaction between the primary oscillation mode and one of its harmonics. For purely four-lobed initial oscillations with $Q_r=\sqrt{(32\pi/3)}< Q_c$ the drop assumes an almost six-lobed shape within two periods of the primary oscillation. Initial disturbances with combinations of four- and six-lobed shapes lead to a continuous exchange of energy between these two modes. In general, the shape evolution in these cases is highly depended on the initial conditions. This effect is shown in Figure VII.1 as a deviation of the quadratic decrease of the frequency for high enough values of the initial disturbance.

The evolution of the drop shape for charge values close to the Rayleigh stability limit is shown in the plane of static shapes ($\omega = 0$). It corresponds to a transcritical bifurcation point between the families of static spherical shapes and oblate ($\varepsilon < 0$) and prolate ($\varepsilon > 0$) axisymmetric forms. Prolate forms exist for lower values of charge and are unstable to small amplitude perturbations. Finite amplitude oscillations destabilize the spherical drops at values of charge below Q_c, with a decrease proportional to the amplitude of the initial disturbance. Oblate static shapes exist for $Q > Q_c$ and are stable to small axisymmetric perturbations.

A possible extension to the present studies is to determine the nonlinear effects of viscous forces. For gas-liquid interfaces the effect of viscosity on the frequency of oscillation and the shape of the drop is small when the product $(\sigma R/\nu^2\rho)^{1/2}$ is large, where n is the kinematic viscosity. Then the timescale for viscous dissipation, or equivalently of vorticity diffusion from the interface is much longer than the characteristic time for the inviscid oscillation. This condition is satisfied for water drops, but not for fluids of interest in materials processing with higher viscosity. For two fluids of comparable density, the inviscid analysis leads to a discontinuity between the components of tangential velocity on the two phases, which is avoided by introducing a viscous boundary layer on both sides of the interface.

Finally, nonlinear analysis of the three-dimensional shapes and oscillations of charged, conducting drops near the Rayleigh limit has been carried out by Natarajan and Brown[32]. It was shown that the oblate spheroids that exist for greater amounts of charge are unstable to non-axisymmetric disturbances and cannot be observed. However, the bifurcating prolate spheroids that exist for values of charge less than the Rayleigh limit are unstable to only axisymmetric perturbations.

VIII. REFERENCES

1. J. R. Melcher & G. I. Taylor, *Ann. Rev. Fluid Mech.*, **1**, 111, (1969).
2. C. M. Elliott & Ockendon, Jr., **Weak and variational methods for moving boundary problems**, Pitman, Boston (1982).
3. J. J. Stoker, **Water Waves**, Interscience, New York (1957).
4. D. D. Joseph, *Arch. Rat. Mech. Anal.*, **51**, 295 (1973).
5. L. Landau, *Doklady Acad. Sci. USSR*, **XLIV**, 311 (1944).
6. J. T. Stuart, *Ann. Rev. Fluid Mech.*, **3**, 347 (1971).

7. C. M. Bender & S. A. Orszag, **Advanced Mathematical Methods for Scientists and Engineers**, McGraw-Hill (1978).
8. O. M. Phillips, *J. Fluid Mech.*, **106**, 215 (1981).
9. T. R. Akylas & D. J. Benney, *Stud. Appl. Math.*, **63**, 209, 1980.
10. H. M. Ettouney & R. A. Brown, *J. Comput. Phys.*, **49**, 118 (1983).
11. A. N. Beris, J. A. Tsamopoulos, R. C. Armstrong & R. A. Brown, *J. Fluid Mech.*, **158**, 219 (1985).
12. H. Saito & L. E. Scriven, *J. Comp. Phys.*, **42**, 53 (1981).
13. T. W. Patzek, R. E. Benner, & L. E. Scriven, *Bull. Amer. Phys. Soc.*, **27**, 1168 (1982).
14. O. A. Basaran, K. R. Amudson, T. W. Patzek, & L. E. Scriven, *Bull. Am. Phys. Soc.*, **27**, 1168 (1982).
15. F. H. Harlow & J. E. Welch, *Phys. Fluids*, **8**, 2182 (1965).
16. M. J. Miksis, J. M. Vanden-Broeck, & J. B. Keller, *J. Fluid Mech.*, **23**, 31 (1982).
17. C. A. Miller & L. E. Scriven, *J. Fluid Mech.*, **32**, 417 (1968).
18. J. W. S. Rayleigh, *Proc. R. Soc.*, **XXIX**, 71 (1879).
19. J. A. Tsamopoulos & R. A. Brown, *J. Fluid Mech.*, **127**, 519 (1983).
20. G. B. Foote, *J. Comp. Phys.*, **11**, 507 (1973).
21. C. T. Alonso, **Proc. (1st)Int. Colloq. on Drops and Bubbles**. (Ed. D. J. Collins, M. S. Plesset & M. M. Saffren.) Jet Propulsion Laboratory (1974).
22. H. Lamb, **Hydrodynamics**, 6th Edition. Cambridge University Press, (1932).
23. E. Trinh & T. G. Wang, *J. Fluid Mech.*, **122**, 315 (1982).
24. E. Trinh, A. Zwern, & T. G. Wang, *J. Fluid Mech.*, **115**, 453 (1982).
25. J. A. Tsamopoulos & R. A. Brown, *J. Fluid Mech.*, **147**, 373 (1984).
26. L. F. McGoldrick,*J. Fluid Mech*, **21**, 305 (1965).
27. J. W. S. Rayleigh, *Phil. Mag.*, **14**, 184 (1882).
28. G. I. Taylor, *Proc. R. Soc. London, A* **280**, 383 (1964).
29. N. Bohr & J. A. Wheeler, *Phys. Rev.*, **56**, 426 (1939).
30. P. M. Adornato & R. A. Brown, *Proc. R. Soc. London, A* **389**, 101, 1983.
31. J. A. Tsamopoulos, T. R. Akylas, & R. A. Brown, *Proc. R. Soc. London, A***401**, 67 (1985).
32. R. Natarajan & R. A. Brown, *Proc. R. Soc.*, *A***410**, 209 (1987).

FIGURE CAPTIONS

(1) IV.1 Mode Coupling for first three spherical harmonics

(2) IV.2 The percentage of each period that the drop in $n = 2$ oscillation spends in a prolate form as a function of the amplitude of the oscillation measured by the maximum ratio of the major to minor axis L/W. Asymptotic results (—), numerical calculations of Foote (---) and experimental results (□) of Trinh and Wang are shown.

(3) IV.3 The change in $n = 2$ oscillation frequency with increasing amplitude of oscillation as measured by L/W. Asymptotic results(—), numerical calculations of Foote (Δ) and Alonso (+), and experimental results of Trinh and Wang (•), R=0.49cm; □, R = 0.62cm) are shown.

(4) V.1 Phase-plane plots of the fundamental (a) and the resonating mode harmonic (b) with $Q = \sqrt{(32\pi/3)}$.

(5) V.2 Shapes of drops initially perturbed by an $n = 4$ mode with $Q = \sqrt{(32\pi/3)}$ and $\varepsilon = 0.2$.

(6) VI.1 Phase-plane diagrams for the amplitude modulation function $A(T_{1/2})$ for (a) $\kappa>0$ and (b)$\kappa<0$.

(7) VI.2 Evolution of amplitude function A for stable drop oscillation plotted as a function of the slow time scale $T_{1/2}$ for three values of κ.

SIMULATION OF THE THREE-DIMENSIONAL BEHAVIOR OF AN UNSTEADY LARGE BUBBLE NEAR A STRUCTURE

Georges L. Chahine[a] and Thomas O. Perdue
DYNAFLOW, Inc.
10422 Mountain Quail Road
Silver Spring, Maryland 20901

ABSTRACT

In most practical applications existing bubble dynamics models, either spherical or axisymmetric, are only more or less appropriate approximations. In this paper we will describe an on-going project which considers the fully three-dimensional bubble dynamics problem. The interaction between a growing, deforming and collapsing bubble near a boundary is simulated numerically using a Boundary Integral Method. The example of large bubble dynamics near a solid flat plate in a gravity field is considered. The plate orientation significantly influences the 3-D bubble shape and behavior. The flow field due to the bubble dynamics is considered to be potential. To initialize the computations, the bubble is taken to be very small and spherical. From there on, no additional assumptions are imposed and the bubble surface is free to move under the influence of the pressure field, inertia forces, and the presence of body forces and a nearby wall. The presence of gas inside the bubble is accounted for using a polytropic law of behavior and surface tension is included in the model. This paper presents the method, addresses the numerical difficulties and shows the influence of the problem geometry on the bubble dynamics.

INTRODUCTION

Bubble dynamics has been the subject of much attention since the end of the last century. However, nonspherical bubble dynamics, as well as interaction between bubbles, and between bubbles and neighboring complex boundaries have received much less attention due to the complexity of the nonspherical free boundary problem involved. In recent years the study of bubble dynamics for cavitation erosion problems has focused on bubble behavior near solid boundaries. Blake and Gibson[1] give a very fine overview of all facets of this work. Concerning the modelling of the phenomena, several analytical and numerical methods have been used. To quote a few: a Finite Difference Method was used by Plesset and Chapman[2]; a Variational Principle Method, by Shima and Nakajima[3]; an Asymptotic Expansion Method by Chahine and Bovis[4]; and Chahine and Liu[5]; a Boundary Element Method by Guerri, Luca and Prosperetti[6], etc... these studies were, however, limited to an axisymmetric configuration.

One of the numerical methods that has proven to be very efficient in solving this type of free boundary problem is the Boundary Integral Method. Guerri, et al. and Blake, et al.[7] used this method in the solution of axisymmetric problems of bubble growth and/or collapse near rigid boundaries. However, there are many cases which cannot be handled with an axisymmetric model and require a three-dimensional treatment. In this paper, such a three dimensional numerical model is developed to account for the growth and collapse of a buoyant vapor and gas bubble near boundaries. Buoyancy effects are included to allow the simulation of large bubbles such as those generated by underwater explosions. The three-dimensionality of

[a] also Research Professor, Department of Mechanical Engineering, Johns Hopkins University, Baltimore MD.

the problem arises from orienting the rigid boundary arbitrarily relative to the direction of gravity. In general, there is no axis of symmetry, and the entire bubble surface must be considered.

In this paper we present an outline of the model, the problem formulation, the numerical difficulties and the method of solution. We then present the results of a systematic study of bubble behavior in the presence of a solid plane wall at different orientations relative to gravity, and at different distances from the wall. The detailed mathematical expressions involved are very cumbersome and are kept to a minimum. The interested reader is referred to Perdue[8] and Chahine, Perdue and Tucker[9] for more detailed descriptions and expansions.

PROBLEM FORMULATION

It has been shown in earlier studies that due to the relatively large velocities associated with underwater explosion and cavitation bubbles (high Reynolds numbers) viscosity has no appreciable effects on the growth and collapse of bubbles in water. Also, since throughout most of the life of the bubble, the motion of the bubble wall is relatively slow compared to the speed of sound in water, compressibility effects can be ignored. This is valid shortly after bubble inception and until its latest collapse phase. These assumptions, classical in bubble dynamics studies, result in a flow that is potential and satisfies the Laplace equation,

$$\nabla^2 \Phi = 0, \tag{1}$$

where Φ is the velocity potential.

The solution must in addition satisfy boundary conditions at infinity, at the bubble wall and at the boundaries of the submerged body in contact with the fluid. In a fixed reference system of axes centered, for instance, at the initial center of the bubble, the velocity potential at infinity is zero. This condition can be written

$$\lim_{\overline{X} \to \infty} \Phi = 0, \tag{2}$$

where $\overline{X}$ is the vector position for a field point P. At all moving surfaces (such as the bubble surface or a moving nearby boundary) an identity between fluid velocities normal to the boundary and the normal velocity of the boundary is to be satisfied. At the bubble-liquid interface, the normal velocity of the moving bubble wall must equal the normal velocity of the fluid, or,

$$\nabla \Phi \cdot \mathbf{n} = \mathbf{V}_s \cdot \mathbf{n}, \tag{3}$$

where $\mathbf{n}$ is the local unit vector normal to the bubble surface and $\mathbf{V}_s$ is the local velocity vector of the moving surface. For a solid stationary boundary, as considered in the examples presented below,

$$\nabla \Phi \cdot \mathbf{n} = 0 \tag{4}$$

For a plane wall, this condition is met exactly by including an image bubble in the computation of the potential and its normal derivative over the bubble surface.

The bubble is assumed to contain noncondensible gas as well as a vapor of the surrounding liquid. The pressure within the bubble at any given time is considered to be the sum of the partial pressures of the noncondensible gases, P_g, and the vapor, P_v, inside the bubble. Vaporization of the liquid is assumed to be fast enough so that the vapor pressure remains constant throughout the simulation and equal to the equilibrium vapor pressure at the ambient liquid temperature. To the contrary, gas diffusion does not have time to occur and the noncondensible gases are assumed to have a polytropic behavior, PV^k = constant, where V is the bubble volume and k is the polytropic constant varying from $k = 1$ for isothermal behavior to $k = 1.4$ for adiabatic conditions.

The pressure at the exterior of the bubble surface, P_L, can be obtained at any time from the following pressure balance expression:

$$P_L = P_v = P_{g_o} \left(\frac{V_o}{V}\right)^k - C\sigma \tag{5}$$

where P_{g_o} and V_o are the initial gas pressure and volume respectively, σ is the surface tension, C is the local curvature of the bubble, and V is the instantaneous value of the bubble volume. P_{g_o} and V_o are known quantities at t=0.

At the beginning of the simulation, the bubble is assumed to be spherical. This approximation is valid as long as the chosen initial bubble size is sufficiently small and the amount of time elapsed since the bubble inception is very short so as to neglect any interaction with the wall. The initial velocity potential of the flow field is then due to a point source of intensity $4\pi R^2 R$ The potential at the bubble wall then becomes

$$\phi_o = -R_o R_o \tag{6}$$

where R_o and R_o are the initial radius and wall velocity of the spherical bubble. The initial wall velocity can be obtained from integrating the Rayleigh-Plesset equation with respect to time (see for instance Chahine[10] or Blake and Taib[7]), using $R = 0$ when $R = R_{max}$. In doing so, one obtains two solutions depending on whether the polytropic constant is greater than or equal to one. For the case where $k > 1$, the following equation is obtained for R_o,

$$\dot{R}_0^2 = \frac{2}{\rho} \frac{P_g}{3\,(1-k)} \left\langle 1-\alpha_m^{3k-3} \right\rangle + \frac{P_v + P_a}{3} \left\langle 1-\alpha_m^{-3} \right\rangle - \frac{\sigma}{R} \left\langle 1-\alpha_m^{-2} \right\rangle \tag{7}$$

where ρ is the liquid density, $\alpha_m = R_o / R_{max}$, R_{max} being the maximum radius attainable by the bubble in an infinite medium: $k = 1.4$ was used in the results given below.

METHOD OF SOLUTION

The three-dimensional Boundary Integral Method chosen for this problem is based on Green's equation which effectively reduces by one the dimension of the problem. If the velocity potential, Φ, and its normal derivatives are known on the fluid boundaries (points M) and Φ satisfies the Laplace equation as specified in Equation (1), then Φ can be determined anywhere in the domain of the fluid (field points P). This can be written:

$$\iint_s \left[\frac{-\delta\phi}{\delta n} \cdot \frac{1}{|MP|} + \phi \frac{\delta}{\delta n} \left(\frac{1}{|MP|} \right) \right] ds = a\,\pi\phi\,(P) \tag{8}$$

where a is positive and defined as follows:

$a = 4$, if P is a point in the fluid,
$a = 2$, if P is a point on a smooth surface, and
$a < 4$, if P is a point at a sharp corner of the surface.

If the field point P is selected to be on the surface of the bubble or its image, then a closed set of equations can be obtained and used at each timestep to solve for values of $\partial\Phi/\partial$ (or Φ) assuming that all values of Φ (or $\partial\Phi/\partial n$) are known at the preceding step.

To solve Equation (8) numerically, it is necessary to discretize the bubble into panels, perform the integration over each panel, and then sum up the contributions to complete the integration over the entire bubble surface. To do this, the initially spherical bubble is discretized into a geodesic shape with flat, triangular panels. This allows for a rather even initial distribution of nearly equal-sized panels. In the numerical code developed, the fineness of the grid can be controlled relatively easily by prescribing the frequency of subdivision of basic triangles of an icosahedron. With the discretized surface, Equation (8) becomes a set of N equations (N is the number of discretization nodes) of index i of the type:

$$\sum_j \left(\frac{\delta\phi}{\delta n} \cdot A_{ij} \right) = \sum_j \left(\delta_j\, B_{ij} \right) - a\pi\phi_i \tag{9}$$

where A_{ij} and B_{ij} are elements of matrices which correspond numerically to the integrals given in Equation (8). For this discretized problem, the solution is more accurate, and stability is improved, when $a\pi$ is taken at each node to be the solid angle under which the fluid is "seen" from the particular node. The computation of this solid angle is performed at each node and at each time step.

To evaluate the integrals in Equation (8) over any single panel, a transformation from cartesian coordinates to an oblique coordinate system (ζ, η) is made (see Figure 1). Then, by assuming a linear variation of the potential and its normal derivative over the panel, the following expressions can be written for Φ and $\partial\Phi/\partial n$ at any point M on the panel ABC:

$$\Phi(M) = (1-\zeta-\eta)\,\Phi(A) + \zeta\Phi(B) + \eta\Phi(C),$$

$$\frac{\delta\phi}{\delta n}(M) = (1-\zeta-\eta)\frac{\delta\phi}{\delta n}(A) + \zeta\frac{\delta\phi}{\delta n}(B) + \eta\frac{\delta\phi}{\delta n}(C) \qquad (10).$$

In this manner, both Φ and $\partial\Phi/\partial n$ are continuous over the bubble surface and are expressed as a function of the values at the nodal points which define the particular panel. Obviously higher order expansions in powers of ζ and η are conceivable, and would improve accuracy at the expense of additional analytical and numerical computation times. These expansions were not considered in this study. The introduction of the oblique coordinates significantly reduce the complexity of the integrations.

The two integrals in Equation (8) can be evaluated analytically and the resulting expressions, very long and complicated warrant a publication on their own. The complete derivations and results can be found in Perdue[8] and Chahine et al.[9], and are available upon request from the authors. These expressions are generally valid for any panel ABC relative to any node P where P is not on the panel. When the field point P is on the panel over which the integration is being performed, the integrals become singular and warrant special treatment, but can still be evaluated analytically. These singular integrals can be evaluated numerically, but we have instead selected to obtain exact analytical expressions.

In order to proceed with the computation of the bubble dynamics several quantities appearing in the above boundary conditions need to be evaluated at each time. The bubble volume presents no particular difficulties, while the unit normal vector, the local surface curvature and the local tangential velocity at the bubble interface need further development. The curvature of the bubble surface is obtained by first computing a local bubble surface three-dimensional fit, $f(x,y,z) = 0$. The unit normal at a node can then be expressed as:

$$\mathbf{n} = \pm\, \nabla f / |\nabla f| \qquad (11).$$

The appropriate sign is chosen to insure that the normal is always directed toward the fluid. The local curvature is then computed by

$$C = \nabla \cdot \mathbf{n} \qquad (12).$$

To obtain the total fluid velocity at any point on the surface of the bubble, the tangential velocity, $\mathbf{V}_t$, must be computed at each node in addition to the normal velocity, $\partial\Phi/\partial n$. This is also done using a local surface fit to $\Phi_l = g(x,y,z)$. Taking the gradient of this function at the particular node in question and eliminating any normal component of velocity appearing in this gradient gives a good approximation for the tangential velocity. Practically, a second degree equation for the local surface around node N at time t was selected. For instance, for an equation such as:

$$a_1 z^2 + a_2 z + a_3 y^2 + a_4 y + a_5 x^2 + a_6 x + a_7 = 0 \qquad (13)$$

seven parameters a_i have to be determined to obtain the local surface fit equation. This was obtained by using the coordinates of the node of interest, N, and six selected nodes around it and solving the resulting set of 7x7 linear equations. The six extra nodes were selected randomly from a matrix associating to each node all the nodes located on the two immediate rows of nodes surrounding it. A random number generating function was used to select the six needed nodes. In order to reduce errors, the procedure was repeated at least five times, eliminating choices that produced a singular or degenerated system of linear equations. The results were then averaged using the five answers. A similar approach was used to generate the function $g(x,y,z)$ at each node N and time t. Once f and g are determined, the sought variables, C, $\mathbf{n}$ and $\mathbf{V}_t$, are determined using the analytical expressions (11) and (12), and the following expression for $\mathbf{V}_t$:

$$\mathbf{V}_t = \nabla\Phi_t \, x \, \mathbf{n} \qquad (14).$$

With the problem initialized and the velocity potential known over the surface of the bubble, an updated value of $\partial\Phi/\partial n$ can be obtained by performing the integrations outlined above and solving the corresponding matrix equation. The unsteady Bernoulli equation can then be used to solve for $D\Phi/Dt$, the total material derivative of Φ, (using $\mathbf{V}_S = \nabla\Phi$),

$$\frac{D\phi}{Dn} = \frac{\delta\phi}{\delta n} + |\nabla\phi|^2 = \frac{P_a - P_L}{\rho} - \rho g z + \frac{1}{2}|\nabla\phi|^2 \qquad (15).$$

$D\Phi/Dt$ provides the time variations of Φ at any node which is followed during its motion with the fluid. Using an appropriate timestep all values of Φ on the bubble surface can be updated using Φ at the preceding time step and $D\Phi/Dt$.

From our experience, the best choice was based on the ratio between the minimum size of panel sides, l_{min}, and the maximum node velocity, V_{max}:

$$dt = l_{min} / V_{max} \qquad (16)$$

This choice has the great advantage of constantly adapting the timestep, by refining it at the end of the collapse, and increasing it during the slow bubble size variation period. An additional advantage is the prevention of surface "folding" in the latest phases of the collapse and instability enhancement. New coordinate positions of the nodes can then be obtained using the position of the previous time step and the displacement $\partial\Phi/\partial n.\mathbf{n} + V_t.\mathbf{e}_t$ where $\mathbf{e}_t$ is the unit tangent vector, $\mathbf{e}_t = \mathbf{V}_t / |\mathbf{V}_t|$. This time stepping procedure is repeated throughout the bubble growth and collapse, resulting in a fairly complete shape history of the bubble.

The major emphasis of this paper is on the growth and collapse of a bubble near an infinite flat rigid boundary. The algorithm is, however, capable of handling the bubble interaction with nearly any structure geometry. For the infinite flat plate, an image bubble is added to satisfy the boundary condition (4). Thus, the integrations must be carried out over two identical bubble shapes, though the symmetry between them allows for some significant reduction in calculations and computation times. For other cases, the adjacent structure is discretized as well and the integrations are performed over all boundaries with solutions obtained for all nodes (see examples in Chahine et al.[9]).

COMPUTATIONAL RESULTS AND DISCUSSION

Before considering the complex cases of interest to this study the numerical code was tested against existing numerical codes in the literature which are relatively simple. The collapse of a spherical bubble has been extensively studied, and an "exact" analytical-numerical solution exists by numerical integration of the well-known second degree differential equation; the Rayleigh-Plesset equation. In Figure 2, we compare the results of the present numerical code with the "exact" solution for increased degrees of mesh refinement, or discretization frequency. Only the first frequency discretization (12 nodes) shows a significant error on the bubble period (more than 7 percent). The second frequency (42 nodes) and third frequency (92 nodes) show errors of 1.8 and 0.6 percent respectively, while frequencies four (162 nodes) and five (252 nodes) show errors less than 0.14 percent and 0.05 percent respectively. The four frequency discretization was selected for most of the runs shown below.

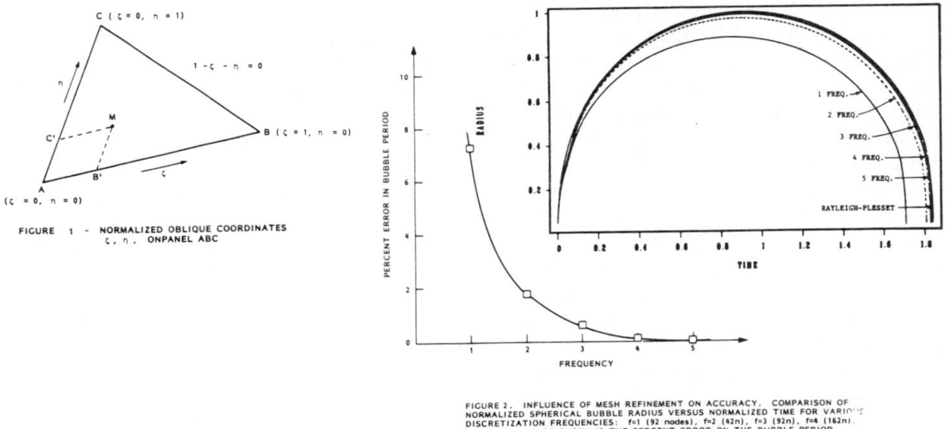

FIGURE 1 - NORMALIZED OBLIQUE COORDINATES ζ, η, ONPANEL ABC

FIGURE 2. INFLUENCE OF MESH REFINEMENT ON ACCURACY. COMPARISON OF NORMALIZED SPHERICAL BUBBLE RADIUS VERSUS NORMALIZED TIME FOR VARIOUS DISCRETIZATION FREQUENCIES: f=1 (92 nodes), f=2 (42n), f=3 (92n), f=4 (162n), f=5 (252n). ALSO SHOWN IS THE PERCENT ERROR ON THE BUBBLE PERIOD.

The second series of comparisons was made with the previously studied axisymmetric cases available in the open literature, and have shown, as described below, differences with these studies of the order of 0.1 percent on the bubble period. Finally, comparison with actual test results of the complex three-dimensional case presented here, Figure 8, shows strikingly similar complex bubble shapes with a systematic error of 12 percent on the times which was proven to be related to the bubble period lengthening influence of the container walls on the experimental results (see Chahine et al.[9]).

Figure 3 shows the results of an axisymmetric case of bubble growth and collapse computed using the above described algorithm. It shows some general 3-D shapes and bubble profiles depicting the growth and collapse in a gravity field near an infinite horizontal plate above the bubble at a standoff ratio of L/R_{max} = 1.50, where L is the perpendicular distance from the initial bubble center to the plate. Here, the initial spherical bubble was discretized using 162 nodes resulting in 320 triangular panels. All coordinates have been normalized by R_{max} = 17cm. The figure shows how the bubble grows nearly spherically, then, during the collapse phase, flattens on the side which is opposite to the plate and forms a reentrant jet in a direction perpendicular to the plate. Note that for this case, the plate is located directly above the bubble, so both the buoyancy effects and the effects of the nearby boundary are in the same direction, vertically upward. The total period of the bubble, scaled by the Rayleigh time, $R_{max}/\sqrt{P/\rho}$ was about 2.084. P is the difference between the ambient pressure and the minimum bubble pressure. This agrees very well with the bubble period given by Blake, et al.[7] who report a value of approximately 2.097 for a similar case computed in the absence of gravity with their axisymmetric scheme and a larger number of panels. The maximum velocity attained by the jet in this simulation was $11.1\sqrt{P/\rho}$. This, too, compares very well with the results of other studies.

Figures 4 and 5 show fully the three-dimensional bubble growth and collapse when the infinite plate was moved to a vertical position adjacent to the bubble. The two cases were ran when the plate was at standoff ratios of L/R_{max} = 1.50, and 1.00. With gravity acting vertically downward, the two competing forces (gravity, presence of plate) are perpendicular, rather than parallel to each other. The same initial discretization is used.

Both of these figures show very clearly the formation of the reentrant jet moving at an angle upward toward the plate. Given that the two competing forces no longer act along the same line, one would expect that the jet would form in some resulting direction depending on the proximity of the plate. Comparison of the two figures shows that indeed this is the case. In Figure 4, where the bubble is initially closer to the plate, the reentrant jet appears to act along a line that is more nearly perpendicular to the plate. In Figure 4, the angle of the reentrant jet is more acute and the jet penetrates the bubble and touches the other side at a point much closer to the top of the bubble than for the bubble in Figure 5. Also one notices that, during the collapse, once the jet has touched the other side of the bubble, the bubble retains a larger volume for the case where the plate is closer. This has been demonstrated in the axisymmetric models and is very evident here as well. It is also apparent from the figures that the bubble growth is an important consideration. Deviation from sphericity, for instance, at maximum bubble size is observed and is larger when the bubble is initially closer to the plate. Bubble periods are noticeably changed by the wall configuration relative to gravity. Placing the boundary vertically rather than horizontally results in an increased normalized period from 2.084 to 2.101. Moving the bubble closer to the wall lengthens the period still more. This is consistent with the known lengthening effect of the boundary and the reduction in period caused by the buoyancy force.

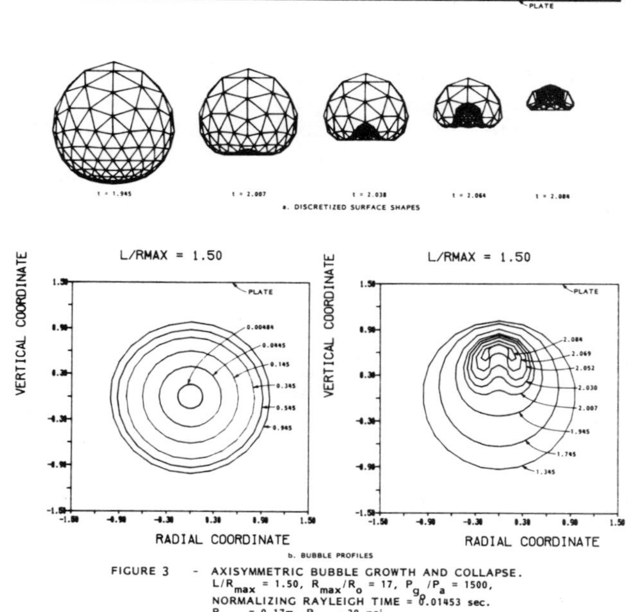

FIGURE 3 - AXISYMMETRIC BUBBLE GROWTH AND COLLAPSE. L/R_{max} = 1.50, R_{max}/R_o = 17, P_g/P_a = 1500, NORMALIZING RAYLEIGH TIME = 0.01453 sec. R_{max} = 0.17m, P_{amb} = 20 psi

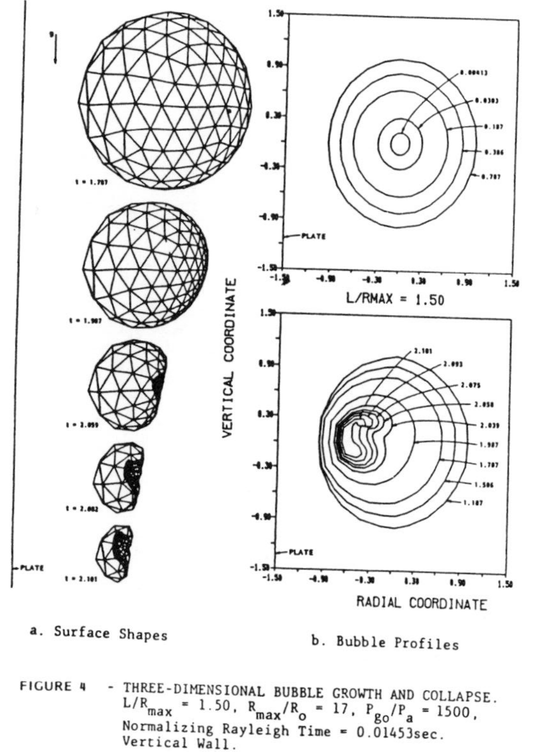

a. Surface Shapes

b. Bubble Profiles

FIGURE 4 - THREE-DIMENSIONAL BUBBLE GROWTH AND COLLAPSE. L/R_{max} = 1.50, R_{max}/R_o = 17, P_{go}/P_a = 1500, Normalizing Rayleigh Time = 0.01453sec. Vertical Wall.

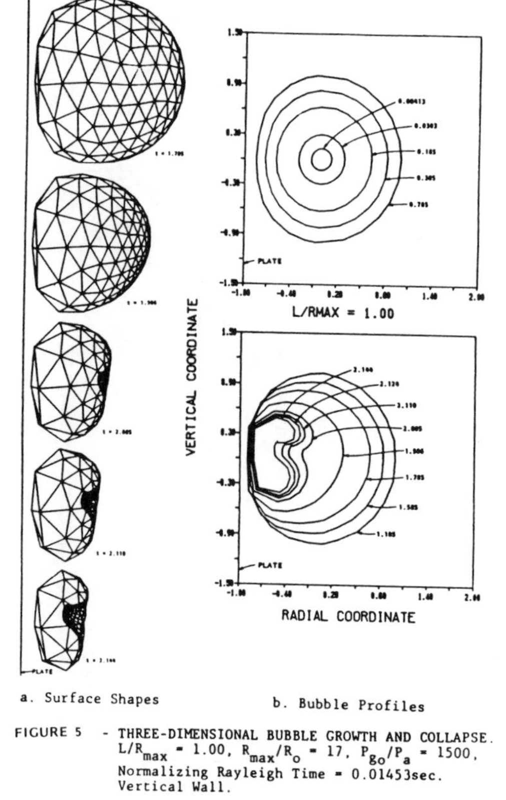

a. Surface Shapes

b. Bubble Profiles

FIGURE 5 - THREE-DIMENSIONAL BUBBLE GROWTH AND COLLAPSE. L/R_{max} = 1.00, R_{max}/R_o = 17, P_{go}/P_a = 1500, Normalizing Rayleigh Time = 0.01453sec. Vertical Wall.

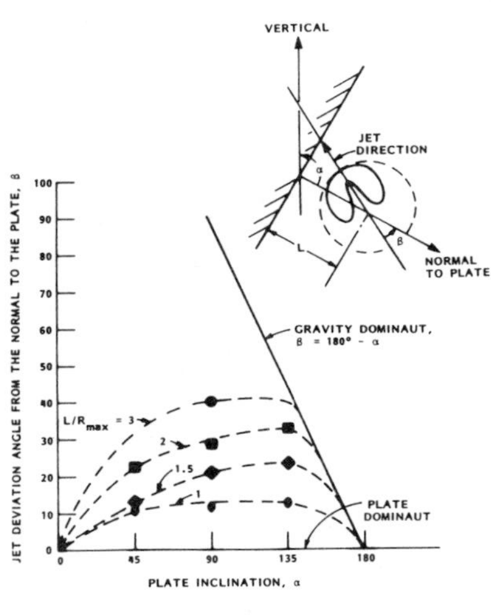

FIGURE 6 - RELATIVE EFFECTS OF GRAVITY FORCE AND PLATE PRESENCE ON REENTRANT JET ANGLE

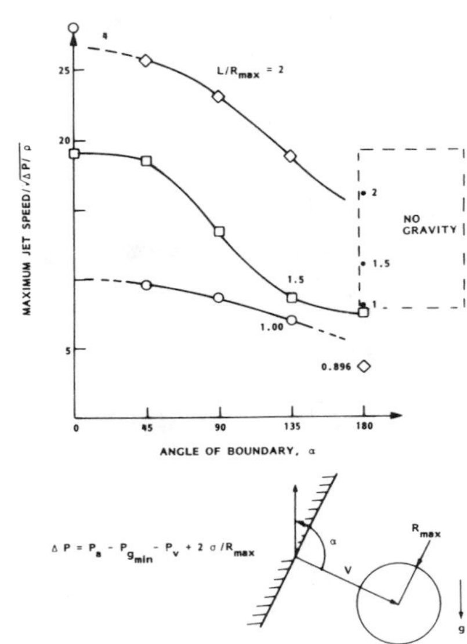

FIGURE 7 - INFLUENCE OF GRAVITY FORCE AND FLAT PLATE PRESENCE ON REENTRANT JET SPEED

Figures 6 and 7 summarize the results of a series of runs where the plate angle relative to gravity field and the plate standoff distance to the center of explosion have been varied. Figure 6 shows the deviation angle from the normal to the plate, β, versus the angle α between the plate normal and the vertical axis. The set of curves in the figure is for various values of $\bar{L} = L/R_{max}$. When $\bar{L}$ is very small, wall effect is predominant and the jet is practically normal to the plate, $\beta = 0$. β increases with $\bar{L}$ and attains a maximum when $\bar{L}$ goes to infinity. In that case, gravity forces are predominant and the reentrant jet is vertical directed upward independent of the plate angle α. However, with our definitions of α and β, the limit value of β depends on α as follows:

$$\beta_{limit} = 180° - \alpha \tag{17}$$

For intermediate values of $\overline{L}$, the deviation angle from normal increases with α to attain a maximum for values of α between 90° and 135°. This result should not be generalized and is probably dependent on a Froude number based on the maximum bubble size, its period (or energy of the explosion) and the acceleration of gravity. The curves show a skewness toward higher α's, which reflects the fact that for $\alpha > 40°$, both gravity and plate presence act in the same direction, while for $\alpha < 90°$, these two forces act in opposite directions. (Note that $\alpha = 0°$ corresponds to a horizontal plate below the explosion center while $\alpha = 180°$ corresponds to a horizontal plate above the explosion center).

Figure 7 shows the influence of gravity and the proximity of the flat plate on the reentrant jet speed. Several important observations can be made form this figure.

a. The jet speed increases with distance from the plate.

b. The jet speed is higher when buoyancy forces and the plate presence do not act in the same direction.

c. The influence of plate angle on the jet speed is negligible both for very large and small values of $\overline{L}$.

d. The influence of α on the jet speed is the sharpest for $\overline{L} \sim 1.5$, limit value for which the jet direction changes sides for $\alpha = 180°$.

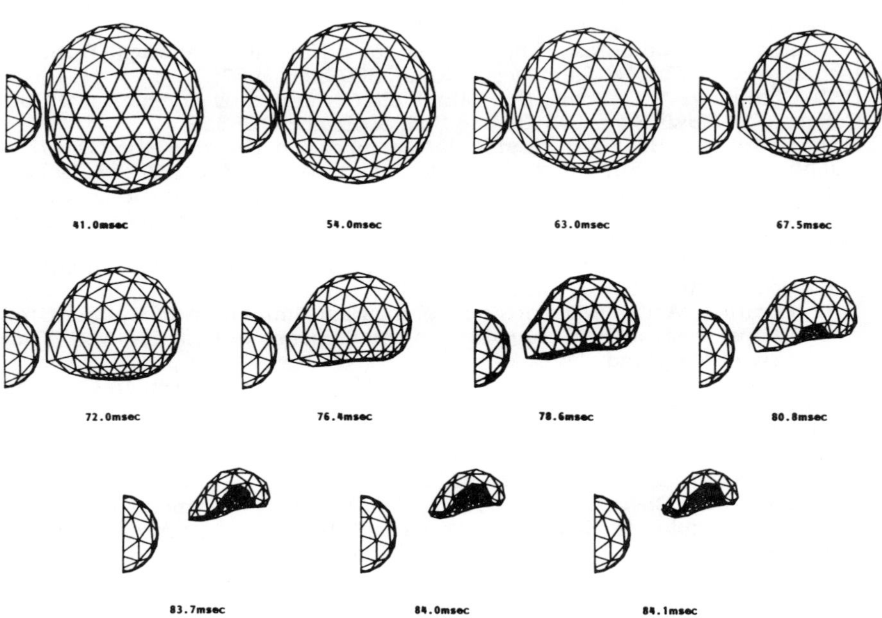

FIGURE 8 - BUBBLE COLLAPSE NEAR AXISYMMETRIC BODY
4 FREQUENCY DISCRETIZATION

Figure 8 shows the interaction between an unsteady bubble and an axisymmetric finite-size body. The center of the submerged body was placed at the same depth as the initial center of the bubble and at a distance to the axisymmetric body of 1.0 maximum bubble radii. Gravity was also included, so the two forces (buoyancy and body attraction) were acting perpendicularly to each other. The parameters for this case were chosen to match the experimental case shown in Figure 5 in Snay, Goertner and Price[11]. The computed bubble shapes closely resembles those actually recorded in the small scale underwater explosion test. A portion of the bubble is seen to adhere to the nearby body while the remainder behaves as if only gravity is the influencing factor. The one major discrepancy between the numerical and experimental results is the period of the growth and collapse of the bubble. The measured period is about 12 percent longer than the computed bubble period. This difference is mainly due to the fact that the experiments were conducted in a finite size cylindrical tank (diameter three times the maximum bubble diameter). The presence of the tank walls was shown by the numerical simulations to have a lengthening effect of the same order of magnitude on an otherwise isolated bubble.

This study is presently being extended to more accurately describe the latest phase of the bubble collapse when the reentrant jet approaches the opposite side of the bubble. Numerical instabilities occur in that last phase and can be reduced by smoothing techniques. Similarly, the pressures generated on the nearby plate become unsteady and need to be more accurately determined.

REFERENCES

1. J. R. Blake & D. C. Gibson, "Cavitation Bubbles Near Boundaries," *Ann. Rev. of Fl.Mech.*, **19**, 99-123, 1987.
2. M. S. Plesset & R. B. Chapman, "Collapse of a Vapor Cavity in the Neighborhood of a Solid Wall," *Cal. Tech. Rep.* 85-48, 1969.
3. A. Shima & K. Nakajima, "The Collapse of a Non-Hemispherical Bubble Attached to Solid Wall," *J. Fluid Mech.*, **80** (2), 369-391, 1977.
4. G. L. Chahine & A. G. Bovis, "Pressure Field Generated by Nonspherical Bubble Collapse," *J. Fluids Eng.*, **105** (3), 356-364, September 1983.
5. G. L. Chahine & H. L. Liu, "A singular perturbation Theory of the Growth of a Bubble Cluster in a Superheated Liquid," *J. Fluid Mech.*, **156**, 257-279, 1985.
6. L. Guerri, G. Lucca & A. Prosperetti, "A Numerical Method for the Dynamics of Non-Spherical Cavitation Bubbles," **Proc. 2nd Int. Coll. on Drops and Bubbles**, JPL Publication 82-7, Monterey CA, Nov 1981.
7. J. R. Blake, B. B. Taib, & G. Doherty, "Transient Cavities Near Boundaries," *J. Fluid Mech.*, **170**, 479-497, 1986.
8. T. O. Perdue, "A Three-Dimensional Numerical Simulation of Bubble Growth and Collapse Near a Rigid Boundary in the Presence of Gravity," M.S. Thesis, University of Maryland, 1988.
9. G. L. Chahine, T. O. Perdue, & C. B. Tucker, "Interaction Between an Underwater Explosion Bubble and a Solid Submerged Structure," Tracor Hydronautics Technical Report 86029-1, April 1988.
10. G. L. Chahine, "Experimental and Asymptotic Study of Nonspherical Bubble Collapse," *Appl. Sci. Res.*, **38**, 187-197, 1982.
11. H. G. Snay, J. F. Goertner & R. S. Price, "Small Scale Experiments to Determine Migration of Explosion Gas Globes Towards Submarines," Navord Report 2280, July 1952.

FIGURE CAPTIONS

Figure 1. Normalized oblique coordinates ζ, η on panel ABC.

Figure 2. Influence of mesh refinement on accuracy. Comparison of normalized spherical bubble radius versus normalized time for various discretization frequencies: $f = 1$ (92 nodes), $f = 2$ (42n), $f = 3$ (92n), $f = 4$ (162n), $f = 5$ (252n). Also shown is the percent error on the bubble period.

Figure 3. Axisymmetric bubble growth and collapse. $L/R_{max} = 1.50$, $R_{max}/R_o = 17$, $Pg_o/P_a = 1500$. Normalizing Rayleigh time = 0.01453 sec. $R_{max} = 0.17$m, $P_{amb} \cong 20$ psi.

3a. Discretized surface shapes

3b Bubble profiles

Figure 4. Three-dimensional bubble growth and collapse, $L/R_{max} = 1.50$, $R_{max}/R_o = 17$, $Pg_o/P_a = 1500$. Normalizing Rayleigh time = 0.01453 sec. Vertical wall.

Figure 5. Three-dimensional bubble growth and collapse, $L/R_{max} = 1.00$, $R_{max}/R_o = 17$, $Pg_o/P_a = 1500$. Normalizing Rayleigh time = 0.01453 sec. Vertical wall.

Figure 6 Relative effects of gravity force and plate presence on re-entrant jet angle.

Figure 7. Influence of gravity force and flat plate presence on re-entrant jet speed.

Figure 8. Bubble collapse near axisymmetric body 4-frequency discretization.

COMPUTER MODELING OF THE DYNAMICS OF BUBBLE ON ROTATING FLUIDS IN LOW AND MICROGRAVITY ENVIRONMENTS

J. Hung and Y. D. Tsao
The University of Alabama in Huntsville
Huntsville, Alabama 35899

Fred W. Leslie
NASA/Marshall Space Flight Center
Huntsville, Alabama 35812

ABSTRACT

Time dependent evolutions of the profile of free surface (bubble shapes) for a cylindrical container partially filled with a Newtonian fluid of constant density, rotating about its axis of symmetry, have been studied. Numerical computations of the dynamics of bubble shapes have been carried out with the sinusoidal function vibration of gravity environment in high and low rotating cylinder speeds. The initial condition of bubble profiles was adopted from the steady-state formulations in which the computer algorithms have been developed by Hung and Leslie[3], and Hung *et al.*[4].

I. INTRODUCTION

Vibration of microgravity environment affects the design concepts of the Gravity Probe-B Spacecraft, the Space Station, and of all other kinds of spacecraft. Dynamics of fluid in a cylindrical cavity plays an important role in determining stability and motion of liquid-filled containers in spacecraft, in particular, under the effect of the vibration of microgravity environment. Surface tension is an important factor in a large variety of fluid flows in low gravity. Free surface shapes of liquids are vitally important in spacecraft fuel-tank design, fluid-shell systems, and also in fluid management systems.

Experiments undertaken abroad orbiting spacecraft are subject to a variety of residual accelerations over a broad range of frequencies. Certain classes of experiments may be more sensitive than others to the residual accelerations. These include fluid experiments which need to avoid buoyancy-driven convection, critical point phenomena, and growth of both organic and inorganic crystals. The sources of the residual accelerations range from the effects of the Earth's gravity gradient, atmospheric drag on the spacecraft, and spacecraft altitude motions to the higher frequency g-jitter arising from machinery vibrations, thruster rings, crew motions, etc. The effect of the high frequency accelerations on fluid motions is not completely understood. Recent work[1,2] suggests that they may be unimportant in comparison to the residual motions caused by low frequency accelerations.

In this study, time-dependent dynamical behavior of surface tension on partially-filled rotating fluids in high and low rotating speeds under a vibrating microgravity environment have been carried out by numerically computing the Navier-Stokes equations subjected to initial and boundary conditions. At the interface between the liquid and the gaseous fluids, both the kinematic surface boundary condition, and the interface stresses conditions for components tangential and normal to the interface, have been applied. The initial condition of bubble profiles was adopted from the steady-state formulations in which the computer algorithms have been developed by Hung and Leslie[3] and Hung *et al.*[4] in rotating cylinder tank; and also Hung et. al.[5] in the coordinate of the Gravity Probe-B Spacecraft[6].

II. MATHEMATICAL FORMULATION

Consider a circular cylinder of radius, a, with length, L, which is partially filled with a Newtonian fluid of constant density ρ and kinematic viscosity ν. The cylinder rotates about its axis of symmetry with angular velocity, $\omega(t)$, which is a function of time, t. Let us use cylindrical coordinates (r, θ, z), with corresponding velocity components (u, v, w). The gravitational acceleration, g, is along the z-axis. For the case of axial symmetry, the θ-dependency vanishes. The governing equations are shown as follows:

(A) Continuity Equation

$$\frac{1}{r}\frac{\partial}{\partial r}(ru)+\frac{\partial w}{\partial z}=0 \tag{1}$$

(B) Momentum Equations

$$\frac{Du}{Dt}-\frac{v^2}{r}=-\frac{1}{\rho}\frac{\partial P}{\partial r}+\nu\left(\nabla^2 u+\frac{u}{r^2}\right) \tag{2}$$

$$\frac{Dv}{Dt}-\frac{uv}{r}=\nu\left(\nabla^2 u+\frac{v}{r^2}\right) \tag{3}$$

$$\frac{DW}{Dt}=-\frac{1}{\rho}\frac{\partial P}{\partial z}-g+\nu\nabla^2 w \tag{4}$$

$$\frac{D}{Dt}=\frac{\partial}{\partial t}+u\frac{\partial}{\partial r}+w\frac{\partial}{\partial z} \tag{5}$$

$$\nabla^2=\frac{1}{r}\frac{\partial}{\partial r}+\left(r\frac{\partial}{\partial r}\right)+\frac{\partial^2}{\partial z^2} \tag{6}$$

Let the profile of the interface between gaseous and liquid fluids be given by

$$\eta=\eta(t, r, z) \quad . \tag{7}$$

The initial condition of the profile of interface between gaseous and liquid fluids at $t=t_o$ is assigned explicitly, and is given by

$$\eta=\eta(t=t_o, r, z) \quad . \tag{8}$$

A set of boundary conditions has to supply for solving the equations. These boundary conditions are as follows:

(1) At the container wall, no-penetration and no-slip conditions assure that both the tangential and the normal components of the velocity along the solid walls will vanish. A constant contact angle is present when the free surface of liquid intersects the container wall.

(2) Along the interface between the liquid and gaseous fluids, the following two conditions apply:

(a) Kinematic surface boundary condition: The liquid (or gaseous) surface moves with the liquid (or gas) which implies

$$\frac{D\eta}{Dt}=0, \text{ or } \frac{\partial\eta}{\partial t}+u\frac{\partial\eta}{\partial r}+w\frac{\partial\eta}{\partial z}=0, \text{ on } \eta=\eta(t=t_i, r, z) \quad . \tag{9}$$

(b) Interface stresses condition: At the interface, the stresses must be continuous. These can be decomposed to the component normal to the interface, n_i, and the components tangential to the interface, n_j. For the components tangential to the interface between liquid and gaseous fluids

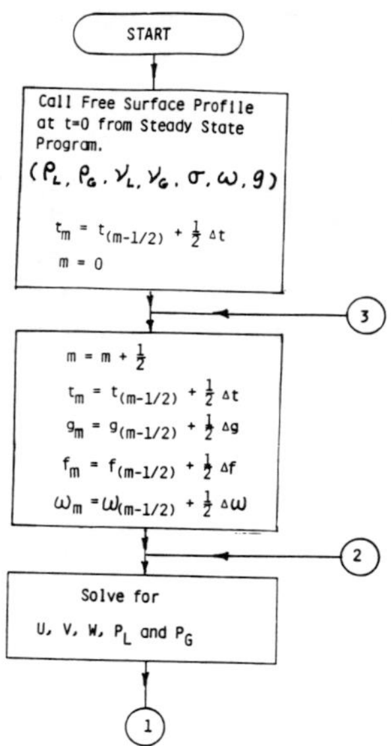

Fig. 1(A)

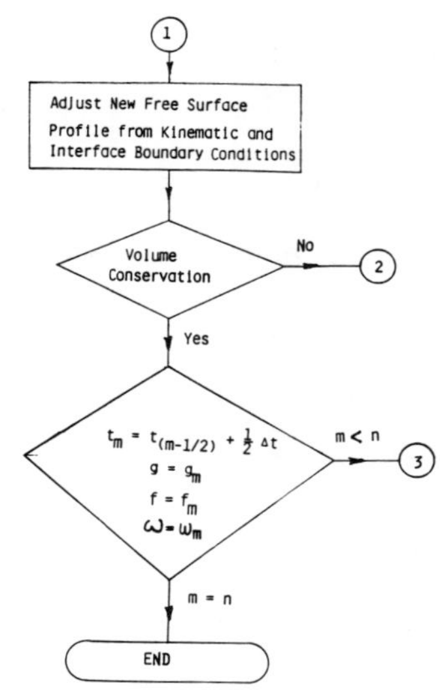

Fig. 1(B)

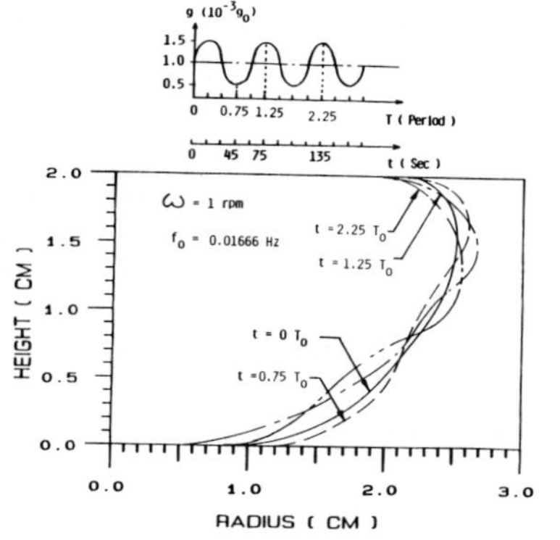

Fig. 2

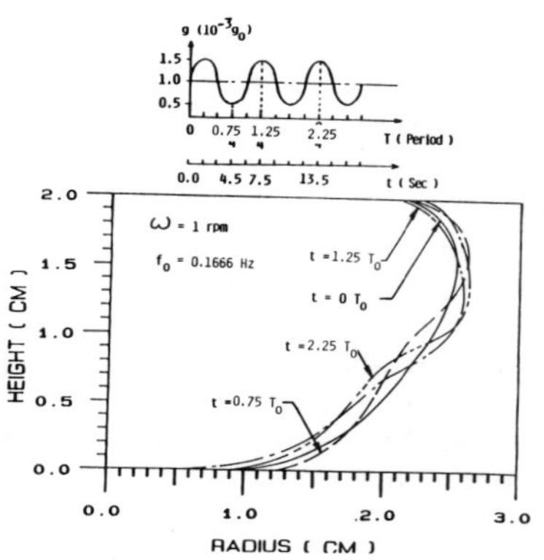

Fig. 3

$$\nu\left(\frac{\partial u_i}{\partial n_j}+\frac{\partial u_j}{\partial n_i}\right)n_i \mid_{Liquid} = \nu\left(\frac{\partial u_i}{\partial n_j}+\frac{\partial u_j}{\partial n_i}\right)n_i \mid_{Gas} \tag{10}$$

must hold. For the component normal to the interface between the liquid and gaseous fluids, neglecting viscous stresses which are small in comparison to the pressure difference across the interface, the expression becomes Laplace's formula which is:

$$P_L - P_G = -\frac{\sigma}{r}\frac{d}{dr}\left[\frac{r\,\psi}{\left(1+\psi^2\right)^{1/2}}\right] \tag{11}$$

where P_L denotes the liquid pressure at the interface; P_G, the gaseous pressure at the interface; σ, the surface tension of the interface; and ψ, the tangent of the interface which is defined by

$$\psi = \frac{dz}{dr} \text{ on } \eta_i = \eta(t_i, r, z) \tag{12}$$

A computer algorithm was developed to integrate Equations (1) to (4) numerically, subjected to the following conditions: (A) initial condition, Equation (8); (B) boundary conditions which include no-penetration and no-slip conditions at the container wall; (C) kinematic surface boundary condition, shown in Equation (9); and (D) interface stresses conditions, shown in Equations (10) to (12).

Vibration of gravity environment is governed by the following formula:

$$g = g_o[1 + \frac{1}{2}\sin(2\pi f_o t)] \tag{13}$$

where g_o denotes the background gravity environment, and $f_o(Hz)$ stands for the vibration frequency. $T\,(1/f_o)$ is the period of the vibration.

III. NUMERICAL CALCULATION OF TIME DEPENDENT BUBBLE PROFILES IN VIBRATION OF MICROGRAVITY ENVIRONMENT

The present study examined time-dependent rotating bubbles under the influence of vibrating microgravity environment. Initial profiles of bubbles were determined from the computations based on steady state formulations in which the computer algorithms have been developed by Hung and Leslie[3], and Hung *et al.*[4]. In other words, the initial profiles of bubbles can be determined from the following parameters: liquid density (ρ_L) and its kinematic viscosity (ν_L), gas density (ρ_G) and its kinematic viscosity (ν_G), surface tension coefficient (σ), angular velocity of rotating cylinder (ω), vibrating frequency (f) of gravity environment and background gravity environment (g). Thus, the initial condition of the interface profile between gaseous and liquid fluids at $t = t_o$ can be assigned explicitly at Equation (8).

A computer algorithm was developed to integrate Equations (1) to (4) numerically, subjected to the initial and boundary conditions described in Equations (8), (9), (10), (11) and (13). Figures 1(A) and 1(B) show the flow charts for the procedures of computation for numerically solving these equations.

For the purpose of showing the computation results easier to compare with the experimental measurements [rotating equilibrium, free surfaces in the low and microgravity environments of a free-falling aircraft (KC-135), carried out by Leslie[8]], the size of the container is assumed to be a radius of 3 cm and a height of 2 cm. The container is partially-filled with ethanol, and the volume of air is 30 cm^3. The contact angles between ethanol and the solid walls for all the cases are $\theta = 10°$.

Figures 2, 3, and 4 show the evolution of bubble shapes with sinusoidal function vibration of gravity environment with vibration frequencies of 0.01666, 0.1666, and 1.666 *Hz*, respectively, in the background gravity environment of 10^{-3}g, and low rotating speed of 1 rpm. There are four profiles of bubble shapes at the vibrating gravity environment corresponding to the time of 0.0, 0.75, 1.25, and 2.25 period of vibration, shown in the figures. The following conclusions can be drawn based on

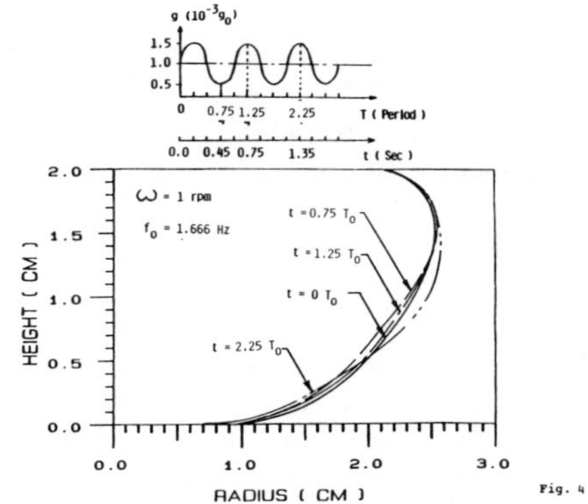

Fig. 4

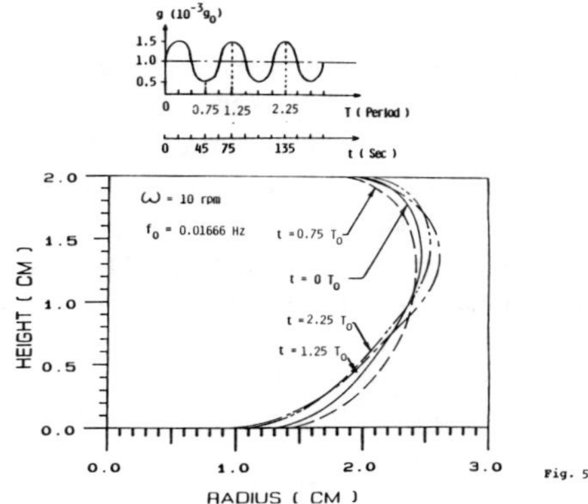

Fig. 5

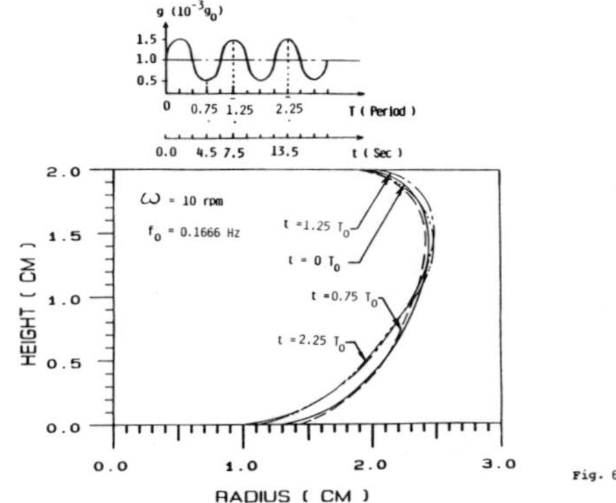

Fig. 6

these three figures: (1) Larger amplitudes of vibrations are shown in the profiles below the maximum radius of the bubble than that of the profiles above the maximum radius of the bubble; (2) Greater back and forth vibrations of the radius of bubble intersecting the bottom wall of the cylinder are shown than that of the radius of bubble intersecting the top wall of the cylinder; and (3) The lower the vibration frequency of gravity environment, the greater the amplitude of the vibration of the bubble profiles.

Figures 5, 6, and 7 show the evolution of bubble shapes with sinusoidal function vibration of gravity environment with vibration frequencies of 0.01666, 0.1666, and 1.666 *Hz*, respectively, in the background gravity environment of 10^{-3}g, and medium rotating speed of 10 rpm. Similar to Figures 2 to 4, these figures also show four profiles of bubble shapes at the vibrating gravity environment corresponding to the time of 0.0 0.75, 1.25, and 2.25 periods of vibration. In addition to the conclusions drawn earlier, the following additional results can be shown: (1) Amplitudes of the vibration of bubble shape decrease as the rotating speed of the cylinder increases; and (2) As the rotating speed increases, maximum radius of the bubble decreases, and shapes of the bubble become flatter with respect to the axis of rotation.

Similar calculations for the evolution of bubble shapes with sinusoidal function vibration of gravity environment with high rotating speed of 100 rpm are made. It is shown in Figure 8 that the amplitudes of the vibration due to the vibrating gravity environment become negligibly small, and practically no vibration of the bubble shape is viewed when the rotating speed of the cylinder is on the order of, or exceeds 100 rpm.

IV. DISCUSSIONS AND CONCLUSIONS

Time-dependent dynamical behavior of surface tension on partially-filled rotating fluids in both low gravity and microgravity environments have been carried out by numerically computing the Navier-Stokes equations subjected to the initial and the boundary conditions. At the interface between the liquid and gaseous fluids, both the kinematic surface boundary condition, and the interface stresses conditions for components tangential and normal to the interface, have been applied. The initial condition of bubble profiles was adopted from the steady-state formulations in which the computer algorithms have been developed by Hung and Leslie[3], and Hung *et al.*[4].

In conclusion, for sinusoidal function vibration of gravity environment in low and medium rotating speeds of the cylinder, it is shown that the back and forth vibrations of the radius of bubble intersecting the bottom wall of the cylinder are greater than that of the radius of bubble intersecting the top wall of the cylinder. Furthermore, as the vibration frequency decreases, the vibration amplitudes of the bubble profiles increase. As the rotating speed of the cylinder increases to the order of 100 rpm and higher, the amplitude of vibration due to the vibrating gravity environment becomes insignificant because the centrifugal force dominates over the fluctuation of the microgravity force.

ACKNOWLEDGEMENT

The authors appreciate the support received from the NASA/Marshall Space Flight Center through the NASA Grant NAG8-035. The authors would like to express their gratitude to Richard A. Potter of NASA/Marshall Space Flight Center for the stimulative discussions during the course of the present study.

REFERENCES

1. Y. Kamotani, A. Prasad, & S. Ostrach, "Thermal Convections in an Enclosure due to Vibrations Aboard a Spacecraft", *AIAA J.*, **19**, 1981, pp. 511-516.
2. V. S. Avduyevsky, (ed.), **Scientific Foundations of Space Manufacturing**, MIR, Moscow, 1984.

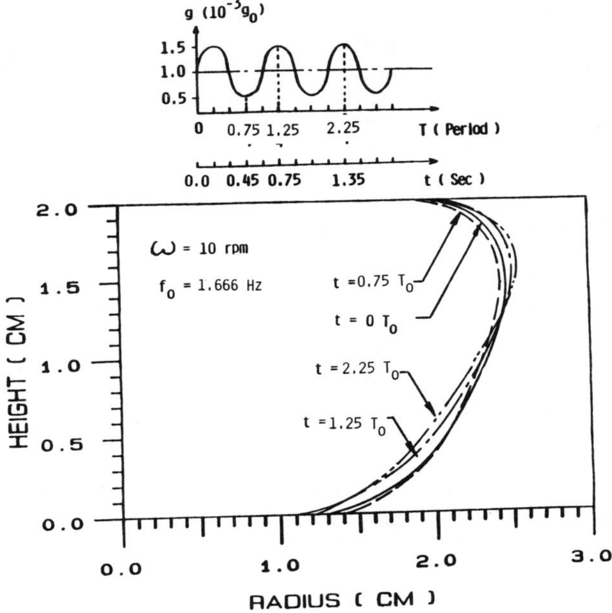
g $(10^{-3}g_0)$
1.5
1.0
0.5
0 0.75 1.25 2.25 T (Period)
0.0 0.45 0.75 1.35 t (Sec)
2.0
1.5
1.0
0.5
0.0
ω = 10 rpm
f_0 = 1.666 Hz
t = 0.75 T_0
t = 0 T_0
t = 2.25 T_0
t = 1.25 T_0
HEIGHT (CM)
0.0 1.0 2.0 3.0
RADIUS (CM)

Fig. 7

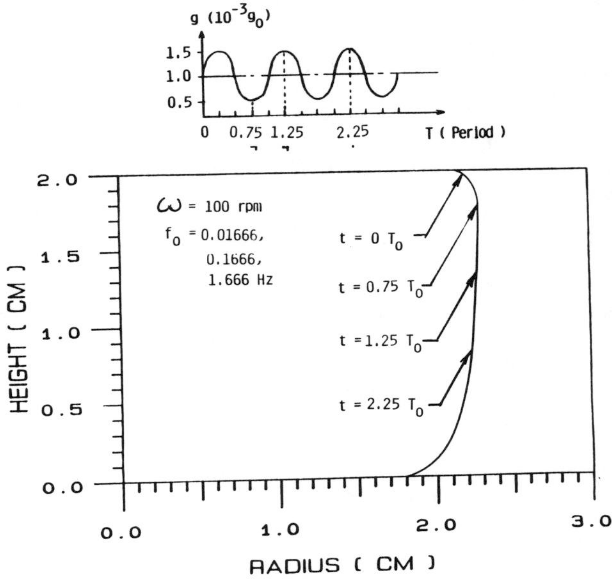
g $(10^{-3}g_0)$
1.5
1.0
0.5
0 0.75 1.25 2.25 T (Period)
2.0
1.5
1.0
0.5
0.0
ω = 100 rpm
f_0 = 0.01666,
0.1666,
1.666 Hz
t = 0 T_0
t = 0.75 T_0
t = 1.25 T_0
t = 2.25 T_0
HEIGHT (CM)
0.0 1.0 2.0 3.0
RADIUS (CM)

Fig. 8

3. R. J. Hung & F. W. Leslie, "Bubble Shapes in a Liquid-Filled Rotating Container Under Low Gravity", *J. Spacecr. Rockets*, **25**, 1988, pp. 70-74.
4. R. J. Hung, Y. D. Tsao, B. B. Hong, & F. W. Leslie, "Dynamical Behavior of Surface Tension on Rotating Fluids in Low and Microgravity Environments", *Int. J. Micrograv. Res. App.*, 1988, in press.
5. R. J. Hung, Y. D. Tsao, B. B. Hong, & F. W. Leslie, "Bubble Behaviors in a Slowly Rotating Helium Dewar in Gravity Probe-B Spacecraft Experiment",*J. Spacecr. Rockets*, **25**, 1988, in press (JSR Log A8236).
6. "Stanford Relativity Gyroscope Experiment (NASA Gravity Probe B)", *Proc. Soc. Photo-Opt. Instrum. Eng.*, **619**, 1986, pp. 1-165.
7. R. J. Hung, Y. D. Tsao, B. B. Hong, & F. W. Leslie, "Effect of Surface Tension on the Dynamical Behavior of Bubble in Rotating Fluids Under Low Gravity Environment", *Adv.Space Res.*, **27**, 1988, in press.
8. F. W. Leslie, "Measurements of Rotating Bubble Shapes in a Low Gravity Environment, *J. Fluid Mech.*, **161**, 1985, pp. 269-279.

FIGURE CAPTIONS

1A. Flow chart for the procedures of computation for numerically determining time-dependent evolution of the bubble shapes.

1B. Continuation of Flow chart for the procedures of computation for numerically determining time-dependent evolution of the bubble shapes.

2. Time-dependent evolution of bubble shapes with sinusoidal function vibration of gravity environment with vibration frequency of 0.01666 *Hz*, rotating speed of cylinder at 1 rpm and background gravity environment of 10^{-3}g.
3. Time-dependent evolution of bubble shapes with sinusoidal function vibration of gravity environment with vibration frequency of 0.1666 *Hz*, rotating speed of cylinder at 1 rpm and background gravity environment 10^{-3}g.
4. Time-dependent evolution of bubble shapes with sinusoidal function vibration of gravity environment with vibration frequency of 1.666 *Hz*, rotating speed of cylinder at 1 rpm and background gravity environment 10^{-3}g
5. Time-dependent evolution of bubble shapes with sinusoidal function vibration of gravity environment with vibration frequency of 0.01666 *Hz*, rotating speed of cylinder at 10 rpm and background gravity environment 10^{-3}g.
6. Time-dependent evolution of bubble shapes with sinusoidal function vibration of gravity environment with vibration frequency of 0.1666 *Hz*, rotating speed of cylinder at 10 rpm and background gravity environment 10^{-3}g.
7. Time-dependent evolution of bubble shapes with sinusoidal function vibration of gravity environment with vibration frequency of 1.666 *Hz*, rotating speed of cylinder at 10 rpm and background gravity environment 10^{-3}g.
8. Time-dependent evolution of bubble shapes with sinusoidal function vibration of gravity environment with vibration frequency of 0.01666, 0.1666, and 1.666 *Hz*, rotating speed of cylinder at 100 rpm and background gravity environment 10^{-3}g.

COMPUTATION OF DROP PINCH-OFF AND OSCILLATION

T.S. Lundgren
Department of Aerospace Engineering and Mechanics
University of Minnesota, Minneapolis, MN 55455

N.N.Mansour
Ames Research Center, Moffett Field, CA 94035

ABSTRACT

Computations of large amplitude motion of free drops in zero gravity have been performed by an inviscid boundary integral method. Impulsive forces were applied to a body of liquid causing it to disintegrate into smaller drops. The break-up process is studied in detail. We find that a narrow throat occurs between the main body of liquid and the developing droplet. When this throat is small we numerically cut off the droplet and follow its motion. It is found that fairly large oscillations of the droplet are caused by a jet of fluid which squirts into the droplet from the main body.

1. INTRODUCTION

Drops created by acoustic focusing (Elrod *et al.*[1]) or by forced vibration of jets (Chaudhary & Redekopp[2]) are observed to oscillate after pinch-off. Little is known about the nature of these oscillations. The objective of this work is to study the details of both the pinch-off process and the resulting drop oscillations by numerical means. The numerical method we will use is an inviscid boundary integral vortex method which treats surface tension accurately. This was described in detail by Lundgren and Mansour[3] for drop oscillation problems, and is summarized in Section 2. The specific drop production mechanism, which will be studied, is previewed in Figure 1, in which a free spherical drop is given an initial impulse in such a way that the ends are driven in opposite directions. Following through the frames in Figure 1, we see that the impulse causes the drop to become very elongated and jet-like, and small daughter drops form at the ends. We cut these off numerically as suggested by Fromm[4], and continue with the computation. Two more drops begin to form at the ends. When a larger initial impulse is applied, the jets become longer and thinner, the droplets which form are smaller, and more drops can be formed before this process is exhausted.

In this paper we wish to focus attention on a single drop as it forms, on the details of the necking down of the throat, and on the resulting oscillations of the drop.

Lamb[5] (Sect. 275) presents Rayleigh's[6] linearized solution for small "vibrations of a globule". The axially symmetric form of the solution is the superposition of modes of the following form,

$$r = a + \varepsilon_n P_n(\cos\theta)\sin(\omega_n t + \eta) \tag{1.1}$$

for the surface shape, and

$$\varphi = -\frac{\omega_n a}{n}\left(\frac{r}{a}\right)^n \varepsilon_n P_n(\cos\theta)\cos(\omega_n t + \eta) \tag{1.2}$$

for the velocity potential inside the drop, with the frequencies ω_n given by

$$\omega_n = n(n-1)(n-2)\frac{T}{\rho a^3} \tag{1.3}$$

where a is the unperturbed radius of the drop, T is the surface tension and $P_n(\cos\theta)$ is a Legendre polynomial (θ is the polar angle). Decomposition of the drop motion into its linear modes will be employed in section 3 as a diagnostic tool.

2. NUMERICAL METHOD

We will treat the liquid motion as an incompressible, irrotational, axially-symmetric flow constrained by surface tension. The density of the outside air will be completely neglected compared to the density of the liquid.

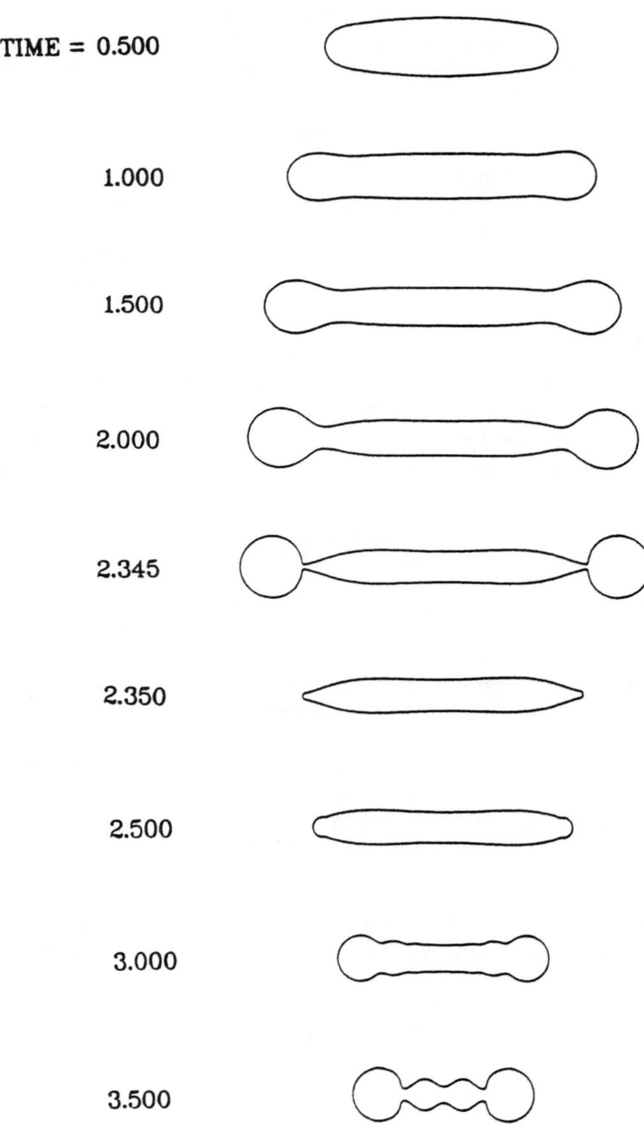

Figure 1

The flow can be represented by singular dipole solutions of Laplace's equation distributed over the deforming interface. This is equivalent to generating the motion by a smooth distribution of vortex rings on the surface. This boundary integral vortex method, due to Baker, Meiron, and Orszag[7,8,9], will be described briefly below. More details may be found in Lundgren and Mansour[3].

A numerical grid of Lagrangian tracer points is introduced on the moving surface. Because of the dipole integral representation the entire computation takes place on this surface (actually on a meridional line on the surface because of axial symmetry). One can compute the velocity **u** at each grid point **r** on the surface from knowledge of the dipole density at all of the grid points by evaluating surface integrals. Each point may be evolved in time by

$$\frac{d\mathbf{r}}{dt} = \mathbf{u} \tag{2.1}$$

to obtain the changing shape of the interface. As the shape changes, the dipole density must change in such a way that the pressure difference across the interface is balanced by surface tension. Upon using the Bernoulli equation for the pressure, and the Laplace formula for the surface tension force, we find an evolution equation for the velocity potential ϕ on the surface following the motion of a grid point

$$\frac{d\phi}{dt} = \frac{1}{2}\mathbf{u}\cdot\mathbf{u} - \frac{T}{\rho}\left(\frac{1}{R_1} + \frac{1}{R_2}\right) \tag{2.2}$$

where R_1 and R_2 are the principal radii of curvature of the surface. These are computed from the local shape of the surface. The remaining parameters are the surface tension T, and the mass density ρ. This equation describes the evolution of ϕ as a function of the surface shape. The dipole density μ may be determined from ϕ by solving an integral equation:

$$\phi(\mathbf{r}) = \frac{1}{2}\mu(\mathbf{r}) + P.V. \int_S \mu' \frac{\partial g(\mathbf{r},\mathbf{r}')}{\partial n'} dS' \tag{2.3}$$

where $g(\mathbf{r},\mathbf{r}')$ is the Green function solution of Laplace's equation in an infinite domain.

To solve for the shape of the free surface one must solve Equations (2.1) and (2.2) for each point on the surface. The numerical strategy is as follows. Given μ and the shape of the surface at the beginning of a time step, one first computes the velocity on the surface. Then (2.1) and (2.2) are used to update the surface points and the potential. In order to determine μ to begin the next time step, one must solve Equation (2.3) with the just determined φ as input. This is a Fredholm integral equation of the second kind and may be solved by iteration.

For the impulsive problem we will study in this paper the jet-like motion of the liquid tends to separate the nodes, giving poor resolution to the evolving drop. Therefore we remesh the integration range after each time step, using cubic splines to interpolate equally spaced node positions and the velocity potential values at these nodes. When regions of high curvature appear we modify this procedure so that the distance between nodes is inversely proportional to curvature in these regions.

All lengths have been made dimensionless with l_0, a characteristic dimension of the drop. Velocities have been made dimensionless with

$$\upsilon_0 = (2T/\rho l_0)^{1/2} \tag{2.5}$$

time by l_0/υ_0.

3. RESULTS

The initial conditions for Figure 1 were taken from the Rayleigh linear mode functions with $r = 1$ and velocity potential $\varphi = \varepsilon P_n(\cos\theta)$ with $\varepsilon = 1.5$. As the column of fluid expands outward droplets begin to form at the ends. Figure 1 shows the expanding jet at the dimensionless times 2.345 and 2.350 just before and just after cut off of two daughter drops. We continue the computation with the mother drop by remeshing the node positions and the velocity potential values, adding additional nodes to keep the total number the same as before cut off. In Figure 1, we have followed a sequence of views of the main body of the drop, discarding the separated droplets. In Figure 2 we repeat that computation up to the time of cut off but then we discard the mother drop and follow the motion of one of the daughters. In this example, the cut off was forced when the throat radius was 0.6 times the mean node spacing, but 0.5 or 0.7 give almost identical results. At cut off, we displaced the node next outward from the throat to the axis along with the value of the velocity potential and discarded all the nodes to the left of this. All the remaining droplet nodes and potential values remained the same. The computation was started with 121 nodes. At the cut off time, only 25 remained on the droplet so we increased the number to 61 by using cubic spline interpolation. Then we simply restarted the computation with these node and velocity potential values as initial conditions.

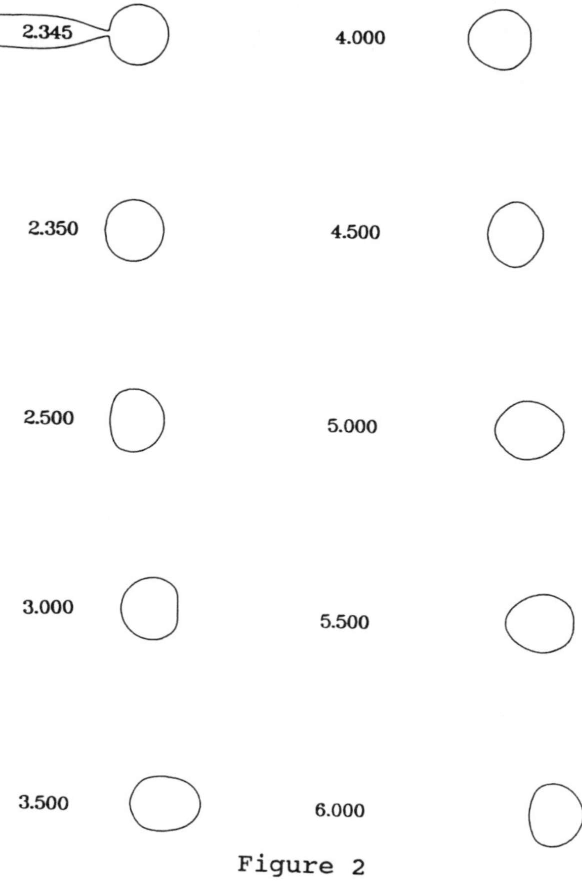

Figure 2

Lamb[5] shows that any potential flow can be regarded as starting from rest under the action of an impulsive pressure applied on the boundary. This impulsive pressure is the negative of the velocity potential times the mass density. We can therefore regard the disconnected droplet as being impulsively set into motion with the impulsive pressures which existed just before the cut off.

In Figure 2, one should note that the drop looks very spherical just before and just after cut off, but still develops fairly large amplitude oscillations. In order to explain this, we have computed vector plots to show the velocity field before and after cut off. In Figure 3,at two time steps before cut off, we see that there is actually a flow through the throat during the collapse, with a maximum velocity just above the throat which is about four times the velocity in the upper part of the droplet. This is partly a Bernoulli effect, due to the flow through the throat, and partly because liquid is being squeezed out of the throat in both directions. In the second part of Figure 3, in the daughter drop, comparably high velocities are seen at points near the bottom of the drop. The conclusion to be drawn is that during pinch-off, a jet of fluid is squirted into the droplet, and it is the jet-momentum which causes the drop to oscillate, as if a spherical drop were started from rest by a fairly concentrated impulsive pressure applied at one end. We also computed the pressure in the throat to be sure that it did not become large as the throat diameter tended to zero. In fact, it was always slightly smaller than the outside pressure.

In order to explore the oscillations in some detail, we have decomposed the drop shape into its linear modes according to

$$r = r_0 + \sum_{n=0}^{\infty} C_n(t) P_n(\cos\theta) \tag{3.1}$$

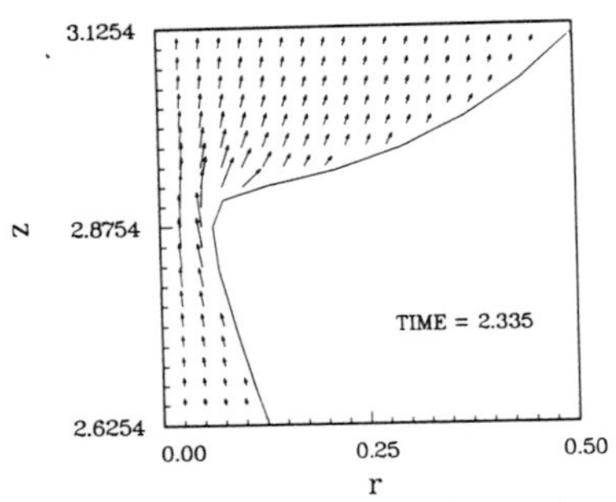

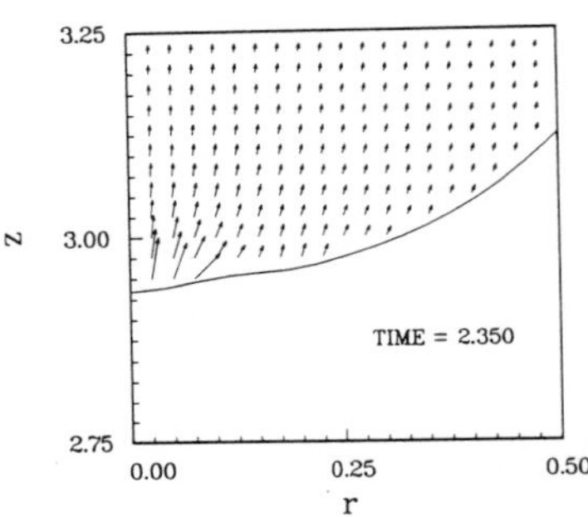

Figure 3

where r is the radius measured from the center of gravity of the drop to a surface point with polar angle θ and r_0 is the radius of a sphere of the same volume. Using orthogonality of the Legendre polynomials, the mode amplitudes C_n are given by

$$C_n = (n + \tfrac{1}{2}) \int_{-1}^{1} (r - r_0)\, P_n(\cos\theta)\, d\cos\theta \tag{3.2}$$

These quantities have been calculated as functions of time by computing the integrals.

Figure 4 shows a time plot of the displacement D of the point on the right end of the drop from the equilibrium radius r_0. Figure 5 shows the amplitudes of modes one to six, normed by D_{max}. For this computation r_0 = 0.662 and D_{max} = 0.151. The relative amplitude is D_{max}/r_0 = 0.228. Clearly the oscillations are dominated by the second mode, but the third and fourth modes also make a considerable contribution.

In Figure 6, the maximum amplitudes of modes one to ten are plotted vs. mode number to form a mode signature. This gives a concise description of the oscillations. Any oscillations with the same signature and the same relative amplitude would look about the same. We have computed drop oscillations initiated with the impulse parameter e between 1.25 and 2.5, all of which result in nearly identical signatures. As the impulse parameter increases, r_0 and D_{max} decrease. At ε=2.5, for instance, r_0=0.531, D_{max}= 0.103 and the relative amplitude is 0.194.

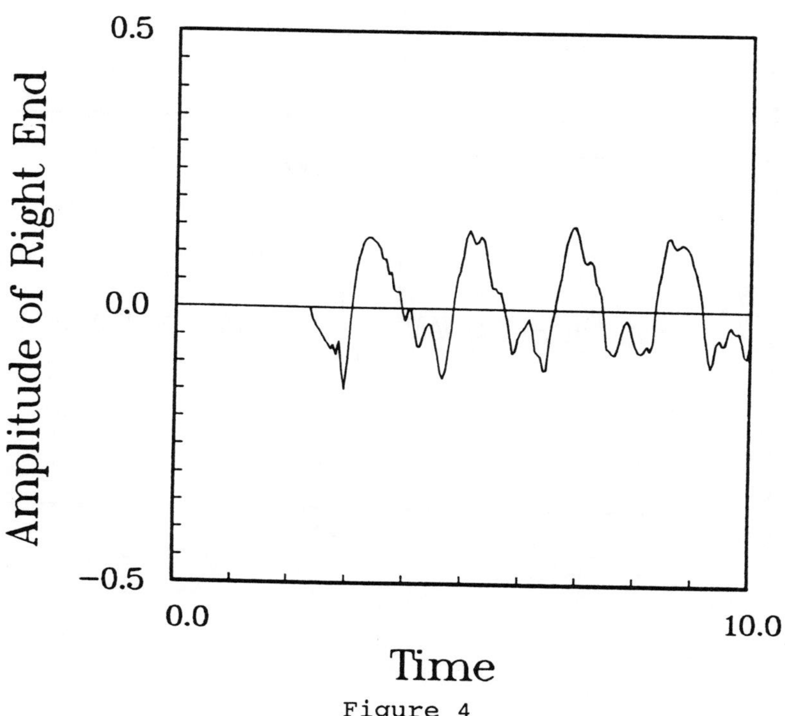

Figure 4

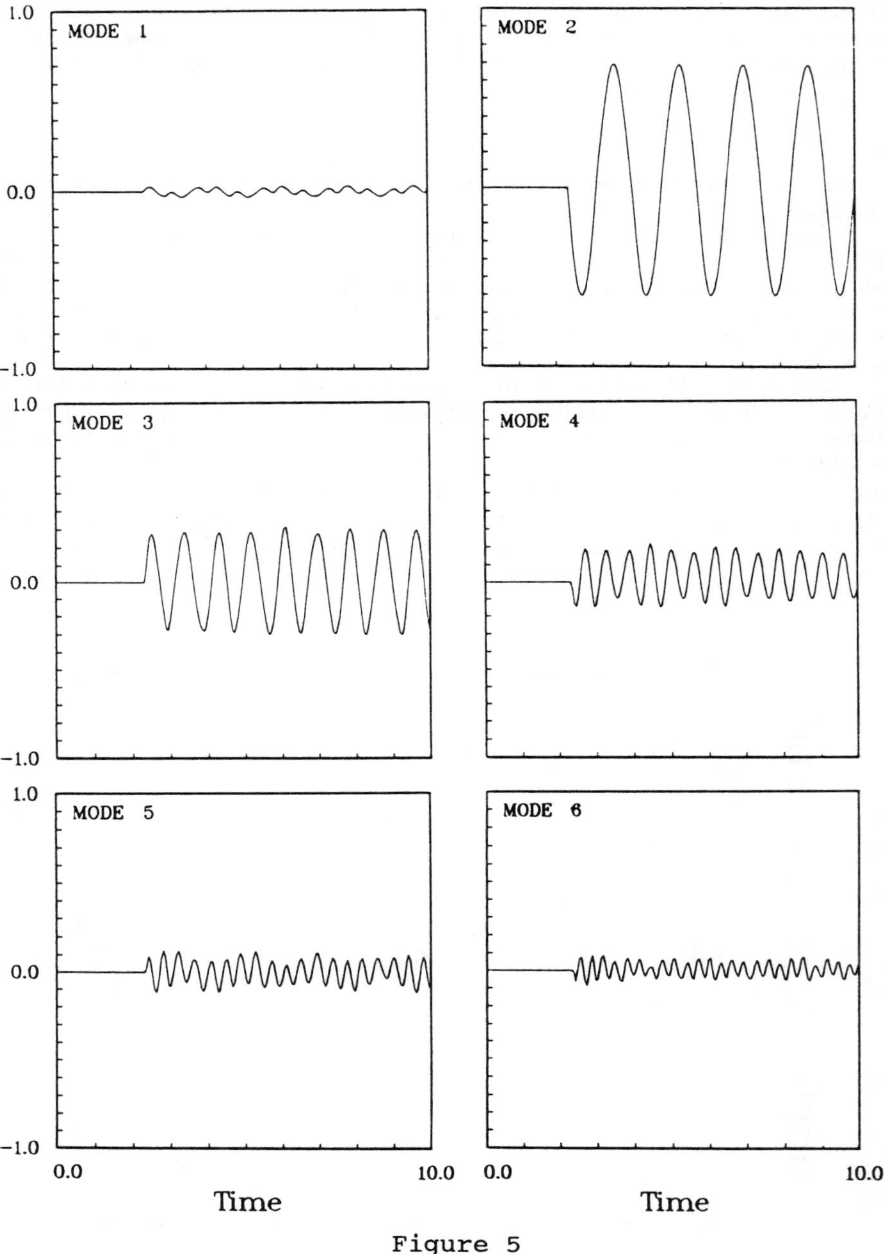

Figure 5

A typical run long enough to produce Figures 4, 5 and 6 takes about three minutes on the CRAY XMP.

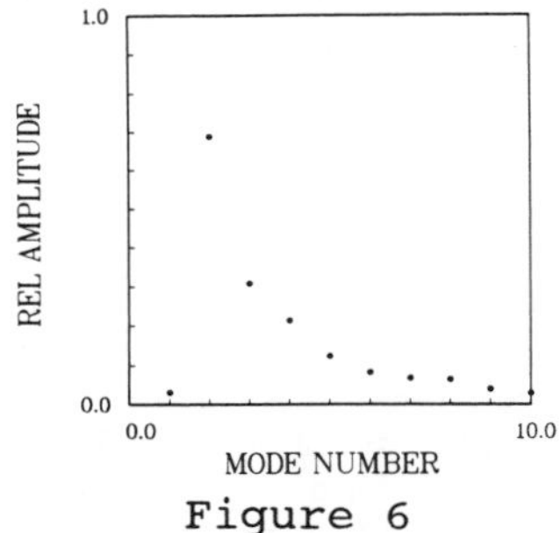

Figure 6

ACKNOWLEDGEMENT

This work was supported by NASA-Ames Research Center under the Directors Discretionary Fund, Contract Number NCA2-330.

REFERENCES

1. S. A. Elrod, B. Hadimioglu, B.T. Khuri-Yakub, E. G. Rawson, E. Richley, C. F. Quate, N. N. Mansour, & T. S. Lundgren, "Nozzleless Droplet Formation with Acoustic Beams", **these Proceedings** , chapter 1 (1989).
2. K. C. Chaudhary & L. G. Redekopp, "The Nonlinear Capillary Instability of a Liquid Jet, *J. Fluid Mech.* **96**, 257-274 (1980).
3. T. S. Lundgren & N. N. Mansour, "Oscillations of Drops in Zero Gravity with Weak Viscous Effects", *J. Fluid Mech.* **194**, 479–510 (1988).
4. J. E. Fromm, "Numerical Calculation of the Fluid Dynamics of Drop-on-Demand Jets", *IBM Journ. Res. Dev.* **28**, 322-333 (1984).
5. H. Lamb, **Hydrodynamics**, 6th ed. (Cambridge Univ. Pr., Cambridge, Eng., 1932)
6. J. W. S. Rayleigh, "On the capillary phenomena of jets", *Proc. R. Soc. London* **29**, 71-97 (1879).
7. G. R. Baker, D. I. Meiron, & S. A. Orszag, "Vortex simulations of the Rayleigh-Taylor instability", *Phys. Fluids* **23**, 1485-1490 (1980).
8. G. R. Baker, D. I. Meiron, & S. A. Orszag, "Generalized vortex methods for free-surface flow problems, *J. Fluid Mech.* **123**, 477-501 (1982).
9. G. R. Baker, D. I. Meiron, & S. A. Orszag, "Boundary Integral Methods for Axisymmetric and Three-Dimensional Rayleigh-Taylor Instability Problems, *Physica* **12D**, 19-31 (1984).

FIGURE CAPTIONS

Figure 1. Shapes of drop impulsively set into motion from spherical with initial velocity potential $\varphi = 1.5\, P_2\,(\cos\theta)$. Daughter drops which are formed at the ends are numerically cut off and discarded.

Figure 2. Motion of one of the daughter drops produced by the impulsive flow of Figure 1.

Figure 3. Velocity field in the drop before and after cut off.

Figure 4. Time trace of the amplitude of the oscillatory motion of the outward end of the daughter drop.

Figure 5. Time traces of six Legendre modes, normed by the maximum amplitude from Figure 4.

Figure 6. Mode signature of the daughter drop.

GEOMETRIC STATISTICAL MECHANICS FOR NON-SPHERICAL BUBBLES AND DROPLETS

S.P. Marsh and M.E. Glicksman
Materials Engineering Department,
Rensselaer Polytechnic Institute, Troy, New York 12180

ABSTRACT

The behavior of many dispersed two-phase systems, such as bubbly flows and droplet mists, are strongly affected by interfacial effects. A statistical approach is presented here which may be used to model the global behavior of such geometrically complex systems. This approach describes the interfaces as a distribution of differential surface patches, each having two local, independent principal curvatures. Statistically-averaged (mean-field) physics describing the appropriate phenomena are formulated for the individual patches. These physics can include the volume-fraction effects of the dispersed phase, as well as flow conditions and interfacial phenomena. The resulting local patch dynamics are then used in a distributional analysis to predict the ensemble behavior and evolution of the dispersed system. Application of this approach to diffusive coarsening in solid-liquid systems will be described briefly, along with suggestions of applications to other two-phase systems of interest.

INTRODUCTION

There is much ongoing interest, both commercial and scientific, in the behavior of finely-divided multiphase systems. The nature of interfacial effects and the geometry of the phases are essential in determining the macroscopic properties and responses of these systems. Examples of physical phenomena in multiphase systems include heat transfer in gas-liquid mixtures, droplet dispersion and evaporation/condensation in aerosols, and flows in porous media.

Many studies have been done to explore the nature and behavior of single bubbles and droplets relative to their surroundings. Rallison[1] has reviewed observations and analysis of droplet and bubble deformation in shear flows, while a discussion of bubble dynamics and cavitation is provided by Plesset and Prosperetti[2]. The goal of such investigations is to relate the observed behavior of a bubble or droplet to the various physical factors present in its environment, such as flow fields, viscosities, and variations in temperature, pressure or concentrations.

The behavior of large collections of bubbles or droplets, such as in aerosols, emulsions, and bubbly and foamy flows, is further complicated by interactions among the dispersed elements. A simple weighting of the phase properties of interest by mass- or volume-fraction cannot account for the local dynamics of phase changes and interfacial effects, yet it is these phenomena which often govern the observed macroscopic properties and behavior of dispersed systems. This is particularly true in foam flows (Kraynik[3]), where the foamy structure is made up of thin liquid films and the volume fraction of this continuous phase is small.

The purpose of this paper is not to present formulations for specific phenomena, but instead to provide the framework for a tractable description of dispersed systems that bridges the gap between the hydrodynamic and macroscopic levels. The approach used is statistical in nature and provides macroscopic, "average" properties that are self-consistently derived from local effects occurring on the hydrodynamic level. This situation is analogous to that of the classical "ideal gas," the mean pressure of which is determined through the velocity (momentum) distribution of the individual particles. In the present case, however, the systems are not necessarily at equilibrium, and the geometry of the interfaces (their curvature and shapes) can influence the local interactions.

THEORY

The present work represents a statistical mean-field approach which is not limited to spherical or near-spherical bodies. The fundamental unit of this model is a small patch of surface area which has two independent principal curvatures. The geometric variables associated with a general patch are defined in the following section, and the relationships among these variables as the surface patch undergoes a displacement is presented. Formulation of the mean dynamical interaction of an interfacial patch relative to its environment is then discussed. Finally, the application of distributional analysis to these local results yields ensemble properties and dynamical behavior of the system of interest.

GEOMETRIC VARIABLES

A general surface patch is shown by the dark-shaded area in Figure 1. The geometry of the patch is characterized by the principal radii of curvature R_1 and R_2. These principal radii are always orthogonal at every point on a general surface or interface. The patch is small enough so that R_1 and R_2 remain essentially constant over the surface. The extensive size of the patch is described by the angles θ and ϕ through which R_1 and R_2 are rotated to form the patch. These angles will in general be differential angles, and their product is defined to be ω, the differential solid angle associated with the patch. Thus, R_1 and R_2 are two independent geometric labels describing the intrinsic character of the patch, whereas ω is the extensive variable which quantifies the "amount" of the patch.

It is more convenient for analysis to describe the geometry of the patch in terms of a dimensionless shape parameter and a single lengthscale. We thus define the shape factor s as

$$s \equiv \frac{R_2 - R_1}{R_2 + R_1}, \tag{1}$$

where $R_1 < R_2$. The shape factor ranges from $s = 0$ for spheres ($R_1 = R_2$) to $s = 1$ for cylinders ($R_2 >> R_1$). Note that this shape factor bears some resemblance to the dimensionless deformation, D, used by Taylor[4] to characterize deformed droplets. However, s is a <u>local</u> descriptor, whereas D describes the shape of an entire deformed droplet in terms of the major and minor axes.

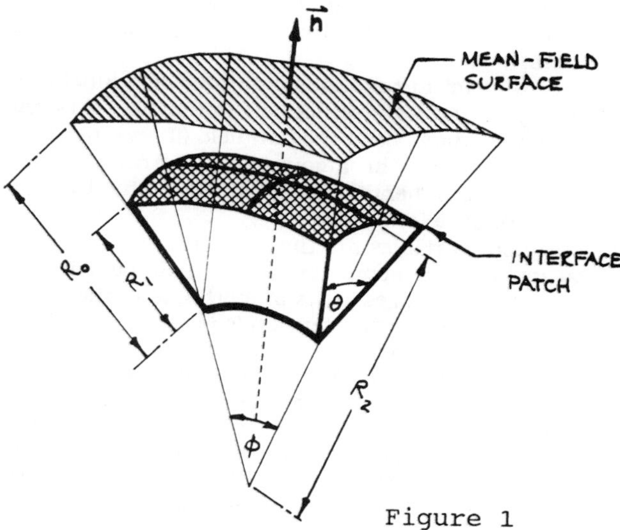

Figure 1

The reciprocal lengthscale used to characterize the patch is the mean curvature, H, defined as

$$H = \frac{1}{2}\left(\frac{1}{R_1} + \frac{1}{R_2}\right) = \frac{1}{R_1(s+1)} \; . \tag{2}$$

The area of a patch, A, and the volume of the associated curved wedge, V (see Figure 1), can be written as

$$A = \frac{1+s}{1-s} R_1^2, \qquad \text{and} \qquad V = \frac{1+2s}{3(1-s)} R_1^3 \tag{3}$$

The curved wedge in Figure 1 is chosen as the volume associated with the surface patch because it reduces to the exact associated volume for the limiting cases of spheres ($s = 0$) and cylinders ($s = 1$).

The variable R_1 represents a single length scale within each class of geometrically similar surface patches. Note from the above equations that, for a fixed value of s, the mean curvature, length, area, and volume of a patch vary as $1/R_1$, R_1, R_1^2, and R_1^3, respectively. The differential changes in these quantities for a given change in R_1 (at constant s) can be written simply as

$$dH = \frac{-1}{1+s} R_1^{-2} \, dR_1 , \tag{4}$$

$$dA = 2 \, \frac{1+s}{1-s} R_1 \, dR_1 , \qquad \text{and} \qquad dV = \frac{1+2s}{1-s} R_1^2 \, dR_1 , \tag{5}$$

where dR_1 is an arbitrary differential change in R_1.

In deriving equations (4) and (5), the assumption is made that each surface patch retains its s-value as it evolves, and only the lengthscale R_1 (or H) changes. Although the s-value may not remain constant locally as the surface evolves, this approximation simplifies the resulting analysis greatly and accounts for the dynamical differences related to the size (or curvature) scale of patches having different shapes. A more general description can be made by using a total derivative of Equations (2) and (3). This would then require simultaneous specification of the variations ds and dR_1, which may or may not be correlated in real dynamical processes.

DYNAMICS

The formulation of an appropriate statistically-averaged dynamical expression describing the behavior of an interface patch, which interacts with its surroundings, will vary with the particular physical system and conditions under consideration. For this reason, specific formulations will not be presented here; instead, the factors which must be considered in forming the appropriate dynamical relations will be discussed.

The purpose of the dynamical equations is to provide a kinetic relationship between the interfacial region and its "average" environment. This relationship may involve viscous effects and turbulence in the flow field (if present), heat and/or mass transfer, gravity, and, of course, the amount and distribution of the dispersed phase(s) present. These environmental effects will determine the rate at which the interface will grow, shrink or deform locally.

The primary dependent variables are the geometric labels of the patch, H and s. The mean curvature, H, represents the differential coefficient dA/dV, which relates local area and volume changes. It is proportional to the pressure difference across the curved interface if the surface tension is isotropic, which is generally true in fluid-fluid systems. Related phenomena include the local supersaturation in binary systems, vapor pressure elevation in liquid-gas systems, and melting-point depression in solid-liquid systems; the magnitude of these deviations from equilibrium values are all proportional to H. The mean curvature thus accounts primarily for the surface tension (or energy) effects relative to the bulk phases.

The shape factor, s, indicates the deviation from sphericity of the patch. As a spherical interface minimizes the surface area and associated energy for a given volume, a shape factor greater than zero indicates the presence of other forces (viscous, gravitational, etc.) supporting the deviation. This introduces a possible orientation factor which relates to the geometry of the dispersed phase. For instance, droplets in a shear flow would tend to deform along the flow lines; surface patches with normals parallel to the flow direction would tend to be more spherical, while the interface patches along the sides of the elongated droplet, with normals perpendicular to the flow, would tend to be more cylindrical, with s-factors approaching unity. The presence of orientation effects can be included through appropriate consideration of the system of interest.

The purpose of the dynamical relations is to quantify the mean response of a geometric "class" of interface (having given values of H and s) to the "average" environment that the patch interacts with. This may involve the self-consistent definition of a "mean-field" which represents an ensemble property of the system, such as pressure or concentration. An example involving a "mean-field" temperature in a coarsening solid-liquid system will be presented later.

DISTRIBUTIONAL ANALYSIS

The continuity equation describes how a distribution of patches will evolve in time. It relates changes in the interfacial curvature distribution to the dynamical behavior of the patches described in the previous section. For a given s-value, the surface elements may have any H-value. The density function $\Omega(H,t)$ is defined for each s-value as the total solid angle, per unit volume of the microstructure, of all patches having a mean curvature between H and $H+dH$, at time t. It can also be related to the surface-area distribution through the parameters H and s, if desired. $\Omega(H,t)$ represents the time-dependent density function for the geometric variables H and s corresponding to the physical interfaces. Note that the number density function for a collection of spheres equals the total solid angle Ω divided by 4π (since there are 4 steradians per sphere); however, Ω is the appropriate dimensionless "counting" variable for arbitrary phase shapes.

The distribution function $\Omega(H,t)$ for a fixed s-value satisfies the continuity equation

$$\frac{\partial \Omega}{\partial t} + \frac{\partial}{\partial H}(\dot{H}\Omega) = 0, \tag{6}$$

with $\dot{H} \equiv dH/dt$. The $\dot{H}$-term is supplied by the appropriate dynamical expression, and may be a function of time. Scaling relations and/or appropriate initial conditions should then be included in the continuity equation. The resulting equation will describe the curvature evolution at a fixed s-value. One can also allow for variations in s, leading to a partial differential equation describing the simultaneous evolution of both the curvature and the shape factor.

APPLICATION TO SOLID-LIQUID COARSENING

Coarsening in finely-divided solid-liquid systems is driven by a reduction in the interfacial area between the phases and the associated excess free energy. Locally, the curved interface causes a depression in the melting point (pure materials) or a supersaturation of solute (alloys). The extent of these effects is proportional to the mean curvature of the interface. The resulting gradients between the various interfaces excite fluxes of heat or solute, which lead to a reduction in the average interfacial curvature and total surface area.

Statistical modeling of diffusion-controlled coarsening systems was first formulated by Lifshitz and Slyozov[5], who considered the case of spherical bodies in the limit of zero volume fraction. Each sphere interacts with a "mean field," which represents the average effect of the rest of the system and is determined self-consistently to maintain a constant total volume of solid. The dynamical equation was obtained by solving the Laplace equation between the surface of a sphere and the "mean field," located an infinite distance away (the dilute limit) to determine the flux between a sphere and the "average" environment. Applying the continuity equation (6) to the distribution of radii leads to a time-independent size distribution when each radius is scaled by the average radius. The average radius (scale factor) was shown to grow as the cube-root of time, at a scaled rate of 4/9. Numerous attempts have been made to modify the mean-field dynamics (generally by modifying the interface-to-mean-field distance) to account for the finite volume fractions which occur in real systems. These theoretical approaches have been reviewed recently by Voorhees[6], again for spherical bodies.

Marsh *et al.*[7] have used the present geometric approach to extend this analysis to non-spherical bodies in non-dilute systems. The results indicate that the average mean curvature, H^*, decays as the cube-root of time (analogous to the growth of the average spherical radius). The curvature distributions become time-independent when scaled by this average curvature, and were found to be virtually independent of the shape factor, *s*. They do broaden somewhat with increasing volume fraction of the dispersed phase. These asymptotic curvature distributions are plotted in Figure 2 (as lengthscales, or reciprocal curvatures) for different volume fractions of the solid phase. Each distribution has a maximum lengthscale equal to 1.5 times the critical lengthscale, as was found in the Lifshitz-Slyozov analysis.

The results of this analysis indicate that, in diffusion-controlled coarsening, the mean undercooling in a solid-liquid mixture of a pure material (which is proportional to the average curvature) will decay as the cube-root of time. The scaled results also predict that the total surface area in alloys will also decay as the cube-root of time, even in geometrically complex systems. Both of these predictions have been confirmed by direct experimental measures (Marsh *et al.*[8], Boettinger *et al.*[9]). Further work relating coarsening rates to material properties is still in progress.

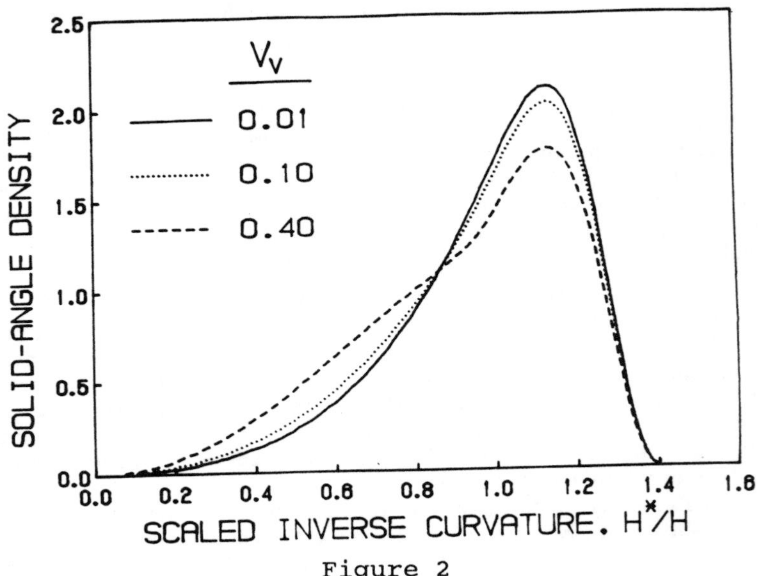

Figure 2

SUMMARY

A statistical approach has been presented which relates interfacial geometry and interactions on the hydrodynamic scale to macroscopic average behavior. This formulation, which is not limited to discrete spherical bodies, can be modified to include orientation effects, viscosity, convection, gravity and other appropriate factors and their interaction with the interface. The resulting dynamics represent the interaction of a local patch of interface with the averaged environment. The continuity equation for the curvature distribution function then describes how a distribution of interfacial curvatures will evolve. For the case of solid-liquid phase coarsening, mean-field interactions between geometrically similar surface patches lead to asymptotic lengthscale distributions which are relatively insensitive to the geometry. The coarsening rate for these classes of surface patches, as measured by the decay of the specific surface area, Sv, show a cube-root of time dependence. This behavior has been verified experimentally.

ACKNOWLEDGEMENT

We wish to acknowledge financial support for this work by the National Science Foundation, Division of Materials Research, under grant DMR86-11302.

REFERENCES

1. J. M. Rallison, *Ann. Rev. Fluid Mech.* **16**, 45 (1984).
2. M. S. Plesset & A. Prosperetti, *Ann. Rev. Fluid Mech.* **9**, 145 (1977).
3. A. M. Kraynik, *Ann. Rev. Fluid Mech.* **20**, 325 (1988).
4. G. I. Taylor, *Proc. R. Soc. London* A**138**, 41 (1932).
5. I. M. Lifshitz & V. V. Slyozov, *J. Phys. Chem. Solids* **19 (1/2)**, 33 (1961).
6. P. W. Voorhees, *J. Stat. Phys.* **38**, 231 (1985).
7. S. P. Marsh, M. E. Glicksman, & D. I. Zwillinger, in **Modeling of Casting and Welding Processes IV** (ed. T. Piwonka & A. F. Giamei, Metallurgical Socy., Warrendale, PA, 1988), in press.
8. S. P. Marsh, M. E. Glicksman, L. Meloro, & K. Tsutsumi, in **Modeling of Casting and Welding Processes IV** (ed. T. Piwonka & A. F. Giamei, Metallurgical Socy., Warrendale, PA, 1988), in press.
9. W. J. Boettinger, P. W. Voorhees, R. C. Dobbyn, & H. E. Burdette, *Met. Trans. A* **18A**, 487 (1987).

FIGURE CAPTIONS

1. Sketch of a general interfacial patch and the associated mean-field surface. The principal radii of curvature, R_1 and R_2, are constant and orthogonal at each point on the patch.
2. Asymptotic, self-similar inverse curvature (lengthscale) distribution obtained from the statistical analysis of solid-liquid phase coarsening. This distribution is virtually independent of the shape factor, s, and broadens with increasing volume fraction of the solid phase, Vv.

AXISYMMETRIC CREEPING MOTION OF DROPS THROUGH A PERIODICALLY CONSTRICTED TUBE

M. J. Martinez[a] and K. S. Udell
Department of Mechanical Engineering
University of California, Berkeley, Berkeley, CA 94720

ABSTRACT

The axisymmetric creeping motion of a neutrally buoyant deformable drop flowing in a tube with periodically varying diameter is analysed with a boundary integral equation method. The undeformed drop radius is comparable to the average tube radius. The fluids are immiscible, incompressible and the suspension flows at constant volume flux. Two tubes are considered. Tube I has a maximum to minimum radius ratio of 1.8 and a maximum radius to wavelength ratio of 0.3. Tube II has a contraction ratio of 3 and the same maximum radius to wavelength ratio. The effects of the capillary number, the drop to suspending fluid viscosity ratio, drop size and the contraction ratio on the extra pressure drop and drop speed are examined.

INTRODUCTION

In this paper we consider the axisymmetric creeping motion of a neutrally buoyant deformable drop flowing in a liquid-filled tube with periodically varying cross-sectional area. Our interest in this problem stems from its usefulness as a model of two-phase flow in a porous material. The rationale behind the use of a sinusoidal tube is to include the effects of strain in the flow which is an essential element of the flow field experienced by a fluid parcel as it follows the tortuous path of the pore space in a porous material. Although an improvement in formulating an appropriate pore model of two-phase immiscible displacement in porous media, the sinusoidal tube is still a highly simplified geometrical representation. It does, however, include important features of the geometry in a real pore space, in particular the converging/diverging geometry prevalent in granular as well as consolidated materials.

The pressure drop requirement for constant volume flux will vary according to the position and shape assumed by the drop. This feature of a porous material is thought to be responsible for trapping of blobs when unable to deform sufficiently to slip through the constricted space. Due to the interconnected nature of the pore space, a feature lacking in the present model, the local conditions in a porous material are probably closer to constant pressure gradient rather than constant volume flux. Hence capillary forces may cause drops to be retained in pores with a large change in cross-sectional area. In the present constant volume flux problem we will not detect entrapment as such, but can only infer its occurrence in systems which develop a large pressure gradient in deforming the drop sufficiently to squeeze through the throat. If the local pressure gradient in an equivalent pore structure is substantially less, the drop will become trapped.

Oh & Slattery (1979) conducted a static analysis to predict the pressure drop, due solely to a difference in leading and trailing curvatures of the drop, for various configurations in a periodic tube. Advancing and receding contact angles were independently chosen based apparently on experimental data. They show large variations in pressure drop for various drop configurations in the tube and associate the maximum value with the minimum pressure drop required to prevent trapping. Similar studies were performed previously by Gardescu (1930) and by Slattery (1974).

[a]*current address:* Fluid & Thermal Sciences Department, Sandia National Labs, Albuquerque, New Mexico 87185

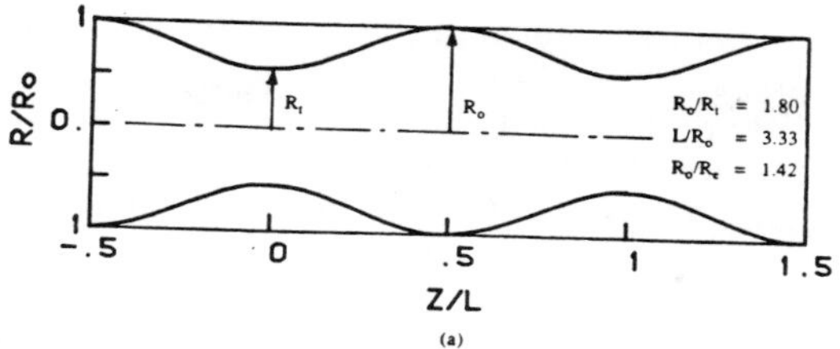

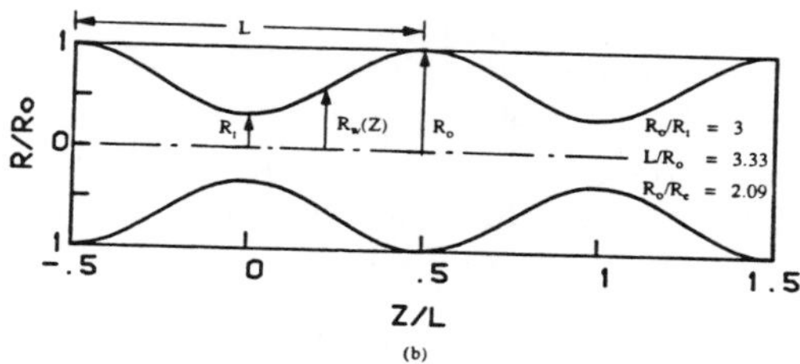

Figure 1

The experimental investigation of most relevance to the present study is that conducted by Olbricht & Leal (1983) for the same flow system considered here. They considered the flow and breakup of drops for Newtonian and viscoelastic fluids in a sinusoidal tube (Tube I of the present study). Other pertinent investigations include the experiments of Han & Funatsu (1978) who considered the deformation and breakup of drops in an abrupt contraction including both Newtonian and viscoelastic fluids. Chin & Han (1979, 1980) considered the corresponding problem, both theoretically and experimentally, in a funnel preceding a uniform tube section. Neither study considered the effects of walls.

In the present study, we consider the axisymmetric creeping motion of a drop with viscosity $\lambda\mu$, translating in a sinusoidal tube filled with another liquid of viscosity μ. The fluids are immiscible and a constant interfacial tension, γ, is exerted along their common interface. The volume of the drop is measured in terms of its undeformed radius, a_o. Both fluids are incompressible and the suspension (including both drop and suspending fluid) flows at constant volume flowrate, Q. The drop speed is not steady, but varies with the location of the drop in the tube. In order to compare the effect of contraction ratio, defined here as the ratio of maximum to minimum tube diameter, two tubes were considered, see Figure 1. The axial variation in wall radius, $R_w(Z)$, of each tube is described by the functional form.

$$R_w = \overline{R} + A \sin 2\pi\frac{Z}{L}, \text{ where, } \overline{R} = \frac{1}{2}(R_o + R_t),\ A = \frac{1}{2}(R_o + R_t),$$

L is the tube wavelength, and R_o and R_t are the maximum and minimum tube radii, respectively. Tube I is the tube used in the experimental investigation of Olbricht & Leal and has a contraction ratio of 1.8 and maximum radius to wavelength ratio of 0.3. Tube II has a contraction ratio of 3 and the same R_o/L.

PROBLEM STATEMENT AND NUMERICAL ANALYSIS

The motion of suspending fluid is governed by the equations of conservation of mass and the Stokes approximation for balance of momentum,

$$\nabla\cdot\mathbf{v} = 0,\quad \mathbf{x}\in\Omega+\Gamma \quad\text{and}\quad -\nabla p + \mu\nabla^2\mathbf{v} = 0,\quad \mathbf{x}\in\Omega, \tag{1}$$

where $\mathbf{v}$ and p are the velocity vector and pressure and Ω denotes the suspending fluid domain with boundary Γ (= $\Gamma_T + \Gamma_B$), including the tube wall and an inlet and outlet plane located in the undisturbed fluid (Γ_T), and the fluid interface (Γ_B). Similar equations govern the drop fluid if we replace velocity, pressure, and viscosity in Equation (1) by $\mathbf{u}$, p_b, and $\lambda\mu$, respectively. These are valid in Ω_B with boundary Γ_B.

The boundary conditions include the no-slip condition on the tube wall and specification of the undisturbed flow field, denoted $V_\infty e_z$, far ahead and behind the drop. Three conditions must be satisfied on the interface. One is again the no-slip condition $\mathbf{v} = \mathbf{u}$. The tangential stress is also continuous on the interface, however, the normal component suffers a jump proportional to the interface curvature modified by the local buoyancy due to the introduction of the dynamic pressure,

$$[\sigma(\mathbf{v}) - \sigma(\mathbf{u})] \cdot \mathbf{n} = \frac{1}{Ca}(\kappa - Bo\ z)\mathbf{n} \qquad (2)$$

where σ is the Newtonian stress tensor, $\mathbf{n}$ is the unit normal to the interface pointing outward from the suspending fluid domain, and κ is the interface curvature. There is also a kinematic condition on the fluid interface requiring that it move in the direction of its normal with the local normal fluid velocity, $(\partial\mathbf{Y}(\mathbf{X},t)/\partial t) \cdot \mathbf{n} = \mathbf{v} \cdot \mathbf{n}$, where $\mathbf{Y}(\mathbf{X},t)$ is the position vector of a point on the interface and $\mathbf{X}$ represents some reference configuration for the drop. The three conditions on the interface do not overspecify the problem since the evolution of the drop shape must also be determined as part of the solution.

In the preceding, the nondimensionalization was accomplished by using an effective radius, R_e, as characteristic length, the average bulk velocity, $V = Q/\pi R_e^2$, as characteristic velocity, $\mu V/R_e$ as characteristic stress (and pressure), and R_e/V as characteristic time. This effective radius was proposed by Franzen (1979) and used by Olbricht & Leal (1983),

$$\left(\frac{\overline{R}}{R_c}\right)^4 = \int_0^1 \frac{d\left(\frac{Z}{L}\right)}{\left(1 + \frac{A}{\overline{R}}\sin 2\pi\frac{Z}{L}\right)^4}$$

Using this definition of effective radius, Olbricht & Leal found that friction factor data collected in tube I coincide precisely with the Hagen-Poiseuille law for laminar flow in a tube of constant diameter. Pertinent parameters for describing the geometry of tubes I and II are included in Figure 1.

The dimensionless parameters appearing in the problem include the capillary number, $Ca=\mu V/\gamma$, which measures the ratio of viscous forces to interfacial tension forces, a Bond number, $Bo = (\rho - \rho_b)gR_e^2/\gamma$, the ratio of buoyancy to interfacial tension forces, the drop to suspending fluid viscosity ratio, λ and the ratio of undeformed drop radius to characteristic tube radius, $a = a_o/R_e$. Finally, the tube geometry specified in terms of the contraction ratio, R_o/R_t, and nondimensional wavelength, L/R_e completes the list of nondimensional parameters in the problem. The list is extensive and would require a prohibitive number of combinations to study the effects of all parameters. Nevertheless, we consider a few variations of relevance to the present investigation of this system as a pore scale model of two-phase flow in porous media.

The problem is time-dependent and nonlinear as a consequence of the boundary conditions on the interface. A major difficulty in the solution of problems involving interfacial phenomena is that the interface shape itself is to be determined as part of the solution. It has been demonstrated that an effective solution method is to reformulate the problem in terms of a boundary integral equation for the unknown velocity and/or stresses on the boundaries of the fluid domain (*e.g.*, Rallison & Acrivos 1978, Lee & Leal 1982). For axisymmetric problems the azimuthal integrations can be performed analytically (Youngren & Acrivos 1975) and the method reduces the problem to solving a pair of integral equations along the generating curve in a meridional plane of the axisymmetric boundary. The integral equation represents the superposition of a distribution of "Stokeslets" on the boundary in such a way as to satisfy the boundary conditions commensurate with the boundary shape.

Details of the construction of the boundary integral equation in terms of the hydrodynamic potentials given by Ladyzhenskaya (1969) follows the development of Rallison & Acrivos (1978) (see also Rizzo & Shippy 1968) and will not be reiterated here, only the resulting boundary integral equations are given for completeness (Martinez 1987):

$$\int_{\Gamma_T}[U_{ki}t_i(v) - T_{ki}V_i]\,d\Gamma + \frac{1}{Ca}\int_{\Gamma_B}U_{ki}n_i(\kappa - Bo\ z))\,d\Gamma - (1-\lambda)\int_{\Gamma_B}T_{ki}v_i\,d\Gamma = \begin{cases} C_{ki}V_i(\mathbf{x}), & \mathbf{x}\in\Gamma_T \\ (1+\lambda)C_{ki}V_i(\mathbf{x}), & \mathbf{x}\in\Gamma_B \end{cases} \tag{3}$$

where $\mathbf{U}(\mathbf{x},\mathbf{y})$ and $\mathbf{T}(\mathbf{x},\mathbf{y})$ are velocity and traction kernals derived from the Stokeslet, and $\mathbf{C}$ includes the principal value of $\mathbf{T}$ as $\mathbf{x}\rightarrow\Gamma(\Gamma_B)$ from $\Omega(\Omega_B)$. $\mathbf{t}(\mathbf{v})$ are boundary tractions, and $\mathbf{V}$ denotes the no-slip condition on the tube wall and undisturbed flow field far from the drop. The boundary integral equations include all the boundary conditions except for the kinematic condition which is left to determine the evolution of the drop shape. The unknowns are the stresses on the tube and surface velocity on the interface.

The numerical solution of the boundary integral equation involves discretizing the integral operators, using finite element techniques, into a linear system to be solved for the unknown velocities and tractions at node points on the boundary for a given drop shape. The boundary geometry and the boundary velocity and tractions are approximated by a finite-dimensional set of basis functions, each composed of polynomials defined on the boundary. The full moving-boundary problem is solved by specifying an initial drop shape, thereby allowing the specification of Γ_B, and solving the linear system for the corresponding stresses on the tube wall and surface velocities on the interface. Repeated substitution of the interface velocities obtained from the integral equation into the kinematic condition generates an initial value problem for the history of the drop shape. The resulting system of ordinary differential equations for the evolution of the coordinates on the interface is solved with a variable-step algorithm which automatically selects the timestep such that a user-specified local truncation error tolerance is maintained. In order to maintain a proper discretization, the mesh of nodes for the interface was periodically updated by fitting a cubic spline to the coordinates and redistributing the nodes with spacing inversely proportional to the interface curvature. The drop volume error was monitored at each step and never allowed to exceed 0.75% of the initially specified value. More complete details of the numerical solution can be found in Martinez (1987).

The undisturbed flow field is required in the specification of the far field boundary conditions, $V_\infty(\mathbf{x})$. This function was determined by solving the steady, single phase flow problem. The problem was solved for both tubes I and II by specifying the no-slip condition on the tube wall and a parabolic radial distribution of axial velocity, with volume flowrate Q, on the inlet and outlet faces of a tube of length $2L$. The function V_∞ was determined by computing the interior velocities at various radii for axial locations distant from the inlet and outlet using integral equations similar to Equation (3), but valid in the fluid domain (Martinez 1987). The single phase solution was recomputed, using the previously determined undisturbed velocity field as inlet and outlet boundary conditions, to determine the pressure drop over one tube period for later use in computing the extra pressure drop, ΔP^+, in the two-phase calculations. ΔP^+ is the instantaneous pressure drop in the tube minus the pressure drop that would obtain in the absence of the drop, for the same volume flowrate. In this manner the nondimensional pressure drop per tube wavelength, $\Delta PR_e/\mu V$, was determined to be 41.6 and 64.9 for tubes I and II, respectively. These compare favorably with the corresponding values 42.0 and 64.6 obtained by linear interpolation in Table 2 of Tilton & Payatakes (1984).

DISCUSSION OF RESULTS

For definiteness, the simulations of drop motion were begun with the drop residing in a section of uniform diameter, equal $2R_o$, preceding the periodically constricted section. In this way the drop attains the equilibrium shape corresponding to a tube of uniform diameter before encountering the sinusoid.

The results separate logically according to the magnitude of the capillary number. For small capillary number ($Ca \geq 0.08$), the extra pressure drop, ΔP^+, varies periodically with the streamwise position of the drop center of mass beginning with the first period of the constricted tube. However, for large capillary number ($Ca \leq 0.5$) the periodicity is attained only after the drop has traversed several periods of the sinusoidal tube, *i. e.*, the drop continues to be deformed in passing through the sequence of constrictions after starting in a tube of uniform diameter. Thus, for the parameter range considered, the capillary number is central to determining the degree of deformation and the corresponding pressure drop response.

Our examination of the parameter effects, though not exhaustive, covers a significant range in capillary number ($0.043 \geq Ca \geq 0.5$) and viscosity ratio ($0.4 \geq \lambda \geq 10$). All cases assume $Bo = 0$. In order to investigate the effects of tube geometry, both tubes I and II were considered (see Figure 1). The use of the characteristic radius for length scale, if appropriate, enables comparison of the effect of contraction ratio. Various parameter values were also chosen for comparison with the experimental measurements of Olbricht & Leal (1983).

The results of most interest are the drop deformation and extra pressure drop as a function of the position of the drop in the tube. The axial location of the drop in the tube is defined by the axial coordinate of its center of mass. The pressure drop signatures are shown along with the corresponding drop shape assumed in flowing through the periodically constricted tube, thereby illustrating the correlation of ΔP^+ with drop shape. Average values of extra pressure drop, denoted $\langle\Delta P^+\rangle$, and of drop speed ratio, U/V, are also given for those systems where periodic response was observed (small Ca). The average extra pressure drop was computed by numerical integration of ΔP^+ with respect to Z/L over one tube period. It is also possible to average ΔP^+ with respect to time; this is related to the spatial average through the drop velocity. The two averages are equivalent only if the drop velocity is constant, however, neither average appears to offer a distinct advantage. The average drop speed was determined by dividing the computed time increment to traverse one tube period into the tube wavelength. This is consistent with the average drop speed presented by Olbricht & Leal thereby allowing comparison between numerical and experimental results. The average values are compiled in Table I. The entry corresponding to $Ca = 0.5$ in tube I is an average over the first period of the sinusoidal tube following a section of uniform diameter.

SMALL *Ca* SYSTEMS

Figure 2a shows the extra pressure drop as a function of the axial position of the drop in tube I for $a = 0.76$ and 0.90, thereby illustrating the effect of drop size on extra pressure drop. The abscissa represents the axial coordinate of the center of mass, scaled with the tube wavelength. The tube throat (minimum cross-section) is at $Z/L = 0$ and 1 on the figure. The capillary number and viscosity ratio are 0.083 and 0.40, respectively. The corresponding drop shapes at various locations in the tube are shown in Figure 2b and illustrate the correlation between shape and the pressure drop required to maintain constant flowrate. These shapes and extra pressure drop signatures are typical for small capillary number and moderate drop size.

Table I
Average drop speed and pressure drop

a	Ca	λ	$\frac{U}{V}$	$\frac{(\Delta P^{+}) R_e}{\mu V}$
		Tube I $(R_o/R_t) = 1.8$		
0.76	0.043	0.40	1.42	2.26
0.76	0.083	0.40	1.46	0.23
0.76	0.083	1.0	1.40	2.56
0.76	0.083	7.5	1.31	7.21
0.76	0.51	10.0	1.23	5.37
0.90	0.083	0.40	1.33	-1.47
		Tube II $(R_o/R_t) = 3$		
0.90	0.083	0.40	1.05	-2.57
0.90	0.083	1.0	1.01	3.22
0.90	0.083	10.0	0.95	19.4
0.90	0.50	1.0	1.02	0.10

In the contraction section the flow is partially extensional in the streamwise direction and results in streamwise elongation of the drop in order to fit through the throat section. This effect yields an asymmetric shape with higher curvature at the leading end for either drop size. The required extra pressure drop increases in response to the deformation necessary to squeeze the drop through the contraction. Working of the drop in the contraction produces the drop deformation and also increases the level of drop pressure. Once the head has passed the throat, however, ΔP^{+} decreases dramatically as the drop's midsection and tail flow easily through the throat. Here the motion is assisted by the interfacial tension exerted along the fluid interface and by the relaxation of drop pressure built up in the contraction. The drops in Figure 2b have fore/aft symmetry as the center of mass passes through $Z/L = 0$.

There are two mechanisms responsible for producing a minimum ΔP^{+} in the throat. One is the fluid replacement mechanism whereby when $\lambda<1$ the drop volume is being replaced by fluid of lower viscosity than in single phase flow. This effect becomes important as the drop size increases since a larger fraction of the volume is occupied by lower viscosity fluid. This mechanism is active in the present case and is readily seen in Table I to result in $\langle \Delta P^{+} \rangle R_e/\mu V = 0.23$ and -1.47 for $a=$ 0.76 and 0.90, respectively. This is not the sole cause, however. The second mechanism is the relaxation, in the divergent section, of drop pressure that is built up on the contraction side by deforming the drop sufficiently to flow through the constriction. In single phase flow, the majority of the dissipation in the sinusoidal tube occurs in the vicinity of the throat. In the present case, the high pressure drop raises the level of pressure in the film of suspending fluid surrounding the drop as it passes through the throat. The drop fluid in the portion past the throat is in the expansion section and the radial drop expansion is assisted by the high pressure in the drop, thereby reducing the pressure gradient required to maintain the volume flux. This effect is clearly evident, although resulting only in a local minimum, for relatively viscous drops as can be seen in Figure 3 showing the effect of λ on the pressure drop signature in tube I. In particular, the highly viscous drop, with $\lambda = 7.5$, always has $\Delta P^{+} > 0$, but a local minimum still occurs as the drop straddles the throat section.

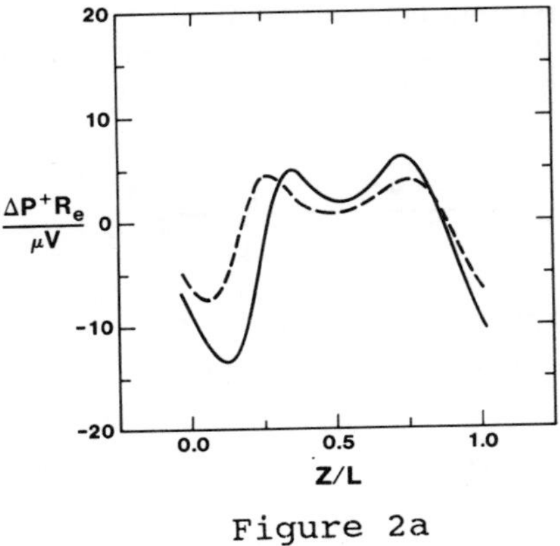

Figure 2a

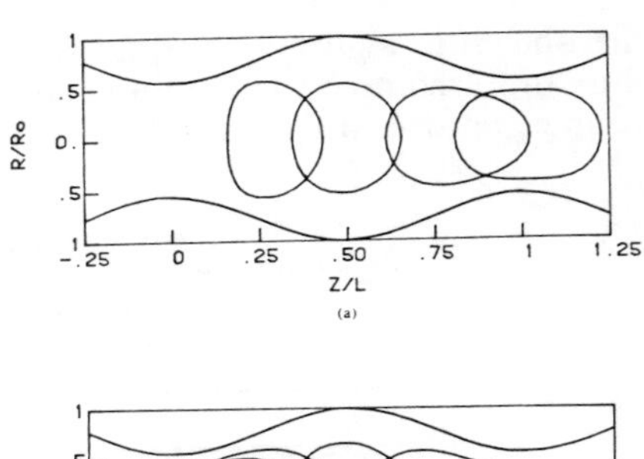

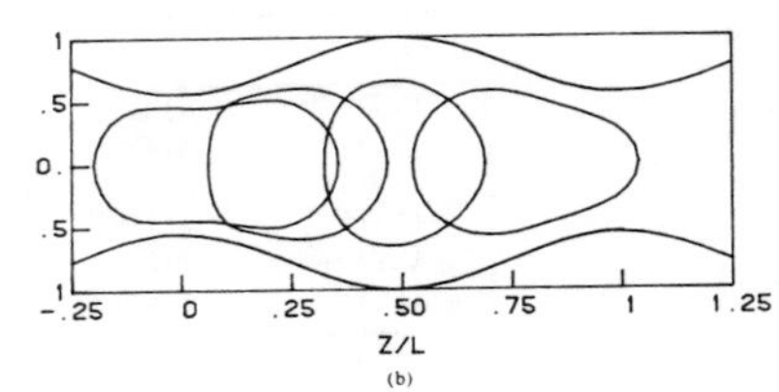

Figure 2b

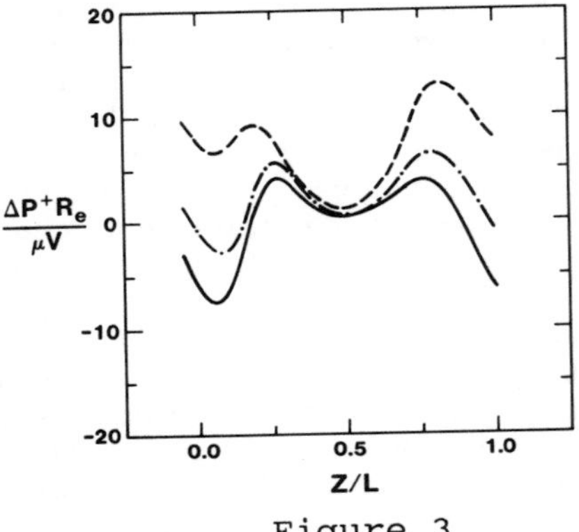

Figure 3

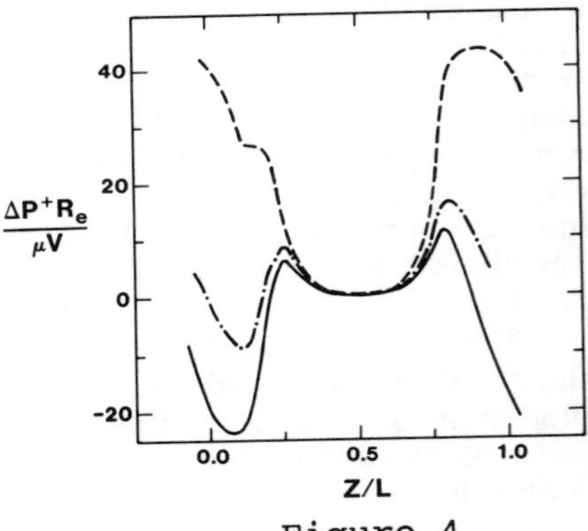

Figure 4

In this case the minimum pressure drop occurs when the viscous drop is in the widest section of the constricted tube, Z/L=0.5.

It is at first surprising that there is a local maximum in ΔP^+ as the drop occupies the expansion section of the tube as shown in Figure 2. This effect is part of the relaxation of drop pressure in the divergent section. The drop is seen to recoil in the biaxial expansion region to such an extent that it temporarily blocks the tube. Furthermore, the head of the drop resides in a region of slow moving fluid and the relatively inviscid drop finds it easier to expand rather than proceed down the tube. The large pressure behind the drop is indicated in Figure 2b by the flat shape of the interface at the trailing end of the drop. This phenomena is a consequence of the constant volume flux condition as well as the drop viscosity. The magnitude of pressure recovery in the expansion is much less for the viscous drop in Figure 3.

The parameter values for the present discussion were chosen to allow comparison of results with values reported in the experimental investigation of Olbricht & Leal (1983). They report average drop speed ratios, U/V, of 1.45 and 1.36 for drop sizes of 0.76 and 0.91, respectively, for $Ca = 0.083$ and $\lambda = 0.40$. The corresponding values computed in the present study are 1.46 and 1.33. The average extra pressure drop results reported by Olbricht & Leal are the arithmetic average of the maximum and minimum values over one tube period and so are not directly comparable to the integrated average values given in Table I. The reported values of average extra pressure drop are 0.9 and 0 for drop sizes of 0.76 and 0.91, respectively. The corresponding average values for the present study are 0.23 and -1.47. Olbricht & Leal also show tracings of drop shapes at various locations in the tube along with the corresponding pressure signatures. The drop shapes of the present investigation compare satisfactorily with the experimental shapes. The comparison between the pressure signatures is less satisfactory, the present results showing enhanced values for the peaks and valleys in the profiles. It is noted that the abscissae on the extra pressure loss signatures presented by Olbricht & Leal are not labeled and it is not clear whether the extra pressure drop is plotted against time or axial position, rendering a direct comparison with the numerical results impossible. Nevertheless, Olbricht & Leal also observed many fine details in the extra pressure loss signal for this drop size, although the magnitudes of extremal values are less. They generally associate the maximum ΔP^+ with the drop traversing the cross-section of minimum area whereas the present results indicate a minimum value at the same location when $\lambda = 0.4$ and a local minimum when λ=7.5. Inspection of the numerical results indicates the Reynolds number based on the drop velocity, which varies with axial position, can be as much as 2.5Re in Tube I and 5Re in Tube II, where $Re = \rho V R_e/\mu$ is the Reynolds number based on the imposed flow. Given that Re is about 0.05 in the experiment, effects of inertia cannot be entirely discounted and are a plausible reason for discrepancies with the present results.

The effect of viscosity ratio on the pressure drop signature is shown in Figures 3 and 4 for tubes I and II, respectively. The extra pressure drop response is similar for both tubes when $\lambda \geq 1$ with two local extrema upon entering and exiting the throat region and a minimum when the center of mass is slightly past the throat. The extrema are significantly enhanced fot the highly constricted tube, however. The pressure recovery in the expansion section is complete in tube I, even for λ=1. The pressure drop is always greater in the convergent section for tube II. The pressure relaxation is inhibited as the drop viscosity increases until the minimum pressure drop no longer is found in the throat but rather occurs when the drop is in the widest section of the tube. In tube I (λ=7.5) the pressure drop continues to experience a local minimum near the throat. The minimum all but disappears in tube II for λ = 10 and the extra pressure drop remains high in traversing the throat region. The

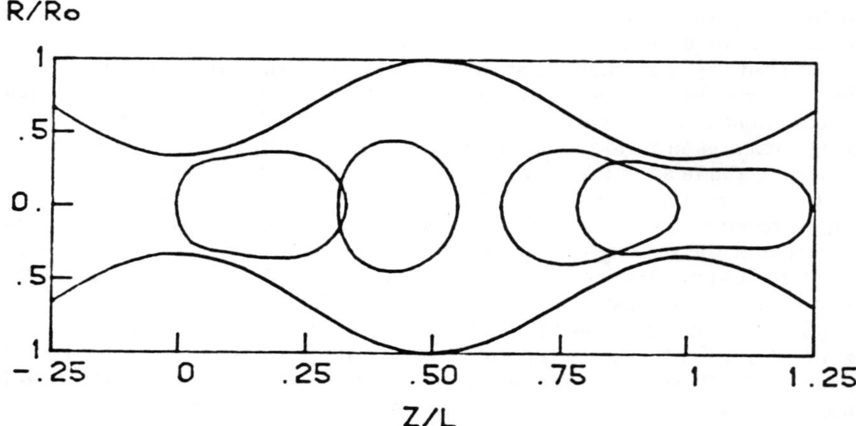

Figure 5

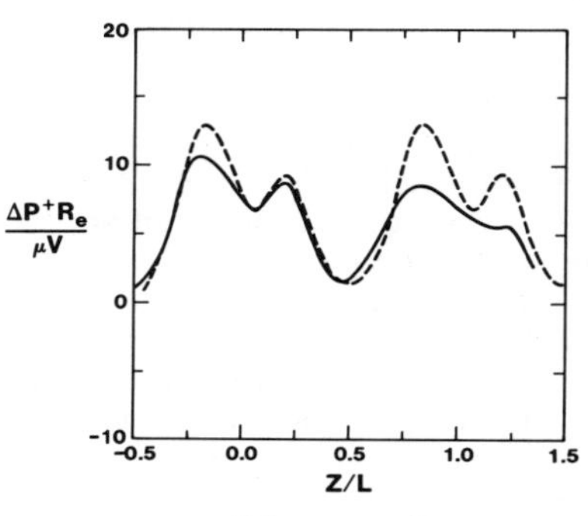

Figure 6a

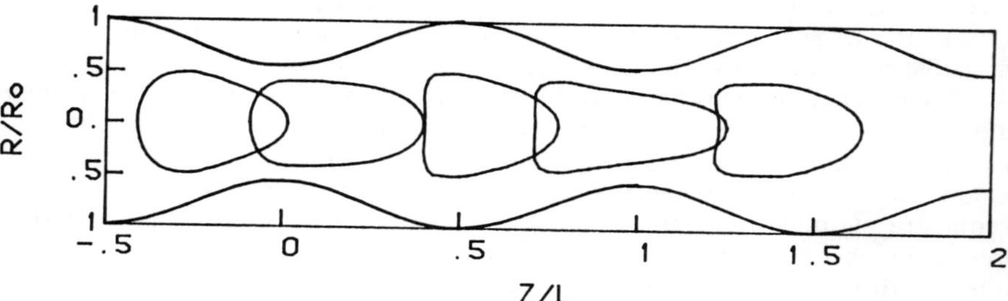

Figure 6b

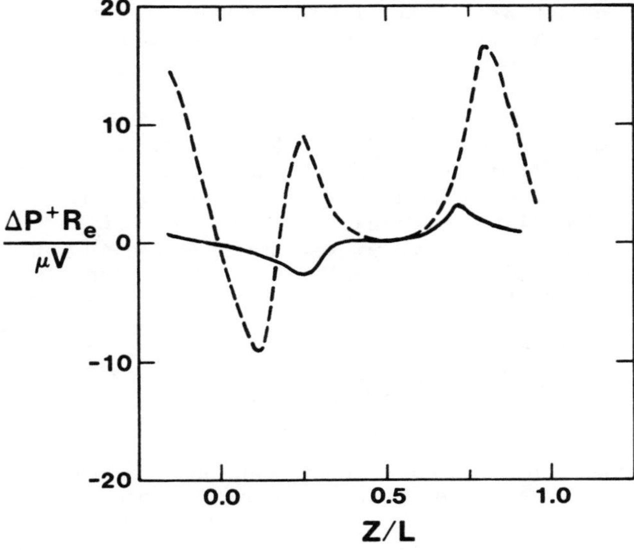

Figure 7a

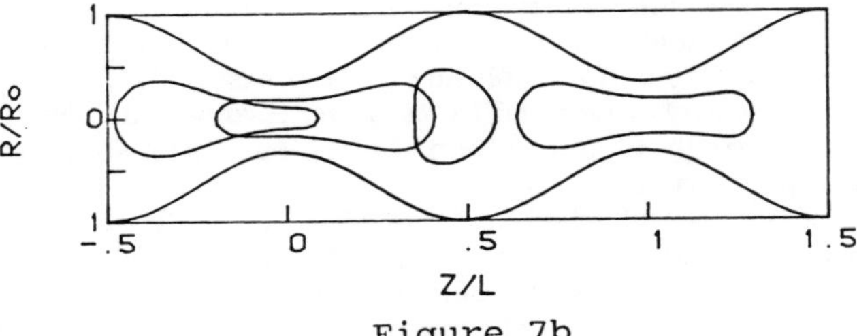

Figure 7b

drop shapes assumed in traversing tube II for these values of capillary number are similar to those shown previously for tube I (Figure 2b) when $\lambda = O(1)$. As λ increases the radial expansion in the diverging section is increasingly inhibited and the drop protrudes further into the wide region of the tube before relaxing to a nominally spherical shape as illustrated in Figure 5. The recoil is assured for this drop size in tube II since the drop, though sizeable compared to the throat radius, is still small compared to the maximum tube radius. This absence of wall effects is evident in the pressure drop signatures for tube II shown in Figure 4 as they all return to the same value of extra pressure drop for Z/L approx 0.5.

The average extra pressure drop increases with λ (see Table I) as it does for the flow of drops in tubes of uniform diameter (Ho & Leal 1975, Martinez 1987). This comparison with straight tubes does not hold true for average drop speed, however. The average drop speed decreases with λ for tube I but is almost independent of λ in tube II. Thus, in a highly constricted tube, the drop's motion is greatly impeded in squeezing through the constriction and the viscosity ratio plays little part in determining the average speed relative to the bulk velocity.

LARGE *Ca* SYSTEMS

In this section we consider a few examples illustrating the essential effects of large capillary numbers (Ca = 0.5). Figure 6a shows the extra pressure drop as a function of the position of the drop centroid in tube I for the parameter set (a, Ca, λ) = (0.76, 0.5, 10). The simulation was performed for 2.5 sinusoid periods downstream of the uniform diameter section. The corresponding shape evolution is shown in Figure 6b. The interesting feature of this system is that the drop does not recoil in the expansion sections as in the previous cases even though the drop is "small" in the wide sections of the tube. Here the drop enters the second throat in an initially deformed state only to be deformed further. The drop develops negative curvature at the trailing end upon exiting the second throat resulting in a heart-shaped drop similar to those observed by Han & Funatsu (1978) for drops on the downstream end of an abrupt contraction. The maximum ΔP^{+} is lower than for the case with Ca = 0.083 shown for comparison in Figure 6a, even though the latter drop is less viscous. Decay of the amplitude of the pressure drop signature is evident as the drop is progressively deformed in its travel through the series of sinusoids. The drop should eventually produce a periodic pressure drop signature after having traversed several tube periods.

An analysis was also performed for Ca = 0.5 in tube II for a= 0.90 and for equal drop and suspending fluid viscosities, λ = 1. The evolution of extra pressure drop and shape as the drop translates though the tube is shown in Figure 7a and 7b, respectively. Although very large deformation of the drop is seen to occur for this large capillary number, the drop recoils in the wide region of the tube and hence the extra pressure drop is nominally periodic beginning with the first period of wavy tube. The recoil is due mainly to the fact that the drop is small, indicating that a permanent deformation requires the drop to be sizeable not only with respect to the characteristic radius, but also with respect to the maximum tube radius. The permanent elongation observed in the previous case was due to the inability of the drop to recover from the deformation experienced in the first throat before entering the second. This behavior suggests that permanent deformation depends on the ratio of the timescale for deformation to the timescale for flow through a tube period. Rallison (1984) (see also Rallison & Acrivos 1978) suggests a deformation timescale, $a_o(\lambda + 1)\,\mu/\gamma$, based on surface tension relaxation in the absence of wall effects. Permanent deformation will occur if the deformation timescale is greater than the flow time, L/U, which leads to a scale for capillary number above which permanent deformation is expected (*c.f.* Olbricht & Leal 1983),

$$Ca > Ca_c = \frac{L}{a_d(\lambda+1)} \frac{V}{U}$$

Thus, large capillary numbers are needed to permanently deform small drops, however more viscous drops suffer permanent deformation for lower values of capillary number. For example, in tube I for $(a, Ca, \lambda) = (0.76, 0.50, 10)$, the formula gives $Ca_c = 0.46$, indicating permanent deformation can be expected, in agreement with computed results. In tube II the case $(a, Ca, \lambda) = (0.90, 0.50, 1)$ results in $Ca_c = 3.8$, here indicating permanent deformation only for much larger values of capillary number.

Even with the recovery of drop shape in the present case, the shape is grossly elongated in the throat sections where the drop is pulled out into a long strand without rupture of the drop. This case shows effects of large capillary number which can result in the formation of "strands", even when the drops are small compared to the maximum tube radius. The corresponding extra pressure drop is seen to be quite small compared to previous values for tube II (see Figure 4). The average value of extra pressure drop is much lower than for $Ca = 0.083$, all other parameters being equal (Table I).

CONCLUSIONS

The creeping flow of a viscous drop through a periodically constricted tube filled with a second viscous fluid was analyzed using a boundary integral equation method. The effects of the drop size, capillary number, and viscosity ratio, on the pressure gradient and drop shape were examined. Two tubes were considered. The maximum to minimum radius ratio was 1.8 for tube I, as compared to a ratio of 3 for tube II. The maximum tube radius to tube wavelength was 0.3 for both tubes.

The drops undergo uniaxial extension in the contraction sections and biaxial expansion in the divergent sections of the tube and consequently the drop speed and extra pressure drop vary with the drop's location in the tube. For small to moderate capillary number ($Ca < 0.5$) drops recover a nearly spherical shape in the wide sections of the periodic tube. For large capillary number, and/or large drop viscosity, the drops are permanently deformed in traversing the narrow throat and do not recoil in the wide sections. Large capillary numbers also promote the formation of strands as drops squeeze through the throat of the sinusoidal tube. The average drop speed varies with capillary number and viscosity ratio in tube I, however the drop speed is nearly independent of these parameters in tube II.

Comparison of results in tube I with the experimental data of Olbricht & Leal (1983) show satisfactory agreement for drop speed but only qualitative agreement for extra pressure drop.

ACKNOWLEDGEMENT

This work performed in part at Sandia National Laboratories supported by the U.S. Department of Energy under contract number DE-AC04-76DP00789.

REFERENCES

H. B. Chin & C. D. Han, 1979, "Studies on Droplet Deformation and Breakup. I. Droplet Deformation in Extensional Flow," *J. Rheol.* **23**, 557-590.

H. B. Chin & C. D. Han, 1980, "Studies on Droplet Deformation and Breakup. II. Breakup of a Droplet in Non-uniform Shear Flow," *J. Rheol.* **24**, 1-39.

P. Franzen, 1979, "Zum Einflus der Porengeometrie auf den Druckverlust bei der Durchstromung von Porensystemen. I. Versuche an Modellkanalen mit variablem Querschnitt," *Rheol. Acta*, **18**, 392-423.

I. I. Gardescu, 1930, "Behavior of Gas Bubbles in Capillary Spaces," *Trans. AIME*, **86**, 351-370.

C. D. Han & K. Funatsu, 1978, "An Experimental Study of Droplet Deformation and Breakup in Pressure-Driven Flows through Converging and Uniform Channels," *J. Rheol.*, **22**(2), 113-133.

B. P. Ho & L. G. Leal, 1975, "The Creeping Motion of Liquid Drops Through a Circular Tube of Comparable Diameter," *J. Fluid Mech.* **71**, 361-384.

O. A. Ladyzhenskaya, 1969, **The Mathematical Theory of Viscous Incompressible Flow**, Gordon & Breach, New York, 2nd. ed.

S. H. Lee & L. G. Leal, 1982, "The Motion of a Sphere in the Presence of a Deformable Interface," *J. Coll. Interf. Sci.*, **87**, 81-106.

M. J. Martinez, 1987, "Viscous Flow of Drops and Bubbles in Straight and Constricted Tubes," Ph.D. Thesis, University of California, Berkeley, CA.

S. G. Oh & J. C. Slattery, 1979, "Interfacial Tension Required for Significant Displacement of Residual Oil," *Soc. Pet. Eng. J.*, April, 83-96.

W. L. Olbricht & L. G. Leal, 1983, "The Creeping Motion of Immiscible Drops through a Converging/Diverging Tube," *J. Fluid Mech.*, **134**, 329-355.

J. M. Rallison & A. Acrivos, 1978, "A Numerical Study of the Deformation and Burst of a Viscous Drop in an Extensional Flow," *J. Fluid Mech.*, **89**, 191-200.

J. M. Rallison, 1984, "The Deformation of Small Viscous Drops and Bubbles in Shear Flows," *Ann. Rev. Fluid Mech.*, **16**, 45-66.

F. J. Rizzo & D. J. Shippy, 1968, "A Formulation and Solution Procedure for the General Non-Homogeneous Elastic Inclusion Problem," *Int. J. Solids. Struct.*, **4**, 1161-1179.

J. C. Slattery, 1974, "Interfacial Effects in the Entrapment and Displacement of Residual Oil," *AIChE J.*, **20**(6), 1145-1154.

J. N. Tilton & A. C. Payatakes, 1984, "Collocation Solution to Creeping Newtonian Flow through Sinusoidal Tubes: a Correction," *AIChE J.*, **30** (6), 1016-1021.

G. K. Youngren & A. Acrivos, 1975, " Stokes Flow Past a Particle of Arbitrary Shape: A Numerical Method of Solution," *J. Fluid Mech.*, **69**, 377-403.

LIST OF FIGURES

Figure 1. Geometry of the periodically constricted tubes; (a) tube I, (b) tube II.

Figure 2a. Effect of drop size on the extra pressure drop in tube I, Ca=0.083, λ=0.40, (dashed line) a=0.76, (solid line) a=0.90.

Figure 2b. Effect of drop size on the drop shape at various locations in tube I, flow is from left to right, Ca=0.083, λ=0.40, (a) a=0.76, (b) a=0.90.

Figure 3. Effect of viscosity ratio on extra pressure drop in tube I, a=0.76, Ca=0.083, (solid line) λ=0.40, (dot/dashed line) λ=1, (dashed line) λ=7.5.

Figure 4. Effect of viscosity ratio on extra pressure drop in tube I, a=0.90, Ca=0.083, (solid line) λ=0.40, (dot/dashed line) λ=1, (dashed line) λ=10.

Figure 5. Drop shapes for a highly viscous drop in tube II, a=0.90, Ca=0.083, and λ=10.

Figure 6a. Extra pressure drop at high capillary number in tube I; (solid line) a=0.76, Ca=0.5, λ=10; (dashed line)a=0.76, Ca=0.083, λ=7.5.

Figure 6b. Drop shape evolution at high capillary number in tube I, a=0.76, Ca=0.5, and λ=10.

Figure 7a. Effect of large capillary number on extra pressure drop in tube II, a=0.90, λ= 1, (solid line) Ca=0.5, (dashed line)Ca=0.083.

Figure 7b. Effect of large capillary number on drop shape in tube II, a=0.90, Ca=0.5, and λ=1.

NUMERICAL SIMULATION OF CAPILLARY DRIVEN VISCOUS FLOWS IN LIQUID DROPS AND FILMS BY AN INTERFACE RECONSTRUCTION SCHEME

J. Y. Poo and N. Ashgriz
Department of Mechanical and Aerospace Engineering
State University of New York at Buffalo, Buffalo, NY 14260

ABSTRACT

The numerical modeling of viscous drops and films have been restrictive due to the difficulty in solving the full transport equations where the location of the boundary is not known *a priori*, and also due to the perplexity in calculating an accurate surface curvature. In this paper we combine an advanced numerical algorithm (FLAIR) with a newly presented and accurate curvature calculation scheme, to study the dynamics of viscous drops and films. It is demonstrated that this technique can resolve the complex flow fields and large surface deformations of the type which are common in studies of viscous liquid drops and films. The evolution of the shape and internal flow field of an initially elliptic, infinitely long, viscous liquid cylinder and a liquid film with both symmetric and asymmetric large amplitude sinusoidal surface disturbances are calculated numerically.

INTRODUCTION

Compared to the analytical studies, the full numerical treatment of fluid dynamics of liquid drops and films are rare. This is due to the great difficulties in handling the free surface boundaries. Some numerical techniques, such as boundary-integral technique[1,2], have been commonly used for the analysis of inviscid or zero Reynolds number drop dynamics. However, boundary-integral methods have to be improved in order to be utilized in finite Reynolds number flow analysis. Yet another powerful technique for managing the free surface flows is the boundary-fitted coordinate method. Although, this is an old technique, its application in drops and bubbles studies has only been recent. Kang and Leal[3] used this technique to study the deformation of a bubble in a uniaxial straining flow at finite Reynolds numbers. The difficulty of this technique lies in the determination of the criterion for the coordinate transformation. For instance, for axisymmetric problems, Ryskin and Leal[4] have shown that the boundary-fitted coordinates at any instant can be connected with the common cylindrical coordinates via a pair of mapping functions, which satisfy the covariant Laplace equations. However, there are no generalized functions for transformation of the coordinates into any arbitrary and vigorously deformed surfaces. Another limitation of these techniques is that they can not handle free surface flows where there are surface folding and merging.

Among the different computational techniques available for solving problems with interfaces and free surfaces, volume tracking methods have shown a great promise for detailed numerical study of large surface deformations and surface merging. Volume tracking techniques rely on some kind of volumetric progress variable, such as marker particles in the Marker and Cell (MAC) technique[5,6], and fraction of cell volume in Volume of Fluid (VOF) technique[7,8]. Marker-and-cell method involves Eulerian flow-field calculations and Lagrangian marker movements. The velocity of a marker is found by taking the average of the Eulerian velocities in its vicinity; this is the proper approach to use because a marker is only a point with no mass. For differential liquid particles there is no difference between Eulerian and Lagrangian velocities; and therefore, the liquid particles can be translated directly according to the local velocity. However, due to the irregularity of the flow field, the markers might accumulate more at some spots or scatter sparsely at others. For example, it is very probable that a liquid interior cell can temporarily have no marker in it and need some way to assure its status. The no-mass markers do make the free surface movement observable; however, the liquid fractions in surface cells can never be determined by counting the number of markers. They must be found after some type of curve fitting has been performed.

The Volume Of Fluid (VOF) method is much easier to use and less computationally intensive. Two major problems arise when the interface is represented by a fractional volume parameter. First is how to identify the exact surface location, and second is how to advect the surface. Several techniques have been introduced for moving the volume fraction field[8-11]. A common one is the so-called Donor-Acceptor technique[8,9]. This technique is based on describing a surface orientation and then moving the surface with the velocity normal to that orientation. In the Donor-Acceptor technique, the surface cell is assumed to be either horizontal or vertical. The decision regarding the orientation is made based on studying the neighboring cells. After deciding the orientation of the surface in the cell some criterion is used to move the cell volume fraction field in space and time. The Donor-Acceptor technique which is used in VOF method, emphasizes control of interface diffusion rather than control of the liquid fraction in a cell. Therefore, *ad hoc* techniques must be designed to remove "bad" points. For example, cells having liquid fractions either less than zero or greater than unity are corrected by redistributing liquid around them.

Another major difficulty in using volume tracking methods is calculation of the surface curvature, which is a necessary parameter for determination of the effect of surface tension forces in drops and bubbles studies. Intuitively we can simply add up the partial volumes in the cells to represent the so-called liquid height and utilize a central finite difference scheme to find the first and second derivatives of the interfaces. This approach can result in substantial errors when there are large variations in the interface curvature. Nichols et al.[8] improved this technique by considering the liquid heights in three neighboring cells to find the slope of a line-segment in the middle cell which represents the first derivative of the fluid surface. The second derivative was then calculated by the change in slopes of two neighboring surface cell line-segments. Smith[12] used the same method to calculate the slop of the surface line-segment, but a different approach to find the second derivative of the surface. He approximated the surface curvature by finding the radius of a circle which passes through midpoints of neighboring surface line-segments. In the aforementioned methods the accuracy of the interface curvature is limited because the second derivatives are obtained after the first derivatives are calculated. Chorin[13] circumvented the calculation of the first derivative and directly obtained the surface curvature based on the cell volume fractions. He found the curvature of an osculating-circle which resulted in the correct partial-volume field in the neighborhood of the cell under consideration. This method was demonstrated to be satisfactory for circles but not as good for sine waves. In this paper we will introduce a more accurate technique for calculating the surface curvature.

As mentioned previously, there are two major difficulties in utilizing the volume of fluid method. One is the advection of the surfaces; the other is the reconstruction and surface curvature calculation based on a given set of surface volume fractions. Because, in drops and bubbles studies, the accuracy of the results is crucially dependent on the accuracy of the technique used for handling the above problems, it is important to improve the accuracy of the presently available methods. It is the purpose of this paper to present a new numerical technique for the surface curvature calculation and implementing that in an advanced numerical technique

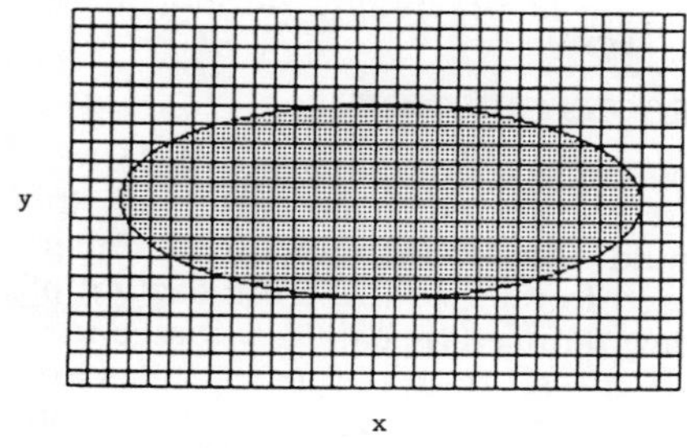

Figure 1

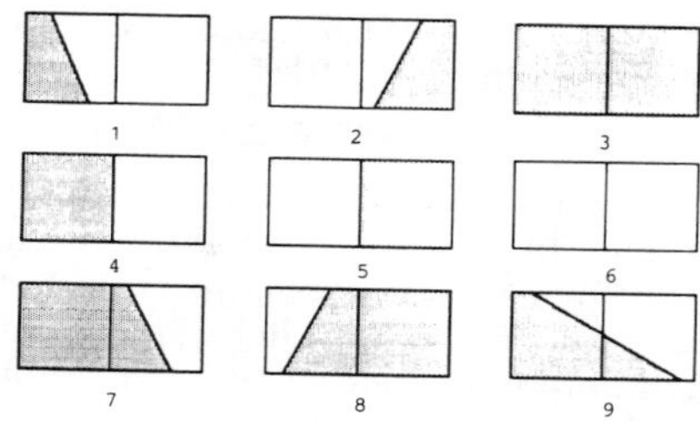

Figure 2

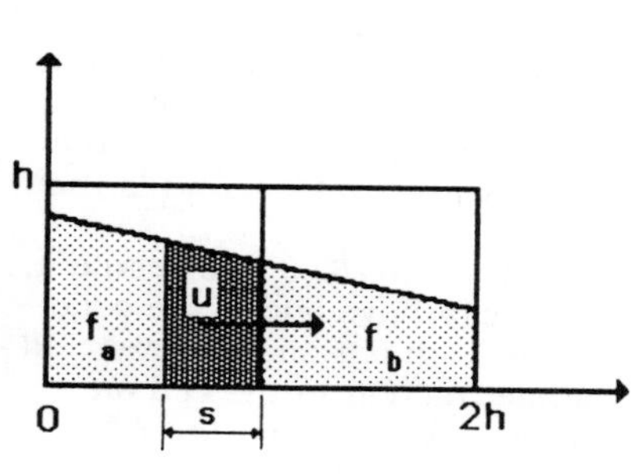

Figure 3

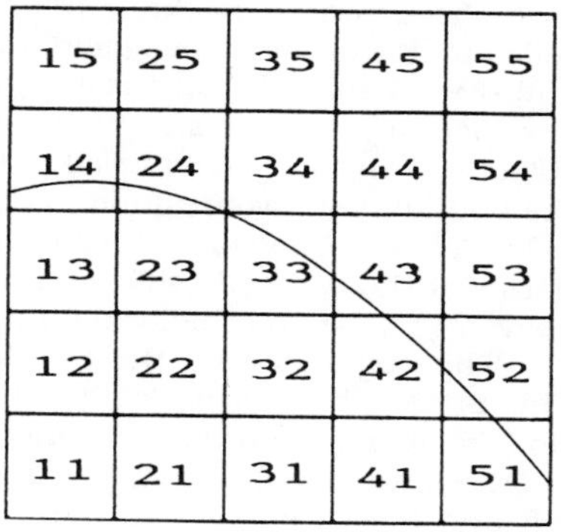

15	25	35	45	55
14	24	34	44	54
13	23	33	43	53
12	22	32	42	52
11	21	31	41	51

Figure 4

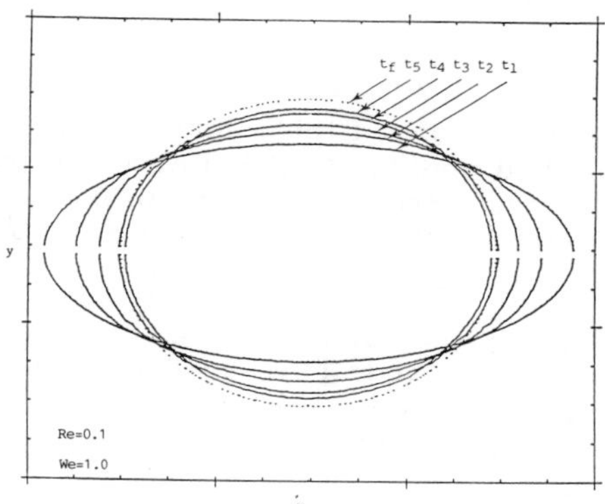

Figure 5

for surface advection and reconstruction which was previously developed by Ashgriz and Poo[14].

SURFACE ADVECTION AND RECONSTRUCTION

In this paper we will only study the liquid motion in a two dimensional Cartesian coordinates. Two different configurations are considered. One is the flow induced by capillary forces in an infinitely long liquid cylinder with an elliptic cross-sectional area. From here on "liquid drop" is used to refer to the cross-section of such liquid cylinder. The other problem is the flow induced by capillary forces in a two dimensional liquid film due to surface disturbances. Both of these studies are carried out in an Eulerian system. We will first describe the numerical method and then present the results.

Consider an elliptic liquid region in an uniform numerical mesh system as shown in Figure 1. The circumference of the elliptic drop will cross through the cells in this system. We will call the cells which contain part of the ellipse periphery the surface cells. A partial volume (or area) parameter, f, can be introduced which represents the fractional volume (or area) of the surface cell that is filled with fluid. Therefore, the volume fraction for each cell can be defined such that $f = 0$ for cells outside of the ellipse (empty cells), $f = 1$ for cells inside of the ellipse (full cells), and $0 \leq f \leq 1$ for cells on the surface (wet cells). We will use the FLAIR algorithm, developed by Ashgriz and Poo[14], to advect the volume fraction field and to reconstruct the surface. Very briefly, this technique seeks the slope of a line-segment at the boundary of two neighboring surface cells. The slope of the line-segment is calculated based on volume fractions of the two cells. This line-segment and the boundaries of the two cells will form a trapezoid. The fluid inside the trapezoid is then convected based on the local fluid velocities once.

The major task in the FLAIR algorithm is to find the interface line-segment. This is done by first noticing that two neighboring cells can have one of the conditions shown in Figure 2. For instance, the surface orientation is said to correspond to Case 1 when $0<f_a<1$ and $f_b = 0$, where, f_a is the volume fraction in the cell on the left and f_b is that on the right. Once the case for the two neighboring cells is identified, the coefficients of the surface line-segment (i.e., a and b of $y = ax + b$) can be calculated based on f_a and f_b. In order to obtain a and b, the area underneath the line-segment is calculated and it is set equal to the corresponding cell volume fraction. This will result in two equations, one for f_a and the other for f_b, to solve for the two unknowns, a and b. For instance, the coefficients for a surface resembling that shown in Figure 3 are:

$$a = f_b - f_a \tag{1}$$

$$b = \frac{h}{2}(3f_a - f_b) \tag{2}$$

After the surface is reconstructed the volume fraction field is advected by calculating the volume of fluid underneath the surface line-segment that is transported from one cell to the neighboring cell (Figure 3). For instance, the fluid flux moving from left cell in Figure 3 to the right cell can be calculated from:

$$\delta f = \left(\frac{u\delta t}{h}\right)\left[a + \frac{b}{h} - \frac{a u \delta t}{2}\right] \tag{3}$$

where u is the liquid velocity at the boundary of the two cells, and δt is the time step in calculations. Ashgriz and Poo[14] have carried out such calculations for all possible line-segment orientations and have developed a case distinction diagram which will identify the surface orientation based only on the two neighboring volume fractions (e.g. f_a and f_b).

CURVATURE CALCULATION

Once the new volume fraction field is obtained, the surface curvature at each cell is calculated based on a technique presented below. This technique is a modification of a previous technique developed by Poo and Ashgriz[16]. However, the present technique reduces the numerically generated surface curvature fluctuations. Such curvature fluctuations were observed to produce large errors in surface shape and flow field calculations.

Consider a 5 x 5 cell unit and an arbitrary curve which passes through it. As before, we introduce an area fraction, f_{ij}, for each cell, c_{ij}, where i increases from left to right and j increases from bottom to top (see Figure 4). Our objective is to solve the following inverse problem: Given the area fractions in a 5 x 5 cell unit, what is the equation for the curve that cuts through the unit? The curvature can then be calculated for the center cell knowing the equation that best describes the interface curve. Basic assumptions of the method are: (1) In a 5 x 5 cell unit there can at most be one connected interface; (2) the grid spacing is small enough to resolve the interface to such an extent that a second order polynomial can be a good approximation of the interface.

In order to find the interface curve, first a standard base for the 5 x 5 cell unit is chosen by comparing the amount of liquid at each side. The heaviest side is then placed at the bottom with the lightest side on top. Adding the values of volume fraction of liquid in each column, we then get five liquid heights, which are denoted by f_A, f_B, f_C, f_D and f_E. For example, in Figure 4, $f_A = f_{11}+f_{12}+f_{13}+f_{14}+f_{15}$. A second-order polynomial (i.e. $y = ax^2 + bx + c$) is assumed to pass through the 5 x 5 cell unit. There are also five vertical fluid columns under this curve, which can be calculated by integrating the volume under the curve. The criterion for choosing the best second-order polynomial is based on a least-square error analysis. When the second-order polynomial is checked against the real fluid columns (i.e. f_A to f_E) the best coefficients (i.e. a, b, and c) can be found using standard least square error method. These coefficients are:

$$a = \frac{1}{36h}(5f_A - 2f_B - 6f_C - 2f_D + 5f_E) \tag{4}$$

$$b = \frac{1}{126}(-110f_A + 17f_B + 105f_C + 53f_D - 65f_E) \tag{5}$$

$$c = \frac{h}{756}(925f_A + 251f_B - 381f_C - 289f_D + 250f_E) \tag{6}$$

After the coefficients are found the first derivative, $y_x = 5ah+b$, and the second derivative, $y_{xx} = 2a$, can be obtained immediately from:

$$y_x = \frac{1}{28}(-5f_A - 4f_B + 4f_D + 5f_E) \tag{7}$$

$$y_{xx} = \frac{1}{18h}(5f_A - 2f_B - 6f_C - 2f_D + 5f_E) \tag{8}$$

Derivatives are evaluated at the center of the center cell, c_{33}, where $x = 5h/2$. Finally, the curvature of the curve in the center cell is computed from:

$$\frac{1}{R} = -\frac{-y_{xx}}{\left(1 + y_x^2\right)^{1.5}} \tag{9}$$

where R is the surface curvature.

PROBLEM FORMULATION

The equations that are solved are the two-dimensional, incompressible, constant properties Navier-Stokes equations with free boundaries. The equations for the conservation of mass and momentum, and the Poisson's equation are:

$$\frac{\partial u}{\partial x} + \frac{\partial v}{\partial y} = 0 \tag{10}$$

$$\frac{\partial u}{\partial t} + \frac{\partial uu}{\partial x} + \frac{\partial uv}{\partial y} = -\frac{\partial p}{\partial x} + \frac{1}{Re}\left(\frac{\partial^2 u}{\partial x^2} + \frac{\partial^2 u}{\partial y^2}\right) \tag{11}$$

$$\frac{\partial v}{\partial t}+\frac{\partial vu}{\partial x}+\frac{\partial vv}{\partial y}=-\frac{\partial p}{\partial y}+\frac{1}{Re}\left(\frac{\partial^2 v}{\partial x^2}+\frac{\partial^2 v}{\partial y^2}\right) \tag{12}$$

$$\frac{\partial^2 p}{\partial x^2}+\frac{\partial^2 p}{\partial y^2}=2\left(\frac{\partial u}{\partial x}\frac{\partial v}{\partial y}-\frac{\partial u}{\partial y}\frac{\partial v}{\partial x}\right) \tag{13}$$

At the present time we are only considering the motion of a liquid drop in vacuum. Therefore, no shear forces at the liquid surface are considered. Also the effect of velocity gradients on the surface pressure are neglected. Hence, the boundary equation is reduced only to the pressure balance at the surface:

$$p=\frac{\sigma}{R}=\frac{-\sigma y_{xx}}{\left(1+y_x^2\right)^{1.5}} \tag{14}$$

where σ is the surface tension. The equation of motion is solved by a fully implicit scheme using staggered grids. However special care has to be taken in writing the finite difference equations for the surface cells. At the free surface, the momentum equation can not be evaluated in the regular way as it is done in the interior of the liquid because the cells at the free surface have at least one cell boundary facing vacuum, which makes the evaluation of the neighboring velocities in that direction impossible. As in MAC method[5], continuity equation is to be used for the surface cells to calculate the velocity at the free surface. Details of velocity field calculation at the free surfaces are given in Ashgriz and Poo[14].

Based on the above discussions, the solution procedure can be summarized as follows: (1) Specify the initial conditions for the surface geometry, and velocities; (2) Move the surface based on the given velocities and the FLAIR method; (3) Find the surface curvature based on the cell volume fractions; (4) Calculate the surface pressure based on the local surface curvatures, and equation (14); (5) Calculate the pressure throughout the drop using the Poisson's equation; (6) Solve for the velocity field throughout the drop or the film using the momentum equations; (7) Iterate between steps 6 and 7 to get convergence; (8) Increment the time and repeat.

CAPILLARY DRIVEN DROP MOTION

The results for the motion of a initially elliptic drop of Figure 1 are presented in Figures 5 and 6. The velocities are initially zero everywhere. However, because there is a non-uniform curvature distribution on the surface of the ellipse, the drop starts a motion based only on surface tension forces. Note that there is no surface forces in the direction normal to the cross section of the liquid cylinder. The damping motion of the drop is shown in Figure 5. The Reynolds number used for this case is 0.1, and Weber number is 1.o. The Weber number is defined based on the initial larger diameter of the ellipse, *D*. Because drop's initial velocity is zero, the velocity scale is obtained by setting the *We* number equal to 1 and calculating the velocity scale from $u=\sqrt{\sigma/\rho D}$. In Figure 5 the drop shape was displayed by plotting the points where the FLAIR line segments intercept the cell boundaries. It is important to emphasis that in presenting the surface shapes no artificial surface smoothing is used. Several drop shapes at different times are shown in Figure 5. It is observed that at such a low Reynolds number the drop damps to a circular shape with a few oscillations. The velocity fields inside the drop at different times are shown in Figures 6a-e. The oscillation of the drop generates a recirculation zone inside the drop, which also damps out after the drop ceases motion. The smoothness of the surface after long computational times, indicates the accuracy of this technique for surface

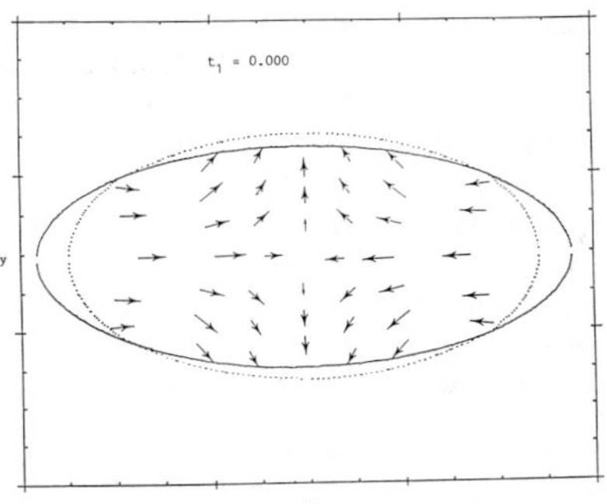

Figure 6a

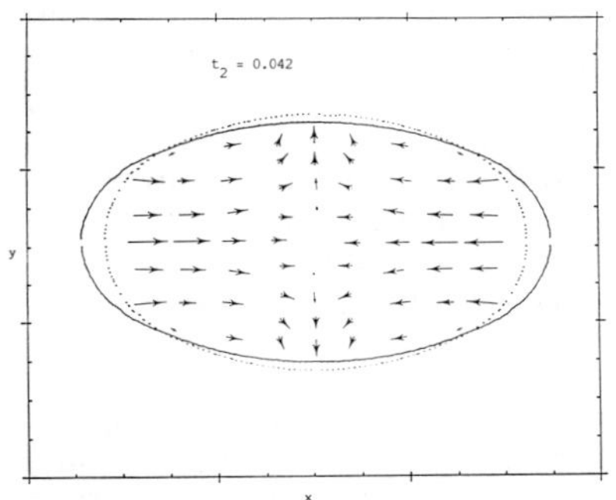

Figure 6b

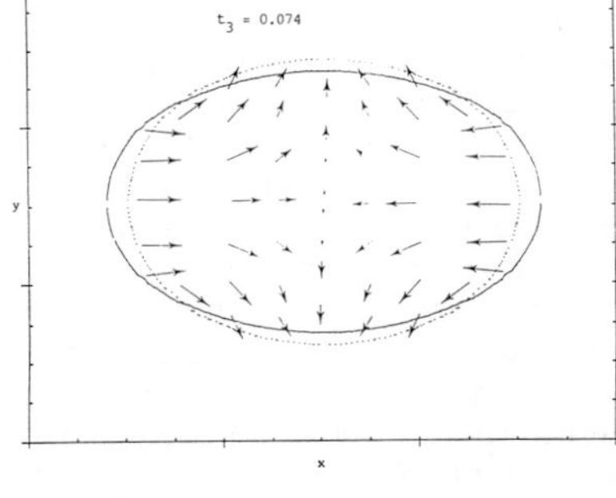

Figure 6c

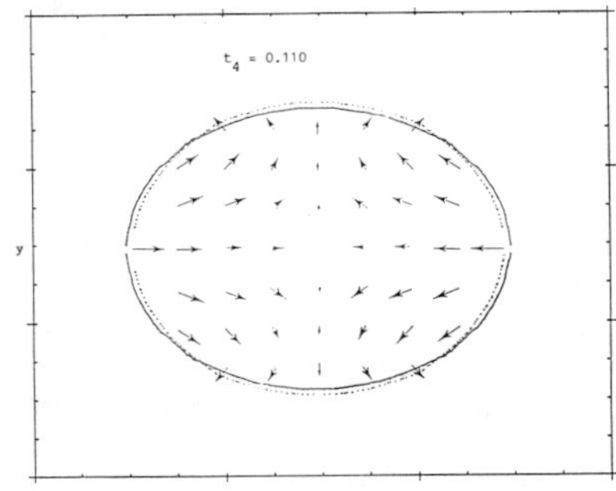

Figure 6d

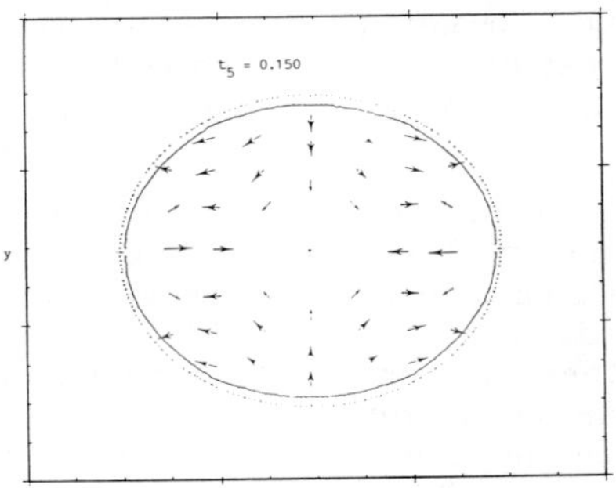

Figure 6e

movement and therefore better curvature calculations. However, as it is seen in Figure 6-e, there is a 4% loss in the drop mass. This error is mainly due to the utilization of a crude numerical scheme for momentum calculation and is being improved at the present time. Another error is because the actual shape of the surface cell is not used when the continuity equation is applied at the surface to find the surface velocities. It is also noteworthy that the area loss can be made equal to zero, if during each time step the liquid is always re-allocated to make sure all the fluid quantities are accounted for. This means that a zero mass loss is not a sufficient condition for an accurate computation but only a necessary one.

CAPILLARY DRIVEN LIQUID FILM DYNAMICS

The surface curvature and advection scheme is also tested on an infinitely long liquid film. In the two problems describe below the Reynolds number is 500 and the Weber number is 6.5. The *Re* and *We* numbers are defined the same way as for the drop case. The first problem is calculation of the shape and internal flow field of a liquid film due to an initially symmetric large amplitude perturbation. This problem is solved using Navier-Stokes equations for an incompressible fluid as described by Equations (10) to (14). The computation procedure is the same as that in the drop case. However, the boundary conditions are changed to model a spatially periodic and temporary varying two dimensional liquid film. An initially stationary liquid film is perturbed by similar sinusoidal surface disturbances on both top and bottom surfaces. A two dimensional liquid film is analytically predicted to be stable to symmetric disturbances. The computational results, presented in Figure 7, also show that the sinusoidal wave at the surface of the liquid film damps out in a few time steps as expected. It is interesting to note that the viscous effects cause that the surface shape is not sinusoidal after short time. Also it is observed that the valleys of the initial sinusoidal disturbance last longer than the peaks. This is thought to be due to the persistence of the internal flow towards the center of the film.

The second problem which is considered here is calculation of shape and internal flow fields in a liquid film subject to an initially asymmetrical sinusoidal disturbances. The purpose of this asymmetrical disturbance is to see the effect of the interaction between the wave on the top surface and the wave on the bottom surface. The test results are shown in Figure 8. The wave at the bottom surface, which has a higher frequency, damps out much faster than the lower frequency wave at the top surface. And while the higher frequency wave is damping out, it prevents the diminishing of the peaks of the lower frequency wave. The effect comes from the flow field within the liquid film. The liquid pushed up by the peaks at the bottom surface is being transferred to the peaks at the top surface while the wave at the bottom surface is damping out. Figure 8-c shows that the complicated flow field inside the liquid film generates several other surface perturbations which did not exist initially. Detailed computational study of dynamics of viscous drops and films are being carried out at the present time.

CONCLUSIONS

A new technique of interface transport and reconstruction is applied for the numerical modeling of liquid drops and films. This technique, called Flux Line-segment model for Advection and Interface Reconstruction (FLAIR), is more accurate in advection of the fraction of volume field than the previous algorithms. Therefore, the details of drop deformation and flow field at finite Reynolds numbers can be determined. Also a more accurate method for calculating the curvature is developed which will allow long time computations provided an accurate momentum calculation scheme is attached. Sample calculations are carried out for shape and internal flow field of perturbed viscous drops and films. High viscosity elliptic drops will commence to be circular without going through typical prolate-oblate

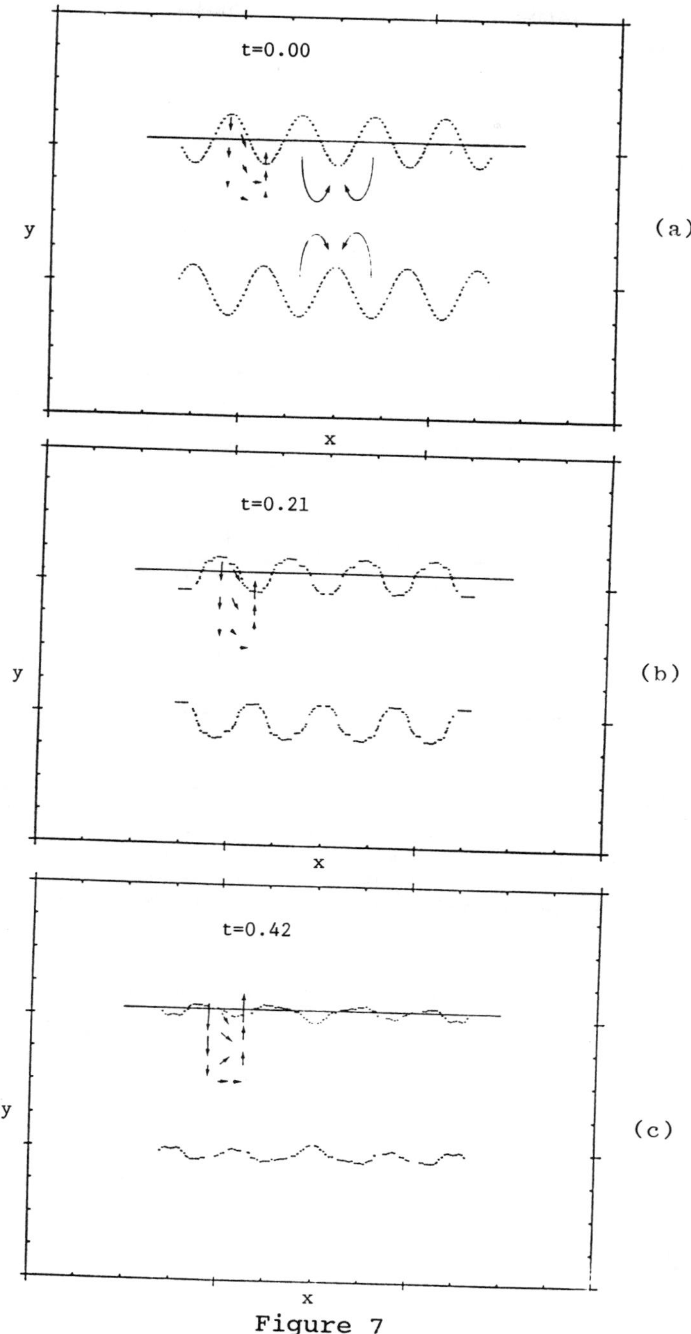

Figure 7

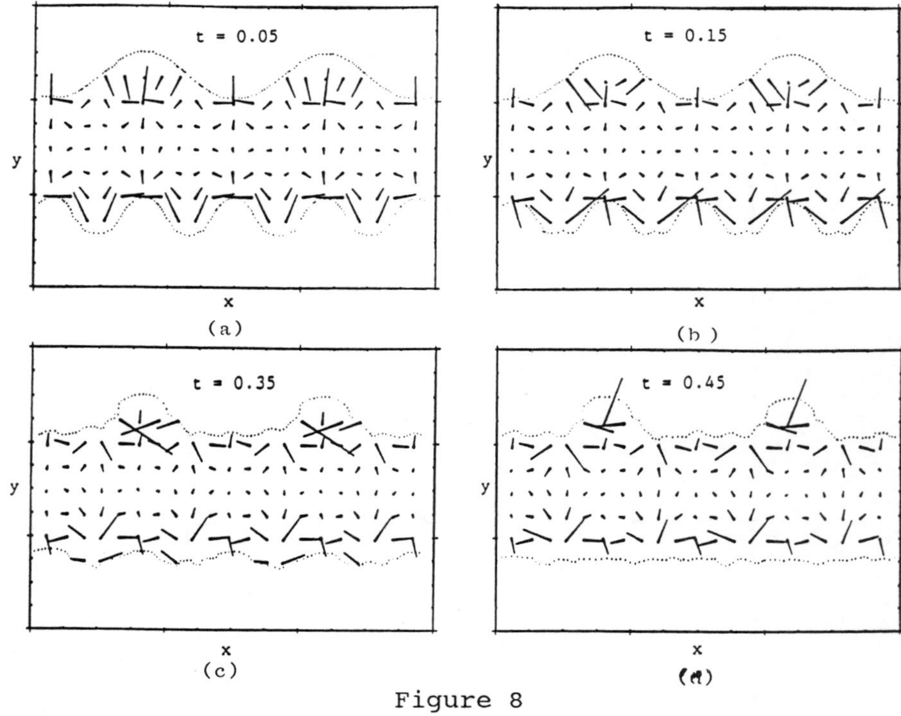

Figure 8

oscillations of the type observed in inviscid drop oscillations. Viscous liquid films subject to large amplitude symmetric and asymmetric sinusoidal surface disturbances exhibit interesting surface shape variation with time. The surfaces do not stay sinusoidal during the damping period and the surface with higher frequency disturbance damps out faster.

REFERENCE

1 R. W. Yeung (1982), *Ann. Rev. Fluid Mech.*, **14**:395-442.
2 A. S. Geller, S. H. Lee, & L. G. Leal (1986), *J. Fluid Mech.*, **169**, 27.
3 I. S. Kang & L. G. Leal (1987), *Phys. Fluids*, **30**(7), pp. 1929-1940.
4 G. Ryskin & L. G. Leal(1983), *J. Comput. Phys.*, **50**, 71.
5 F. H. Harlow & J.F. Welch (1965), *Phys. Fluids*, 8:2182-2189.
6 G. R. Foot (1971), Ph. D. Thesis, Dept. of Atm. Sci., University of Arizona.
7 C. W. Hirt & B. D. Nichols (1981), *J. Comput. Phys.*, 39:201-225.
8 B. D. Nichols, C. W. Hirt, & R. S. Hotchkiss (1980), Los Alamos Scientific Laboratory, LA8355.
9 W. F. Noh (1964), in B. Adler, S. Fernbach & M. Rotenberg, eds., **Methods in Computational Physics**, **3**:117-179.
10 W. F. Noh & P. Woodward (1976), in A.I. van de Vooren & P.J. Zandbergen, eds., *Lecture Notes in Physics, Vol.* **59**: **Proc. of the Fifth Int. Conf. on Numerical Methods in Fluid Dynamics**, Springer-Verlag, NY, 330-340.
11 A. J. Chorin (1980), *J. Comput. Phys.*, 35:1-11.
12 Smith, J., (1981),*J. Comput. Phys.*,. 39, 112.
13 A. J. Chorin (1985), *J. Comput. Phys.*, 57, 472.
14 N. Ashgriz & J. Y. Poo (1989), submitted to the *J. Comput. Phys.*,,.
15 J. Y. Poo & N. Ashgriz (1989), *J. Comput. Phys.*, in press.

FIGURE CAPTIONS

4. Astrophysics

Norman R. Lebovitz
Session Chair

GRAVITATIONAL INSTABILITIES IN ASTROPHYSICAL FLUIDS

Joel E. Tohline
Department of Physics and Astronomy, Louisiana State University, Baton Rouge, LA 70803-4001 U.S.A.

ABSTRACT

Over the past decade, the significant advancements that have been made in the development of computational tools and numerical techniques have allowed astrophysicists to begin to model accurately the nonlinear growth of gravitational instabilities in a variety of physical systems. The fragmentation or rotationally driven fission of dynamically evolving, self-gravitating "drops and bubbles" is now routinely modeled in full three-dimensional generality as we attempt to understand the behavior of protostellar clouds, rotating stars, galaxies, and even the primordial soup that defined the birth of the universe. A brief review is presented here of the general insights that have been gained from studies of this type, followed by a somewhat more detailed description of work, currently underway, that is designed to explain the process of binary star formation. A short video animation sequence, developed in conjunction with some of the research being reviewed, illustrates the basic nature of the fission instability in rotating stars and of an instability that can arise in a massive disk that forms in a protostellar cloud.

1. INTRODUCTION

The past decade has seen an enormous enhancement in our ability to study the dynamics of self-gravitating fluids through the techniques of computer modeling. Instabilities that have been known to exist through linear stability analyses for most of this century are now being studied and analyzed as they develop into nonlinear structures. We have now begun to get a handle on just how efficient dynamical fragmentation in the expanding, early universe must have been and how effective fragmentation is at producing clusters of young stars in interstellar clouds. The concept that binary fission can occur via a dynamical instability in rapidly rotating stars has been cast aside as a result of nonlinear models, and we have begun to study the breakup of massive accretion disks around protostars as a possible avenue to binary star formation. We begin the review of these topics with a quick overview of the general physical concepts that provide a common thread through all of these studies.

2. GENERAL PHYSICAL CONCEPTS

The physical ingredient that makes studies of astrophysical systems unique, when compared to most other investigations of the properties of drops and bubbles, is gràvity. Each of the four areas of investigation described in this review examine the nonlinear development of a gravitationally driven, dynamical instability in some type of astrophysical fluid: the first depicts the growth of structure in a dynamically expanding flow that is generally thought to be representative of our evolving universe; the second examines fragmentation in dynamically contracting flows in an attempt to understand how clusters of stars form in our Galaxy; and the last two examine the growth of distortions in rapidly rotating, equilibrium structures in an attempt to understand how planetary systems and/or short-period binary star systems form.

Other than the inclusion of self-gravity, the mathematical technique used to describe the time-evolution of these physical systems is familiar—standard equations representing the three-dimensional (3-D) flow of inviscid, compressible fluids (or gases) are employed. Generally, an explicit time-integration of the continuity equation,

$$\frac{\delta\rho}{\delta t} + \nabla \cdot (\rho \mathbf{v}) = 0 \quad (1)$$

an energy equation representing the first law of thermodynamics,

$$\frac{\delta u}{\delta t} + \nabla \cdot (u \mathbf{v}) = -P \;\; \nabla \cdot \mathbf{v} + c_1 u \frac{ds}{dt} \quad (2)$$

and the three components of the equation of motion,

$$\frac{\delta S}{\delta t} + \nabla \cdot (S \mathbf{v}) = -\rho \frac{\delta \phi}{\delta r} - \frac{\delta P}{\delta r} + \frac{A^2}{\rho r^3} \quad (3)$$

$$\frac{\delta T}{\delta t} + \nabla \cdot (T \mathbf{v}) = -\rho \frac{\delta \phi}{\delta z} - \frac{\delta P}{\delta z} \quad (4)$$

$$\frac{\delta A}{\delta t} + \nabla \cdot (A \mathbf{v}) = -\rho \frac{\delta \phi}{\delta \theta} - \frac{\delta P}{\delta \theta} \quad (5)$$

—written here in cylindrical polar coordinates—is performed in conjunction with an implicit solution of the Poisson equation,

$$\nabla^2 \Phi = 4\pi G\rho \quad (6)$$

This group of equations specifies, as a function of the time t and the Eulerian coordinate positions (r, z, θ), the evolution of five principal dynamical variables—the mass density ρ, the internal energy density u, the angular momentum density A, and the linear momentum densities S and T in the r- and z-directions, respectively. Besides the gravitational constant G and the constant c_1, which is definable in terms of the gas constant $\Re$ and the ratio of specific heats of the gas γ, the standard equations are written here in terms of: the fluid velocity $\mathbf{v} \equiv (v_r, v_z, v_\theta)$, where $v_r \equiv S/\rho$, $v_z \equiv T/\rho$, $v_\theta \equiv A/(r\,\rho)$; the gas pressure P; the gravitational potential Φ, and the specific entropy of the gas s. In the studies being reviewed here, the system of equations is closed by adopting a perfect gas relation between P and u, *i.e.*,

$$P = (\gamma - 1)u, \quad (7)$$

and (usually) by adopting either an isothermal ($P = c^2\rho$ where c is the chosen isothermal sound speed) or adiabatic ($P = K\rho^\gamma$, where K specifies the specific entropy of the homentropic gas) equation of state, in which case the time-variation of u is given trivially in terms of the time-variation of ρ. Whenever such a simple equation of state is used, neither an explicit specification of ds/dt nor numerical integration of Equation (2) is required.

In order to get a feeling as to how gravity can totally dominate the global evolution of an astrophysical system, consider the behavior of a nonrotating, spherically symmetric gas cloud that initially sits at rest with a uniform gas density ρ_o but in which the pressure is everywhere zero, *i.e.*, consider the collapse of a "pressure-free" sphere. As explained, for example, by Tohline (1982), the time-evolution of this system can be described very simply through an *analytical*, Lagrangian integration of the one-dimensional (spherically symmetric) equation of

---

Although the studies described here deal with inviscid fluids, one should not conclude that viscous effects are always unimportant in astrophysical fluids, nor that they are always ignored. A great deal is known from linear stability analyses, for example, about the importance of viscosity in self-gravitating systems [see Chapters 10-11 of Tassoul 1978, and references cited therein]. Nonlinear studies, however, have not yet come to grips with this tougher problem owing, primarily, to the differing time scales that arise and to the necessity for implicit integration of the dynamical equations when the effects of viscosity are considered.

motion in which pressure gradients are ignored. The analytic solution shows a uniform-density, spherically symmetric collapse in which the cloud density ρ remains uniform but it increases steadily with time (formally becoming infinitely large at a time when the cloud radius shrinks to zero) according to the expression

$$\frac{\rho}{\rho_o} = \sec^6 \zeta \tag{8}$$

where ζ specifies the time through the parametric equation

$$\zeta + \frac{1}{2}\sin 2\zeta = \frac{\pi}{2}\frac{t}{\tau_d}, \tag{9}$$

$$\tau_d \equiv [3\pi/32G\rho_o]^{1/2} . \tag{10}$$

In this example problem, the gas cloud dynamically evolves to a drastically altered state on a time scale $\tau_d \sim 1/\sqrt{G\rho_o}$..

A very important lesson can be learned from this simple example: if gravity exerts a dominating influence on the evolution of a physical system, the governing dynamical time scale will be $1/\sqrt{G\rho_o}$ where ρ_o is the mean density of the system. This statement holds true in much more complicated classical systems as well as in expanding general relativistic gas flows. For this reason τ_d, as defined above, is the characteristic time scale that will show up in every study covered by this review.

Now consider a situation in which gas pressure exerts a *non*negligible influence on a self-gravitating fluid. Through a linear perturbation analysis of the one-dimensional (1-D) equations governing fluid flow (*cf.*, Chandrasekhar 1961, §119) one identifies a critical perturbation wavelength,

$$\lambda_J \equiv [\pi c^2 / G\rho_o]^{1/2} \tag{11}$$

below which perturbations do not amplify but, instead, propagate through the fluid as sound waves. A perturbation having a wavelength $\lambda > \lambda_J$, or, equivalently, an enclosed mass

$$M > M_J \sim \rho_o \lambda_J^3 \tag{12}$$

however, encounters the so-called "Jeans instability" (named after the British astronomer who, in the late 1800's, first discovered it) and grows exponentially on a time scale τ_d. It is the nonlinear development of this general type of instability—associated with gas masses $M > M_J$—that will be described as I review what we have learned in recent years about fragmentation in expanding cosmologies and in collapsing protostellar clouds.

Fluid systems in which pressure gradients everywhere balance the force of gravity and which, at the same time, are stable against a Jeans-type collapse and fragmentation, do exist in Nature. Our Sun, the Earth, or any one of the multitude of stars and planets that we see in the sky is an example. In the absence of rotation, such a fluid system will exist as a spherically symmetric object, having a structure describable by 1-D equilibrium equations. If angular momentum is added to the system, the object will flatten into an equilibrium structure approximately represented by an axisymmetric, oblate spheroid and its equatorial-to-polar axis ratio $a/b > 1$ will get larger and larger as the system is spun faster and faster. At some critical degree of flattening ($a/b \sim 3$), a rotationally driven dynamical instability will set in, permitting the object to develop large-amplitude, nonaxisymmetric structure. (In rotating *viscous* fluids, an important nonaxisymmetric instability is known to arise in systems having an axis ratio somewhat smaller than the one required for the dynamical instability to set in. As has been reviewed, for example, by Tassoul [1978] and Durisen and Tohline [1986], this viscously driven instability is also believed to play an important role in certain stages of stellar evolution. Because this instability is generally expected to develop on a time scale that is much longer than the dynamical time scale and because, to date, its nonlinear development has not been satisfactorily modeled in realistic astrophysical systems, it will not be discussed further here.)

Because centrifugal forces are comparable to gravity in a rotationally flattened system, the time scale governing the development of a rotationally driven instability necessarily will be on the order of both τ_d and the rotation period of the system. Linear stability analyses have indicated that, when this instability first sets in, the nonaxisymmetric distortion that grows most rapidly has an ellipsoidal, or barlike, geometric shape—*i.e.*, considering distortions in the azimuthal coordinate θ of the form $\frac{\delta\rho}{\rho_0} \propto \cos(m, \theta)$, the fastest growing mode has $m = 2$. It is the nonlinear development of this $m = 2$, or "bar-mode," instability that will be described as I review what recent numerical experiments have taught us about instabilities in rapidly rotating stars and in massive accretion disks.

Both the Jeans-type instability and the rotationally driven, dynamical instability just described can only be adequately studied by integrating forward in time the fully three-dimensional (3-D) nonlinear hydrodynamic equations that govern the behavior of compressible, self-gravitating fluids. In situations where gas pressure is thought to play a negligible role in the dynamics, noninteracting Lagrangian particles can be used to represent the fluid and N-body integration schemes have been used to study gravitational fragmentation. In situations where gas pressure is thought to be important, both "finite-difference" representations of the governing partial differential equations and "smoothed particle hydrodynamics" techniques have been used to study various aspects of these problems. In what follows, examples of all three numerical integration techniques will be shown, although the concentration will be on recent results produced with "finite-difference" codes.

3. EXPANDING COSMOLOGIES

At the present time, the universe exhibits a great deal of structure. Nonlinear density enhancements are evident in the form of luminous galaxies, clusters of galaxies, and superclusters of galaxies. At the longest length scales presently observable, ordered structures resembling huge voids (or bubbles), filaments and pancakes are seen. On the other hand, observational data reflecting the properties of matter in the very early universe, such as data obtained from measurements of the global isotropy of the 3° background radiation, indicates that the universe was initially very homogeneous. Recently, numerical simulations have been used to demonstrate how, during the first ten billion years of its life, nonlinear-amplitude structures have been able to grow spontaneously from initially very low amplitude perturbations in the dynamically expanding universe.

One of the best demonstrations of how structures developed during the early evolution of the expanding universe has been presented by Centrella and Melott (1983, 1984). Centrella and Melott used N-body integration techniques to study the growth of gravitational instabilities in a cubic cell of pressure-free gas whose volume is steadily undergoing radial expansion as it participates in the global, general relativistic expansion of the universe. The cubic cell contained a great deal of matter, but it was assumed to occupy a small enough fraction of the total volume of the universe that Newtonian gravitational interactions were sufficient to describe the evolution of fluid structures within the cell itself. What they found was that gravitational instabilities can grow on a dynamical time scale in a relativistically expanding, pressure-free fluid. However, they found that structures of differing shapes, differing length scales, and differing degrees of complexity can arise, depending on what spectrum of initial perturbations is introduced into the

See, also, the nice account by Centrella [1988] of how computer graphics and imaging can be very useful in scientific studies of this sort.

fluid initially, and depending on how rapid the global expansion of the universe is assumed to be (*i.e.*, whether the universe is assumed to be very open, or only marginally so).

Centrella and Melott's "best fit" to currently observed, large-scale structures in the universe is illustrated nicely in the opening color figure of Centrella (1988). Large bubbles, filaments and pancakes are seen to have developed in the fluid in ample supply within the first ten billion years of the expansion of the universe. Looking ahead to the results of the next few sections of this review, what Centrella and Melott have also shown is that gravitational fragmentation is a very efficient process in expanding, pressure-free fluids. As we shall see, structures develop much less readily in fluids that undergo global *collapse*, and they develop much less efficiently in systems where thermal pressure plays a role in governing the dynamical motions of a fluid.

4. COLLAPSING PROTOSTELLAR CLOUDS

New stars are observed to be forming at a fairly steady rate from the interstellar medium in our Galaxy and in other galaxies. It is generally thought that the Jeans instability initiates the star formation process in most instances. The interstellar medium is composed, in large part, of molecular hydrogen gas clouds embedded in a more diffuse, neutral hydrogen gas having a mean density $\rho_o\ 10^{-23} \mathrm{gcm}^{-3}$. This medium has a density that is more than twenty decades smaller than the mean density of the gas found in normal stars like the Sun. Following the onset of a Jeans instability, therefore, the volume of gas which encloses the mass $M>M_J$ must undergo an enormous amount of compression on its way to forming a star.

After encountering an initial Jeans instability, can a single, massive gas cloud spontaneously fragment into multiple pieces, thereby giving birth to a cluster of moderate- or low-mass stars instead of forming one massive, single star? Apparently Nature's answer to this question is "Yes," because multiple star systems are the rule rather than the exception in our Galaxy. At the previous gathering of this audience, Silk (1981) summarized our, then current, understanding of fragmentation in this type of setting as it was understood from linear stability analyses. In order to honestly test our physical models of star formation and to fully understand the process of cloud fragmentation, however, we must follow the nonlinear development of gravitational instabilities using 3-D hydrodynamics. As you may well imagine, this is a difficult task. Given the huge compressions involved, it is a challenge to model the dynamical transition of the interstellar medium from its initially diffuse state to its final, compact stellar state even under the assumption of spherically symmetric (1-D) compressions. To adequately resolve an evolution in three dimensions is impossible, even given current supercomputer technology. Nevertheless, over the past decade some strides toward understanding the process of gravitational fragmentation in dynamically collapsing clouds have been made.

I will summarize for you the works of Boss (1986) and Miyama (1988) on this subject since they have published, by far, the most complete results based on numerical simulations of the cloud fragmentation process. Examining a wide range of initial parameters, they have exhaustively studied the tendency for *rotating* gas clouds to fragment while undergoing either isothermal or adiabatic collapse. The results can be summarized most easily by examining a two-dimensional parameter space: one parameter,

$$\beta_o \equiv \frac{\textit{Kinetic Energy}}{|\textit{Gravitational Potential Energy}|}\ , \tag{13}$$

identifies the relative importance of rotation in the initial cloud ($\beta_o = 0$ means no rotation and β_o=1/3 means centrifugal forces initially balance gravity in the equatorial plane of the cloud); the second parameter,

$$\alpha_o \equiv \frac{\textit{Thermal Energy}}{|\textit{Gravitational Potential Energy}|} , \qquad (14)$$

specifies how many Jeans masses the cloud encompasses (α_o= 1/2 means $M=M_J$ so the cloud is only marginally Jeans unstable initially, and α_o<<1 means $M>>M_J$.

For strictly isothermal collapses and for adiabatic collapses in which $\gamma \leq 4/3$, cloud fragmentation occurs, at least to some degree, as long as α_o<1/2. Although some amplification of initial density perturbations in clouds can occur *during* the dynamical collapse phase—particularly for those clouds that enclose many Jeans masses initially—the fragmentation process seldom goes to completion until after one dynamical time when the entire cloud has collapsed to a rotationally flattened disk. For adiabatic collapses in which γ>4/3, fragmentation tends not to occur unless α_o is initially very small. Even large initial nonaxisymmetric perturbations in the cloud density die away if α_o is initially close to 1/2, basically reflecting the fact that the cloud encompasses only one Jeans mass of material.

Several figures presented by Miyama (1988) illustrate to what degree fragmentation takes place as a function of α_o and β_o in clouds that collapse along adiabats having different adiabatic exponents. His figures show that, as the parameter α_o drops, a cloud is able to fragment into a larger number of pieces. Also, as promised, gravitational fragmentation proceeds much more efficiently when a cloud collapses along an adiabat that has a small index. If $\gamma \leq 4/3$, a cloud will also fragment into a larger number of pieces if β_o is smaller.

Instead of studying the evolution of clouds that are forced to follow a fixed adiabat, Boss (1986) has examined the more physically realistic situation of gas clouds, initially having different masses, that begin their collapse from interstellar cloud densities along an isothermal path but that, later, heat up, effectively moving onto adiabats having γ>4/3. Figure 2 in Boss (1986) displays the geometric structure of four clouds near the end of their dynamical phase of collapse; unlike the examples shown in Miyama's work, in Boss's study all four clouds began their collapse with identical values of α_o and β_o— 0.25 and 0.04, respectively. Because the models had different total masses, however, nonisothermal effects set in at different rates (the least massive cloud *d* heated up much sooner during its collapse than did the most massive cloud *a*) and the fragmentation efficiency changed drastically from one model to the next. As both Boss (1986) and Miyama (1988) have explained, the relative ease with which different clouds fragment can be explained fairly readily in terms of the relative competition between thermal pressure, gravity, and rotation (see also Tohline 1982).

5. RAPIDLY ROTATING STARS

Early this century, mathematicians and astronomers discovered via linear perturbation analyses that a rapidly rotating star can become susceptible to a nonaxisymmetric, dynamical "bar-mode" instability if rotation is sufficiently important that β_o> 0.27 in the star. Based on the discovery of this instability in inviscid fluids, and the discovery of a similar type of gravitational instability in viscous fluids, speculation arose that, in the nonlinear growth regime, a rotationally driven instability might provide an avenue through which single stars can fission into close binary systems. Working with Richard Durisen (Indiana University), and a host of other collaborators, I have worked over the past several years to try to examine the nonlinear development of the dynamical "fission" instability and determine how useful it is in creating binary star systems. In summary, what we have found is that the instability does not lead to binary fission, at least in the sense that binary fission was originally conceived.

A videotape we have made illustrates graphically the outcome of a typical simulation which starts with a centrally condensed and differentially rotating star that is rotating sufficiently rapidly that linear theory would predict it is unstable toward the development of the heretofore mentioned "bar-mode" instability. As has been described in detail by Durisen and Tohline (1986), Durisen *et al.* (1986) and Williams and Tohline (1988), independent of the exact type of nonaxisymmetric perturbation that is introduced into the otherwise initially axisymmetric structure, an $m = 2$ barlike distortion grows spontaneously with an exponential growth rate and an azimuthal pattern speed in the fluid that matches the predictions of linear theory. As soon as the exponentially growing pattern becomes well-defined, however, it is clear that the natural eigenmode of this type of system is somewhat different than was expected from linear analyses. The growing mode exhibits a two-armed spiral pattern, rather than a pure barlike shape. As a result of the spiral pattern, the nonlinear development of the pattern does not result in the appearance of a rotating ellipsoid which pinches off, or bifurcates, through its center into a binary system. Instead, gravitational torques acting between the central regions of the object and the outer, trailing spiral arms efficiently redistribute angular momentum from the central regions, outward. High specific angular momentum material in the spiral arms then gets shed in the equatorial plane of the system and a less flattened, central object is left behind.

As a result of this work, the notion that the classical, dynamical nonaxisymmetric instability in rapidly rotating stars can lead to binary fission must be abandoned. As Williams and Tohline (1988) discuss, however, the information that has been learned in these studies about the efficiency with which gravitational torques can redistribute angular momentum in such systems will help us understand more fully the process by which circumstellar disks form around objects as diverse as protostars and neutron stars.

6. ACCRETION DISKS AND TORI

There are many situations in astronomy in which an extended, rapidly rotating disk of material is observed to surround a central, "point-like" object of mass M_c. The ring system around the planet Saturn is one example that is familiar to most people, but gaseous disks are also observed around many compact stellar objects, such as white dwarf stars and neutron stars, and disks from which material accretes onto massive black holes are believed to play a fundamental role in fueling the tremendously energetic radiation phenomena that emanate from the nuclei of galaxies. The disk, usually having a mass $M_d \ll M_c$, rotates differentially under the influence of the $1/r$ potential defined by the gravitational field of the central object. When material flows from the disk onto the central object, the system is often referred to as an accretion disk system and the flow is called the accretion flow. If gas pressure is, structurally, relatively unimportant in the disk, then the disk will be geometrically thin and its circularized orbits will exhibit a "Keplerian" velocity profile, *i.e.*, as a function of radius r, the angular velocity Ω of orbiting fluid elements will be

$$\Omega \propto r^{-\ell/2}, \tag{15}$$

with $\ell = 3$. If gas pressure is structurally important, then the disk can be geometrically thick, looking more like a torus than a thin annulus, and it can exhibit an angular velocity profile much different from the Keplerian one. Understanding the dynamical behavior of accretion disks (or tori), particularly the dynamical development of nonaxisymmetric structures in the disks, is fundamental to understanding many astrophysical phenomena. It is the dynamical response of an accretion disk to either internally or externally generated disturbances that ultimately controls the rate and level at which material flows from the disk onto the central object about which it orbits.

It has been known for some time that fluid systems which rotate differentially in the presence of a central force field encounter an *axisymmetric* dynamical instability if the specific angular momentum of the fluid $j=r^2\Omega$ decreases outward, *i.e.*, if the value of ℓ in (6.1) is $\ell< 4$. The criterion $\ell>4$ demanded for axisymmetric stability was first derived for inviscid, incompressible fluids by Rayleigh, but it generalizes to homentropic, compressible fluids as well (*cf.*, Tassoul 1978). This particular instability has not aroused much interest in accretion disk studies, primarily because it does not arise in thin, Keplerian disks; it is of even less interest to us here because it does not require the influence of gravity in the disk.

Over the past few years, however, the study of dynamical instabilities in accretion disks has received a great deal of attention because astronomers have begun to realize that:

1. Fat accretion tori can exist with nonKeplerian angular velocity profiles;
2. Nonaxisymmetric, dynamical instabilities can arise in disks whose angular velocity profiles are stable according to the axisymmetric, Rayleigh criterion;
3. In certain systems, the accretion disk may have a mass $M_d \geq M_c$, in which case instabilities driven by the self-gravity of the disk may develop.

Much of this renewed interest has been stimulated by the pioneering work of Papaloizou and Pringle (1984) who demonstrated, through a local linear stability analysis, that the Rayleigh criterion is not sufficient to stabilize nonaxisymmetric modes in incompressible fluids. In their initial investigation, Papaloizou and Pringle identified *sonic* modes that are able to grow exponentially on, roughly, the sound-crossing time in zero mass tori having power-law angular velocity profiles in the range $4>\ell> 3.5$. More recently, Goodman and Narayan (1988) have extended this type of study to include self-gravitating disks and have found, at least in very slim tori, that a Jeans-type, nonaxisymmetric instability can also arise if, effectively, the ratio M_d/M_c is large enough.

In collaboration with Izumi Hachisu (Kyoto University), I have embarked on a long-term project to study the linear and nonlinear development of gravitationally driven nonaxisymmetric instabilities in accretion disks. Our study is motivated primarily by the desire to find out if, in protostellar gas clouds, massive accretion disks can fragment into a few large pieces, giving rise to the formation of binary or small multiple star systems. It is evident from the works of Papaloizou and Pringle (1984) and Goodman and Narayan (1988), however, that the results of this project should influence studies of accretion disks in a wide range of other areas of astrophysical research. I will share with you here some of the earliest findings of this numerical study (see also Tohline and Hachisu 1989).

While examining the stability of fully self-gravitating tori, *i.e.*, systems in which there is no central mass ($M_d/M_c= \infty$), we have discovered that a global $m= 2$ instability grows dynamically in systems having β_o as small as 0.17. That is, the instability sets in when the rotational kinetic energy of the system is substantially smaller, relative to the system's gravitational potential energy, than it must be for a dynamical instability to arise in centrally condensed stars. In addition, by following the nonlinear development of this instability, we have found that the $m= 2$ mode does not grow without bound. Once the $m= 2$ mode reaches a certain finite amplitude, the torus settles into a steady-state structure that has an ellipsoidal, or "barlike," shape and a correspondingly deformed central hole. Actually, the final structure should not be thought of as an ellipsoidal torus that is simply resting in inertial space, nor should it be visualized as a bar that is tumbling about its shortest axis as a solid body. Instead, the steady-state structure is a geometrically thick, differentially rotating fluid disk in which a large-amplitude, ellipsoidal wave sloshes around the disk with a pattern speed that is slow compared to all fluid elements in the disk. Therefore, this gravitational instability, when it first sets in, does not cause the disk to break into two well-defined pieces. A small, binary density enhancement is observed to develop at the ends of the steady-state bar, but unlike a well-defined binary system, fluid continues to flow around the distorted torus and no Lagrangian fluid elements become permanently assigned to either enhanced region.

As β_o is increased more and more above the level where the instability first sets in, the $m = 2$ distortion grows to a larger and larger limiting amplitude. Finally, at $\beta_o \sim 0.23$, the instability appears to grow without bound. Apparently, at this higher rotation rate, azimuthal modes having $m > 2$ begin to influence the nonlinear evolution. Within one or two rotation periods, the ellipsoidal torus pinches together along its narrowest dimension, and the fluid is segmented into two pieces.

Through this investigation, we have identified new, dynamically important phenomena that can arise in self-gravitating accretion disks. We are extending the analysis to include systems having mass ratios in the entire range $0 < M_d / M_c < \infty$ so that our study can contribute to a broad spectrum of astrophysical fluid systems.

7. SUMMARY

Large scale computing facilities have tremendously enhanced our ability to study the nonlinear growth of gravitational instabilities in astrophysical fluid systems. Some of the studies described in this review are still in their infancy, but already it is clear that models of dynamical fission and fragmentation are destined to play a permanent, important role in constraining or guiding the development of astrophysical theories in a number of diverse arenas. It is also clear that techniques in computer graphics/imaging and video animation are intimately tied to computational studies of this magnitude and that the continued development of these techniques will benefit astrophysical research as well. The next decade promises a great deal of excitement in the study of gravitational instabilities in "drops and bubbles."

ACKNOWLEDGMENT

This research has been supported in part by grant number AST-8701503 from the U. S. National Science Foundation.

REFERENCES

A. P. Boss, 1986 "Protostellar Formation in Rotating Interstellar Clouds. V. Nonisothermal Collapse and Fragmentation." *Astrophys. J. Suppl. Ser.* **62**, 519-552.

J.M. Centrella, 1988 "Using the Computer as a Camera." *Computers in Physics* **2**, 34-39.

J. M. Centrella & A. L. Melott, 1984 "The Large-scale Structure of the Universe: Three-dimensional Numerical Models." *in* **Numerical Astrophysics** (ed. J. M. Centrella, J. M. LeBlanc, & R. L. Bowers), 334-360. Jones and Bartlet.

J. M. Centrella & A. L. Melott, 1983 "Three-dimensional Simulations of Large-scale Structure in the Universe." *Nature* **305**, 196-198.

S. Chandrasekhar, 1981 **Hydrodynamic and Hydromagnetic Stability.** Dover.

R. H. Durisen, R. A. Gingold, J. E. Tohline, & A. P. Boss, 1986 "Dynamic Fission Instability in Rapidly Rotating $n = 3/2$ Polytropes: A Comparison of Results from Finite-Difference and Smoothed Particle Hydrodynamics Codes." *Astrophys. J.* **305**, 281-308.

R. H. Durisen & J. E. Tohline, 1985 "Fission of Rapidly Rotating Fluid Systems." *in* **Protostars and Planets II** (ed. D. C. Black & M. S. Matthews) pp. 534-575. Univ. of Arizona Pr.

J. Goodman & R. Narayan, 1988 "The Stability of Accretion Tori — III. The Effect of Self-Gravity." *Mon. Not. R. Astron. Soc.* **231**, 97-114.

S. Miyama, 1988. *Astrophys. J.*, submitted.

J.C. B. Papaloizou, J. E. Pringle, 1984 "The Dynamical Stability of Differentially Rotating Discs with Constant Specific Angular Momentum." *Mon. Not. R. Astron. Soc.* **208**, 721.
J. Silk, 1981 "Fragmentation of Interstellar Clouds and Star Formation." *in* **Proceedings of the Second International Colloquium on Drops and Bubbles** (ed. D. H. LeCroissette), pp. 214-221. Jet Propulsion Laboratory.
J.-L. Tassoul, 1978 **Theory of Rotating Stars**, Princeton University Press.
J. E. Tohline,. 1982 Hydrodynamic Collapse. *Fund. Cosmic Phys.* **8**, 1-82.
J. E. Tohline & I. Hachisu, 1989, *Astrophys. J.* submitted.
H. A. Williams & J. E. Tohline, 1988 "Circumstellar Ring Formation in Rapidly Rotating Protostars." *Astrophys. J.* **344**, 449-464.

PION PRODUCTION IN RELATIVISTIC COLLISIONS OF NUCLEAR DROPS

C. T. Alonso, J. R. Wilson, T. L. McAbee, and J. A. Zingman
Lawrence Livermore National Laboratory,
Livermore, California 94550

ABSTRACT

In a continuation of the long-standing effort of the nuclear physics community to model atomic nuclei as droplets of a specialized nuclear fluid, we have developed a hydrodynamic model for simulating the collisions of heavy nuclei at relativistic speeds. Our model couples ideal relativistic hydrodynamics with a new Monte Carlo treatment of dynamic pion production and tracking. The collective flow for low-energy (200 MeV/N) collisions predicted by this model compares favorably with results from earlier hydrodynamic calculations which used quite different numerical techniques. Our pion predictions at these lower energies appear to differ, however, from the experimental data on pion multiplicities. In the case of ultra-relativistic (200 GeV/N) collisions, our hydrodynamic model has produced baryonic matter distributions which are in reasonable agreement with recent experimental data. These results may shed some light on the sensitivity of relativistic collision data to the nuclear equation of state.

1. INTRODUCTION

Ever since the discovery of nuclear fission, scientists have attempted, with remarkable success, to model the dynamics of atomic nuclei by assuming they behave like tiny drops of a nuclear fluid. In this fluid the strong nuclear force is analogous to the Van der Waals force of an ordinary liquid. Two main assumptions are made when a hydrodynamic model is applied to nuclear dynamics: (1) there are enough particles for the implied statistical nature of hydrodynamics to be valid (two fused uranium nuclei, for example, comprise 480 nucleons); and (2) the interaction distances are short compared to nuclear dimensions. This latter assumption is not well understood at present. Generally the nuclear force interaction length is approximately 1-2 fm (a fm or "fermi" is the unit of nuclear distance, equal to 10^{-13} cm) while the diameter of a heavy element nucleus is about 15 fm.

Since there is as yet no complete quantum description of relativistic heavy-ion collisions, we use macroscopic phenomenology to model the events that occur in these reactions. Many models have been proposed and investigated over the past decade, ranging from single particle models which attempt to mock up quantum effects, such as quantum molecular dynamics[1] and VUU theories,[2] to collective models such as the hydrodynamics reported here.[3,4,13]

Relativistic nuclear collisions can be studied at several new and proposed facilities. Low energy (2 GeV/N, or lower, in the laboratory frame) collision data have been available since 1975 at the Bevalac accelerator in Berkeley. Mid-energy beams (15 GeV/N, lab) have recently become available at Brookhaven's AGS. Also during the last year data from the SPS at CERN have been available at 200 GeV/N (lab) or 60 GeV/N (lab). Proposed for the future is RHIC, a U.S.-based accelerator capable of 100 GeV/N in the center-of-mass frame. The push toward higher energies reflects the desire to create and explore the quark gluon plasma predicted [5] at energy densities of several GeV/fm^3.

2. THREE-DIMENSIONAL RELATIVISTIC HYDRODYNAMICS

The relativistic ideal-fluid hydrodynamics problem may be cast as a set of coupled nonlinear conservation equations: a continuity equation for baryon number density, an energy equation, and a momentum equation. The specific forms[6] which we use are

$$\delta_t (D) + \frac{1}{G}\, \delta_i (DGV^i) = 0 \,, \tag{1}$$

$$\delta_t (E) + \frac{1}{G}\, \delta_i (EGV^i) + P\, \delta_t(\gamma) + \frac{P}{G}\, \delta_i (GU^i) = 0 \,, \tag{2}$$

$$\delta_t (S_j) + \frac{1}{G}\, \delta_i(GS_jV^i) + \delta_j P = 0 \,. \tag{3}$$

Here, γ is the Lorentz factor, $D = \rho\gamma$ is the coordinate density, $E = \rho\, \varepsilon\, \gamma$ is the internal energy density, $S_i = (D+E+P\gamma)\, U_i$ is the momentum density, ε is the specific internal energy and ρ is the density. P is the pressure, which is related to E and D through an equation of state. V^i is the transport velocity with respect to the coordinate grid, defined from the proper velocity U^i as $V^i = U^i/\gamma$. G is the square root of the determinant of the 3-metric. These forms differ slightly from those of other workers.[7] We solve these coupled, nonlinear equations using a monotonic continuous-fluid differencing scheme developed at Lawrence Livermore National Laboratory.[6,8,9] A second-order van Leer scheme is used for the advection terms. Momentum and baryon number are explicitly conserved. While energy is not ex-plicitly conserved, it has been rigorously tested to accuracies of a few percent as described below.

Hydrodynamic modeling of a nuclear collision involves three physical processes. First, the lighter nucleus and the center of the heavier nucleus are stopped by a strong shock wave. Second, a weaker shock wave propagates outward from the shocked central region into the tangential surface region of the heavy nucleus. Third, the system expands adiabatically. To test whether our numerical algorithm is sufficiently accurate under similar conditions we calculated wall shocks, shock tubes, and rarefactions and compared the results to analytic solutions.[9,10,11]

To obtain accurate results at very high velocities (for example 200 GeV/N, with a relativistic factor γ=10, corresponds to 0.995c in the equal speed frame) requires extremely careful treatment and testing of the solution scheme. We subjected our code to analytic shock tube tests, wall shock tests, and adiabatic expansion tests. Our typical accuracy was 1% to 3% in both the compression ratio and total energy conservation for a range of Lorentz factors from 1 to 10.

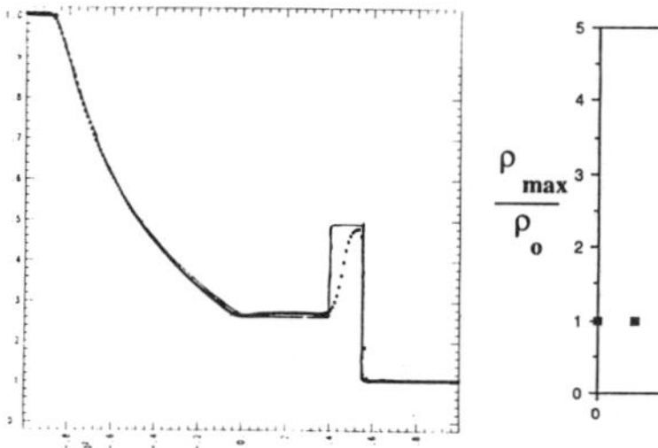
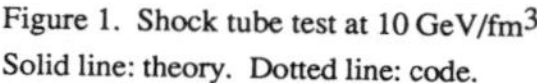

Figure 1. Shock tube test at 10 GeV/fm^3. Solid line: theory. Dotted line: code.

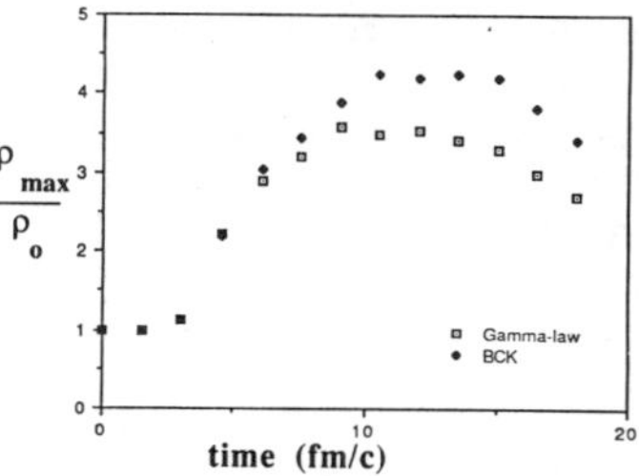

Figure 2. Maximum compression vs time for Au on Au at 200 MeV/N with two EOS's.

In Figure 1 we present the results of the shock tube test at a specific energy of 10 GeV/fm^3 in the equal speed frame. For a rarefaction with this initial specific energy we obtain an error of less than 1% in the proper velocity down to an expansion ratio of 10 in density. The shock velocity is also found to better than 1% accuracy. In the absence of shock compression our code remains on an adiabat to one part in several thousand for compression ratios of order 10.

3. THE NUCLEAR EQUATION OF STATE

The equation of state (EOS) of nuclear matter has importance in both nuclear physics and astrophysics. In the latter, whether nuclear matter is soft or hard greatly affects the bounce of collapsing stellar cores, and can determine whether or not a supernova explosion occurs. In the former, whether nuclear matter is soft or hard is believed to greatly affect pion and quark-gluon plasma production in high energy collisions. As the internal energy in a nuclear system increases, it has been predicted that nuclear matter might undergo a phase transition resulting in a plasma consisting of bound quarks and the gluons which hold the nucleons together. The latent heat of vaporization from nucleons to quarks is estimated[5] to be about 1-2 GeV/fm^3.

At lower energies a quasi-empirical nuclear EOS parameterization has developed over the years. Most researchers use the so-called Skyrme[12] formula. We have used a very similar EOS which basically describes a relativistic baryonic Fermi gas with delta resonances included.[13] Our EOS was constrained to fit the known data, namely the nuclear binding energy and the compressibility and pressure of normal nuclear matter.

At higher energies we chose to use the gamma-law equation of state (EOS)

$$P = (\Gamma-1)\, E/\gamma\ , \qquad (4)$$

covering the range from moderately soft to very stiff EOS's. This choice seems appropriate because real sound speeds require $\Gamma>1$ and causality limits require $\Gamma\leq2$ at very high energies. One expects $\Gamma=4/3$ for a quark gluon plasma at high energy, so a choice of $\Gamma=4/3$ in particular seems well justified.

We used our code to study heavy ion collisions in both the low and high energy regimes.[13] At Bevalac energies (200 and 1350 MeV/N, $\gamma= 1.3$) nuclear compressions appear to be only marginally sensitive to changes in the nuclear matter EOS. In Figure 2 we show the maximum compression as a function of time for Au on Au at 200 Mev/N using a hard and a soft EOS. The maximum compressions differ by about 15-20%. Given the large uncertainties in the experimental data, it is hard to discriminate between such small differences.

At CERN energies (200 GeV/N, $\gamma= 10$), nuclear compressions during high-energy collisions appear to be more sensitive to changes in the nuclear EOS. While wall shock compression ratios are limited theoretically to the value of 4 for nonrelativistic $\Gamma = 5/3$ fluids, for relativistic fluids the maximum wall shock compression for $\gamma= 10$ and $\Gamma= 5/3$ is 27. Figure 3 shows the maximum compressions calculated during the collision of ^{16}O and ^{208}Pb at 200 GeV/N as a function of time.

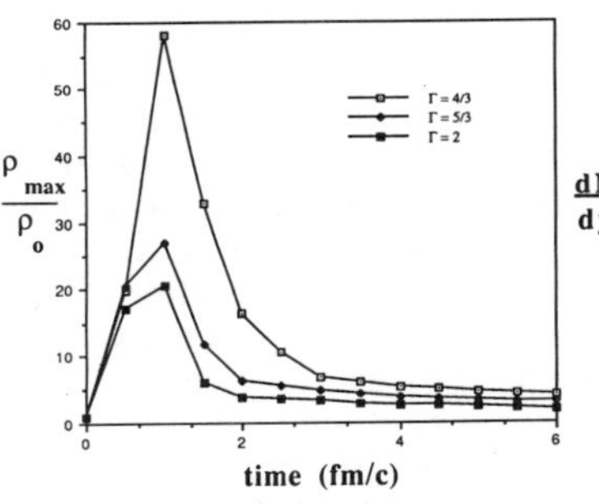

Figure 3. Maximum compression vs time for O on Pb at 200 GeV/N with Γ= 4/3, 5/3 and 2.

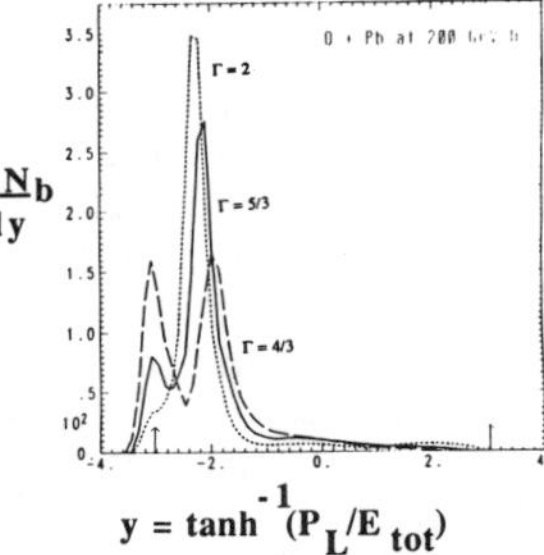

Figure 4. Differential baryon number vs rapidity for O on Pb at 200 GeV/N with Γ = 4/3, 5/3 and 2.

In Figure 4 we show the quantity dN_b/dy vs. y for three different gamma-law EOS's (Γ= 2, 5/3, and 4/3). Here N_b is the number of baryons present and the "rapidity" y is defined by

$$y = \tanh^{-1}(p_L/E_{tot}) \ , \tag{5}$$

where p_L is the longitudinal momentum of a particle (fluid element) and E_{tot} is its total energy. Rapidity is a favored quantity for describing nuclear collisions because its transfer is an indication of the stopping power (as opposed to transparency) of colliding nuclear fluids. The three curves in Figure 4 show a marked dependence upon the choice of nuclear EOS. There is a trend toward more baryons remaining at the initial rapidity for softer equations of state. For the Γ= 4/3 EOS we obtain baryon number densities greater than $20\rho_O$ for a time of 0.8 fm/c. A maximum total energy density at peak compression of about 60 GeV/fm^3 is attained.

4. PIONS AS A NUCLEAR FLUID SIGNATURE

In the search for a sensitive indicator of the nuclear EOS, much interest has fallen upon the pions which are generated and propagated in hot dense nuclear matter. It is our hope that they will act as little thermometers. Pion generation in nuclear matter depends upon attaining a sufficiently high local energy density. The details are not very well known, but the requirement is typically 0.2 GeV/fm^3. The scattering, absorption, and emission of pions depend upon the details of their interactions in hot dense nuclear matter and upon the equilibrium achieved. These details are not known in nuclear science today. We have been able to model these processes with the use of our pion code. This Monte Carlo pion overlay is now described below.

Pion production has heretofore only been investigated statically in chemical fluid models,[14,15] but our model treats the pions as dynamic Monte Carlo particles interacting with the baryonic fluid by exchanging momentum and energy during production, scattering, and absorption.[13] We define a master equation for the pion momentum distribution function,

$$\begin{aligned}\frac{dN_\pi(p_\pi)}{dt} = & \, \upsilon_\pi \rho \sigma_a(\rho, p_\pi) f_{BE}(p_\pi, T) - \upsilon_\pi \rho \sigma_a(\rho, p_\pi) N_\pi(p_\pi) \\ & - \rho N_\pi(p_\pi) \int \upsilon_{\pi N} \sigma_S(\rho, p_{\pi N}) D_B\,(p_{\underline{N}}, T) d^3 p_N + \rho \int \langle \upsilon_\pi \sigma_S N'_\pi \rangle d^3 p'_\pi\end{aligned} \tag{6}$$

Here $N(p)$ is the distribution function, σ_a and σ_S the absorption and scattering cross-sections for pions from nucleons, respectively, and < > indicate a thermal average over the nucleon velocity in the fluid rest frame. Subscripts p denote pion quantities, while N indicates a nucleon. ρ is the baryon density, f_{BE} is the Bose-Einstein distribution, which we take to be the equilibrium solution for the model, and D_B is a relativistic Boltzmann distribution for the nucleons. Our model does not require equilibrium conditions; in fact, we have used the code to study approaches to equilibrium. The first term in the equation represents the production of pions by the thermal motion of the baryonic fluid. The second represents absorption of pions on pairs of nucleons. Note that these terms are taken so that detailed balance is guaranteed. The third term describes pion-nucleon scattering out of a given momentum state; the last term describes the scattering of pions with momentum p'_p into the state p_π by a thermalized baryon distribution.

In studying dynamic pions in this manner for the first time, we soon discovered that pions pose many interesting questions. For example, what are the details of pion generation in nuclear matter? Does scattering occur via the bare pion-nucleon cross section or the pions dressed? Will the pions scatter on each other? How soon is equilibrium reached? How transparent is the nuclear matter? What is the duration of the pion-baryonic matter interaction? At what energy do pions dominate the energy balance? Do other particles like kaons and delta resonances matter? How will phase changes in the nuclear matter affect pion production and transport?

While we have been able to explore answers to some of these questions with our code, many other questions remain unanswered. Some of our results are given below for the low energy and high energy regimes.

5. LOW ENERGY (200 MEV/N AND 1350 MEV/N) SIMULATIONS

Many experiments have been done at the Bevalac in the 200 MeV/N and 1350 Mev/N energy ranges. Analysis of the data shows that at these energies the colliding nuclear drops stop each other and impart considerable transverse momentum which results in "side splashing". We simulated such side splashing for Au on Au at 200 MeV/N with our code, as can be seen in Figure 7b.

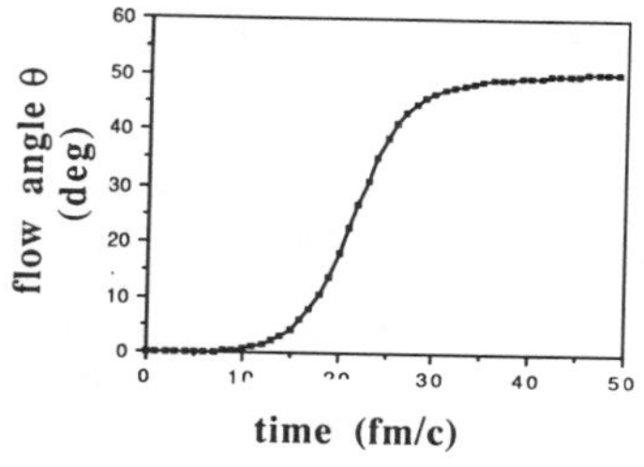

Figure 5. Flow angle vs time for Au on Au at 200 MeV/N.

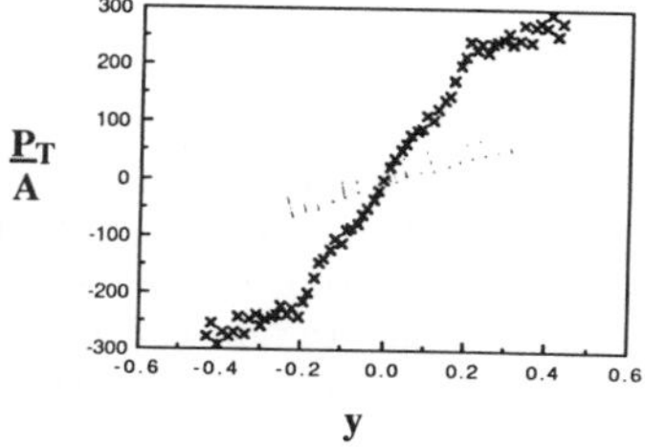

Figure 6. Transverse momentum vs rapidity for Au on Au at 200 MeV/N.

In Figure 5 we show our prediction, as a function of time, for the flow angle for this collision. Since rapidity is a measure of stopping power, Figure 6 shows the transverse momentum as a function of rapidity. The shaded line corresponds to deductions from experimental data. From experimental data the slope at zero rapidity has been deduced to be about 120 MeV/c.[16] Our hydrodynamic models, and those of other researchers, predict this slope to be more like 300 MeV/c. That is, in the low energy regime hydrodynamic models predict too much splash.

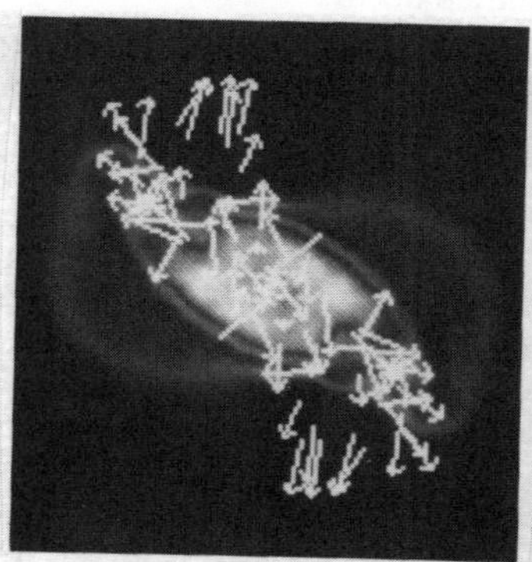

Figure 7a. Density plot for La on La at MeV/N at time 10 fm/c. The arrows show pion directions.

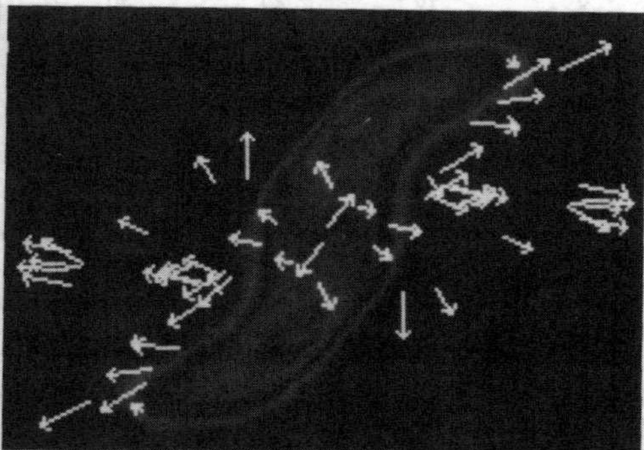

Figure 7b. Density plot for La on La at 1350 MeV/N at time 15 fm/c. The arrows show pion directions.

We also calculated the dynamic pion production and tracking for La on La at 1350 MeV/N (γ= 1.3). We show in Figure 7a and 7b our visual simulations, shaded in density, of the above reaction at times 10 fm/c (maximum compression) and 15 fm/c (separation).

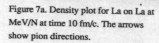

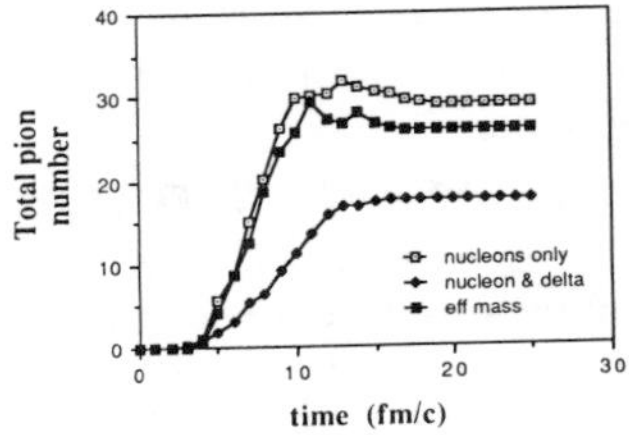

Figure 8. Pions vs time for La on La at 1350 MeV/N, for three different EOS's.

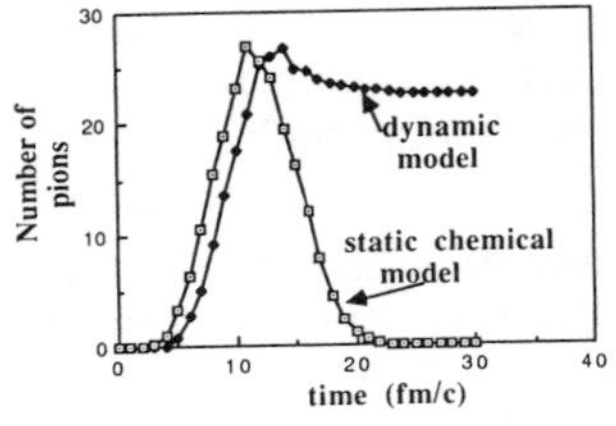

Figure 9. Pions vs time for La on La at 1350 MeV/N. Dynamic model compared to a static chemical model.

Figure 8 shows the number of pions present in the calculation as a function of time for three different EOS's. Most of the pions are generated during the transit of the initial shock. The most pions we generated was around 30, whereas experimental deductions indicate about 50 in the actual collision. Thus our dynamic pion hydro model seems to underpredict the pion multiplicity by almost a factor of two while it overpredicts hydrodynamic splash by a factor of two.

We see a clue as to where the missing pions may come from in Figure 9. We present here a time history of our pion production along with a static chemical model calculation in which the equilibrium number of pions was calculated in each cell and summed over the grid.[13] No pions from resonances were included. We observe that even though we do not force our system to assume an equilibrium solution, the number of pions at late times is very nearly that from the chemical model taken at maximum compression. We also observe a drop in the number of pions chemically produced at late times when the system expands and cools. The drop in the dynamic model comes from reabsorption of produced pions. We therefore conclude that we may be underpredicting with respect to the observed pion number[17] because we do not have a channel for pion production from resonance decay.

Inclusion of a mean field such that the pion energy E_π (ρ, p_π) is a function of the baryon density ρ as well as the pion momentum p_π is also an important improvement which is in progress.

6. HIGH ENERGY (200 GeV/N) SIMULATIONS

Newly available heavy-ion beams at the CERN SPS[18,19] have sparked much interest in the behaviour of hot hadronic matter at high compression. Predictions of a transition to a new phase of baryonic matter, the quark-gluon plasma, require that energy densities reach several GeV/fm^3. It has also been suggested that nuclei may be partially to completely transparent to each other in very high energy collisions.[20,5] Early experimental results[18,19] for the collisions of ^{16}O with ^{208}Pb however, point to a high degree of stopping at lab energies of 200 GeV/N. This large stopping power may be indicative of hydrodynamic-like behavior in the collision systems.

Our calculations were performed in the equal speed system, so target and projectile rapidities were initially symmetric about zero. In this system, for the collision of ^{16}O (200 GeV/N in the lab) with ^{208}Pb, the Lorentz factor is 10.41. To obtain accurate results with such high velocities (0.995c in the equal speed frame) requires extremely careful treatment and testing of the solution scheme.

In Figure 10 we show the final-state baryon rapidity distribution for the collision of ^{16}O with ^{208}Pb at an energy of 200 GeV/N assuming a $\Gamma = 4/3$ EOS, for a calculation in which no pions were included. For the problem studied here y is initially 3.03 in the equal speed system. The numerical grid was 124 x 62 cells with dimension 0.067 x 0.167 fm^2 at the origin. Along the collision axis, geometric increases in cell size, with a ratio of 1.015, were employed to extend the grid. We ended our calculation at a time of 7 fm/c. The principal features of the rapidity distribution were not changing appreciably past 4.5 fm/c. Also shown in Figure 10 is the experimental result[18] for ^{16}O collisions with a ^{197}Au target.

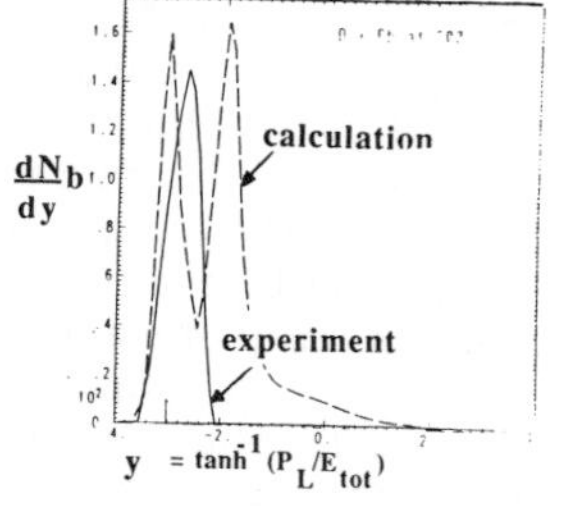

Figure 10. Differential baryon number vs rapidity for O on Pb at 200 GeV/N; our calculation compared to experiment.

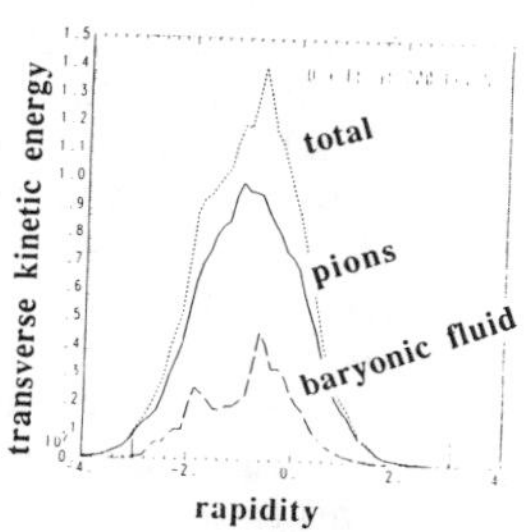

Figure 11. Transverse energy partition (GeV) between pions and baryons as a function of time for O on Pb at 200 GeV/N

A striking feature of Figure 10 is that about 60 nucleons remain at the initial Pb target rapidity. These nucleons represent a Pb corona (or "spectator fragment") which extends well beyond the original oxygen nucleus. The oxygen literally punches a hole in the lead, with formation of a weak radial shock. Thus, it is not surprising that an appreciable fraction of the Pb target remains at rest in the lab frame. Experimentally, about 65% of the total system mass is found to be localized (i.e. within one unit) about the initial target rapidity, compared with 70% for $\Gamma = 4/3$, 75% for $\Gamma = 5/3$, and 85% for $\Gamma = 2$ in our calculations. The remaining 35% of the experimental mass was not accounted for.[18]

We recently completed simulations that explicitly include the pion component. These preliminary runs, which have not been carried out far enough in time for pion generation to cease, indicate that about 300 pions are generated in ^{16}O with ^{208}Pb at 200 GeV/N. Since an experimental number of 310 has been reported, we find this to be a remarkable preliminary agreement. We note that adding the pion component substantially changes the whole nature of the collision, as seen in the transverse energy plot shown in Figure 11 and the rapidity curves in Figure 12. Note the big difference in stopping power between Figure 10 (pure hydro with no pions) and Figure 12 (pions included). Our calculations indicate that the pions have captured fully 25% of the initial available energy. This is not surprising since the pion energies (1-2 GeV) are comparable to the baryon energies (10 GeV) in this frame. Indeed the pion number (300) at these high energy densities is greater than the nucleon number (224)!

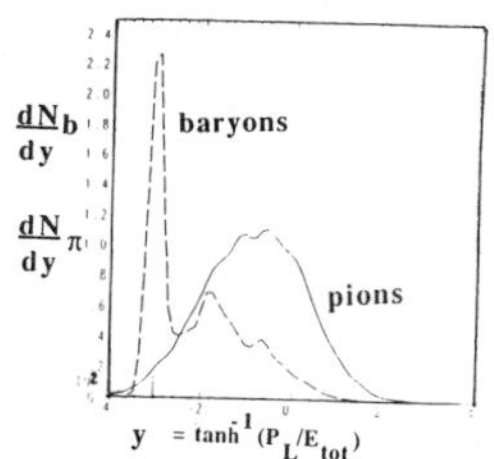

Figure 12. Differential baryon and pion numbers vs rapidity for O on Pb at 200 GeV/N. Solid line: pions.

Our runs have indicated that at these high energies pions are still generated at times well after shock transit has completed. This is in contrast to Bevalac runs where most of the pion generation occurs in the shocks. The reason is that at CERN energies the average energy density and compression remain high enough at late times for bulk pion production to continue (see Figure 3). We have not yet been able to carry our calculations out long enough in time (past 10 fm/c) to see the end of the pion generation. In fact at their rather comparable speeds it may be almost impossible for the pion fluid to disengage from the nuclear fluid. At late times the CERN composite appears to be expanding at a Bevalac-like velocity.

The question arises whether the pions measured in the experimental detectors are providing information about the hot dense interior of the nuclear composite, where quark-gluon plasma may have formed, or whether the detectors only observe pions that last interacted on the surface and carry little information about quark-gluon plasma. The answer is not clear yet, but many of our pions appear to have had their last recorded interaction in regions of low density, presumably the surface regions. If so, then the pions may not be such useful little thermometers of the nuclear interior as we had hoped. It may be that kaons, which have a much smaller scattering cross section because they are strange, will prove to be more efficient carriers of the signature of the hot dense nuclear interior. We plan eventually to include kaons in our model.

SUMMARY

We have made the following conclusions about hydro model simulations of low-energy (200 MeV/N) heavy ion collisions: (1) hydrodynamic flow predictions, using very different numerical approaches, are similar; (2) the hydrodynamic simulations overpredict the magnitude of the collective flow by about a factor of two in slope at zero rapidity; (3) including dynamic pion tracking with hydrodynamics underpredicts the pion multiplicity by almost a factor of two, 30 (theory) to 50 (expe-iment); and (4) including viscosity and a pion decay channel for delta resonances may correct these discrepancies.

Concerning high-energy (200 GeV/N) collisions, our hydrodynamic model appears to give qualitative agreement with the data. Hydrodynamic simulations reproduced the general features of the measured rapidity shift and pion multiplicity. These preliminary results indicate that the stopping power in such high energy collisions remains significantly high. We found that highly relativistic calculations require extremely careful treatment and testing. The relativistic composite appears to generate pions over long periods of time, long after the shock transit time. The energy balance is almost dominated by the pions, which have captured 25% of the available energy from the baryonic fluid.

While ideally the stopping power should be theoretically derivable directly from the nuclear force, in practice experience has shown that only experimental data can provide real insight into the behavior of such complex systems. Our hydrodynamic model provides a base for direct comparison with experimental data. Such physical processes as partial transparency and phase changes still need to be included. However, gross features such as baryon rapidity distributions and transverse kinetic energy have in fact been qualitatively described by our model. We conclude that our approach of overlaying Monte Carlo pion dynamics on a hydrodynamics base appears to have some utility.

ACKNOWLEDGMENT

This research was performed at Lawrence Livermore National Laboratory under the auspices of the United States Department of Energy, contract No. W-7405-Eng-48.

REFERENCES

1. J. Aichelin, H. Stöcker, *Phys. Lett.* **176B**,14 (1986).
2. H. Kruse, B. V. Jacak, J. J. Molitoris, G. D. Westfall, H. Stöcker, *Phys. Rev. C* **31**, 1770 (1985); H. Kruse, B.V. Jacak, H. Stöcker, *Phys. Rev. Lett.* **54**, 289 (1985); J.J. Molitoris, H. Stöcker, *Phys. Rev. C* **32**, 346 (1985).
3. A. A. Amsden, G. F. Bertsch, F. H. Harlow, J. R. Nix, *Phys. Rev. Lett.* **35**, 905 (1975); G. Buchwald, G. Graebner, J. Theis, J. Maruhn, W. Greiner, H. Stöcker, *Phys. Rev. Lett.* **52**, 1594 (1984).
4. J. Zingman, T. McAbee, J. Wilson, C. Alonso, **LLNL Report UCRL-97153** (1987); J. Zingman, T. McAbee, J. Wilson, C. Alonso, **LLNL Report UCID-21126** (1987).
5. L. McLerran, *Rev. Mod. Phys.* **58**, 1021 (1986).
6. J. F. Hawley, L. L. Smarr & J. R. Wilson, *Astrophys. J.* 277, 296 (1984).
7. H. Stöcker & W. Greiner, *Phys. Rep.* **137**, 277 (1986); R.B. Clare & D. Strottman, *Phys. Rep.* **144**, 177 (1986).
8. J. R. Wilson & G. J. Mathews, in Conference Proceedings, "Frontiers in Numerical Relativity", Champaign, Illinois, 1988, to be published.
9. J. Centrella & J. R. Wilson, *Astrophys. J. Suppl. Ser.* **54**, 229 (1984).
10. K. W. Thompson, *J. Fluid Mech.* **171**, 365 (1986).
11. J. F. Hawley, L. L. Smarr & J. R. Wilson, *Astrophys. J. Suppl. Ser.* **55**, 211 (1984).
12. H. Stöcker, W. Greiner, *Phys. Rep.* **137**, 277 (1986).
13. J. A. Zingman, T. L. McAbee, J. R. Wilson & C. T. Alonso, *Phys. Rev. C* **38**, 760, (1988).
14. H. Stöcker, W. Greiner, W. Scheid, *Z. Phys.* **A286**, 121 (1978); P. Danielewicz, *Nucl. Phys.* *A***314**, 465 (1979).
15. D. Hahn, H. Stöcker, *Nucl. Phys.* *A***452**, 723 (1986).
16. J.W. Harris *et al.*, LBL-23476 (1987).
17. J.W. Harris *et al.*, *Phys. Rev. Lett.* 58, 463 (1987).
18. H. Schmidt *et al.* (WA80 Collaboration), in **Proc. of the 27th Int'l Summer School for Theoretical Physics**, Zakopane, Poland, 1986, to be published.
19. A. Bamberger *et al.* (NA35 Collaboration), *Phys. Lett.* *B***184**, 271 (1987), R. Albrecht *et al.* (WA80 Collaboration), *Phys. Lett.* *B***199**, 297 (1987).
20. J.D. Bjorken, *Phys. Rev.* **D27**, 140 (1983).

FIGURE CAPTIONS

Figure 1. Shock tube test at 10 GeV/fm^3. Solid line: theory; Dotted line: code.

Figure 2. Maximum compression vs time for Au on Au at 200 MeV/N with two EOS's.

Figure 3. Maximum compression vs time for O on Pb at 200 GeV/N with Γ=4/3, 5/3 and 2.

Figure 4. Differential baryon number vs rapidity for O on Pb at 200 GeV/N with Γ=4/3, 5/3 and 2.

Figure 5. Flow angle vs time for Au on Au at 200 MeV/N.

Figure 6. Transverse momentum vs rapidity for Au on Au at 200 MeV/N.

Figure 7a. Density plot for La on La at MeV/N at time 10 fm/c. The arrows show pion directions.

Figure 7b. Density plot for La on La at 1350 MeV/N at time 15 fm/c. The arrows show pion directions.

Figure 8. Pions vs time for La on La at 1350 MeV/N, for three different EOS's.

Figure 9. Pions vs time for La on La at 1350 MeV/N. Dynamic model compared to a static chemical model.

Figure 10 Differential baryon number vs. rapidity for O on Pb at 200 GeV/N; our calculation compared to experiment.

Figure 11 Transverse energy partition (GeV) between pions and baryons as a function of time for O on Pb at 200 GeV/N.

Figure 12. Differential baryon and pion numbers vs rapidity for O on Pb at 200 GeV/N. Solid line: pions.

THE DYNAMICS OF SELF-GRAVITATING FLUID MASSES

N. R. Lebovitz
Department of Mathematics, The University of Chicago, Chicago IL 60637

ABSTRACT

The equations describing the dynamics of rotating, self-gravitating fluid masses are formulated, with particular emphasis on quasi-steady approximations needed to resolve the problem of "stiffness" due to the presence of widely separated timescales. The solutions of the quasi-steady system need to be tested separately for stability on the shortest timescale, and the essentially Lagrangian, or Hamiltonian, nature of that stability problem is indicated. Application is made to the classical ellipsoidal figures, and recent progress toward understanding the fission process of binary-star formation is described in this context.

I. INTRODUCTION

A star begins its life with a stage of dynamical collapse characterized by a single timescale, the so-called dynamical timescale T_d. Integrating the equations governing the star's evolution during this phase is, in principle if not in practice, a mathematically and numerically straightforward initial-value problem. When the collapse phase, which begins with an enormously rarified protostar, comes to a close, the star finds itself in a state of rough hydrodynamic equilibrium, continuing to contract slowly in order to make up in gravitational potential energy what is being lost through radiation. The timescale T_c of this contraction is governed by the energy loss, or luminosity, of the star, and it exceeds the dynamical timescale by a large factor, 10^8 or more. When internal temperatures rise to the level where nuclear reactions can release energy, these take over the rôle of supplying the energy lost through radiation, and have the effect of introducing a third timescale T_n determined by the rate of release of nuclear energy.

Most of the star's life following the initial phase of dynamical collapse is spent in one of those phases governed by a longer timescale (the sun is in that phase governed by the nuclear timescale T_n, the "main-sequence" phase). From the mathematical standpoint, and particularly from the computational standpoint, following the evolution during these phases is no longer a straightforward initial-value problem. The latter entails solving the equations "on all timescales," including the shortest one T_d. Computationally this means that if one wishes to follow the evolution for a length of time T_c, 10^8 or more timesteps are needed. This computational inefficiency must be overcome by some approximation allowing a direct evolutionary calculation on the longer timescale. Once this is achieved, the evolutionary calculation on the long timescale must be supplemented by considering the stability, on the short timescale (or timescales), of the solutions so found. This is necessary because such (possibly unstable) motions have been suppressed by the approximation method.

Astrophysicists learned long ago how to do these evolution and stability problems in the (most important!) case of spherical stars without fluid motions beyond those associated with a slow change of radius. However, for stars with a large amount of rotation, or with more complicated velocity fields, there is no general and systematic procedure for following evolution on a long timescale, and, for the most part, calculations on the long timescale and with general velocity fields have been done as standard initial-value problems, despite the inefficiencies due to the stiffness (Lucy 1977). On the other hand, under certain simplifying assumptions of a kind now commonly made in the earliest phases (the collapse and contraction phases), there *is* a systematic procedure (Lebovitz 1981). We describe it briefly in the following section.

Insofar as stability on the shortest timescale is concerned, there is no difficulty of principle, since the underlying problem is again a standard initial-value problem. We show in §III how to formulate an approximation to the stability problem in the form of the Lagrangian dynamics of a system with a finite number n of degrees of freedom. Here we have in mind the presence of only two timescales, T_d and T_c, in order not to have to consider stability on several timescales.

In §IV we describe the ellipsoidal figures of uniform density, which represent exact solutions of the fully dynamical system of equations, and whose steady counterparts represent exact solutions of the slow-evolution equations described in §II. The approximation to the stability problem, described in §III, becomes exact in this context.

Finally, in §V, we indicate the applications of these methods to a problem of long standing in astrophysics, that of the formation of a binary star from a single star during the contraction phase by a process of fission.

II. EVOLUTION ON A LONG TIMESCALE

The general equations believed to govern the dynamics of a star are those of inviscid fluid dynamics; the neglect of viscosity appears justified by virtue of the enormous Reynolds numbers estimated for stars. Following this assumption, we find for the basic governing equations the Euler equations of fluid dynamics (written below with respect to a rotating reference frame, for later convenience) expressing momentum conservation, the continuity equation expressing mass conservation, and an equation expressing energy conservation. The "natural", i.e. shortest, timescale is $T_d = (G\rho)^{-1/2}$ where ρ is a typical density and G is the universal gravitational constant; this timescale is to be suppressed. The long timescale is $T_c = GM^2/RL$, where M is the star's mass, R a typical radius, and L is the luminosity, or rate of energy loss due to radiation. If the equations are nondimensionalized using, not the natural timescale T_d, but the contraction timescale T_c, as the unit of time, they take the forms

$$\varepsilon\frac{\partial \mathbf{u}}{\partial t} + \mathbf{u}\cdot\nabla\mathbf{u} + 2\omega\times\mathbf{u} + \omega\times(\omega\times\mathbf{x}) + \varepsilon\frac{d\omega}{dt}\times\mathbf{x} = -\rho^{-1}\nabla p + \nabla V, \tag{1}$$

$$\varepsilon\frac{\partial \rho}{\partial t} + \mathrm{div}\,(\rho\mathbf{u}) = 0\ , \tag{2}$$

$$\varepsilon\frac{\partial s}{\partial t} + \mathbf{u}\cdot\nabla s = -\varepsilon Q\ , \tag{3}$$

where $\epsilon = T_d/T_c$ is a small parameter, $\mathbf{x}$ is the position vector, $\mathbf{u}$ the fluid velocity with respect to the rotating frame, ω the angular velocity relative to inertial space, p the pressure, V the gravitational potential, s the entropy per unit mass, and Q represents the effects of radiative transport of energy. The most complicated physics and chemistry of stellar-evolution theory are buried in this Q, which is a functional of s and ρ. In our treatment, which will follow recent convention in simplifying the physics while attending to the complicated dynamics, they will remain buried.

These differential equations must be supplemented by constitutive equations

$$p = P(\rho, s) \quad \text{and} \quad V(\mathbf{x},t) = G\int_D \frac{\rho(\mathbf{y},t)}{|\mathbf{x}-\mathbf{y}|}d\mathbf{y}\ . \tag{4}$$

The first is a thermodynamic equation of state and the second relates the self-gravitational potential V to the density (and to the domain D as well unless the latter is understood to be implicitly defined by the density). The physical simplification that permits us to "integrate" the energy equation (3) is simply to replace s by a function $\phi(F,t)$, where $F(\mathbf{x},t)$ is defined by the conditions

$$\frac{DF}{Dt} = \varepsilon\frac{\partial F}{\partial t} + \mathbf{u}\cdot\nabla F = 0, \quad F(\mathbf{x},0) = F_0(\mathbf{x}) \tag{5}$$

and F_0 is a function vanishing on the boundary of D at time $t = 0$ and (say) positive in the interior; in consequence of equation (5), the boundary of D at any time t is given by $F = 0$, and $F > 0$ in the interior. The function ϕ is chosen so as to ensure that the total energy of the figure is decreasing. A fuller explanation is given in Lebovitz (1981). The upshot is that p becomes a function of ρ, F *and* t, and the energy equation (3) is supplanted by equation (5).

Suppressing the rapid time variations looks easy: set $\varepsilon = 0$ in equations (1), (2) and (5). This does embody the assumption that the dynamics are taking place on the slow timescale, and the resulting "reduced" equations indeed form some of the conditions determining the slow evolution. It is easy to see, however, that further conditions must be imposed as well. For example, the total mass of the figure is conserved. This follows from equation (2), but is no longer a consequence of the (steady-state) equation obtained from (2) by setting $\varepsilon = 0$: mass conservation must be restored to the reduced system "by hand". It is likewise for angular-momentum conservation. But this is not all: there is an infinite family of circulation integrals that is conserved exactly by the full system, but whose conservation cannot be inferred from the reduced system. What to do?

In general it is not clear what to do, but when the initial velocity field has a sufficiently simple structure, one can proceed as follows. Suppose that the streamlines of the initial velocity field are closed and can be represented as the intersections of level sets of the function $F_0(\mathbf{x})$ with level sets of a second scalar function $G_0(\mathbf{x})$. This holds in the important special case when the motion is one of pure rotation (not necessarily rigid rotation). Let $G(\mathbf{x},t)$ satisfy the same equation as F and reduce to G_0 at $t = 0$. Then the circulation integrals over the contours with $F =$ constant and $G =$ constant are conserved, as are also certain mass integrals (e.g. the mass enclosed inside the subdomain with boundary $F =$ constant or $G =$ constant). This leads to a system of equations, believed to be well-set, for the slow evolutions of the figure and its large-scale velocity field, as follows (Lebovitz 1981):

$$\mathbf{u}\cdot\nabla\mathbf{u} + 2\omega\times\mathbf{u} + \omega\times(\omega\times\mathbf{x}) = -\rho^{-1}\nabla p + \nabla V, \tag{6}$$

$$\operatorname{div}(\rho\mathbf{u}) = 0 \tag{7}$$

$$\mathbf{u}\cdot\nabla F = 0 \tag{8}$$

$$\mathbf{u}\cdot\nabla G = 0 \tag{9}$$

$$\int_{F>0} \rho(x_1u_2 - x_2u_1 + \omega\,[x_1^2 + x_2^2])\,d\mathbf{x} = \text{const.}, \tag{10}$$

$$\int_{F>\alpha,\,G>\beta} \rho\, d\mathbf{x} = \text{const.} = M_{\alpha\beta}, \tag{11}$$

$$\int_{F=\alpha,\ G=\beta} (\mathbf{u} + \omega\times\mathbf{r})\cdot d\mathbf{r} = \text{const.} = C_{\alpha\beta}\,; \tag{12}$$

$$p = P(\rho, F, t) \quad \text{and} \quad V(\mathbf{x},t) = G\int_{F>0} \frac{\rho(\mathbf{y},t)}{|\mathbf{x}-\mathbf{y}|}d\mathbf{y}\,. \tag{13}$$

Equations (10)-(12) represent conservation of angular momentum, mass and circulation, respectively. We have assumed that the angular velocity is directed along the x_3 axis. Note that the parameter t, representing time measured on the long timescale, remains in the system of equations in virtue of its explicit presence in the pressure.

III. STABILITY ON THE SHORT TIMESCALE

An evolutionary path inferred from these equations can only be possible if each of the steady-state figures of that path is stable on the short timescale. This has to be checked separately from the evolutionary calculation since the short timescale has been suppressed from the latter. On the short timescale the star represents a conservative system: each mass element conserves its entropy. Such a fluid system is derivable from a Lagrangian and, as Kulsrud (1968) has shown, the linearized stability equations then take a simple form when expressed in terms of the Lagrangian displacement ξ:

$$\xi_{tt} + A\xi_t + B\xi = 0 \tag{14}$$

Here A and B are linear operators (for their forms, see Lynden-Bell and Ostriker 1967, Lebovitz 1988). On the natural inner product, A is antisymmetric and B is symmetric:

$$(\xi,\eta) = \int_{F>0} \rho\,\xi\cdot\eta\, d\mathbf{x}, \qquad (\xi, A\eta) = -(A\xi, \eta) \qquad \text{and} \qquad (\xi, B\eta) = -(B\xi, \eta)\,.$$

The system (14) can be reduced to finite-dimensional form conveniently by means of a Galerkin procedure. Suppose a basis ξ_i has been chosen for the domain D (where $F > 0$). Then the truncation

$$\xi = \sum_{i=1}^{n} c_i(t)\xi_i$$

leads to the finite-dimensional system

$$\mu\ddot{c} + \alpha\dot{c} + \beta c = 0 \tag{15}$$

Here c is the n-vector of coefficients in the truncated expansion, and the matrices μ, α, and β are respectively symmetric positive-definite, antisymmetric, and symmetric.

Equation (15) is therefore identical with an n-degree-of-freedom system in Lagrangian particle dynamics, linearized about an equilibrium point. A fair amount of qualitative information about such systems is available, and can be a useful guide in interpreting and checking numerical solutions. Moreover, Equation (15) can easily be generalized to the nonlinear case. This can be done either by first generalizing Equation (14) to the nonlinear case and then performing the Galerkin truncation, or by performing the truncation within the Lagrangian. The action of latter, when expressed for the perturbed flow described by ξ, has the form

$$S = \int_{t_1}^{t_2}\int_{F>0} \rho\left\{|\xi_t|^2 + L\,\xi_t + \Phi[\xi]\right\} d\mathbf{x}\,dt\ . \tag{16}$$

Substituting the Galerkin truncation as above and varying the Lagrangian now leads to

$$\mu\ddot{c} + \alpha\dot{c} + \nabla f(c) = 0, \quad \text{where } f(c) = \int_{F>0} \Phi\Big[\sum_1^n c_i\xi_i\Big] d\mathbf{x}\ . \tag{17}$$

The gradient appearing in equation (17) is of course with respect to c.

This can be an effective method for dealing with the stability problem, but is quite sensitive to the choice of basis.

IV ELLIPSOIDAL FIGURES OF UNIFORM DENSITY

Equations (1) and (2) are exactly satisfied if D is an ellipsoid with (time variable) semiaxes a_1, a_2 and a_3, uniform density $\rho(t)$, and velocity and pressure given by

$$u_1 = \lambda a_1 x_2/a_2,\ u_2 = -\lambda a_2 x_1/a_1,\ u_3 = 0 \quad \text{and} \quad p = p_c(t)(1 - \sum_1^3 x_k^2/a_k^2) \ . \tag{18}$$

Hence, if one thinks of a self-gravitating fluid of uniform density as a drop held together by gravity rather than by surface tension, it has an analytically very simple form. The result of substituting these expressions in equations (1) and (2) is a system of six nonlinear ordinary differential equations for the six variables a_1, a_2, a_3, ρ, ω and l. The equations contain p_c, which must be specified in terms of the other variables or (in the incompressible case) eliminated via a further condition; for details, see Chandrasekhar (1969).

This has the effect of reducing the partial differential equations of the original problem to a system of ordinary differential equations.

By observing the exact integrals of this system, setting the remaining time derivatives equal to zero, and specifying the pressure as an explicit function of the time t (measured on the long timescale), one obtains from these equations six nonlinear equations in the six unknowns for determining their solution as a function of t. If Equations (18) are substituted in the slow-evolution Equations (6)-(13), one gets the same system (Lebovitz 1981, Equations 8.3 and 8.4). In either way, one can follow the slow evolution of the ellipsoidal figures. If, as seems reasonable in astrophysical applications, one supposes that the initial figure is axisymmetric or nearly so, the shapes of the evolving figures are severely circumscribed by one of the constants of the motion (actually a linear combination of the angular momentum and circulation integrals):

$$(a_1 - a_2)^2\ (\omega + \lambda) = \text{constant} \tag{19}$$

This constant is zero if the initial figure is axisymmetric. Hence the evolving figure either remains axisymmetric or must follow a path on which $\omega + \lambda = 0$. There is such a path of nonaxisymmetric figures, and the axisymmetric ellipsoids (the Maclaurin spheroids) become dynamically unstable at the point where the two paths cross (this was discovered by Riemann). It is evident that further evolution must take place along the figures for which $\omega + \lambda = 0$. This explains the evolutionary paths found by Lebovitz (1972, 1974).

Riemann was able to prove that the Maclaurin spheroids become unstable at the point where the family with $\omega + \lambda = 0$ bifurcates because the perturbation takes a spheroid into an ellipsoid, and could therefore be described within his system of ordinary differential equations governing the motion of ellipsoids. To investigate the stability of fluid ellipsoids to non-ellipsoidal perturbations, one needs a more general approach. There are many, including the use of the virial equations (Chandrasekhar 1969). The Galerkin method described in the preceding section is also appropriate in this case, and the choice of a suitable basis quite practical.

This is due to the following circumstance. Let the operators A and B of Equation (14) above be those appropriate to the case when the unperturbed configuration is one of the ellipsoidal figures just described. If ζ is a vector polynomial , i.e. if each of its components is a polynomial in the cartesian coordinates of maximum degree k, then $A\,\zeta$ and $B\,\zeta$ are likewise vector polynomials of maximum degree k. Hence in the Galerkin approximations discussed above, if one forms a basis of vector polynomials of degree not greater than k, there will be an exact solution involving only finitely many basis functions, and the truncation gives an exact solution.

The evolutionary and stability properties of these ellipsoidal figures have played an important rôle in the fission theory of binary stars, described below (see also Lebovitz 1989).

V THE FISSION THEORY OF BINARY STARS

The word "fission" refers to the formation of a binary star from a single, rotating star during a phase of slow, or secular contraction. Two timescales (the dynamical and the contraction timescales) characterize the pre-main-sequence phase when fission is contemplated. The quasistatic approximation described above has so far been implemented only in the context of idealized models of uniform density.

The traditional, quasistatic approach toward fission envisages the following sequence of events (or preconception). Initially, the figure is axisymmetric, or nearly so. In consequence of a slow loss of energy through radiation, the axisymmetric figure becomes progressively more flattened, resulting in a progressively larger fraction of its energy in the form of kinetic energy of rotation. When the ratio of kinetic to gravitational energy reaches a critical value, the axisymmetric figure becomes unstable to a symmetry-breaking perturbation carrying the circular cross-section into an elliptical one. Further evolution takes place quasistatically along a nonaxisymmetric (or bar-shaped) family on which the equatorial cross-section lengthens progressively in one direction. This continues until this nonaxisymmetric family itself becomes unstable. The nature of this secondary instability and of the family bifurcating at this point is crucial for the theory. One imagines that instability is initiated by a perturbation tending to concentrate matter toward the ends of the bar, resulting in a dumbbell shape. This process continues as evolution proceeds until the matter connecting the two ends of the dumbbell has thinned to the vanishing point, and one is left with a pair of binary stars orbiting each other.

There is nothing in this preconception of binary-star formation that would preclude the use of realistic models in checking the various assertions. However, the quasistatic method described above has not yet been attempted for realistically stratified models, and the only models in which attempts at verification have been made are ellipsoidal figures of uniform density and perturbations therefrom. Realistic models *have* been employed in another approach to the formation of binary stars, also referred to as "the fission theory." This other approach is discussed in the article by Tohline in this volume (see also Durisen *et. al.* 1986). To avoid confusion we refer below to the original conception of fission as "the classical fission theory."

We turn now to a brief description of the classical fission theory and of its revised form, wherein one attempts to verify the preconception of fission described above in the ellipsoidal context. A fuller description of these topics is given by Lebovitz (1987).

V.1 THE CLASSICAL VERSION

Here the assumption is made that viscosity is important. This has the effect of enforcing a uniform rotation and influences the stability properties of the figures as well. The axisymmetric family consists of Maclaurin spheroids, which become unstable when the ratio of kinetic to gravitational energy reaches a certain critical value. The family of Jacobi ellipsoids (those described above for which $\lambda = 0$) bifurcates from this critical point, and becomes the stable, bar-shaped family along which further evolution takes place. The Jacobi family in turn becomes unstable when the bar reaches a critical elongation. Up to this point of instability along the Jacobi family, all the assertions of the preconception above are verified (in fact, they were originally abstracted from this model). Beyond this point, however, the facts fail to agree with the preconception. The perturbation initiating the instability of the critical Jacobi figure does not shape it into a dumbbell. More seriously, the family of figures bifurcating from this critical point is itself unstable: there are no stable, quasistatic figures just beyond the critical Jacobi figure. Hence quasistatic evolution ceases at this point and motion must take place (at least for a while) on the dynamical timescale. The precise behavior of the figures at this point has never been fully resolved.

V.2 THE REVISED VERSION

Here viscosity is neglected, on the ground that this assumption is nearer to the truth than the opposite assumption of the classical version. A wider variety of ellipsoidal configurations is now possible (the Riemann ellipsoids, described by Equation 18 above, in which there are fluid motions beyond those of rigid-body rotation). The initial evolution again takes place along the Maclaurin family (approximately). Again an instability along this family signals a shift in the evolution onto a family of ellipsoids with three unequal axes. This family itself subsequently becomes unstable. While details differ, there is a rather complete analogy with the classical version up to this point. However, there is an important difference beyond this point. On the revised version, the perturbation initiating the instability carries the ellipsoid into a dumbbell shape. Moreover, the family of figures bifurcating at the critical point is stable beyond the critical point, allowing quasistatic evolution to continue.

V.3 CONCLUSIONS

The principal conclusion is that, by removing the artificial assumption of a dominating viscosity from the theory, one finds a pattern of behavior supporting the preconception of fission as described above. There remain a number of mathematical details to confirm before this picture is mathematically solid (Lebovitz [1987]), but it appears at this writing that the impasse that affected the classical version of the fission theory is overcome on the revised version.

The theory suffers from an obvious defect: the models are uniformly dense whereas stars are centrally condensed. One may wonder whether the qualitative conclusions drawn from these idealized models would persist in the context of realistically stratified models. I am hopeful that this matter can be addressed with the aid of the formulation of slow evolution for stars with nontrivial fluid motions, as described in §II above.

REFERENCES

S. Chandrasekhar (1969) **Ellipsoidal Figures of Equilibrium** (New Haven: Yale Univ. Pr.).

R.H. Durisen, R.A. Gingold, J.E. Tohline, & A.P. Boss (1986) *Astrophys. J.* **305**, 281.

R. Kulsrud, (1968) *Astrophys. J..* **152**,1121.

N. Lebovitz (1972) *Astrophys. J.* **175**, 171.

N. Lebovitz (1974) *Proc. Int. Coll. Drops and Bubbles* (eds. D. J. Collins, M. S. Plesset, M. M. Saffren), **1**.

N. Lebovitz (1981) *Proc. R. Soc. London A* **375**, 249

N. Lebovitz (1987) *Geophys. Astrophys. Fluid Dyn.*, **38**, 15.

N. Lebovitz (1988) *Geophys. Astrophys. Fluid Dyn.*, to appear.

N. Lebovitz (1989) **Proc. 20th Gen. Assembly International Astronomical Union**, to appear.

L. Lucy (1977) *Astron. J.* **82**, 1013.

D. Lynden-Bell & J. Ostriker (1967) *Mon. Not. R. Astron. Soc.* **136**, 293.

J. Tohline, (1989) this volume [and chapter].

LIGHT SCATTERING FROM SPHEROIDAL DROPS: EXPLORING OPTICAL CATASTROPHES AND GENERALIZED RAINBOWS

Philip L. Marston, Cleon E. Dean, and Harry J. Simpson
Department of Physics, Washington State University
Pullman, WA 99164-2814

ABSTRACT

Light scattered from spheroidal drops produces a variety of complicated diffraction patterns which are examples of optical catastrophes. These patterns decorate caustics in the outgoing rays which have been refracted by (and internally reflected from) the drop's surface. Near the rainbow scattering angle, hyperbolic umbilic, cusp, and other catastrophes have been observed with laser and white light illumination. Lips caustics in the backscattering pattern have also been observed. Calculations of caustic parameters (from generalized ray tracing) and wavefields are summarized.

1. INTRODUCTION AND REVIEW

Observations of light scattered from spheroidal drops lead to the discovery of unexpected diffraction catastrophes not present in scattering from spheres.[1,2] The present section reviews patterns observed and relevant results from catastrophe theory.[1-3] All of the experiments described in the present paper, are with oblate drops having a vertical axis of rotational symmetry and illuminated by a horizontally directed light. The drops (usually water) are levitated acoustically in air; they have an oblate shape whose aspect ratio is determined by the amplitude of the sound.[4] Varying both the aspect ratio of the drop, and the scattering angle, facilitates the observation of diffraction catastrophes for a range of control parameter space[5] not as easily explored by other methods. In all the experiments, the scattering patterns were obtained by photography with a camera focused on infinity: such patterns are the same as those cast on a ground glass screen placed at a large distance from the drop. A second camera was usually focused on the drop to record the drop's shape and (sometimes) the locations of intense rays leaving the drop. The intersection of a horizontal plane with the drop gives a circle. Figure 1 shows rays through the drop when the plane cuts the equator. The rays labeled 1 and 2 are for a *horizontal scattering angle* θ of 152°. A distant observer viewing the drop from that θ sees two bright rays in that plane. As θ is decreased to the rainbow angle $\theta_R \approx 138°$, the rays merge and they disappear for $\theta < \theta_R$. The aforementioned rays exist even if the drop is a sphere; however, when the drop is oblate, other rays can contribute to the horizontal scattering which do not lie in the equatorial plane. The merging of those rays gives rise to new caustics, more complicated than the rainbow, which can be classified with catastrophe theory.[5] As rays to a distant observer merge, the Gaussian curvature vanishes for the associated outgoing wavefront (i.e. the wavefront directed towards a distant caustic is locally flat). Section 2 considers the local shapes of relevant outgoing wavefronts.

Figure 2 shows scattering patterns in the rainbow region from the original study[1,2] arranged in order of increasing aspect ratio $q=D/H$, defined in Figure 3(a). Visible on the left of Figure 2(a) is an arc associated with the usual rainbow caustic. On the other side of Figure 2(a) is a cusp diffraction catastrophe, which is not present for small values of q. The θ of the rainbow caustic remains at θ_R and does not depend on q. As q is increased, the cusp shifts to the left towards a smaller θ until it merges with the rainbow caustic in Figure 2(b). The resulting pattern is

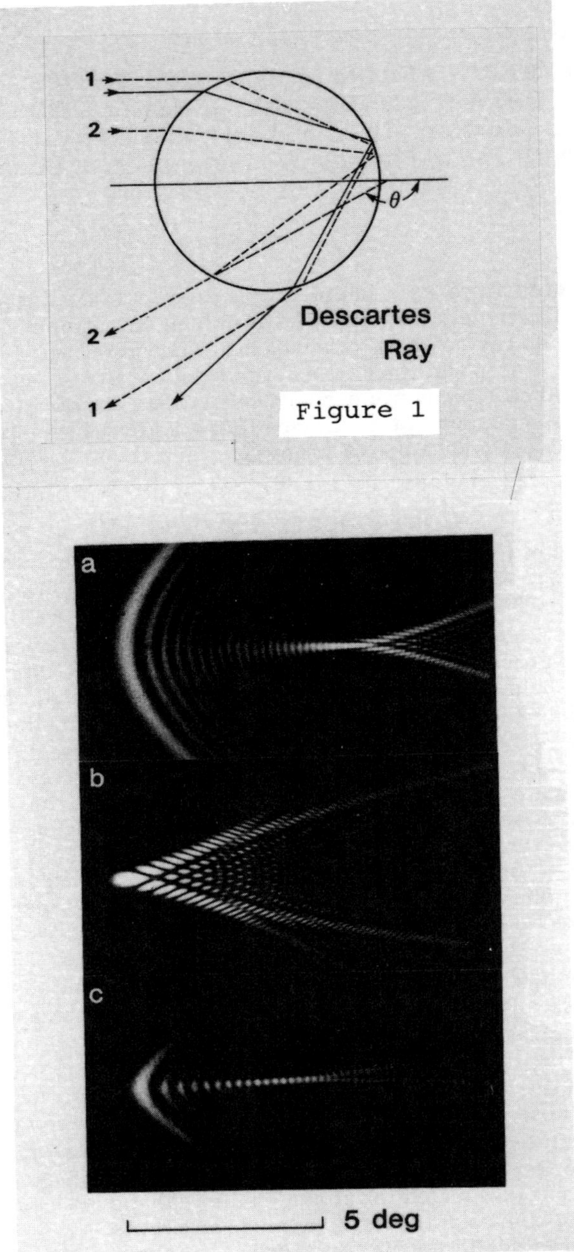

Figure 1

Figure 2

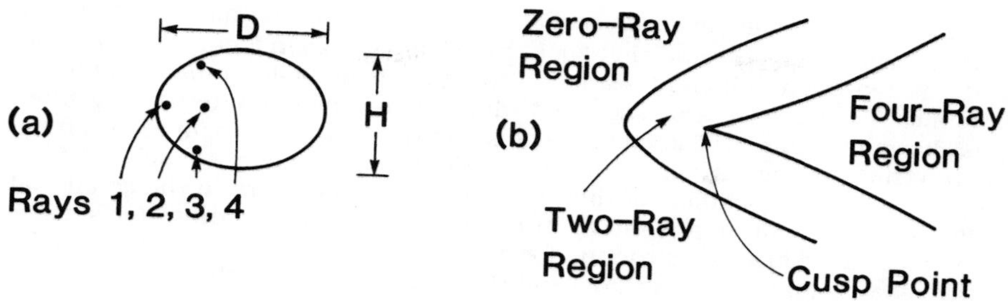

Figure 3
(a) & (b)

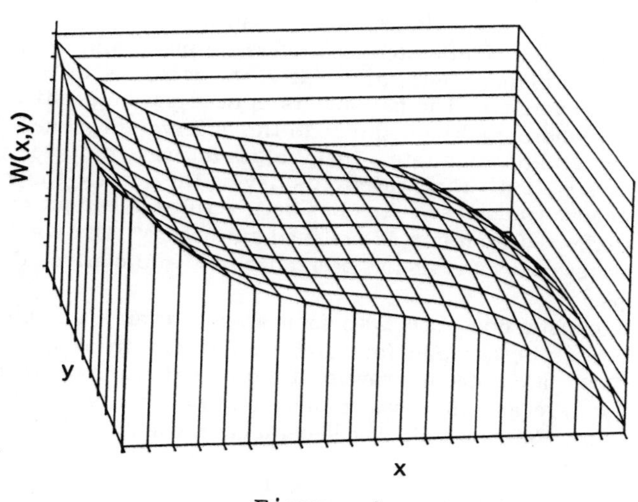

Figure 4

that of a *hyperbolic umbilic focal section (HUFS)*. An additional increase in q causes the cusp to reappear and shift back toward larger θ, as in Figure 2(c). This set of pat-terns represents different sections through a hyperbolic umbilic diffraction catastrophe. The caustics are associated with rays having one internal reflection, such as shown in Figure 1. The number of distinct rays of that class, depends on the direction to the observer, as shown in Figure 3(b) when the cusp is displaced from the rainbow arc. The locations of the rays, as seen by a camera in the equatorial plane focused on the drop, are shown in Figure 3(a). For larger drops (or smaller wavelengths l) the diffraction structure for patterns like those in Figure 2 should be finer. The shapes and general locations of the brightest regions should remain the same for a spheroid with a given θ.

Nye[3] calculated that the aspect ratio giving a HUFS is $q_4=[3\mu^2/4(\mu^2-1)]^{1/2} \approx 1.311$, where for water the refractive index $\mu=1.332$. This agrees with the measured condition.[1] The critical condition is that four rays merge. Marston[2] used the condition that three rays merge in the direction of the cusp point, as in Figure 3(b), to calculate the angular location of the cusp point. The angular separation of the cusp point from the rainbow ray was measured for patterns like those shown in Figure 2(a) and (c) for various q. The measurements support the theory. The theory may be used to invert scattering data for q.

2. SHAPES OF WAVEFRONTS GIVING CUSPS AND HYPERBOLIC UMBILIC FOCAL SECTIONS

Though catastrophe theory is useful for classifying the generic shape of wavefront producing a given type of diffraction pattern, the detailed shape may vary by a smooth transformation. Consequently the local shapes, which propagate to produce cusps (such as in Figure 2(a)) and HUFS (Figure 2(b)), were not known. The usual example of a cusp caustic considered in catastrophe optics is a longitudinal cusp in which the cusp opens up along (or opposite to) the general direction of propagation.[6] The cusps produced by scattering from oblate drops open up perpendicular to the general direction of propagation and may be described as *transverse cusps*. The generic shape of wavefront which propagates to produce such cusps was found by Marston to have the local form[6-8]

$$W(x,y) = -(a_1 x^2 + a_2 y^2 x + a_3 y^2), \qquad (1)$$

where x and y are cartesian coordinates in a reference plane and W specifies the displacement from that plane. Here, and below, the x axis is in the equatorial plane and the y axis is vertical. The parameters a_j need not be positive; however, $a_2 \neq 0$ and $a_2 \neq (-2z)^{-1}$ where z is the distance to the observation plane. The a_j affect the cusp location and opening rate. Their values for wavefronts leaving oblate drops are under investigation. That wavefronts having the generic shape of Equation (1) propagate to give transverse cusps has been verified by simple experiments in which light was reflected from curved polished surfaces.[6,8] The ways in which rays merge at transverse cusp caustics has been analyzed[7] and clarifies the merging of rays leaving drops.[2]

In a HUFS, two caustic lines meet with some angle Ψ at an apex. Inspection of Figure 2(b) and other photographs indicates that water drops give $\Psi \approx 42 \pm 2°$. Description of the associate outgoing wavefront requires that parameters be introduced which affect Ψ. Previous discussions of generic wavefronts were limited to Ψ of 60° and 90° (e.g., see Reference 5). The general case, considered by Marston,[6,8] shows that the wavefront

$$W(x,y) = -(\alpha x^3 + 3\gamma y^2 x)/6, \qquad \Psi = 2\tan^{-1}\beta \qquad \beta = (\gamma/\alpha) \qquad (2, 3, 4)$$

produces a far-field HUFS where (3) and (4) relate the caustic parameter Ψ to the shape parameters α and γ. This wavefront shape is shown in Figure 4. The ray from $x = y = 0$ propagates to the apex of the "V" shaped caustic where the irradiance diverges most rapidly as $\lambda \rightarrow 0$. This ray is the Descartes ray in the equatorial plane. The calculation of α and γ is complicated by the three *spheroidal* reflecting or refracting surfaces which the initial plane wave meets. The key to calculation of α and γ is to notice that the *principal curvatures* κ_j of this wavefront are

$$\kappa_1 (x,y=0) = -\alpha x, \qquad \kappa_1 (x,y=0) = -\gamma x \tag{5, 6}$$

for points on the wavefront with infinitesimal x in the equatorial plane (where $y=0$). Consequently it is only necessary to determine the κ_j along a ray adjacent to the Descartes ray. To achieve this, Marston[9] made use of generalized ray tracing[11] which is a method of tracking the κ_j of wavefronts through optical systems. The method is used in the design of instrumentation[10] but is not widely used in scattering theory. This was done for a spheroidal drop having Nye's critical aspect ratio $D/H = q_4$. The analysis is tedious but yields the simple results[9]

$$\alpha = \frac{18}{D^2} \frac{\left(4-\mu^2\right)^{1/2}}{\left(\mu^2-1\right)^{3/2}}, \quad \beta = \frac{\mu}{\sqrt{12}} = \tan(\psi/2) \tag{7, 8}$$

where μ is the refractive index and D is the drop diameter. Inspection of (2) shows that $W(x,y=0) = -(\alpha x^3)/6$. Standard rainbow theory[11] for spheres yields a cubic wavefront with the same expression for α.

The important and new result is Equation (8). For water drops, $\mu = 1.332$, and Equation (8) gives $\psi = 42.1°$, in agreement with the observed $42 \pm 2°$. As an additional test, HUFS were observed for levitated drops of microscope immersion oil for which $\mu = 1.515$ and Equation (8) gives $47.2°$. Technical limitations made it necessary to determine ψ in real time with a reticle instead of photographically. The observations give $46 \pm 2°$ and support the theory.

3. DIFFRACTION PATTERN IN THE HYPERBOLIC UMBILIC FOCAL SELECTION

When the incident light is polarized with the electric field vertical, the amplitude of the field E_o in the outgoing wave is fairly uniform near the Descartes ray. The scattered field at a distance r from the origin $x = y = 0$ of the reference plane (used in specifying W) may be approximated[7] by the diffraction integral

$$E(U, V, r) \approx \frac{k}{2\pi i r} E_o e^{ikr} \iint_{\infty}^{\infty} \exp\{-ik[W(x,y) + Ux + Vy]\} dx dy, \tag{9}$$

where the implied time dependence is $\exp(-ikct)$, $k = 2\pi/\lambda$, $U = u/z$, $V = v/z$, $r = (u^2 + v^2 + z^2)$ and u and v are cartesian coordinates in an observation plane at a large distance z from the reference plane. For W given by Equation (2), Marston[9] evaluated Equation (9) by carrying out a shearing of coordinates with the result

$$E \approx E_o \frac{k^{1/3} 2^{4/3} \pi}{i r \alpha^{2/3} \beta} e^{ikr} Ai(w_1) Ai(w_2), \quad Ai(w_j) = \frac{1}{\pi} \int_0^\infty \cos\left(\frac{1}{3}s^3 + w_j s\right) ds \tag{10, 11}$$

$$w_j = -k^{2/3} (U \pm V\beta^{-1})/(2\alpha)^{1/3}, \ j = 1,2 \tag{12}$$

where $Ai(w_j)$ is the standard Airy function for the real argument w_j. The arguments U and V are essentially horizontal and vertical scattering angles (in radians) *relative to* the apex point. At the apex point $w_1 = w_2 = 0$, $Ai(w_j) = 0.355$ so that $E \propto k^{1/3}$ and the scattered irradiance $\propto (D/r)(kD)^{2/3}$. The factor $(kD)^{2/3}$ is a consequence of a focusing of light in the apex direction; for a sphere, the corresponding factor in the rainbow direction is weaker being[11] $(kD)^{1/3}$. Typically

$kD>>1$ and the irradiance at a hyperbolic umbilic focus greatly exceeds that for rainbow scattering.

The above calculation was confirmed by comparison with observations, Figure 2(b) for a drop with D=1.42 mm. The field amplitude, Equation (10), is proportional to $S(U,V)=Ai(w_1)\, Ai(w_2)$ and the irradiance is proportional to S^2. Figure 5 compares the observed irradiance pattern with a contour plot of the function S. The solid contours are for regions where $S>0$ while the dashed contours are where $S<0$. Though the irradiance pattern would be better described by contours of S^2, Figure 5(b) is useful for locating the bright regions of the pattern while also revealing the sign of S. Since 5(a) and (b) are shown on the same angular scale, there are no adjustable scaling parameters in this comparison.

4. WHITE LIGHT OBSERVATIONS OF GENERALIZED RAINBOWS

The conditions that produce the cusp on the left of Figure 2(a), may exist in raindrops of natural origin illuminated by sunlight.[2] This is because freely falling drops take on an oblate shape. It is worth viewing such patterns from levitated drops lit by white light to see what colors might be anticipated in nature. A rainbow arc can be produced by white-light illumination of a single drop viewed through a camera focused on infinity. For monochromatic illumination, the supernumerary bows appear as bands adjacent to the primary arc[1]; see e.g. the left side of Figure 2(a). In Figure 2(b) the structure described by the function $S^2(U,V)$ should replace the primary and supernumerary bows.

Observations were made of levitated water drops having diameter D from 1.2 to 2.3 mm, illuminated with collimated white light from a xenon arc lamp. When the drops were nearly spherical (corresponding to low acoustic drive) the pattern had the general appearance of a natural bow with reddish hue appearing on the outside of the bow. As with natural bows, the hues corresponding to longer wavelengths λ tend to be the most distinct as a consequence of dispersion: the scattering angle θ_R of the Descartes ray for violet *exceeds* θ_R for red by nearly 2°. Hence, the region where the short-λ scattering is most intense, lies in the two-ray region for long-λ scattering. The hue in that region is somewhat dependent on drop size since the angular width of a monochromatic supernumerary bow[11] varies as $(\lambda/D)^{2/3}$. When the drop was flattened, the cusp comes into view in the region of large θ. It is noteworthy that the cusp appears bright and white, suggesting that dispersion does not significantly affect the cusp point location. In nature, it appears even less likely that the cusp caustic would be colored since the pattern there would be associated with scattering from many drops, not all having the same D/H. When the critical D/H of q_4 is reached the cusp merges with the colored rainbow arc to give the HUFS. As with the arc, the colors in the HUFS are most distinct near the transition to the zero-ray region where θ is smallest. As a consequence of dispersion, the small-θ side of the apex and the caustic lines appear reddish. With a subsequent increase in D/H, the rainbow arc separates from the cusp as in Figure 2(c). The bow is colored while the cusp is white. Since the novel cusp and hyperbolic umbilic caustics are a generalization of the rainbow caustics, these patterns for scattering from drops are named *generalized rainbows*.[2] The patterns had the same features whether observed with unpolarized white light or with light with the E-field *vertically* polarized since the vertical component is expected to dominate the scattering in this region.[11]

Subsequent increase in $q=D/H$ above ~1.5 gives rise to a caustic which has yet to be classified. The first hint that this caustic is being approached occurs when the bow straightens up and bends over backwards [i.e. with a curvature opposite to that of Figures 2(a) and (c)]. Increasing q to close to ~1.7 causes a band extending to large

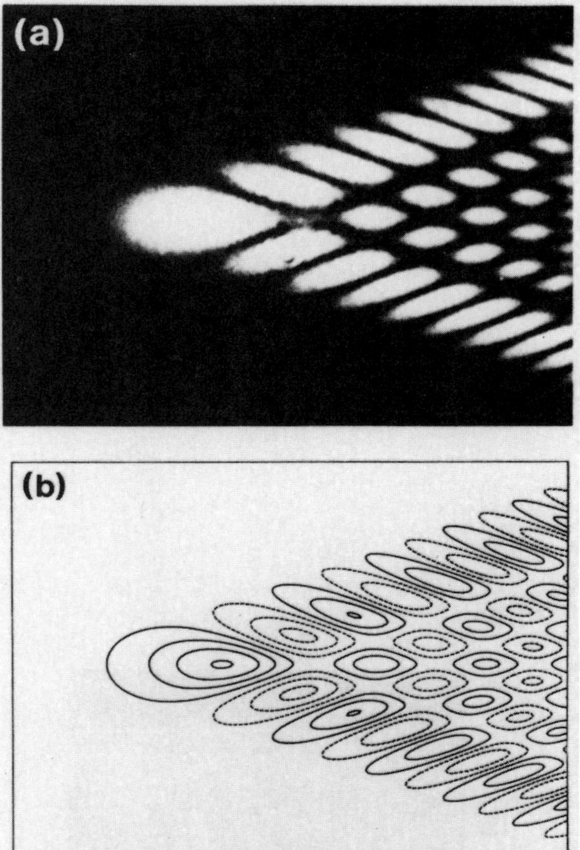

Figure 5

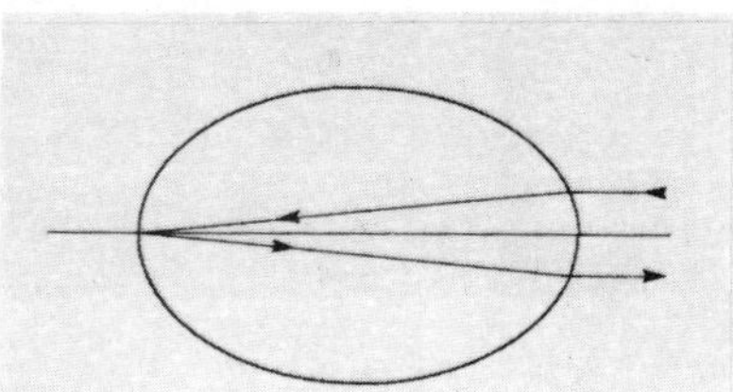

Figure 6

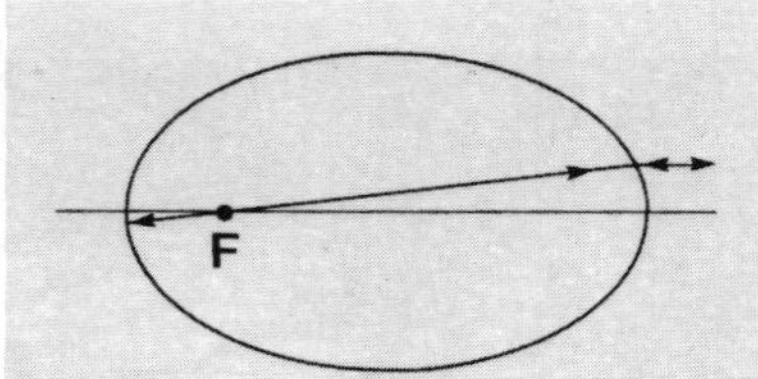

Figure 7

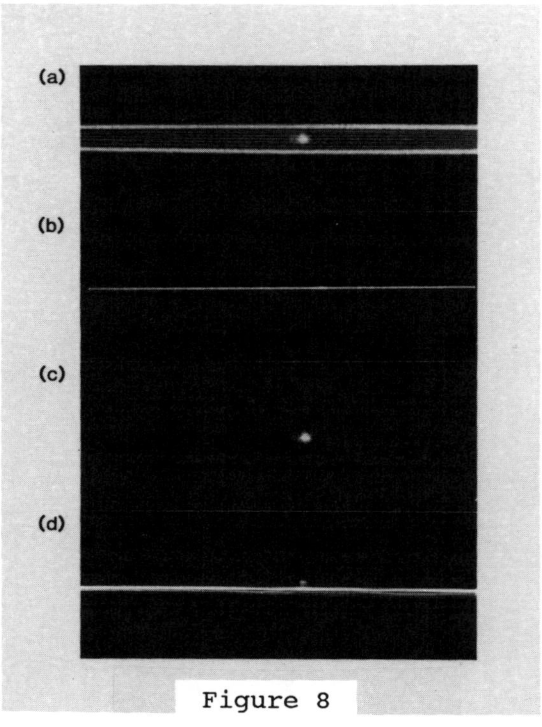

Figure 8

θ (but with small vertical scattering angles) to appear a bright white. Where that band meets the (reversed) bow is most bright, and is evidently part of a higher catastrophe. Increasing q results in the reappearance of a bow with the usual orientation but as part of a *swallowtail caustic* [see e.g. Figure 2.5(c) of Reference 5]. In the above sequence, reddish hues are present in the region of the pattern illuminated by the Descartes ray from the equatorial plane. This is a consequence of dispersion. With laser illumination, the sequence reveals rich diffraction patterns which decorate the caustics. The above phenomena must result from flattened outgoing wavefronts which differ from those considered in Section 2. Ray tracing suggests they are associated with an internal focus within the drop. The focus is positioned so that refraction of the outgoing rays flattens the wavefront as in Figure 6 discussed below.

5. LIPS CAUSTICS IN LIGHT BACKSCATTERED FROM OBLATE DROPS

Catastrophe theory allows the caustics leaving two cusp points to join smoothly.[5] The resulting bounded caustics have the appearance of a pair of lips. With an additional flattening of outgoing wavefront, the lips-like caustics are drawn together and the two cusp points gradually merge. The merging of cusp points is known as a "lips event" in catastrophe terminology.[5] Nye[3] predicted two such events occur in light backscattered from spheroidal drops having critical D/H of

$$q_{L1} = [\mu/(2\mu-2)]^{1/2} \approx 1.416, \qquad q_{L2} = [(2\mu-1)/(2\mu-2)]^{1/2} \approx 1.584 \qquad (13, 14)$$

where μ for water is 1.332. For $q_{L1}<q<q_{L2}$, there are no caustics in the near backward direction. Marston's analysis[2] of the cusp point location predicts that the horizontal scattering angle θ_3 to the cusp point $\rightarrow 180°$ as $q \rightarrow q_{L1}$. That analysis indicates e.g. that if q is reduced below q_{L1} to 1.40, then $|\theta_3 - 180°| \approx 26°$.

Nye's derivations are brief so a discussion is merited. The condition for the event at q_{L1} is that paraxial rays *in the vertical plane of symmetry* are focused on the back of the drop (Figure 6). The focal length in this plane is given by the lens maker's formula $f = \mu\rho/(\mu-1)$ where $\rho = H^2/2D$ is the vertical radius of curvature at the equator. Setting $f = D$ and solving for D/H yields Equation (13). The condition for the event at q_{L2} is shown in Figure 7. The internal paraxial focus F is now a distance ρ from the back side of the drop so that the reflected rays are also focused at F. Setting $\rho = D - f$ yields Equation (14).

We have confirmed these events exist by viewing laser light backscattered from levitated drops of water.[12] The ultrasonic field for our levitator is excited by a pair of vibrating rods which face each other. This facilitates mounting a camera (focused on infinity) closer to the drop than previously possible. The incident beam reflects from a beam splitter (mounted adjacent to the rods) onto the drop. The backscattered light passes through the splitter to the camera. A second camera records the drop's profile. Figure 8 shows backscattering patterns recorded for various D/H. (The bright point visible in these is light from a distant beam dump illuminated by that part of the beam not scattered by the drop. It is useful as a fiducial mark of the backward direction.) Drops flattened by this amount from acoustic radiation pressure may have significant deviations from the perfect spheroidal shape assumed in the derivations. Hence it is appropriate to consider an optically equivalent *spheroid* (with aspect ratio q_s) where the vertical curvature at the equator ρ^{-1} matches that of the actual drop. Let q_{sj} denote the apparent q_s for the jth part of Figure 8. *The patterns in Figure 8 may be understood as follows*: q_{sa} is sufficiently less than q_{L1} that the lips appear as two horizontal caustic lines. These are the distinct bright lines which are displaced in (a) by vertical angle ~1.2°. The drop is flatter in (b) so that q_{sb} is much closer to q_{L1} and the horizontal caustics are nondistinct. Nevertheless, as in (a), θ_3 is not close enough to 180° for the cusp points (where the caustics terminate) to be in the field of view. (A somewhat wider field of view than the present 17° would be needed to photograph symmetric cusp points.) In (c), a slight increase of the acoustic amplitude raises q_s to $q_{sc}>q_{L1}$ so that no caustics are visible. The reason why the actual merging of symmetric cusp points at 180° would be difficult to photograph is evident by inspection of Figure 4 of Reference 2: the rate at which $q_3 \rightarrow 180°$ diverges as $q_s \rightarrow q_{L1}$. An additional flattening of the drop gives rise to a new horizontal caustic at $q_{sd}>q_{L2}$ because of the second lips event.

To facilitate a quantitative check of the theory, several drops were photographed for patterns equivalent to (a) - (d) of Figure 8. For each photo, q_s was determined by measuring the equatorial radius of curvature ρ and averaged with other measurements of the *j*th class. The resulting *averages* are q_{sa}=1.398 ±0.085, q_{sb}=1.408±0.052, q_{sc}=1.461±0.048, and q_{sd}=1.587±0.093. These results are consistent with the predictions that $q_{sa}<q_{L1}$, $q_{sb}\lesssim q_{L1}$,$q_{L1}\lesssim q_{sc}<q_{L2}$ and $q_{sd}\gtrsim q_{L1}$. The size of the uncertainties is indicative of the difficulty in determining ρ from photographs. The D ranged from 1.3 to 2.3 mm. In spite of the aforementioned uncertainties, the predicted sequence of lips caustics is confirmed.

ACKNOWLEDGEMENT

This research was supported by the Office of Naval Research. The photographs in Figure 2 were obtained at Jet Propulsion Laboratory with the assistance of Dr. E. H. Trinh.

REFERENCES

1. P. L. Marston & E. H. Trinh, "Hyperbolic umbilic diffraction catastrophe and rainbow scattering from spheroidal drops," *Nature (London)* **312**, 529-531 (1984).
2. P. L. Marston, "Cusp diffraction catastrophe from spheroids: generalized rainbows and inverse scattering," *Opt. Lett.* **10**, 588-590 (1985).
3. J. F. Nye, "Rainbow scattering from spheroidal drops—an explanation of the hyperbolic umbilic foci," *Nature (London)* **312**, 531-532 (1984).
4. E. H. Trinh & C. J. Hsu, "Equilibrium shapes of acoustically levitated drops," *J. Acoust. Soc. Am.* **79**, 1335-8 (1986); P. L. Marston, S. E. LoPorto-Arione, & G. L. Pullen, "Quadrupole projection of the radiation pressure on a compressible sphere,", *ibid.* **69**, 1499-1501 (1981); **71**, 511 (1982).
5. M. V. Berry & C. Upstill, "Catastrophe optics: Morphologies of caustics and their diffraction patterns," *Prog. Opt.* **18**, 257-346 (1980).
6. P. L. Marston, "Wavefront geometries giving transverse cusp and hyperbolic umbilic foci in acoustic shocks," in **Shock Waves in Condensed Matter 1987,** S. C. Schmidt & N. C. Holmes, eds. (Elsevier Science Pubs., Amsterdam, 1988) pp. 203-206.
7. P. L. Marston, "Transverse cusp diffraction catastrophes: Some pertinent wavefronts and a Pearcey approximation to the wavefield," *J. Acoust. Soc. Am.* **81**, 226-232 (1987).
8. P. L. Marston, "Surface shapes giving transverse cusp catastrophes in acoustic or seismic echoes," in **Acoustical Imaging Vol. 16,** ed. by L. W. Kessler (Plenum, New York, 1988) pp. 579-588.
9. P. L. Marston, **Annual Reports** for 1986 & 1987 (available from Defense Tech. Info. Ctr., Cameron Stn., Alexandria, VA, Accn. Nos. AD-A174401 & A185785).
10. J. A. Kneisly, "Local curvature of wavefronts in an optical system," *J. Opt. Soc. Am.* **54**, 229-235 (1964).
11. H. C. van de Hulst, **Light Scattering by Small Particles** (Wiley, New York, 1957).
12. H. J. Simpson, M.S. degree project report, Washington State University 1988 (unpublished). Available by contacting author.

FIGURE CAPTIONS

1. Rays through a spheroidal water drop in the horizontal equitorial plane.
2. Scattering of light with λ=632 nm into the rainbow region of an oblate water drop with a diameter D and aspect ratio D/H of: (a) 1.39 mm and 1.23; (b) 1.42 mm and 1.31; and (c) 1.40 mm and 1.37. The horizontal scattering angle θ increases from left to right.

3a. Locations of outgoing rays relative to the drop's profile.

3b. Rainbow and cusp caustics partition a scattering pattern.

4. A hyperbolic umbilic focal section is produced at large distances from wave-fronts of this shape.
5. Comparison of the observed pattern (a) with a contour plot (b) of the theoretical amplitude function $S(U,V)$. THe angular scales of (a) and (b) are the same for a drop diameter of 1.42 mm such that the angular width of (b) is 3.37°. The contours begin at -0.2 and have an interval of 0.08.
6. Rays are focused at the back of the drop for the first lips event.
7. Focusing of the rays for the second lips event.
8. Backscattering patterns for oblate drops with D and D/H of: (a) 1.40 mm and 1.418; (b) 1.69 mm and 1.437; (c) 1.71 mm and 1.484; and (d) 1.62 mm and 1.596 The lips caustics are distinct in (a) but not in (b). The field of view is 17°.

5. Undercooling & Solidification

Robert Bayuzick
Session Chair

PROCESSING OF UNDERCOOLED MELTS

J. H. Perepezko and W. P. Allen
Dept. of Metallurgical and Mineral Engineering,
University of Wisconsin-Madison,
1509 University Avenue, Madison, WI 53706

ABSTRACT

The undercooling of liquids is observed often, but is restricted by the catalysis of heterogeneous nucleation sites in contact with the liquid. Containerless liquid processing and liquid dispersal into fine droplets yields an effective nucleant isolation which permits a deep undercooling approaching 0.3-0.4 T_m before the onset of solidification. At high undercooling the nucleation of an equilibrium phase may be superseded by metastable product structures. Often processing variables such as melt superheat, liquid surface films, powder size and cooling rate can be used to control the undercooling and produce a transition in phase selection kinetics. In this case the use of metastable phase diagrams is important for the interpretation and prediction of solidification products such as supersaturated solutions, metastable intermediate phases and glasses. At the same time a consideration of the competitive nucleation and growth kinetics and solidification thermal history is essential in the analysis of solidification microstructure. These features highlight a number of solidification reaction paths and new opportunities for the development of microstructure control.

1. INTRODUCTION

The generation of non-equilibrium solid phases has been observed frequently during solidification processing of undercooled melts (Anantharaman and Suryanarayana 1971 and Giessen 1976). It is recognized that the level of liquid undercooling at the onset of crystallization is a principal factor in bringing about rapid solidification conditions (Jackson and Chalmers 1956). Since the free energy available for the formation of nonequilibrium structures is directly related to the undercooling, the level of undercooling at the onset of crystallization must be considered in order to understand and characterize metastable phase formation during solidification processing (Perepezko and Anderson 1980). With increasing undercooling, product phase selection during nucleation is expanded to an increasing variety of phases and microstructural morphologies. Therefore, in order to maximize the undercooling potential of the liquid, it is important to identify key processing variables and analyze their effect on the undercooling response.

Since solidification of a large continuous liquid sample is catalyzed by the most potent nucleation site present, a means of circumventing the effect of catalytic sites such as oxides or container walls is necessary in order to observe extensive undercooling. Although several experimental techniques have been developed to achieve high levels of liquid undercooling with slow cooling rates (Turnbull and Cech 1950, Kattamis and Flemings 1966, and Wang and Smith 1950), the most successful method for minimizing the action of catalytic sites is based upon the dispersal of a liquid sample into a collection of fine droplets (Vonnegut 1948, Turnbull 1952 and Perepezko 1980). The high purity liquid sample is dispersed within a suitable carrier fluid, and droplet independence is maintained by the formation of thin, inert surface coatings which are not catalytic to crystallization. Thermal analysis techniques can then be used to evaluate the effect of various processing parameters on the attainable undercooling under controlled experimental conditions.

In fact, the experience derived from the application of the droplet method has provided a basis to interpret the solidification response during splat quenching, melt spinning and containerless processing where direct and accurate measurement of the initial phase selection and solidification temperature is difficult.

During solidification processing of an undercooled melt, nucleation and/or growth kinetics may favor the formation of a non-equilibrium phase rather than the thermodynamically stable phase. The reaction path followed during the initial stages of crystallization appears to be determined by the relative nucleation kinetics of the competing phases. From this basis the relationships between undercooling, cooling rate and particle size during the solidification of a distribution of liquid droplets (that are revealed experimentally) can be examined in further analysis using numerical nucleation rate calculations to reveal the range of processing conditions that expose different phase selection options.

2. PROCESSING VARIABLES

An effective approach that has been applied to obtain large melt undercoolings involves the cooling of a collection of fine liquid droplets (Perepezko 1980). Past experience with the droplet method has identified a number of processing parameters that are likely to govern the optimization of undercooling in powder samples. These processing variables include droplet size refinement, sample purity, droplet surface coating, uniformity of coating, cooling rate and melt superheat as well as alloy composition and applied pressure. The influence of the processing variables can be most effectively demonstrated with illustrations from droplet experiments, but the results are general and are expected to apply to other processing methods where direct measurement and control of solidification is more difficult.

The importance of liquid subdivision in realizing a large undercooling potential is indicated by the series of crystallization thermograms given in Figure 1 for samples of high purity tin (99.999%) (Perepezko 1980). In each case the processing conditions were identical except for the degree of liquid dispersal. The coarsest dispersion with a most frequently occurring size, $\overline{D}$, of about 275 μm displays a major crystallization exotherm centered at 184°C which corresponds to an undercooling of 48°C. There are also several additional exotherms apparent which extend about 50°C below the crystallization peak at 184°C. As $\overline{D}$ decreases with an increasing degree of droplet refinement, the crystallization exotherm at 184°C is observed to decrease in magnitude. A similar behavior is exhibited by the other exotherms in Figure 1. In addition, accompanying each level of droplet refinement is the development of further crystallization exotherms at decreasing temperatures. The shift in exotherms to lower temperatures with increasing refinement indicates that even for dispersions which may be considered to be relatively fine (i.e. $\overline{D}$= 30 μm), several nucleants are likely to be present within many of the droplets.

The average particle size is not the only distribution parameter relating to the level of refinement since the range of sizes within the droplet dispersion can vary considerably. For example, the undercooling behavior of a tin droplet sample having a well characterized size distribution is shown in Figure 2. The fit given in Figure 2 indicates that the droplet distribution produced by mechanical shearing is log-normal with unimodal behavior. The solidification exotherm corresponding to this droplet sample is relatively broad and exhibits multiple crystallization peaks. This example illustrates that for a given droplet surface coating, the greatest level of undercooling and the most uniform crystallization behavior are obtained with a fine and narrowly distributed size range.

The above observations illustrate the influence of droplet size on the undercooling and crystallization behavior of liquids. The dispersal of a bulk sample of liquid into a collection of fine droplets effectively isolates potent nucleation sites

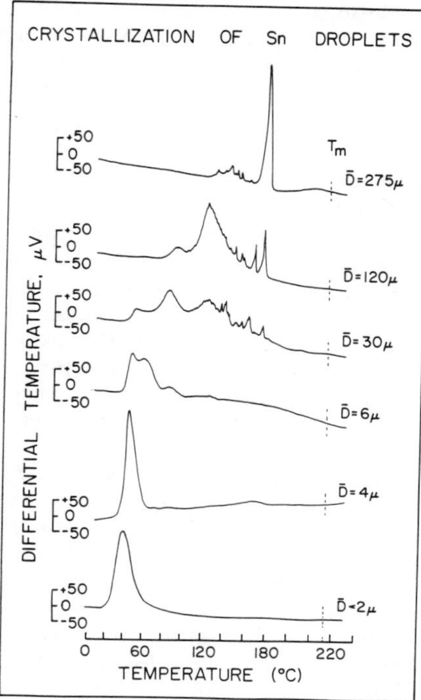

Figure 1

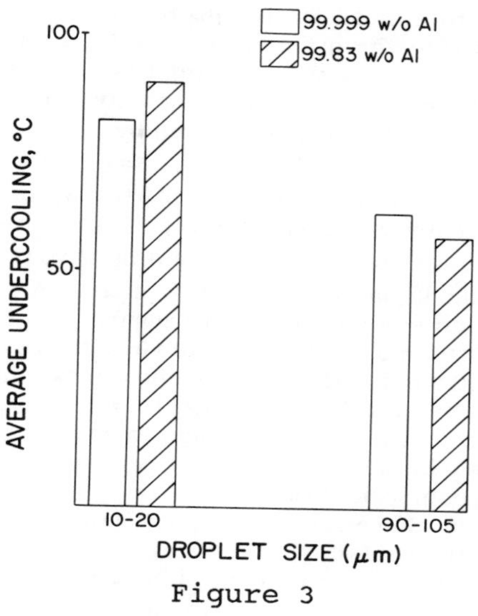

Figure 3

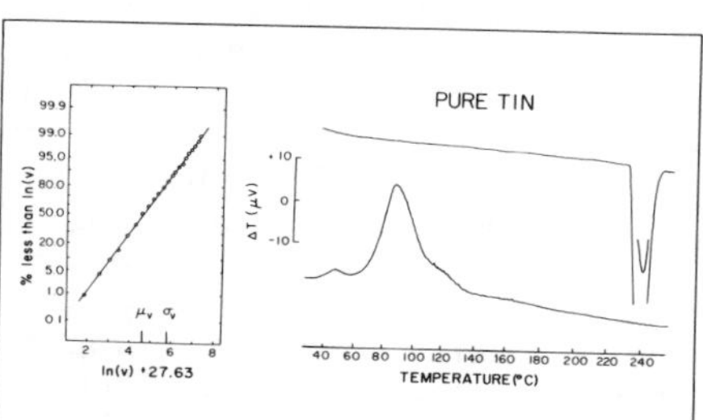

Figure 2

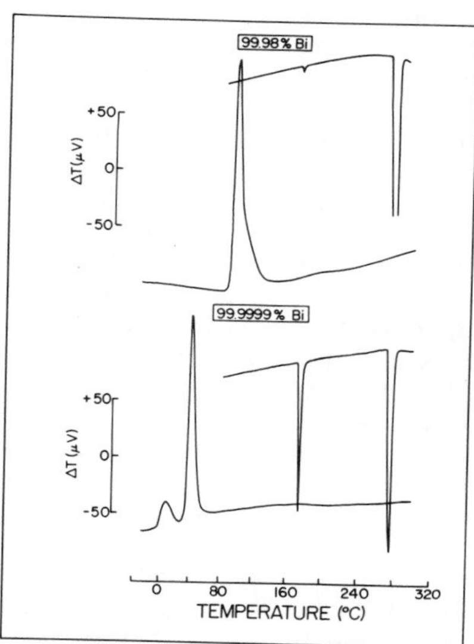

Figure 4

into a small fraction of the droplet population. If the nucleating sites contained within the original liquid volume are distributed randomly, the arrangement of nucleants among the droplets for a high degree of dispersal may be described by a Poisson distribution. For a log-normal droplet size distribution the nucleant free droplet fraction, X, is given by

$$X=\sum_i P(v_i)X_i = \sum_i P(v_i)\exp\,(-M_i v_i) \qquad (1)$$

where $P(v_i)$ is the log-normal distribution function (Box *et al.* 1978), X_i is the nucleant free fraction in the *i*th size group, M_i is the concentration of nucleants per unit volume and v_i is the droplet volume. In the case of monodisperse droplet sample, Equation (1) can be written as $X = \exp(-Mv)$. With such a distribution of nucleation sites, it is only necessary for a sample to contain an active catalyst concentration of $7\mathrm{x}10^{-13}$ m^{-3} to yield crystallization of 90% of droplets with a diameter of 40 μm at a temperature above maximum undercooling. However, reducing the droplet size to 14 μm will yield $X = 0.9$. While this example emphasizes the importance of particle size refinement in achieving large undercooling, it also suggests that very small active catalyst concentrations can have a significant effect on the undercooling behavior.

Since Mv in Equation (1) is proportional to the total number of active nucleants contained in the starting liquid, the fraction of nucleant-free droplets in some cases will be influenced by the purity of the starting material. Figure 3 illustrates the effect of melt purity on the crystallization behavior of aluminum droplets. For the two size ranges investigated, it is observed that changing the purity from 99.999% to 99.83% has only a minor effect on the liquid undercooling level. Although specific impurity levels do not seem to be crucial for aluminum, other examples have been noted where the impurity content can be an important factor. Thermograms presented in Figure 4 show the effect of purity on the undercooling of two bismuth droplet samples having the same size range and treated using the same surface coating procedure. A change in purity from 99.9999% to 99.98% yields a measurable decrease in undercooling with samples prepared from 99.98% pure material crystallizing over a broader temperature range. These examples illustrate that although the largest undercooling level is usually achieved with high purity liquids, melts of nominal purity can also exhibit substantial undercoolings.

Based upon the results generated by theoretical and experimental analysis (Fletcher 1958 and Maurer 1959), it appears that the nucleating efficiency of foreign particles becomes significant for impurity clusters approximately 100Å in size or larger. This information coupled with bulk impurity concentration data can be used to analyze the effect of impurities on the nucleation behavior (Hoffmeyer 1985). Assuming a monodisperse droplet population and that all impurity atoms combine to form spherical clusters approximately 100Å in size, the number of impurity clusters contained within droplets of a given size can be calculated as a function of melt purity. If this calculation is performed for a series of monosized droplet populations, the overall influence of impurity clusters on the nucleation response can be estimated. In Figure 5, the log of the overall impurity concentration, C_I, is plotted versus Z which is the log of the average number of impurity clusters per droplet. The number of impurity clusters present in monosized droplets with diameters of 0.1, 1.0 and 10 μm is indicated as a function of bulk impurity concentration. The intersection of the curve corresponding to a given droplet size with the line Z=0 defines the purity level necessary to observe an average of one impurity cluster per droplet. Examination of the curve corresponding to 1 μm droplets shows that for an overall purity of 99.9999%, the possibility exists for a significant fraction of the droplets to nucleate in the absence of residual impurity clusters.

The operation of droplet dispersal isolates the liquid from interaction with container walls by the formation of a stable surface coating which is required to maintain droplet independence. When internal nucleants are effectively isolated into a small fraction of the droplet population, the droplet surface coating is believed to be the limiting factor for determining the liquid undercooling level (Perepezko 1980). This aspect of undercooling behavior is illustrated in Figure 6 by a series of DTA thermograms corresponding to bismuth droplet samples prepared with different surface treatments. Each of the three droplet emulsions displays a narrow size distribution with an average diameter of less than 10 μm. Although all of the samples exhibit relatively uniform cooling to a single well defined crystallization exotherm, the level of undercooling achieved in each case varies in response to changes in surface coating. The potency of a surface in catalyzing the nucleation of a crystalline phase from an undercooled liquid has been characterized by a contact angle, θ (Pound and LaMer 1952 and Turnbull 1950). In order to observe a large amount of undercooling, the surface coating should wet the liquid phase to a greater extent than the nucleating crystalline phase. Surface treatments resulting in the formation of droplet coatings with a complex structure that is dissimilar to the crystalline phase are expected to be desirable for obtaining large undercooling.

The effect of surface coating on the undercooling behavior of aluminum is shown in Figure 7 (Perepezko, Graves, and Mueller 1987). Modification of the surface oxide was achieved by dispersing the bulk liquid within various inorganic molten salt systems. Powders produced in each of the salts do attain a distinct undercooling level with powders prepared in the sulfate salt achieving the largest undercooling level. It should be noted that not all of the powders produced in the sulfate salt attained the maximum undercooling level since a smaller crystallization exotherm was also present at approximately the same undercooling as observed for powders prepared in the chloride salt. As a means of investigating the ability of the chemical environment in which the droplets are dispersed to alter the chemistry of the surface coating, powders were characterized by Auger Electron Spectroscopy (AES). Surface chemistry analysis of powders produced in the different salts indicates that while an Al_2O_3 oxide coating is present on each of the three powder samples, differences in coating chemistry may be induced by varying the chemistry of the dispersal media (Figure 8). Although the phosphate salt induces a surface concentration of phosphorus and the sulfate salt results in the presence of residual sulfur, droplet dispersal in the chloride salt seems to produce a relatively pure Al_2O_3 coating. These observations indicate that changes in the droplet surface chemistry can be induced by changes in the chemical environment in which the powders are produced and that even slight coating modifications can have a significant influence on the attainable undercooling.

While surface catalysis can yield a well defined crystallization in droplets over a narrow temperature range, this characteristic is not universal. In some cases there is a continuous crystallization over a wide range of temperatures. This type of undercooling behavior is illustrated in Figure 9 for an indium droplet sample with a fine average size of 8 μm and a narrow size range (5 to 12 μm). The continuous crystallization is represented by a baseline offset in the thermogram and accounts for the solidification of about one-half of the sample. When crystallization is controlled by a single nucleation event, the entire sample can be frozen by holding at one temperature within the crystallization range (Turnbull 1952). The narrow and symmetric crystallization at the maximum undercooling limit in trace A of Figure 9 appears to be controlled by single nucleation kinetics. In trace B the completion of crystallization at maximum undercooling by an isothermal holding treatment is clearly demonstrated. Similarly, separate holding treatments at high temperature in Figure 9 provide a basis for understanding the solidification behavior prior to maximum undercooling in terms of an overlap of many separate nucleation events.

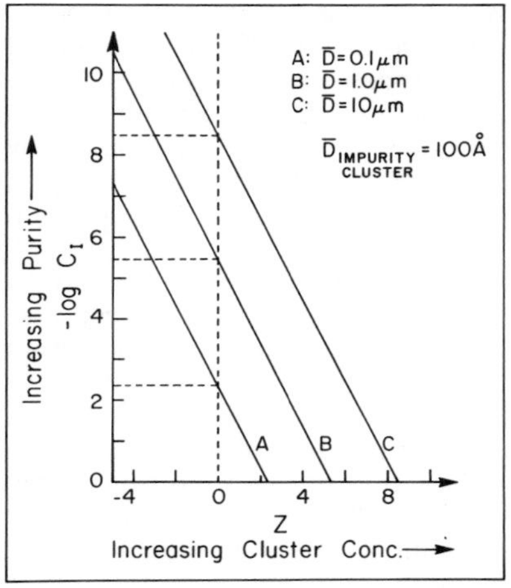

Figure 5

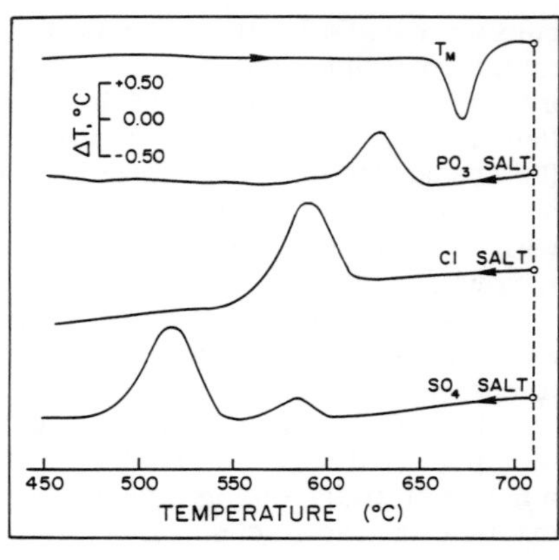

Figure 7

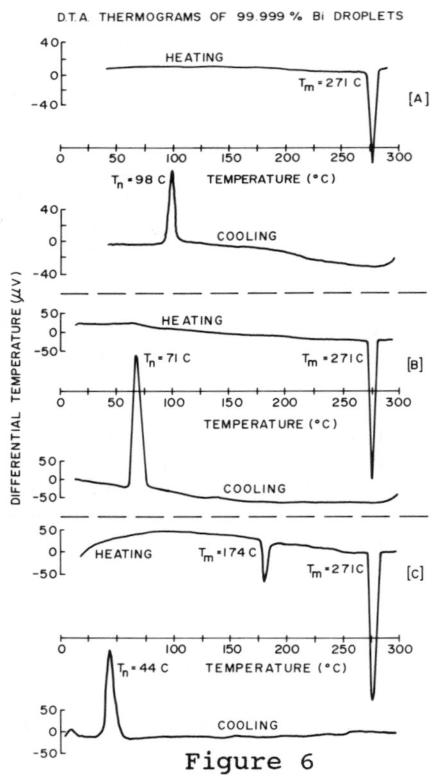

Figure 6

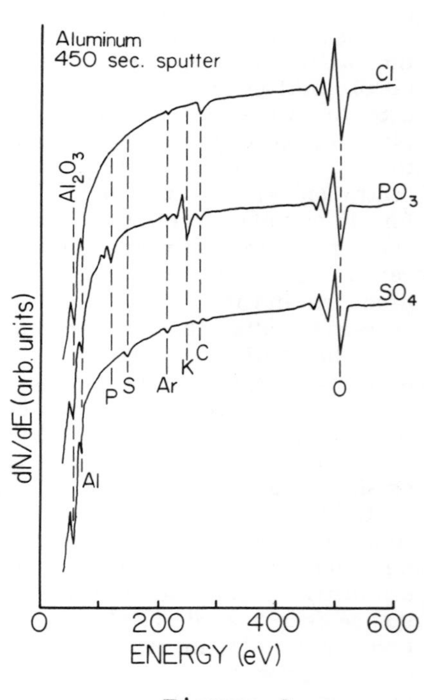

Figure 8

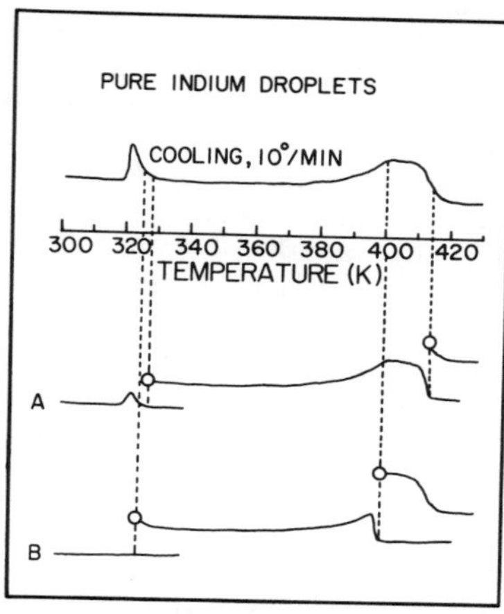

Figure 9

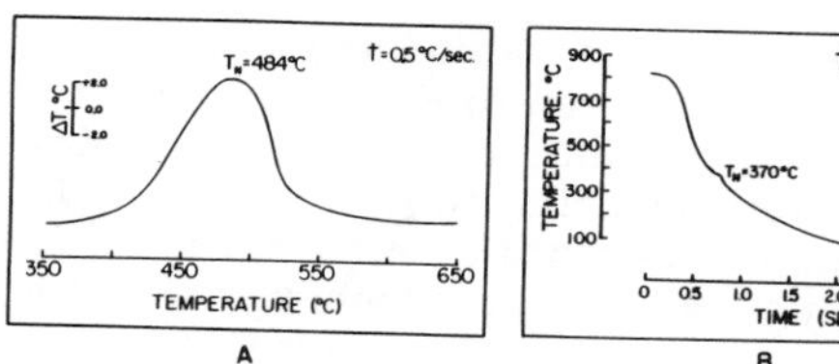

Figure 10

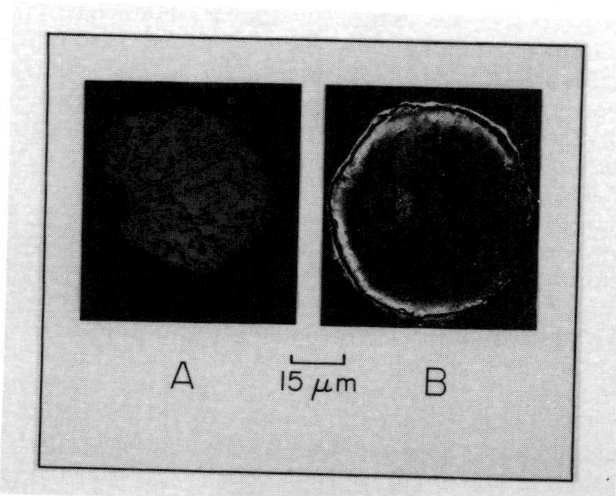

Figure 11

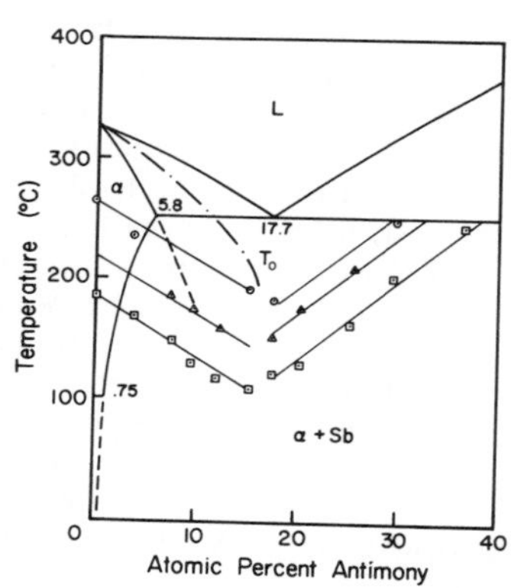

Figure 12

Isothermal holding within the high temperature range results in a flat baseline at each holding temperature and an onset for the crystallization peak with a sharp leading edge. A sharp leading edge indicates that each nucleation event is restricted to a narrow temperature range. The high temperature crystallization characteristics can be described in terms of the activity of a collection of droplet surface sites (of varying size) distributed among the droplets (Turnbull 1953). Other experiments involving x-ray diffraction indicate that the surface sites are likely to be small patches of In_2O_3 (Perepezko and Paik 1982). These features demonstrate that even in a finely dispersed liquid sample with a specific droplet surface treatment intended to provide a controlled undercooling behavior, other variations in surface coating catalysis can result in non-uniform undercooling with a wide crystallization range.

Another factor which can influence the undercooling potential of the liquid is the applied cooling rate. In the Al-Si binary system the nucleation temperature of hypereutectic and eutectic alloys appears to lie on a plateau indicating that a common catalyst is limiting the undercooling (Mueller, Richmond, and Perepezko 1985). In this case, the common nucleant appears to be primary silicon, which forms just below the liquidus and then catalyzes the remaining liquid to freeze at a temperature of 484°C. When eutectic droplets are quenched at about 500°C/sec, the nucleation temperature of the sample is depressed to 370°C as is shown in Figure 10. This increase in undercooling manifests itself in the microstructural morphology of the powders following crystallization. Although the solidification microstructure of eutectic droplets slowly cooled at 0.5°C/sec exhibits primary silicon particles, those quenched from the liquid state do not usually show evidence indicating primary silicon formed from the melt (Figures 11a and 11b respectively). Thus, the increase in cooling rate suppresses formation of primary silicon yielding a transition to an alternate nucleation kinetics operating at lower temperatures, which results in a modified microstructure containing primary aluminum.

Other factors controlling the undercooling potential of a liquid include melt superheat, alloy composition and pressure. The temperature above the melting point to which the molten sample is heated before cooling has been reported to influence the level of liquid undercooling achieved prior to solidification (Glicksman and Childs 1962). The retention of "crystalline adsorbates" above the melting temperature has been proposed as a possible explanation for the observation that the thermal history of the sample affects the undercooling response (Turnbull 1950). These adsorbates are actually embryos of the crystalline phase which are retained within cavities on the surface coating and act as preferred sites for nucleation of the solid during cooling. As the superheating is increased these adsorbates become less stable and eventually melt allowing for greater undercooling. It should be noted that excessive levels of superheating can have undesirable consequences such as oxidation of the sample, alteration of the surface coating and vaporization of volatile constituents.

The effect of alloy concentration on the undercooling behavior of Pb-Sb binary alloys is summarized in Figure 12 by a plot of crystallization temperatures on the lead-rich portion of the phase diagram. The application of different surface treatments was used to produce the three distinct undercooling levels indicated. As observed in a number of other alloy systems the composition dependence of the nucleation temperature can exhibit a pattern which tends to follow the trend of the alloy liquidus and may reflect a common nucleus composition (Perepezko and Paik 1982). However, in other systems the trend of crystallization temperatures does not appear to be correlated with the liquidus curve and in some cases appears to be due to the intervention of metastable phase nucleation.

The influence of applied pressure on the undercooling response is illustrated in Figure 13 (Yoon and Paik *et al.* 1986). The nucleation temperatures of pure bismuth droplet samples exhibiting two distinct undercooling levels are denoted on the temperature-pressure diagram (Klement *et al.* 1963) by T_{N1} and T_{N2}. Thermal cycling was used to identify the relationship between crystallization temperature and formation of the metastable Bi(II) phase. It was determined that solidification of droplets at the higher undercooling level, T_{N2}, resulted on the formation of Bi(II) while nucleation at T_{N1} was associated with solidification of Bi(I). Comparison of the pressure dependent crystallization temperatures with the measured melting points of Bi(I) and metastable Bi(II) as a function of pressure indicates that the change in nucleation temperature with pressure tends to follow the trend of the corresponding melting temperature.

The effect of the various processing parameters on the undercooling response is summarized in Table I. Even when these processing conditions are satisfied to produce maximum undercooling, experience suggests that solidification is still initiated by a heterogeneous nucleation site associated with the sample surface (Perepezko *et al.* 1986). Therefore, it appears that close attention to the nature of the powder surface coating is of prime importance in achieving reproducible, large undercooling values in fine powders. The maximum undercooling limits for several pure metals measured using thermal analysis are listed in Table II. Based upon the comparison presented in Table II, current studies have demonstrated clearly that the previous maximum undercooling values actually correspond to heterogeneous nucleation conditions. Indeed, the present findings emphasize that further nucleation kinetics studies are necessary before a complete characterization of the revised undercooling limits in terms of heterogeneous or homogeneous kinetics is possible.

Table I:
Effect of processing parameters on undercooling

Parameter	Undercooling Response	Remarks
Droplet Size	Increased ΔT with size refinement at constant $\dot{T}$	Nucleant isolation follows Poisson statistics
Droplet Coating	Function of coating structure and chemistry; major effect in limiting ΔT	Most effective coating is catalytically inert
Cooling rate	ΔT generally increases with increasing $\dot{T}$	Changing $\dot{T}$ can alter the nucleation kinetics
Melt Superheat	System specific	Appears to be related to coating catalysis
Composition	T_n follows trend of T_L	Melt purity not usually critical
Pressure	T_n parallels melting curve trend	Change in response can signal alternate phase formation

Table II:
Maximum undercooling limits

Element	Previous Studies			Current Studies	
	ΔT (°C)	$\Delta T/T_m$	Reference	ΔT (°C)	$\Delta T/T_m$
Al	130	0.14	Turnbull and Cech (1950)	175	0.19
Sb	135	0.15	Turnbull and Cech (1950)	210	0.23
Bi	90	0.16	Turnbull and Cech (1950)	227	0.41
Cd	---	---	---	110	0.19
Ga	150	0.50	Bosio et al. (1966)	174	0.58
In	---	---	---	110	0.26
Pb	80	0.13	Turnbull and Cech (1950)	153	0.26
Hg	80	0.34	Turnbull (1952)	88	0.38
Te	---	---	---	236	0.32
Sn	117	0.23	Pound and Lamer (1952)	191	0.38

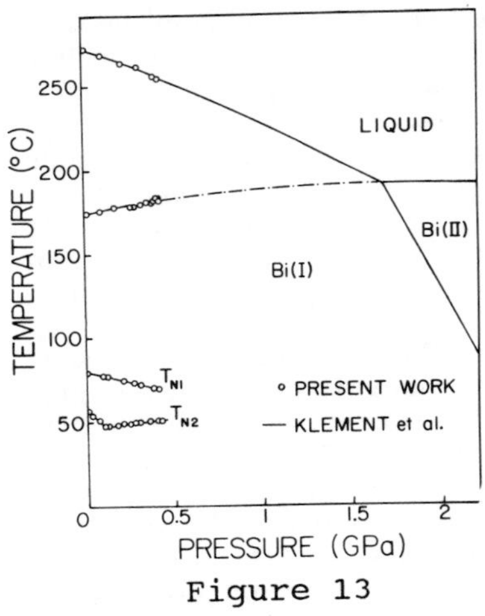

Figure 13

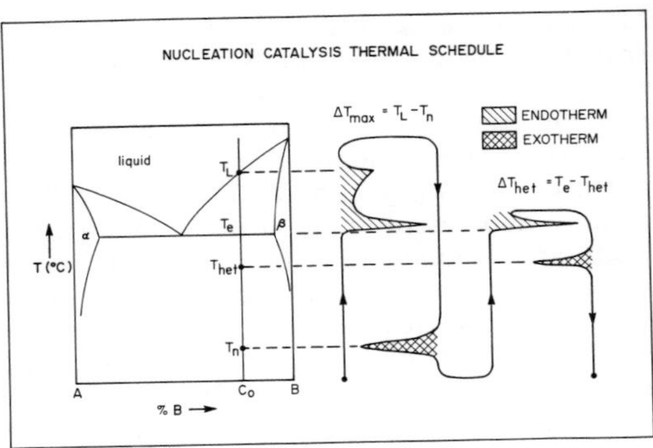

Figure 15

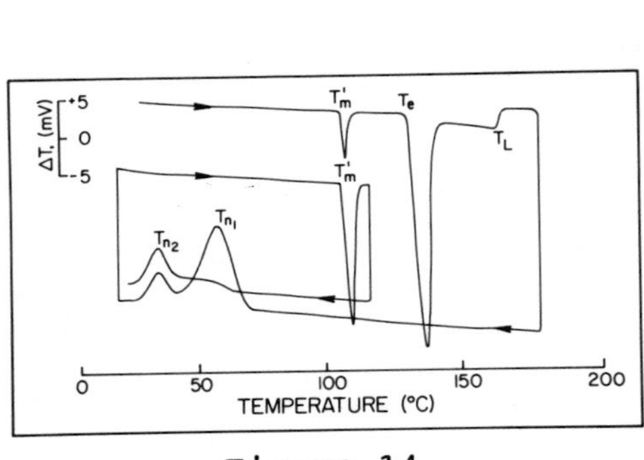

Figure 14

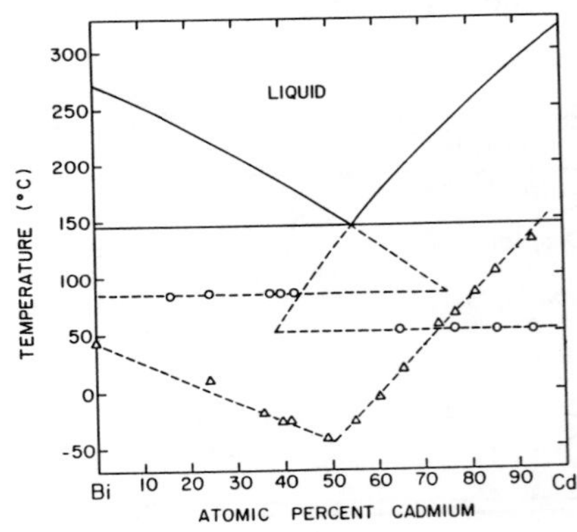

Figure 16

3. THERMAL ANALYSIS

The application of thermal analysis techniques such as Differential Thermal Analysis (DTA) and Differential Scanning Calorimetry (DSC) is an effective method for evaluating metastable phase equilibria. Thermal cycling experiments can be used to determine the relationship of a peak on the thermal analysis trace to the appropriate stable or metastable phase diagram. In addition, thermal analysis may be applied to measure the catalytic potency of a primary solid phase for nucleation of the remaining liquid.

It has been observed that a metastable phase can be produced in highly undercooled Bi-48.6 at% Sn droplet samples (Yoon and Perepezko 1986). The relationship between this metastable phase formed during rapid solidification at ambient pressure and a high-pressure phase in the same alloy system was considered by previous investigators (Gordon and Deaton 1972). An example of the application of thermal cycling to determine the correspondence of a given thermal analysis peak to metastable phase formation is presented in Figure 14. In addition to the equilibrium eutectic and liquidus signals, denoted by T_e and T_L respectively, a metastable endothermic event, T_m', is also observed during heating. During cooling from above the liquidus temperature two distinct exothermic peaks (T_{n1} and T_{n2}) corresponding to crystallization of the liquid are present. While the first cycle represents a complete heating and cooling trace, heating during the second cycle is halted just after T_m'. Cooling from this temperature results in the presence of only the lower temperature nucleation exotherm at T_{n2}. This thermal behavior indicates that metastable phase formation is primarily associated with that fraction of droplets which crystallize at the higher undercooling level, T_{n2}.

Within the framework of the droplet technique, the identity of possible heterogeneous sites can be examined in a controlled nucleation catalysis experiment. The approach which has been developed to permit the evaluation of alloy phase catalysis is summarized in Figure 15 (Sundquist and Mondolfo 1961). By annealing droplet samples above the eutectic temperature, T_e, a two-phase mixture is generated consisting of primary solid solution and liquid. Provided that other nucleation sites can be shown not to participate in the catalysis, the catalytic potency of the primary phase for nucleation of the contacting liquid can then measured on cooling. The first thermal cycle in Figure 15 indicates that the sample exhibits an undercooling ΔT_{max} when cooled from a temperature above the liquidus. When cooled from a temperature between T_e and T_L as in the second cycle, a level of undercooling ΔT_{het} below T_e is observed which can be used as a measure of the catalytic potency of the β for nucleation of the remaining liquid. An indication of the generality of the observed T_{het} reaction temperatures is given in Figure 16 for the Bi-Cd system (Anderson 1982). Maximum undercoolings (triangles) are indicated as well as T_{het} values (circles) representing catalysis by each of the primary phases. Liquidus extrapolations are also shown which can be used to estimate the composition of the contacting liquid at the corresponding reaction temperature. Thermal analysis of the solidification catalysis reaction involving bismuth primary phase and liquid reveals the presence of two distinct crystallization exotherms as is shown in Figure 17 (Ohsaka 1986). Repeated thermal cycles performed between a temperature just above the eutectic and the catalysis exotherm demonstrates that the relative magnitude of the low temperature nucleation peak increases significantly with cycling. Microstructural analysis indicates that the bismuth primary phase develops growth facets as a result of this thermal cycling, and the transition from rough (non-faceted) to faceted bismuth primary is illustrated by the two micrographs shown in Figures 18a and 18b (Anderson 1982). Based upon the thermal analysis results and microstructural observations it is believed that the higher and lower temperature exotherms correspond to nucleation of the liquid on non-faceted and faceted bismuth substrates respectively.

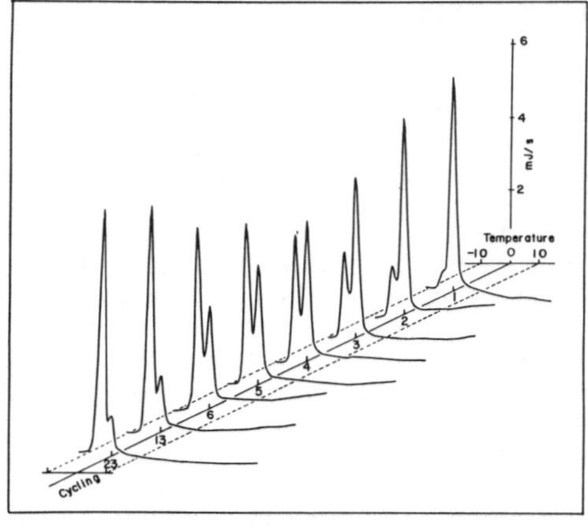

Figure 17

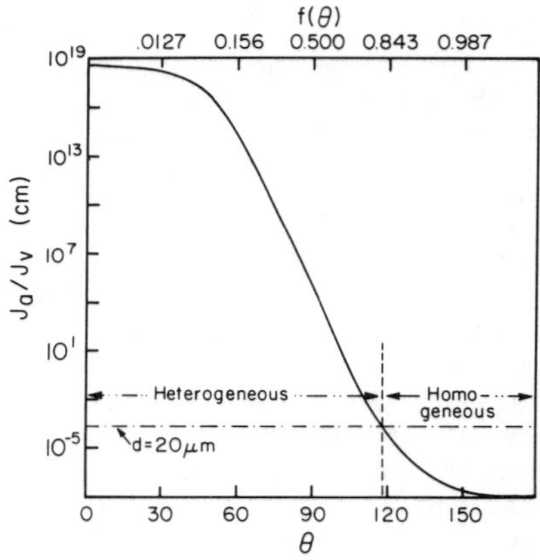

Figure 19

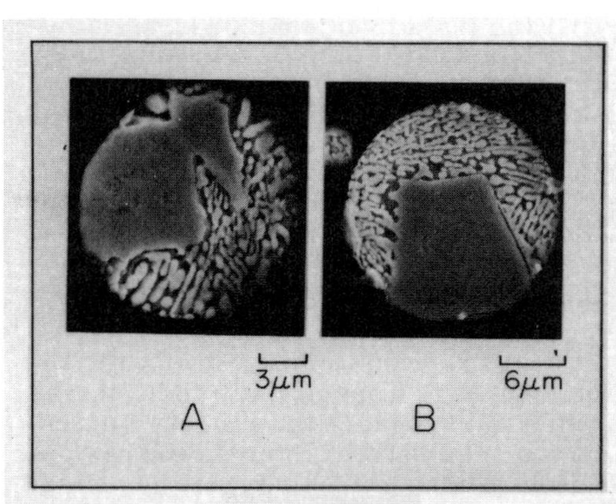

Figure 18

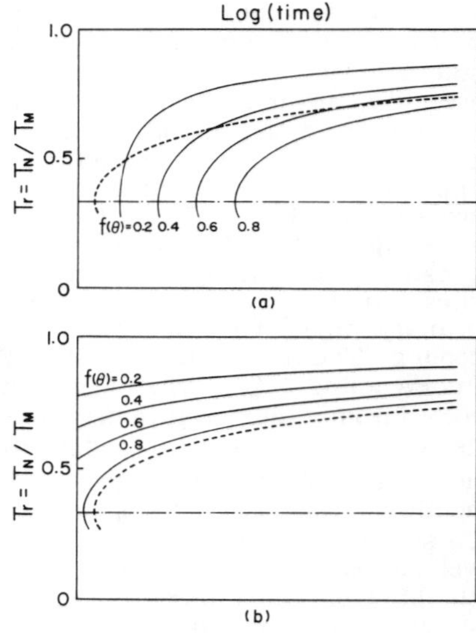

Figure 20

4. DROPLET NUCLEATION KINETICS

A more quantitative determination of the effect of melt cooling rate and droplet size on the undercooling level can be achieved through an investigation of the crystallization kinetics. It should be noted that the results of numerical nucleation rate calculations are highly dependent upon the values assigned to the kinetic parameters, and accurate, independent experimental determination of several of these parameters (such as the nucleation site potency and density and the interfacial energies) is not possible at the present time. However, a consideration of several cases involving limiting conditions of crystal nucleation can be used to develop a number of useful guiding relationships.

4A. Isothermal Conditions

The nucleation of a solid phase from the liquid is a fluctuation process involving small clusters of atoms. These clusters vary in size and each can be considered to be a potential nucleus for continued growth of the solid. Clusters of the critical size necessary to nucleate the new phase are produced by the addition and dissolution of individual atoms. At steady state the distribution of cluster sizes is stationary, and the nucleation rate is constant. The mechanism through which nucleation of the crystalline phase occurs is usually classified into two general types. Homogeneous nucleation refers to the case in which nucleation occurs throughout the volume of liquid without an association with catalytic sites. For the majority of processing conditions, nucleation is initiated at some catalytic site and is referred to as heterogeneous nucleation. These sites may be distributed within the volume of liquid or on the droplet surface. Based upon classical theory (Turnbull 1952) a general expression for the steady state nucleation rate, J_i, can be represented as

$$J_i = \Omega_i \exp[-\Delta G^* f(\theta)/kT] \qquad (2)$$

where J_i relates to either heterogeneous surface nucleation, J_a; homogeneous nucleation, J_v; or heterogeneous nucleation on catalytic sites distributed within the volume, J_s. Appropriate values for the prefactor, Ω_i, activation energy barrier, ΔG^*, and contact angle function, $f(\theta)$, are used in Equation (2) and kT has the usual meaning. The terms for Ω_i involve a product of the nucleation site density on a sample surface or volume basis, the number of atoms on a nucleus surface and a liquid jump frequency. For many cases $\Omega_v = 10^{30}\eta^{-1}\text{cm}^{-3}\text{s}^{-1}$ and $\Omega_a = 10^{22}\eta^{-1}\text{cm}^{-2}\text{s}^{-1}$ with the liquid shear viscosity in poise, η, given by

$$\eta = 10^{-3.3} \exp[3.34T_l/(T-T_g)] \qquad (3)$$

where T_l is the liquidus temperature and T_g is the glass transition temperature (Turnbull 1969). The term ϕ in the expression for Ω_a is defined as the fraction of surface sites which are active. For J_v and J_a, ΔG^* is given by $\Delta G^* = b\sigma^3/\Delta G_v{}^2$ where σ is the liquid-solid interfacial energy, ΔG_v is the driving free energy for nucleation of a unit volume of product phase and $b = 16\pi/3$ for spherical nuclei with $f(\theta) = 0.25[2\text{-}3\cos\theta + \cos^3\theta]$. For heterogeneous volume nucleation the value of Ω_s depends on the specific catalyst concentration and size as well as the form of ΔG^* which is related to the details of the catalyst-nucleus interaction (Turnbull 1981).

The nucleation rate is a relatively steep function of temperature with a magnitude determined principally by the exponential term involving ΔG^* at the nucleation temperature , T_n and to a lesser degree by the prefactor term. In evaluating the temperature dependence of J_i, usually a constant value is taken for Ω_v, but for Ω_a and Ω_s it is necessary to consider that the catalytic site density may vary for different conditions. Similarly, since little information is available to estimate the catalytic potency of a particular site, a range of $f(\theta)$ values is normally used in calculations. However, the most important parameters in determining J_i and hence the maximum undercooling level are ΔG_v and which have received continued experimental and theoretical study (Perepezko and Paik 1984, Skapski 1956 and Spaepen 1975).

In general, the critical condition to observe nucleation experimentally during isothermal holding over a time t (usually one nucleus per sample) is $J_v vt = 1$, $J_a at = 1$ or $J_s st = 1$ depending on the nucleation kinetics controlling solidification. Therefore, using Equation (2) the time for experimentally observable nucleation under isothermal conditions, t_i, (assuming heterogeneous surface kinetics) can be written as

$$ln(t_i) = -ln(a\Omega_a) + \Delta G^* f(\theta)/kT \qquad (4)$$

or

$$ln(t_i) = -ln(a\Omega_a) + 16\pi\sigma^3 f(\theta)/3k\Delta H^2 T_m T_r(1-T_r)^2 \qquad (5)$$

where a is the droplet surface area, ΔH is the heat of fusion, T_m is the melting temperature and T_r is the "reduced undercooling" defined as $T_r = T_n/T_m$ where T_n is the nucleation temperature. It should also be noted that a lower bound for the onset of nucleation of a crystalline phase has been estimated to be $T_r = 1/3$ (Perepezko and Paik 1982). Below this temperature the undercooled liquid can no longer be maintained metastably and crystallization will occur unless the glass transition intervenes.

The effect of droplet size on the nature of the nucleation process can be estimated through an evaluation of Equation (2). A consideration of the conditions necessary for ($J_v vt$) and ($J_a at$) to be of comparable magnitude (for a given sample size) can be used to determine the transition point between heterogeneous surface and homogeneous volume dependent nucleation (Boettinger and Perepezko 1985). For a monodisperse droplet size one can write the ratio J_a/J_v as

$$J_a/J_v = [\Omega_a/\Omega_v \exp \Delta G^*(1-f(\theta))/kT] = v/a \qquad (6)$$

where v and a correspond to the droplet volume and area respectively. Assuming $\Omega_v = 10^{35} cm^{-3}s^{-1}$ and $\Omega_a = 10^{27} cm^{-2}s^{-1}$ and using the limiting value for the onset of sensible nucleation ($\Delta G^* = 60kT$), the ratio J_a/J_v can be written as a function of $f(\theta)$ (Figure 19). Figure 19 illustrates that for a 20 μm droplet ($v/a = 3.3 \times 10^{-4}$ cm), at a critical contact angle, θ_t, of 118° there is an equal probability for the heterogeneous surface and homogeneous volume nucleation processes. Homogeneous nucleation would be favored over heterogeneous nucleation for catalysts exhibiting a potency which can be described by contact angles in excess of 118°. It should be noted that this analysis also assumes a constant active surface site fraction of $\phi = 1$.

Another feature to consider when examining the conditions necessary to observe a transition in the nucleation kinetics is the influence of a possible range of catalytic site densities represented by variation of Ω_a in Equation (5) as well as changing catalytic potencies reflected by variation of $f(\theta)$. The effect of these variables on the appearance of a kinetic transition is illustrated by the series of isothermal time-temperature-transformation curves shown in Figure 20. In both cases the dashed curve represents homogeneous nucleation. For a relatively low catalyst site density ($\Omega_a/\Omega_v \cong 10^{-20}$), a kinetic transition from heterogeneous surface to homogeneous volume nucleation is possible depending on the potency of the active catalyst (Figure 20a). On the other hand a relatively high site density ($\Omega_a/\Omega_v \cong 10^{-10}$) would, in general, preclude the observation homogeneous nucleation as is illustrated in Figure 20b. The conditions examined in Figure 20 demonstrate that a given undercooled volume does not have to be free of all nucleants in order to achieve a transition to homogeneous kinetics. It is only necessary that the degree of dispersal is sufficient to mitigate the influence of the most potent nucleants.

When a liquid sample is solidified in a highly undercooled condition, the nucleation process becomes the rate determining step for solidification. Therefore, the nucleation rate is equivalent to the solidification rate. In a fine droplet dispersion, the nucleation event in each droplet is an independent occurrence. On this basis a first order reaction law can be applied to describe the rate of solidification in a distribution of droplets (Turnbull 1952) as

$$dN(v_i,t)/dt = -k_i N(v_i,t) \tag{7}$$

or

$$N(v_i,t) = N(v_i,0) \exp(-k_i t) \tag{8}$$

where $N(v_i,t)$ is the number of liquid droplets belonging to the ith size group at time t with volume v_i and k_i is the operative nucleation frequency. For homogeneous nucleation, $k_i = J_v v_i$, and for the more commonly observed surface area dependent heterogeneous nucleation, $k_i = J_a a_i$ where a_i is the surface area of a droplet with volume v_i.

The time dependence of the liquid droplet distribution or the freezing rate is illustrated schematically in Figure 21 for isothermal conditions. Initially at $\tau = 0$ the liquid droplet size distribution is log-normal. As time passes, larger droplets freeze before smaller sizes at a rate given by Equation (7). As a result, the initial distribution characteristics of liquid droplets changes, as shown in Figure 21. The rate of this size distribution change will be different for volume dependent and surface area dependent kinetics and can be measured experimentally to determine the operative nucleation mechanism (Turnbull 1952, and Paik and Perepezko).

Most often a droplet sample contains a distribution of nucleants either internal or on the surface which can become active during cooling. As noted previously the droplet fraction which is free of such nucleants can be evaluated from the Poisson distribution function. Under these conditions the number of liquid droplets of volume v_i that remain in a distribution after cooling from T_m to T_n can be obtained as

$$N(v_i,0) = N_T P(v_i) \exp(-m_i a_i) \tag{9}$$

where N_T is the total number of droplets, $P(v_i)$ is the log-normal distribution function (Box *et al.* 1978) and m_i is the number of surface catalysts per unit area. The isothermal solidification of a droplet dispersion is then described by the total number of unfrozen droplets at time t, $N_T(t)$, as the summation of nucleation events over all size classes

$$N_T(t) = N_T \Sigma_i P(v_i) \exp(-m_i a_i)) \exp(-J_a a_i t) \tag{10}$$

In practice, $N_T(t)$ can be monitored by calorimetric[1], dilatometric (Turnbull 1952) or direct observation measurements (Wood and Walton 1970).

4b. Continuous Cooling

During continuous cooling the critical nucleation condition (e.g. $J_a at = 1$) requires some modification to account for the total number of nuclei, N, formed while cooling from T_m to T_n (Hirth 1978). Assuming a constant cooling rate, $\dot{T}$, N can be written as

$$N = -(1/\dot{T}) \int_{T_m}^{T_n} J_i(T)\, dT \tag{11}$$

where $J_i(T)$ is the nucleation rate at temperature T. The integration involved in Equation (11) is the area under the curve in Figure 22. Although this area can be evaluated numerically, it can be approximated by the shaded region as

$$-\int_{T_m}^{T_n} J_i(T)\, dT \cong (T' - T_n)\, J_i(T_n) \tag{12}$$

where T' is the intersection between the tangent to $J_i(T)$ at T_n and the T axis. Assuming that σ and Ω_i are independent of temperature over the narrow range of interest, differentiation of Equation (2) with respect to temperature gives

[1]Interested parties are invited to write J. H. Perepezko regarding this research conducted by J. S. Paik and J. H. Perepezko.

$$T' = T_n - \frac{\Delta G_v{}^3 T_n{}^2}{A\left(2T_n d\, \Delta G_v / dT + \Delta G_v\right)} \quad (13)$$

where $A = 16\pi\sigma^3 f(\theta)/3k$. For the case of heterogeneous surface nucleation, the critical condition to observe nucleation during continuous cooling can now be written as

$$J_a(T_n)\ \text{at} = \frac{-A\ \Delta T\ \left(2T_n d\ \Delta G_v / dT + \Delta G_v\right)}{\Delta G_v{}^3 T_n{}^2} = K(T_n) \quad (14)$$

where $t = \Delta T/\dot{T}$. Typically, $K(T_n)$ is on the order of 10^2 for $0.6 \leq T_r \leq 0.7$ (Boettinger and Perepezko 1985). From this analysis, the time required to observe heterogeneous nucleation during cooling, t_c, is given by

$$\ln(t_c) = -\ln[a\Omega_a/K(T_n) + 16\pi\sigma^3 f(\theta)/3k\Delta H^2 T_m T_r(1-T_r)]^2. \quad (15)$$

The comparison of Equations (5) and (15) indicates that the ratio $t_c/t_i \cong 100$. In other words, the transformation curve for continuous cooling will be shifted to longer times with respect to the corresponding isothermal curve. A central assumption of this analysis is the maintenance of steady state conditions over a range of nucleation temperatures where the approximation of Equation (12) is reasonable. It should be noted that for large ΔT and high $\dot{T}$ transient conditions are likely to become important in determining the rate of kinetic responses.

For a given Ω_a/Ω_v ratio, variation of the $f(\theta)$ term in Equation (15) can be used to generate a series of transformation curves identical to those calculated from Equation (5) for the isothermal case (Figure 20). These curves may be used to explain the difference in undercooling behavior displayed by tin and indium droplet samples (Figures 1 and 9). The sequence of crystallization exotherms exhibited by the tin sample suggests that a spectrum of nucleants of varying catalytic potency is present within the liquid. It is clear that the active nucleants are not all equally effective in catalyzing solidification and are not distributed in equal abundance. The occurrence of crystallization exotherms at similar temperatures in different droplet dispersions would suggest that in each case a common nucleation site initiated solidification. The development of exotherms at distinct undercooling levels implies that each of these sites may be characterized by discrete values of $f(\theta)$. Melts containing catalytic sites with potencies described by a continuum of $f(\theta)$ values would be expected to exhibit an undercooling response like that of indium.

A series of schematic continuous cooling transformation curves is given in Figure 23 to illustrate the application of transformation diagrams for analyzing crystallization kinetics (Boettinger and Perepezko 1985). Curve A corresponds to formation of a metastable phase, and the curves B, C and D refer to nucleation of the equilibrium phase. Transformation curve C represents the most catalytic sites. As the value of $f(\theta)$ increases, *i.e.*, the potency decreases, the curve breadth tends to decrease, and the nose shifts to shorter times. The dashed lines in Figure 23 represent typical continuous cooling paths. The undercooling limit for a cooling rate of $\dot{T}_1$ would be given by the intersection of the cooling path with transformation curve C. However, if the cooling rate is increased to $\dot{T}_2$, the effect of catalyst *C* can be negated. Therefore, the undercooling level achieved can be increased dramatically by increasing the cooling rate. The breadth of the transformation curve associated with metastable phase formation, A, is larger than that for the equilibrium phase, B. If nucleants of higher catalytic potency are not present, i.e. curves C and D are absent, nucleation of the metastable phase will dominate regardless of the applied cooling rate. It is also possible for the nucleation of the equilibrium phase to be favored with increasing cooling rate. This can occur if the nucleation site density for the equilibrium phase, curve D, is greater than that of the metastable phase, A. Under these conditions, an increase in the cooling rate would result in a reduced amount of metastable phase formation.

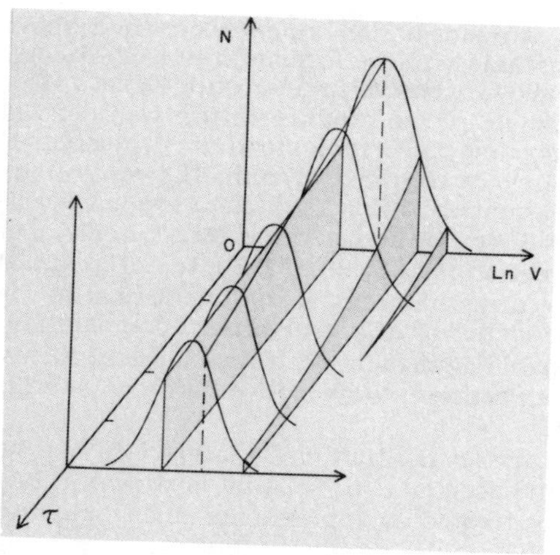

Figure 21

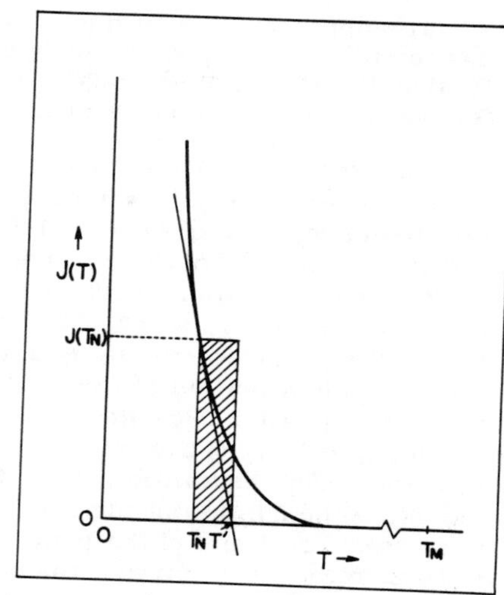

Figure 22

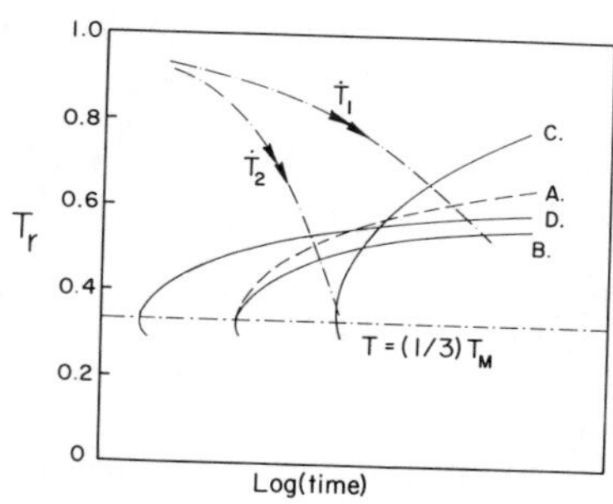

Figure 23

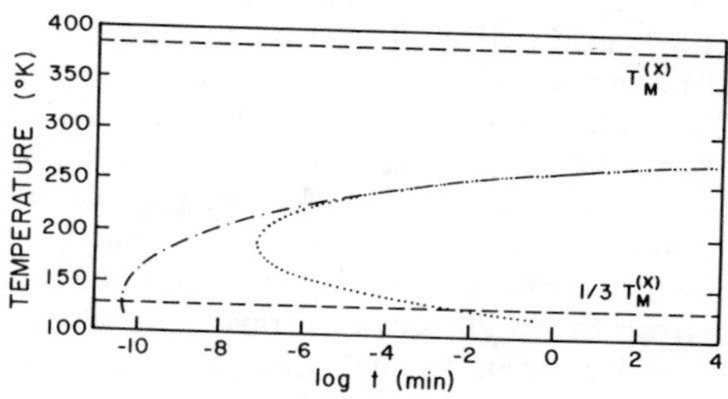

Figure 24

The approach to continuous cooling transformation kinetics outlined above has recently been applied to analyze metastable x-phase formation in a Pb-56 at% Bi alloy (Perepezko *et al.* 1985). In this alloy extensive undercooling occurs (T_r = 0.7) yielding nucleation of a metastable single phase product with a well defined crystallization onset, and measurements using DSC indicate that T_n decreases from 265°K to 255°K over a cooling rate range of 10 to 320°K/min. The continuous cooling nucleation results can be fitted to Equation (15) in terms of a single nucleation frequency with a slope of 1.74x10^7 KJ^{2}cm^{-6} and a value for $\Omega_a a/K(T_n)$ of 3.3x10^{22}s^{-1}. Instead of the usual approximation for $\Delta G_v = \Delta H(1-T_r)$, the ΔG_v values used were derived from direct calorimetric measurements of the latent heat of the metastable x-phase, the temperature dependence of ΔC_p and reported molar volumes. The prefactor term, Ω_a is in reasonable agreement with the classical theory result if only a portion of the surface area represents an active catalytic site for metastable phase nucleation (*i.e.*, $\phi \cong 0.01$).

If Ω_a, σ and $f(\theta)$ are treated as constant, evaluation of Equation (15) can be used to develop the transformation diagram based on $J_a at$ = 1 that is shown in Figure 24. Although the calculation should be treated as approximate since estimates were used for several of the parameters, the results do illustrate some interesting features regarding the transformation kinetics. For example, if the linear approximation for ΔG_v had been applied instead of the measured values, the nose of the transformation diagram would be located at T_r = 1/3 or 128°K rather than 187°K. Also, the cooling rates required to reach the nose of the curve differ by a factor of 10^3 for the different ΔG_v values. This observation reveals the importance of using measured values for the nucleation parameters even if they represent a limited temperature range. It is also of interest to compare the droplet kinetics behavior with rapid quenching results. In this regard the calculated transformation diagram is consistent with the observation that suppression of the metastable phase by splat quenching is difficult (Barromee-Gautier *et al.* 1968). According to Figure 24, a critical cooling rate in excess of 5 x 10^7°K/s is required to bypass x-phase nucleation. Further, other droplet results indicate that another nucleation site for the metastable phase is active at a greater undercooling than that shown in Figure 24. As a result, at high cooling rates a transition from one type of nucleation site to another for formation of the same phase is possible.

An extension of the droplet emulsion technique which provides conditions favorable for attaining large liquid undercoolings in high melting point materials is drop tube processing (Shong *et al.* 1987). Powder samples are allowed to free fall through a cylindrical quartz chamber approximately three meters in height. The powders melt as they pass through a hot zone at the top of the chamber and then solidify during free fall. Unlike the droplet emulsion method direct measurement of the undercooling level is not possible during drop tube processing. However, a microstructural comparison of highly undercooled InSb-Sb eutectic droplets produced with the droplet method and samples of the same alloy processed in the drop tube has demonstrated that high undercooling levels can be achieved with this method. As in the droplet emulsion technique, processing parameters such as droplet size, droplet surface coating, cooling rate and melt superheat can be varied to analyze the effect of these parameters on the undercooling level achieved (Perepezko, Graves and Mueller 1987). Currently powder samples having melting temperatures approaching 2000°C may be processed, and estimated maximum cooling rates are in excess of 10^4°C/s.

This experimental technique has been recently applied to investigate the kinetic competition between α (hcp) and β (bcc) crystallization in a Ti-40 at% Al alloy (Graves 1987). An observed change in phase selection from an equilibrium β solidification product in coarse powder to a formation from the melt in fine powder is consistent with an increase in undercooling with decreasing droplet size. Using image analysis and SEM to quantitatively assess the microstructural abundance of α and β as a function of droplet diameter, a comparison can be made between powders processed in a He gas environment and those processed in Ar to examine the effect of the higher conductivity He on the phase selection (Figure 25). To provide a basis for the comparison an expression for the cooling rate of a falling particle as a function of droplet size and gas environment can be written as

$$\frac{dT}{dt} = -\frac{\epsilon A\,\sigma}{mc}(T^4 - T_\infty^4) - \frac{hA}{mc}(T - T_g) \tag{16}$$

where ϵ is the emissivity of the liquid, A is the surface area of the droplet, σ is the Stefan-Boltzmann constant, m is the mass of the droplet, c is the specific heat, h is the heat transfer coefficient, t is time, T is room temperature and T_g is the gas temperature (Geiger and Poirier 1973). The first and second terms in Equation (16) represent the contribution of radiative and convective cooling respectively to the overall cooling rate. Assuming Newtonian cooling conditions during free fall, an observed shift in the microstructural abundance curves of α and β with a change in gas atmosphere (Figure 25) may be described by a single translation in cooling rate.

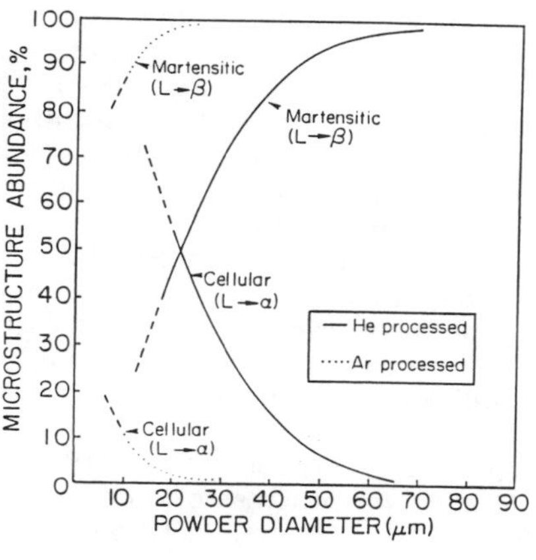

Figure 25

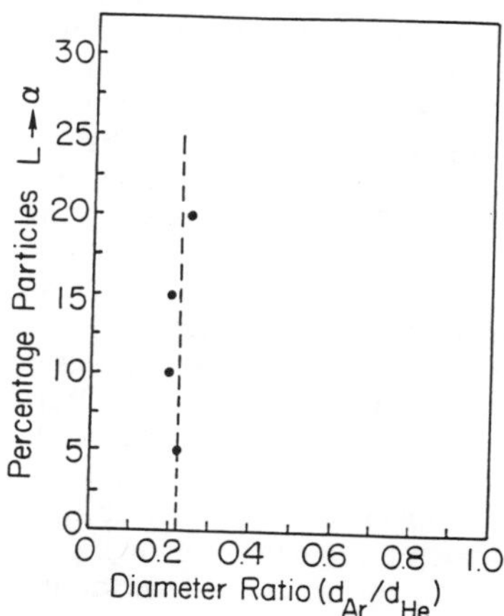

Figure 26

In other words, the difference in droplet size required to produce an equivalent volume fraction of metastable product under Ar and He atmospheres is approximately equal to that which would be required to produce equivalent cooling rates during free fall. For example, comparing the cooling rates of powders yielding 10% α, the calculated cooling rate of the 10 μm particle required in an Ar atmosphere is approximately equal to that of the 45 μm particle required in a He atmosphere (8 x 10^5°K/s vs. 1 x 10^6°K/s respectively). Since a number of assumptions are required for these calculations (Clyne *et al.* 1985), a comparison of the relative change in cooling rate resulting from the change in gas atmosphere can be used to minimize the influence of errors arising from the estimate of various parameters in the heat flow model. An alternative means of expressing the equivalence in thermal conditions for drop tube processed samples is through a simple comparison of the droplet diameters required to produce a given percentage of the population solidifying as a under He and Ar environments. A summary plot of this diameter comparison is presented in Figure 26 which shows that the ratio of the diameters for Ar and He processed powder remains approximately constant for an increasing percentage of metastable solidification product.

The observed shift in microstructure abundance curves by a simple translation in cooling rate implies the operation of a single nucleation kinetics for the formation of the metastable α-phase in the Ti-40 at% Al alloy. If nucleation of equilibrium β-phase also occurs by single mechanism, that fraction of the droplet population which does not contain this site will then undercool to form α. Therefore, assuming heterogeneous surface nucleation, Poisson statistics (Equation (9)) can be used to interpret the microstructural abundance of α as a function of droplet size for the two gas environments[2].

5. CONCLUSIONS

The observations of the undercooling behavior of fine droplet dispersions highlight some of the important thermodynamic and kinetic features of rapid solidification processing. A number of important processing variables have been identified in terms of their influence on the optimization of the undercooling potential. These parameters include droplet size, cooling rate, droplet surface coating, melt superheat, alloy composition and applied pressure. While control of these variables has allowed for the attainment of new maximum undercooling limits for a number of metals, the observed crystallization kinetics emphasizes the importance of heterogeneous nucleation in dominating the initial product formation in most cases. The possibilities for nucleation controlled structure modification are numerous and are closely related to a phase selection promoted by the attainment of a high undercooling level. In fact, variation of the liquid undercooling brought about by changes in droplet size, droplet surface coating and/or cooling rate can result in a change in the phase selection kinetics allowing for some level of control over the solidification product. Alternatively, a control over melt undercooling variables can be used to scale up sample sizes to approach bulk levels with selected phases and microstructures characteristic of undercooled solidification.

Overall, it appears that experimental techniques and methods of nucleation kinetics analysis are evolving to a level that should allow for progress beyond the rationalization of post solidification structural observations, to a realization of some predictive capability in the application of modified surface catalysis for control of the phase selection.

ACKNOWLEDGEMENT

The support of the DARPA (DAALL03-86-K-0164), ARO (DAAL03-86-K-0114) and NASA (NAG-3-436) is gratefully acknowledged. It is a pleasure to acknowledge many stimulating discussions with Dr. W.J. Boettinger of NIST.

REFERENCES

T. R. Anantharaman & C. Suryanarayana, "Review: a Decade of Quenching from the Melt." *J. Mat. Sci.*, **6**, (1971), 1111-1135.

I. E. Anderson, "Solidification Catalysis in Metals and Alloys." Diss., University of Wisconsin 1982.

C. Barromee-Gautier, B. C. Giessen, & N. J. Grant, "Metastable Phases in the Pb-Sb and Pb-Bi Systems." *J. Chem. Phys.*, **48**, (1968), 1905-1911.

[2]Interested parties are invited to write W. P. Allen regarding this research conducted by D. J.Thoma, W. P. Allen, and J. H. Perepezko.

W. J. Boettinger & J. H. Perepezko, "Fundamentals of Rapid Solidification," in **Rapidly Solidified Crystalline Alloys**, ed. S.K. Das, B.H. Kear & C.M. Adam. Warrendale, PA: TMS-AIME, 1985, 21-58.

L. Bosio, A. Defrain, & I. Epelboin, *J. Phys.* (Paris), **27**, (1966), 61.

G. E. P. Box, W. G. Hunter, & J. S. Hunter, **Statistics for Experimenters**, New York: J. Wiley, 1978.

T. W. Clyne, R. A. Ricks, & P. J. Goodhew, "The Production of Rapidly-Solidified Aluminum Powder by Ultrasonic Gas Atomization Part I: Heat and Fluid Flow." *Int. J. Rapid Solid.*, **1**, (1985), 59-80.

N. H. Fletcher, "Size Effect in Heterogeneous Nucleation." *J. Chem. Phys.*, **29**, (1958), 572-576.

G. H. Geiger & D. R. Poirier, **Transport Phenomena in Metallurgy**, Reading, MA: Addison-Wesley Pub., 1973.

B. C. Giessen, "Classification and Crystal Chemistry of Ordered Metastable Alloy Phases," in **Rapidly Quenched Metals**: **Proc. of the 2nd Int. Conf. on Rapidly Quenched Metals**, ed. N.J. Grant & B.C. Giessen. Cambridge, MA: MIT Press, 1976, 119-134

M. E. Glicksman & W. J. Childs, "Nucleation Catalysis in Supercooled Tin." *Acta Metall.*, **10**, (1962), 925-933.

D. E. Gordon & B. C. Deaton, "Induced High-Pressure Phases in the Bi-In, Bi-Sn and Bi-Tl Alloy Systems." *Phys. Rev. B*, **6**, (1972), 2982-2984.

J. A. Graves, "Rapid Solidification of High Temperature Aluminide Compounds." Diss. ,University of Wisconsin 1987.

M. Hillert, "Nuclear Composition-A Factor of Interest in Nucleation." *Acta Metall.*, **1**, (1953), 764-766.

J. P. Hirth, "Nucleation, Undercooling and Homogeneous Structures in Rapidly Solidified Powders." *Metall. Trans. A*, **9A**, (1978), 401-404.

M. K. Hoffmeyer, "Nucleation Catalysis by Dispersed Particles." M.S. Thesis, University of Wisconsin 1985.

K. A. Jackson & B. Chalmers, "Kinetics of Solidification." *Can. J. Phys.*, **34**, (1956), 473-490.

T. Z. Kattamis & M. C. Flemings, "Dendrite Structure and Grain Size of Undercooled Melts." *Trans. AIME*, **236**, (1966), 1523-1532.

W. Klement, A. Jayaraman, & G. C. Kennedy, "Phase Diagrams of Arsenic, Antimony, and Bismuth at Pressures up to 70 kbars." *Phys. Rev.*, **131**, (1963), 632-637.

R. D. Maurer, "Effect of Catalyst Size in Heterogeneous Nucleation." *J. Chem. Phys.*, **31**, (1959), 444-448.

B. A. Mueller, J. J. Richmond, & J. H. Perepezko, "Solidification Structures Developed in Undercooled Al-Si and Pb-Sb Alloys," in **Rapidly Quenched Metals**: **Proc. of the 5th Int. Conf. on Rapidly Quenched Metals**, ed. S. Steeb & H. Warlimont. North Holland, NY: Elsevier Sci. Pub., 1985, 47-50.

K. O. Ohsaka, "Nucleation Kinetics of Solidifying Alloys." Diss., University of Wisconsin 1986.

J. H. Perepezko, "Crystallization of Undercooled Liquid Droplets." in **Rapid Solidification Processing**: **Proc. of the Second Int. Conf. on Rapid Solidification Processing**, ed. R. Mehrabian, B.H. Kear & M. Cohen. Baton Rouge, LA: Claitors Pub., 1980, 56-67.

J. H. Perepezko & I. E. Anderson, "Metastable Phase Formation in Undercooled Liquids" in **Synthesis and Properties of Metastable Phases**, ed. T.J. Rowland & E.S. Machlin. Warrendale, PA: TMS-AIME, 1980, 31-63.

J. H. Perepezko, J. A. Graves, & B. A. Mueller, "Rapid Solidification of Highly Undercooled Liquids," in **Processing of Structural Metals by Rapid Solidification**, ed. F.H. Froes & S.J. Savage. Metals Park, OH: ASM, 1987, 13.

J. H. Perepezko, B. A. Mueller, & K. O. Ohsaka, "Solidification of Undercooled Alloys," in **Undercooled Alloy Phases**, ed. E.W. Collings & C.C. Koch. Warrendale, PA: TMS-AIME, 1986, 289-320.

J. H. Perepezko, B. A. Mueller, J. J. Richmond, & K. P. Cooper, "Crystallization Kinetics in Undercooled Liquids," in **Rapidly Quenched Metals: Proc. of the 5th Int. Conf. on Rapidly Quenched Metals.** Ed. S. Steeb & H. Warlimont. North Holland, NY: Elsevier Sci. Pub., 1985, 43-46.

J. H. Perepezko & J. S. Paik, "Undercooling Behavior of Liquid Metals," in **Rapidly Solidified Amorphous and Crystalline Alloys**: Proc. of Mat. Res. Soc. Symp., ed. B.H. Kear, B.C. Giessen & M. Cohen. North Holland, NY: Elsevier Sci. Pub., 1982, 49-63.

J. H. Perepezko & J. S. Paik, "Thermodynamic Properties of Undercooled Liquid Metals." *J. Non-Cryst. Solids*, **61**, (1984), 113-118.

G. M. Pound & V. K. LaMer "Kinetics of Crystalline Nucleus Formation in Supercooled Liquid Tin." *J. Am. Chem. Soc.*, **74**, (1952), 2323-2332.

W. E. Ranz & W. R. Marshall, *Chem. Eng. Prog.*, **48**, (1952), 141-146, 173-180.

D. S. Shong, J. A. Graves, Y. Ujiie, & J. H. Perepezko "Containerless Processing of Undercooled Melts," in **Materials Processing in the Reduced Gravity Environment of Space**: Proc. of Mat. Res. Soc. Symp., ed. R.H. Doremus & P.C. Nordine. Pittsburgh, PA: MRS, 1987, 17-27.

Skapski, A.S. "A Theory of Surface Tension of Solids--I. Application to Metals." *Acta Metall.*, **4**, (1956), 576-582.

F. A. Spaepen, "Structural Model for the Solid-Liquid Interface in Monatomic Systems." *Acta Metall.*, **23**, (1975), 729-743.

B. E. Sundquist & L. F. Mondolfo, "Heterogeneous Nucleation in the Liquid-to-Solid Transformation in Alloys." *Trans. AIME*, **221**, (1961), 157-164.

D. Turnbull, "Kinetics of Heterogeneous Nucleation." *J. Chem. Phys.*, 18, (1950), 198-203.

D. Turnbull, "Kinetics of Solidification of Supercooled Liquid Mercury Droplets." *J. Chem. Phys.*, **20**, (1952), 411-424.

D. Turnbull, "Theory of Catalysis of Nucleation by Surface Patches." *Acta Metall.*, **1**, (1953), 8-14.

D. Turnbull, "Under What Conditions Can a Glass be Formed?" *Contemp. Phys.*, **10**, (1969) 473-488.

D. Turnbull, "On Anomalous Prefactors from Analysis of Nucleation Rates," in **Prog. Mat. Sci.-Chalmers Ann. Vol.**, ed. J.W. Christian, P. Haasen & T.B. Massalski. Oxford: Pergamon, 1981, 269-275.

D. Turnbull & R. E. Cech, "Microscopic Observation of the Solidification of Small Metal Droplets." *J. Appl. Phys.*, **21**, (1950), 804-810.

B. Vonnegut, "Variation with Temperature of the Nucleation Rate of Supercooled Liquid Tin and Water Drops." *J. Colloid Sci.*, **3**, (1948), 563-569.

C. C. Wang, & C. S. Smith, "Undercooling of Minor Liquid Phases in Binary Alloys." *Trans. AIME*, **188**, (1950), 136-138.

G. R. Wood & A. G. Walton, "Homogeneous Nucleation Kinetics of Ice from Water." *J. Appl. Phys.*, 41, (1970), 3027-3036.

W. Yoon, J. S. Paik, D. LaCourt & J. H. Perepezko "The Effect of Pressure on Phase Selection During Nucleation in Undercooled Bismuth." *J. Appl. Phys.*, **60**, (1986), 3489-3494.

W. Yoon & J. H. Perepezko "The Effect of Pressure on Metastable Phase Formation in the Undercooled Bi-Sn System." *J. Mat. Sci.*, **23** (1988), 4300-4306.

FIGURE CAPTIONS

Figure 1. Undercooling and crystallization of tin droplets at different levels of size refinement.

Figure 2. Log-normal size distribution of a tin droplet sample and the corresponding undercooling behavior.

Figure 3. Undercooling behavior of Al powders at different levels of purity and size refinement.

Figure 4. The influence of changes in total impurity content on undercooling of Bi droplets prepared with similar size distribution and coating treatment.

Figure 5. Plot of the relationship between melt purity and cluster concentration for various droplet sizes.

Figure 6. Effect of different coating treatments on the undercooling behavior of pure Bi droplets.

Figure 7 . DTA thermograms showing the variation in undercooling which results from producing Al powders in different inorganic salts.

Figure 8. AES spectra of the external surface of Al powder made in the various salts.

Figure 9. Undercooling behavior of pure In droplets.

Figure 10. Undercooling behavior of Al-Si eutectic alloy droplets. (a) DTA trace at 0.5°C/sec; (b) continuous cooling curve at 500°C/sec.

Figure 11. Solidification microstructures developed in undercooled Al-Si eutectic alloy droplets. (a) T_N=484°C at $\dot{T}$=0.5°C/sec; (b) T_N=370°C at $\dot{T}$=500°C/sec.

Figure 12. Summary of undercooling behavior in Pb-Sb alloy droplets: circles, triangles and squares indicate the composition dependence of T_N for three different surface treatments.

Figure 13. Effect of pressure on the undercooling response of pure Bi droplets. T_{N1} and T_{N2} represent two distinct undercooling levels within the same droplet sample. The measured pressure dependence of the melting points of Bi(I) and metastable Bi(II) is also indicated.

Figure 14. Thermal cycling behavior of Bi-48.6 at% Sn droplet sample. The metastable endothermic behavior is associated with droplets exhibiting the largest undercooling.

Figure 15. Thermal treatment schedule for the study of nucleation catalysis.

Figure 16. Summary of primary phase catalysis results for Bi-Cd system. Includes maximum undercoolings (triangles) and T_{het} values (circles).

Figure 17. Effect of thermal cycling on relative magnitudes of two solidification exotherms in Bi-40 at% Cd droplet sample.

Figure 18. SEM micrographs of Bi-38 at% Cd droplets. (a) after completing one repetition of catalysis reaction thermal cycle; (b) after completing eight repetitions of catalysis reaction thermal cycle.

Figure 19. Relative magnitudes of heterogeneous surface to homogeneous nucleation in a 20 µm droplet as a function of contact angle.

Figure 20. Time-Temperature-transformation diagram representation of the transitions in nucleation kinetics that may occur in undercooled liquids.

Figure 21. Illustration of the decay of the number of unfrozen droplets during isothermal nucleation at different intervals of $\tau = J_v t$ for a volume dependent process.

Figure 22. Graphical representation of the approximate evaluation of droplet nucleation during continuous cooling.

Figure 23. Transformation diagram representing different nucleation kinetics that may occur during continuous cooling of undercooled liquids.

Figure 24. Calculated heterogeneous nucleation kinetics for x-phase in Pb-56 at% Bi droplets. The dotted curve is based on measured ΔG_v and the dotted and dashed curve is derived from the linear ΔG_v approximation.

Figure 25. Summary of microstructural analysis for drop tube processed Ti-40 at% Al powder.

Figure 26. Summary of the diameter ratio for Ar and He drop tube processed powder as a function of metastable product structure abundance in Ti-40 at% Al.

CONTAINERLESS PROCESSING OF NICKEL ALUMINIDES AND THEIR RESULTING MECHANICAL PROPERTIES

G. Carro and W.F. Flanagan
Vanderbilt University, Nashville, TN 37235

ABSTRACT

By utilizing electromagnetic levitation, it is possible to control both the degree of superheating and, more importantly, the degree of supercooling of a melt prior to splat-quenching. This allows some control over the resulting microstructure. For the case of Ni_3Al-Cr alloys, this enabled us to control the antiphase domain size relative to the grain size. Subsequent annealing treatments were used to modify these microstructural features. Miniature tensile tests were performed on the resulting material in order to test current theories of ductility in these ordered alloys. Conducive to the testing of such theories is the ability to solidify the various alloys at different degrees of undercooling, which is conveniently done with containerless processing.

INTRODUCTION

We are particularly interested in improving the properties of Ni_3Al alloys, which have important high-temperature applications based on the stability of their ordered gamma prime phase. Their main attribute is that their strength increases with increasing temperature, but ductility is limited, particularly at room temperature.

It is known[1] that boron, in small percentages and equilibrated so that it is preferentially located in grain boundaries, provides ductility in the sub-stoichiometric alloys. The mechanism of how this is accomplished is not fully understood, but most theories focus on its effect on "grain-boundary" strength. R. W. Cahn and co-workers, on the other hand, have proposed that anti-phase-domains (APDs) are responsible for the ductility in these alloys, especially when small amounts of disordered phase are precipitated on their boundaries[2,3]. Because of the high-temperature stability of the ordered phase, there is a composition range over which the ordered phase forms directly from the liquid. Since the occurrence of anti-phase domain boundaries requires the prior presence of a disordered gamma phase, this therefore restricts the range of composition and temperature over which ductility is predicted. By controlling the alloy composition and the quenching rate, microstructural features, such as the occurrence and size of the APDs can be manipulated, allowing such ideas to be experimentally tested. In particular, metastable gamma-prime phase, having the same composition as the disordered gamma phase, is desirable as an intermediary state, and this can be obtained with a sufficiently high quenching rate.

Levitation processing allows us to study the effect of such microstructural control on the properties of these alloys. In particular, such an approach does not involve container, thereby precluding a major source of impurities and, more to the point, extrinsic heterogeneous nucleation sites. Lack of the latter allows the possibility of undercooling, which in turn offers a means by which the microstructure can be refined, because the driving force for solidification is greatly enhanced and the sensible heat that must be removed during the process is reduced. Microstructural dimensions vary reciprocally with the velocity of the liquid/solid interface, which increases with undercooling[4]. If the cooling rate is sufficiently high, the possibility of forming the metastable phases also exists. Splat quenching can be used to increase the cooling rate once the specimen has been undercooled a

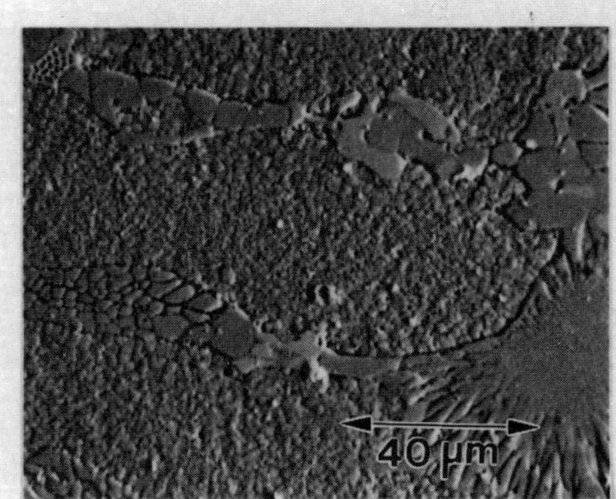

Fig. 1. SEM micrograph of slow-cooled (i.e., as cast) IC-218 alloy.

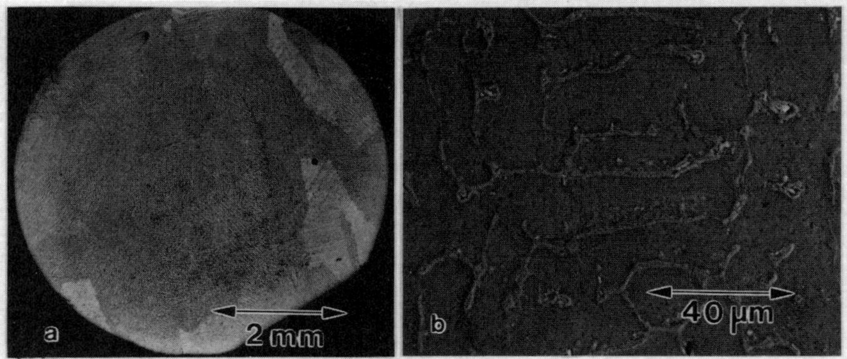

Fig. 2. Drop-solidified IC-218 alloy: (a) optical micrograph showing the macrostructure; and (b) higher-magnification SEM micrograph of (a).

Fig. 3. Optical micrograph of cross-section of a splat-quenched foil of IC-218 alloy.

certain degree. However, because the technique involves contact between the liquid and the anvil plates, heterogeneous nucleation accompanies such an increase in cooling rate, and would necessarily result in a different microstructure than would be obtained by drop-solidification.

One Ni_3Al alloy which solidifies without forming APDs (and contains boron), and two which do (one of which contains boron) were investigated to see if some improvement in room-temperature ductility could be obtained by microstructural modification using an approach based on the above.

EXPERIMENTAL DETAIL

Tensile curves and microstructural analysis were performed on several alloys. One alloy, IC-50 ($Ni_{76}Al_{23}Zr_{0.5}B_{0.2}$), which solidifies directly to the stable gamma-prime ordered phase, and two chromium-bearing alloys, IC-218 ($Ni_{74}Al_{17}Cr_8Zr_{0.5}B_{0.1}$), and A-1 ($Ni_{74}Al_{17}Cr_9$) which offer the possibility of solidifying initially to the disordered gamma phase, were prepared by levitation melting followed by "splat-quenching" in a vacuum system at the Center for Space Processing of Engineering Materials at Vanderbilt University. In the latter tests, different degrees of super-heating or undercooling were accomplished by adjusting the gas flow and/or the power to the levitating coil prior to quenching.

The tensile tests were done on miniature samples prepared from the solidified alloys. The technique is described in the literature[5], and has been shown to be accurate, provided that the grain size is sufficiently small to assure at least 4 grains through the thickness[6]. In our case, for all of the samples, there were at least 12 through the thickness, which certainly meets this criterion. Standard TEM and SEM techniques were used for metallographic analysis.

RESULTS

The effect of different solidification processes on microstructure is readily shown using IC-218. The normal as-cast microstructure is shown in Figure 1, where primary gamma dendrites form with interdendritic (divorced) eutectic of gamma and gamma-prime phases. The gamma-prime phase, in this instance, has a very "blocky" shape. On continued cooling in the solid-state, additional gamma-prime phase precipitates in the super-saturated primary gamma phase. If the same alloy is solidified during free-fall, the same microconstituents are present but the structure is more refined; however, drops which are solidified in this fashion often exhibit shrinkage porosity and a relatively large grain size, both of which invalidate mechanical test results for such small specimens. This is seen in Figures 2a and b. Also, the recalescence which accompanies the initial solidification process heats the drop back up close to the melting point, so that a large fraction of the drop solidifies at a relatively slow rate. On the other hand, when this alloy is splat-quenched, the resulting microstructure is highly refined and appears to be a single homogeneous phase. The grain size is equiaxed, even through the thickness of the splat foil, as seen in Figure 3.

When splat quenched, both alloy A-1, which contains no boron, and alloy IC-218, which does, had similar microstructures which exhibited very fine APDs, formed by the nucleation of the ordered gamma-prime phase from the disordered gamma phase on cooling, as seen in Figures 4a for alloy A-1 (Figure 5a shows the same for alloy IC-218); but alloy IC-50, which contains boron, has a composition adjusted so that it solidifies directly to the ordered gamma-prime phase, with the result that no APDs form, as seen in Figure 4b. All the alloys had roughly the same small grain size.

If the degree of supercooling before splat quenching is varied, or even if the temperature remains above the equilibrium melting temperature, it is found that the

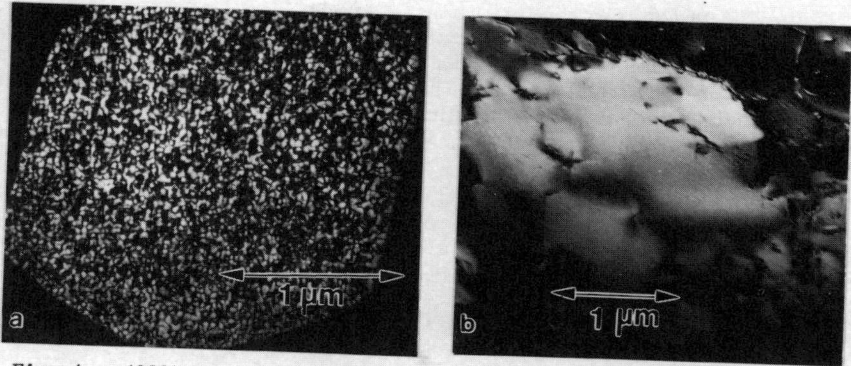

Fig. 4. (001) dark-field image of (a) A-1 and (b) IC-50 alloys in the as-splat-quenched condition.

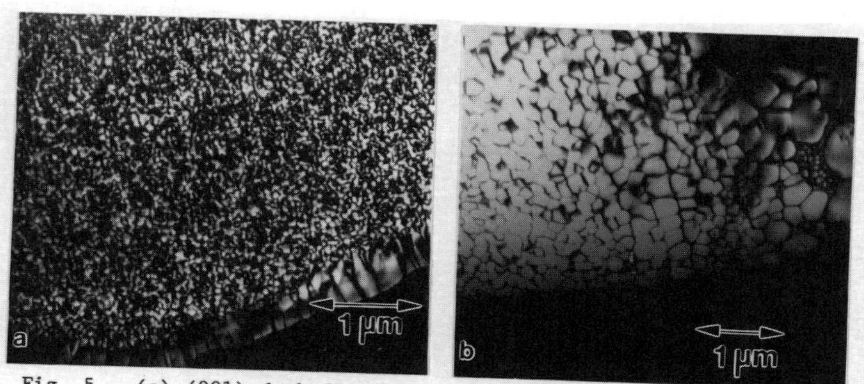

Fig. 5. (a) (001) dark-field image of IC-218 alloy splat-quenched with 7% undercooling. (b) (011) dark-field image of IC-218 alloy splat-quenched with 28% superheating.

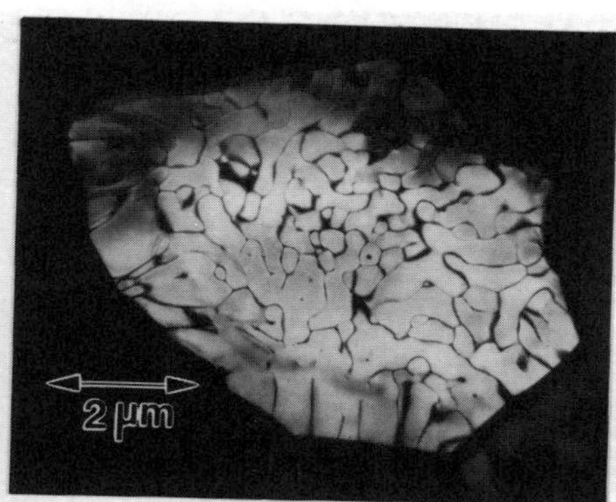

Fig. 6. (011) dark-field image of splat-quenched IC-218 alloy annealed for 2 h @ 900°C.

resulting grain size does not vary significantly. On the other hand, the APD size is markedly affected, decreasing with the degree of undercooling. This is seen in Figures 5a and b for alloy IC-218, where TEM micrographs are shown of samples quenched from below and above the melting temperature, i.e., undercooled and overheated respectively. Subsequent annealing treatments cause APD boundary growth, as seen in Figure 6.

The effect of such treatments on the tensile properties of these alloys is summarized in Figure 7. The grain size, APD boundary size, ductility, and the experimental parameters leading to these, are summarized in Table I. It should be emphasized that specimens of alloy IC-50 contain no APDs and alloy A-1 contains no boron.

Table I
APD size and grain size of treated samples

Alloy	Thermal History	Grain Size (m)	APD Size (nm)
IC-218	Undercooled 7%, SQ	6.5 ± 0.6	40 ± 4
	Undercooled 7%, SQ + 2h ann @ 900°C	6.8 ± 0.3	543 ± 100
	Overheat 28 %	7.0 ± 0.3	133 ± 17
IC-50	Undercooled 8%, SQ	7.0 ± 0.2	----
	Undercooled 8%, SQ + 2h ann @ 900°C	7.0 ± 0.8	----
	Overheat 4%, SQ	10 ± 1.0	----
A-1	Undercooled 4%, SQ	12 ± 0.9	30 ± 5
	Undercooled 4%, SQ + 2h ann @ 900°C	18 ± 2	328 ± 25

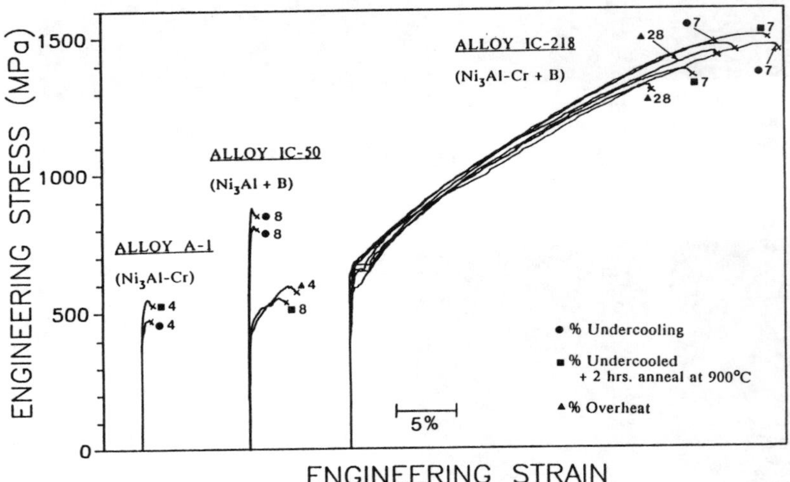

Fig. 7

DISCUSSION

By adjusting the composition and processing variables, it is possible to vary the grain size and APD size independently, which allows us to evaluate the contribution of each to the ductility of these high-temperature alloys. In Figure 7 it is seen that both boron and APDs are present in the highly ductile alloy (IC-218), whereas the absence of either boron (alloy A-1) or APDs (alloy IC-50) characterize the low ductility alloys. In addition, the APD size seems to have no effect on the magnitude of the strength or ductility, even though it seems to be a prerequisite for the latter. The APD boundaries may act as obstacles to dislocation motion, particularly if gamma phase is precipitated on them. If they are small enough, they will prevent the occurrence of dislocation pile-ups in the same way that fine precipitates do in dispersion-hardened materials[7]. This would prevent large concentrations of stress from building up at the grain boundaries. Boron, on the other hand, segregates to the grain boundaries[1], and could facilitate the production of slip into the adjoining grains[8]. This would essentially "weaken" the boundaries in that the stress at which slip would be transmitted through them would be less, and this in effect would make them much tougher. The role of the APD boundaries then would be to prevent stresses from locally exceeding the grain boundary fracture stress.

There is reported in the literature evidence of ductility for boron-free $Ni_3Al(X)$ (where (X) represents alloying with Cr, Mn, Fe, Co, or Si) when they are very rapidly quenched[9]. Their microstructure is similar to what we have reported here, in that APDs are present and have the same size. Their samples were very fine wires formed by quenching directly from the liquid state into water. We can find no explanation for why their results differ from ours.

We have tested samples with and without APDs, we have varied the size of the APDs, and have done annealing treatments to assure the presence of gamma phase on these boundaries, and we find that boron seems to be a corequisite with the requirement of APDs for extensive ductility in these nickel-aluminides. This does not support the thesis of Cahn et.al., but in fact their suggestion may still apply, in that APDs may supplement the role of boron, and vice-versa. The ability to test these ideas is made possible through containerless processing, which allows for control of the undercooling.

CONCLUSIONS

Levitation melting, followed by splat quenching, provides a means of obtaining very small APDs, in the nanometer range, as well as small grain size, 6 microns.

The hypothesis of Cahn et.al. that APDs invoke ductility in nickel-aluminides does not seem to be wholly correct, but rather they seem to act in conjunction with the effect of boron. Thus the effect of boron must involve a modification of grain-boundary strength as well as the formation of APDs.

ACKNOWLEDGEMENTS

The authors wish to thank ARMCO Inc. for providing some of the alloys studied. This work was supported by the NASA Office of Commercial Programs.

REFERENCES

1. C.T.Liu, C.L.White, & J.A.Horton, *Acta Metall.* **33**, 2753 (1985).
2. R.W.Cahn, P.A.Siemers, J.E.Geiger, & P.Bardhan, *Acta Metall.* **35**, 2737 (1987).
3. R.W.Cahn, P.A.Siemers, & E.L.Hall, *Acta Metall.* **35**, 2753 (1987).
4. R.Trivedi, *Metall. Trans.* **15A**, 977 (1984).
5. G.Carro & W.F.Flanagan, Scripta *Metall.* **22**, 903 (1988).
6. N.Igata, K.Miyahara, K.Ohna, & T.Uda, *J. Nucl. Mat.* 122 & 123, 354 (1984).
7. M.Ashby, Oxide Dispersion Strengthening (G.S Ansell, T.D.Cooper, & F.V.Lenel, eds., Gordon and Breach, N.Y., 1968), p. 159.

8. H.J.Frost, *Acta Metall.* **36**, 2199 (1988).
9. A.Inoue, H.Tamioka, & T.Masumoto, *Metall. Trans.* **14A**, 1367 (1983).

FIGURE CAPTIONS

Figure 1. SEM micrograph of slow-cooled (i.e., as cast) IC-218 alloy.

Figure 2. Drop-solidified IC-218 alloy:
(a) optical micrograph showing the macrostructure; and
(b) higher-magnification SEM micrograph of (a).

Figure 3 Optical micrograph of cross-section of a splat-quenched foil of IC-218 alloy.

Figure 4. (001) dark-field image of
(a) A-1 and
(b) IC-50 alloys in the as-splat-quenched condition.

Figure 5. (a) (001) dark-field image of IC-218 alloy splat-quenched with 7% undercooling;
(b) (011) dark-field image of IC-218 alloy splat-quenched with 28% superheating.

Figure 6. (011) dark-field image of splat-quenched IC-218 alloy annealed for 2 h @ 900°C.

Figure 7. Engineering stress/strain curves for the splat-quenched alloys. APDs are not present in IC-50, but are in A-1 and IC-218. Only IC-218 contains both boron and APDs.

6. Combustion

Jack Salzman
Session Chair

SOME RECENT ADVANCES IN DROPLET COMBUSTION

C. K. Law
Department of Mechanical and Aerospace Engineering, Princeton University Princeton, NJ 08544

ABSTRACT

This paper reviews the theoretical and experimental advances in droplet combustion since the 1982 Second International Colloquium on Drops and Bubbles. Specific topics discussed include multicomponent droplet combustion and microexplosion, convective droplet combustion, the combustion of slurries, propellants and hazardous wastes, soot formation in droplet burning, and several miscellaneous subjects. Areas of further research are suggested.

1. INTRODUCTION

At the Second International Colloquium on Drops and Bubbles held in 1982, the state of understanding on the mechanisms of droplet combustion was reviewed.[1] Specific topics discussed include the classical d^2-law solution and its subsequent modification, finite-rate kinetics and the flame structure, combustion and microexplosion of multicomponent fuel blends, near- and super-critical combustion, and droplet dynamics and interaction. Since the time of the Second Colloquium, advances in the understanding of droplet combustion have been made along two directions, namely a more mature understanding of the mechanisms of multicomponent droplet combustion, and applications of droplet combustion to several problems of practical interest. The present review summarizes these advances and suggests areas for further research.

Because of the diverse background of the intended audience, in Section 2 we shall provide a brief discussion of the basic phenomenology of single-component droplet combustion. The combustion and microexplosion of multicomponent fuels are then discussed in Sections 3 and 4 respectively, which are followed by a review on convective droplet combustion in Section 5. In Sections 6, 7, and 8 we discuss the practical problems involving droplet combustion of slurry fuels, liquid propellants and hazardous wastes, and soot formation in droplet burning. Some miscellaneous studies are covered in Section 9. The final section, Section 10, presents suggestions for further research.

2. COMBUSTION OF SINGLE-COMPONENT DROPLETS

Studies on droplet combustion are of relevance to the understanding and optimization of processes within liquid-fueled combustors in which the fuel is introduced in the form of a spray of droplets. In the simplest possible mode of droplet combustion, a single-component droplet is motionless in a stagnant, gravity-free, oxidizing environment of infinite extent. Droplet gasification is effected by the continuous heat transfer from the flame to the droplet surface and the simultaneous transport of fuel vapor from the droplet surface to the flame. The rate-limiting processes governing droplet gasification are heat and mass diffusion due to the existence of temperature and concentration gradients, and Stefan convection due to the continuous transfer of mass from the droplet surface to the ambience. Reaction between the outwardly-transported fuel vapor and inwardly-transported oxidizing gas can be considered to occur at an infinite rate such that the flame, within which reaction takes place, can be approximated to be an infinitely thin surface located where the diffusive transport of fuel and oxidizer are in stochiometric proportion.

The bulk combustion parameters of interest are the instantaneous values of the

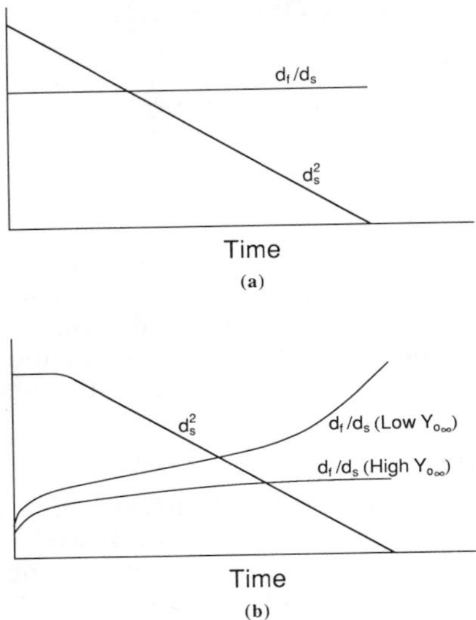

Figure 1

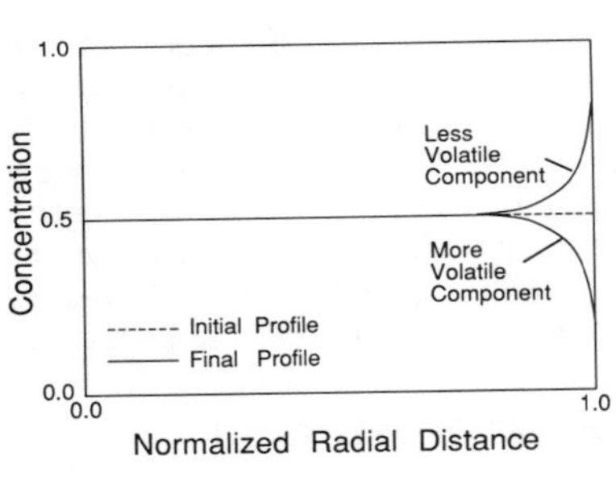

Figure 2

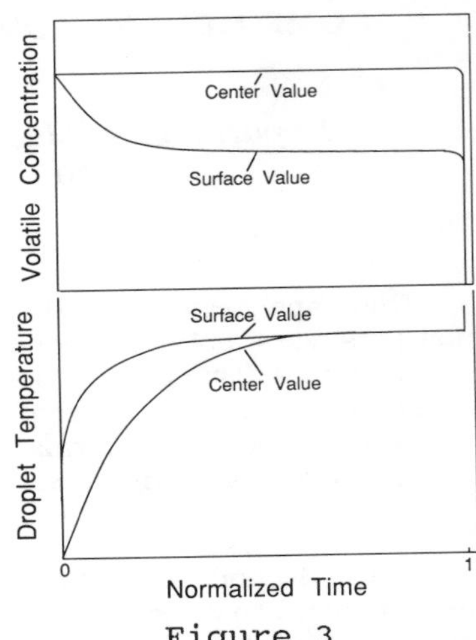

Figure 3

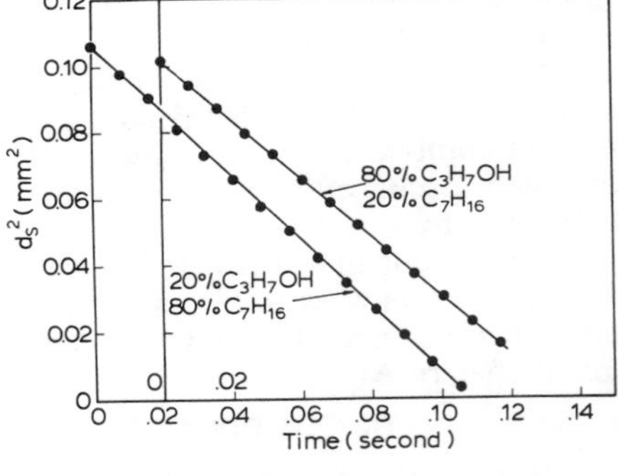

Figure 4

regression rate of the droplet diameter (d_s), the flame diameter (d_f), and the flame temperature (T_f). If we further assume that the droplet surface temperature (T_s) is constant, that the gas density is much smaller than the liquid density such that the gas-phase processes are quasi-steady, and that the instantaneous fuel gasification rate at the droplet surface (m_s) is equal to the instantaneous fuel consumption rate at the flame (m_f), then the classical d^2 law states[2,3] that d_s^2 decreases linearly with time and s is characterized by a burning rate constant $K = -d(d_s^2)/dt$, that the ratio (d_f/d_s) is a constant, and that T_f is simply the adiabatic flame temperature which is also a constant. These behavior are shown in Figure 1a.

The d^2-law model has since been revised to include the two major transient processes of droplet heating[4,5] and fuel vapor accumulation[6], which are consequences of the influence of initial conditions neglected in the d^2-law. Droplet heating accounts for the fact that the droplet is initially cold and therefore a substantial amount of the transferred heat goes into the sensible heating of the droplet rather than fuel gasification. This results in an initially slower gasification rate and a smaller flame size, as shown in Figure 1b. The flame temperature is also lowered. Fuel vapor accumulation accounts for the fact that upon ignition the flame diameter is smaller than that predicted by the d^2-law because initially there does not exist sufficient fuel vapor in the inner region between the droplet and the flame to support such a large-diameter flame. Thus the amount of fuel gasified is partly consumed at the flame and partly used to build up the fuel vapor profile in the inner region. This implies that $m_s \neq m_f$, while the scaled flame diameter (d_f/d_s) also continuously increases. The increase is without bound when the ambient oxidizer concentration ($Y_{O\infty}$) is low and the steady-state flame size large, but the scaled flame diameter approaches a constant value for higher values of $Y_{O\infty}$ (Figure 1b). The physical flame size (d_f) first increases and then decreases.

The above behavior have been verified experimentally for low- and micro-buoyancy situations[6-9] in which the diameter of a (spherical) flame can be meaningfully defined. Thus the qualitative combustion characteristics of a single-component fuel droplet under sufficiently subcritical pressures are reasonably well understood.

3. COMBUSTION OF MULTICOMPONENT DROPLETS

Practical fuels are frequently multicomponent in nature. Compared with single-component fuels, we now also wish to determine the sequence and rates with which fuel components of different physical and chemical properties are gasified and thereby depleted from the droplet interior. It is clear that the nature of the gas-phase reaction, both in the droplet vicinity as well as in the bulk, depends on what has been gasified, while the nature of the liquid-phase reaction, if there is any, correspondingly depends on what has been left behind.

The earlier viewpoint that multicomponent droplet gasification follows that of batch distillation, in that the sequence of gasification is controlled only by the volatility differentials among the different components, has been found to be incorrect. Rather, it is now generally accepted that, because of the extremely slow rate of liquid-phase mass diffusion as compared with those of liquid-phase heat diffusion, gas-phase heat and mass diffusion, and droplet surface regression, liquid-phase mass diffusion together with the volatility differentials are crucial factors in determining the gasification behavior of a multicomponent droplet. Indeed, the influence of this strong diffusional resistance persists even in the presence of rapid internal circulatory motion generated by external flow[10].

A unique feature of this diffusion-dominated droplet gasification mechanism is the possible attainment of approximately steady-state temperature and concentration profiles within the droplet, and consequently a steady gasification rate and constant average droplet composition.

Figure 2 shows typical calculated droplet concentration profiles of the volatile component of an initially equal-concentration binary mixture, with an extremely slow rate of liquid-phase mass diffusion[11]. Figure 3 shows the temporal variations of the center and surface values of the volatile component concentration and temperature within the droplet. It is seen that after gasification is initiated, the concentration of the volatile component in the surface layer is rapidly reduced, although its value in the inner core is hardly affected because of diffusional resistance. The droplet temperature, however, attains a somewhat uniform and constant value fairly early. Towards the end of the droplet lifetime, diffusion becomes efficient because of the small droplet size. This leads to the rapid depletion of the volatile component from the droplet composition and consequently to an increase of the droplet temperature to a value close to the boiling point of the less volatile component.

There are two situations under which liquid-phase diffusional resistance may not be sufficiently strong to effect the steady-state behavior discussed above. The first situation is for liquid mixtures whose mass diffusivity is only moderately, but not excessively, small. The second situation is for slow rates of droplet gasification such that, relative to the droplet lifetime, the mass diffusion process can be considered to be quite efficient. Under these two situations, the more volatile component can be preferentially gasified in a continuous manner, although at a rate smaller than if batch distillation is assumed. The droplet composition will therefore continuously change, causing the droplet gasification process to resemble a mixed-mode behavior intermediate between those of batch distillation and diffusion-limited steady state.

Experiments have been conducted[7] for binary mixtures with either equal or differing volatilities. Figure 4 shows that for mixtures of heptane and propanol, whose boiling points are both very close to 100°C, the effect of volatility differential is eliminated such that d_s^2 varies linearly with time in the manner of a single-component fuel. The burning rate constant K still depends on the mixture composition and has been found[7] to vary somewhat linearly with the initial volume fraction of either component.

Figure 5 shows the temporal variations of d_s^2 and d_f/d_s for a mixture of heptane and hexadecane, which have vastly different volatilities because the latter has a boiling point of 256°C. It is clear that the combustion behaviors here are qualitatively different from those of a single-component fuel shown in Figure 1. Specifically, Figure 5 shows that the d_s^2-t variation consists of two fairly straight segments separated by a short, almost horizontal transition segment. Visually, during the transition period the flame becomes fainter and its size also reduces. This flame shrinkage phenomenon corresponds to the local minima region in the (d_f/d_s)-t plot in Figure 5.

The above results agree with the anticipated behavior of liquid-phase diffusion-limited combustion. That is, the first segment of Figure 5 corresponds to the establishment of the surface concentration layer, during which the volatile component in the surface region is preferentially gasified while the droplet temperature is also relatively low, being controlled by the boiling point of the more volatile component. As the surface layer is gradually established, it becomes more concentrated with the less volatile, higher-boiling-point component. Thus, there must exist a short, transitional, droplet heating period to bring up the droplet temperature. The diversion of heat for sensible droplet heating slows the gasification rate and consequently causes a reduction in the flame size and temperature. These effects account for the observed transition and slope change in the second segment of the d_s^2-t curve. Finally, after the droplet is mostly heated up, it assumes the somewhat

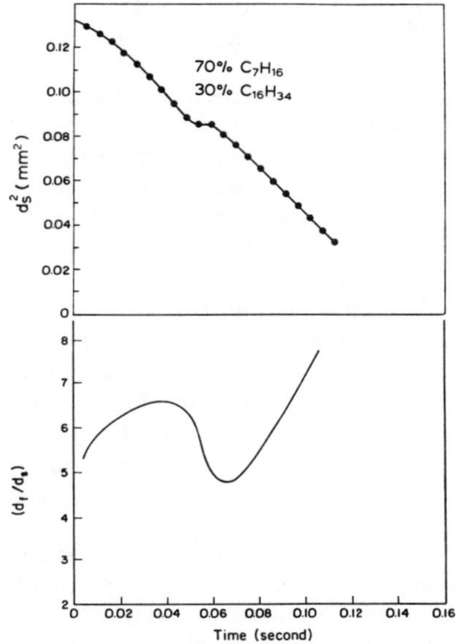

Figure 5

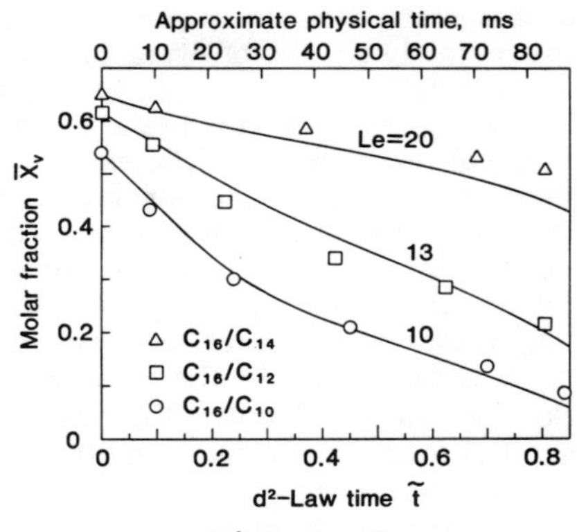

Figure 7

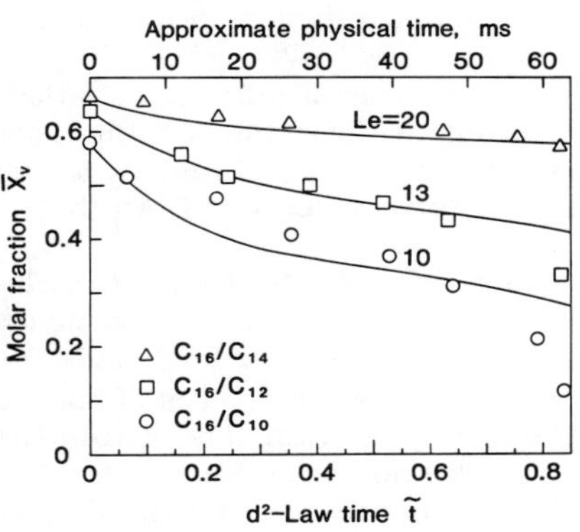

Figure 6

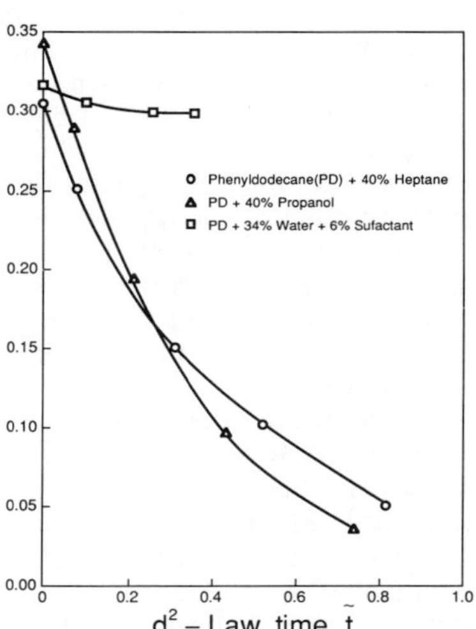

Figure 8

steady-state behavior represented by the last segment of the curves. The recent microgravity results of Avedisian and Yang,[12] and Yang and Avedisian[13] also seem to indicate this three-staged behavior.

While the above results are interesting, they do not indicate the intensity of diffusional resistance and the extent the volatile component is trapped in the droplet interior. In fact, these observed behaviors can also be qualitatively explained on the basis of batch distillation, in which case the more volatile component will have been mostly depleted from the droplet interior during the third segment. Thus definitive statements regarding the effectiveness of diffusional resistance can be made only based on temporal variations of the droplet composition.

Figure 6 shows[14] the variation of the spatially-averaged volatile molar fractions within the droplet, $\overline{X}_v$, for mixtures of hexadecane (designated simply by the symbol C_{16}) with tetradecane (C_{14}), dodecane (C_{12}), and decane (C_{10}) undergoing combustion in an air environment of about 1300°K. The initial, zero-time datum corresponds to the state when ignition has just been achieved. The lower abscissa is $\tau = 1-(d_s/d_{so})^2$, where d_{so} is the initial droplet diameter. It is an approximate, nondimensional indication of time, and can be called the d^2-law time because it is linearly proportional to time when the d^2-law holds.

It is seen from Figure 6 that while the C_{16}/C_{14} droplet behaves in a nearly steady-state manner, for the C_{16}/C_{10} droplet the concentration of the more volatile component, decane, continuously decreases. This behavior clearly demonstrates the influence of the liquid-phase mass diffusivity. That is, since the diffusion coefficient D depends on the reduced mass of the molecules of the two components, the C_{16}/C_{14} mixture has a relatively small diffusivity and therefore offers stronger diffusional resistance as evidenced by the nearly steady-state behavior. The opposite holds for the C_{16}/C_{10} mixture. While a greater mass diffusivity is expected of this mixture, what is especially significant is the complete absence of steady-state for the C_{16}/C_{10} droplet, indicating that mass diffusion in fuel blends is more efficient than previously recognized.

Figure 7 shows the temporal variations of $\overline{X}_v$ in the case of pure vaporization, with the gas temperature T_∞ being about 1000°K. Compared with the burning cases of Figure 6, it is seen that with a slower surface gasification rate and thereby longer time available for diffusion, the gasification behavior further deviates from that of steady state.

The above experimental results have been compared with theoretical calculations[11]. An important parameter which emerges from the model is the liquid-phase Lewis number, $Le = \alpha/D$, which is much greater than unity because for liquids the mass diffusivity D is much smaller than the thermal diffusivity α. It is of special interest to note that while the estimated Le are indeed large numbers, within the range of 10 to 20 for the mixtures tested herein, they are not excessively large. The consequence of these moderately large values of Lewis number is that the droplet exhibits a gasification behavior intermediate of those of batch distillation and diffusion-limited steady state, as experimentally observed. Figures 6 and 7 demonstrate the satisfactory comparison between theory and experiment.

To further demonstrate the influence of liquid-phase diffusional resistance, Figure 8 compares $\overline{X}_v$ for a phenyldodecane/water macro-emulsion with miscible mixtures of phenyldodecane with heptane and with propanol[15]. Since the diffusivity of the water micro-droplets within the emulsion can be considered to be vanishingly small, the diffusional resistance for this emulsion is infinitely large. Figure 8 therefore shows that while the concentrations of heptane and propanol in the miscible mixtures continuously decrease, the water concentration in the emulsion remains basically at its initial value such that the gasification process can be considered to truly resemble that of the diffusion-limited steady-state.

In conclusion, there now exist convincing experimental and theoretical evidence that, for conventional fuel blends, liquid-phase diffusional resistance is only moderately strong such that the gasification behavior is intermediate of those of distillation and diffusion-limited steady state. Its effectiveness is further reduced with increasing diffusivities and volatility differentials among the mixture constituents, and with decreasing droplet gasification rate.

4. MICROEXPLOSION OF MULTICOMPONENT DROPLETS

An interesting phenomenon frequently accompanying multicomponent droplet combustion is microexplosion[16] which involves an almost instantaneous, violent rupturing of the droplet as it undergoes combustion. The cause of microexplosion is believed to be the following[11]: as the surface concentration layer is being established after the droplet started gasification, the droplet temperature rapidly increases because of the increase in the concentration of the less volatile component near the droplet surface. However, since the droplet inner core retains a relatively high concentration of the more volatile component because of diffusional resistance, the volatile component can be heated to the state at which homogeneous nucleation is initiated. The subsequent rapid internal gasification then leads to the violent rupturing observed.

This postulated microexplosion mechanism has two distinctive global properties[11]. First, it can occur only if the volatilities of the components are sufficiently different and their initial concentrations lie within an optimum range. The reason being that microexplosion requires both the less volatile component to drive up the droplet temperature and the volatile component to facilitate homogeneous nucleation. This property has been experimentally confirmed[7,17]. Figure 9 shows the normalized droplet explosion size (d_{se}/d_{so}) and volume $(d_{se}/d_{so})^3$ as a function of concentration for heptane/hexadecane mixtures, where d_{se} is the droplet diameter just before the onset of rapid internal gasification and microexplosion. It is seen that, in agreement with theoretical predictions, the dependence of microexplosion on concentration is parabola-like, with the optimum being around equal concentration.

The second property of microexplosion mechanism is that the occurrence of microexplosion is enhanced with increasing pressure[18]. This is because the droplet temperature increases with increasing pressure as a result of the elevation of boiling points, although the homogeneous nucleation temperature is mostly insensitive to pressure variations when the pressure is not close to that of the critical state. This property is demonstrated in Figure 10, in which (d_{se}/d_{so}) and $(d_{se}/d_{so})^3$ are plotted as functions of pressure for 50/50 mixtures of propanol/octadecane, butanol/octadecane, and butanol/hexadecane. It is clear that, with increasing pressure, the droplet microexplodes with larger diameters and thereby earlier in its lifetime.

Microexplosion has also been found to depend sensitively on the stability of droplet generation. To demonstrate this point, Figure 11 shows three modes of droplet generation with different stabilities. In Figure 11a the generation mode is optimal in that stable, spherical, monosized droplets are readily formed at a short distance away from the nozzle. For this mode microexplosion is not readily observed. In Figures 11b and 11c, the generation mode is less than optimal in that a small ligament is seen to be attached to the main droplet. This ligament may (Figure 11b) or may not (Figure 11c) be absorbed by the main droplet. If it is not absorbed, then microexplosion is again not readily observed. However, if absorption occurs, then occurrence of microexplosion is greatly facilitated. The cause of this dependence is so far unclear.

5. CONVECTIVE DROPLET COMBUSTION

A comprehensive review on convective droplet combustion has been given by Fernandez-Pello[19]. Most of the studies since the reviews of Law[1,20] and of Sirignano[10]

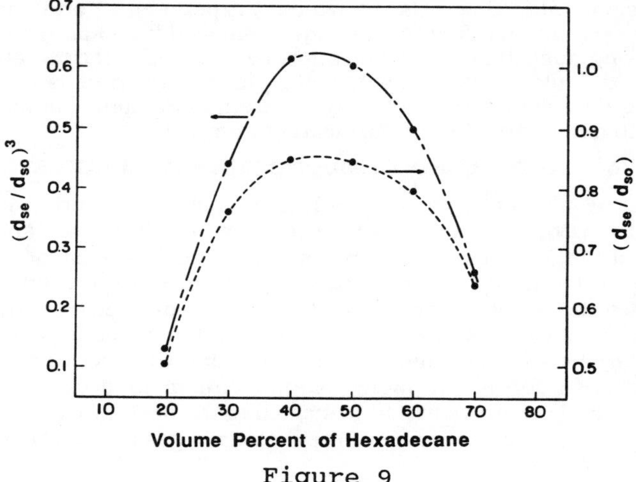

Figure 9

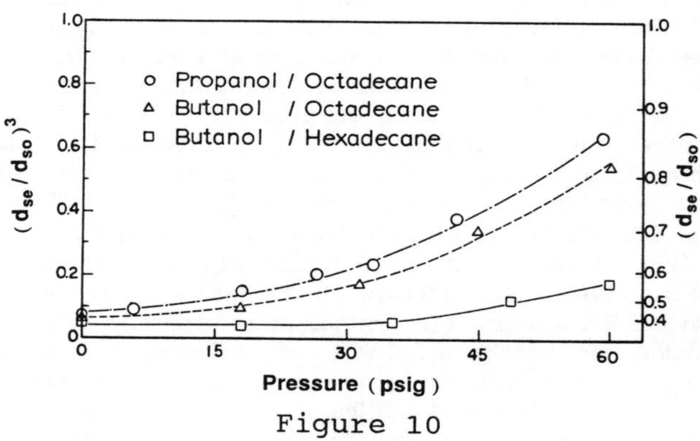

Figure 10

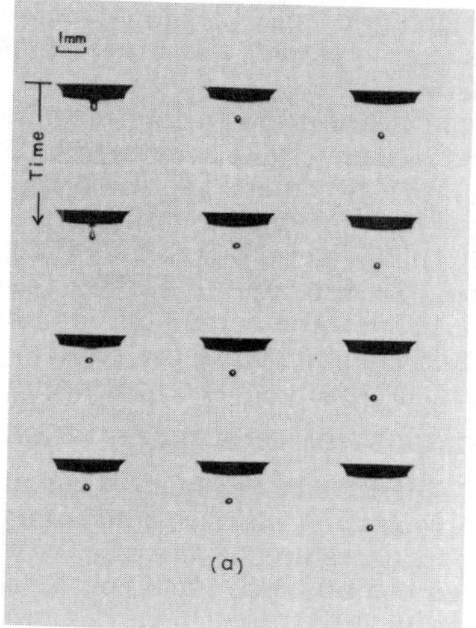

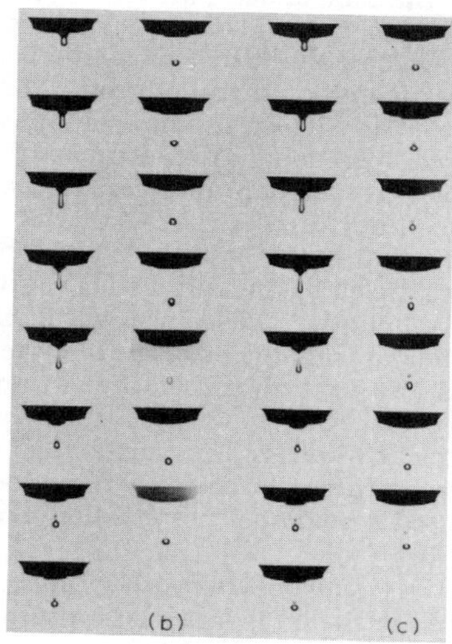

Figure 11
(a), (b), (c)

have been theoretical in nature. Specifically, Tong and Sirignano,[21] and Rangel and Fernandez-Pello[22] have studied the high-Reynolds number flow situations, with boundary layers developed on both the liquid and gas side of the interface. Series expansion solutions were obtained and expressions for the integrated combustion parameters were derived.

Complete numerical solutions for low to intermediate Reynolds number flows have also been obtained and parametrically correlated by Renksizbulut and Yuen,[23] Renksizbulut and Haywood,[24] Haywood *et al.*,[25] and Dwyer and Sanders.[26-28] These studies yield the following insights. First, upon injection the high Reynolds number droplet (say $Re \approx 100$) quickly slows down, indicating that the boundary layer analysis would break down. Second, due to surface gasification, the drag coefficients are very much reduced from the corresponding hard sphere values. Third, presentation, correlation, and comparison of the results depend sensitively on the selection of the averaged property values used in nondimensionalization.

6. COMBUSTION OF SLURRY DROPLETS

The recent interest in slurry fuels has been stimulated by two quite different applications. Coal-oil and coal-water slurries consisting of 200 to 400 mesh coal particles have been used in furnaces and boilers as a means of direct coal utilization in conventional liquid-fueled combustors, while boron and carbon slurries consisting of micronized particles in JP-10 fuel have been formulated as high-energy propellants for tactical use.

Earlier studies[29-31] have shown that during slurry droplet combustion, the total droplet lifetime consists of a relatively short initial period during which the liquid fuel vaporizes while the suspended particles agglomerate, which is followed by a very long period of agglomerate burning. Further experiments[32] reveal that under most situations the agglomerate is hollow, with a porous shell structure.

Experiments have been conducted to determine the instantaneous size and liquid content of the droplet[33-35]. Figure 12 shows a typical set of data for the time-resolved droplet size and composition. From studies of this nature the following mechanism of slurry droplet combustion and shell formation emerges (Figure 13). During the initial stage the slurry droplet gasifies as if it were a pure liquid, with a burning rate constant being that of the liquid. The continuously regressing droplet surface concentrates the particles in the surface layer until a porous, rigid shell is formed. Gasification subsequently takes place within the shell, which therefore continuously thickens. The outer diameter of the shell remains constant. A continuously growing bubble is also formed in the droplet interior because of liquid depletion. This rigid shell gasification mode is eventually followed by microexplosion. At present it is not clear what are the exact cause of microexplosion and the amount of liquid remaining in the inner core when microexplosion occurs.

Antaki[36] has formulated a rigid shell combustion model, which shows that the rate of liquid mass loss is given by

$$\dot{m} = (\pi/4)\rho_\ell d_s K$$

while the inner diameter of the shell, d_i, varies with time according to

$$d_i^3 = d_s^3 - \left(\frac{3d_s}{2c_o}\right)K_t$$

where ρ_ℓ is the liquid density, c_o the initial liquid volume fraction in the slurry, and d_s the outer shell diameter. Since d_i^3 varies linearly with time, the above model has been termed the d^3-law. The validity of the expression for $\dot{m}$ has been experimentally verified by Lee and Law[35]. An analysis of their data also indicates that the shell porosity is about 0.5, and that the shell becomes rigid when it is two to four particles thick. Based on these information, it is now possible to describe the gasification history of a slurry droplet until the liquid fuel is almost depleted.

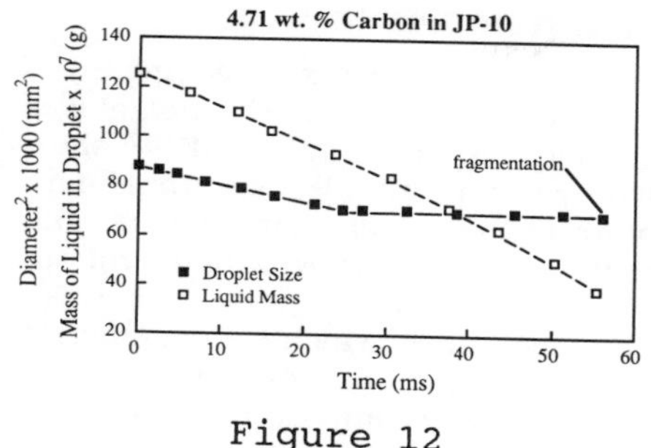

Figure 12

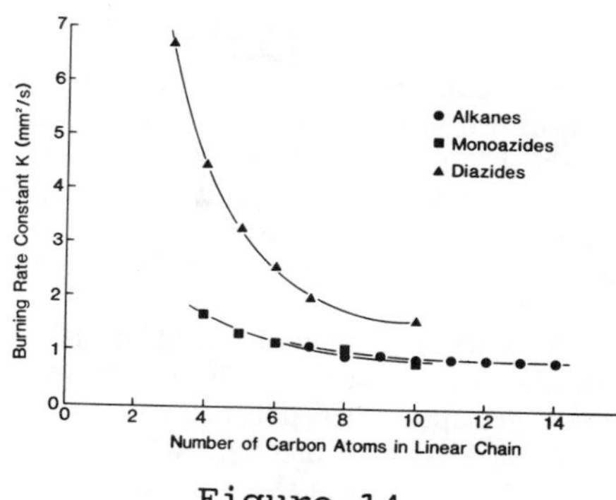

Figure 14

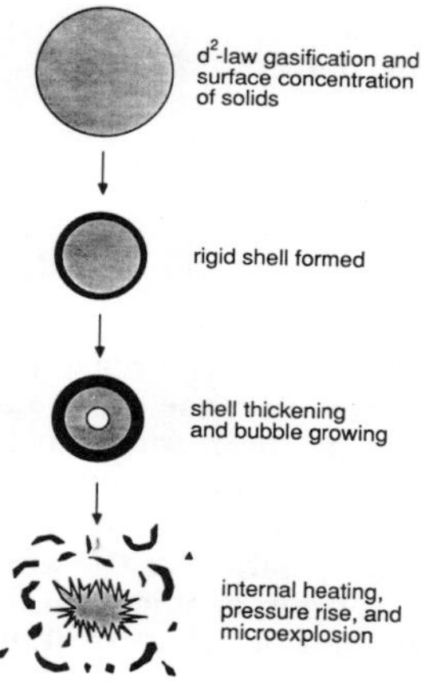

Figure 13

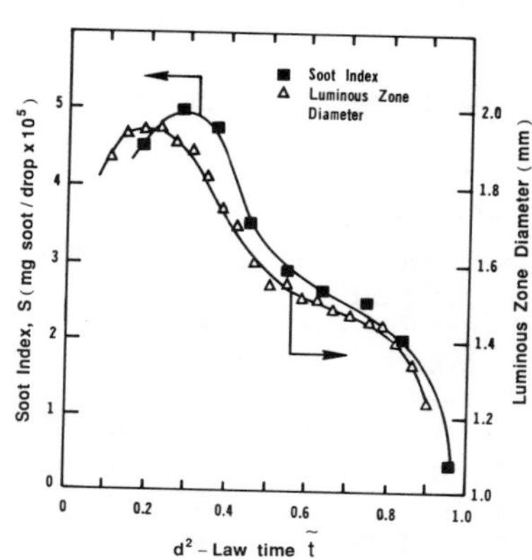

Figure 15

7. COMBUSTION OF PROPELLANT DROPLETS

Two classes of liquid propellants have been studied. The first is composed of ammonium salts dissolved in water, and is for application in guns. The advantages are the high energy density and safety in handling. Experimental results[37] show that the initial droplet gasification is dominated by water. This leads to rapid concentration of the salt in the propellant, elevation of the droplet temperature, and eventually runaway in the liquid-phase reaction, and thereby droplet micro-explosion.

The second class of liquid propellants are organic azides, which are compounds obtained by substituting one or more of the hydrogen atom in an organic molecule by N_3 groups. The potential of the azides is that they decompose upon heating, releasing heat and nitrogen which can greatly enhance the droplet gasification rate. Figure 14 compares the burning rate constant K for n-alkanes, mono-substituted alkyl azides, and di-substituted alkyl azides as function of the number of carbons in the alkane chain[38]. The significant increase in the burning rate of the diazides as compared with the corresponding alkanes is quite impressive. For example, $K \approx 7$ mm^2/s for diazidopropane while $K \approx 1$ mm/s for n-heptane. Even larger values of K for other organic azides have been achieved. The significance of these results can be further appreciated by recognizing that for conventional hydrocarbon droplet combustion an increase of 10 to 20% in K through, say, increases in the gas temperature or oxygen concentration is already quite difficult to achieve.

To explain such a large increase in K, we first note that from d^2-law,

$$K \sim \ln(1+B)$$

where the heat transfer number B is

$$B = \frac{c_p\,(T_\infty - T_s) + Y_{O\infty}\, q/\nu}{\ell}$$

c_p is the specific heat, q the heat of combustion per unit mass of fuel reacted, ν the stochiometric oxidizer-to-fuel mass ratio, and ℓ an effective specific latent heat of vaporization. Since B>1, it is clear that a factor of, say, seven, increase in K would require an exponential increase in B. Such a large increase in B cannot be achieved by the gas-phase heat production term given by the numerator in the expression of B, especially the heat of decomposition is not large as compared to the heat of oxidation. The magnitude of B, however, can be significantly increased by reducing ℓ to very small values. In other words, decomposition of the azide has to take place either in the droplet interior or at its surface upon gasification, but not in the gas phase. The decomposition heat release can then be used for liquid gasification, and this effectively reduces ℓ

An adequate description of the combustion behavior of these propellant droplets requires understanding of the associated liquid-phase reactions. Very little work has been done in this area.

8. DROPLET COMBUSTION APPLIED TO ENVIRONMENTAL PROBLEMS

Knowledge of the fundamental mechanisms of droplet combustion has been applied to gain understanding of two environmental-related issues.

The first problem is soot formation in droplet burning. Based on previous findings in diffusion flames,[39] the phenomenology of soot formation in droplet burning is the following. First, soot precursors form from pyrolyzed fuel molecules in the high temperature region on the fuel side of the diffusion flame. The high molecular weight species are then transferred back into the fuel rich region between the droplet and the flame via thermophoresis, where they undergo further growth. This inward transport of the soot particles is eventually stopped by the outwardly-directed Stefan flow of the fuel vapor at a location where the two forces balance, resulting in the formation of a layer of soot particles. Some recent microgravity results[40] also seem to indicate that, in the complete absence of external convection, there may exist situations under which large amount of soot particles can be accumulated this way. Furthermore, if these hot particles can reach the droplet surface, their contact with the cold droplet will lead to immediate boiling and consequent microexplosion of the droplet.

The size of the soot layer is found to be proportional to that of the flame, as expected[41]. Figure 15 shows that the instantaneous amount of soot present around the droplet closely follows the size of the luminous zone, which in turn has been found to vary with the size of the primary oxidation zone.

Thus a strategy to minimize soot formation in droplet burning is to maintain small flames so that the total volume of the soot layer is correspondingly small. The presence of weak external convection can also cause oxidation of the soot particles by convecting them through the flame, although excessive blowing can lead to flame extinction and thereby exhaust of the soot in the wake region. Alternatively, in the absence of external convection, the soot particles can also be oxidized eventually as the flame regresses towards the droplet, provided that early extinction does not occur. It has also been suggested[41] that the blending of a sooty fuel with a less sooty fuel is more effective in reducing soot formation if the latter is also the more volatile component. The reason being that since the flame size is the largest during the initial phase of the droplet lifetime, and since it is the more volatile component that dominates gasification during this phase, it is desirable that this component be the less sooty one. The occurrence of flame shrinkage due to droplet heating, as discussed earlier, can also reduce the soot.

The second environmental problem of interest is the incineration of liquid hazardous wastes (HWs). A major difficulty here is that most hazardous wastes are chlorinated hydrocarbons, which are incineration resistant due to their low heats of combustion and due to the scavenging of the crucial hydrogen radicals needed to propagate the oxidation reactions by the chlorine atoms.

Blending of an HW with a regular hydrocarbon fuel has been suggested as an approach to enhance incinerability.[42] Here it has been found that the fuel additive needed should be the less volatile component such that the HW, being more volatile, can be vaporized earlier in the droplet lifetime. This allows a longer gas-phase reaction time for its destruction.

9. MISCELLANEOUS STUDIES

Advances in high pressure combustion have been made by Niioka and Sato[43] and Sato *et al.*,[44] who experimentally studied the combustion of a suspended multicomponent droplet under elevated pressures up to 2 MPa. The three-staged combustion behavior of Wang *et al.*[7], as discussed previously, was again observed. Distinction between the different stages, however, diminishes with increasing pressure. The influence of strong natural convection and the suspension fiber on their results, especially those on microexplosion, are not clear.

Dryer and co-workers[45,46] have studied coke formation in the combustion of heavy oil droplets. Experimental results show that coke is formed only during the last 10% of the droplet lifetime, and that the mass of coke formed is about 3% of the original mass of the heavy oil. Dilution by a lighter oil or changing the initial droplet size do not seem to have much effect on this ratio. Active development has also been underway to measure the temperature distribution within the droplet by nonintrusive optical means. A promising technique[47] is to add two organic compounds to the hydrocarbon fuel, a fluorescent monomer M and an appropriately chosen ground state reaction partner G. Electronically excited M may then react with G to form an exciplex (excited state complex) E^*, whose emission is red-shifted from the M^* emission. The temperature dependence of the M^*/E^* ratio is then used to determine the droplet temperature.

10. SUGGESTIONS FOR FURTHER STUDIES

While our understanding of the fundamental mechanisms governing the combustion of single- and multi-component droplets is reasonably satisfactory, our ability to describe their combustion behavior quantitatively is quite inadequate. In fact, we have not been able to quantitatively predict any of the bulk combustion parameters (*e.g.* K, d_f/d_s, T_f) pertaining to the spherically-symmetric, steady combustion of a single- component droplet.

The above state of quantitative inadequacy actually prevails for practically all problems in combustion. Since (spherically-symmetric) droplet combustion probably has the simplest geometry and flow field, it is ideally suited to be the first candidate for quantitative numerical simulation.

Such a numerical simulation can be conducted at several levels. At the lowest level we should be able to predict the histories of d_s, d_f, and T_f for given initial values of d_{so} and d_{fo}; the latter can be obtained from reported experimental data. Droplet heating can be suppressed by using a volatile fuel, say heptane, while flame chemistry can be simplified by assuming dissociation equilibrium among the several major species.

At the next level, droplet heating and multicomponent effects can be included. Finally flame chemistry can be incorporated in order to describe droplet ignition and extinction. It is important to recognize that there now exist reasonably accurate experimental data on the bulk combustion parameters to provide at least partial validation of the numerical solutions. A complete validation, however, requires spatially-resolved temperature and concentration information, whose determination necessitates sophisticated development of optical techniques. Such developments should be encouraged.

The numerical solutions of Dwyer and Sanders[26-28] have shown a significant reduction in the droplet drag coefficient due to surface gasification. A well-controlled experimental effort is therefore needed to accurately determine the drag of droplets undergoing rapid gasification. Furthermore, since great uncertainties exist in the selection of transport properties needed for nondimensionalization, therefore raw, instead of processed, experimental data should be reported.

Recently there have been many international activities on microgravity droplet combustion. Interesting results such as the role of soot formation have already emerged. Results on high-pressure, near- and super-critical combustion under microgravity conditions would also be particularly useful.

Very little work on the turbulent dispersion of vaporizing droplets has been conducted. The problem is a challenging and important one. Studies on droplet collision and coalescence for dense spray applications should be encouraged.

Finally, it is also clear that present efforts on the droplet combustion of slurries, propellants, and hazardous wastes will continue to be important topics of droplet research.

ACKNOWLEDGEMENTS

The author's research on fundamental combustion including those on droplet processes have been supported by the National Science Foundation, the Department of Energy, the Air Force Office of Scientific Research, the Army Research Office, and the Office of Naval Research. This review was prepared with support from NASA-Lewis.

REFERENCES

1. C. K. Law (1982), "Mechanisms of droplet combustion," **Proc. Second Intl. Colloquium on Drops and Bubbles**, JPL Pubn. 82-7 (Ed.: D.H. Le Croissette).
2. G. A. E. Godsave (1953), "Studies of the combustion of drops in a fuel spray: The burning of single drops of fuel," **Proc. 4th Symposium (Int.) on Combustion**,

Williams & Wilkins, 818.
3. F. A. Williams (1985), **Combustion Theory**, Benjamin-Cummins, Palo Alto.
4. C. K. Law (1976), "Unsteady droplet combustion with droplet heating," *Combust. Flame* **26**, 17.
5. C. K. Law & W. A. Sirignano (1977), "Unsteady droplet combustion with droplet heating. II: Conduction limit," *Combust. Flame* **28**, 175.
6. C. K. Law, S. H. Chung, & N. Srinivasan (1980), "Gas-phase quasi-steadiness and fuel vapor accumulation effects in droplet burning," *Combust. Flame* **38**, 173.
7. C. H. Wang, X. Q. Liu, & C. K. Law (1984), "Combustion and microexplosion of freely falling multicomponent droplets," *Combust. Flame* **56**, 175.
8. I. Gökalp, C. Chauveau, & G. Monsallier, (1988), "Experiments on combustion in reduced gravity and turbulent vaporization of n-heptane droplets," **Proc. 3rd Intl. Colloq. on Drops and Bubbles** [*this vol. and chapter.*]
9. I. Gökalp, C. Chauveau, J. R. Richard, M. Kramer, & W. Leuckel (1989), "Observations on the low temperature vaporization and envelop or wake flame burning of n-heptane droplets at reduced gravity during parabolic flights," **Proc. 22nd Symp. (Int.) on Combustion**, The Combustion Institute, in press.
10. W. A. Sirignano (1983), "Fuel droplet vaporization and spray combustion theory," *Prog. Energy Combust. Sci.* **9**, 291.
11. C. K. Law (1978), "Internal boiling and superheating in vaporizing multicomponent droplets," *A1ChE J.* **24**, 626.
12. C. T. Avedisian & J. C. Yang (1988), "Some experiments on free droplet combustion at low-gravity," **Proc. 3rd Intl. Colloq. on Drops and Bubbles** [*this vol. and chapter.*]
13. J. C. Yang & C. T. Avedisian, "The combustion of unsupported heptane/hexadecane mixture droplets at low gravity," **Proc. 22nd Symp. (Int.) on Combustion**, The Combustion Institute, in press.
14. A. L. Randolph, A. Makino, & C. K. Law (1988), "Liquid-phase diffusional resistance in multicomponent droplet gasification," **Proc. 21st Symp. (Int.) on Combustion**, The Combustion Institute, 601.
15. A.L. Randolph & C.K. Law (1988), "Time-resolved gasification and sooting characteristics of droplets of alcohol/oil blends and water/oil emulsions," **Proc. 21st Symp. (Int.) on Combustion**, The Combustion Institute, 1125.
16. F. L. Dryer (1977), "Water addition to practical combustion systems: Concepts and applications," **Proc. 16th Symp. (Int.) on Combustion**, Combustion Institute, 279.
17. J. C. Lasheras, A. C. Fernandez-Pello, & F. L. Dryer (1981), "On the disruptive burning of free droplets of alcohol/n-paraffin solutions and emulsions," **Proc. 18th Symp. (Int.) on Combustion**, The Combustion Institute, 293.
18. C. H. Wang & C. K. Law (1985), "Microexplosion of fuel droplets under high pressure," *Combust. Flame* **59**, 53.
19. A. C. Fernandez-Pello (1986), "Convective droplet combustion," Fall Tech. Meeting, Eastern Section, Combustion Institute, San Juan, Puerto Rico.
20. C. K. Law (1982), "Recent advances in droplet vaporization and combustion," *Prog. Energy Combust. Sci,* **8**, 169.
21. A. Y. Tong & W. A. Sirignano (1986), "Multicomponent droplet vaporization in a high temperature gas," *Combust. Flame* **66**, 221.
22. R. H. Rangel & A. C. Fernandez-Pello (1984), "Mixed-convective droplet combustion with internal circulation," *Combust. Sci. Technol.* **42**, 47.
23. M. Renksizbulut & M. C. Yuen (1983), "Numerical study of droplet evaporation in a high temperature stream," *J. Heat Transfer* **105**, 389.
24. M. Renksizbulut & R. J. Haywood (1988), "Transient convective droplet evaporation with variable properties and internal circulation at intermediate Reynolds numbers," *Int. J. Multiphase Flow*, in press.
25. R. J. Haywood, R. Nafziger, & M. Renksizbulut (1988), "A detailed examination of gas and liquid phase transient processes in convective droplet evaporation," *J. Heat Transfer*, in press.
26. H. A. Dwyer & B. R. Sanders (1985), "Detailed computation of unsteady droplet

dynamics," **Proc. 20th Symp. (Int.) on Combustion**, Combustion Institute, 1743.
27. H. A. Dwyer & B. R. Sanders (1988), "A detailed study of burning fuel droplets," **Proc. 21st Symp. (Int.) on Combustion**, The Combustion Institute, 633.
28. H. A. Dwyer & B. R. Sanders (1989), "Calculation of unsteady reacting droplet flows," **Proc. 22nd Symp. (Int.) on Combustion**, Combustion Institute, in press.
29. K. Miyasaka& C. K. Law (1980) "Combustion and agglomeration of coal-oil mixtures in furnace environments," *Combust. Sci. Technol.* **24**, 71.
30. G. A. Szekely. & G. M. Faeth, (1983), "Reaction of carbon black slurry agglomerates in combustion gases," **Proc. 19th Symp. (Int.) on Combustion**, Combustion Institute, 1077.
31. G. E. Liu & C. K. Law (1986), "Combustion of coal-water slurry droplets," *Fuel* **65**, 171.
32. S. C. Yao & L. Liu (1983), "Behavior of suspended coal-water slurry droplets in a combustion environment," *Combust. Flame* **51**, 335.
33. P. Antaki & F. A. Williams (1987), "Observations on the combustion of boron slurry droplets in air," *Combust. Flame* **67**, 1.
34. F. Takahashi, F. L. Dryer, & F. A. Williams (1988), "Combustion behavior of free boron slurry droplets," **Proc. 21st Symp. (Int.) on Combustion**, Combustion Institute, 1983.
35. A. Lee & C. K. Law (1988), "Experiments on the gasification and shell characteristics in slurry droplet burning," Paper No. 88-40, Spring Tech. Meeting, Western States Section, Combustion Institute, Salt Lake City, Utah.
36. P. Antaki (1986), "Transient processes in a rigid slurry droplet during liquid vaporization and combustion," *Combust. Sci. Technol.* **46**, 113.
37. D. L. Zhu & C. K. Law (1987), "Aerothermochemical studies of energetic liquid materials: 1. Combustion of HAN-based liquid gun propellants under atmospheric pressure," *Combust. Flame* **70**, 333.
38. A. Lee, C. K. Law, & A. L. Randolph (1988), "Aerothermochemical studies of energetic liquid materials: 2, Combustion and microexplosion of droplets of organic azides," *Combust. Flame* **71**, 123.
39. B. S. Haynes & H. G. Wagner (1981), "Soot formation," *Prog. Energy Combust. Sci.* **7**, 229.
40. B. D. Shaw, F. L. Dryer, F. A. Williams, & J. B. Haggard, (1987), "Sooting and disruption in spherically symmetric combustion of decane droplets in air," **Proc. 38th Cong. of the International Astronautical Federation**, Brighton, U.K.
41. A. L. Randolph & C. K. Law (1986), "Influence of physical mechanisms on soot formation and destruction in droplet burning," *Combust. Flame* **64**, 267.
42. N. W. Sorbo, R. R. Steeper, C. K. Law, & D. P. Y. Chang (1989), "An experimental investigation on the incineration and incinerability of chlorinated alkane droplets," **Proc. 22nd Symp. (Int.) on Combustion**, Combustion Inst., in press.
43. T. Niioka & J. Sato (1988), "Combustion and microexplosion behavior of miscible fuel droplets under high pressure," **Proc. 21st Symp. (Int.) on Combustion**, Combustion Institute, 625.
44. J. Sato, M. Tsue, M. Niwa, & M. Kono (1988), "Microgravity droplet combustion in high pressures near critical pressures of fuels," **Proc. 3rd Intl. Colloq. on Drops and Bubbles** [*this vol. and chapter*].
45. N. J. Marrone, I. Kennedy, & F. L. Dryer (1984), "Coke formation in the combustion of isolated heavy oil droplets," *Combust. Sci. Technol.* **36**, 149.
46. C. B. Katz, F. L. Dryer, & F. Takahashi (1987), "Some further observations on the relations of heavy fuel oil properties to coke particulate emissions," Fall Tech. Meeting, Western States Section, Combustion Institute, Honolulu, HI.
47. A. M. Murray & L. A. Melton (1985), "Fluorescence methods for determination of temperature in fuel sprays," *Appl. Opt.* **24**, 2783.

FIGURE CAPTIONS

Figure 1. Qualitative behavior of the droplet diameter (d_s) and flame diameter (d_f): (a) for d^2-law and (b) allowing for the transient processes of droplet heating and fuel vapor accumulation.

Figure 2. Qualitative behavior of the concentration profiles of a two-component droplet with strong liquid-phase diffusional resistance.

Figure 3. Qualitative behavior of the temporal variations of the surface and center values of the temperature and volatile concentration of a two-component droplet with strong diffusional resistance.

Figure 4. Experimental data exhibiting d^2-law behavior for a two-component mixture with nearly equal volatilities.

Figure 5. Experimental data of droplet and flame diameters for a two-component mixture with very different volatilities.

Figure 6. Temporal variation of the average molar fraction of the volatile component for two-component mixture droplets undergoing combustion. Solid lines are calculated values.

Figure 7. Temporal variation of the average molar fraction of the volatile component for two-component mixture droplets undergoing vaporization. Solid lines are calculated values.

Figure 8. Temporal variation of the average droplet molar fraction of the volatile component for mixtures of phenyldodecane with water, heptane, and propanol.

Figure 9. Droplet size and volume at microexplosion for different initial concentrations of heptane/hexadecane mixtures, where d_{so} is the initial diameter and d_{se} the diameter at microexplosion.

Figure 10. Droplet size and volume at microexplosion for 50-50% alcohol/alkane mixtures under elevated pressures, where d_{so} is the initial diameter and d_{se} the diameter at microexplosion.

Figure 11. Sequence of different droplet generation modes: (a) stable; (b) unstable, with absorption of ligament; (c) unstable, with separation of ligament.

Figure 12. Droplet size and liquid mass history for carbon slurry droplet burning in air.

Figure 13. Schematic showing the sequence of gasification, shell formation, and microexplosion during slurry droplet combustion.

Figure 14. Influence of number of carbon atoms for n-alkanes, monoazides and diazides on the burning rate constant K.

Figure 15. Variation of the luminous zone diameter and the amount of soot present during droplet burning.

FURTHER OBSERVATIONS ON MICROGRAVITY DROPLET COMBUSTION IN THE NASA-LEWIS DROP TOWER FACILITIES: A DIGITAL PROCESSING TECHNIQUE FOR DROPLET BURNING DATA

Mun Young Choi and Frederick L. Dryer
Mechanical and Aerospace Engineering, School of Engineering and Applied Sciences, Princeton University, Princeton, NJ 08544

John B. Haggard, Jr. and Mike H. Brace
NASA-Lewis Research Center
Cleveland, OH

ABSTRACT

This paper presents recent results from a continuing study on the burning characteristics of pure, single component fuel droplets in a microgravity environment. A new, economical, microprocessor-based, computer imaging and data analysis method for determining the diameter and the position of the burning droplets from photographic records is described and demonstrated using experimental data previously obtained on the combustion of n-decane droplets in air. Initial observations on the combustion of methanol droplets in air are also reported.

1.0 INTRODUCTION

The combustion of a pure, single component liquid droplet provides an ideal problem from which to obtain valuable information for both basic and applied scientific purposes. The problem involves a chemically reacting, two-phase flow with phase change and thus represents one of the simplest systems in which to study the complex coupling of these phenomena. However, the presence of either forced or natural convective effects considerably complicates the problem, not only by invoking asymmetrical considerations, but by introducing additional coupled transport effects both inside and outside the droplet[1,2]. Thus, the fundamental physical and chemical theories governing the burning of spherically-symmetric fuel droplets are strictly valid (without correction) only under conditions where the effects of buoyancy and forced convection have been eliminated[3,4].

Forced convective effects can be minimized in unsupported, isolated, free droplet burning, by using levitation methods[5], or by minimizing surrounding gas and droplet relative motion[6]. However, the elimination of natural convective effects is much more difficult to achieve. The use of very small fuel droplet diameters similar to those found in practical combustors would minimize these effects, but the physical size and reduced burning time limit the ability to photographically resolve physical dimensions of the droplet and flame. Thus the ability to study the transient effects of droplet heating, combustion extinction, and dissolution of combustion products at the liquid surface is hindered. The study of relatively large droplets (1200 μm) is desirable in order to reduce inaccuracies resulting from optical resolution and to enhance the duration of transient burning phenomena. Reduced environmental pressure is another means of limiting natural convective effects[7], and pressure reduction appears to permit attainment of flame structure which is nearly symmetric to the droplet, even for droplets suspended on filaments. However, reduced pressure introduces modifications in the governing chemical kinetics which are difficult to assess, and perturbations from the presence of suspending filaments are well recognized[8,9].

Reduced gravity appears to be the most appropriate means to achieve a buoyancy-free environment without affecting other issues. By releasing a test combustion vessel into free fall, a burning droplet appropriately positioned and dispensed will remain stationary with respect to the surrounding gas. Therefore, it will not experience any forced convective air currents or buoyancy induced flows, and the analysis of the conservation and energy equations is reduced to considering a single spatial dimension. This simplification provides the theoretical means to determine values such as the surface regression rate of the droplet and the stand-off distance of the flame as a function of burning time, and theory may then be directly compared with experimental measurements are recorded as photographic data.

Kumagai and Isoda[10], pioneered the use of drop towers to study the burning of suspended fuel droplets under reduced gravity conditions, and later, Kumagai *et al.*[11], first described a single-filament-withdrawal technique to study the combustion of isolated, unsupported droplets under conditions at 10^{-3} to 10^{-4} of earth's gravity. Yet, these and other studies which have further exploited this technique have been incapable of observing more than 45% of the (projected) burning time for the fuel droplet diameters (on the order of 800 to 1200 μm) which can be successfully deployed with low residual velocity.

Avedisian and his colleagues[12,13] have reported another approach which yields droplets small enough to be completely burned in drop towers, and Hara and Kumagai[14] have recently adapted an opposed needle configuration (similar to that described below) to yield similar capabilities. However, many of the effects of interest would be best characterized through the study of large droplets over a range[15] of diameters. Space-based experiments offer the optimal conditions in which to conduct such experiments. The major objectives of the present project, which was initiated[16,4,17,18] in the early 1980's have been: to provide information necessary for the successful design of buoyancy-free droplet-burning experiments for space-based platforms[17,19]; to develop data acquisition and analysis techniques to determine geometrical characteristics as a function of burning time; and to develop theoretical models for comparison with the experimental observations. Burning rate, and extinction properties are of particular interest in the study.

The results reported by Shaw *et al.*[18], point to the need for a more precise and accurate method to determine droplet diameter and flame position as a function of time from photographic records, and the need for selection of a liquid fuel which might yield less sooting under microgravity environments characteristic of space-based platforms. In what follows, a new economical approach for analyzing photographic droplet images is described, and results are compared to Vanguard analysis of some of the n-decane data obtained earlier[18]. In addition, some recent observations on the burning characteristics of methanol in air are also reported.

2.0 EXPERIMENTAL FACILITY DESCRIPTION

In order to develop suitable hardware for the project[19], initial experimental package design and operation have been tested in the 2.2 second drop tower at NASA Lewis Research Center. The drop tower package, with its associated drag shield is capable of achieving gravity levels of 10^{-6} g, comparable to or below those available on space-based platforms, and considerably lower than many of the drop tower approaches used in other low gravity droplet combustion studies. Yet, the available observation time is too short to observe the full combustion history of droplets in the size range of interest for eventual space-based experimentation (1200 μm).

An opposed, hollow needle retraction scheme has been chosen to dispense and deploy the droplet to be studied[17]. A droplet is dispensed through and suspended on a matched pair of horizontal, opposed, 0.2 mm, hollow needles (Figure 1a). To reduce interfacial contact area of the liquid with the external surface of the needles, the needles are continually separated during the dispensing of liquid to form a

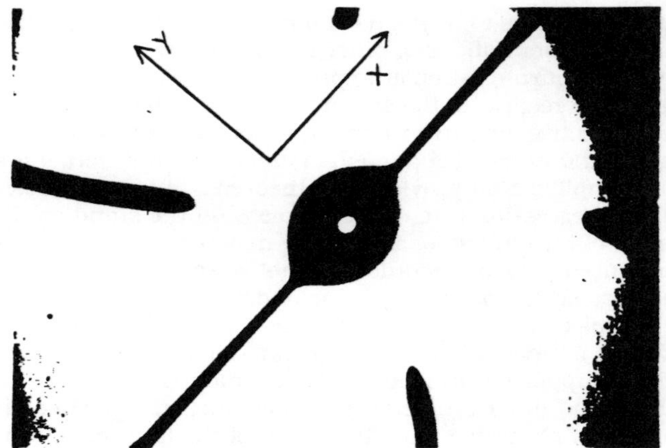

FIGURE 1a

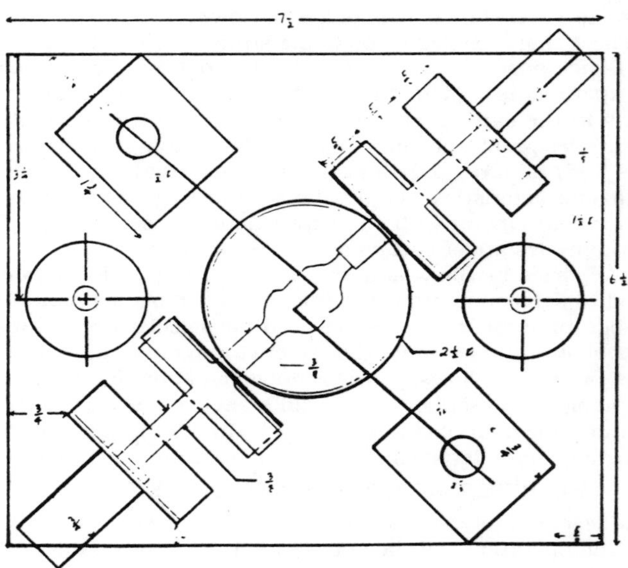

Figure 1b

"stretched droplet". Droplets may be partially or fully dispensed previous to drop package release, with approximately 0.5 seconds being required to dampen droplet motion due to the package release and g-level transition. After the droplet is fully dispensed, and stabilized in microgravity conditions, the needles are symmetrically extracted at a defined accelerative rate along the needle axes, to deploy the droplet. The conceptual advantages of this approach are its simplicity, as well as the increased likelihood of producing a symmetric vapor cloud surrounding deployed droplets of more volatile fuels, such as simple alcohols and small-carbon-number hydrocarbons.

Approximately 20 to 30 milliseconds after the droplet is deployed, the droplet is ignited by two spark discharges symmetrically positioned (in the horizontal plane) relative to the deployment position. This configuration is chosen to minimize the net impulse delivered to the droplet by the spark discharges[17,18], and this procedure also produces a more symmetric energy deposition to the droplet surface (vapor generation) and vapor cloud ignition than that produced by a single spark. After ignition, all hardware is retracted to more than 20 diameters away from the burning droplet. The combustion volume (8000 cc) surrounding the burning droplet can be pressurized up to two atmospheres with air or selected gas mixtures. Back-lighted droplet images and flame diameters are recorded using 16 mm cine-photography. The experimental design is described in further detail elsewhere[17,20].

3.0 IMAGE PROCESSING TECHNIQUES

While simple Vanguard Analyzer processing represents a mechanical improvement over manual measurement of photographic images to obtain droplet and flame diameter, the techniques are similar in that both depend on the ability of the human eye to discriminate edges of the specific image from the background. Small movements in the droplet, variations in background and foreground lighting, and the film grain cause the intensity of the image to vary slightly relative to the background. Frame-by-frame visual image discrimination of cine-photography film is thus very difficult to perform with any precision. Furthermore, it remains difficult with both techniques to obtain several dimensional measurements from which to deduce a mean diameter from the photographic image of a burning droplet.

As part of the present work, an economical micro-processor-based imaging and analysis technique has been developed at Princeton specifically for processing photographic images of burning droplets. The image processing technique used is described below.

3.1 Digital Image Acquisition and Preprocessing

The photographic image of a droplet (in this case, a single frame of 16 mm Kodak Tri-X, cine-photography film) is projected onto a high-resolution, ground glass screen using a 16 mm frame-projector. This image is captured using a black and white CID video camera connected to a DATA TRANSLATION 2851 frame grabber board. The frame grabber board segments the video camera image into a 480 by 512 screen pixel array. The brightness level for each pixel (screen array element) is converted to graylevel values between 0 (white) and 255 (black). The frame grabber board is installed in a Compaq 386, 16 MHz microprocessor equipped with a 40 megabyte fixed disk, 3 megabytes of memory, a 30387 math co-processor, and EGA graphics display. The graylevel values generated by the frame grabber board are stored as a 480 x 512 two dimensional screen pixel array.

To reduce processing time and eliminate unnecessary background regions, a simple pre-processing algorithm has been developed to identify the active screen region of interest. The algorithm searches the screen pixel array for the approximate extreme row and column locations of the envelope separating the image of the droplet from the background. A histogram-count of the graylevels for the pixels extreme

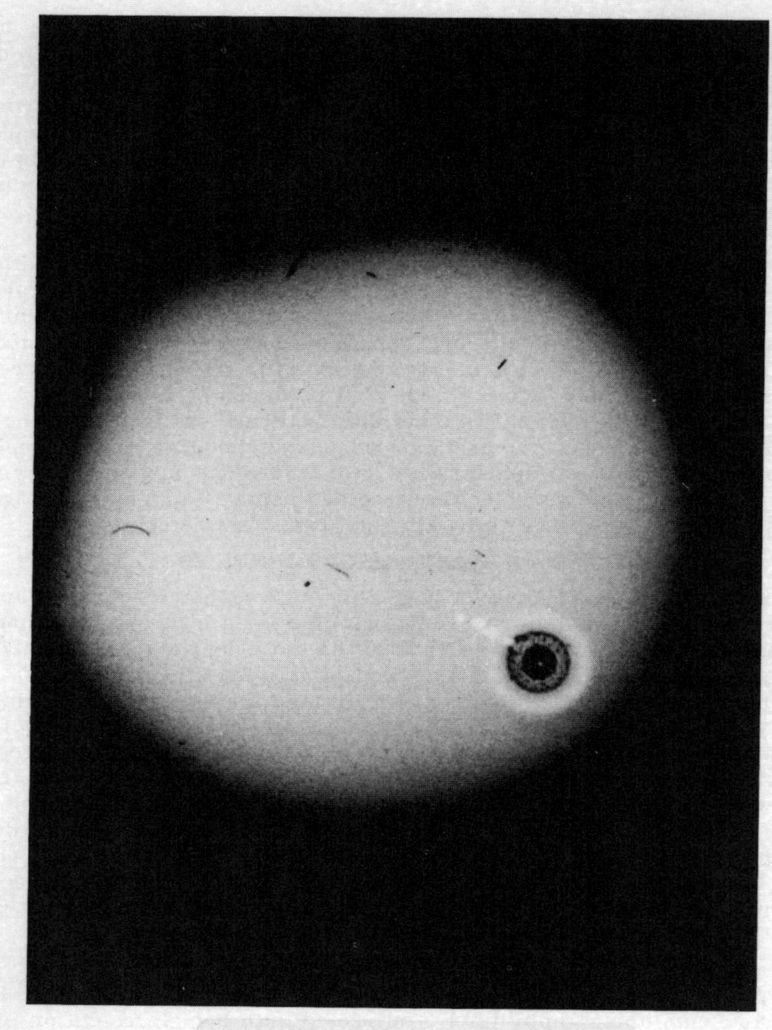

Fig. 2

row and column locations of the envelope separating the image of the droplet from the background. A histogram-count of the graylevels for the pixels contained within the specified window is performed and the histogram is searched for graylevel values greater than or equal to a preliminary threshold determined from the dynamic range of graylevels found for the entire screen array. If no pixels with at least this threshold graylevel are present, the window slides to the next row or column location until the edge can be located. This iterative process is repeated for windows sliding from each extremity of the screen pixel array until the four extreme-most threshold locations are determined.

As seen in Figure 2, in some circumstances, a dark soot region (shell) surrounds the droplet during combustion. Thus, the screen pixel array region of interest determined above might also include the soot shell image. It is difficult to distinguish between the soot shell/background interfaces and the droplet/background interface, based on graylevel gradients alone, since the droplet and soot ring graylevel dynamic ranges overlap. The aspect ratio of the active region defined above is compared with those calculated from the previous photographic frame analysis or with values input by the operator. If the aspect ratio of the new active region is outside the predetermined limits or greater than that from the previous analysis, then further processing is utilized to eliminate the soot shell from the active region.

3.2 Image Enhancements Methods

Methods used to enhance images vary with different applications. For instance, imaging techniques may be used to optimize visual perception. However, for the application of machine detection, visual appearance is not as critical as machine perception and computational efficiency[23]. The issues important to machine discrimination of the "edge" or interface between the background and an object (so that the object area may be computed) are best envisioned from a histogram distribution of the graylevels of all pixels within the active region.

For images with bi-modal distributions in graylevel, this histogram provides a useful global description of the graylevel content associated with the object of interest and the background[24]. Figure 3a, 3b compares such a histogram distribution for an "ideal" image (generated by the micro-processor), with that found from analysis of an actual photographic image of a droplet similar to that in Figure 2. While the ideal image consists of very narrow dynamic ranges for background and object with no pixels possessing intermediate graylevel values, the histogram for the real image is degraded by diffractional and electronically-induced sampling effects, as well as through spatial variations in (background and foreground) lighting, optical resolution and the film grain. Thus, the pixels near the droplet-background boundary will assume gray level values intermediate to the individual droplet and background dynamic ranges[25]. The boundary between the object and background can be easily obtained for the ideal case by selecting any graylevel threshold between points A and B to define the contour of this ideal image since there is an infinite graylevel gradient at the boundary. The choice of appropriate graylevel threshold separating background and object in the real case is much less clear. This fact becomes apparent by inspecting the ideal and real image line profiles, *i.e.*, a display of the graylevel values at each pixel along a row of the screen pixel array (Figure 4a, 4b).

For the actual image, the choice of the threshold graylevel affects the measured diameters due to the finite number of pixels in the intermediate graylevel region between the background dynamic range and that of the object. Any algorithm which reproducibly enhances the graylevel discontinuity at the interface between the regions of distinctly different brightness ranges should apparently improve the ability to locate the contour edge of the object.

For example, spatial convolution filters may be used to perform this enhancement[26]. Such filters operate on screen pixel array neighborhoods of chosen

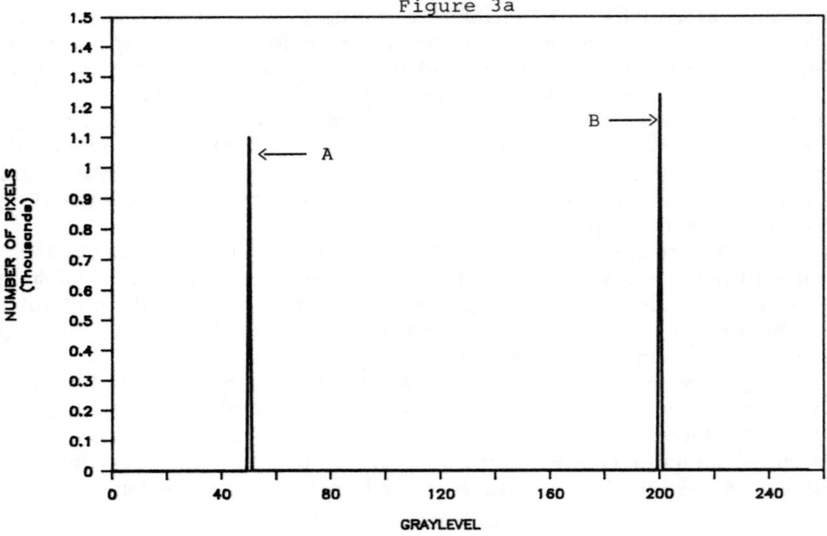
Figure 3a
NUMBER OF PIXELS (Thousands)
1.5
1.4
1.3
1.2
1.1
1
0.9
0.8
0.7
0.6
0.5
0.4
0.3
0.2
0.1
0
A
B
0
40
80
120
160
200
240
GRAYLEVEL

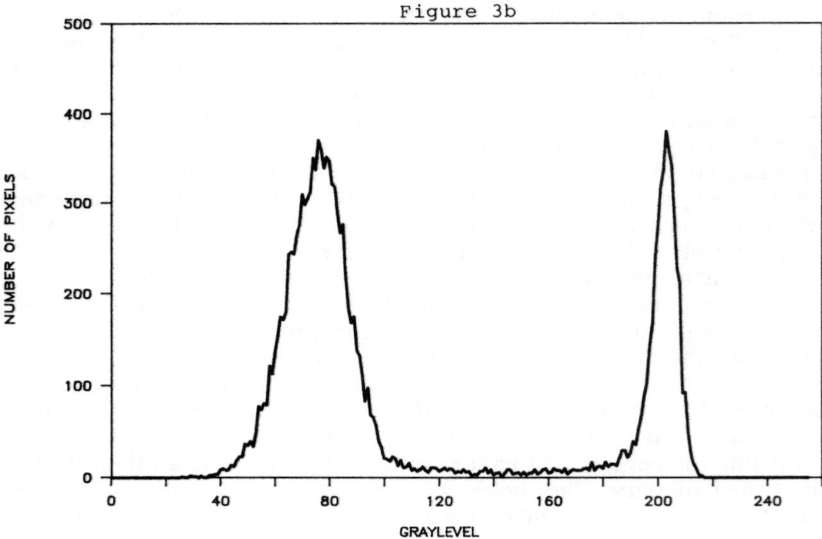
Figure 3b
NUMBER OF PIXELS
500
400
300
200
100
0
0
40
80
120
160
200
240
GRAYLEVEL

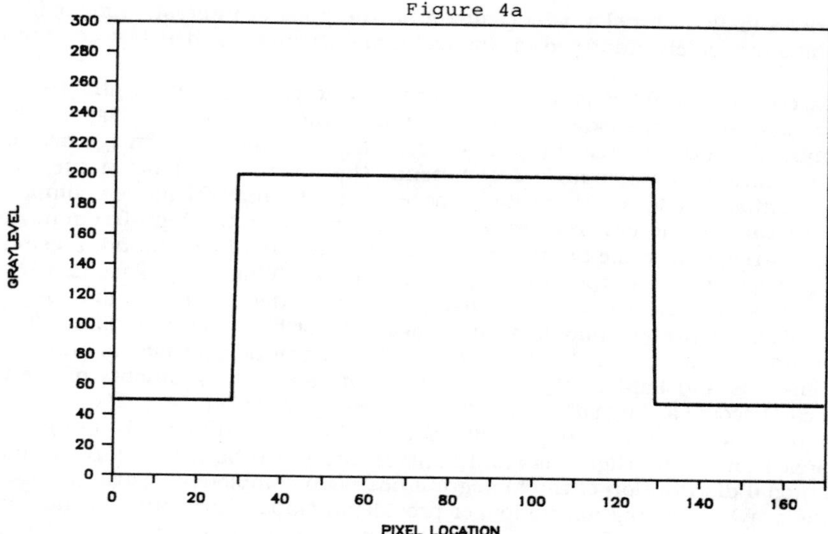
Figure 4a
GRAYLEVEL
PIXEL LOCATION
0
20
40
60
80
100
120
140
160
180
200
220
240
260
280
300

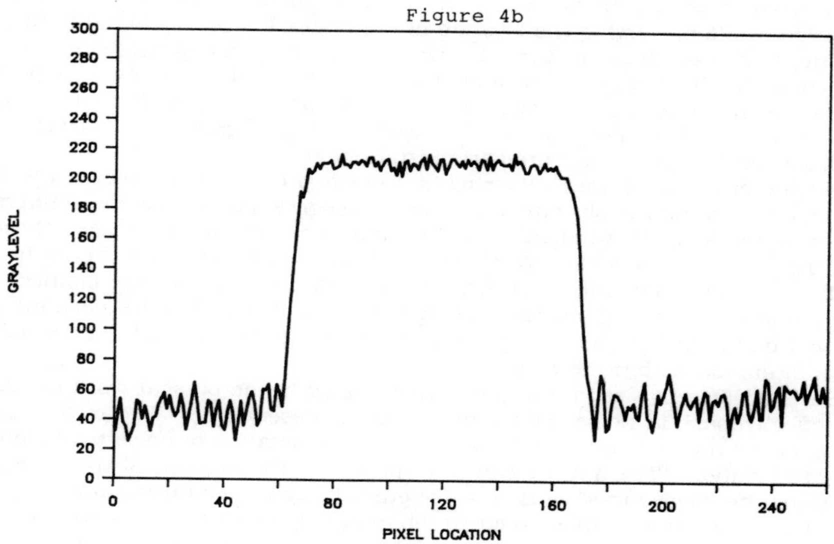
Figure 4b
GRAYLEVEL
PIXEL LOCATION
0
40
80
120
160
200
240
20
60
100
140
180
220
260
280
300

dimensions (called kernels), which surround the pixel of interest. Linear filtering algorithms are widely employed in this technique because of their fast computation rate.

The convolution filtering process can be used to pass selected spatial frequency components by simply modifying the elements of the filter kernel. Figure 5 shows the graylevel frequency distribution of a real droplet photograph image before any filter application. Castleman[24] recommends the use of High-Pass or the Laplace filters to enhance edges. These filters accentuate the high-frequency components while attenuating the low-frequency of the image. Since the high-frequency components correspond to the boundaries between object and background, these filters would be expected to sharpen the edges of the object. While High-Pass and Laplace filters appear to perform well for images with well defined edges and negligible visual noise, the present photographic images do not have these qualities. The additive random noise in graylevel at the object background interface is amplified by the High-Pass and Laplace filters, and, in such cases, the gradients of the noise spikes can become as large as the gradient of the edge itself[23].

Figure 6 shows the histogram distribution for the real image before and after the application of the High-Pass and Laplace filters. In both post-filtered images, the bi-modal distribution of the histogram has been completely destroyed. As a result, the graylevel histogram no longer provides a global description of the appearance of the image. While these same filtering processes will improve the visual quality of the image, the process degrades the ability of the computer to detect edges.

Yang and Huang[27], suggest that median or mean filters can be used to increase the signal to noise ratio by decreasing the noise components of the graylevel dynamic range, while preserving the location of the original edge. Median filtering is a non-linear process which replaces the center pixel of a defined neighborhood with the median graylevel value determined from the pixel matrix. Thus, it is particularly useful in reducing "salt and pepper" noise associated with digital images. However, the median filter can attenuate the signal, if extreme graylevel values within the pixel neighborhood are part of true signal[23].

On the other hand, mean filtering is a linear process which reduces additive random noise within the pixel matrix. Baxes[26], suggests that while mean filters are effective in reducing the random noise, they may also "smear" the image. However, smearing is less severe for an edge with nearly constant gradient if the mean filter matrix is smaller than the pixel distance in which the ramp edge connects the background and droplet graylevels. Using the mean filter within the constant ramp edge will maintain the identical slope as long as the filter kernel uses local gray level information within the slope.

Several different sized mean filters were tested in the present work to observe their effectiveness in noise reduction and edge preservation. The smallest filter matrix, of 3 x 3 dimension, displayed the best combination of noise reduction and edge preservation. The 3 x 3 mean filter preserves the location of the edge while smoothing the "jaggedness" caused in the gradient by random noise and the fractal character of the pixel. While a median filter could perform the same function with small improvements, the mean filter is much more computationally efficient. The mean filter required only 0.5 second to process an entire screen pixel array, while the non-linear median filter required 85 seconds to complete the same processing.

Figure 7 shows line profiles of an actual droplet processed through two successive applications of the 3 x 3 mean filters. Each application of the mean filter shows a reduction of the noise within the background and droplet dynamic ranges without distortion of the edge location. The histogram distribution after the successive applications of mean filters are shown in Figure 8. The individual dynamic ranges of the droplet and background become more uniform due to the reduction of the random

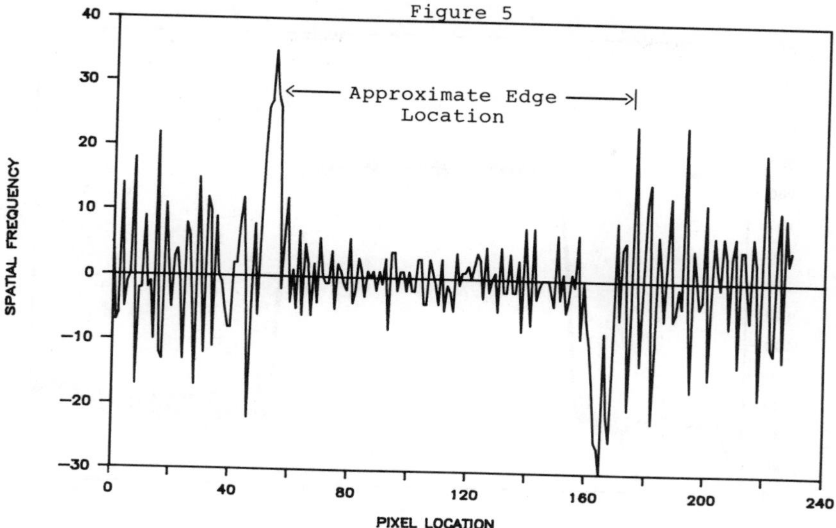
Figure 5
Approximate Edge Location
SPATIAL FREQUENCY
PIXEL LOCATION
40
30
20
10
0
−10
−20
−30
0
40
80
120
160
200
240

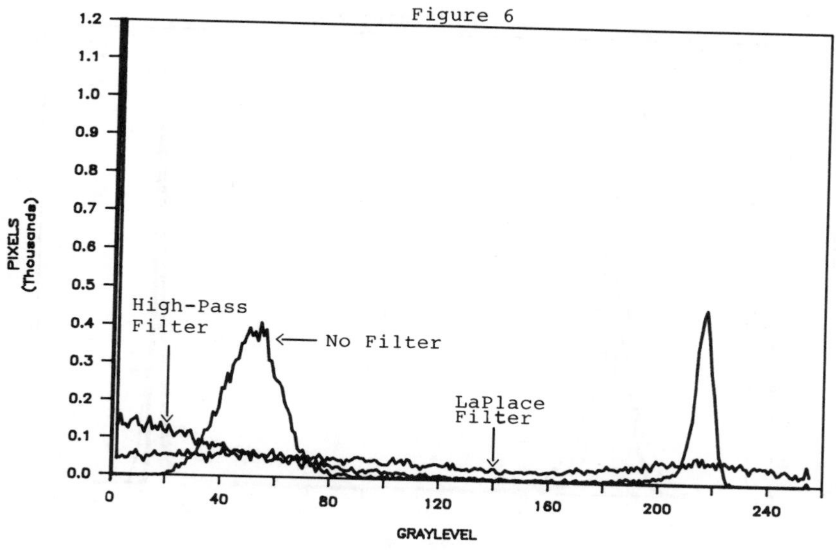
Figure 6
High-Pass Filter
No Filter
LaPlace Filter
PIXELS (Thousands)
GRAYLEVEL
1.2
1.1
1.0
0.9
0.8
0.7
0.6
0.5
0.4
0.3
0.2
0.1
0.0
0
40
80
120
160
200
240

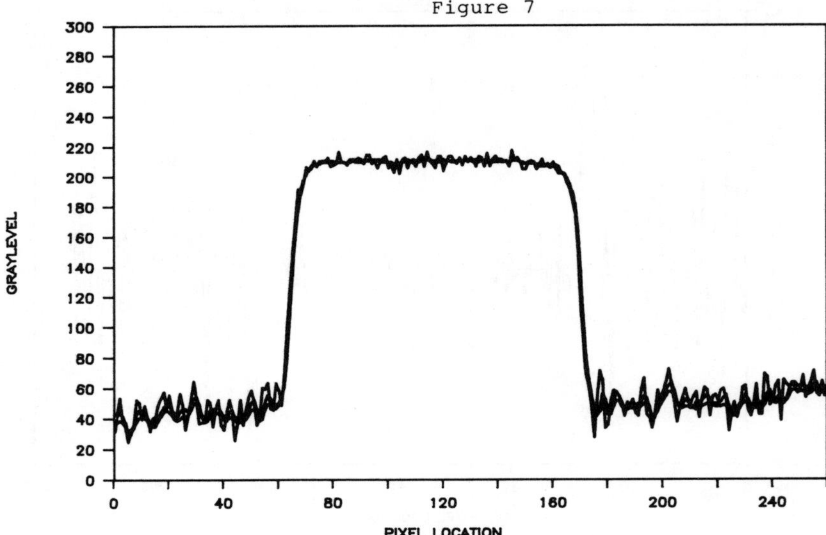
Figure 7
GRAYLEVEL
PIXEL LOCATION
0
20
40
60
80
100
120
140
160
180
200
220
240
260
280
300
0
40
80
120
160
200
240

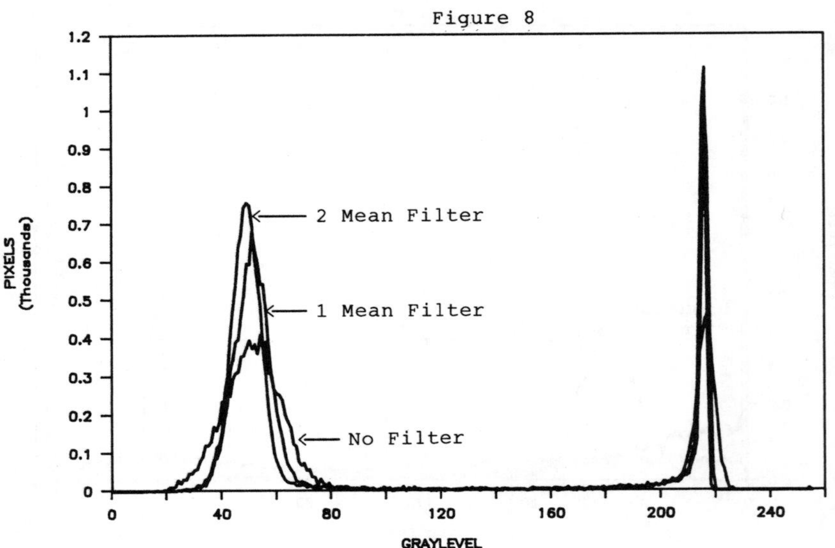
Figure 8
PIXELS (Thousands)
GRAYLEVEL
2 Mean Filter
1 Mean Filter
No Filter
0
0.1
0.2
0.3
0.4
0.5
0.6
0.7
0.8
0.9
1
1.1
1.2
0
40
80
120
160
200
240

noise and tend to emulate the bi-modal distribution of the ideal image depicted earlier.

Upon selection of an edge enhancement scheme, the image enhancement technique further needs an algorithm to partition the processed screen pixel array between background and object reproducibly. For an image with a bi-modal graylevel distribution, the global threshold between the object and the background is the gray level located at the valley between the two peaks[32]. However, the remaining random noise obscures the true location of this point in a real image. Castleman[24] recommends choosing the threshold as the mean graylevel between the two peaks of the histogram. Although this method produces a reasonably accurate boundary, it is limited in that it assumes that the width of the graylevel dynamic ranges for the object and the background are equal.

A more accurate method is to approximate each of the distributions as Gaussian[28], from which one might then determine the threshold using a standard-deviation criterion. A 3-standard-deviation probability for a one-dimensional Gaussian distribution such as the dynamic ranges of the background and droplet encompasses 99.9% of all gray levels associated with each peak. Theoretically, only a small number of pixels should be present in the region bounded by points C and D in Figure 9. Since this region occurs at the interface between the droplet and background and the number of pixels under both dynamic ranges are nearly equal, the pixel distribution in this region should be equally divided between the droplet and the background dynamic ranges. To verify this assumption, an ideal image was degraded by the application of five 7 x 7 mean filters (Figure 10). The processing caused equal numbers of pixels from the background and droplet dynamic ranges to be distributed into the interfacial zone. Thus, the expected value of the global boundary threshold should be the mean graylevel value between points C and D. The local statistics including the mean and the standard deviation of the histogram distribution are determined from the individual dynamic ranges.

Global thresholding was selected instead of local thresholding in the present work for two reasons. First, the histogram displayed narrow individual dynamic ranges of the background and droplet. The contrast between the two were nearly uniform in the regions surrounding the droplet, thus either method would define similar thresholds. Second, the global thresholding method provides greater computational efficiency. For a single frame analysis, the choice between the two thresholding methods is not critical, however, in an automated system which may analyze 500 frames or more of cine-photography, the total reduction of analysis time can be very significant.

An investigation of the sensitivity of measured diameter to the selection of threshold showed a maximum discrepancy of 1.27% with respect to the average value for graylevel values varied from 125 to 170 (See Figure 11). This relative insensitivity of the measured diameter as a function of the graylevel threshold in the interfacial region is expected because the number of pixels in this region is negligible compared to the number of pixels within the droplet dynamic range.

Kittler *et al.*[29], suggest that for real images with blurred edges, the optimum threshold is better defined as the point of the highest spatial frequency. To compare results of the above method with this approach, an algorithm which calculates the spatial frequency of an image was developed to measure the rate of change in the graylevel along a row of pixels. The two peaks corresponding to the highest frequency components in the image define the edges of the droplet. While this is indeed a more accurate method of thresholding, it is also computationally-intensive. For a droplet with a diameter of 100 pixels, 100 line profiles have to be calculated in order to determine the frequency distribution used to calculate the boundary of the droplet. Figure 12 displays the line profile with the corresponding frequency analysis.

Figure 9

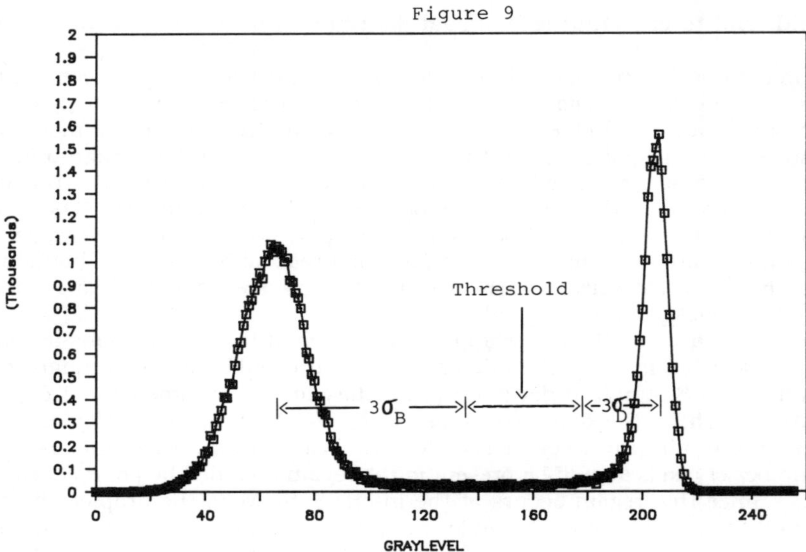

Figure 10

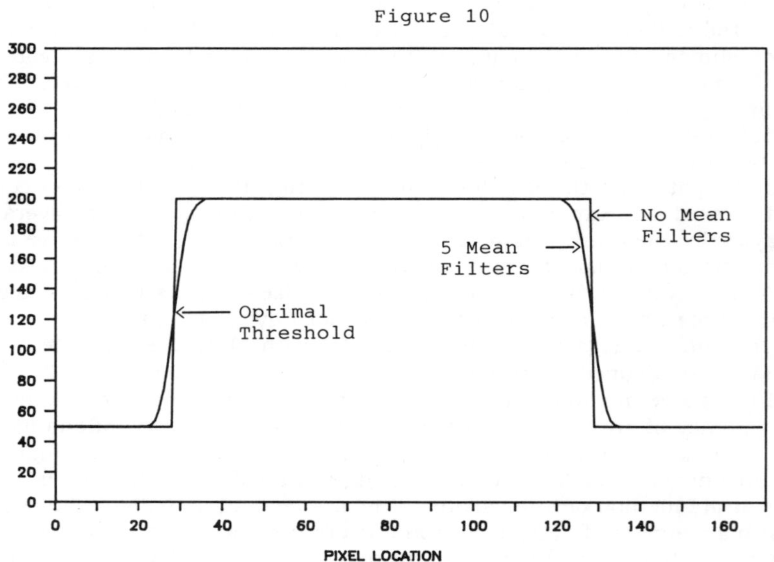

Figure 11

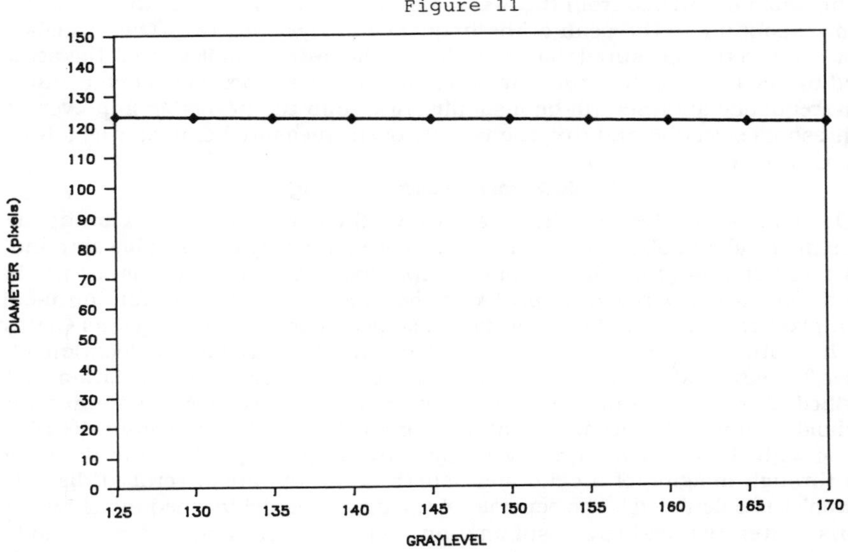

Figure 12

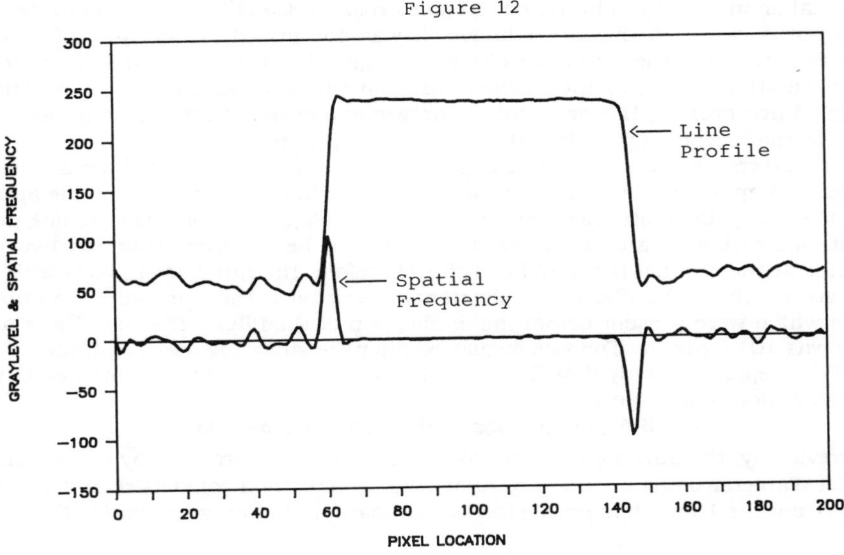

The threshold calculated from the 3-standard deviation method is 159. The threshold corresponding to the peak of the frequency analysis is 160. This translates to 0.06% error in the measured diameter. Thus, the authors believe that the accuracy gained by determining the "true" threshold by extensive processing such as the maximum frequency analysis will be insignificant due to the negligible improvement in the threshold selection and the insensitivity of the measured diameter as a function of the threshold.

3.3 Image Postprocessing

On the basis of the above results, two 3 x 3 mean spatial filters are applied to reduce the random noise and to smooth the edge boundary between background and droplet. After this processing, a subroutine finds the two locations for the background and droplet graylevels peaks in the graylevel histogram for the modified screen pixel array graylevel distribution. The peaks are approximated as Gaussian, and the optimum graylevel threshold value for identifying the location of the droplet/background interface is determined by the 3-standard deviation method described above. The number of pixels within the active region bounded by the threshold contour (the droplet area) are summed from the histogram distribution starting with the threshold graylevel. By converting the pixel size to dimensional units through imaging of a calibrated length scale, the area-averaged diameter of the droplet is calculated. The centroid of the droplet is determined by using combinations of iterative histogram summation and the increasing-region method. For comparison with Vanguard analysis, two perpendicular axes are drawn through the centroid to calculate the horizontal and vertical diameters.

3.4 Image Processing Demonstrations

To ascertain the absolute accuracy of the method, an object of known size with a projected dimension comparable to the droplet's projected image of 85—100 pixels was selected. The average diameter of eight computer measurements of the object to three significant digits was found to be 1.11 cm. The actual diameter measured using a micrometer was 1.11 ± 0.005 cm.

To determine the effect of pixel spatial resolution on the measurement, the projected image of the washer was doubled to 170 pixels by utilizing a larger video lens. The average diameter of eight computer measurements was again 1.11 cm. In the smaller image, the simulated droplet cross-sectional area was comprised of 7500 pixels, while in the larger image, it was comprised of nearly 28000 pixels. Thus in both cases the spatial resolution is higher than the optimal level such that no information is lost in the translation from the original image to the digitized image. Furthermore, the negligible improvement in using the larger projection is offset by the longer computation time required to process it.

To determine the algorithm's capability of defining the edge from a smeared image, a computer-generated circle was measured after being blurred by the application of a 7 x 7 mean filter five times successively. Figure 10 displays the line profile of this image before and after the filters. It can be observed that the two plots coincide at the optimal threshold of 125. Therefore, the number of pixels with gray level greater than 125 after the application of the filter is the same as the number of pixels which were present before the application of the filter. The initial size of the circle was 101.5 pixels. The size measured after the smearing was 101.3 pixels. The small discrepancy of only 0.19% is most likely due to the fractal properties of pixel representation of an object.

3.5 Comparison with Vanguard Analysis

Previously, the burning rates of droplets have been determined by measuring the droplet diameter using a Vanguard Analyzer. This method consists of projecting the image from the 16mm film onto a digitizing board and manually locating the edge

Figure 13a

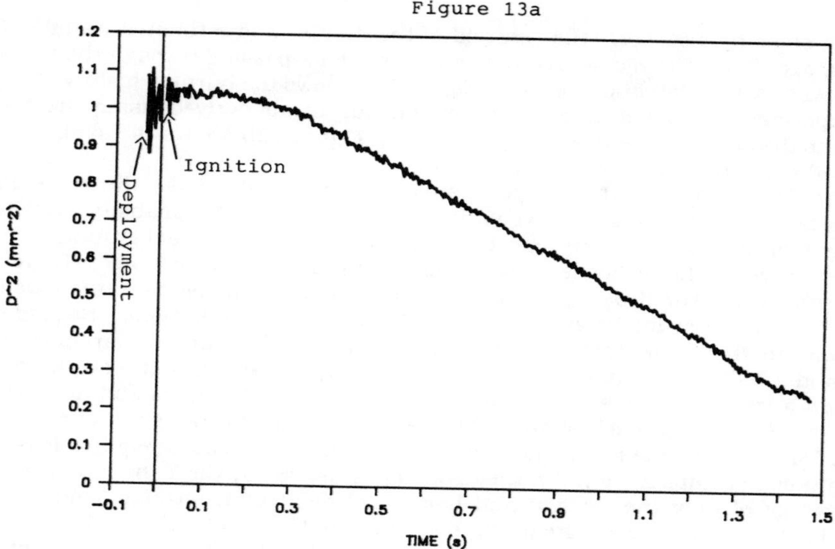

Figure 13b

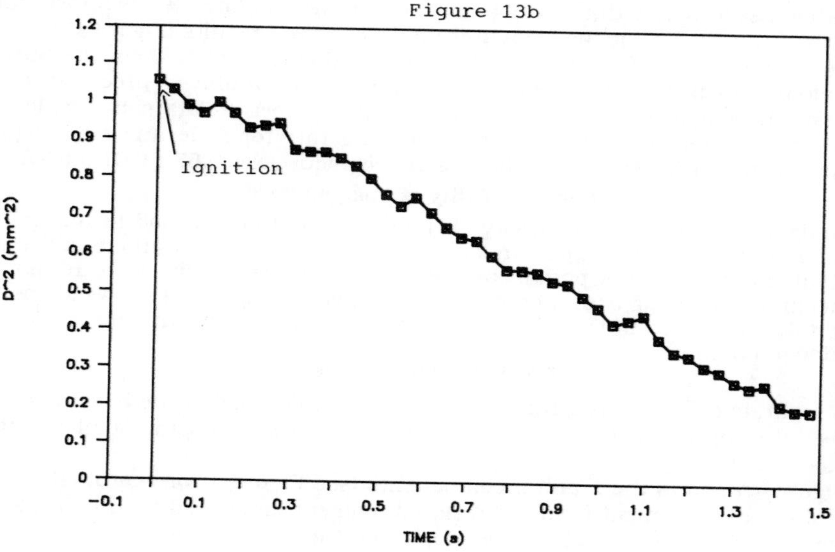

using a cross-hair cursor. The centroid of the droplet was estimated visually, and the average of the orthogonal diameter measurements passing through the centroid was used to define the drop diameter. Figure 13a shows the burning history of a typical experiment for an n-decane droplet burning in air derived using the digital image analysis technique described above, and Figure 13b shows the same history analyzed using the Vanguard analysis.

The Vanguard method has three distinct disadvantages. First, the method has low precision due to the subjective nature of the human eye's capability to systematically identify the boundary between droplet and (varying) background in each frame analyzed. The response of the human eye is actually logarithmic. Therefore, the diameter measured using human perception will be smaller than the true diameter[24]. Computer-image analysis has neither of these limitations. Second, the Vanguard method is very time-consuming. It requires approximately an hour to an hour and a half for an analysis of 50 frames. A typical analysis of a single experiment on a frame-by-frame basis would involve nearly 500 frames and at least ten hours to complete (no such analysis has been performed previously). The analysis of 500 frames using the technique and hardware described above requires less than 75 minutes to complete. Finally, the diameter measured by the Vanguard system is the average of only the vertical and horizontal diameters, although much more intensive efforts could generate a crude estimate of the droplet area. On the other hand, the digital imaging approach used here can easily produce an area averaged diameter for the droplet. The computer analyzed data presented in Figure 13a displays a constant burning rate after the initial heat-up period. The random "d^2" fluctuation associated with the Vanguard film analysis (Figure 13b) is substantially larger than that found for the computer-image analyzed results (Figure 13a). Furthermore, the effects of imprecision in the determination of "d^2" on the burning rate calculation is essentially eliminated due to the improved precision of the computer image analysis technique and the large number of data points which can be determined efficiently. The reported burning rate for n-decane by Vanguard analysis was 0.63 ± 0.09 mm^2/s. The present technique gives 0.67 ± 0.01 mm^2/s.

3.6 Reproducibility of Imaging Results

To demonstrate the insensitivity of the image analysis method to the the film projected image size, and other random sources of potential variations in the data reduction apparatus, two separate analyses were performed on the same 16 mm film of a fuel droplet burning in air at atmospheric pressure. Figure 14 shows almost imperceptible differences in the image analysis results for the square of the diameter as a function of time.

3.7 Velocity Measurements

A separate computer program which locates the centroid of the burning droplet has been developed to determine the displacement in the photographic plane during the experiment.

Figure 15 shows the x and y components (see Figure 1 for coordinate definitions) of the displacement from the residual momentum caused by the droplet deployment, and that imparted to the droplet by the ignition process[18] for an experiment conducted on a 1100 μm diameter methanol droplet burning in air. The magnitude of the residual velocity in this plane is 9.0 mm/s, and it is observed to decrease with increasing initial diameter of the deployed droplet.

For deployed droplets with diameters larger than 1500 μm, the residual velocity induced from the deployment is consistently less than 3.0 mm/s. The residual velocity component produced by the ignition process is strongly influenced by the symmetry of the droplet position and the spark locations at the time of ignition. The precision with which the spark locations are defined appears to be responsible for most of the variation of the residual velocity from experiment to experiment

Figure 14

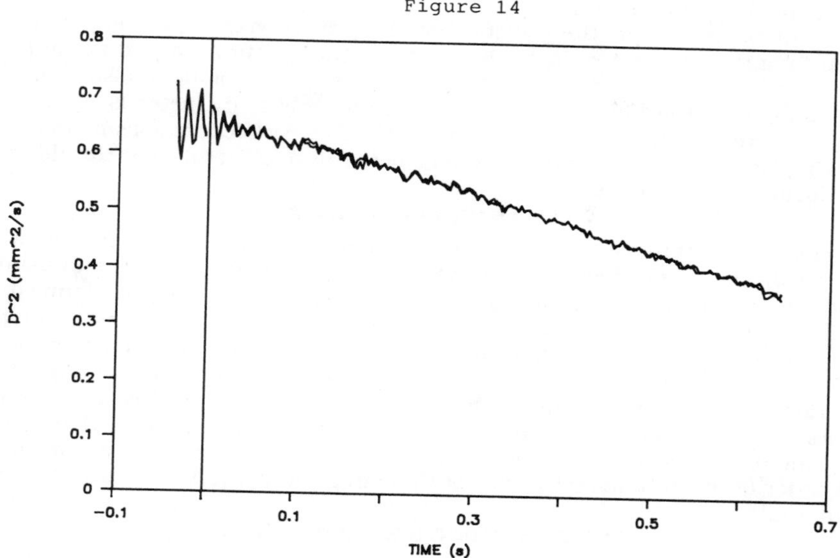

Figure 15

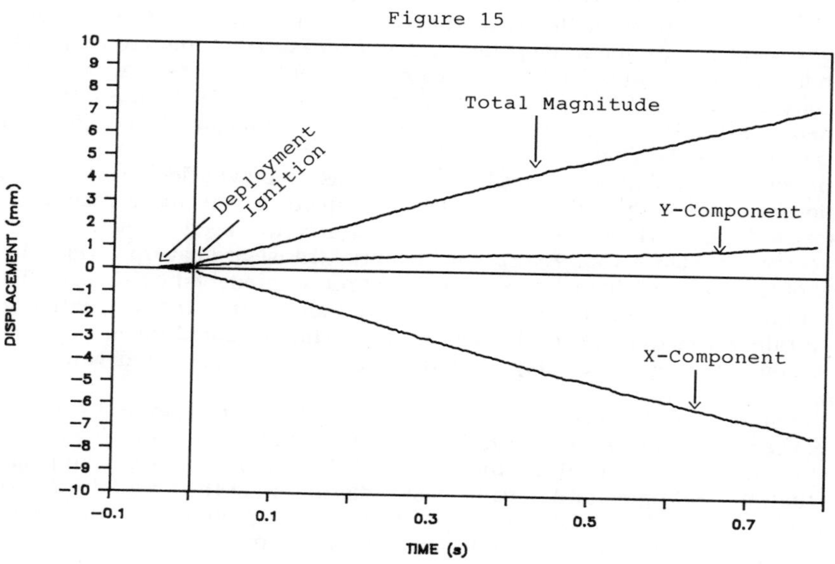

with the the present apparatus.

The displacement in the direction of the gravity vector (-z component, not shown) commonly translates to a velocity of less than 1 mm/s, and this component is not significantly affected by the ignition process. This plane is also that of the depth of field for the film image from which the droplet diameter is typically derived. This small drift and the optics design of the experiment causes no more than a 1% change in the magnification of the camera image used for diameter determination during the experiment.

3.9 Oscillation Damping

From the computer-analyzed data presented in Figure 13a, it is evident that both the droplet deployment and droplet ignition processes produce substantial oscillations in the droplet geometry, however, these oscillations are damped very rapidly relative to the estimated total burning time (0.1 sec). A primary assumption made in calculating the burning rate is that the oscillation is completely damped before the onset of the steady- state burning period. However, this assumption was not verified until the present data reduction was used. The frame-by-frame analysis permits a much improved, quantitative measure of this oscillatory behavior. The small fluctuations which are evident throughout the burning history are on the same order of magnitude as the size of the film grain (10 μm) and thus cannot be eliminated.

4.0 EXPERIMENTS WITH n-DECANE

From early results obtained during development of the deployment and ignition technique in the 2.2 second drop tower which was described previously, Shaw *et al.*[20] have reported observations for 1000 μm n-decane droplets burning in air. The test time available prevented the complete burning of the droplet to be observed for droplets of this size range, but droplet diameter and flame positions were determined for the early burning history using a Vanguard film analyzer. Differences in the burning rate constants determined ranged from 1.36 mm^2/s to 0.3 mm^2/s with an average least square among six experiments of 0.63 $mm^2/s \pm 0.09\ mm^2/s$. The scatter in determined values was suggested to result from both experimental sources and Vanguard data reduction methods. However, the least square value of 0.63 mm^2/s is to be compared with a theoretical value of 0.66 mm^2/s in room temperature, quiescent air at one atmosphere[18], and a measured value of 0.81 mm^2/s in suspended droplet combustion[21] and 0.93 in free droplet studies[21] at earth gravity conditions. Reduction from values at one-g appear to be similar to those experienced for n-heptane droplets in experiments performed by Kumagai and Okajima[22] (0.97 mm^2/s at one-g, and 0.78 mm^2/s at "zero" g).

The low residual velocity coupled with negligible gravity levels allows the observation of spherically-symmetric burning droplets in the more recent experiments with decane. The square of the computer area-averaged diameter plotted vs. time is shown in Figure 13a. Except for the initial transient heat-up period, the rate change of the squared diameter remains constant as predicted by theory. The regression analysis started at different initial times shows that the fluctuation in the burning rate is less than 1% of the mean value. The measured mean value of 0.67 mm^2/s compares very favorably with the theoretically calculated value of 0.66 mm^2/s.

In addition, it was noted by Shaw *et al.*[20] that the formation and collection of soot between the droplet surface and the flame (Figure 2) became so severe as to lead (through some undetermined mechanism) to shrinkage of the flame diameter relative to that of the droplet, and in some cases, disruption of the liquid droplet at less than 25% of its total estimated burning time. These phenomena have not been reported elsewhere in the literature in any previous combustion studies of pure n-alkanes or other fuels. Although the sooting characteristics is an interesting phenomena which deserves further study, it interferes with the major objective of this research, which is to obtain steady-state burning characteristics of fuel droplets.

5.0 INITIAL EXPERIMENTS WITH METHANOL

Experiments have been initiated on droplet combustion of methanol, a fuel with a very low sooting tendency. Figure 16 shows the behavior of the square of the area-averaged diameter of the burning methanol droplet as a function of time for a large droplet (1100 μm initial diameter). After the initial heat-up period and the decay of the droplet oscillation caused by deployment and ignition, the burning rate first accelerates and then continues to decrease with time until observations could no longer be made. The burning rate of methanol was calculated using the computer method. The linearly fitted burning rate calculated from the regression starting at $t = 0.34$ seconds to $t = 0.56$ seconds is 0.66 mm^2/s. The burning rate calculated from the regression analysis starting at $t = .66$ seconds to $t = 0.83$ seconds is 0.52 mm^2/s. Thus, the burning rate has decreased by nearly 21%, within a time period which is less than half the droplet burning lifetime.

Earlier work[30] has shown that simple alcohol droplet vaporization (combustion was not studied) is affected by the surrounding humidity. Water, one of the major combustion products created in the surrounding diffusion flame may be transported to the liquid droplet surface, condense and dissolve at the liquid-gas phase boundary. This phenomena would significantly modify the gas phase boundary conditions and the energy transport mechanism to the droplet surface both through the release of its latent heat of vaporization upon condensation, and through mass diffusive transport effects in a thin layer within the liquid droplet (prolonging droplet heat-up). The magnitude of these effects should be a function of the droplet burning time (affected either by initial droplet size or increased oxygen concentration). These coupling characteristics will produce an environment which is a more definitive test of droplet combustion models and their transient behavior. As the solubility of water in an alcohol rapidly decreases with increasing carbon number, the later study of higher carbon number alcohols could experimentally identify the magnitude of this effect on the transient burning characteristics of droplets and on droplet burning extinction.

From these initial experiments, it appears that a fuel as volatile as methanol can be studied successfully using the present experimental methods for droplet dispensing, deployment and ignition. Methanol is a particularly interesting fuel in that under microgravity conditions, transient burning conditions are accentuated by combustion product dissolution, enriching the data characteristics to be compared with theoretical models. Finally, the detailed chemical kinetics of methanol oxidation are well understood and considerably simpler when compared with those of most liquid hydrocarbon fuels[31]. Additional experiments and detailed numerical modelling efforts are currently underway.

6.0 CONCLUSION

An economical, precise, and rapid image processing approach has been developed for analyzing droplet burning rate information from photographic records, and a similar approach is under development for determining flame and soot shell diameter. The method uses a digital processing technique for droplet edge enhancement by mean filtering and for graylevel threshold selection between droplet and background by optimization. Analyses of droplet burning photographic data obtained using the droplet combustion test apparatus and the 2.2 sec NASA-Lewis drop tower show the ability to generate very reproducible burning rate data at microgravity levels comparable to those obtained in space environments. Analysis of droplet burning data for n-decane in air show that the burning rate constant reported earlier

Figure 16

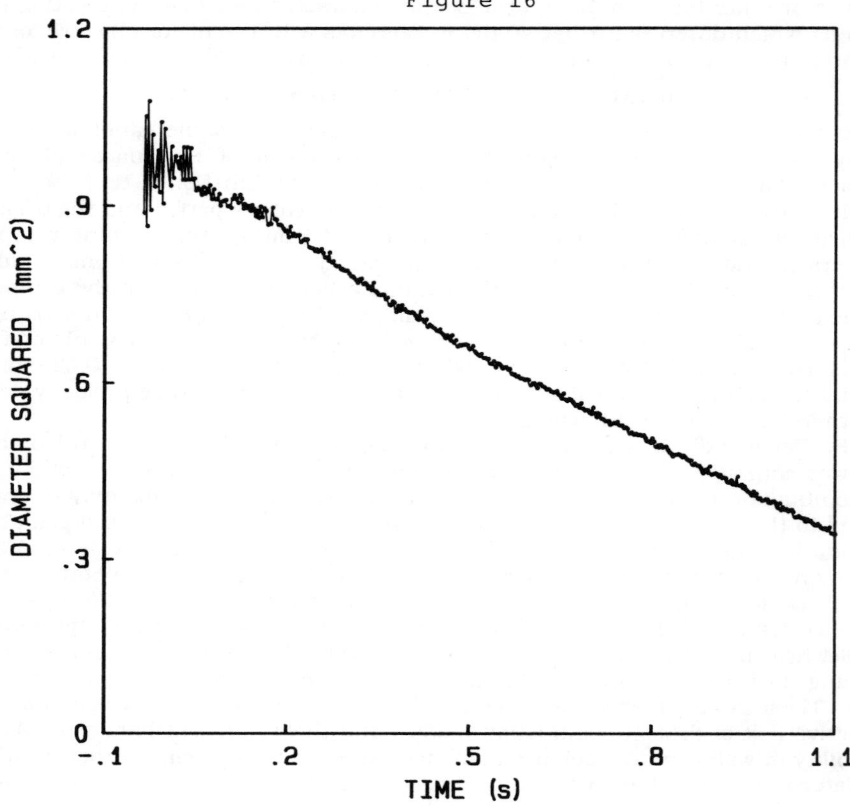

of 0.63 ± 0.09 mm^2/s should be revised to a value closer to 0.67 ± 0.01 mm^2/s, and that the low precision of earlier measurements was primarily due to the use of Vanguard data reduction method employed. While the duration of test time prevents the observation of complete burning history for the larger droplet diameters (800 μm), recent experiments show that complete burning characteristics should be able to be obtained for droplets smaller than 800 μm with low residual velocity. The rate of success of achieving these conditions for each drop tower experiment decreases as the initial droplet size is decreased.

While the combustion of n-decane at higher oxygen indices eliminates the dense soot shell found for burning in air, soot formation continues to be observed. Preliminary experiments using methanol as fuel show that soot formation does not occur with this fuel. However, the burning rate constant does not achieve a quasi-steady value, and, after the initial heat-up transient, decreases with burning time. It is hypothesized that this departure from the "d^2-law" is caused by the condensation and dissolution of water vapor at the liquid droplet surface. Additional experiments on methanol are planned. Initial experiments on methanol droplets burning in air at 10^{-6} g suggest that diffusive transport and subsequent condensation of water on the droplet surface modifies the gasification rate of the liquid fuel droplet. The majority of the water vapor is created at the flame front as the product of combustion and may also be present in the surrounding air. The energy release due to the condensation of water vapor provides an additional source of heat (in addition to the heat diffusing from the flame front) which may initially enhance the burning rate of the methanol droplet. As the burning process progresses, a thin liquid shell containing water forms at the droplet surface inhibiting the rate of methanol vaporization. The net result of the water condensation and the reduction of methanol gasification is a reduced rate of droplet diameter decay with time.

ACKNOWLEDGEMENT

The portions of this research which were conducted at Princeton University was sponsored by NASA through contract # NAS3-24640. M. Y. Choi is a NASA GRADUATE STUDENT RESEARCHER FELLOW.

REFERENCES

1. C. K. Law, 1982, "Mechanisms of Droplet Combustion", **Proc. of the 2nd Int. Colloquium on Drops and Bubbles**, ed. D. H. le Croissette, JPL Pubn. 82-7, Jet Propulsion Laboratory, Pasadena, pp. 39-53.
2. C. K. Law, 1982,"Recent Advances in Droplet Vaporization and Combustion", *Prog. Energy Comb. Sci.* **8**, pp. 171-201.
3. F.A. Williams, 1985, "Diffusion Flames and Droplet Burning", Ch. 3, **Combustion Theory**, 2nd edn., Benjamin/Cummings Publ., Menlo Park, CA, pp. 52- 91.
4. F. A. Williams, 1985, "Ignition and Burning of Single Liquid Droplets", *Acta Astron.* **12**, pp. 547-553.
5. D. J. Maloney & J. F. Spann, 1988, "Evaporation, Agglomeration, and Explosive Boiling Characteristics of Coal-Water Fuels under Intense Heating Conditions", presented at the **22nd Symposium (Int.) on Combustion**, Seattle WA, August. Paper No. 211. Proc. in press.
6. J. D. Lasheras, A. C. Fernandez-Pello, & F. L. Dryer, 1979, "Initial Observations on the Free Droplet Combustion Characteristics of Water-in-Fuel Emulsions", *Comb. Sci. Tech.*, **21**, pp. 1-14.
7. K. Miyasaka & C. K. Law 1981, "Combustion of Strongly Interacting Linear Droplet Arrays", **Proc. of the 18th Symposium (Int.) on Combustion**, The Combustion Inst., Pittsburgh, PA, p 283.
8. A. Williams, 1976, "Fundamentals of Oil Combustion", *Prog. Energy Comb. Sci.*, **2**, pp. 167-179.
9. G. M. Faeth, 1977, "Current Status of Droplet and Liquid Combustion", *Prog. Energy Comb. Sci.*, 3, pp. 191-224.

10. S. Kumagai & H. Isoda, 1957, "Combustion of Fuel Droplets in a Falling Chamber", **6th Symposium (Int.) on Combustion**, Reinhold Publ., NY, pp. 726-731.
11. S. Kumagai, T. Sakai, & S. Okajima, 1971, Combustion of Free Fuel Droplets in a Freely Falling Chamber, **13th Symposium (Int.) on Combustion**, The Combustion Institute, Pittsburgh, PA, pp. 779-785.
12. J. C. Yang, G. S. Jackson, & C. T. Avedisian, 1988, "Some Experiments on Free Droplet Low Gravity Combustion", *This volume, this chapter.*
13. J. C. Yang & C. T. Avedisian, 1988, "The Combustion of Unsupported Heptane/Hexadecane Mixture Droplets at Low Gravity", presented at the **22nd Symposium (Int.) on Combustion**, Seattle WA, August. Paper No. 215. Proc. in press.
14. H. Hara & S. Kumagai, 1988, "A New Apparatus for Free Droplet Combustion under Microgravity", poster presented at the **22nd Symposium (Int.) on Combustion**, Seattle WA, August. Poster No. 257.
15. F. A. Williams & F.L. Dryer, 1988, "A Science Requirement Document for the Microgravity Droplet Combustion Experiment, Revision B", In preparation: interested parties are invited to write the authors.
16. F. A. Williams, 1981, "Droplet Burning", Chapter 3 of **Combustion Experiments in a Zero-Gravity Laboratory**, T. H. Cochran, ed., ***Progress in Astronautics and Aeronautics*, 73**, AIAA, New York, pp. 31-60.
17. J. B. Haggard & J. L. Kropp, 1987, "Development of a Droplet Combustion Experiment for Low Gravity Operations", AIAA Paper No. 87-0576.
18. B. D. Shaw, F. L. Dryer, F. A. Williams, & J. B. Haggard, jr., 1987, "Sooting and Disruption is Spherically Symmetrical Combustion of Decane Droplets in Air", Presented at the Int. Astron. Federation Conf., Brighton, UK, October 10-17, 1987. Paper No. 87-403. To appear in *Acta Astron.*.
19. J. B. Haggard & J. L. Kropp, 1989, "Droplet Combustion Experiment Drop Tower Tests Using Models of Space Flight Apparatus", to be presented at the 1989 Annual Meeting of the AIAA. Interested parties are invited to write the authors.
20. B. D. Shaw, F. L. Dryer, F. A. Williams, & N. Gat, 1988, "Interactions between Gaseous Electrical Discharges and Single Liquid Droplets", *Comb. & Flame*, to appear. Interested parties are invited to write the authors.
21. A. R. Hall & J. D. Diederichsen, 1953, "Experimental Study of the Burning of Single Drops of Fuel in Air at Pressures up to Twenty Atmospheres", **4th Symp. (Int.) on Combustion**. Williams & Wilkins, Co., Baltimore, pp. 837-846.
22. S. Okajima & S. Kumagai, 1975, "Further Investigations of Combustion of Free Droplets in a Freely Falling Chamber Including Moving Droplets", **15th Symp. (Int.) on Combustion**, The Combustion Institute, Pittsburgh, pp. 401-407.
23. Jae S. Lim, 1984 "Image Enhancement", **Digital Image Processing Techniques**, Academic Pr., Orlando, FL, 1-51.
24. Kenneth R. Castleman, 1979 "Digital Image Processing", Prentice-Hall, Englewood Cliffs, NJ.
25. Nobuyuki Otsu, 1979 "A Threshold Selection Method From Graylevel Histograms", *IEEE Trans. on Sys., Man Cybern.*, **SMC-9**, no. 1, 62-66.
26. Gregory A. Baxes, 1984 "Digital Image Processing: a practical primer", Prentice-Hall, Englewood Cliffs, NJ.
27. J.G. Yang & T.S. Huang, 1981 "The Effect of Median Filtering on Edge Location Estimation", *Comp. Graph. Image Process.*, **15**, p 224-245.
28. Jong-Sen Lee, 1986 "Edge Detection by Partitioning", **Statistical Image Processing and Graphics**, Marcel Dekker, N.Y., 59-69.
29. J. Kittler, J. Illingworth, & J. Foglein, 1985 "Threshold Selection Based on Simple Image Statistics", *Comp. Vision, Graph., Image Process.*, **30**, p 125-147.
30. C. K. Law, T. Y. Ziong, & C.H. Wang, 1987 "Alcohol Droplet Vaporization in Humid Air", *Int. J. Heat Mass Trans.*, **30**, No. 7, p 1435-1442.
31. T. S. Norton & F.L. Dryer, 1988, "Some New Observations on Methanol Oxidation Chemistry", *Comb. Sci. Tech.*, to appear. Interested parties are invited to write the authors.
32 Joan S. Weszka, 1978 "A Survey of Threshold Selection Method", *Comp. Graph. Image Process.*, **7**, p 259-265.

FIGURE CAPTIONS

Figure 1a Photograph of Drop Configuration. The z-direction corresponds to the direction of the gravity vector.

Figure 1b Schematic of Apparatus.

Figure 2 Photograph of Sooting Droplet of n-Decane Burning in Air at 10^{-6} of Earth's Gravity.

Figure 3a Histogram of Ideal Object and Background.

Figure 3b Histogram of a Burning Droplet of n-decane (in Air at Atmospheric Pressure) Recorded on Kodak Tri-X 16mm Film (Backlighted).

Figure 4a Line Profile of Ideal Object and Background used to Generate Figure 3a.

Figure 4b Line Profile of a Burning Droplet of n-decane (in Air at Atmospheric Pressure) Recorded on Kodak Tri-X 16mm Film (Backlighted). The Experimental Data are the Same as Used in Figure 3b.

Figure 5 Spatial Frequency of a Burning Droplet and Background Before Filter Applications.

Figure 6 Histogram of the Graylevel Distribution for a Burning Droplet Real Image Before and After High Pass and Laplace Filter Processing.

Figure 7 Line Profile of Burning Droplet and Background After Mean Filter Applications.

Figure 8 Histogram of Droplet and Background After Mean Filter Applications.

Figure 9 Histogram Distribution with Statistics.

Figure 10 Line Profile of Ideal Object Before and After Mean Filtering.

Figure 11 Measured Diameter as a Function of the Choice of Graylevel Distribution used in the Analysis.

Figure 12 Line Profile and Spatial Frequency After Mean Filter Applications.

Figure 13a Computer Analysis of Photographic Data from a Decane Droplet Burning in Air at Atmospheric Pressure. Initial droplet diameter was 1100 μm.

Figure 13b Vanguard Analysis of the same photographic data analyzed in Figure 3a.

Figure 14 Comparison of Two Separate Analyses of the Same 16 mm Film Record of a Burning Droplet of n-Decane in Air at Atmospheric Pressure.

Figure 15 Position History of a Burning Droplet Relative to the Location of the Deployment.

Figure 16 Computer Analysis of a Methanol Droplet Burning in Air at 1 atm. Initial Droplet Diameter, 1100 μm.

EXPERIMENTS ON COMBUSTION IN REDUCED GRAVITY AND TURBULENT VAPORIZATION OF n-HEPTANE DROPLETS

I. Gökalp, C. Chauveau and G. Monsallier

Centre National de la Recherche Scientifique
Centre de Recherches sur la Chimie de la Combustion et des Hautes Temperatures
45071 Orléans Cédex 2, France

ABSTRACT

A Droplet Burning Facility which allows the investigation of droplet vaporization and burning under various dynamic and thermal conditions has been constructed. The system has been operated during the parabolic flights of the NASA KC135 aircraft. An important set of data has been collected on the low temperature turbulent vaporization and envelope burning of suspended n-heptane droplets at ground and reduced gravity conditions. From digitized images obtained by a rapid video camera, the time evolution of the droplet and the flame dimensions are determined with great accuracy. The information is used to deduce the vaporization rate constant, the flame diameter and the flame standoff ratio under various conditions.

1. INTRODUCTION

In studies on the vaporization and combustion of fuel droplets and their application to the modeling of spray combustors, the classical d^2-law model has been widely used. One of the basic assumptions of this quasi-steady and diffusive-convective model is the approximation of spherical symmetry. However, even in a stagnant environment, the hot flame in a gravitational field causes gases to rise buoyantly, establishing an axisymmetric flow under natural convection, so that it is no longer meaningful to identify a flame diameter[1]. During droplet combustion with forced convection, depending on the magnitude of the droplet Reynolds number, the system may also easily depart from spherical symmetry. In fact, these two sources of departure from spherical symmetry, natural and forced convection, are almost always simultaneously present in practical systems. This observation has recently led to the introduction of a unified mixed convection parameter in the analysis of the governing equations of the system[2].

Several attempts have been made to reduce the effects of asymmetrical unsteady buoyancy. One method is to investigate small droplets projects into a hot gas stream[3], and another is to investigate burning droplets in reduced pressure[4]. In this second method, suspended droplet have been employed and the entire droplet history has been photographed. Another way to decrease buoyancy is to reduce or eliminate gravity. Elimination of gravitational effects on droplet combustion has been pioneered by Kumagai *et al.*[5] They studied the phenomenon occurring in a free-falling chamber. In their experiments, zero gravity was achieved for less than 1 second, so that burning to completion in spherical symmetry never occurred. More recent attempts to study droplet combustion in a gravity-free environment also used a drop frame[6,7]. Purely stationary free droplets were difficult to observe, but the data obtained from a high-speed cine-camera gave interesting results.

In spite of the generally turbulent nature of reacting two-phase systems (i.e. spray combustion), very little is known about the influence of turbulence on the vaporizing and burning characteristics of fuel droplets[2,8]. In order to contribute to this area, we have initiated the investigation of the influence of a turbulent convective flow on the vaporization and burning characteristics of single fuel droplets. The issues which are addressed are the effects of varying the turbulence structure in terms of its intensity and length and time scales on the mass and heat transfer characteristics of fuel droplets and the transition velocity between the envelope and wake flame regimes.

In the work described below, the parabolic flights of an aircraft are used to create a reduced gravity environment of the order of 10^{-3} g. As will appear from the results presented below, with regards to droplet vaporization and combustion, the most important feature of the parabolic flight technique is an operational low-gravity time longer than 15 seconds. In this paper we first briefly describe the experimental set-up and then present and discuss some of the results concerning the envelope flame burning of n-heptane droplets under reduced gravity. We also present some preliminary results from ground experiments on low temperature vaporization of n-heptane droplets to address the study of the influence of a turbulent flow on the droplet vaporization.

2. DESCRIPTION OF THE EXPERIMENTAL SET-UP

The experiments are performed in a facility specially designed for this purpose: the Droplet Burning Facility (DBF) which is fully described in Reference 9. It is briefly presented here for completeness. The DBF consists of a closed loop flow channel (Figure 1) composed of an electrical axial fan, an electrical heater, a set of laminarization ducts and a test section of square cross-sectional area (side: 15 cm; length: 50 cm). A by-pass channel and a return channel of circular cross-section (diameter: 20 cm) close the loop to the electrical fan. Turbulence grids, used for turbulent convective flow runs, can be positioned at the inlet of the test section. Two circular glass windows (diameter: 10 cm; thickness: 8 mm) are mounted on the parallel sides of the test section to allow optical access. The test section also supports the injection and the ignition systems.

The air flow can be generated by the axial fan at a maximum velocity of approximately 5 m/s in the test section, and the flow can be heated by the electrical heater up to 600°K. The flow rate is adjusted in order to obtain a nominal pressure of 1 atm in the test section. The characteristics of the forced flows were measured previously during ground experiments with laser Doppler anemometry, and the flow monitoring system was calibrated against the mean longitudinal velocity at the location of the droplet.

A droplet is formed by using a syringe located at the top of the test section. The delivery of 0.5 µl of n-heptane allows the formation of a droplet of approximately 1 mm diameter. A quartz fiber (diameter: 0.2 mm) is tilted toward the syringe tip allowing the transfer of the droplet to the fiber. The ignition of the droplet is provided by two tungsten electrodes positioned under the droplet. This sequence of operations is controlled by a micro-computer.

The recording equipment is composed of two systems: A high speed video camera, Kodak-Ektapro 1000, which can record up to 1000 full frames per second, and a normal video camera, Sony 8 mm PAL. For more information about the experimental procedure, Reference 10 may be consulted.

3. EXPERIMENTAL RESULTS AND DISCUSSION

3.1. Envelope Flame Burning of n-Heptane Droplets at Reduced Gravity

The burning of n-heptane droplets has been observed during parabolic flights with the KC 135 aircraft at NASA Johnson Space Center, Ellington Air Force Base. The time evolution of the burning droplets suspended on the silica filament and of the surrounding flame is recorded **in silhouette**, with the high speed video camera.

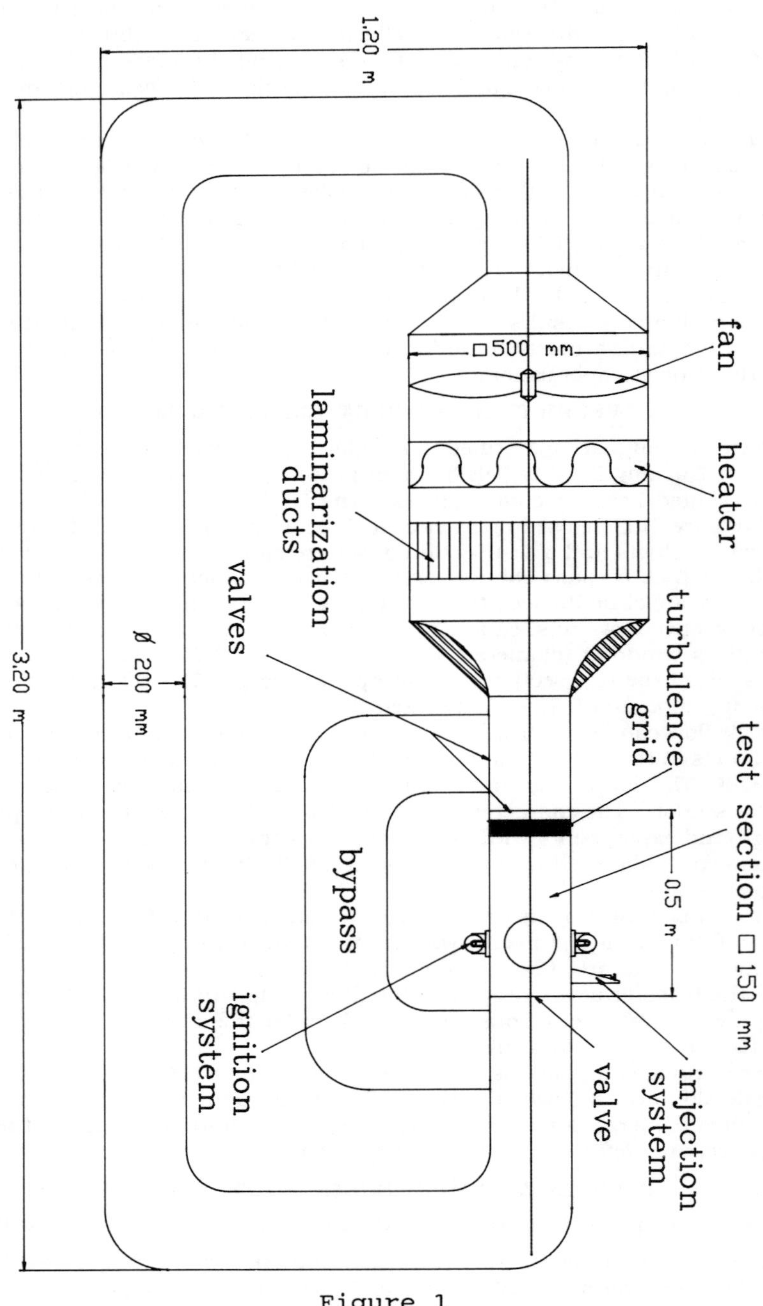
1.20 m
3.20 m
□500 mm
fan
heater
laminarization
ducts
turbulence
grid
valves
ø 200 mm
test section □150 mm
0.5 m
bypass
ignition
system
injection
system
valve

Figure 1

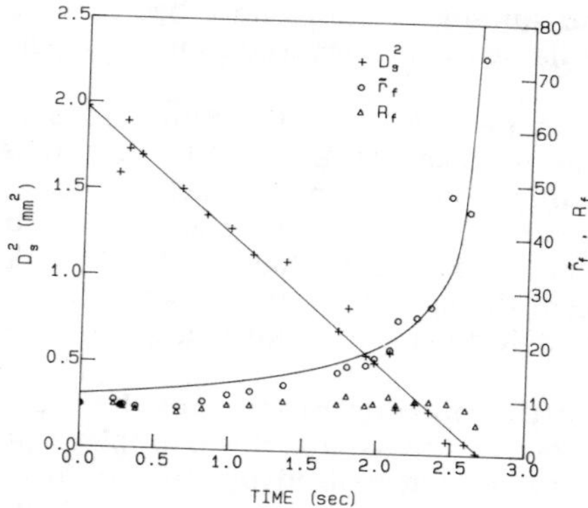

Figure 2

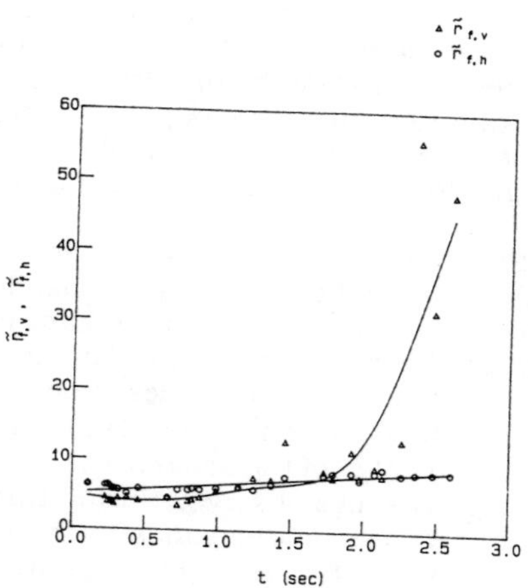

Figure 3

The digitized video film records of the burning droplets are measured with a reticule on the video monitor, and suitable readings have been taken to enable the dimensions of the drop and the flame to be expressed in terms of the diameter of a sphere of equivalent surface area.

For all the runs, color pictures from the Sony video camera show a yellow flame zone, indicating intense soot formation. The brightness of this yellow zone does not completely obscure the droplet and for each run it is possible to deduce the time evolution of both the liquid droplet and the flame. The burning sequences from the rapid video camera also indicate that the droplet is less spherical than the shape of the flame. Indeed, the sphericity degree of the flame, in terms of the ratio between its horizontal and vertical dimensions, lies around 0.8 during the entire droplet lifetime.

In the discussion of the burning droplet results we will present the time evolutions of three parameters: the square of the droplet equivalent diameter, D_s^2, the flame equivalent diameter normalized by the initial droplet diameter, $R_f = D_f/D_o$, and the flame standoff ratio, $\tilde{r}_f = r_f/r_s$. The initial droplet diameter is 1.41 mm. A typical droplet burning run is presented on Figure 2. This figure shows that the surface regression rate of the droplet follows closely the d^2-law until its complete consumption, which lasts 2.7 s. A least-square fit gives an average slope of 0.73 mm^2/s for the surface regression rate. This value of the burning rate constant agrees quite well with others obtained from the burning of n-heptane at reduced gravity, such as the value of 0.68 mm^2/s for an initial diameter[7] of 0.5 mm or the value of 0.78 mm^2/s for an initial diameter[5] of 1 mm. It is also worth noting that, during these runs, while the total lifetime of the droplet at reduced gravity (burnout time) is recorded, extinction due to finite-rate gas-phase combustion kinetics was not seen. These observations have been made for smaller initial diameter unsupported droplets at reduced gravity[7], and also for suspended droplets at reduced pressure[11].

The time evolution of the normalized flame diameter is shown in the same figure. This ratio (R_f) increases gradually up to approximately 10 during the first second of the droplet combustion and then remains constant for over 90-percent of the droplet lifetime. The range of variation of this parameter corresponds to that observed by Kumagai *et al.*[5], for an observation time less than 1 sec.

The data concerning the time evolution of the flame front standoff ratio, $\tilde{r}_f$, display a constant behaviour up to 0.7 s. of the data collection time. This behaviour, indicative of stationary burning, was also observed during Kumagai's experiments. The order of magnitude of the ratio also corresponds to the values determined by Kumagai and co-workers. However, after this quasi-stationary behaviour, an increase of the $\tilde{r}_f$ parameter is observed. The global shape of the $\tilde{r}_f$ evolution may be approached by a parabolic form. In fact, one can note that $\tilde{r}_f = D_o \times R_f/D_s$, where D_o = 1.4 mm, $D_s = (1.91 - 0.73\ t)^{0.5}$ from the experimental regression rate, and $R_f \simeq 10$ for at least t = 1 s to 2.5 s. The curve corresponding to the evolution of $\tilde{r}_f$ calculated from the above approximation is also plotted on Figure 2. During the last part of the droplet lifetime $\tilde{r}_f$ reaches values above 40. It is noteworthy that the global shape of the $\tilde{r}_f$ behaviour is remarkably close to the predicted behaviour when the effects of the fuel vapor accumulation is included in the formulation[4].

As clearly shown on burning sequences obtained by the high speed video camera, during the second half of the droplet burning time, the silica filament strongly influences the accuracy of the evaluation of the equivalent droplet diameter. On Figure 2, $\tilde{r}_f$ values are calculated from flame dimensions based on the equivalent sphere surface area method. We have also evaluated $\tilde{r}_f$ from two other definitions: $\tilde{r}_{f,h}$ which is the ratio between the horizontal dimensions of the flame and the droplet, and $\tilde{r}_{f,v}$ which is the ratio between their respective vertical dimensions. Figure 3. shows that the time evolution of $\tilde{r}_{f,v}$ is similar to that of $\tilde{r}_f$, but $\tilde{r}_{f,h}$ is nearly constant and increases linearly with time. This comparison emphasizes the influence of the suspending filament on the shape of the droplet and flame and on the evaluation of some of the parameters in droplet burning.

The global tendencies of these three parameters as observed during several burning runs, where the same droplet initial diameter is used, have been presented in Reference 9. The reported data show very similar patterns to those of the individual case presented here, except that the mean value of the vaporization rate K is found to be equal to 0.69 mm^2/s.

3.2. Low Temperature Vaporization of n-Heptane Droplets in a Grid Induced Turbulence

To investigate the vaporization characteristics of fuel droplets in a turbulent medium, a turbulent flow is generated by grids in the DBF, and the intensity and scales of turbulence are varied independently. All the relevant characteristics of the dynamic field are measured with Laser Doppler Anemometry. The influence of the turbulence intensity and time and length scales on the low temperature vaporization rate of n-heptane droplets (d_o = 1.5 mm) is investigated here. Two grids of square mesh, 10 mm and 5 mm are used. The kinetic energy of the grid induced turbulence, shown in Figure 4, decays rapidly with axial distance for Reynolds numbers based on mesh size Re_m>1700. For this range of the Reynolds number, an almost isotropic and homogeneous turbulent velocity field is obtained ($u'/v' \approx 1$).

Figure 5 presents the time evolution of the square of the droplet diameter (based on the equivalent surface area method) normalized by the initial droplet diameter, for four different Reynolds numbers based on droplet diameter and axial mean velocity at ambient temperature. From these linear evolutions the vaporization rate constant, K, can be deduced by a least square fit. The dependency of K/K_o, the vaporization rate constant normalized by its value under stagnant conditions, on the square root of the Reynolds number, is shown on Figure 6. In this relation the value of K_o is equal to 0.014 mm^2/s. A Frossling type linear relation, with a slope of 0.220, is observed. This slope increases to 0.248 if the one-third power of the Schmidt number is included in the correlation. This rate compares well with the existing results[12]. However, the variations between different values of this rate may be related to the differences in the turbulence structure in each experiment.

Figure 7 presents the evolution of the K/K_o ratio with increasing distance from the turbulence grid, i.e. with decreasing turbulence intensity, and for four values of the Reynolds number based on the grid mesh (M = 10 mm). These results are obtained, by varying the distance between the grid and the droplet, for each mean flow velocity (or each Re_m). As the mean velocity downstream of the grid remains constant for the x/M values explored here, this experimental procedure allows the isolation of the influence of the turbulence intensity on the mass transfer rate. In spite of the threefold variation of the turbulent kinetic energy between the extreme values of the x/M (see Figure 4), the normalized mass transfer rate constant does not vary with x/M, for any of the Re_m values explored here. The same observation is made for the evolution of K/K_o in smaller scale turbulence produced with the grid of 5 mm mesh size. In a mixed heat and mass transfer process, when the turbulent flow produced by the small scale grid is heated up to 338°K, the only difference from the isothermal turbulent mass transfer case is the global increase of the K/K_o ratio, where K_o increases also to 0.037 mm^2/s, from 0.014 mm^2/s.

The insensitivity observed here of the mass transfer rate to the turbulence intensity is intriguing, for several experimental studies have reported significant

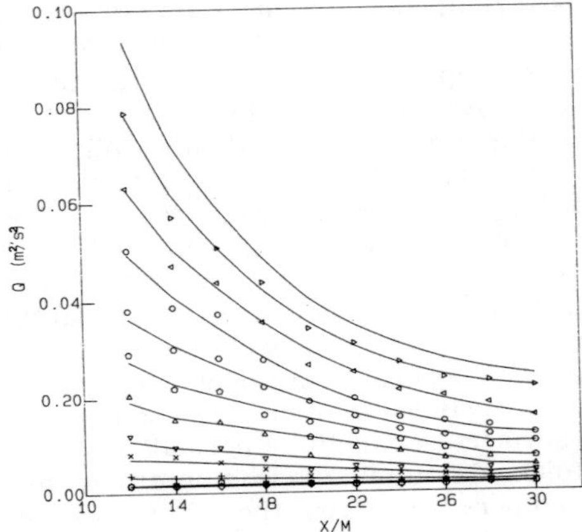

Figure 4

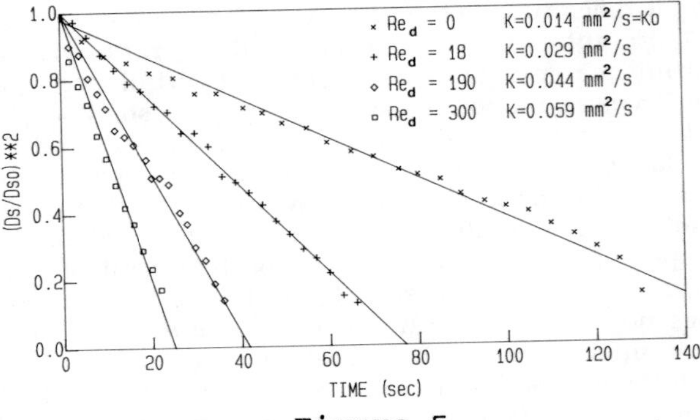

Figure 5

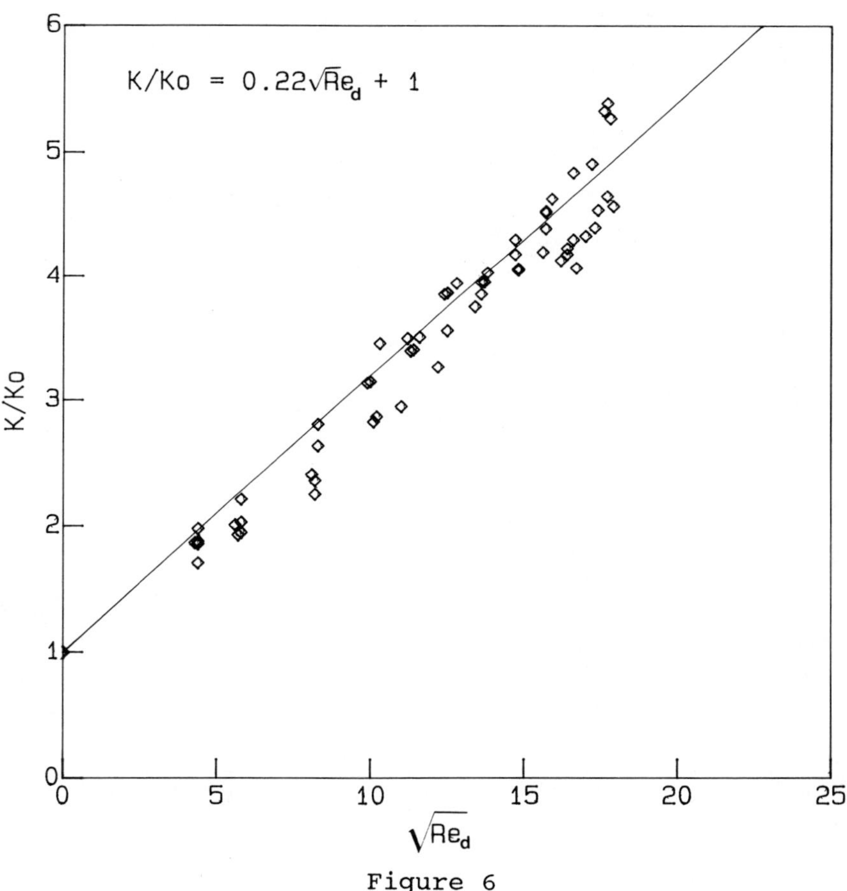

Figure 6

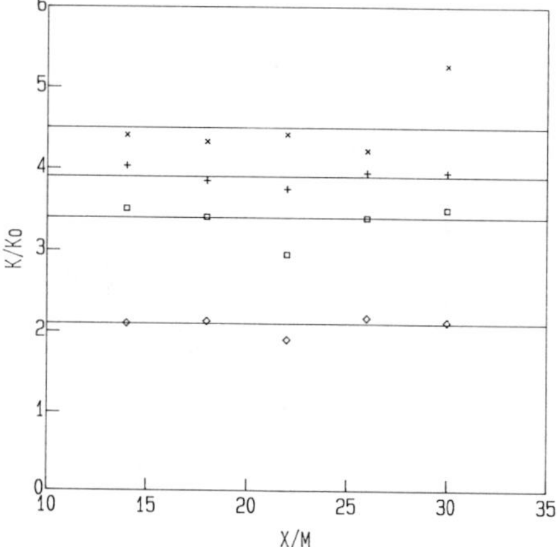

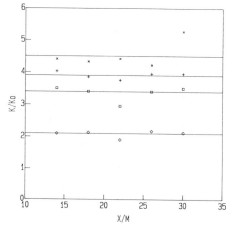

Figure 7

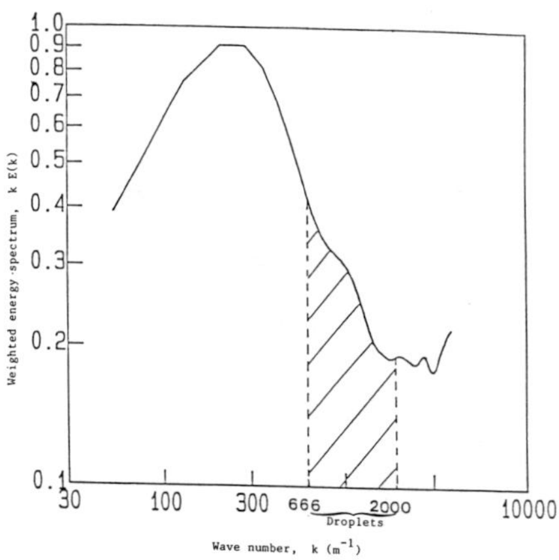

Figure 8

variations in the heat and mass transfer rates from or to drops or solid particles, when the turbulent intensity of the flow is changed[12]. Some studies have also shown a dependence on the integral length scale of the turbulence field[14,15].

In order to understand this observation, a spectral analysis of the turbulence energy distribution has been conducted. The one-dimensional wave-number spectrum weighted by the wave-number is shown on Figure 8 for the large scale grid (M= 10 mm). The wave-numbers corresponding to the droplet diameters investigated here are also indicated. It is clear that the interaction between the droplet and the turbulence is only confined to smallest scales, well below the most energetic scales, located around k = 200 m^{-1}. In fact, the spectral measurements at different x/M values have shown that the length scale of the most energetic eddies, l_e, varies between 4 and 6 mm for the M=10 mm grid and between 2 and 5 mm for the M=5 mm grid[13]. Consequently, the initial droplet sizes investigated here do not allow the interaction with the most energetic turbulent eddies, but only with the most dissipative ones. The weak energy content of the turbulent eddies which actually interact with the droplet may then explain the insensitivity of the mass transfer rate to the variations in the turbulence intensity.

4. CONCLUDING REMARKS

Concerning the reduced gravity experiments with the use of parabolic flights, the results presented here show both their utility and their limits. The (approximately) 15 seconds of operational reduced gravity allowed, for the first time, the simultaneous observation of the droplet and flame diameters for large fuel droplets, which, more easily than with small droplets, permit the differences between the buoyant and non-buoyant droplet combustion to be observed. The presence of an operator during the effective reduced gravity period is also an important factor in the conduction of the experiments. The limits of the parabolic flights are related to the relatively high level of the residual gravity. But this limit may be relaxed by allowing the free floating of the whole experimental set-up during the reduced gravity period of the parabolic flight.

Concerning the investigation of the influence of turbulence on the droplet vaporization, the preliminary results presented here suggest that the mass and heat transfer rates of liquid drops may depend strongly on the ratio between the initial droplet diameter and the characteristic scales of the most energetic eddies.

ACKNOWLEDGMENTS

This work is supported by the Microgravity Office of the European Space Agency and by the Centre National d'Etudes Spatiales (C.N.E.S.). We wish to acknowledge the efficient monitoring of these supports by H. WALTER (E.S.A.) and R. BONNEVILLE (C.N.E.S.). Many thanks also to the ESTEC/ESA staff at Noordwijk, Netherlands (A. GONFALONE, V. PLETSER and F. BAUD) for their continuous support and assistance. It is also a pleasure to acknowledge the hospitality of the NASA JSC. One of the authors (C. C.) is supported by a joint CNRS-CNES grant.

REFERENCES

1 C. K. Law (1982) *Prog. Energy Combust. Sci.* **8**:171.

2 A. C. Fernandez-Pello (1986) "Convective Droplet Combustion", Invited Paper, 1986 Fall Technical Meeting, Eastern Sections of the Combustion Institute, San Juan, Puerto Rico, December.

3 J. J. Sangiovanni & A. S. Kesten (1977) *Combust. Sci. Tech.* **16**:59.

4 C. K. Law, S. H. Chung & N. Srinivasan (1980) *Combust. Flame*, **38**: 173.

5 S. Kumagai, T. Sakai & S. Okajima (1971) **Proc. of the 13th Symp. (Int.) on Combustion**, The Combustion Institute, Pittsburgh, p.779.

6 F. A. Williams (1981) *Prog. Astron. Aeron.* **73**:31.

7 J. C. Yang, C. T. Avedisian & C. H. Wang (1987) "An Experimental Method for Studying Combustion of an Unsupported Fuel Droplet at Reduced Gravity", paper presented at the 1987 Fall Eastern States Section Meeting of the Combustion Institute, November.

8 W. A. Sirignano (1983) *Prog. Energy Combust. Sci.* **9**:291.
9 I. Gökalp, C. Chauveau, J. R. Richard, M. Kramer & W. Leuckel (1988) "Observations on the low temperature vaporization and envelope or wake burning of n-heptane droplets at reduced gravity during parabolic flights", paper presented at the 22nd Symp. (Int.) on Combustion, Seattle, Washington, August 14-19.
10 I. Gökalp, C. Chauveau, J. R. Richard, M. Kramer & W. Leuckel (1987) "Droplet vaporization and combustion in microgravity", ESA/CNES 1st Progress Rept., September, 24p.
11 Chung S.H. and Law C.K. (1986) Combust. Flame 64:237.
12 R. Clift, J. R. Grace & M. E. Weber (1978) **Bubbles, Drops, and Particles**, NY: Academic Press.
13 I. Gökalp, C. Chauveau, & G. Monsallier (1988) "Some experimental observations on the influence of turbulence on the vaporization rate of fuel droplets", paper presented at EUROMECH 234 on Turbulent Two Phase Systems, Toulouse, France, May 9-11.
14 G. D. Raithby & E.R.G. Eckert (1968) *Int. J. Heat Mass Transfer* **11**: 1233.
15 B. G. Van der Hegge Zijnen (1958) *Appl. Sci. Res. A* **7**: 205.

FIGURE CAPTIONS

1. Schematics of the Droplet Burning Facility
2. Time evolution of the squared droplet diameter (D_s^2), the normalized flame radius (R_f) and the flame standoff ratio ($\tilde{r}_f$) at reduced gravity for a burning n-heptane droplet.
3. Time evolution of the flame standoff ratio defined from horizontal ($\tilde{r}_{f,h}$) and vertical ($\tilde{r}_{f,v}$) dimensions for a n-heptane droplet burning at reduced gravity.
4. Centerline decay of turbulent kinetic energy, $Q = 1/2(u'^2 + 2v'^2)$, for various mesh Reynolds numbers Re_m. M=10 mm, M/d = 3.3.
5. Time evolution of the normalized squared droplet diameter of a vaporizing n-heptane droplet in a room temperature turbulent flow, for different droplet Reynolds numbers, Re_d.
6. Reynolds number dependency of the normalized mass transfer rate constant, K/K_o.
7. Behaviour of the normalized mass transfer rate constant with down-stream distance x/M, for various mesh Reynolds numbers. M= 10 mm.
8. Comparison of the droplet diameter range with the energetic scales in the one-dimensional wave-number energy spectrum. M =10 mm, Re_m= 3500.

MICROGRAVITY VAPORIZATION OF LIQUID DROPLETS UNDER SUPERCRITICAL CONDITIONS

E.W. Curtis, J.P. Hartfield and P.V. Farrell
Department of Mechanical Engineering
University of Wisconsin-Madison, Madison, WI 53706

ABSTRACT

A computer model is proposed for single droplet vaporization at high pressure and temperature. The model is one-dimensional in space, with no gas or liquid motion. The model is used to calculate the transient heat-up and subsequent vaporization of the droplet. The numerical results indicate a rapid heat up period during which the droplet temperature surpasses its thermodynamic critical point under some relatively high pressure and temperature conditions. The current model does rely on some provisional methods for describing near-critical point transport properties. Thus, the model should be improved by experimental data. An experimental procedure is described to verify the model results. Microgravity drop tower studies using a compression device with a droplet placed at a position of compressive symmetry will be pursued. The photographic results obtained from experiment should provide vaporization rate data throughout the droplet lifetime.

INTRODUCTION

Research on liquid fuel injection dynamics occurring in combustion engines is important to the progress of the combustion engineering field. The efficiency and emissions of combustion engines are directly related to the fuel injection event, particularly spray dynamics. Several fundamental factors that contribute to the liquid jet and droplet break-up and their subsequent vaporization remain unspecified. These factors include liquid jet break-up effects, droplet dynamics, near-critical effects and turbulent effects.

In the particular case of the compression ignition internal combustion engine, a relatively cold liquid fuel is sprayed into a gaseous environment which may be at a pressure greater than the critical pressure of the fuel and at a temperature nearly that or above the fuel critical temperature. During this process, the spray jet has been observed to break up into individual, vaporizing droplets. During their lifetime, these droplets may undergo a transient heat-up such that the liquid temperature approaches the fuel critical temperature.

Previous experimental and numerical studies in this field have dealt primarily with low pressure, low temperature environments[1,2]. These investigations have not taken into account the variation in transport properties, loss of surface tension, or general transient behavior of high pressure, high temperature vaporization and combustion. The current project is directed at studying the fundamental processes of droplet vaporization at near-critical conditions. Major efforts include development of a numerical model to simulate droplet vaporization and experimental investigation of actual near-critical vaporization.

The problem to be numerically modeled is the vaporization of a single, spherical liquid droplet in an infinite gaseous field at near or supercritical conditions. The objective of the model is to accurately describe the transient, variable property, near-critical vaporization process based on fundamental conservation principles and available transport property information. This model may be used to predict liquid fuel vaporization for the variety of fuels and environments encountered in combustion engines.

It is desirable to compare the predictions of the numerical model with experimental data, particularly in the region of the critical point. However, near-critical vaporization and property data are very limited. In the current project, the size and vaporization history of a single droplet evaporating in a hot, high pressure gas field is to be investigated experimentally. It is hoped that results will allow validation of the computer model and elicit a deeper understanding of one of the mechanisms important in spray injection.

COMPUTER MODEL

The problem modeled is that of a single, spherical, liquid droplet heating up and vaporizing in a hot, quiescent, high-pressure gas atmosphere in zero gravity. The zero gravity atmosphere prevents any convection in the gas or droplet due to buoyancy forces. Eliminating both natural and forced convection leaves diffusion as the dominant form of mass transport.

The initially cold (300°K) droplet heats up and vaporizes when placed in the hot, high pressure gas. To simplify the evaluation of the transport properties, the droplet is a single component liquid vaporizing into a single component gas field. Due to the lack of convection and the spherical geometry of the droplet, one-dimensional behavior is assumed. The gas environment is assumed to be sufficiently large that the temperature and composition far from the droplet are not affected by its vaporization; therefore the temperature and pressure far from the droplet are always known. At supercritical pressures a fluid is above its vapor dome and a distinct phase change is not observed when it is heated beyond its critical temperature. Due to the lack of a distinct phase change, we define any "liquid" that has diffused into the "gas field" as vapor.

The model simultaneously solves the equations for conservation of mass, energy and species along with all temperature and composition dependent properties such as thermal conductivity, enthalpy, density, and mass diffusivity. The equation for conservation of mixture mass for a spherical droplet is

$$\frac{\partial \rho_m}{\partial t} + \frac{1}{r^2}\frac{\partial (r^2 \rho_m V_m)}{\partial r} = 0 \qquad (1)$$

where ρ_m is the mixture density, V_m is the radial diffusion velocity and r is the radial distance from the droplet center. The conservation of energy equation is

$$\frac{\partial(\rho_m h_m)}{\partial t} + \frac{1}{r^2}\frac{\partial(r^2 \rho_m V_m h_m)}{\partial r} = \frac{1}{r^2}\frac{\partial\left(r^2 k \frac{\partial T}{\partial r}\right)}{\partial r} - \frac{\partial\left(\rho_m D_{AB}\frac{\partial(w_A h_m)}{\partial r}\right)}{\partial r} - \frac{\partial\left(\rho_m D_{BA}\frac{\partial(w_B h_m)}{\partial r}\right)}{\partial r} - \frac{\partial P}{\partial t} \qquad (2)$$

where h_m is the mixture enthalpy, k is the thermal conductivity, w_A and w_B are the mass fractions of components A and B, P is pressure and T is temperature. The mass diffusivity of component A into component B is given by D_{AB}. D_{BA} is the mass diffusivity of component B into component A; (D_{AB} is assumed equal to D_{BA}). The equation for conservation of species A is

$$\rho_m \frac{\partial w_A}{\partial t} + \frac{1}{r^2}\frac{\partial(r^2 \rho_m V_m w_A)}{\partial r} = \frac{1}{r^2}\frac{\partial\left(r^2 \rho_m D_{AB}\frac{\partial w_A}{\partial r}\right)}{\partial r} \qquad (3).$$

The equations for the transport properties of the mixture are:,

$$k_m = w_A k_A + w_B k_B \; ; \; k_A, k_B = f\,(T)$$

$$h_m = \int Cp_m \, dT$$
$$\rho_m = w_a \rho_a = w_b \rho_b$$
$$\rho_m D_{AB} = \rho_m D_0 \frac{\mu}{\mu_0}; \ \mu = f(T)$$

where μ is the dynamic viscosity, C_p is the specific heat, and the subscript 0 refers to the property value at the reference temperature (T_o = 300°K).

The initial conditions for the problem are:

Liquid:

$$h(r,0) = h_l, \qquad w_A(r,0) = 1, \qquad \rho(r,0) = \rho_A;$$

Gas:

$$h(r,0) = h_g, \qquad w_A(r,0) = 0, \qquad \rho(r,0) = \rho_B.$$

The boundary conditions for the problem are:

Liquid:

$$\frac{\partial h_l}{\partial r}(0,t) = 0, \ h_l(R,t) = h_g(R,t)$$
$$\frac{\partial w_A}{\partial r}(0,t) = 0, \ w_A(R,t) = w_{AEl}$$
$$\frac{\partial \rho_l}{\partial r}(0,t) = 0, \ \rho(R,t) = \rho_{ml}$$

Gas:

$$h(\infty,t) = h_\infty$$
$$w_A(\infty,t) = 0, \qquad w_A(R,t) = w_{AEg}$$
$$\rho(\infty,t) = \rho B, \qquad \rho(R,t) = \rho_{mg}.$$

W_{AEi} is the equilibrium mass fraction of species A in phase i at the droplet-atmosphere interface. The subscripts, l and g, refer to the liquid and gas phases, and the subscript m refers to the mixture. R is the instantaneous droplet radius. The final boundary condition is provided by an energy balance at the droplet interface.

$$-r^2 k_l \left[\frac{\partial T}{\partial r}\right]_l + r^2 k_g \left[\frac{\partial T}{\partial r}\right]_g - \left[\frac{\partial m}{\partial t}\right]_i L = 0 \tag{4}$$

The $\left[\frac{\partial m}{\partial t}\right]_i L$ term in Equation (4) refers to the energy transferred through the interface due to vaporization. The latent heat of vaporization, L, of the liquid mass m is a function of pressure and goes to zero above the critical pressure of the fluid.

The droplet-atmosphere interface is assumed to be in thermodynamic equilibrium at all times. For thermodynamic equilibrium[3] $dG_{T,P} = 0$, where $G_{T,P}$ is the Gibbs function. For a multicomponent system, the above equation is equivalent to

$$x_A \hat{f}_A^{\,l} = y_A \hat{f}_A^{\,g}$$
$$x_B \hat{f}_B^{\,l} = y_B \hat{f}_B^{\,g} \tag{5}$$

where $\hat{f}_j^{\,i}$ is the fugacity of component j in the mixture in phase i. The mole fractions of component h in the liquid and gas phases are given by x_h and y_h. The vapor phase fugacity for component i in a mixture can be found from the thermodynamic relationship[4,5,6],

$$\ln \frac{\hat{f}_i}{y_i P} = \int_0^v \left[\frac{1}{v} - \frac{1}{\overline{R}T}\left(\frac{dP}{dn_i}\right)_{T,P,n_j}\right] dv - \ln Z \tag{6}$$

with a suitable equation of state (e-o-s) to describe the pressure-volume-temperature (P-V-T) behavior of the vapor. In Equation (6), v is the specific volume of the mixture, P is the pressure, T is the temperature, Z is the compressibility factor, $\overline{R}$ is the universal gas constant, and n_i refers to the number of moles of component i. In Equation (6) for a pure gas, the fugacity equation is simplified to

$$\ln \frac{f}{P} = \int_0^P \left(\frac{v}{\overline{R}T} - \frac{1}{P}\right) dP. \tag{7}$$

At low pressures the ideal gas equation adequately describes the P-V-T behavior of most gases. At high pressures, especially near the critical point of the vapor, the ideal gas relations are no longer satisfactory.[7,8] The Peng-Robinson equation is a modification of the Redlich-Kwong two-constant e-o-s. The ideal gas equation is given by $Pv = \overline{R}\,T$. The Redlich-Kwong equation of state is given by $P = \frac{\overline{R}\,T}{v-b} - \frac{a\,T^{-0.5}}{v\,(v+b)}$ where a and b are constants dependent on the critical temperature and critical pressure of the gas. The Peng-Robinson e-o-s is designed to give more accurate vapor-liquid equilibrium calculations than Redlich-Kwong under a wide range of conditions, but especially under high pressure conditions.[4] The Peng-Robinson e-o-s is given by

$$P = \frac{\overline{R}\,T}{v-b} - \frac{a\,(T)}{v\,(v+b) + b\,(v-b)}$$

where a is now a function of temperature as well as the critical temperature and critical pressure of the gas.

An ideal solution is often assumed to simplify the calculation of the fugacities of the gas constituents. For an ideal solution, the fugacity of each component in the mixture is assumed equal to the fugacity of the pure component if it were at the same temperature and pressure as the mixture:[9]

$$\hat{f}_j^{\,g} = y_j f_j^{\,g}$$

For the current set of calculations, the Peng-Robinson e-o-s is used to model the droplet-atmosphere interface and the Redlich-Kwong e-o-s with the ideal solution assumption is used for comparison. The two interface equations behave almost identically until pressures become very high and the saturated vapor approaches its critical point. The liquid phase fugacities are calculated under the assumption that very little gas will be dissolved into the liquid. Henry's law for dilute non-ideal solutions is used to calculate the fugacity of the solute using the equation $\hat{f}_B^{\,l} = x_B H_B$; $H_B = f\,(T)$, where x_B is the mole fraction of B in the liquid, and H_B is the Henry's constant for the binary mixture.[9] The fugacity of the solvent in the liquid mixture is calculated using Raoult's law, and the equation $\hat{f}_A^{\,l} = x_A P_A^{sat}$; $P_A^{sat} = f(T)$, where P_A^{sat} is the saturation pressure of component A at temperature T.[3]

The model is solved by applying an implicit finite difference scheme to a one-dimensional, variable spaced, radial grid in both the liquid and gas systems. The grid points are concentrated near the droplet-atmosphere interface because the temperature and species gradients may be large near the droplet. The general solution procedure is to first guess an interface temperature and determine the equilibrium species concentrations at the interface. The liquid and gas systems are solved separately using the interface boundary condition. Once the liquid and gas systems are solved, the guess for the interface temperature is checked with the energy balance given in Equation (4). If energy is not conserved in Equation (4), a new interface temperature is guessed and another system iteration takes place. If Equation (4) is satisfied the estimated temperature is assumed correct and the model proceeds to the next time step. The initial heat-up of the droplet interface is rapid, and up to 25 system iterations per time step may be required for the early stages of the droplet vaporization. After approximately five to ten time steps, the number of system iterations usually reduces to two or three per time step, depending on vaporizing conditions. Typical CPU run times on a VAX 8600 vary from 5 to 25 minutes depending on vaporization conditions and run parameters.

MODEL PREDICTIONS FOR N-OCTANE DROPLET VAPORIZATION

The computer model was run under two different high-pressure vaporizing conditions. For both cases, the droplet liquid was n-octane and the environment gas was nitrogen. Our choice of working fluids was based on a desire to closely approximate actual fuels vaporizing in air without the threat of combustion. In future work, refrigerant R-113 and helium will also be used as working fluids. The critical temperature and pressure of R-113, as well as its low ambient-temperature vapor pressure, make it an attractive fluid to use in our near critical vaporization experiments. The large ratio of specific heats for helium makes it easier for our experimental compression device to reach supercritical temperatures.

The well known d^2 law for droplet vaporization predicts a short period of heat-up followed by quasi-steady vaporization at the saturation temperature of the fluid[1]. For both vaporizing conditions the pressure was above the critical pressure of n-octane, so the supercritical saturation temperature is not defined. Hence, steady state vaporization at a saturation temperature is not expected and the d^2 law was not used for comparison.

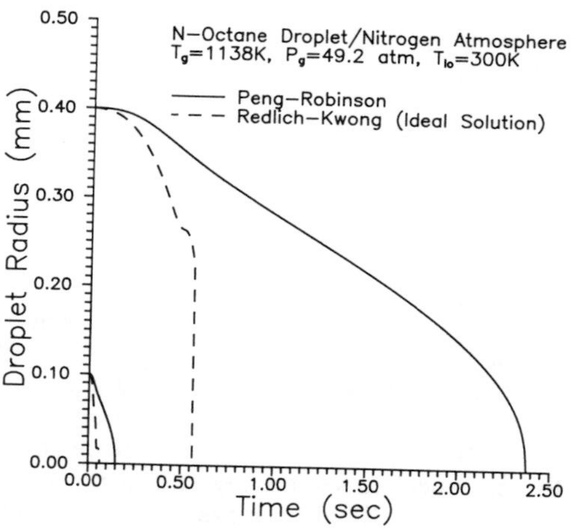

Figure 1

The model was for three droplet sizes using the Peng-Robinson and the Redlich-Kwong ideal solution interface modeling equations for both of the vaporizing conditions. The first vaporizing condition was at a pressure of 4.98 MPa (49.2 atm) and a temperature of 1138°K, about twice the critical values for n-octane. Figure 1 shows calculated droplet radius vs. time for this condition, and an initial droplet radius of 0.1 and 0.4 mm. Calculations were also performed for an initial droplet radius of 0.2 mm with similar results. A fairly short heat-up period followed by rapid vaporization is predicted. The curves calculated using the Redlich-Kwong e-o-s show a similar heat up period to the Peng-Robinson calculations but much more rapid vaporization.

Figure 2 shows comparable heat up times for the two interface models, but while the Peng-Robinson calculations show the interface temperature will exceed the critical temperature of n-octane, the Redlich-Kwong ideal solution model will not. It is not clear why the interface temperature for the Peng-Robinson calculations falls back below T_c after the initial heat-up period. It is believed that changes in the transport properties of the gas field near the droplet caused by the increasing n-octane mole fraction may have an effect on the interface temperature. Figures 3 and 4 show the profiles of temperature and mass fraction of n-octane near the droplet for the same condition as Figures 1 and 2, calculated for a droplet initial radius of 0.1mm using the Peng-Robinson e-o-s. The calculations show the steep temperature and species gradients near the droplet surface.

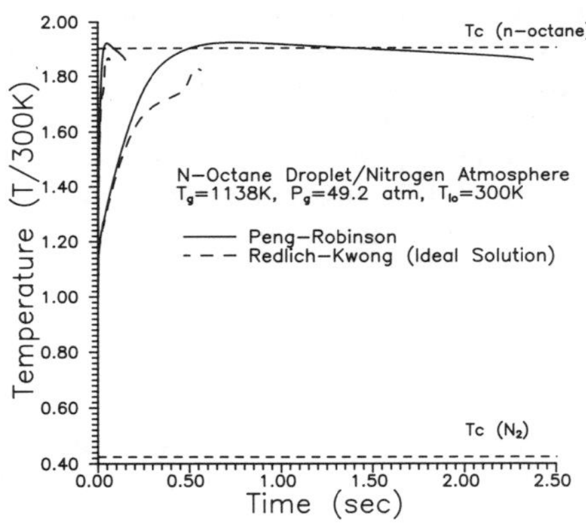

Figure 2

The calculations of Figures 5 to 8 were done for the second vaporizing condition of a pressure of 2.53 MPa (25 atm) and a temperature of 550°K. This condition is $1.02P_c$ and $.967T_c$ of n-octane. These values are similar to those in a compression ignition engine where the pressure can be above the critical pressure of the fuel while the temperature may be slightly below it. The pressure is above the critical for n-octane, so we would not expect to see steady state vaporization, but the interface temperature cannot reach the critical temperature of n-octane. Figures 5 and 6 show the droplet continuously heating as it vaporizes, with no steady state reached.

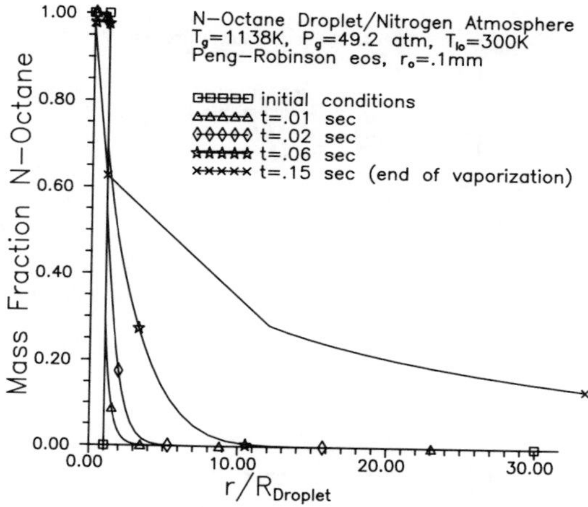

Figure 3

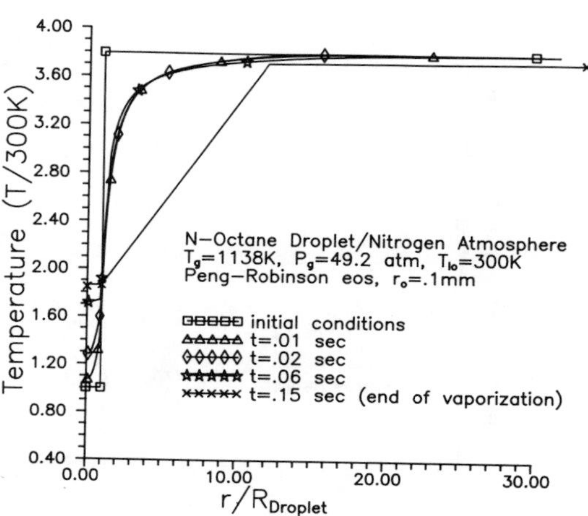

Figure 4

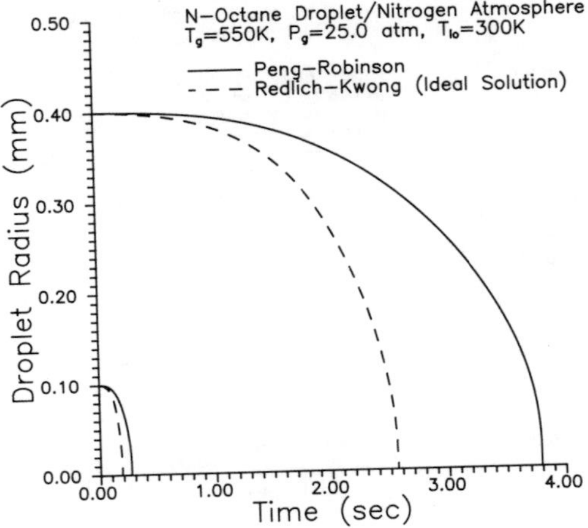

Figure 5

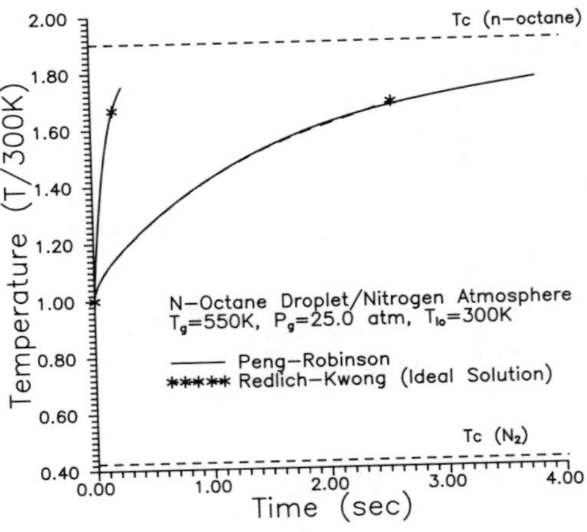

Figure 6

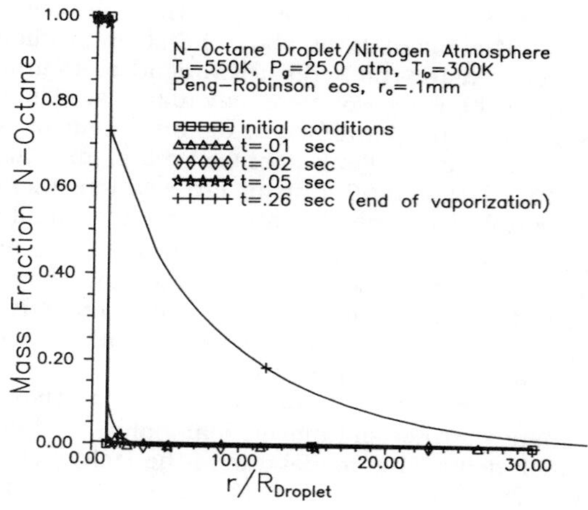

Figure 7

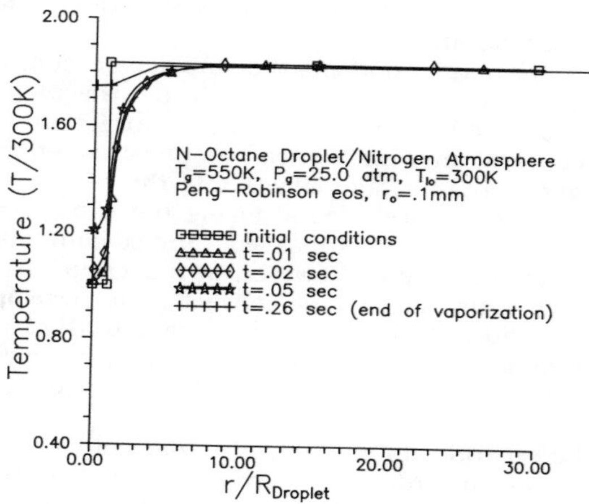

Figure 8

The Redlich-Kwong ideal solution calculations show more rapid vaporization than the Peng-Robinson calculations, as they did for the previous condition. The rate of heating of the droplet interface is the same for both equation-of-state models. Figures 7 and 8 show the temperature and species concentration profiles for a droplet vaporizing under the same condition as Figures 5 and 6 at various stages of its life. The droplet initial radius is 0.1mm, and the interface is modeled using the Peng-Robinson e-o-s. The calculations show the steep gradients near the droplet surface. The calculations using the Redlich-Kwong ideal solution equations showed faster vaporization than the Peng-Robinson calculations. The Redlich-Kwong ideal solution gives a slightly higher equilibrium mole fraction of n-octane in the gas under the same conditions as the Peng-Robinson e-o-s, and this phenomenon leads to faster vaporization. The Peng-Robinson e-o-s is specifically designed to give better vapor-liquid equilibrium performance than the Redlich-Kwong equation, especially at high pressures.[4] The ideal solution assumption represents a simplification of the real problem, which may induce some error in the equilibrium calculations for non-ideal conditions. For the above reasons, we expect that the Peng-Robinson calculations are more accurate, but comparison with actual experiments is necessary for model validation.

There is considerable evidence of anomalies in the transport properties of pure substances near their thermodynamic critical point.[10,11] Transport property anomalies are not included in this model due to the lack of evidence that they exist for mixtures. It is important to obtain experimental data for improved transport properties to replace provisional methods in the model.

EXPERIMENTAL PROBLEM FORMULATION AND APPROACH

The experimental problem at hand is to determine the vaporization rate and qualitative historical character of a single liquid droplet evaporating in a quiescent gas when diffusion controlled, near-critical effects are induced. These effects may include optical opalescence, anomalous vaporization rates, and shape distortion due to a loss of surface tension.

At low (nearly 1) Reynold's numbers, the diffusionally controlled effects may be of the same order of magnitude as convective effects. However, for large Reynold's numbers (greater than 100), it is probable that convection would dominate the vaporization process and may introduce complicating effects such as shape distortion, flow separation and wake effects into the problem. Thus it is highly desirable to suppress convective effects during the experiment so diffusionally controlled near-critical effects are not masked. Forced convection can probably be minimized through careful design. However, introduction of a cold droplet into a hot environment may cause significant convection to develop due to buoyancy effects. A means for reducing spatial and temporal growth of a natural convection boundary layer will be to operate the experiment in micro-gravity conditions. The 2.2 second drop tower at NASA-Lewis will be used to develop microgravity conditions.

The focus of the experiments is to provide sufficiently accurate qualitative and quantitative information in order to resolve the stated problem of near-critical droplet vaporization. A typical experiment involves the introduction of a droplet at room temperature into a gas field at a similar temperature in micro-gravity. Initial conditions are easily identified. A rapid compression of the gas follows, yielding a high-temperature, high-pressure environment where vaporization occurs. Visualization of the evaporation process and tracking of the temperature and pressure of the gas is carried out.

The experiment requires deployment of a small, ~ 1 mm diameter droplet with minimal internal motion and near zero relative velocity with respect to the gas.

One method used with some success has been the double-needle droplet generator developed by NASA-Lewis researchers. This system is shown schematically in Figure 9.

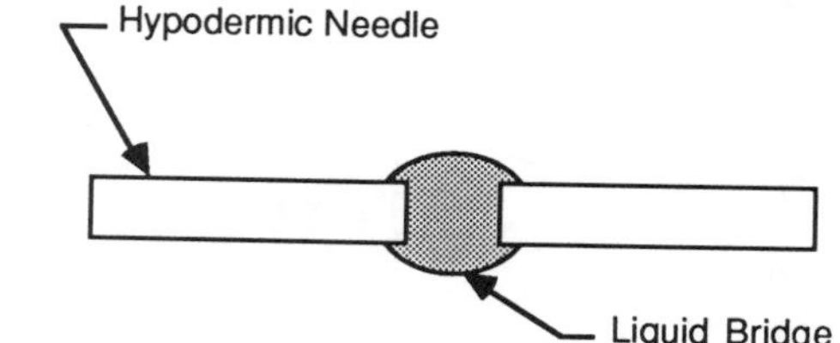

Figure 9a. Fluid dispensed between two hypodermic needles to form a liquid bridge.

Figure 9b. Needles simultaneously retracted and stationary droplet is left behind.

Figure 9
(a) & (b)

In this scheme, two hypodermic needles are brought close together. Fluid is dispensed and a liquid bridge is formed between needle tips. A quasi-stationary droplet is left behind as the needles are simultaneously retracted.

Introduction of a cold droplet directly into a hot, high pressure gas is likely to pose practical problems, especially in identifying the initial conditions of the experiment. An alternative approach of introducing a cold droplet into a cold environment, then rapidly changing the environment to a desired state has been pursued. Initial size, temperature and pressure can be well defined. A rapid compression device with variable initial pressure and a compression ratio of eight-to-one has been designed and tested. In this apparatus, two pistons approach a plane of symmetry simultaneously and from opposite directions. A droplet is positioned centrally on the plane of compressive symmetry. A diagram of the compressor is presented in Figure 10. Compressive time is less than 50 msec and the cylinder leakage time constant for gaseous nitrogen has remained greater than 250 sec. Impact forces resulting from 28 drop tower tests have not resulted in failure of this system.

The droplet generator has been placed at the point of compressive symmetry, thereby minimizing imparted gas motion close to the droplet due to compression. However, some problems in deployment of a droplet under microgravity conditions exist. They are attributed to needle conditioning and alignment, each of which has a substantial effect on performance of the double-needle droplet generator. It is hoped that precise control over needle tip preparation and matching as well as improved needle alignment will yield successful droplet deployment in the future.

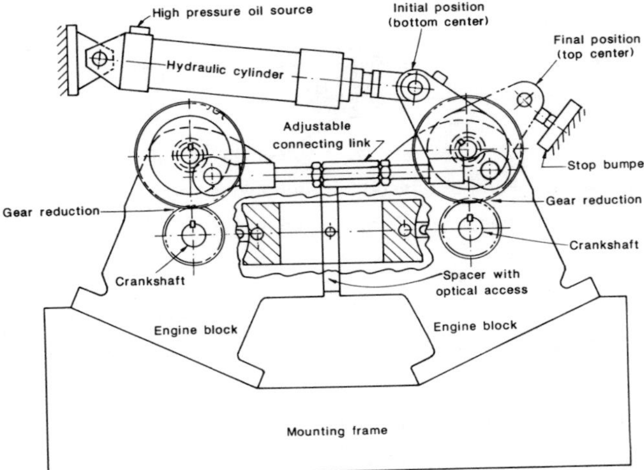

Figure 10

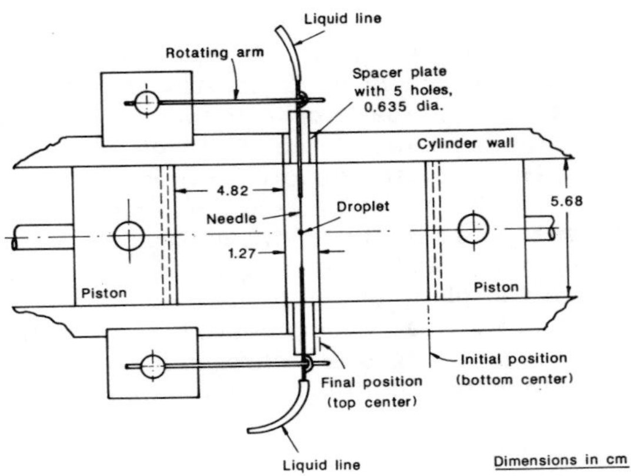

Figure 11

A diagram of the position of the droplet generator with respect to the compressor is presented in Figure 11.

Initial experiments are to be conducted with liquid refrigerant 113 and gaseous nitrogen or helium. The current program outlook is to resume microgravity experiments with refined droplet generator performance in the winter of 1988/89.

ACKNOWLEDGEMENT

This work is supported by a grant from NASA, NAG-3-718.

NOMENCLATURE

a	Equation of state constant
b	Equation of state constant
C_p	Constant pressure specific heat
D_{AB}	Mass diffusivity of component A into component B
f	Fugacity of a pure substance
$\hat{f}$	Fugacity in a mixture
$G_{T,P}$	Gibb's function
h	Specific enthalpy
H	Henry's Law constant
k	Thermal conductivity
L	Latent heat of vaporization
m	Mass
n	Number of moles
P	Pressure
r	Radius r
$\overline{R}$	Universal gas constant
R	Droplet radius
t	Time
T	Temperature
v	Specific volume
V	Radial diffusion velocity
w	Mass fraction
x	Liquid phase mole fraction
y	Gas phase mole fraction
Z	Compressibility factor
μ	Dynamic Viscosity
ρ	Density

SUBSCRIPTS AND SUPERSCRIPTS

A	Component A
B	Component B
c	Critical state
E	Equilibrium state
g	Gas
l	Liquid
m	Mixture
0	Reference property
∞	Ambient

REFERENCES

1. D. B. Spalding, "The Combustion of Liquid Fuels", **Fourth Symposium (International) on Combustion**, The Combustion Institute, Pittsburgh PA, 1953.
2. K. K. Kuo, **Principles of Combustion**, pp. 370-382, John Wiley and Sons, New York, 1986.
3. G. Van Wylen & R. Sonntag, **Fundamentals of Classical Thermodynamics**, 3rd ed., Wiley, 1986.
4. Ding-yu Peng & Donald B. Robinson, *Ind. Eng. Chem., Fundam.*, **15**, No. 1, 1976.
5. O. Redlich & J. N. S. Kwong, *Chem. Rev.*, **44**, 233 (1949).
6. Giorgio Soave, *Chem. Eng. Sci.*, Vol. 27, pp 1197-1203, 1972.
7. J. M. Prausnitz & P. L. Chueh, **Computer Calculations for High Pressure Vapor-Liquid Equilibria**, Prentice-Hall, 1968.
8. J. A. Manrique & G. L. Borman, *Int. J. Heat Mass Transfer* 12, 1081-1095, 1969.
9. Richard E. Balzhiser, Michael R. Samuals, & John D. Eliassen, **Chemical Engineering Thermodynamics**, Prentice-Hall, Inc., 1972.
10. J. V. Sengers, NBS Misc. Pub. 273, 165-178, 1968
11. H. Becker, & U. Griguill, in **Seventh Symposium on Thermophysical Properties**, ASME, 1977.

LIST OF FIGURES

MICROGRAVITY DROPLET COMBUSTION IN HIGH PRESSURES NEAR CRITICAL PRESSURES OF FUELS

Jun'ichi Sato[a]
Research Institute, Ishikawajima-Harima Heavy Industries Co., Ltd.
Toyosu, Koto-ku, Tokyo 135, Japan

Mitsuhiro Tsue, Mario Niwa, and Michikata Kono
Department of Aeronautics, The University of Tokyo,
Hongo, Bunkyo-ku, Tokyo 113, Japan

ABSTRACT

Burning behavior of a fuel droplet in a quiescent high-pressure atmosphere has been studied experimentally to explore the effects of the ambient pressure just around the critical pressure of fuel. Since the natural convection changes the transport field around the burning droplet, microgravity field made by a falling apparatus were used to suppress the natural convection generated by the droplet burning. Experiments showed that spherical flames were obtained for ambient pressures both below and above the critical pressure of fuel. As the ambient pressure is increased, the burning life time decreases and reaches the minimum at the critical pressure of fuel, beyond which the burning life time increases. Flame diameter increases during the burning time to a maximum and then decreases to burnout. This behavior is the same both below and above the critical pressure of the fuel. The maximum flame diameter attained during burning is a function of pressure and decreases with ambient pressure throughout the range investigated. There is no change in slope or minimum at the critical pressure.

INTRODUCTION

The most useful and convenient method to generate chemical energy of liquid fuels is spray combustion. It is widely used not only in boilers and furnaces, but also in internal combustion engines such as a diesel engine, jet engine, and rocket engine. Pressure levels of these engine combustors are being increased to improve thermal efficiency and to reduce their sizes, and, now, their maximum pressures exceed the critical pressures of their fuels. Knowledge of the combustion process in these engine combustors is needed to achieve proper designs of internal combustion engine combustors for stable operation, high efficiency, and low emission levels. However, current understanding on liquid spray combustion in a high-pressure atmosphere, especially, near and above the critical pressure of fuels is far from complete.

In the burning fuel spray, the fuel droplet evaporation, fuel vapor diffusion and mixing, vapor phase combustion, and the fuel droplet combustion may occur simultaneously. Single droplet evaporation and combustion have been studied by many researchers as one of the basic studies to solve these complicated processes. Although research on high-pressure combustion and evaporation of fuel droplets is very important in understanding the engine combustion, most of these experimental and theoretical studies were for fuel droplets at atmospheric pressure levels or relatively low pressure levels[1-3]. This is because there was few useful data compared with the theoretical studies, and high-pressure experiments are very

[a]*Correspondent author*: Jun'ichi Sato, Research Institute, Ishikawajima-Harima Heavy Industries Co., Ltd., 3-1-15 Toyosu, Koto-ku, Tokyo 135, Japan. Telex: IHITOY J23507; Telefax (81) 3-534-3322.

difficult to conduct.

Only a few experimental studies have been performed for high pressures near and/or above the critical pressures of fuels[4-8]. Faeth *et al.* performed the experiments under microgravity condition and measured the burning life time of an n-decane droplet[5]. They have found that the burning life time of fuel droplets decreases with the increase of the ambient pressure below the critical pressure condition but increases with the ambient pressure above the critical pressure condition. Kadota & Hiroyasu and Tsue *et al.* have performed the experiments for many fuels under normal-gravity condition[6,7]. They have found that the variations of the burning life times and the burning rate constants with the ambient pressure arise at the critical pressure of fuel. Recently, Sato *et al.* carried out the experimental study for both normal and microgravity field, and they have explored the effects of the natural convection on the burning rate constant for wide variations of the ambient pressure throughout subcritical and supercritical pressure of fuels[8].

Although these studies have yield much useful information on high-pressure droplet combustion, some unknowns have still remained for combustion near and above the critical pressure of fuel. One of these is the burning behavior just around the critical pressure of fuel, a point where the physical properties of fuels vary remarkably with pressure. Thus, in this study, experiments have been performed to explore the effects of the ambient pressure on the burning process of a fuel droplet near the critical pressure of fuel. Since the natural convection changes the transport field around the burning droplet and natural convection effects become stronger with an increase of the ambient pressure, microgravity technique, which was developed by Kumagai *et al.*[9,10], was used to suppress the natural convection generated by the droplet combustion. The burning life times and the dimensions of the flame have been measured precisely throughout below and above the critical pressure of fuel.

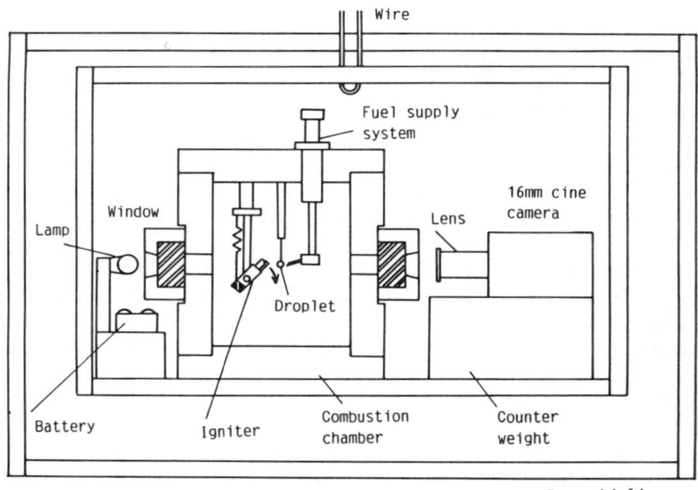

Figure 1

EXPERIMENTAL

The microgravity high pressure droplet combustion experiments have been performed for a filament positioned droplet by using a high pressure combustion chamber and a falling apparatus. The filament-positioning technique is much easier than the free-droplet technique, but how much this technique affects the phenomenon is the problem. However, Kumagai *et al.* have shown by experiment that the fine filament which positions the droplet under microgravity field has little effect on the droplet-burning phenomena.

A schematic of an experimental apparatus used in this study is shown in Figure 1. The high-pressure combustion chamber is a cylinder with an inner diameter of 100 mm and a inner height of 200 mm. Burning behavior can be well observed through four windows on the chamber wall. Experiments can be conducted at ambient air pressure up to 20 MPa.

Single component fuels make the analysis of the burning process of a fuel droplet easier. The fuels employed were n-heptane and n-octane. Critical pressure and critical temperature of n-heptane are 2.74 MPa and 540°K, and those of n-octane are 2.52 MPa and 570 K, respectively. The fuel was forced by a high pressure microplunger pump through a thin pipe to a needle. A fuel droplet was formed at the end of the needle and transferred to the end of a 0.15 mm diameter quartz filament. The initial droplet diameter was about 0.8 mm. After forming the droplet at the end of the quartz filament, the fuel supply system was moved away from the quartz filament. An arc discharge igniter was used for igniting the fuel droplet, which was also moved away from the droplet after a spark discharge.

Microgravity field was made by a falling chamber system with a height of 15 m, which can produce about an effective time of microgravity of 1.6 second. A drag shield, in which the falling apparatus was installed, produced the gravitational acceleration of less than 10^{-5}-10^{-6} g necessary to suppress the natural convection generated by burning of the droplet.

The droplet was formed and attached on the end of the filament under normal gravity condition, and then falling of the apparatus was started. Ignition of a fuel droplet was performed automatically under microgravity field. The burning life time of the droplet and the variations of the flame dimension were measured by direct photography with a 16 mm high speed cinecamera.

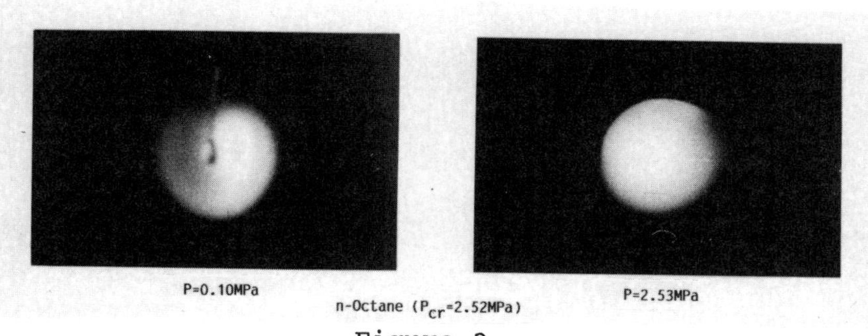

Figure 2

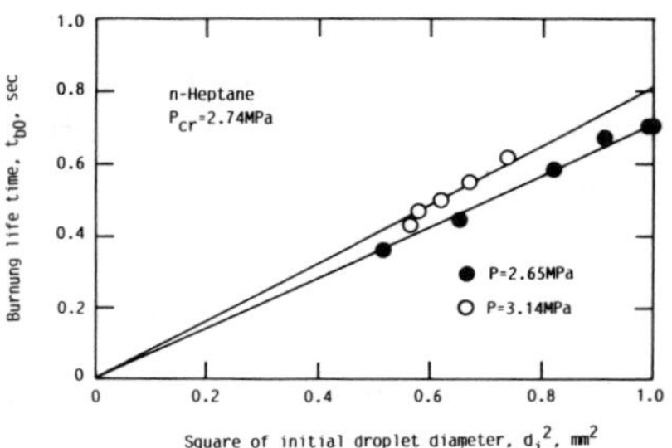

Figure 3

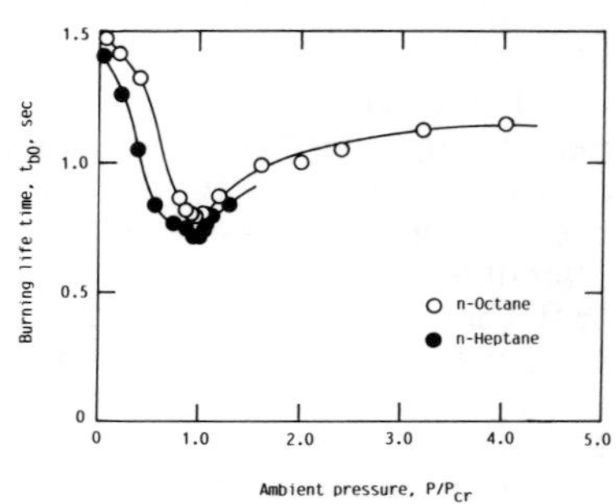

Figure 5

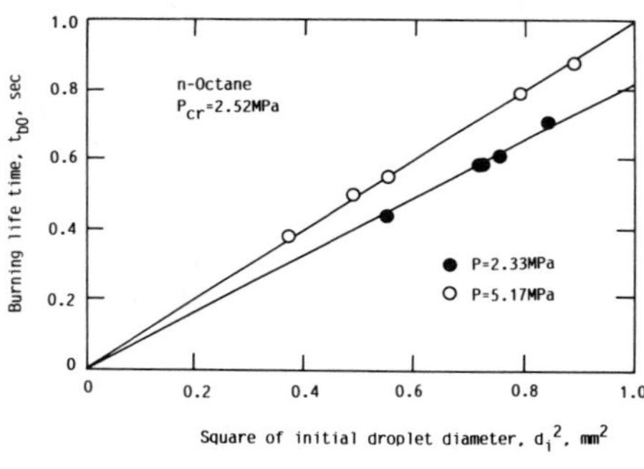

Figure 4

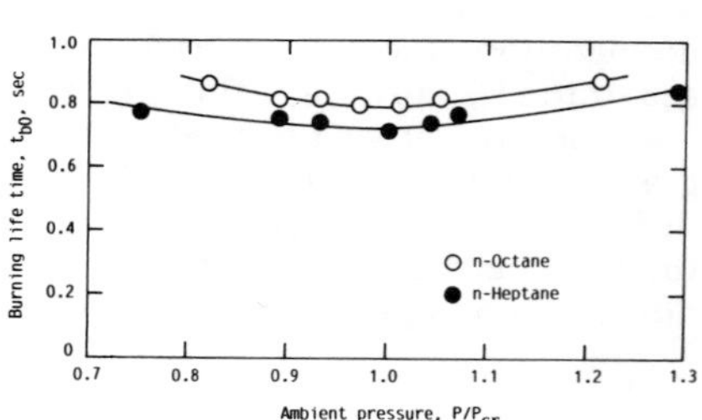

Figure 6

RESULTS AND DISCUSSION

Figure 2 shows the direct photographs of the burning droplet of n-octane. Under microgravity field, spherical flames were formed around the droplet. For the ambient pressures over about 0.5 MPa, the droplet diameter under burning could not be measured. It was because the droplet was surrounded by the bright soot layer, and the flame was too opaque to see the droplet through it. Thus, only the burning life time and the variations of the flame diameter with time were measured by the direct photography.

The burning life time was defined as the time between the ignition of the droplet and the disappearance of the flame. Figures 3 and 4 shows the relations between the square of the initial droplet diameter and the burning life time for n-heptane and n-octane. It has been found that the burning life time is linearly proportional to the square of the droplet diameter both for below and above the critical pressure of fuel. Since Figures 3 and 4 show that the burning life time divided by the square of the initial droplet diameter is constant, the inverse relationship of diameter squared to time also holds, and the d^2 law of the droplet combustion is shown to hold at high pressures. The droplet combustion experiments under normal-gravity field have shown that the liquid droplet was observed under burning up to about 1.5 times of the critical pressure of fuel[6,7]. However, for very high pressures such as over two or three times of the critical pressure of fuel, existence of the liquid droplet under burning is not clear experimentally.

Figure 5 shows the variations of the burning life time with the ambient pressure normalized by critical pressure. The burning life times shown in Figure 5 are for the initial droplet diameter of 1 mm obtained from the plotted values, such as in Figures 3 and 4. The burning life time decreases with the ambient pressure and reaches a minimum value at the critical pressure, beyond which the burning life time increases. This tendency is the same as that previously obtained by Faeth *et al.* for n-decane under microgravity field and by Tsue *et al.* for some paraffin and alcohol fuels under normal gravity field[5,7].

Figure 6 is an expansion of Figure 5 just around the critical pressures of n-heptane and n-octane. This figure shows that the values of the burning life time vary smoothly near the critical pressure of fuel, but they have minimums just at the critical pressures. The droplet under burning could not be observed in this experiment, because the flame was too opaque to see the droplet through it. But our previous experiments under normal-gravity field have observed the liquid droplet in the burning period at pressures up to 1.5 times of the critical pressure of fuel. Therefore, the variations of the burning life time with the ambient pressure around the critical pressure of fuel may be discussed considering the existence of the liquid droplet under burning.

Figures 7 and 8 show the variations of the flame diameter with time for various values of the ambient pressure for n-heptane and n-octane respectively. The definition of the flame diameter d_f is shown in Figure 8 schematically. The flame diameter and the time counted from ignition are normalized by the initial droplet diameter d_i and the burning life time t_{bo}, respectively. The flame diameter increases with time, reaches a maximum, and then decreases. Variations of the maximum value of the flame diameter with the ambient pressure are shown in Figure 9. The pressure is normalized by the critical pressure, and a single relationship applies to both fuels. The maximum flame diameter decreases smoothly with the increase of the ambient pressure from below to above the critical pressure of fuel. The minimum value exhibited by the burning life time does not exist for the maximum flame diameter at the critical pressure.

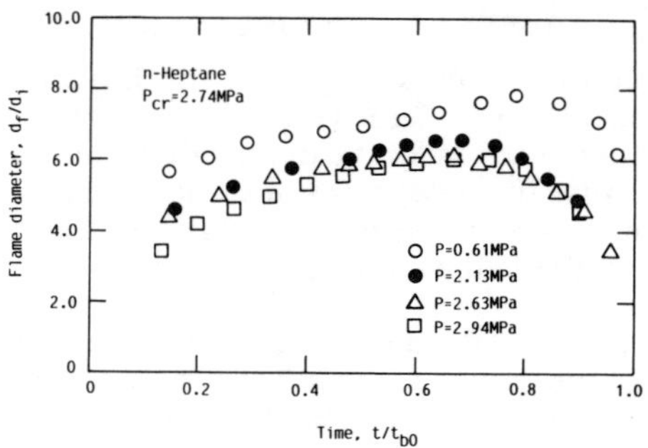

Figure 7

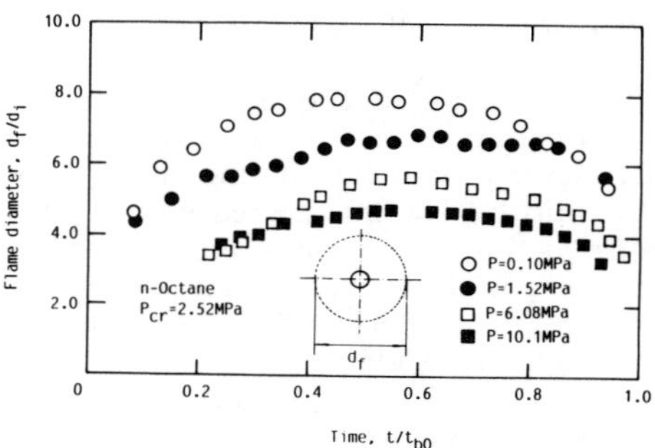

Figure 8

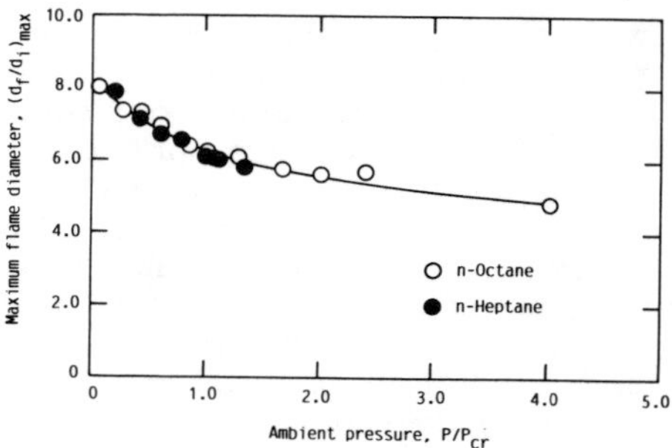

Figure 9

CONCLUSIONS

The burning behavior of a fuel droplet in a quiescent high pressure atmosphere has been studied experimentally to explore the effects of the ambient pressure just around the critical pressure of fuel. Experiments were conducted under a microgravity field to suppress the natural convection generated by the droplet combustion. Spherical flames were obtained both below and above the critical pressure of fuel. Conclusions are as follows:

1). As the ambient pressure is increased, the burning life time decreases smoothly with pressure, reaches a minimum just at the critical pressure of fuel, and increases with pressure above the critical pressure.

2). Flame diameter increases to a maximum during the burning period and then decreases until burnout. The maximum flame diameter attained during burning decreases with the ambient pressure through and above the critical pressure of fuel. There is no change of slope or minimum at the critical pressure.

ACKNOWLEDGMENTS

This work is supported by the Special Coordination Fund for Promoting Science and Technology, through the Science and Technology Agency of the Japanese Government.

The authors are grateful to Dr. Seiichiro Kumagai, Professor Emeritus of University of Tokyo, Noritz Corporation Research Laboratory, for his kind support of microgravity experiments. Finally, the contribution from discussions with Prof. Takashi Niioka of Tohoku University is sincerely appreciated.

REFERENCES

1. A. Williams, 1973. *Combustion and Flame* 21, 1.
2. F. A. Williams, 1985. **Combustion Theory**. 2nd Ed. Benjamin/Cummings.
3. K. K. Kuo, 1986. **Principles of Combustion**. John Wiley & Sons.
4. T. A. Brzustowski & R. Natarajan, 1966. *Can. J. Chem.* Eng. 44, 194.
5. G. M. Faeth, J. F. Dominicis, J. F. Tulpinsky, & D. R. Olson, 1969. **Twelfth Symposium (International) on Combustion.** The Combustion Institute, 9.
6. T. Kadota & H. Hiroyasu, 1981. **Eighteenth Symposium (International) on Combustion**. The Combustion Institute, 275.
7. M. Tsue, H. Miyano, J. Sato & M. Kono, 1987. *J. Jpn. Soc. Aeron. Space Sci.*, 35, 433.
8. J. Sato, M. Tsue, M. Niwa, & M. Kono, (to appear) *Combustion and Flame.*
9. S. Kumagai & H. Isoda, 1957. **Sixth Symposium (International) on Combustion**. Reinhold, 726.
10. S. Kumagai, T. Sakai, & S. Okajima, 1971. **Thirteenth Symposium on Combustion**. The Combustion Institute, 779.

FIGURE CAPTIONS

Figure 1. Schematic of experimental apparatus.
Figure 2. Direct photographs of burning droplet.
Figure 3. Variations of the burning life time with the square of initial droplet diameter for n-heptane.
Figure 4. Variations of the burning life time with the square of initial droplet diameter for n-octane.
Figure 5 Dependence of the burning life time on the ambient pressure.
Figure 6. Dependence of the burning life time on the ambient pressure near critical pressure of fuel.
Figure 7. Variations of the flame diameter with time for n-heptane.
Figure 8. Variations of the flame diameter with time for n-octane.
Figure 9. Dependence of the maximum flame diameter on the ambient pressure.

SOME EXPERIMENTS ON FREE DROPLET COMBUSTION AT LOW GRAVITY

J.C. Yang, G.S. Jackson, and C.T. Avedisian
Sibley School of Mechanical and Aerospace Engineering
Cornell University, Ithaca, N.Y. 14853

ABSTRACT

A small-scale (7.6m) drop tower was used for studying the combustion of unsupported fuel droplets (about 500µm initial diameter) in a stagnant ambience under low gravity. The experimental procedure consisted of generating a droplet in a near vertical trajectory and then releasing the chamber within which the droplet was introduced, as well as associated instrumentation, into free-fall when the droplet reached the apex of its trajectory.

Some results of the burning of n-heptane, toluene, and heptane/hexadecane mixture droplets are reported. The range of the heptane data is discussed in terms of possibly varying ambient conditions around the droplet during burning due to droplet motion. Microexplosions were not observed for the mixture reported.

1. INTRODUCTION

Experiments on the combustion of unsupported droplets that used gravity as the parameter through which buoyancy effects were minimized have been carried out in drop towers in which the test droplet, its enclosed environment (i.e., the combustion chamber containing the air within which the droplet was burned), and associated instrumentation were simultaneously released into free-fall[1-4]. The test droplets were formed by first holding captive the liquid sample by one or more fibers or needles and then freeing the sample from the fiber(s) by a jerking motion of the fiber(s) along its axis. The droplets produced by this method were large (~1000µm initial diameter) by comparison to droplet sizes that are encountered in most practical combustion situations—typically less than 100µm (e.g., burning sprays). In an effort to study droplets of sizes more characteristic of those likely to be encountered in practical applications, we recently reported on a different technique for generating smaller free droplets—in the range of 400µm to 500µm diameter [5]. While it may be argued that droplets in this size range are still too large to yield results representative of those encountered in practical applications, the present effort is a step in that direction.

This paper summarizes some of our recently obtained experimental results on the combustion of free droplets of n-heptane, toluene, and a mixture of n-heptane and n-hexadecane in a low gravity environment. The experiments were conducted in room temperature air at 0.101 MPa.

2. EXPERIMENTAL METHOD

A schematic diagram of the drop tower facility is illustrated in Figure 1. It consisted of: 1) a vertical shaft; 2) an upper work station with movable floor cover to expose the drop shaft; 3) a drop package which was made of an aluminum and steel support framework and which contained the droplet generator, combustion chamber, spark ignition system, high-speed motion picture camera, and light source; 4) a hoist system; 5) an electromagnet for holding and releasing the drop package, 6) a deceleration tank; and 7) a timing control circuit. Further details may be found in Reference 5.

The idea was to first generate droplets at a rate of 2 drops/s in a steady stream of

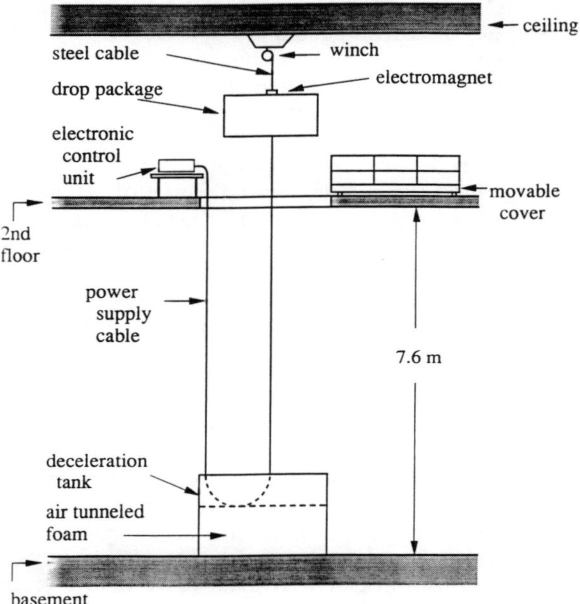

Figure 1

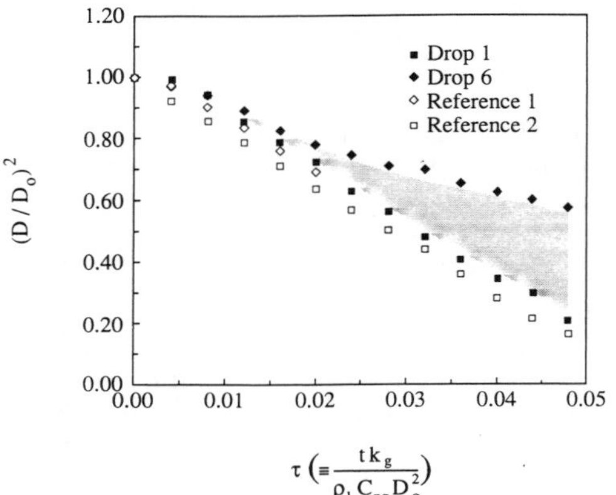

Figure 2

mono-dispersed droplets between 400μm and 500μm diameter; smaller droplets were not studied due to difficulties of optical resolution. The droplet trajectories were almost vertical. The stream was then shut off and the last droplet of this stream was used to perform the experiment. When this last droplet was near the apex of its trajectory, it was ignited by spark discharge across two electrodes positioned around the apex. The drop package was then released into free fall. The droplets studied exhibited a small velocity that was created by inaccuracies in timing the release of the platform with the time of flight of the droplet to its apex, or by the small horizontal velocity component of the droplet due to its non-vertical trajectory. The cumulative effect of these motions yielded Reynolds numbers (*Re* based on droplet diameter) on the order of a maximum of 0.1.

The gravity levels experienced in the moving frame of reference with the present experimental set up were on the order of 10^{-3} that of earth normal gravity. The corresponding Grashof number (based on the initial droplet diameter) was on the order of 10^{-5} (see Appendix). Droplets larger than 1000μm diameter require lower gravity levels to achieve the same dynamic effect. For example, doubling the droplet diameter (from 500μm to 1000μm) would require reducing gravity by about an order of magnitude if all other parameters were held constant. The use of an air drag shield around the drop package to further reduce gravity is an option which has not yet been exploited in the present experimental design.

The primary means of data acquisition was photographic. A 16mm high speed movie camera was operated at 250 frames/s to record the droplet burning history. Direct back lighting was provided by a single tungsten halogen projector lamp. The light intensity used did not permit the flame to be observed for heptane droplets because our initial efforts were directed at obtaining shadow images of the droplets.

Droplet dimensions were obtained directly from the film record by a frame-by-frame analysis. A 16mm film projector was used to vertically project the droplet image onto a horizontal drafting board. The projector lens was positioned approximately 65cm above the board. With this arrangement the operator could focus the image while remaining seated, which facilitated accurate focusing of each frame. Projected droplet images at ignition ranged between 1.5cm and 2 cm. Maximum horizontal and vertical diameter measurements were obtained by marking lines with a pencil with a 0.3mm lead. From these measurements, the equivalent diameter was calculated as the diameter of a sphere with the same volume as the droplet, assuming an ellipsoidal droplet. The greatest source of error in data reduction involved visually identifying the boundary between the droplet and background. In this respect, the present method may be expected to yield comparable accuracy to that obtained from a Vanguard™ analyzer, but is not as rapid as more automated systems[6].

3. RESULTS AND DISCUSSION

3.1 Single Component Hydrocarbons

Results from n-heptane droplets are summarized in Figure 2 in nondimensional form to suppress the effects of initial droplet diameter. The shading indicates the range of six observations, with the evolution of droplet diameter reported for the droplets which burned the fastest and slowest. Figure 3 provides representative shadow photographs from these six heptane droplets (drop 7 was a heptane/ hexadecane mixture to be discussed in Section 3.2); for each set the enlargement is uniform, but different enlargements were used for each of the photographic sets. Drops 2 to 6 were captured on the downward side of their trajectories and were thus drifting downward toward the nozzle exit during the period of low gravity; drop 1 was captured near to the apex and drifted laterally (to the right). The maximum Reynolds numbers and initial diameters are listed in Table 1. Only data in the time

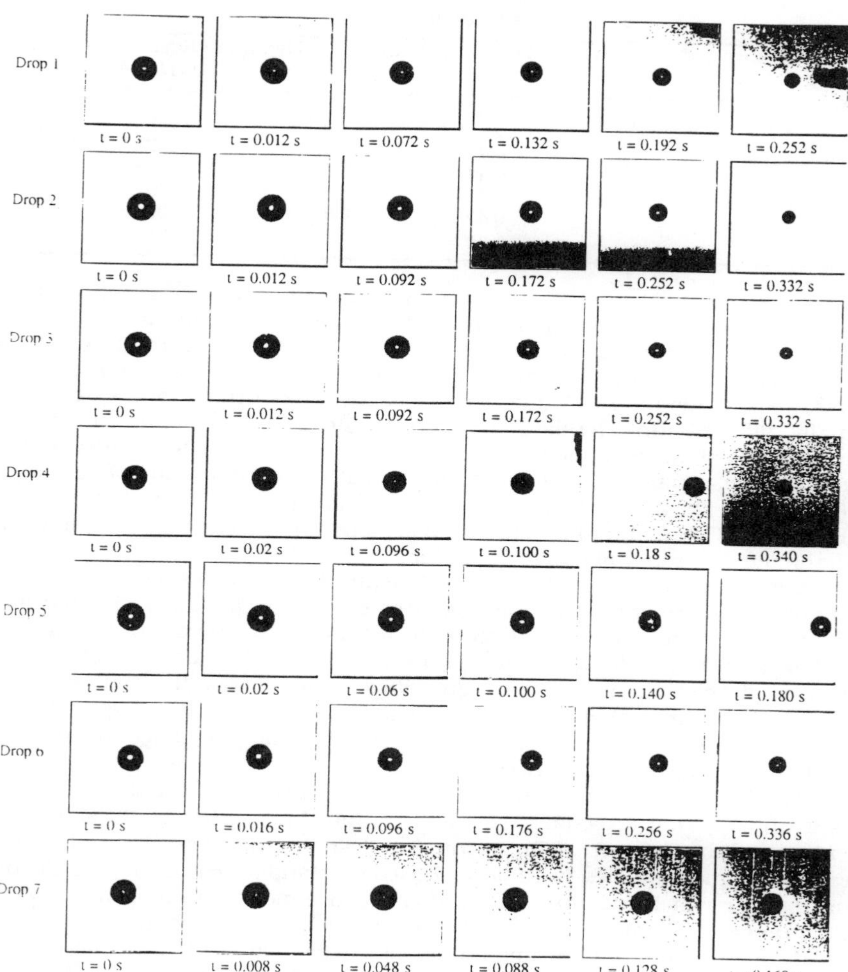
Drop 1
t = 0 s
t = 0.012 s
t = 0.072 s
t = 0.132 s
t = 0.192 s
t = 0.252 s
Drop 2
t = 0 s
t = 0.012 s
t = 0.092 s
t = 0.172 s
t = 0.252 s
t = 0.332 s
Drop 3
t = 0 s
t = 0.012 s
t = 0.092 s
t = 0.172 s
t = 0.252 s
t = 0.332 s
Drop 4
t = 0 s
t = 0.02 s
t = 0.096 s
t = 0.100 s
t = 0.18 s
t = 0.340 s
Drop 5
t = 0 s
t = 0.02 s
t = 0.06 s
t = 0.100 s
t = 0.140 s
t = 0.180 s
Drop 6
t = 0 s
t = 0.016 s
t = 0.096 s
t = 0.176 s
t = 0.256 s
t = 0.336 s
Drop 7
t = 0 s
t = 0.008 s
t = 0.048 s
t = 0.088 s
t = 0.128 s
t = 0.168 s

Figure 3

domain from ignition but prior to extinction and/or vaporization-like phenomenon are included in Figure 2.

Table I
Initial Diameter and Maximum Reynolds Numbers for Droplets Reported in Figure 3

Drop Number	Initial Diameter (mm)	Maximum Reynolds Number ($VD\rho/\mu$)
1	0.50	0.058
2	0.45	0.11
3	0.43	0.078
4	0.44	0.082
5	0.43	0.097
6	0.42	0.096
7	0.49	0.058

Two facts to note are that the flame is not visible in the photographs shown in Figure 3, and that for the six heptane droplets studied (the evolution of diameters of which fell in the shaded region bounded by drops 1 and 6 in Figure 2), the burning rates were not constant and exhibited values during burning which ranged between 0.5 mm^2/s and 0.78 mm^2/s. We note that because the burning rate is obtained by essentially differentiating the data, it is very sensitive to uncertainties in the measurements and the range of data selected to define it, particularly if the evolution of droplet diameter squared is not linear throughout the entire period of burning (cf, drop 6 in Figure 2). If the evolution of D^2 is not linear over the period of burning (which may have been due to effects in the present experiment that are conjectured below), the burning rate will be time dependent.

The maximum burning rate among the heptane droplets studied (drop 1) is similar to previously reported unsupported droplet measurements carried out at low gravity[1,2]. Lower values are considered to be attributed to: 1) variations of the convective flow around the droplets during burning; 2) size-dependent effects from different initial droplet diameters; 3) irregularities in the spark's interaction with the droplet and the surrounding gases; 4) soot shell geometry and its effect on the heat transfer rate to the droplet; and/or 5) variations in the gas concentration, most notably the oxygen concentration, during burning.

The low Reynolds numbers (based on droplet diameter) of the droplets studied makes it unlikely that an axial vapor flow around the droplets could significantly influence the burning rate. The maximum Reynolds numbers were well in the Stokes flow regime as shown in Table 1. Based on standard corrections to the droplet burning rate which account for forced convective motion around burning droplets[7], the ranges in Figure 2 could not be due simply to variations in Re from droplet to droplet. In addition, the burning rate is also dependent on the droplet diameter. However, the range of initial diameters shown in Table 1 is not sufficiently large to explain the range[1] shown in Figure 2.

The spark ignition may influence the burning process because of its inherent transience and its uneven heating of the surrounding gas. In the present method, ignition was thought to be achieved by the spark heating gases into which the droplet moved as it was rising to its apex. The spark did not come in direct contact with the droplet. Droplet ignition was manifested by a visible yellow plume near the droplet which disappeared within 40 msec. However, it is not clear that these ignition characteristics play a significant role in the main portion of the burning. Further investigations are being pursued to assess the effect of spark location and energy, and multiple electrode configurations[4], upon the droplet burning process.

The burning rate increases with ambient oxygen concentration[8]. Pre-existing fuel vapors in the vicinity of the droplet at ignition would tend to lower the initial oxygen concentration there below the far-field ambient concentration. Since the experiment required a continuous stream of droplets to achieve a steady trajectory, fuel vapors could have accumulated during the preparatory stage of an experiment from some pre-vaporization of these droplets, both in the parabolic path traversed by the stream and in the recessed region adjacent to the generator nozzle support as the droplets in the continuous stream entered this region. A droplet ignited in this oxygen lean surrounding and drifting about within it would then burn more slowly than in the far ambience.

If fuel vapor accumulation occurred in the parabolic path of the continuous stream, it was evidently below the lean flammability limit because ignition of these vapors was not observed. A fuel rich zone near the nozzle exit caused by vaporization of liquid fuel accumulating there during preparation of an experiment (i.e., during the period in which the relevant timing parameters are being measured for the stream), could be envisioned to have created a concentration gradient of fuel (and oxygen) vapors such that the fuel (oxygen) concentration decreased (increased) from a peak (minimum) near the nozzle exit toward the apex. A droplet drifting downward toward the nozzle would then experience a decreased oxygen concentration and thus burn progressively more slowly. Drops 2 to 6 were captured just on the downward side of their trajectories and were drifting toward the nozzle, though not necessarily in a straight line. The entrance to the nozzle "well" can be seen in the fourth and fifth photograph for drop 2, and the sixth photograph for drop 4 in Figure 3. The slowest burning rate of the six droplets reported with downward trajectories in Figure 3 was drop 6, and these data are plotted in Figure 2. Unfortunately, it has not yet been possible to sample the gas phase around the droplet to quantify its composition. The effect of gas composition on the droplet burning rate bears on the physical property values and the transfer number.

Drop 1 exhibited a lateral movement after ignition which is considered to have brought it into a zone characterized by the standard ambience after being ignited. The evolution of the diameter for this droplet is similar to prior measurements[1,2] as shown in Figure 2.

Due to the intense background illumination and the camera framing rate used (250 frames per second), neither the particularly "bluish" flame around the n-heptane droplets nor the extent of soot around the heptane droplets could be observed so it is an open question as to whether extensive sooting of small heptane droplets such as studied here occurs. However, toluene droplets were also studied with the existing optical set-up, and it was found that toluene droplets sooted extensively. For this liquid the light emitted from the toluene flame and soot around toluene droplets was sufficient to expose the film as shown in Figure 4, in contrast to heptane (Figure 3).

Figure 4a and 4b show two representative photographic images of a toluene droplet taken just after ignition, and well into burning, respectively. Figures 4c and 4d show computer-generated images of the photographs displayed in Figures 4a and 4b. The images enhance the flame and soot boundaries, and clearly reveal the flame shape. Two zones are revealed: an outer luminous region which probably defined the primary reaction zone of the flame, and an inner carbon or soot ring between the droplet surface and the outer region. Initially, a carbon shell penetrated across the outer luminous zone of the droplet (Figure 4c) and then the shell began to break up as it became spherical (Figure 4d) revealing a perhaps fragile shell structure.

It can be conjectured that the soot shell thickness and the location of the droplet within it should effect the droplet burning rate. One thought is that the soot shell can introduce a thermal resistance for heat transfer between the droplet surface and flame thereby reducing the heat transfer rate and the droplet burning rate.

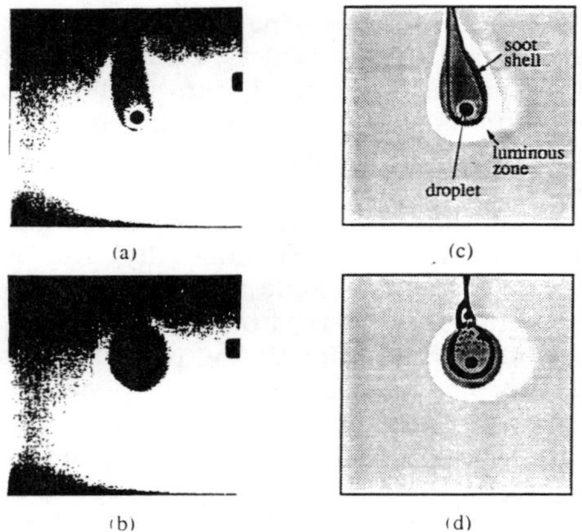

Figure 4

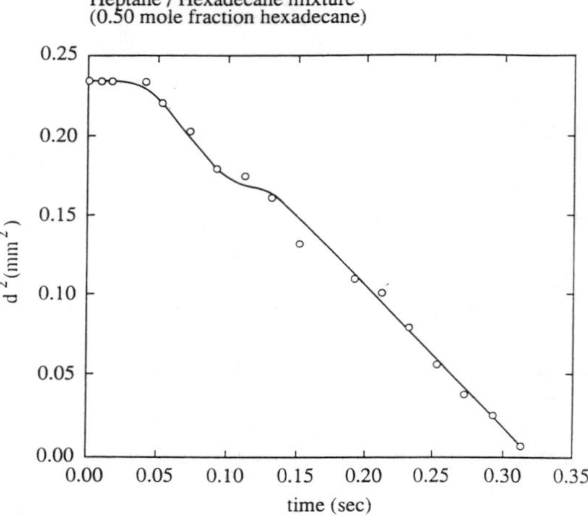

Figure 5

Another perspective is that the hot soot particles can increase the heat transfer rate to the droplet since the soot particles would reside closer to the droplet than the flame. In either event, if the ratio of the soot to the droplet radius is not constant during burning, caused perhaps by varying drifting trajectories of the droplets from run to run and a difference in the relaxation time for the shell to respond to movement changes of the droplet (or vice versa), the burning rate would vary as well from run to run.

Single component unsupported droplets have been observed to explode during their combustion at low gravity. The phenomenon was first identified as "flash extinction" by Knight and Williams[3], and later studied in more detail by Shaw *et al.*[4] using decane droplets about 1 mm initial diameter. The small heptane droplets studied here were not observed to explode (at least the droplet size at which such explosions may have occurred was too small to be optically resolved); no mention of explosions was made in connection with the heptane experiments of Hara and Kumagai[2] which involved a droplet of about 920μm initial diameter. The phenomenon might somehow be sensitively dependent on droplet motion, initial droplet size, method of droplet deployment or ignition, soot shell or flame structure, etc., as these aspects relate to the formation and movement of the soot shell with respect to the droplet.

3.2 Binary mixtures

The mixture components were chosen to yield nearly ideal solutions in order to facilitate physical property predictions which might be required in future model development. The mixture components studied thus far were heptane and hexadecane.

Figure 5 shows the temporal variations of the square of the droplet diameter for one particular 0.5 mole fraction hexadecane in heptane mixture. These data come from drop 7 in Figure 3. Similar trends were also observed for a 0.33 and 0.75 mole fraction hexadecane in heptane mixture. Droplet heating was revealed by an initial period after ignition during which the droplet diameter was approximately constant. Such heating has also been shown to prevail in the combustion of multicomponent droplets at earth normal gravity[9,10].

The solid line in Figure 5 is a curve of best fit, which was drawn to represent a possible trend of droplet diameter squared. The data are not sufficient to conclusively show the typical three-stage burning that has been observed for certain types of miscible mixtures of liquids. One reason could be the resolution of the photographic images, but this is just conjecture. There may be evidence, though, that between 0.1 s and 0.15 s after ignition hexadecane began to dominate the vaporization process. The solid line was drawn to lend support to that possibility. Such a change would indicate a form of preferential vaporization previously observed by others[10].

Microexplosions were not observed for the heptane/hexadecane mixtures (at atmospheric pressure) over the range of compositions studied. Theoretical predictions of this phenomenon are based on the droplet temperature and its superheat limit commensurate with the droplet composition and ambient pressure. If the droplet temperature exceeds the superheat limit of the liquid, microexplosions are possible, though the extent to which the process can shatter the droplet depends on the rate of growth of the initial bubble.

A very simple model considered that the temperature at the edge of the mass diffusive boundary layer within a multicomponent droplet (where the temperature was assumed to vary linearly), commensurate with the liquid phase Lewis number, dictated whether or not the droplet had the potential for microexplosions[11]. If this temperature is compared with the predicted superheat limit for a heptane/hexadecane mixture[12], it is seen to be below the superheat limit for the liquid compositions studied at 0.101MPa. However, if the droplet temperature is taken as spatially uniform and equal to the boiling point of hexadecane, then the occurrence of homogeneous nucleation within the droplet is more likely. The literature has revealed situations in which an unsupported heptane/hexadecane mixture droplet has[13] and has not[10] undergone microexplosion at earth normal gravity. Gravity alone will not effect the superheat limit of a liquid. However, the method of droplet deployment, soot formation around the droplet, etc., could possibly influence the liquid composition within the droplet and thereby the superheat limit, the droplet temperature, and thus the propensity for microexplosions.

ACKNOWLEDGEMENT

This study was supported by the National Science Foundation through Grant No. CBT-8451075 and the New York State Center for Hazardous Waste Management. This support is gratefully acknowledged.

REFERENCES

1. S. Okajima & S. Kumagai, 1975. *15th Symp. (Int.) Comb.*, pp. 401-407.
2. H. Hara & S. Kumagai, 1988. Poster No. P257 *22nd Symp. (Int.) Comb.*
3. B. Knight & F. A. Williams, 1980. *Comb. Flame* **38**, 111-119.
4. B. D. Shaw, F. L. Dryer, F. W. Williams, & J. B. Haggard, 1987. Paper No. IAF-87-403, International Astronautical Congress, Brighton, England, 10-17 October.
5. C. T. Avedisian, J. C. Yang, & C. H. Wang, 1988. *Proc. Roy. Soc. London, A***420**, 183-200.
6. M. Y. Choi, F. L. Dryer & J. B. Haggard, 1989. This volume and chapter.
7. C. K. Law & F. A. Williams, 1972. *Comb. Flame* **19**, 393-405.
8. M. Goldsmith, 1956. *Jet Propulsion* **26**, 172-178.
9. B. J. Wood, H. Wise & S. H. Inami, 1960. *Comb. Flame* **4**, 235-242.
10. C. H. Wang, X. Q. Liu,& C. K. Law, 1984. *Comb. Flame* **56**, 175-197.
11. T. Niioka & J. Sato, 1986. *21st Symp. (Int.) Comb.*, pp. 625-631.
12. C. T. Avedisian & J. R. Sullivan, 1984. *Chem. Eng. Sci.*, **39**, 1033-1041.
13. J. C. Lasheras, A. C. Fernandez-Pello, & F. L. Dryer, 1980. *Comb. Sci. Tech.* **22**, 195-209.
14. Task Group on Fundamental Physics and Chemistry, National Research Council, 1988. **Space Science in the Twenty-First Century.** Chapter 5, National Academy Press, Washington, D.C. .

APPENDIX

The intent of the experimental design was to create a low buoyancy environment. However, gravity alone does not dictate the extent to which buoyancy induced flows will be minimized in our experiment. That is, there is no single acceptable gravity level below which buoyancy induced flows can be neglected[14]. Similarly, there is no single minimum acceptable droplet velocity for the effective free stream flow around the droplet to have a negligible effect on the radial flow field. For a nonzero gravitational level and a nonzero droplet velocity there will be an axial vapor flow around the droplet.

The criterion for neglecting buoyancy induced flows around the droplet is that the Grashof number be small, that is

$$Gr_D \ll 1 \quad (1)$$

where

$$Gr_D = g\beta(T_f - T_\infty)D^3\rho^2/\mu^2 \quad (2)$$

Similarly, effective forced convective flow around the drop may be neglected when

$$Re_D \ll 1 \quad (3)$$

where

$$Re_D = U_\infty Dr/\mu \quad (4)$$

There are several length scales upon which calculation of Gr and Re can be based, for example the droplet diameter D or the flame diameter D_f. D is apparently a convenient length scale for correlating droplet burning data as evidenced by the relatively large number of correlations for the burning rate which have used this length scale in the definition of the Grashof and Reynolds number for burning droplets[7].

Equations 1 and 2 show that lowering gravity is one of *several* ways to minimize buoyancy in a droplet experiment. To further illustrate, consider an ideal gas where $\beta \sim 1/T_\infty$, $\rho \sim P/(RT_\infty)$ and with $\mu \sim T_\infty^{1/2}$. The criterion for neglecting buoyancy induced flows and droplet motion becomes

$$C_1(T_f - T_\infty)\, P^2\, D^3 g / T_\infty^4 << 1 \qquad \text{[to neglect buoyancy]} \qquad (5)$$

and

$$C_2 U_\infty\, D\, P/T_\infty^{3/2} << 1 \qquad \text{[to neglect droplet motion]} \qquad (6)$$

where C_1 and C_2 are constants with the appropriate units. Equation 5 shows that there are four ways by which buoyancy effects can be minimized (all of which have been exploited in the literature): 1) lower g; 2) carry out the experiment at low pressure, 3) reduce the droplet radius; or 4) increase the ambient temperature. Equation 6 shows that not only can U_∞ be reduced, but the droplet radius or pressure can be reduced as well to achieve the same dynamic effect.

The restrictions on droplet velocity are less severe, though perhaps more difficult to satisfy experimentally, than on gravitational level because $Gr_{Rs} \sim D^3 g$ while $Re_{Rs} \sim DU_\infty$. For example, if the initial droplet diameter is doubled from 500μm initial diameter which characterizes our experimental method to, say, 1000μm which is typical of all previously performed low gravity droplet combustion experiments, then the gravity must be reduced by nearly an order of magnitude ($\{0.5/1\}^3$) to yield an equivalent dynamic effect. If droplets in the 2mm to 5mm diameter range are to be studied, then gravity must be reduced yet further from the present 10^{-3} g level to about 10^{-5} g and 10^{-6} g, respectively.

For the conditions of the present experiment, we follow the prescription of Law and Williams[7] to arrive at the following property estimates: $\rho \sim 2.68 \times 10^{-4}$ g-cm^{-3} (air), $\mu \sim 4.96 \times 10^{-4}$ g-cm^{-1}-s^{-1}(air), $T_f \sim 2300$°K, $T_\infty \sim 300$°K, $\beta \sim 7.69 \times 10^{-4}$°K^{-1}, and g~0.98cm/s^2 (for the present experiment). Since $D \sim 0.05$cm, Equation 1 shows that $Gr_D \leq 5.5 \times 10^{-5}$. If the characteristic length scale is chosen as the flame radius, and we further assume that $D_f/D \sim 10$, then $Gr_{Df} \sim 5.5 \times 10^{-2}$.

FIGURE CAPTIONS

Figure 1 Drop tower facility.

Figure 2 Comparison of the variation of non-dimensional droplet diameter squared $(D/D_o)^2$ with non-dimensional time t of the present heptane data with those reported in References 1 and 2: $k_g = 0.1141 \times 10^{-2}$ W/cm K; $\rho_l = 0.6139$ g/cm^3; $C_{pg} = 4.283$ J/g K. Data from References 1 and 2 were interpolated.

Figure 3 Photographic sequences from six heptane runs (Drop 1 to Drop 6), and a 0.5 mole fraction hexadecane in heptane mixture (Drop 7).

Figure 4 A burning toluene droplet. Figures 4a and 4b show direct back-lighted photographs taken 0.056 s and 0.136 s respectively after ignition (total burning time for this droplet was 0.364 s). Figures 4c and 4d are computer-generated images of the photographs shown in Figures 4a and 4b.

Figure 5 Variation of diameter squared with time for a 0.50 mole fraction hexadecane in heptane mixture droplet.

7. Meteorology

Clive Saunders
Session Chair

ON THE CRYSTALLIZATION OF SPHERES AND SHELLS

J. Hallett
Desert Research Institute
Reno, Nevada 89506

ABSTRACT

Isolated drops of many solutions or pure materials readily supercool below their equilibrium crystallization point. In some cases a glass forms; otherwise the crystallization process is usually initiated at a specific point in the drop leading to a complex structure. Under moderate supercooling, in the first stage of crystallization dendrites grow throughout the drop, leading to latent heat release, and temperature rise. The dendrite growth velocity, tip radius and crystal orientation depend critically on the supercooling.

The second stage of crystallization depends on the geometry of the heat loss from the periphery and results in freezing from one side of the drop for asymmetric heat loss or as a thickening shell for symmetrical heat loss. Most solute is rejected at this stage, to nucleate in a geometry determined by the original dendrite distribution and the growth interface, and the internal pressure as it responds to volume changes and cracking of the shell. New crystal orientations appear for nucleation at large supercooling and also in large drops, following reorientation of dendrite arms separated during Ostwald Ripening.

Crystal growth is usually uninfluenced by the interface in a gas or liquid environment but changes habit to thin needle like crystals, and increases growth velocity for a solid interface. Drops of size of the tip radius will no longer be spherical on crystallization. For a solvent which evaporates, crystals grow faceted (as hydrates); surface tension forces may move solvent over the surface to give well formed faceted crystals.

It is suggested that faceted defect free crystals can be grown from drops under low g using controlled nucleation in a controlled vapor and temperature environment in the absence of convective motion.

INTRODUCTION

The crystallization of spheres is of interest in a number of areas of endeavor and raises questions of our understanding of crystal growth, together with latent heat and mass transport in regions of confined geometry. Applications lie in subjects as diverse as atmospheric science and metallurgy. Cloud droplets and raindrops freeze and solution droplets solidify to solute or eutectic; lead shot has been manufactured since the 18th century by free fall and crystallization of molten lead in a tower some 200 ft high; drops solidify to give the raw material for sintered metal. Shell crystallization is of interest in foamed materials, and the production of inertial fusion targets. More fundamental questions arise in understanding how crystals nucleate and grow in a uniformly supercooled or supersaturated environment, how these crystals interact with the drop or shell surfaces, and how the crystallization goes to completion. The shape of the drops may be reconfigured following volume phase change, subject to surface tension effects and solvent evaporation. The redistribution of solute impurities is of considerable importance, for example in determining the mechanical properties of a frozen drop or shell, and also in assessing the possible enhancement of chemical reactions during the freezing of "acid" rain. A further consideration is the solidification of a droplet as it is accreted on surface, as for example in rimed snow or during "splat" cooling of metal drops on a cold surface.

THE CRYSTALLIZATION PROCESS
A. Initial freezing — the adiabatic process.

As a solution is cooled below its point of crystallization or melt below its equilibrium melting point T_m, it becomes metastable with respect to its solid phase, such that if the solid phase be present, it will grow at the expense of the liquid or solute. In the absence of the solid, the liquid becomes progressively supersaturated (supercooled for a melt) until nucleation takes place either by contained impurities (heterogeneous nucleation) or by the formation of a nucleus of the solid sufficiently large to grow beyond the Kelvin equilibrium radius (homogeneous nucleation). See Hobbs 1974; Strickland-Constable, 1968; Pruppacher and Klett, 1980.

In practice, it is found that small quantities of liquid (μl) readily supercool, and only nucleate when supercooled by 0.2 to 0.3 T_m. This is true for metals, organic and inorganic liquids. The only exception to this is highly viscous liquids which supercool to a glass, and only crystallize when they are warmed up to a temperature such that the crystal growth is sufficiently rapid. From the viewpoint of laboratory time scales of ~10 ^{3}s, ml water samples readily supercool by -20°C and in smaller volumes to -40°C, nickel (mp 1455°C) by 400°C. Sodium acetate trihydrate (mp +53°C) on the other hand, readily supercools and forms a glass below about 50°C supercooling, and only crystallizes when it is slowly warmed back above this temperature. Drops freely suspended — either falling at terminal velocity in particle free air or in a electrostatic/acoustical levitation system (Rhim *et al.*, 1982) have a greater chance of not encountering a suitable nucleating particle as drops in contact with a solid surface, and can, in general, be supercooled by a much greater extent. The smaller the volume of a drop, the smaller the probability of it containing a freezing nucleus; homogeneous nucleation can be achieved in principle and practice by producing drops sufficiently small.

The rate of crystallization and the crystal form depends on the material, the supersaturation or supercooling. In the case of a solution both heat transport and mass transport are important in the crystallization process; the latter is usually rate determining in the absence of kinetic effects in as far as thermal diffusivity is usually much greater than solute diffusivity (Ohara and Reid, 1973). The governing equations are:

For heat
$$\frac{dQ}{dt} = 4\,\pi F_h\,KC\,(T_S - T_\infty)$$

For mass
$$\frac{dm}{dt} = 4\,\pi F_m\,DC\,(\rho_S - \rho_\infty)$$

For local balance
$$L\,\frac{dm}{dt} = \frac{dQ}{dt}$$

F represents enhancement caused by fluid motion, in practice orientation dependent.

D, K coefficient of solute diffusion and heat conductivity.

C is a geometry factor equivalent to electrostatic capacitance.

L is the latent heat of fusion

T_∞, ρ_∞ are conditions a long way from the crystal;

T_s, ρ_s are the surface conditions, given simplistically (equilibrium assumption, no effect of kinetics) by the Clausius Clapeyron equation, but in practice modified by growth radius through the Kelvin equation and kinetics on different faces.

The detailed solutions are complex for any real situation. For low viscosity liquids nucleation and crystal growth occurs prior to glass formation; much ingenuity has been expended in achieving high cool rates to form glass in such substances.

In the case of small supercooling and small growth rates, the dislocation structure of the crystal may be important in determining the rate processes on different crystal faces; otherwise the occurrence of preferred crystal growth directions will depend on the occurrence of thermodynamically smooth or rough crystal faces. At larger supercooling, fluids with high viscosity slow the growth as molecules lack sufficient activation to enter preferred sites in the growing lattice; a glass forms, and crystal growth ceases.

For liquids which do not readily form a glass, such as water, the linear growth rate of rough surfaces in the form of dendrites can be represented by a power law

$$V \alpha \Delta T^n;$$

for aqueous systems of low viscosity n~2 (Hallett 1964; Lindenmeyer and Chalmers, 1966; Pruppacher 1967). For defect growth on facets at small ΔT (0.1°C) the growth law is exponential, which becomes a power law for large ΔT as surface nucleation dominates; the faceted surface becomes rough under these conditions.

Ice is an example of a crystal with marked growth rate anisotropy (100 to 1) in both liquid and vapor; in the liquid the anisotropy is a maximum for small supercooling ~ 0.1°C and becomes small at 10°C, where growth rate is a maximum at an angle of ~22°C to the basal plane (Macklin and Ryan, 1966). With passage of the dendrite through a given position, the temperature rises to near T_m, and is given, as equilibrium is approached, by the mean Kelvin radius of the crystals. In an isolated system this changes slowly with time, Ostwald Ripening takes place with high curvature surfaces melting back and low curvature surfaces growing (Vorhees and Glicksman, 1985; Marsh and Glicksman, 1987). This situation might be expected to occur in the centre of a large volume in an environment close to T_m with minimum interaction by heat flow from the periphery. For a solution there is a further complication, as the concentration changes following preferential rejection or incorporation of the solute (DeMicheli and Iribarne, 1962), so that the equilibrium temperature also changes. Ice growth in sodium chloride solution leads to rejection of ions giving a denser solution which convects downwards; sodium sulphate decahydrate rejects water molecules leading to less dense fluid and upward convection (Hallett *et al.*, 1987). It follows that only partial solidification takes place during the initial phase of adiabatic crystallization, leaving complete freezing, and eutectic solidification if appropriate, to take place as further heat is removed at the periphery, by conduction, convection or radiation depending on the system.

The crystal size during growth, as characterized by the tip radius, which is inversely related to supercooling or supersaturation. (Hallett, 1964; Huang and Glicksman, 1981a). This implies that crystallization takes place quite differently in a system of the size of this radius than in a system much greater than this radius. In the latter case, the dendrites propagate throughout the volume to the drop surface. For ice, the tip radius at ΔT -1°C (-10°C) is ~50 μm (10 μm). Larger drops crystallize as a mush, with a fraction

$$\frac{\sigma \Delta T}{L}$$

(σ = liquid specific heat, L = fusion latent heat) solidifying initially. This is ~1/8 for water supercooled by 10°C. Observations are lacking for smaller drops but it would be expected that the drop shape would be dominated by the crystal tip radius, and liquid move over the crystal through gradients of surface tension.

B. Secondary Freezing

Subsequent freezing depends on the geometry of heat removal from the periphery. In a stationary environment, heat loss in low Reynolds, and low Grashof numbers (as in low g) is symmetrical, and freezing proceeds symmetrically inwards.

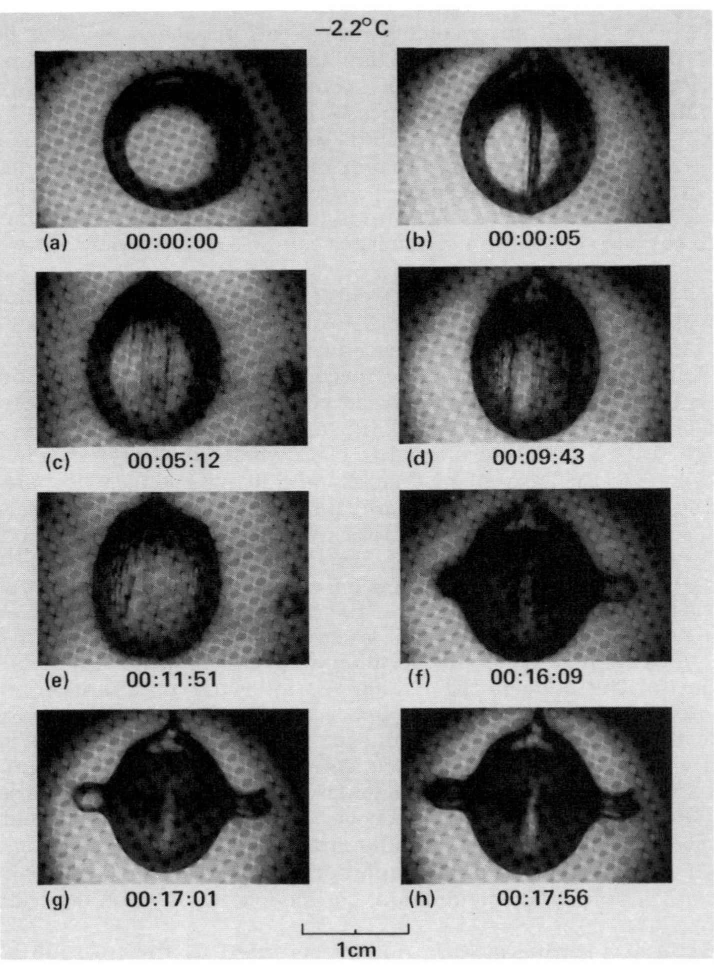

Fig. 1: Spikes produced by water expansion on freezing from a single crystal drop. (a) Water drop suspended in immiscible liquid nucleated by a single crystal inserted at the top of the drop with "a" axis normal to the drop surface. (b) Shows the initial dendrite, which thickens (c, d), changing the drop shape from surface tension effects (e) and producing spikes (f – h) as solidification completes and the water is expelled by expansion. Note air bubble structure in (h) in the central regions of drops and spikes as the inward freezing goes to completion.

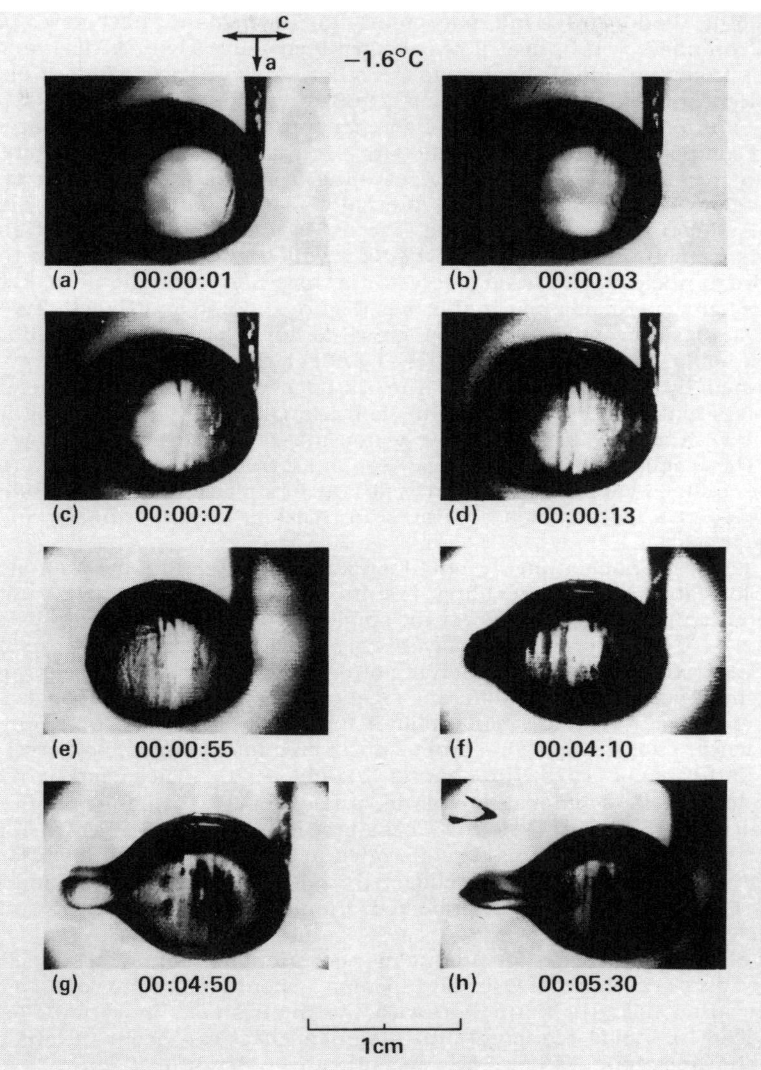

Fig. 2: Similar to Fig. 1, only the drop is nucleated with a crystal having its "a" axis tangential to the drop surface.

On the other hand in free fall, particularly for Reynolds number flow $\gtrsim$100 and in Grashof numbers > 100, heat flow becomes increasing asymmetrical, and freezing proceeds from one side of the drop to the other. This will occur for a drop falling in fixed orientation (Johnson & Hallett 1968). The former case leads to inward freezing and entrapment of rejected components in the middle; the latter pushes rejected components to one side of the drop. Physical entrapment of solute may also occur in the dendrite mush a way resulting from the initial growth of crystals. Volume changes on freezing lead to internal pressure changes and mass transport. Ice and silicon expand on freezing, the excess liquid being ejected as freezing proceeds, sometimes through hollow spikes which grow outwards. In the case of water drops nucleated by a single crystal "a" axis normal to the drop, symmetrical ice spikes grow perpendicular to this axis (Figure 1) (Zhao and Hallett, 1989). In the case of "a" axis tangential to the drop, growth occurs slowly in "c" axis direction, and only one spike forms (Figure 2). It is evident that the orientation of spikes is determined by the orientation of the initial crystals following nucleation. A favorable situation occurs with a threefold triangle of crystals growing in "a" axis direction with "c" axis in the surface; water is ejected outward in a long triangular spike. These spikes can sometimes be seen in ice trays in a refrigerator (Figure 3) or in other confined volumes such as a melt hole in a glacier (Figure 4) (Hallett 1960). In the case of a metal which contracts on freezing, (lead) shrinkage and internal cavities appear.

In general, nucleation at small supercooling results in a single crystal, nucleation at large supercooling results in a polycrystalline structure, the transition temperature being lower for smaller drops; 1mm water drops are single crystal to ~-10 °C (Figure 5), and polycrystal at lower temperature (Figure 6). Several possibilities exist for the production of polycrystalline drops. It appears that a self-epitaxy may occur between different facets in the ice lattice — basal and prism planes are closely related. Some higher index planes give a 70° orientation observed, where supercooled cloud drops impact on a much larger snow crystal surface (Uyeda and Kikuchi, 1978; Hiroshi and Kikuchi, 1976). These can be related to a <30$\bar{3}$4> or <30$\bar{3}$8> composition plane, analogous to twin formation in calcite (Furukawa and Kobayashi, 1978). These processes are more likely at high supercooling. Such reorientation is demonstrated in crystals of new orientation growing from the vapor on cloud drops collected and frozen on a large dendritic snow crystal; these crystals have a preferred orientation (Iwai, 1971; Furukawa and Kobayashi, 1978).

An alternative process for producing new orientations may occur, particularly for larger drops, in terms of Ostwald Ripening. Should melt occur of high curvature dendrite arms near their junction with the main stem, buoyancy effects could readily lead to crystal reorientation. A similar effect can occur in larger systems due to buoyancy induced flows over recently grown dendrites (Figure 7), conditions for convection are dependant on the cavity Rayleigh number and the dendrite size.

When single crystal frozen drops are grown slowly from the vapor they form well developed facets; (Yamashita and Takahashi, 1972); a single crystal of rock salt suspended in solution behaves similarly (Schnorr, 1928). At high supersaturation thin plates, dendrites or hollow columns grow in the same direction as the original crystal orientation giving planar crystals or rosettes depending on whether the frozen drop was a single or polycrystal (Jiusto and Weickmann, 1973).

In the case of rapid external cooling, the spike tip is sealed and internal pressure builds up. When ice freezes from water, air in solution is rejected and becomes supersaturated; eventually bubble nucleation occurs. Should a reduction in pressure occur, air bubble nucleation occurs in the region of maximum supersaturation–

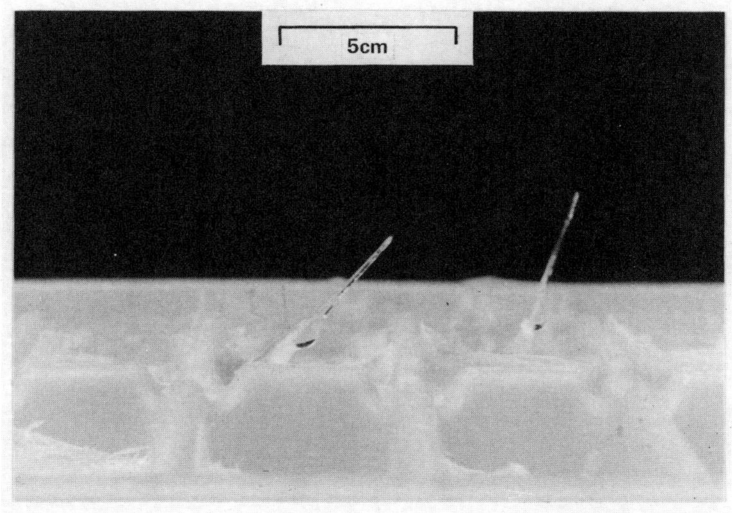

Fig. 3: Ice spikes from a refrigerator plastic tray.

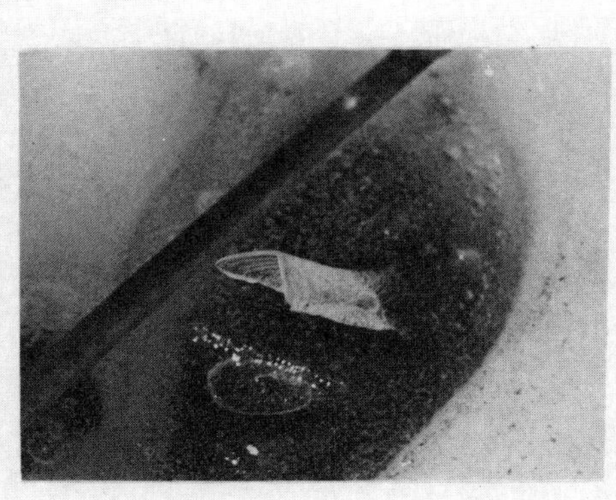

Fig. 4: A tricrystal 7 cm long ice spike from a freezing melt hole in a glacier, showing crystal facets. One face subsequently was melted and the water drained away.

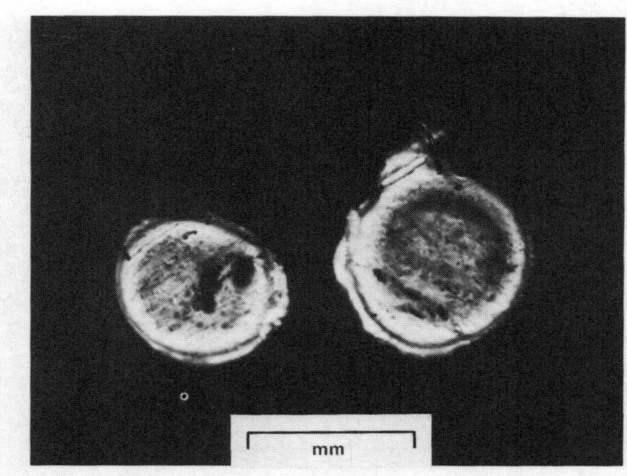

Fig. 5: Single crystal water drops, frozen at small supercooling by contact with ice crystals, viewed in polarized light. The right hand drop has produced a spike. These drops were produced in the spray from Old Faithful Geyser and nucleated by ice crystals formed by dry ice seeding at -78°C.

Fig. 6: Water drops frozen in LN_2 viewed in polarized crystal showing polycrystalline texture. Bubble rings are just visible resulting from nucleation following pressure release on fracture.

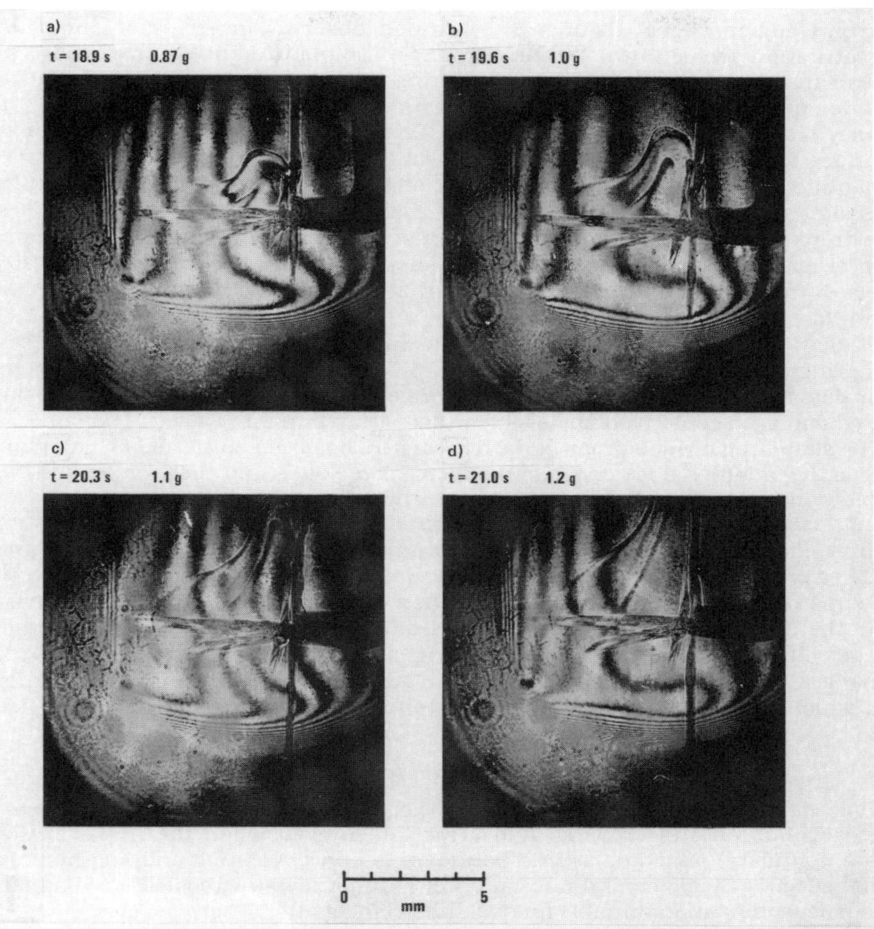

Fig. 7: Initiation of convection from a growing crystal of Na_2SO_4-$10H_2O$ as gravity increases in a KC - 135 aircraft parabolic trajectory viewed in a Mach Zender optical system (From Hallett, et al, 1987).

at the growing interface (Figures 8, 9). Such observations therefore map the supersaturation profile in the freezing drop, as a miniature bubble chamber. The pressure may be released periodically during cracking; in this case concentric nucleation of rejected solute rings can occur (Bari and Hallett, 1974); eventually the drop may shatter catastrophically (Visagie, 1969). A drop freezing upward from an ice surface shows superficially a similar effect (Figure 10), which, however results from a quite separate process. This is a periodic nucleation and growth of bubbles and is similar to the formation of Leisgang rings—this occurs in crystallization of solute drops (Figure 11). Even in the absence of such events, unusual spatial distribution of air bubbles occur because of trapping and subsequent freezing of water pockets during the initial dendrite growth—this gives slabs of bubbles, revealed as lines in thin section (Figure 12).

Similar rejection of ions from solution during freezing of raindrops or dew drops can give enhanced concentration with pH reduction and an enhanced potential for damage in acid rain situations. The role of these phase changes in causing localized damaged could be of major importance (Harrison and Hallett, 1988a.).

The shattering of such drops in the atmosphere has been suggested as important in nucleating additional ice crystals in supercooled clouds and therefore enhancing precipitation. Such a process would apparently only occur with symmetrical drop freezing, associated with a drop tumbling as it falls, subject to oscillatory motion set up by eddy shedding in its wake. It is estimated however that other processes associated with riming may be numerically more important (Mossop, 1985). In this case, supercooled droplets frozen on accretion are subjected to differential strain during the second stage of solidification, leading to occasional shatter of drops diameter >25 μm in the range of supercooling of 3 to 8°C (Dong and Hallett, 1989). Internal stresses in such frozen drops lead to generation of an extensive dislocation array, which is readily revealed by synchroton radiation X-ray topography (Figure 13).

C. Influence of drop boundaries

The relative growth rates along different crystallographic directions discussed earlier depends on supercooling, and are retained even when the initial crystal meets a boundary, providing such a boundary is non nucleating and has minimal thermal effects. Single crystal nucleation of a uniformly supercooled cm diameter drop suspended in an immiscible fluid at supercooling of ~ 1°C gives a crystal which grows in the "a" axis direction as a thin dendrite, fills in to become a circular disc (Figure 1a) and subsequently grows slowly in the "c" axis direction, until the drop has solidified, leading to the spikes in Figure 1e, f). The initial dendrite grows adiabatically; it grows subsequently by heat transfer through the drop and to the drop environment. Uniform growth in the "c" axis direction takes place by heat loss through the drop and periphery until solidification is completed. No preferential growth occurs at the water — liquid interface. This situation is in sharp contrast to drops in contact with a solid such as a metal or glass. The effect is demonstrated by measurement of growth velocity of ice crystals in capillary tubes of different material (Yang & Good 1966); ice growth velocity is much greater in these tubes compared with free growth. Figure 14 shows free growing dendrites and faster thinner crystals, on a glass surface. New crystal orientation may also occur when free growing dendrites meet a glass interface at sufficient supercooling > 2°C. It appears that the dendrites "reflect" on meeting the glass surface at an angle, with approximately equal angles to the normal as in Figure 15. This observation is explicable in part by enhanced heat transport from the growing crystal in contact with the solid compared with its supercooled melt. On the other hand the growth velocity in some plastic tubes is less than in free growth (Kerr *et al.*, 1987). This is

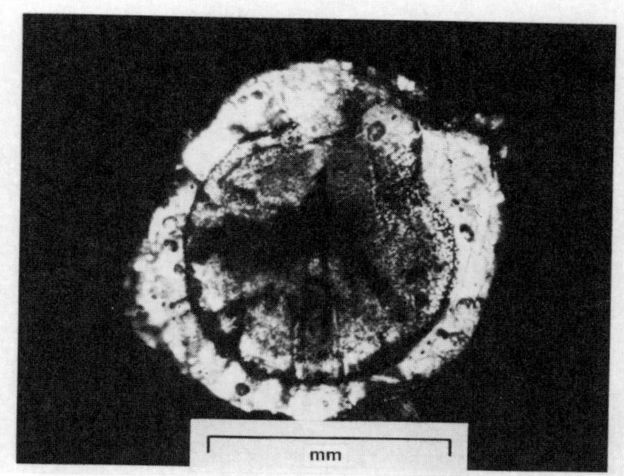

Fig. 8: A bubble ring and internal air bubbles nucleated in a drop supercooled, nucleated and frozen in LN_2. This bubble ring forms as pressure is released as the drop cracks open subject to internal pressure build up. Thin section, in polarized light, showing polycrystal structure.

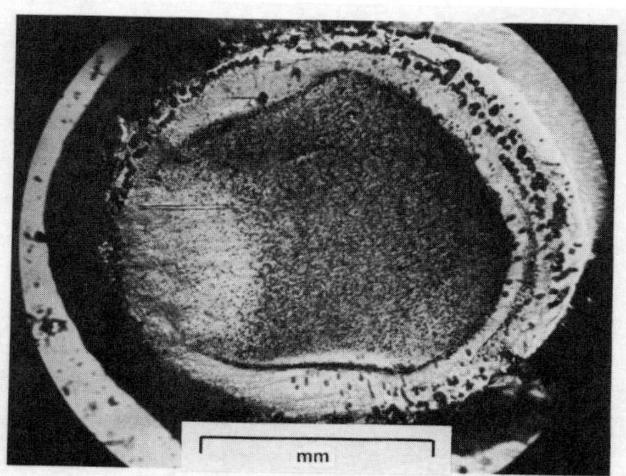

Fig. 9: Supercooled drop nucleated and frozen in silicone oil at -30°C. Note several air bubbles rings in outer regions; the pressure eventually built up to give a major failure leading to near uniform air bubble nucleation in the central region of the drop.

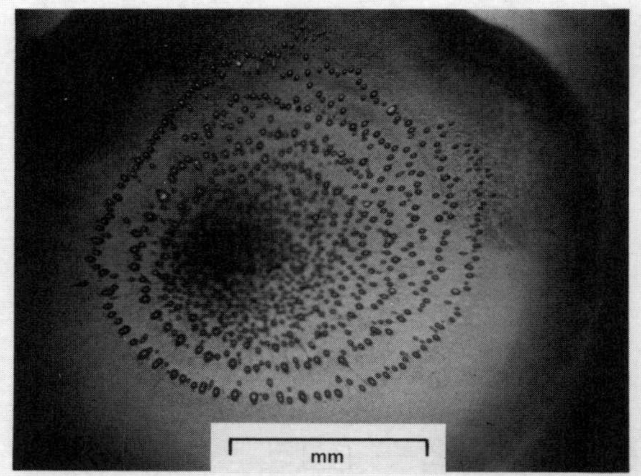

Fig. 10: Bubble rings formed as a drop impacts, spreads and freezes upwards from a cold ice surface. The periodicity is caused by an air supersaturation dependant nucleation; once bubbles from, supersaturation falls and growth ceases; the process then repeats.

Fig. 11: Periodic crystallization of a drop of sodium sulphate solution (as decahydrate) on a glass slide.

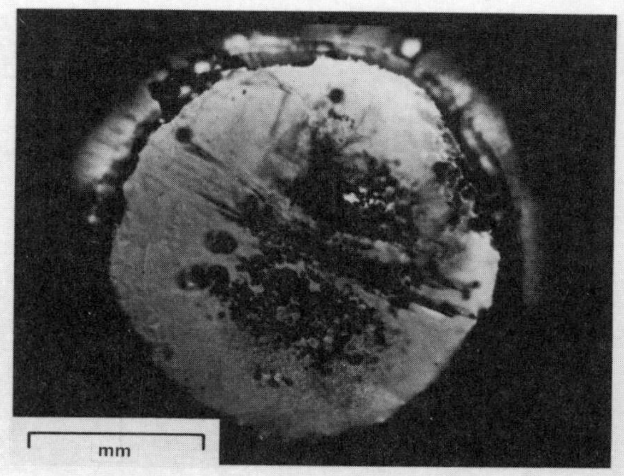

Fig. 12: A drop, in thin section nucleated at 5° supercooling crystallized to a single crystal. The air bubble distribution is determined by dendrite sheets which separated slabs of water during the first stage of freezing.

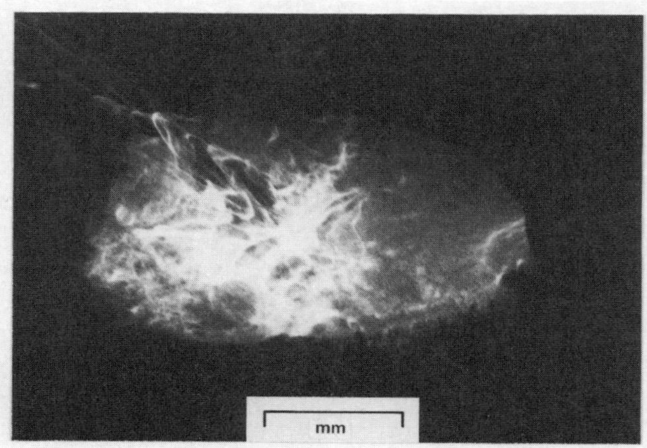

Fig. 13: Defect structure by X ray Synchrotron radiation of a drop supercooled by 2°C frozen, solid on a 0.2 mm glass support.

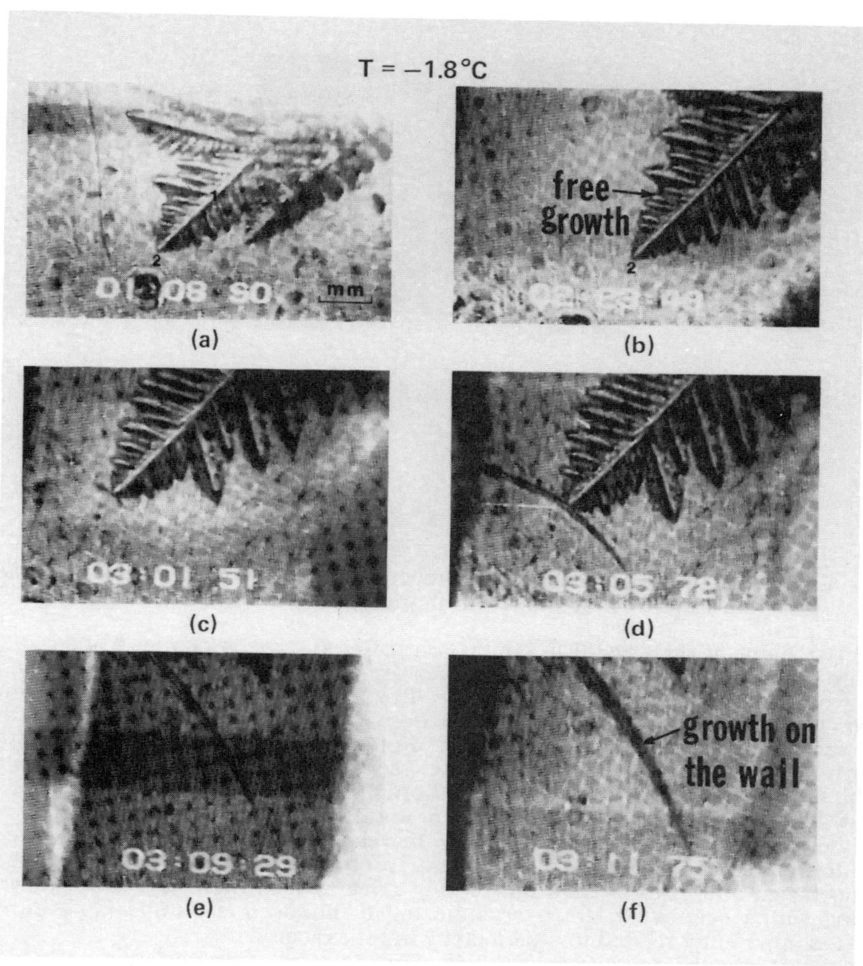

Fig. 14: Comparison between growth rate on the glass wall of a cell (curved, 0.28 cm s^{-1}) and free growth (dendritic, 0.20 cm s^{-1}) in water originally supercooled by 1.8°C.

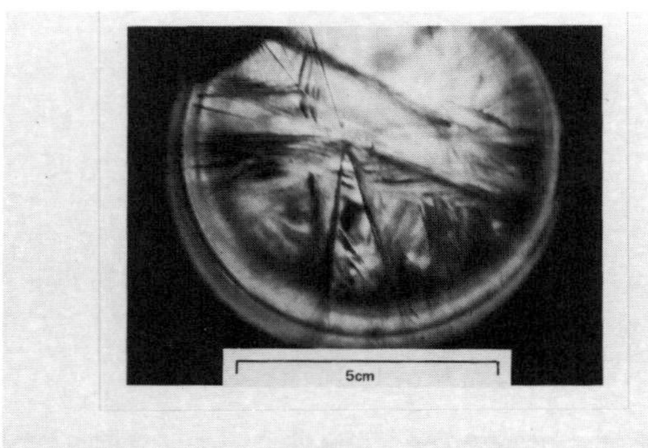

Fig. 15: Polycrystalline growth of ice in water supercooled by 4° in a glass dish nucleated by a single crystal, following secondary ice production on dendrite contact with the walls.

explained in part by reoriented crystals growing around the tube, rather than along the tube, which implies that the solid changes the kinetics of growth at the ice contact point.

The habit of the crystal in contact with the surface is changed, also as can readily be seen in Figure 14; these crystals are similar to the curved patterns of a thin water film crystallization in window "frost". It follows that crystallization of a drop isolated by suspension in air or vacuum is quite different from that of a drop on a surface or in a crucible. In the case of droplets on an ice single crystal substrate growth initially occurs as parallel dendrites, separation depending on supercooling. In the case of a liquid poured into a cold crucible, polynucleation occurs over the cold walls and only dendrites of favored orientation grow inward, slowing down as heat flow to the crucible is reduced with increasing thickness of solid. A second question arises concerning the crystallization of a drop with no rigid boundaries. As discussed above, expansion on freezing gives spikes; there is also an external ridging which occurs for drops crystallizing in immiscible liquids (as Figure 1) which is apparently related to a periodicity in "c" axis growth.

Crystal growth in supercooled shells can easily be examined in a supercooled soap bubble. Figure 16 shows a bubble of soap solution with 15% sucrose to slow down crystallization. The bubble is uniformly supercooled and nucleated at one point. The crystal growth is found to be uniform around the shell. It is inferred that the film stress is continuously deforming the crystal, so that it is bending, presumably by incorporation of defects as it grows. Growth in thin films may also be used to investigate the effect of temperature and solute field around a growing crystal tip. Figure 17 shows the approach of two dendrites in a flat supercooled film. The dendrites do not slow as they approach; it is inferred that the solution rejection field and the thermal field distance is less than the resolution of the photography, ~ 50 μm, which is consistent with simple theory. This technique has potential for studying these fields under a variety of conditions.

D. Crystallization of solutions

In the case of crystallization from solution at modest supersaturation, quite different crystal growth process occur. In the case that solvent is rejected from solution, it may evaporate depending on ambient conditions. At slow growth rates,

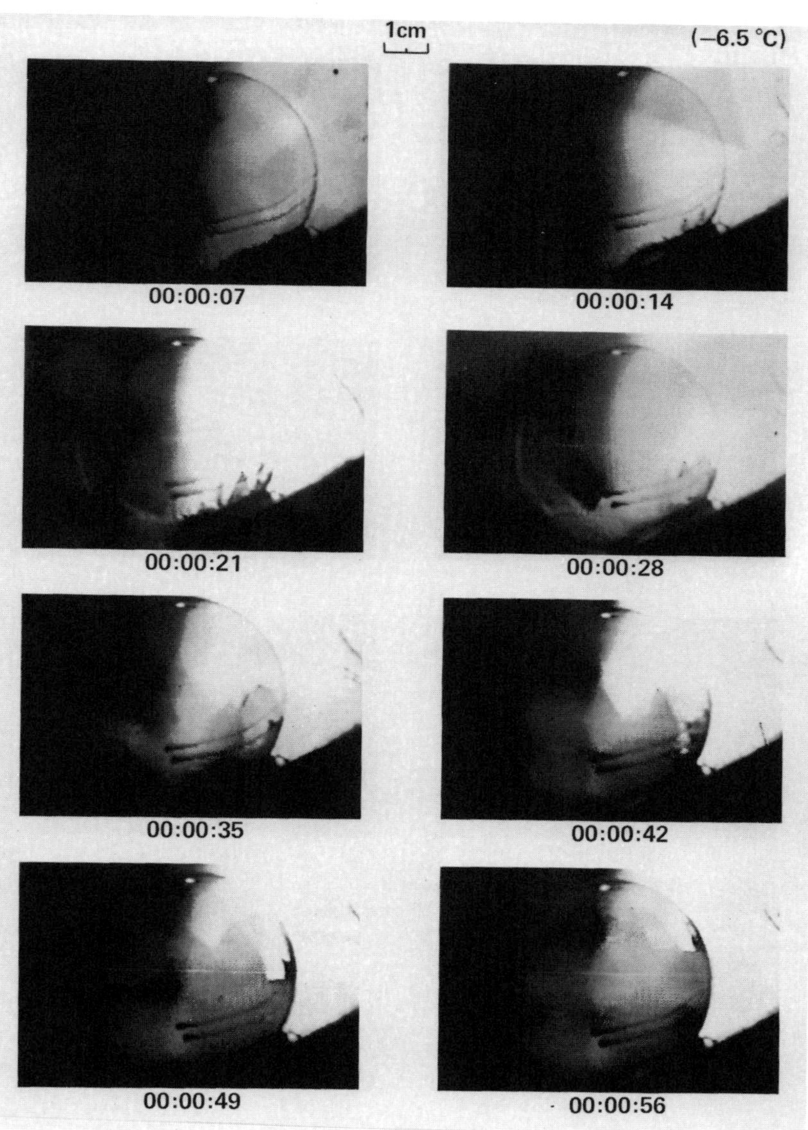

Fig. 16: Dendrite growth in a uniformly supercooled spherical soap bubble with 15% sucrose. The growth rate is constant around the surface of the drop, with continuously changing crystal direction.

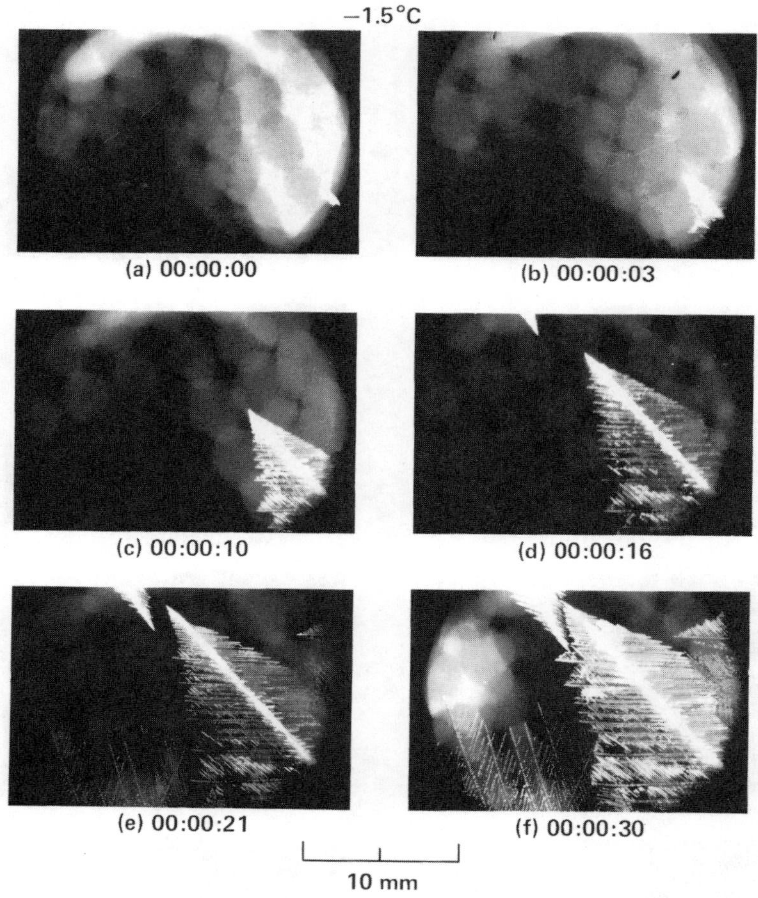

Fig. 17: Approach of two crystals in a flat soap film with 15% sugar. Crystals approach to within ~ 50 μm of each other without slowing down, showing lack of thermal or solute interaction.

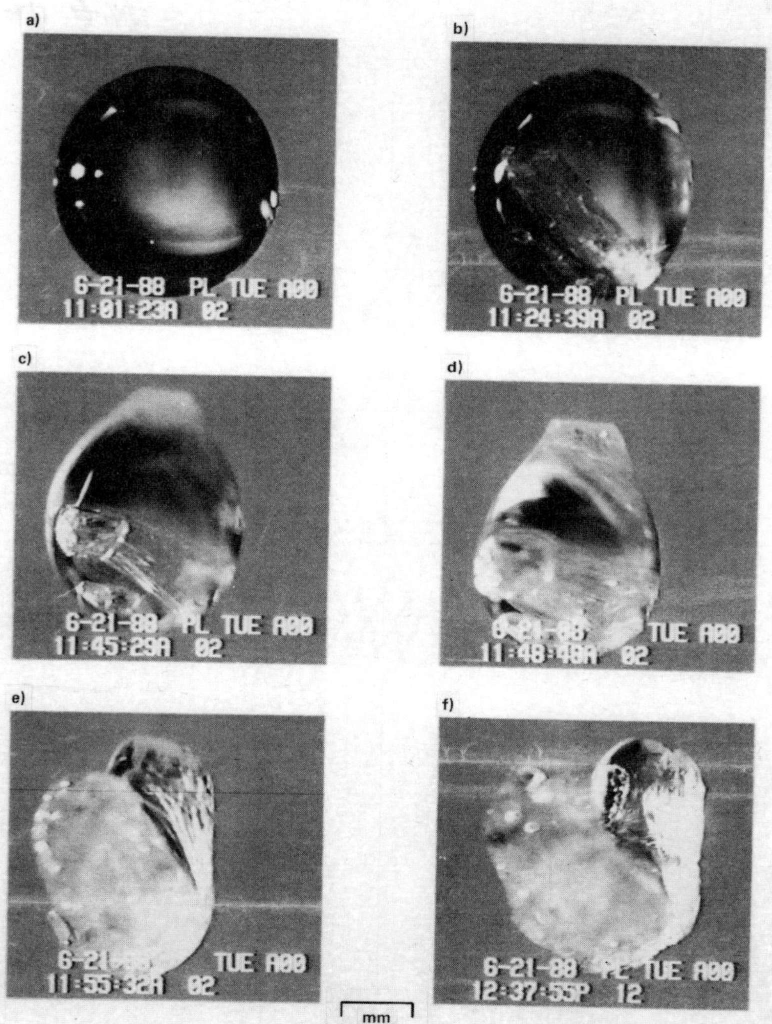

Fig. 18: Evaporation of a drop of sodium sulphate (crystallizing as decahydrate) in the Jet Propulsion Laboratory electrostatic suspension system. Two crystals nucleate and grow outwards by capillary creep of solution. The final photograph is still not completely crystallized; there is still fluid internal to the crystals (from Rhim and Hallett, 1989 to be published).

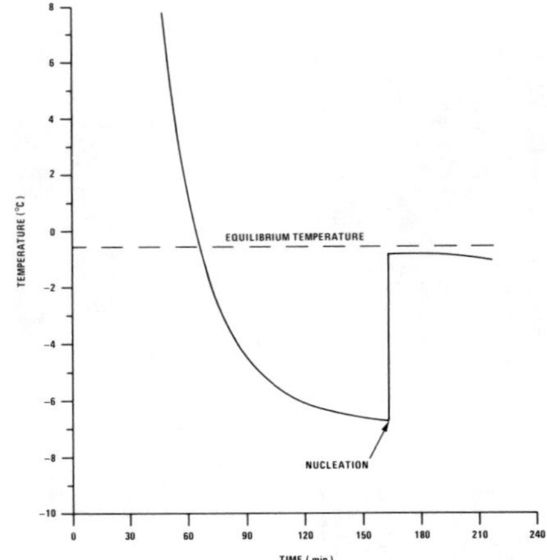

Fig. 19: Increase of temperature following adiabatic crystallization of ice from NaCl solution. Note that the initial temperature of the solution is 0.25°C below that of the equilibrium temperature of the original solution, resulting from solute rejection during ice crystal growth.

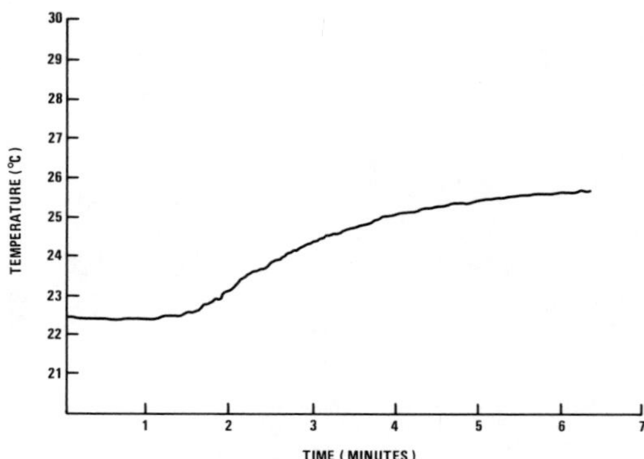

Fig. 20: Increase of temperature with time during adiabatic crystallization of Sodium Sulfate Decahydrate from solution. Crystallization visually appeared complete after 2 minutes, but final temperature was not reached until 4 minutes later, showing that slow growth and latent heat release were still taking place.

crystals showing facets occur, with rate processes determined by their defect structure. Under these circumstances, surface tension driven spreading occurs, the liquid migrating to those regions where more rapid growth is taking place. Hence the original drop changes shape; the original sphere gives way to specific crystal facets (Figure 18), (Rhim and Hallett, 1989). Under these conditions solute may sometimes accumulate inside a shell to give a hollow crystal, with slow internal growth occurring should cooling proceed further or slow evaporation occur.

Temperature changes during the adiabatic and second stage of crystallization demonstrate the different growth processes. For ice dendrites in solution growth (NaCl for example), solute is rejected to solution and temperature rises immediately (10^{-3}s) after nucleation and approaches equilibrium, (adiabatic crystallization) then falls slowly as lateral growth occurs until a new equilibrium is established with the solution of higher concentration (Figure 19). For hydrate crystallization with surface kinetics to give faceted growth, the temperature increases much more slowly (5 minutes) in adiabatic crystallization (Figure 20) and falls back even more slowly to the environmental value as heat is lost from the periphery. This process, together with the tendency of crystals to grow faster on the surface of a solid, gives cause for concern in measurement of temperature of a mush. With the sensing element between crystals, the measurement may be low; with the sensor in contact with a crystal the temperature may be high — that is, closer to the thermodynamic rather than the kinetic value.

A theoretical approach to calculating conditions at the crystal tip in solution growth gives only a small rise in temperature in adiabatic crystallization, in contrast to observation. Transport of heat should be much greater than mass transport which should leave a boundary of high concentration fluid around the growing ice crystal tip and inhibit growth. This apparently does not happen and suggests either a surface transport process at the molecular level away/towards the growing tip, or a modification of the growing tip to a smaller radius of curvature to give a thin diffusion boundary layer at the tip approaching molecular dimension.

Suspended solution drops can become highly supersaturated, in the absence of nucleation sites. These show rapid dendrite growth from solution similar to pure water; particularly, NaCl and $NH_4(SO_4)_2$, and sea salt crystals exhibit this behavior. These solutions commonly occur in haze drops in the atmosphere and crystallize under low relative humidity.

Earlier studies of crystallization of solution drops by evaporation at low humidity have shown, in general, extensive polycrystalline nucleation, with initial crystal growth as a hollow shell, with a hole (Charlesworth & Marshall, 1960). In the former experiments such evaporation took place with drops in a dry turbulent air stream; otherwise at terminal fall velocity in dry air (Leong, 1981; Cheng, *et al.*, 1988). Two considerations are important. First, since the diffusivity of solute in water is low (10^{-5} cgs), a highly concentred shell of solute can form around the periphery of an evaporating drop. Once nucleation occurs, drop internal motion readily redistributes crystals inside the drop leading to a polycrystal structure, and formation of "Fur" — presumably a whisker growth form on the outside fed by capillary flow of solution. This leads to local fluid motion analogous to the position dependent surface tension driven motions of the Marangoni Effect; its origin however lies in concentration gradients caused by differential crystal growth as well as differential evaporation. As gravity decreases, convection resulting from these effects would be expected to dominate over buoyancy driven convection. Crystallization of a suspended solution drop under low gravity and low relative motion with respect to its environment would be expected to proceed quite differently, giving more well defined crystal facets and an absence of periodicity in growth associated with convective fluid flow.

These experiments demonstrate that the possibility exists of growing a single crystal of known shape and size by controlled nucleation and growth under suspension in low g. Growth of crystals under low g will have a further effect in as far as mass and heat transfer to growing crystal surfaces will be inhibited compared with 1 g because of the lack of convection. This effect is important in solution growth at modest supersaturation where convective motion (characterized by a Grashof/Rayleigh number) is faster than the linear crystal growth rate (Hallett *et al.*, 1987, Hallett and Harrison, 1988). This is important conceptually, since growth of more perfect crystals is possible under low ventilation when faceted crystals can grow slowly in the absence of impurity incorporation or nucleation of additional defects in periodicities associated with the convection; the method has the potential for growth of faceted crystals of order of centimeter size.

CONCLUSION

This review has examined problems of crystallization of drops of pure materials and solution both in suspension as well as on a surface. A complex freezing processes results from different supercooling, or supersaturation, and different cooling and evaporation geometry. Figure 21 demonstrates important situations which occur. In the atmosphere, we are concerned with the solidification of pure water cloud drops and rain drops, which are substantially supercooled and have two distinct freezing stages, adiabatic and isothermal as first approximation. For small cloud drops it appears that shape changes may occur at modest supercooling. In the case of solutions in water, as haze droplets in the atmosphere, substantial supersaturation may prevail to give again two stages of growth, with the second stage dominated by evaporation. Many of these considerations apply quite generally to drop crystallization, of metals, organic materials, semiconductor materials and their solutions.

There are distinct differences in crucible or surface crystallization compared with suspended drop crystallization and differences yet again for droplets in suspension under low g. In suspended drops in low g there is no surface nucleation and no surface enhanced growth; in low g there is no internal convection and rearrangement of crystallites produced by Ostwald Ripening. This points to the simplicity of single crystal nucleation and growth in such suspended system. It appears that curved interfaces (positive or negative) fail to influence growth under situations of liquid or gas or vacuum. In the case of solution drop crystallization, the second stage of growth can substantially influence the distribution of rejected impurities, and any associated chemical reactions which take place in the liquid phase; freezing of acid rain or dew, on a vegetation surface may lead to a significant enhancement over effects predicted from initial concentrations, as the solute concentration increases.

An industrial application of these ideas is in freeze drying where initial freezing temperature determines the degree of dispersion of the ice in the secondary freezing stage; the drying rate is improved for crystals of optimum size. If the intent is to produce a crystalline shell of known characteristics— say for inertial fusion— it would appear that single crystal nucleation would only be desirable in a cubic system, as other systems would exhibit stress/strain anisotropy to give non-uniform collapse. In the latter case controlled geometrical nucleation— for example at tetrahedral points in a spherical shell — might be desirable.

ACKNOWLEDGEMENT

This work was supported in part by NSF Grant #8715636 and a NASA Grant #957764 through Jet Propulsion Laboratory, Pasadena, CA.

Figure 21: HOW DROPS CRYSTALLIZE

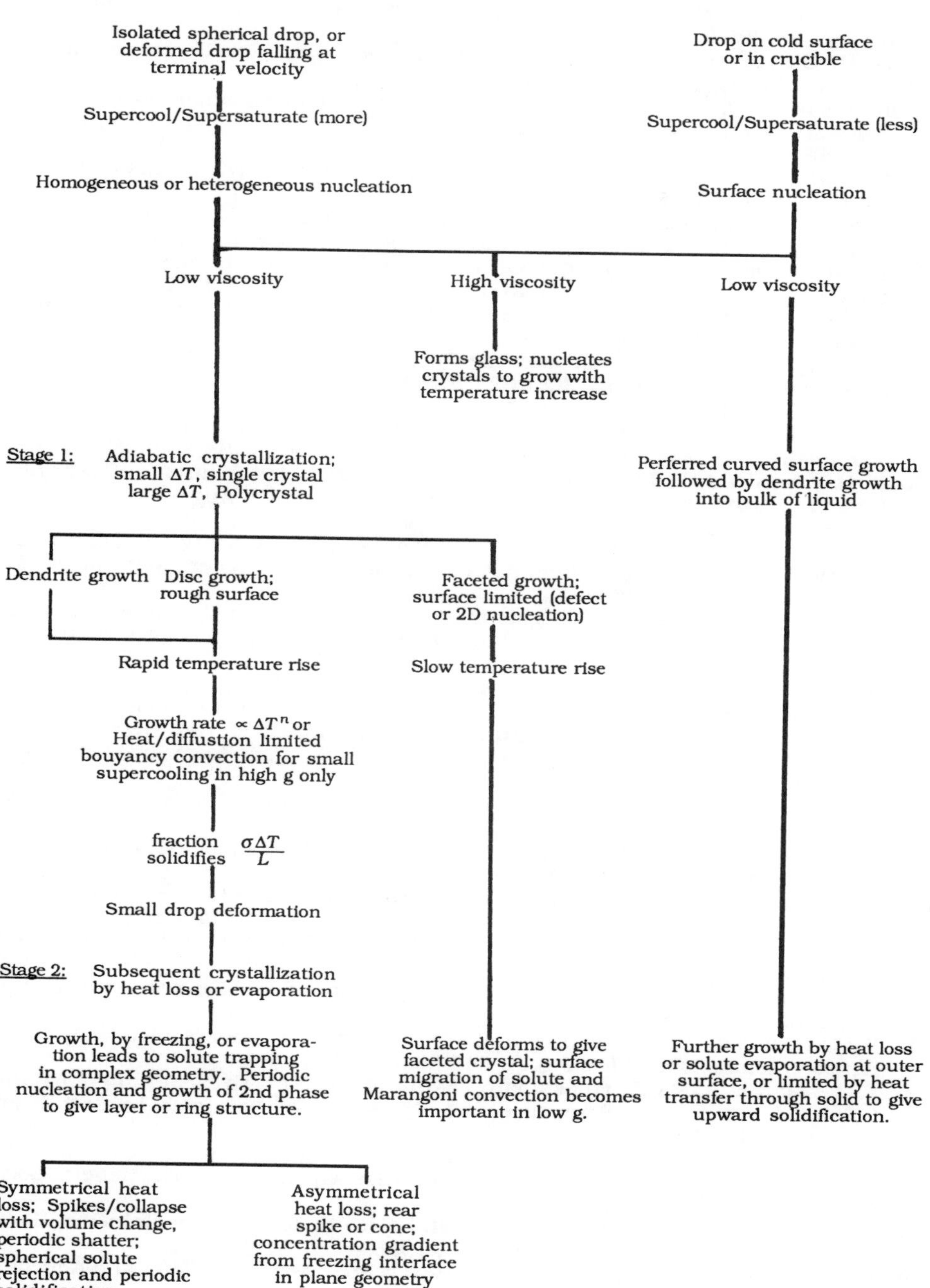

REFERENCES

1. S. Bari & J. Hallett, "Nucleation of Air Bubbles at a Growing Ice-water Interface," *J. Glac.*, **13**, 489-520, 1974.
2. D. H. Charlesworth & W. R. Marshall, "Evaporation from Drops Containing Dissolved Solids" *AIChE J.*, **6**, 9-23, 1960 .
3. R. J. Cheng, D. C. Blanchard & R. J. Cipriano, "The Formation of Hollow Sea-Salt Particles from the Evaporation of Drops of Seawater," *Atmos. Res.*, *22*, 15-25, 1988.
4. S. M. DeMicheli & J. V. Iribarne, "La Solubilitédes Electrolytes Dans La Glace," *J. Chem. Phys.*, **60**, 767-774, 1962 .
5. Y. Y. Dong & J. Hallett, "Droplet Accretion During Rime Growth and the Formation of Secondary Ice Crystals," *Roy. Meteor. Soc., Qtrly. J.* In press, (1989).
6. Y. Furukawa & T. Kobayashi, "On the Growth Mechanism of Polycrystalline Snow Crystals with a Specific Grain Boundary," *J. Cryst.Gr.*, **45**, 57-65, 1978.
7. J. Hallett, "Experimental Studies of the Crystallization of Supercooled Water," *J._Atmos. Sci.*, **21**:6, 671-682, 1964.
8. J. Hallett, "Crystal Growth and the Formation of Spikes in the Surface of Supercooled Water," *J. Glac.*, **3**:28, 698-704, 1960
9. J. Hallett, N. Cho., K. Harrison, A. Lord, E. Wedum, R. Purcell & C.P.R. Saunders, "On the Role of Convective Motion During Dendrite Growth: Experiments Under Variable Gravity, Final Report", NASA Contract Number NAS8-3465, (1987).
10. K. Harrison & J. Hallett, "Crystallization of Supercooled Solutions," Preprints, from **Proc. 10th International Cloud Physics Conference, August 15-20, 1988**, Bad Homburg, Germany (to be published) 1988a.
11. J. Hallett & K. Harrison, "Influence of High and Low Gravity On Convection Around Growing Crystals," **Proc. AIAA 26th Aerospace Sciences Meeting**, Reno, NV, January, pp. 1-11, 1988.
12. U. Hiroshi & K. Kikuchi, "Remeasurement of the Axial Angle between Spatial Branches of Natural Polycrystalline Snow Crystals," *J. Fac. Sci.*, Hokkaido University. *Series VII (Geophysics)*, **5**:1, 21-28, 1976.
13. Hobbs, P. V., **Ice Physics**, Oxford Pr., p. 837, 1974.
14. S. C. Huang & M. E. Glicksman, "Fundamentals of Dendritic Solidification II. Development of Sidebranch Structure," *Acta Metall.*, **29**, 717-734, (1981a).
15. K. Iwai, "Note on Snow Crystals of the Spatial Type," *J. Meteor. Soc. Japan*, **49**, 516-519, 1971.
16. J. E. Jiusto & H. K. Weickman, "Types of Snow Fall," *Bull. Am. Meteor. Soc.*, **54**, 1148-1162, 1973.
17. D. Johnson & J. Hallett, "Freezing and Shattering of Supercooled Water Drops," *O. J. R. Meteor. Soc.*, **94**, 468-482, 1968.
18. W. L. Kerr, D. T. Osuga, R. R. Feeney & Y. Yeh, "Effects of Antifreeze Glycoproteins on Linear Crystallization Velocities of Ice," *J. Cryst. Gr.*, **85**, 449-452, 1987.
19. K. H. Leong, "Morphology of Aerosol Particles Generated from the Evaporation of Solution Drops," *J. Aerosol Sci.*, **12**, 417-436, 1981.
20. C. S. Lindenmeyer & B. Chalmers, "Morphology of Ice Dendrites," *J. Chem. Phys.*, **45**, 2804-2806, 1966.
21. W. C. Macklin & B.F. Ryan, "Habits of Ice Grown in Supercooled Water and Aqueous solutions," *Phil. Mag.*, **14**, 847-860, 1966.
22. S. P. Marsh & M. E. Glicksman, "Evolution of Length Scales in Partially Solidified Systems" in **Structure and Dynamics of Particles in Solidified Systems** (Ed D. E. Loper) Martinus Nijoff, NATO, 1987.
23. S. C. Mossop, "Microphysical Properties of Supercooled Humidities Clouds in which an Ice Particle Multiplication Process Operated," *Roy. Meteor. Soc., Qtrly. J.*, **111**, 183-198, 1985.
24. M. Ohara & R. C. Reid, **Modeling Crystal Growth Rates from Solution**, 1-45 Prentice-Hall, Inc. Englewood Cliffs, NJ, 1973.
25. H. Pruppacher, "Growth Modes of Ice Crystals in Supercooled Water and Aqueous Solutions," *J. Glac.*, **6**, 651-622, 1967.

26. H. R. Pruppacher & J. D. Klett, **Microphysics of Clouds and Precipitation**. D. Reidel, p. 714, 1980.
27. W. K. Rhim, M. M. Saffren & D. D. Elleman, "Development Electrostatic Levitator at JPL," in **Materials Processing in the Reduced Gravity Environment of Space.** Guy E. Rindone, ed. ,115-119, Elsevier Science Publishing Co., Inc. 1982.
28. W. Schnorr, "Growth of Solution Bodies and Spheres of Rock Salt," *Zeitschrift fur Kristallographie*, **68**, 1-14, 1928.
29. R. F. Strickland-Constable, **Kinetics and Mechanism of Crystallization**. Academic Press, London & New York, p. 347, 1968.
30. H. Uyeda & K. Kikuchi, "Freezing Experiment of Supercooled Water Droplets Frozen by Using a Single Ice Crystal." *J. Meteor. Soc. Japan*, **56**, 43-51, 1978.
31. P. W. Voorhees & M. E. Glicksman, "Thermal Measurement of Oswald Ripening Kinetics in Partially Solidified Crystallized Mixtures," *J. Cryst. Gr.*, **72**, 599-615, 1985.
32. P. J. Visagie, "Pressures Inside Freezing Water Drops," *J. Glac.*, **8**, 301-309, 1969.
33. H. K. Weickmann, "Current Understanding of the Physical Processes Associated with Cloud Nucleation," *Beiträge Zur Physik der Atmosphare* **30**, 97-118, 1957.
34. A. Yamashita & C. Takahashi, "Initial Growth Processes of Snow Crystals from Frozen Water Drops," **Proc. Intl. Cloud Physics Conf.**, London, p. 50-51, 1972.
35. L. C. Yang & W. B. Good, "Crystallization of Water in Cylindrical tubes," *J. Geophys. Res.*, **71**, 2465-2496, 1966.

LIST OF FIGURES

Figure 1. Spikes produced by water expansion on freezing from a single crystal drop. (a) Water drop suspended in immiscible liquid nucleated by a single crystal inserted at the top of the drop with "a" axis normal to the drop surface. (b) The initial dendrite, which thickens (c, d), changing the drop shape from surface tension effects (e), and producing spikes (f-h), as solidi-fication completes and the water is expelled by expansion. Note air bubble structure in (h) in the central regions of drops and spikes as the inward freezing goes to completion.

Figure 2. Similar to Figure 1, only the drop is nucleated with a crystal having its "a" axis tangential to the drop surface.

Figure 3. Ice spikes from a refrigerator plastic tray.

Figure 4. A tricrystal 7 cm long ice spike from a freezing melt hole in a glacier, showing crystal facets. One face subsequently was melted and the water drained away.

Figure 5. Single crystal water drops, frozen at small supercooling by contact with ice crystals, viewed in polarized light. The right hand drop has produced a spike. These drops were produced in the spray from Old Faithful Geyser and nucleated by ice crystals formed by dry ice seeding at -78°C.

Figure 6. Water drops frozen in LN_2 viewed in polarized crystal showing polycrystalline texture. Bubble rings are just visible resulting from nucleation following pressure release on fracture.

Figure 7. Initiation of convection from a growing crystal of Na_2SO_4-$1OH_2O$ as gravity increases in a KC-135 aircraft parabolic trajectory viewed in a Mach Zender optical system (From Hallett *et al.* 1987).

Figure 8. A bubble ring and internal air bubbles nucleated in a drop supercooled, nucleated and frozen in LN_2. This bubble ring forms as pressure is released as the drop cracks open subject to internal pressure build up. Thin section, in polarized light, showing polycrystal structure.

Figure 9. Supercooled drop nucleated and frozen in silicone oil at -30°C. Note several air bubbles rings in outer regions; the pressure eventually built up to give a major failure leading to near uniform air bubble nucleation in the central region of the drop.

Figure 10. Bubble rings formed as a drop impacts, spreads and freezes upwards from a cold ice surface. The periodicity is caused by an air supersaturation dependant nucleation; once bubbles from, supersaturation falls and growth ceases; the process then repeats.

Figure 11. Periodic crystallization of a drop of sodium sulphate solution (as decahydrate) on a glass slide.

Figure 12. A drop, in thin section nucleated at 5° supercooling crystallized to a single crystal. The air bubble distribution is determined by dendrite sheets which separated slabs of water during the first stage of freezing.

Figure 13. Defect structure by X-ray Synchrotron radiation of a drop supercooled by 2°C frozen, solid on a 0.2 mm glass support.

Figure 14. Comparison between growth rate on the glass wall of a cell (curved, 0.28 cm s^{-1}) and free growth (dendritic, 0.20 cm s^{-1}) in water originally supercooled by 1.8°C.

Figure 15. Polycrystalline growth of ice in water supercooled by 4° in a glass dish nucleated by a single crystal, following secondary ice production on dendrite contact with the walls.

Figure 16. Dendrite growth in a uniformly supercooled spherical soap bubble with 15% sucrose. The growth rate is constant around the surface of the drop, with continuously changing crystal direction.

Figure 17. Approach of two crystals in a flat soap film with 15% sugar. Crystals approach to with ~ 50 μm of each other without slowing down, showing lack of thermal or solute interaction.

Figure 18. Evaporation of a drop of sodium sulphate (crystallizing as decahydrate) in the Jet Propulsion Laboratory electrostatic suspension system. Two crystals nucleate and grow outwards by capillary creep of solution. The final photograph is still not completely crystallized; there is still fluid internal to the crystals (from Rhim and Hallett, 1989, to be published. Interested parties are welcome to write this author).

Figure 19. Increase of temperature following adiabatic crystallization of ice from NaCl solution. Note that the initial temperature of the solution is 0.25°C below that of the equilibrium temperature of the original solution, resulting from solute rejection during ice crystal growth.

Figure 20. Increase of temperature with time during adiabatic crystallization of Sodium Sulfate Decahydrate from solution. Crystallization visually appeared complete after two minutes, but final temperature was not reached until four minutes later, showing that slow growth and latent heat release were still taking place.

Figure 21. How drops crystallize.

WAKE-EXCITED RAINDROP OSCILLATIONS

Kenneth V. Beard[a] and H. T. Ochs III
Climate and Meteorology Section, Illinois State Water Survey, 2204 Griffith Drive, Champaign, Illinois 61820

ABSTRACT

Interpretation of microwave scattering in rain depends on the shape of raindrops, normally considered to have an "equilibrium" axis ratio from a force balance at each point on the surface between surface tension, hydrostatic and aerodynamic pressure. Data from the Illinois State Water Survey raindrop cameras and also observations using a high resolution, dual-polarization radar suggest a pronounced shift from equilibrium axis ratios for small raindrops. New laboratory measurements have been made for small water drops falling at terminal velocity. The experimental findings agree well with equilibrium axis ratios from perturbation theory for $d \leq 1$ mm, but scatter significantly above the theoretical curve for $d > 1$ mm. The scatter is in the same direction as the shifted axis ratios in the field observations. The one-sided scatter is consistent with wake-excited oscillations, but only for modes having a single nodal meridian through the poles (corresponding to spherical harmonics of degeneracy, $m=1$). These identified modes also best match the forcing pattern associated with eddy shedding.

1. INTRODUCTION

The notion that raindrop oscillations can be excited by eddies detaching from the drop originated with Gunn (1949)[3], who postulated that the lateral drift behavior of 1 mm diameter drops might be caused by a mechanical distortion at resonance between the eddy shedding frequency and the oscillation frequency. In spite of the wide acceptance of this hypothesis as fact, the resonance phenomenon had not been observed. The major difficulty in studying wake-excited oscillations for water drops falling in air is the coincidental drift behavior; such drops cannot be contained within a wind tunnel or an acceleration column. We designed a special experiment to determine the effect of wake forcing on drop shape under free-fall conditions. The experimental approach minimized the lateral drift problem by generating drops at terminal velocity in still air whereby wake-excited oscillations were investigated over relatively short distances.

2. METHOD

Details of the design of our one-story experiment for measuring the axis ratio of small raindrops are given in Figure 1. The upper third of equipment is the system for producing isolated, uniform drops (consisting of pressurized water supply, drop generator assembly, charging electrode, electrostatic deflection chamber and PC controls). The lower two chambers are for measuring drop velocity, using a cathetometer and strobe (dark-field), and for bright-field photography drop silhouettes. The fall columns that extend the generator-to-camera distance from 1.5 to 2.5m are not shown. The chambers and water reservoir are on separate platforms to isolate the experiment from vibrations. The camera, cathetometer and strobes are supported by a metal frame around the experiment that rests on the floor.

---

[a] Also affiliated with Dept. Atmospheric Sciences, Univ. Illinois, Urbana-Champaign

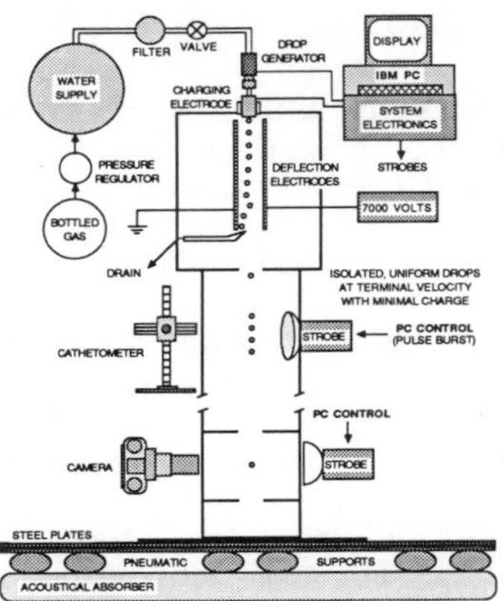

Figure 1

3. RESULTS

The major phase of our experimentation on wake forcing has been completed with results that are summarized by figure 2 ,where the axis ratio for small raindrops is given as a function of drop diameter. For the smaller drops ($d \leq 1$ mm), our observations of axis ratios agree well with the perturbation theory (Pruppacher and Pitter, 1971[4]) based on the numerical pressure distribution for steady state flow around a sphere (curve 1). Error bars show 95% confidence intervals for the axis ratio ,from seven to sixteen measurements at each of the five smallest sizes (57 in total).

Above d=1 mm, the axis ratios scatter above perturbation theory, based on the average pressure distribution for unsteady, separated flow around a sphere (curve 2). Note that there are four experiments: two near d=1.2 mm and two near 1.3 mm. The error bars show ±2 standard deviations based on six to twenty-seven observations of axis ratio for each experiment (58 in total). We discovered that this one-sided scatter above the equilibrium axis ratios is consistent with oscillations only for the transverse mode (m=1), of the two lower harmonics (n=1, 2). We also found that the transverse mode was the only oscillation response matching the forcing pattern expected from pressure changes as eddies detach from alternate sides of the upper pole.

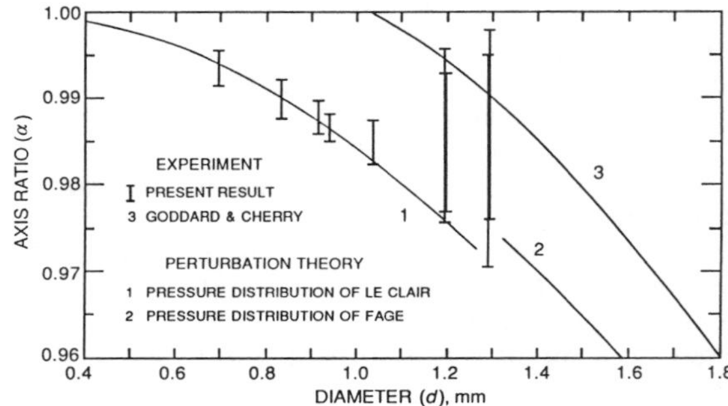

Figure 2

The result of Goddard and Cherry (1984)[2], shown as curve 3, is based on Z_{DR} measurements in rain, using the Chilbolton (UK) high resolution (1/4° beam, 0.5 μs pulse), dual-polarization radar and measurements of raindrops sizes using distrometers. The axis ratios given by curve 3 were used to reconcile calculated values of Z_{DR} from the raindrop size distributions with the measured values of Z_{DR} [an axis ratio shift near d=2 mm similar to Goddard and Cherry (1984), was found by Chandrasekar *et al.* (1988)[1] from aircraft measurements of raindrop shape for $d \geq 2$ mm].

It seems plausible that raindrop oscillations of the type observed in our experiment could have been responsible for the shift from the equilibrium axis ratio postulated by Goddard and Cherry and measured by Chandrasekar *et al.*, if the same type of oscillations also occurs at larger sizes. Since curve 3 is based on data from distributions of sizes extending over several millimeters, a suitable comparison will require additional axis ratio measurements at larger sizes, and averaging of the laboratory axis ratios over the range drop sizes in rain showers. A direct comparison of shapes can be made with the results of Chandrasekar *et al.*, when lab results are obtained for $d \geq 2$ mm.

From our limited data, it is clear we have verified the long-standing hypothesis that small raindrops oscillate because of resonant interaction with eddy shedding—but for larger sizes, and possibly higher harmonics, than originally proposed by Gunn (1949). We plan to make additional measurements to extend the axis ratio measurements to about d=1.5 mm using the present apparatus [experiments on larger drops are underway using a seven-story fall column.] Measurements will also be made near d = 1.1 mm to help establish the transition size for resonant oscillations. We also plan to make frequency measurements to distinguish between the two possible harmonics.

ACKNOWLEDGEMENT

This material is based upon work supported by the National Science Foundation under Grants ATM84-19490, ATM86-01549 and ATM87-22688.

REFERENCES

1. V. Chandrasekar, W. A. Cooper & V. N. Bringi (1988), "Axis ratios and oscillations of raindrops", *J. Atmos. Sci.* 45, 1323-1333.
2. J. W. F. Goddard & S. M. Cherry (1984), "The ability of dual-polarization radar (copolar linear) to predict rainfall rate and microwave attenuation", *Radio Sci.* 19, 201-208.
3. R. Gunn (1949), "Mechanical resonance in freely falling drops", *J. Geophys. Res.* 54, 383-385.
4. H. R. Pruppacher & R. L. Pitter (1971), "A semi-empirical determination of the shape of cloud and rain drops", *J. Atmos. Sci.* 28, 86-94.

FIGURE CAPTIONS

Figure 1 Diagram of experimental apparatus.

Figure 2 Axis ratio as a function of raindrop diameter. Data are shown from present experiment as 95% confidence intervals for the mean axis ratio for smaller drops without oscillations (d=0.7–1.03 mm), and two standard deviations for axis ratios for larger drops with oscillations (d=1.2–1.3 mm).

A NUMERICAL MODEL OF THE ELECTROSTATIC-AERODYNAMIC SHAPE OF RAINDROPS

Catherine Chuang and Kenneth V. Beard[a]
Dept. of Atmospheric Sciences, University of Illinois, Urbana-Champaign, 105 South Gregory Avenue, Urbana, Illinois 61801

ABSTRACT

The model of Beard and Chuang (1987), using the complete form of Laplace's formula and adjustments to the aerodynamic pressure distribution for the effect of drop distortion, has been extended to raindrop shapes under the influence of vertical electric fields and drop charges. A finite volume method with numerically generated transformation to a boundary-fitted coordinate system was used to calculate the shape-dependent electric field. Sufficient constraints (*viz*, drop volume, overall force balance, and shape-dependent surface distributions of aerodynamic and electrostatic stresses) allow the calculation of a unique shape by integration from the upper to lower pole using a multiple iteration scheme. The model has been verified against solutions for a stationary drop in a uniform electric field (Taylor, 1964; Brazier-Smith 1971; Zrnic *et al.* 1984). Numerical shapes of drops falling in electric fields show a pronounced extension of the upper pole. The increased fall speed of electrostatically stretched drops enhances the aerodynamic flattening of the base. The resultant triangular drop profiles are similar to wind tunnel observations (Richards & Dawson 1973; Rasmussen *et al.* 1985).

1. INTRODUCTION

Early theoretical and semiempirical models to predict the shape of raindrops have been subject to varying restrictions, either in physical realism or in applications. A new model based on the equilibrium equation across a curved surface was developed by incorporating an aerodynamic pressure, and electrostatic effects. The method of applying an empirical pressure distribution around a sphere, originally proposed by Savic[11], has been refined to include variations in the pressure distribution with Reynolds number and drop distortion. A finite volume technique was used to calculate the shape-dependent electric effect with a numerically generated boundary-fitted coordinate system.

2. EQUILIBRIUM EQUATION

The equilibrium shape of an interface was determined by Laplace's pressure balance at each point on the surface

$$\sigma(1/R_1 + 1/R_2) = \Delta P \qquad (1)$$

where σ is the surface tension, R_1 and R_2 are the principal curvature radii, and ΔP is the net pressure. The curvatures were evaluated using a tangent angle coordinate system in a vertical plane (Hartland and Hartley 1976)[7], where $1/R_1 = d\phi/ds$ and $1/R_2 = \sin\phi/x$ (see Figure 1).

There are five physical factors that control the shape of raindrops: surface tension, hydrostatic pressure, aerodynamic pressure, internal circulation and electric stress. The initial step in developing the model was a force balance among the first four factors

$$\sigma(d\phi/ds + \sin\phi/x) = 2\sigma/R_t + \Delta\rho g z + P_{ic} + (P_{top} - P)_a \qquad (2)$$

where R_t is the radius of curvature at the top, $\Delta\rho$ is the density difference between water and air, P_{ic} is the pressure from internal circulation and $(P_{top}-P)_a$ is the aerodynamic pressure change from the upper pole.

[a] Also affiliated with the Climate & Meteorology Section, Illinois State Water Survey

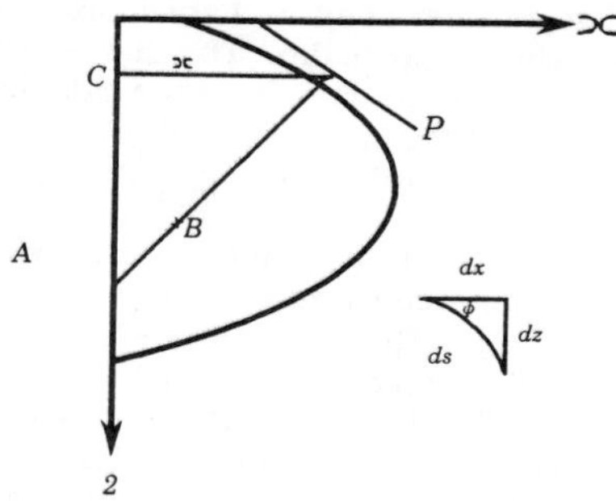

Figure 1. Diagram for the drop surface with R_1 given by BP and R_2 by AP.

The force balance on the raindrop, computed by integration of Equation (2) over the total drop surface A, yields

$$0 = -mg + \int_A P_{ic} \cos\phi \, dA - \int_A P_a \cos\phi \, dA \qquad (3)$$

Equation (3) implies that mechanical equilibrium is satisfied when the total weight is supported by the net normal pressure, $P_{ic} - P_a$, at the drop surface. Since the aerodynamic pressure distribution around a circulating water sphere is essentially the same as rigid sphere (Le Clair, *et al.*, 1972)[8], the integral of P_a is the usual pressure drag, and therefore, the integral of P_{ic} is equivalent to the skin friction.

Unfortunately, there is no information on the distribution of P_{ic} at Reynolds numbers appropriate for raindrops, so approximations have to be made to achieve mechanical equilibrium. These include increasing the amplitude of P_a so that the pressure drag balances the weight (increased drag method), and reducing the drop density to balance the pressure drag (reduced weight method). The first method is equivalent to using a pressure distribution P_{ic} that mimics P_a, as is the case for a circulating liquid sphere in potential flow or Stokes flow. The second method reduces the pressure within the drop, so that the combination $\Delta\rho gz + P_{ic}$ is the same as the creeping flow solution for a sphere (Taylor and Acrivos, 1964)[9]; *i.e.*, P_{ic} is a linear function of z. Additional details on how mechanical equilibrium is achieved are given in Section 4.

The shape was calculated by forward integration of $d\phi/ds$, using the local force balance given by Equation (2), for assumed values of the initial curvature at the top with a boundary condition $d\phi/ds = \sin\phi/x = 1/R_t$, where $x = z = \phi = 0$. An iterative scheme is applied to determine the initial curvature at the top ($1/R_t$) necessary to achieve the proper volume for a particular drop size. A close drop shape was obtained only for an overall force balance (Equation (3)).

3. MODIFYING THE PRESSURE DISTRIBUTION

Before each computation of drop shape, the aerodynamic pressure distribution was adjusted for the effects of drag beginning with an appropriate distribution for a sphere. Measurements of the surface pressure around rigid spheres at very large Reynolds number (Re=10^4-10^5) and numerical results (Re=400) are practically identical in the unseparated region. The distributions for high Reynolds number are rather flat behind the separation point and show only a few degrees variation in the location and breadth of the pressure dip. The distribution of Fage (1937)[10] was used, but instead of applying the data directly to the calculation as in the perturbation model of Savic (1953)[11] and Pruppacher & Pitter (1971)[12], an adjustment was made before each calculation to obtain the appropriate pressure drag for a particular Reynolds number.

In the range of Reynolds number applicable to larger raindrops (d = 2-9mm) the pressure drag (C_{dp}) was approximated by an empirical formula as

$$C_{dp} = C_d\,[1 - 13.4\,Re^{-0.58}] \tag{4}$$

where C_d is the total drag. This interpolation formula was obtained by using end points from Le Clair *et al.* (1970)[13] at Re = 400 and Achenbach (1972)[14] at Re = 75 000.

The adjustment of pressure drag was made by altering the pressure distribution of Fage in the wake region (88° to 180°) and in the intermediate region (72° - 88°) using

$$\kappa'(\psi) = 1 - \Gamma[1 - \kappa_{fage}(\psi)] \qquad 88° \le \psi \le 180°$$

$$\kappa'(\psi) = 1 - \Gamma'[1 - \kappa_{fage}(\psi)] \qquad 72° < \psi < 88° \tag{5}$$

where Γ is a constant and $\Gamma' = 1 - (1 - \Gamma)(\psi - 72°)/16$ provides a linear transition between the unseparated region at 72° and the wake at 88°. The value of Γ was adjusted before the shape calculation so that the pressure drag, obtained by integration $\kappa'(\psi)$ over the surface of a sphere, was the appropriate fraction of the weight as given by Equation (4).

4. AERODYNAMIC SHAPE OF RAINDROPS

To achieve mechanical equilibrium, the amplitude of the pressure distribution κ' was readjusted before every calculation, by using $\kappa = \Lambda\kappa'$, where the constant Λ is determined by the increased drag method or the reduced weight method. This step was needed to readjust the pressure drag for the effect of distortion.

Mechanical equilibrium was achieved by the increased drag method where the entire weight is offset by a pressure drag such that $C_{dp} = C_d$, or the reduced weight method where the drop density is reduced to $(C_{dp}/C_d)\,\Delta\rho$. [Note that the shape is calculated by integration of Equation (2) with P_{ic} = 0, so the effect of internal circulation enters through the adjustments in the drag or weight.] As discussed in Beard & Chuang (1987)[1], the first method increases drop distortion by increasing the aerodynamic drag (increasing C_{dp} to C_d), whereas the second method decreases drop distortion reducing the hydrostatic forcing (reducing $\Delta\rho$ to $(C_{dp}/C_d)\,\Delta\rho$). The axis ratios by these two methods are somewhat different, being 0.690 and 0.710 for d = 5 mm, respectively. As a compromise, a mean forcing method was adopted whereby the drag increase and the weight reduction are halved.

Successive calculations were made using an updated pressure distribution based on the previous drop shape, with κ' obtained from a new value of Γ to satisfy Equation (4), and with κ obtained from a new value of Λ. Iteration was stopped after the raindrop shapes were sufficiently converged (typically 4-8 cycles). The final pressure distributions are similar to those shown in Beard and Chuang where the pressure amplitudes increase with distortion.

The computed aerodynamic shape of raindrops are shown in Figure 2 for d = 1 to 6mm. The drop shapes are placed with the center of mass at the origin with corresponding dashed circles having the radius of the equivalent volume sphere. The flatter base for larger raindrops results from the increasing influence of hydrostatic and aerodynamic effects. Because of the absence of a dimple in the base, the raindrop cross sections in Figure 2 are also profiles.

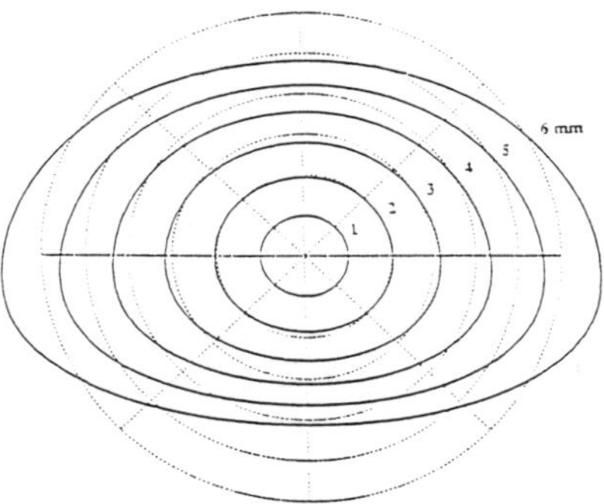

Figure 2. Model shapes for raindrops.

5. ELECTROSTATIC SHAPE OF RAINDROPS IN VERTICAL ELECTRIC FIELDS

In the absence of space charge, the static potential distribution for any conductor configuration can be determined from Laplace's equation. The electric stress from either an applied electric field or surface charges will increase the pressure difference ΔP by an amount $E_s^2/8\pi$ which modifies the equilibrium equation to

$$\sigma(d\phi/ds + \sin\phi/x) = 2\sigma/R_t + \Delta\rho gz + P_{ic} + (P_{top} - P)_a + (1/8\pi)\,[E_s^2/(\phi) - E_s^2/(0)] \qquad (6)$$

where E_s is the magnitude of electric field in electrostatic units (e.s.u.) at the outside surface of drop.

The main difficulty in solving for the electrostatic effect is that the drop shape must be known. A numerical iterative method was used to solve the unknown boundary shape. In order to avoid loss of accuracy, Laplace's equation for the electric potential was solved in a domain with boundaries that are coincident with the drop surface. The boundary-fitted coordinate systems was numerically generated by taking the transformed curvilinear coordinates to be solutions of a suitable elliptic partial differential equation in the physical plane (Thompson *et al.* 1974)[15]. Laplace's equation was solved by a finite volume method to simplify the Neumann boundary condition on the polar axis.

The computed shapes for the pure electrostatic distortion are very close to the theoretical study by Taylor (1964)[2] based on a spheroidal distortion. For predicting drop instability, Taylor found a maximum value of 1.63 for the electrostatic parameter, $X = E_0\,(a_0/\sigma)^{1/2}$, where a_0 is the radius of the equivalent sphere, at an axis ratio of $a = 1.86$. In the numerical model, a stable drop shape could be calculated up to $X = 1.67$ with an the axis ratio of α=1.80. This larger value of X may represent a slight increase for the stability of the natural drop shape.

The effect of a uniform vertical electric field on the drop shape is illustrated in Figure 3 for d = 1-6mm and E = 10 kV/cm. Larger fields produce a pronounce extension of the upper pole. The flattened base for larger drops is a consequence of an increased aerodynamic effect from the higher terminal fall speed of the electrostatic raindrop shapes. The axis ratio (α_e) under different magnitudes of vertical electric field are given in Table 1.

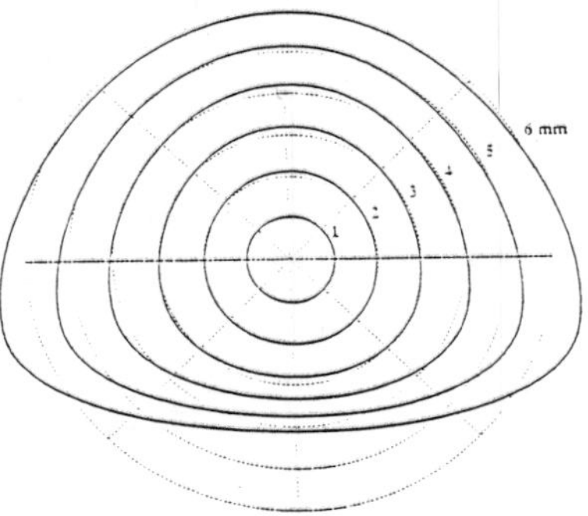

Figure 3. Model shapes for raindrops in vertical electric fields.

Table I. Model axis ratios for raindrops in vertical electric fields.

E(kV/cm)	d=3mm	4	5	6
1.0	0.847	0.769	0.701	0.642
2.0	0.851	0.772	0.703	0.645
3.0	0.856	0.777	0.708	0.649
4.0	0.864	0.785	0.715	0.654
5.0	0.875	0.795	0.724	0.662
6.0	0.889	0.808	0.735	0.673
7.0	0.908	0.825	0.751	0.686
8.0	0.932	0.847	0.770	0.704
9.0	0.964	0.876	0.797	0.728
10.0	1.008	0.918	0.836	0.764

6. DISCUSSION

Model results for aerodynamic shapes when compared with observations in wind tunnels follow the trend in measurements of the equilibrium axis ratios even for extremely large drops. The results for electrostatic-aerodynamic drops are similar to the mean shapes and axis ratios obtained by Richards & Dawson (1973)[5] from photographs of drops in a wind tunnel, but have a distinctly different trend in axis ratio with drop size than the findings of Rasmussen *et al.* (1985)[6]. Although no measurements could be found for the shape of charged drops falling in air, our result differs somewhat from the simple theory of Zrnic *et al.* (1984)[4], based on the method of Green (1975)[17], which neglects aerodynamic effects.

Our findings indicate detectable shape changes in thunderstorms in fields that would initiate lightning. Shape changes are induced by strong electric fields, but modified considerably by alterations in fall speed. The most favorable situation appears to be in vertical fields just before initiation of lightning (about 10 kV/cm, Pruppacher and Klett, 1980[16]) where the differential reflectivity radar signal (Z_{DR}) could be as large as several dB. However, we find that by assuming conduction charging, the maximum charge ($Q_m = 0.5E\, a_0^2$) would have an entirely negligible effect on drop shape, since it is too small compared to the Rayleigh bursting charge (Q_R), *i.e.*, $Q_m^2/Q_R^2 < 0.05$. It is only through changes in fall speed by the QE force that the shape is affected by charge, and then only for larger values of QE found in thunderstorms.

ACKNOWLEDGEMENT

This material is based upon work supported by the National Science Foundation under Grants ATM84-19490 and ATM86-01549.

REFERENCES

1. K. V. Beard & C. Chuang, "A new model for the equilibrium shape of raindrops", *J. Atmos. Sci.* **44**, 1509-1524 (1987).
2. G. Taylor, "Disintegration of water drops in an electric field", *Proc. R. Soc. London* **A280**, 383-397 (1964).
3. P. R. Brazier-Smith, "Stability and shape of isolated and pairs of water drops in an electric field," *Phys. Fluids*, **14**, 1-6 (1971)
4. D. S. Zrnic, R. J. Doviak, & P. R. Mahapatra, "The effect of charge and electric field on the shape of raindrops," *Radio Sci.*, 19, 75-80 (1984).
5. C. N. Richards & G. A. Dawson , "Stress on a raindrop falling at terminal velocity in a vertical electric field--A numerical method", *Phys. Fluids* **16**, 796-800 (1973).
6. R. Rasmussen, C. Walcek, H. R. Pruppacher, S. K. Mitra, J. Lew, V. Levizzani, P. K. Wang, & U. Barth, "A wind tunnel investigation of the effect of an external electric field on the shape of electrically uncharged rain drops," *J. Atmos. Sci*, **42**, 1647-1652 (1985).
7. S. Hartland & R. W. Hartley, **Axisymmetric Fluid-Liquid Interfaces** (Elsevier, NY 1976).
8. B. P. Le Clair, A. E. Hamielec, H. R. Pruppacher & W. D. Hall, "A theoretical and experimental study of the internal circulation in water drops falling at terminal velocity in air", *J. Atmos. Sci.* **29**, 728-740 (1972).
9. T. D. Taylor & A. Acrivos, "On the deformation and drag of a falling viscous drop at low Reynolds number," *J. Fluid. Mech.*, **18**, 466-476 (1964).
10. A. Fage, "Experiments on a sphere at critical Reynolds Numbers", Aero. Res. Comm., England, **Rep. and Memo. No. 1766**, 20 pp. (1937)
11. P. Savic, "Circulation and distortion of liquid drops falling through a viscous medium", Natl. Res. Council (Canada) Rept. NRC-MT-22 (1953).
12. H. R. Pruppacher & R. L. Pitter, "A semi-empirical determination of the shape of cloud and rain drops",*J. Atmos. Sci.* **28**, 86-94 (1971).
13. B. P. Le Clair, A. E. Hamielec & H. R. Pruppacher, "A numerical study of the drag on a sphere at low and intermediate Reynolds number", *J. Atmos. Sci.* **27**, 308-315 (1970).
14. E. Achenbach, "Experiments on the flow past spheres at very high Reynolds numbers", *J. Fluid Mech.* **54**, 565-575 (1972).
15. J. F. Thompson, F. C. Thames & C. W. Mastin, "Automatic numerical generation of body-fitted curvilinear coordinate system for field containing any number of arbitrary two-dimensional bodies",*J. Comp. Phys.* **15**, 299-319 (1974).
16. H. R. Pruppacher & J. D. Klett, **Microphysics of Clouds and Precipitation** (Reidel, Boston, 1978).
17. A. W. Green, "An approximation for the shapes of large raindrops," *J. Appl. Meteor.* **14**, 1578-1583 (1975).

LIST OF FIGURES

RAINBOW-GLORY SCATTERING FROM SPHERES: THEORY AND EXPERIMENTS

Dean S. Langley
Department of Physics, St. John's University, Collegeville, MN 56321

ABSTRACT

The light scattered by a transparent sphere can be made unusually bright in the forward or backward directions by proper choice of the sphere's refractive index *M*. An axial focusing effect in the forward and backward directions gives rise to the strong scattering enhancement known as the glory. For certain *M* values, rainbow scattering coincides with the glory, giving an additional enhancement. Conditions for the existence of rainbow-glory scattering have been explored using ray optics. Mie-theory computations of scattered irradiance show prominent rainbow-glory effects in the predicted *M* ranges. Rainbow-glory backscattering was observed from a glass fiber-coupling sphere in a liquid where *M* was varied by adjusting the liquid temperature.

INTRODUCTION

The light scattered from a transparent sphere can be made unusually bright in certain angular regions by a proper choice of the sphere's refractive index. The spherical symmetry can produce a scattering enhancement in the forward and backward directions known as the glory. The backward glory from water drops is a well known meteorological phenomenon[1,2]; the forward glory is normally masked by other types of scattering[3] but has been demonstrated for air bubbles in liquids[4]. Rainbows are another scattering enhancement familiar from meteorology; reflections and refractions at the surface of the sphere produce a concentration of scattered energy near rainbow angles. Certain values of the sphere's refractive index *M* give a rainbow in the forward or backward direction; such a rainbow-glory combines the two enhancements to produce exceptionally strong scattering. Glory and rainbow scattering are described, and conditions on *M* are given in Section II. Mie scattering computations are used in Section III to demonstrate the presence of rainbow-glories in the predicted *M* ranges. Observations of rainbow-glory backscattering from a glass fiber-coupling sphere are presented in Section IV.

II. GLORY AND RAINBOW SCATTERING

Some features of the glory and rainbow can be understood from the geometry of light rays scattered by a sphere. Ray paths are designated by integers (P,L) where P is the number of chords a ray makes inside the sphere and L is the number of times it crosses the optic axis of the sphere. Glory scattering is associated with rays that emerge in the forward or backward direction but do not lie on the optic axis; Figure 1 illustrates a (2,1) glory ray. Two similar non-glory rays are also pictured to indicate the sphere's effect on an incident plane wave. The geometry of such rays is used to compute the curvature of the emerging wavefront (the curvature is exaggerated in this and subsequent figures for illustrative purposes). Because of rotational symmetry about the optic axis, the wavefront associated with the (2,1) glory is approximately toroidal and weakly focused at infinity along the axis. Diffraction theory applied to wavefronts of this form[5] yields a far-zone scattered irradiance $\propto xI_R$ at small angles from the axis; here x is the dimensionless size parameter (sphere circumference/incident wavelength), and I_R is the scattered irradiance from a perfectly reflecting sphere of the same size. The axial focusing effect, which is intrinsic to

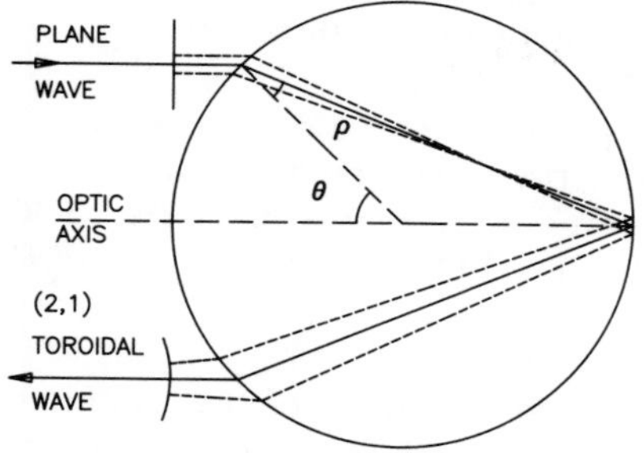

Figure 1

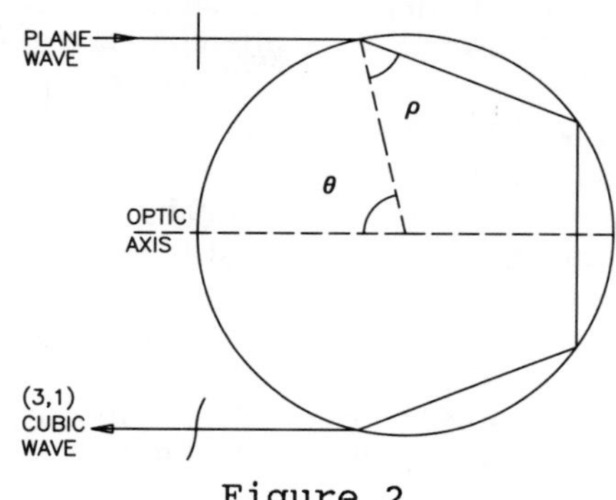

Figure 2

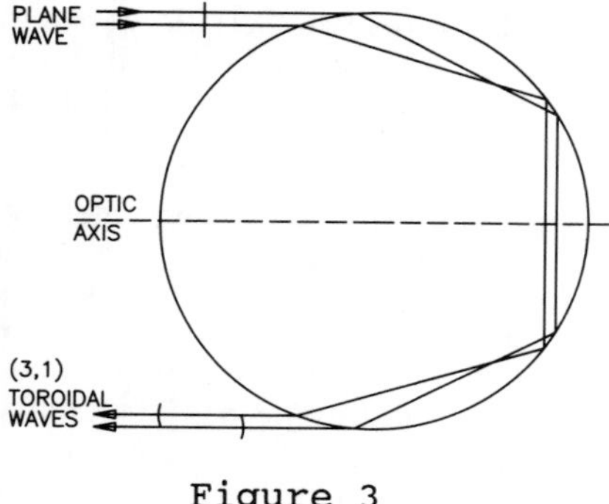

Figure 3

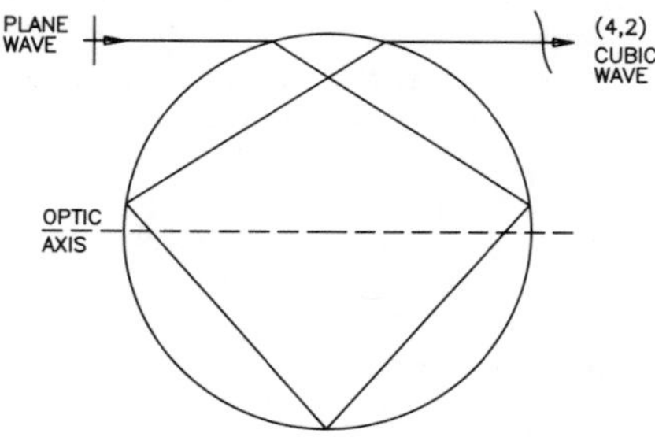

Figure 4

glory scattering, accounts for the enhancement by a factor x.

The angle of a scattered (P,L) ray, measured from its original direction is

$$\phi = (-1)^L \{2P\rho - 2\theta + \pi[L - P + 1/2 + (-1)^L/2]\} \quad (1)$$

where the incidence angle r and refraction angle ρ of any ray are related by Snell's law, $\sin\theta = M \sin\rho$, and M is the refractive index of the sphere relative to the outer medium. For a forward glory ray $\phi = 0$ and L is an even integer, while for a backward glory ray $\phi = \pi$ and L is odd. Using Snell's law and Equation (1), the (P,L) glory condition may be written in terms of M and ρ:

$$M = \cos[P\rho - \pi(P-L)/2]/\sin\rho. \quad (2)$$

Analysis of the geometry of rays gives the approximate shape of the scattered wavefront[6]. The curvature of the wave is expressed by the dimensionless parameter Q = (sphere radius/wavefront radius of curvature):

$$Q = A/(A + M\sin\rho), \quad (3)$$

where

$$A = 2(P\tan\rho - \tan\theta). \quad (4)$$

For a glory wave, $Q>0$ implies a toroidal wavefront diverging from a ringlike virtual source, while $Q<0$ implies a converging wavefront, as in Figure 1.

It is essential to note that a given M may allow 0, 1, or 2 glory rays of type (P,L). The 1-ray ranges are:

$$\sec(L\pi/2P) > M > 0 \qquad (P - 1 > L)$$
$$\csc(\pi/2P) \le M \le P \qquad (P - 1 = L) \quad (5)$$

Thus, a single (2,1) glory exists for $2^{1/2} \le M \le 2$, like the example in Figure 1. The curvature parameter Q is nonzero in the 1-ray ranges, so the glory wavefront remains toroidal. Two distinct (P,L) glory rays occur for M in the range

$$\sec(L\pi/2P) \le M < M_{P,L} \quad (P - 1 > L > 0). \quad (6)$$

The upper bound $M_{P,L}$ is discussed below. Figure 2 illustrates two (3,1) glory rays in a sphere with $1.155 \le M < M_{3,1} = 1.180$. Note that two axially focused toroidal waves are present in the scattering, so prominent interference effects can be anticipated. The curvature parameters Q will have opposite signs for the two waves, though this was difficult to illustrate clearly in Figure 2.

As M approaches the upper bound $M_{P,L}$, the two glory rays lie closer together, and their Q values approach zero from opposite sides; physically, the wavefronts are becoming flatter while retaining opposite curvatures. The two glory rays coalesce when $M = M_{P,L}$, as illustrated in Figure 3 for the case $(P,L) = (3,1)$. The scattered wavefront takes on a cubic form, which is characteristic of rainbow scattering. Geometrically, rainbows occur when the scattering angle ϕ is stationary with respect to variations in the impact parameter of incident rays; i.e, $d\phi/d\theta = 0$. Using Snell's law and Equation (1), the rainbow condition for a (P,L) ray may be written

$$\tan\theta = P\tan\rho. \quad (7)$$

It is clear from Equations (3), (4) and (7) that $Q = 0$ for the wavefront associated with the rainbow ray; the ray passes through the inflection point of a cubic curve that approximates the wavefront locally. Diffraction theory applied to such a wavefront[1] yields a far-zone irradiance $\propto x^{1/3} I_R$, so the rainbow enhancement factor is $x^{1/3}$. However, when the rainbow angle is in the forward or backward direction, the wavefront is axially focused as well, and the resulting enhancement factor is $x^{4/3}$ compared to a perfectly reflecting sphere. Thus it is expected that for large x (spheres much larger than the wavelength of light) the backward or forward scattered irradiance will show a significant increase if M varies through one of the $M_{P,L}$ values. In general, the greatest contributions to the scattering correspond to rays with small values of P and L; energy is lost by the ray at each of the $P-1$ internal reflections, and the reflected fraction tends to decrease as L increases. Rainbow-glory scattering effects should be most evident when $M_{P,L}$ occurs in a range that disallows glory rays with smaller (P,L) values.

$M_{P,L}$ values may be determined by imposing the rainbow and glory conditions simultaneously; manipulation of Equations (2) and (7) yields a simple equation for the refraction angle of the rainbow-glory ray:

$$P \tan\rho = B(\tan P\rho)^B, \tag{8}$$

where $B = (-1)^{P-L-1}$. The resulting ρ may be used in Equation (2) to obtain $M_{P,L}$. It is possible to reduce Equation (8) to a polynomial in $\tan\rho$ of 4th-order or less for 32 cases with (P,L) ranging from (3,1) to (11,8), and obtain analytic solutions. The ρ and $M_{P,L}$ results for the (3,1) (backward) and (4,2) (forward) cases are:

$$(3,1)\ \rho = \arctan[(1 + 3^{1/2}2/3)^{1/2}] \approx 55.7354°, \qquad M_{3,1} = (3^{1/2}6 - 9)^{1/2} \approx 1.17996, \tag{9}$$

$$(4,2)\ \rho = \arctan[(1/3 + 10^{1/2}2/15)^{1/2}] \approx 40.9871°, \qquad M_{4,2} = (10^{1/2}30 - 84)^{1/2}4/9 \approx 1.46521. \tag{10}$$

Figure 4 illustrates the (4,2) forward rainbow-glory ray. The strongest contribution to the total forward scattering comes from what is usually known as forward diffraction; the part of the incident wavefront that is not obstructed by the sphere diffracts into the geometrical shadow region along the forward axis. The enhancement factor associated with the forward-diffracted irradiance[1] is x^2, making it very dominant for large x. To allow the forward rainbow-glory to be observed, polarization properties of the light may be exploited. If the incident beam is plane-polarized, the forward-diffracted light is similarly polarized, while glory-scattered light also contains a perpendicular, or cross- polarized (in the sense of Reference 7) component[1,5,8], designated here as I_{CP}. The forward glory has been observed for air bubbles in liquids[4] using I_{CP}, however no experimental observations of forward rainbow-glories are known.

III. MIE SCATTERING COMPUTATIONS

The Mie theory[9] gives an exact solution for the scattered electromagnetic field from a homogeneous dielectric sphere under plane-wave illumination. The solution takes a series form that converges after approximately x terms, where x is the sphere's size parameter. The theory is presented in a modern form by many authors[1,10] and efficient algorithms are available for computations[11]. The Mie solution is commonly used[5,6] to generate angular patterns of scattered irradiance from a sphere with constant x, or to demonstrate other scattering and polarization effects[3] due to variations in x. Here the Mie solution was employed to compute scattered irradiances in the glory regions as the refractive index M of the sphere was varied. The cross-polarized irradiance I_{CP} was of the most interest for experimental purposes, but it vanishes identically in the exact forward and backward directions. Therefore, the program computed I_{CP} over a range of near-forward and near-backward angles and located the first maximum. For all the computations, the x value corresponded to a glass sphere of diameter 1.3 mm and refractive index 1.84491 immersed in a liquid of variable index such that the relative index was M, and illuminated by light with a vacuum wavelength 633 nm; this corresponded to the experimental conditions of Section IV. Wiscombe's[11] MIEVO algorithm was used for the computations.

The main results of this work are presented in Figures 5 and 6, which demonstrate the presence of rainbow-glory scattering in I_{CP} at the predicted M values. The upper half of each figure marks the 1-ray and 2-ray ranges of M for several

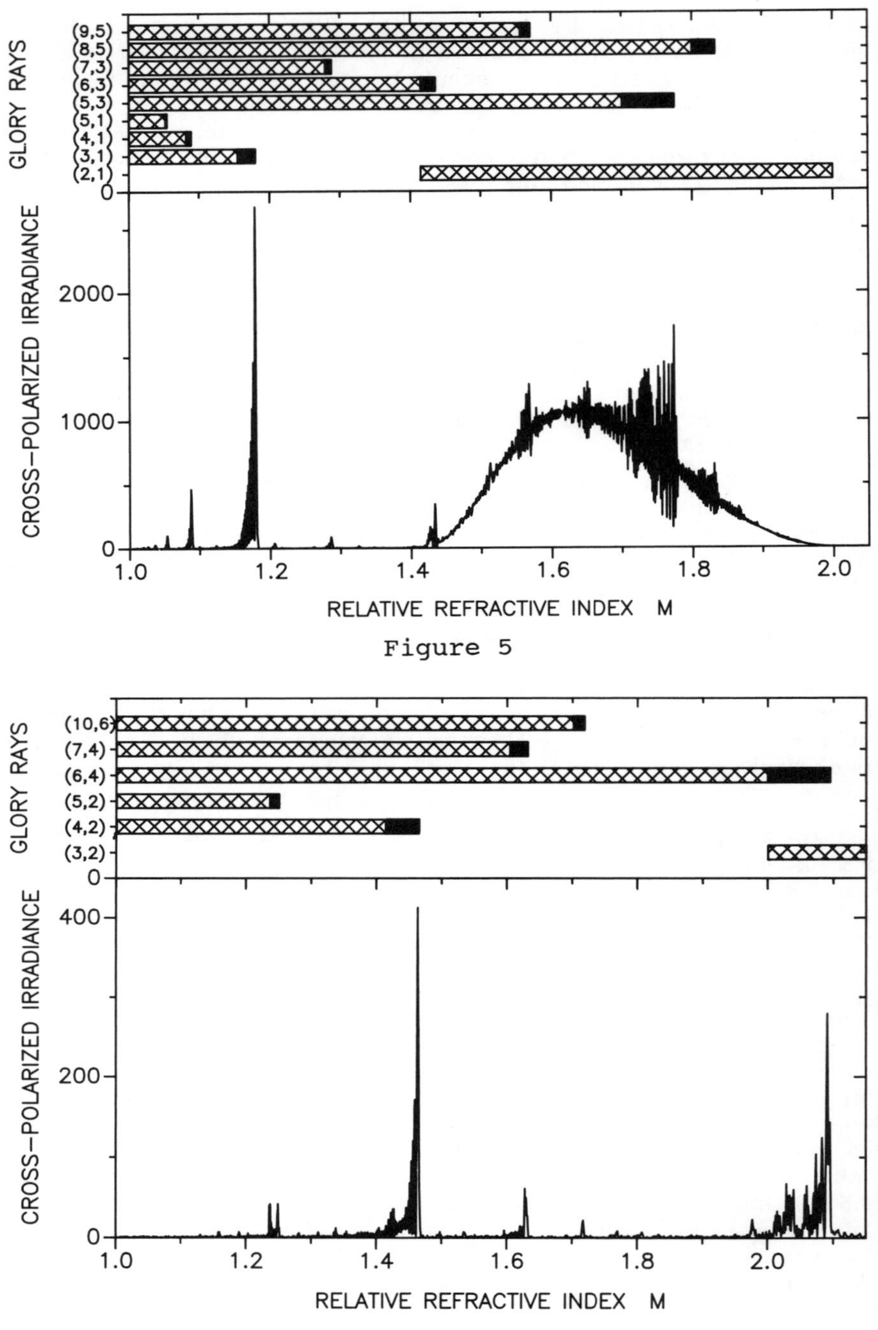

Figure 5

Figure 6

(P,L) glory rays, taken from Equations (5) and (6). The 1-ray ranges are the cross-hatched horizontal bars, and the 2-ray ranges are solid bars; $M_{P,L}$ values lie at the extreme right edge of the 2-ray regions. The lower half of each figure shows the Mie result for I_{CP}, normalized for comparison with the scattered irradiance from a perfectly reflecting sphere ($I_R = 1$).

In Figure 5, one of the most prominent features in the Mie results is the broad peak that fills the 1-ray range of the (2,1) glory ray, $2^{1/2} \leq M \leq 2$. The other main feature is the tall peak that appears at the rainbow-glory index for the (3,1) ray, $M_{3,1} = 1.18$. The enormity of these peaks is significant; the non-rainbow (2,1) glory irradiance exceeds $1000 I_R$, and the (3,1) rainbow-glory irradiance reaches $\sim 2700 I_R$. Note also that the non-rainbow (3,1) glory does not appear at all on this scale; without the rainbow enhancement, the $P = 3$ glory is far weaker than the $P = 2$ glory. Several other rainbow-glories are also evident; the (4,1), (5,1), and (7,3) peaks stand out in otherwise flat regions, while the (5,3), (6,3), (8,5), and (9,5) cases appear superposed on the broad (2,1) peak. Other peaks noticeable in the original plots and at higher M can be aligned with $M_{P,L}$ values as well.

Figure 6 presents similar evidence for forward rainbow-glories at the predicted $M_{P,L}$ values. The (4,2) rainbow-glory peak at $M_{4,2} = 1.465$ is the most outstanding. Other peaks clearly present include the (5,2), (6,4), (7,4) and (10,6) rainbow-glories. A broad non-rainbow (3,2) glory peak occurs in the range $2 \leq M \leq 3$, but this region was not plotted in its entirety since few $M_{P,L}$ having small (P,L) are present. A distinctive feature of rainbow-glory peaks in Figures 5 and 6 is the oscillatory structure that appears in the 2-ray regions. As noted in Section II, the two (P,L) glory rays will produce an angular interference pattern near the axis. As M approaches $M_{P,L}$, the brightness of the first maximum in the angular pattern undulates. Figure 7 gives a detailed look at the 2-ray region of the (3,1) backward glory. The interference effect ceases abruptly at $M_{3,1}$, and I_{CP} drops to a near zero. These features were expected to produce a dramatic effect experimentally.

IV. EXPERIMENTS

The experimental design was to produce the (3,1) backward glory from a sphere illuminated with He-Ne laser light of wavelength 633 nm. To achieve M values in the 2-ray region, $1.155 \leq M \leq 1.180$, a glass (LaSF9) fiber-coupling sphere (1.3 mm diam) of index 1.84491 was immersed in a (Cargille standard) liquid of index ≈ 1.564. The liquid was contained in a glass cell, and its refractive index was adjusted by temperature control, using thermoelectric heat pumps on two faces of the cell; the temperature coefficient for the liquid permitted calibrated M values in the range 1.176 to 1.181 with temperatures 17°C to 33°C. The optical arrangement for observing backscattering is diagrammed in Figure 8. Polarizers placed in the incident and backscattered beams allowed I_{CP} to be selected for viewing by the camera. Figure 9(a) is a close-up photograph of the cell showing the heat pumps on the left and right edges. The sphere itself, suspended by an attached thread at the center of the cell, is not visible. The bright circle of light is the ringlike source of the glory wave, described in Section II; here $M < 1.176$. Figures 9(b) and 9(c) show the effect of increasing M through $M_{3,1}$; the ring reached a maximum brightness near $M = 1.180$, and then grew dimmer as expected. More detailed studies of these near-zone features are underway. It was also desirable to record the far-zone scattering pattern (by focusing the camera on infinity) for comparison with Mie computations, however, temperature inhomogeneities in the liquid prevented an undistorted view. The expected angular structures and polarization features were evident, and the irradiance I_{CP} was been seen to vary in general agreement with the Mie results. Further experiments are in progress to measure the irradiance changes and angular structures.

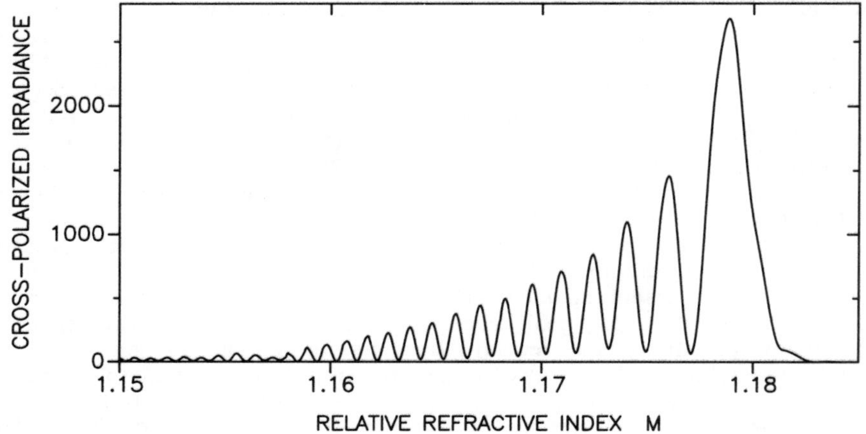

Figure 7

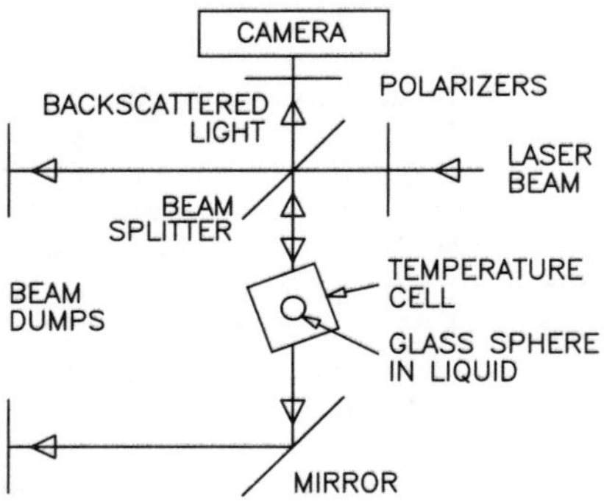

Figure 8

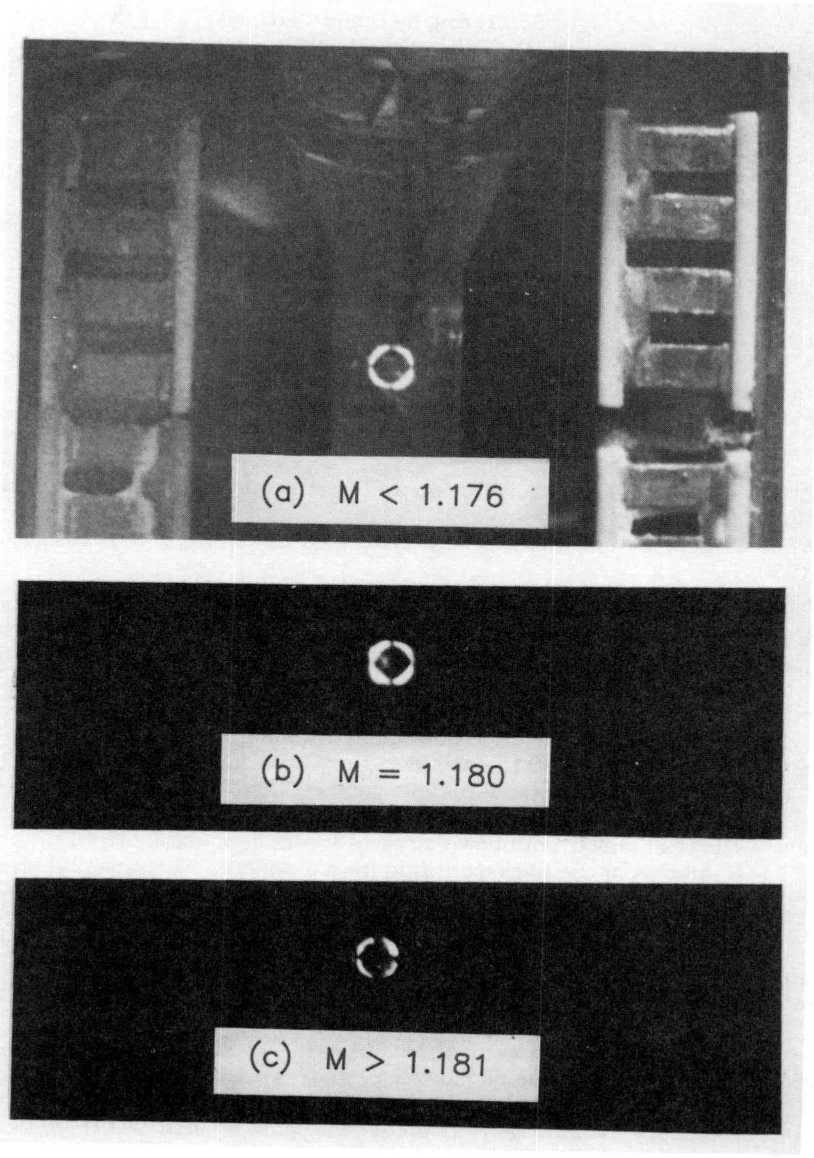

Figure 9

REFERENCES

1 H. C. van de Hulst, **Light Scattering by Small Particles**, (Wiley, New York, 1957).

2 H. C. Bryant & N. Jarmie, The Glory, *Sci. Am.* **231**, 60-71 (July, 1974)

3 H. M. Nussensveig & W. J. Wiscombe, Forward optical glory, *Opt. Lett.* **5**, 455-457 (1980).

4. D. S. Langley, Light Scattering from Bubbles in Liquids, Ph.D. thesis, Washington State University, 1984, Chap. 3.

5. D. S. Langley & P. L. Marston, Glory in Optical Backscattering from Air Bubbles, *Phys. Rev. Lett.* **47**, 913-916 (1981).

6. D. S. Langley & P. L. Marston, Critical-angle scattering of laser light from bubbles in water: measurements, models, and application to sizing of bubbles, *Appl. Opt.* **23**, 1044-1054 (1984).

7. P. L. Marston, Uniform Mie-theoretic analysis of polarized and cross- polarized optical glories, *J. Opt. Soc. Am.* **73**, 1816-1818 (1983).

8. P. L. Marston & D. S. Langley, Strong backscattering and cross polarization from bubbles and glass spheres in water, **Ocean Optics VII**, *Proc. SPIE* **489** 130-141 (1984).

9. G. Mie, Beitrage zur Optik truber Medien, speziell kolloiadaler Metallosungen, *Ann. Phys.* **25**, 377-445 (1908).

10. C. Bohren & D. Huffman, **Absorption and Scattering of Light by Small Particles** (Wiley, New York, 1983).

11. W. J. Wiscombe, Improved Mie scattering algorithms, *Appl. Opt.* **19**, 1505-1509 (1980).

FIGURE CAPTIONS

Figure 1 The simple (2,1) glory ray (2 chords within the sphere and 1 crossing of the optic axis) exists for $2^{1/2} \leq M \leq 2$. Rotating the figure about the axis, the scattered wavefront is approximately toroidal.

Figure 2 Two distinct (3,1) glory rays exist for $1.155 \leq M \leq 1.180$. The two toroidal scattered waves produce prominent interference effects.

Figure 3 At $M = M_{3,1} \approx 1.180$ the two (3,1) glory rays coalesce into a rainbow-glory ray. The wavefront is axially focused and cubic in form, producing strongly enhanced scattering near the backward axis.

Figure 4 The (4,2) forward rainbow-glory ray occurs at $M = M_{4,2} \approx 1.465$.

Figure 5 Rainbows in the backscattering from a sphere. The cross-hatched horizontal bars indicate the 1-ray regions of M for several (P,L) rays, and the solid bars show the 2-ray regions with $M_{P,L}$ values at the right extreme of the bar. Mie computations of I_{CP} vs. M show rainbow-glories.

Figure 6 Rainbows in the forward scattering from a sphere.

Figure 7 Detailed view of the (3,1) rainbow-glory region, showing interference effects in the 2-ray range.

Figure 8 Experimental arrangement for observing backward rainbow-glory.

Figure 9 Photographs of ringlike source of glory wave as M varies in the vicinity of $M_{3,1}$. (a) $M < 1.176$, where two glory waves interfere. (b) $M \approx 1.180$, where the rainbow-glory is predicted. (c) $M > 1.181$, where no (3,1) glory rays occur.

COMPREHENSIVE MODEL RELATING THE MARINE AEROSOL POPULATION OF THE ATMOSPHERIC BOUNDARY LAYER TO THE BUBBLE POPULATION OF THE OCEANIC MIXED LAYER

E. C. Monahan and D. K. Woolf
Marine Sciences Institute, University of Connecticut
Avery Point, Groton, Connecticut 06340

ABSTRACT

The series of sea surface aerosol generation models beginning with that of Monahan, *et al.*[7], which relate the flux of spray droplets up from the interface to the fractional whitecap coverage, have been used successfully by, for example, Burk[19] and Stramska[20], to predict the aerosol population of the MABL. Combining these models with the insights into parent bubble-daughter jet droplet relationships found in Blanchard[6], it has been possible to infer, again in terms of fractional whitecap coverage, the associated bubble flux up to the sea surface. This bubble flux model, when taken together with the information on effective bubble rise velocities provided by Thorpe[29], leads to estimates of mixed layer bubble populations, estimates which are in first order agreement with the limited collection of bubble populations reported for comparable sea states.

INTRODUCTION

Our early field studies of sea spray immediately above the ocean surface[1] convinced us that there was at that time no practical technique available that could be applied to measure this droplet population, or the flux of these droplets up from the sea surface, particularly under the high wind conditions when these droplets play an important role in the sea-air flux of moisture and heat[2]. (While DeLeeuw[3,4] has recently succeeded, even under high wind conditions, in measuring at relatively low elevations the concentration of the larger spray droplets, a technique for the direct measurement of sea surface droplet flux still eludes us.) Now it is known that when an individual bubble rises to the sea surface, its upper, protruding, surface, or film, proceeds to drain, thin, and eventually shatter, producing many film-droplets which are typically less than a micrometer in radius. The remaining bubble cavity then collapses, with the resulting formation of a vertical jet, which usually ejects into the air one or more jet-droplets, whose radii are characteristically greater than one micrometer, as was clearly demonstrated[5,6] a full quarter-century ago. Given that oceanic whitecaps are the surface manifestations of plumes of rising bubbles, it was reasonable to assume that these whitecaps were the immediate source of all but the largest marine aerosol droplets. (The very large spume drops, which are only present at wind speeds of $10 ms^{-1}$ and higher, result from the actual mechanical disruption of the wave crests[7].) This hypothesis led us to attempt to relate the rate of generation at the sea surface of spray droplets (all those with radii less than about 30 μm) to the fraction of the sea surface covered at any instant by whitecaps.

SEA SURFACE AEROSOL GENERATION MODELS

The specific modelling approach adopted to provide estimates of the instantaneous rate of injection of spray droplets into the lower marine atmospheric boundary layer involved estimating the fraction of the sea surface from which whitecaps disappeared per unit time, which is simply proportional to the area of the sea surface covered at any instant by whitecaps since the areas of individual whitecaps decay in an exponential fashion,and combining such field measurements with laboratory determinations of the number and size of the spray droplets produced during the decay of a unit area of whitecap. Equation (1) is a general statement of this model[8].

$$\partial F_o/\partial r = W_B \tau^{-1} \quad \partial E/\partial r \tag{1}$$

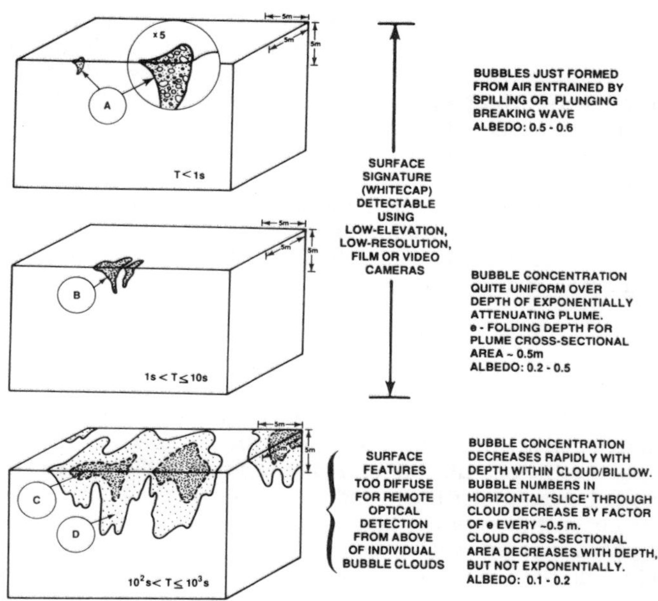

Figure 1

Here, $\partial F_o/\partial r$ represents the number of spray droplets generated per second, per square meter of sea surface, per μm in droplet radius, as a result of the bursting of whitecap bubbles. Here and elsewhere, r, is the radius the spray droplets would have if they were allowed to adjust to their equilibrium size in air characterized by a relative humidity of 80%. The quantity W_B is the estimate of the fraction of the sea surface covered by Stage A and Stage B whitecaps[9,10]. These stages in the evolution of a whitecap, which are illustrated in the upper two panels of Figure (1), are both detectable using the images collected with the ship-mounted photographic system[11] used to record sea surface whitecap coverage during the BOMEX[12], JASIN[13], STREX, and MIZEX 83 experiments. The term in the denominator, τ, is the time constant characterizing the exponential decay of the individual whitecaps. A value of 3.53τ for was obtained from the analysis of cine-film records of the lifetimes of small laboratory whitecaps[5], and recently Nolan[14] obtained a value of 4.27τ from his analysis, with a Hamamatsu Area Analyzer, of the U-matic videotape records of a series of whitecaps collected during the HEXMAX experiment. The final term, $\partial E/\partial r$, is the quantity that is evaluated from measurements made in a hooded whitecap simulation tank[7,8]. Specifically, $\partial E/\partial r$, is a measure of the number of aerosol droplets, per µm increment in droplet radius, introduced into the air within the hood of the tank, normalized by the initial area, in m^2, of the whitecap whose decay resulted in the generation of these droplets.

If the early, laboratory value of is used along with the $\partial E/\partial r$ expression derived from the results of the early experiments carried out in the whitecap simulation tank then in operation at University College, Galway, Ireland, the relationship given in Equation (2) is obtained[7,15].

$$\partial F_o/\partial r = 3.57 \times 10^5\, r^{-3}\,(1 + 0.057r^{1.05}) \times 10^{1.19\,e^{-Y^2}} \times W_B,$$
$$Y = (0.380 - \mathrm{Log}\, r)/0.650 \qquad (2)$$

There are several possible formulations for W_B which can be introduced into Equation (2) to evaluate F_o/r for a given 10m-elevation wind speed[16,17]. Equation (3) was derived from the analysis[16] of the BOMEX and other warm water whitecap photographs described by Monahan[12], and of the photographs of whitecaps taken in the East China Sea and adjacent waters by Toba and Chaen[18]. Here U represents the wind speed, as measured at the standard

$$W_B = 3.86 \times 10^{-6}\, U^{3.41} \qquad (3)$$

anemometer height of 10m, expressed in ms^{-1}.

The model represented by Equations (2) and (3) has been used with some success at the sea surface source function in the modelling of the aerosol population of the marine atmospheric boundary layer carried out by Burk[19], and by Stramska[20]. When the $\partial F_o/\partial r$ expression is divided by v_d, the effective dry deposition velocity as given in the literature[21,22], the resulting low-elevation aerosol population spectrum (when weighted by droplet volume) has a broad peak at about the same radius as that associated with the peak in the JASIN droplet spectra derived from measurements taken at intermediate wind speeds[23]. The model aerosol population spectrum for winds of $7ms^{-1}$ has, in the vicinity of its broad peak, an amplitude that agrees well not only with the amplitude of the JASIN aerosol spectrum for that wind speed[15], but also with the amplitude of the near cloud base aerosol population spectrum for comparable winds deduced from the early findings of Woodcock[24].

Efforts are underway to refine and improve the expression for $\partial F_o/\partial r$. When the $\partial E/\partial r$ formulation derived from the analysis of the last major set of experiments in the Galway Whitecap Simulation Tank[25] is combined with the latest τ value of $4.27s$[14], a new expression for $\partial F_o/\partial r$ results:

$$\partial F_o/\partial r = 0.234 W_B\, \mathrm{Log}_{10}^{-1}\,[6.98 - 1.49\ \mathrm{Log}_{10} r - 1.08\,(\mathrm{Log}_{10} r)^2 + 0.527\,(\mathrm{Log}_{10} r)^3] \qquad (4)$$

SEA SURFACE BUBBLE FLUX MODEL

Given the substantial success attained with the sea surface aerosol generation models in predicting the concentration and spectral shape of the marine aerosol populations for selected wind conditions[15], it seemed appropriate to attempt to "work backward" from the basic aerosol flux model (Equations (2) and (3)) in an effort to arrive at a sea surface bubble flux expression.

This approach, which is given in full elsewhere[26], leads to Equation (5), where

$$\partial K/\partial R = \partial F_o/\partial r \bullet \partial r/\partial R \bullet P(r) \bullet J^{-1} \qquad (5)$$

$\partial K/\partial R$ represents the number of bubbles arriving, and bursting, per second, per square meter of the sea surface, per micrometer increment in bubble radius, R. The first term on the right-hand side is the aerosol flux expression, as given by Equations (2) and (3). The second term, $\partial r/\partial R$, expresses the rate of increase of droplet radius, r, per unit increment of bubble radius, R. Making use of the detailed results obtained by Blanchard[6] from his laboratory investigations of the bursting of individual bubbles, the relationship between the size of the daughter jet droplets and the size of the parent bubble can be expressed as shown in Equation (6).

$$r = 8.77 \times 10^{-2} R + 0.98 \quad (6)$$

The third term, $P(r)$, is the jet-droplet/film-droplet partition function. An expression, based on the results of a series of experiments in the Whitecap Simulation Tank[27], for this fraction of the bubble-generated droplets that is composed of jet-droplets, as opposed to film-droplets, is given in Equation (7).

$$P = 1 - e^{-0.343 r} = 1 - 0.715 e^{-0.030 R} \quad (7)$$

The final term on the right-hand of Equation (5) is the reciprocal of J, where J is a measure of the number of jet-droplets generated as a result of the collapse of a single whitecap bubble. In one previous study[28], it was assumed that each bubble produced five jet-droplets; *i.e.*, that J was equal to 5.

In order to test the bubble flux model against field observations, it is necessary to first deduce the near-surface bubble population, $\partial C/\partial R$, that is associated with a specific bubble flux. This can readily be done by dividing the bubble flux expression of Equation (5) by the radius-dependent terminal rise velocities for air bubbles in sea water, $v_m(R)$. These terminal rise velocities vary markedly, not only with bubble radius but also according to whether the bubbles are hydrodynamically "clean" or "dirty"[29].

Since there are several published observations of the near-surface bubble population for winds of about 13ms^{-1}, it is appropriate to evaluate C/R for that wind speed making use of Equations (2), (3), (5), (6), and (7). The curve labeled $W \times B_D$ on Figure (2) is the predicted spectrum for the near-surface bubble population, where J has been assumed to be equal to one, and the bubbles are taken to be hydrodynamically "dirty". The curve marked $W \times C_C$ on this same figure is the predicted spectrum when J is taken to be five, and the bubbles are assigned the $v_m(R)$ values appropriate for hydrodynamically "clean" bubbles. Thus, the shaded region between $W \times B_D$ and $W \times C_C$ represents the range of near-surface bubble populations compatible with Equation (5) for W_B equals 0.0241; *i.e.*, for winds of 13ms^{-1}.

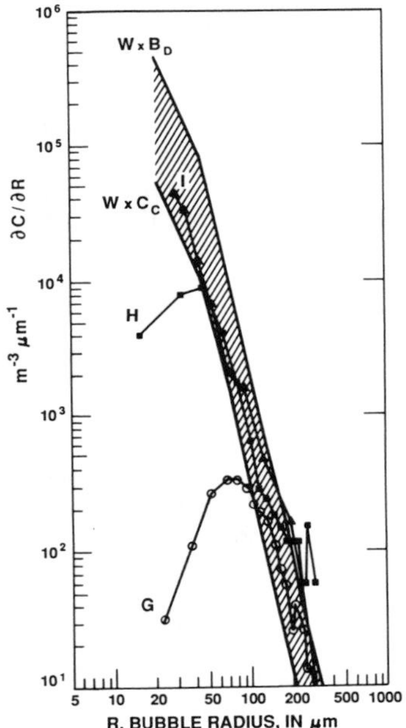

Figure 2

Two near-surface oceanic bubble spectra obtained when the winds were blowing at 11-13ms^{-1} have likewise been plotted on Figure (2). The spectrum describing the population measured by Kolovayev[30] at a depth of 1.5m using a semi-automatic cylindrical bubble trap[31], labeled *G*, falls well below the predicted range, and the other measured spectra, for bubble radii less than 120 μm. This may be due to some dissolution, and some coalescence, of smaller bubbles during their captivity within the bubble trap[32]. Johnson's and Cooke's spectrum[32], labeled *H* on Figure (2), is consistent with the model predictions for all *R* greater than 50 μm. This spectrum, based on photographs taken at 0.7m depth with a camera in a waterproof housing accompanied by three external strobe lamps, drops below the range of predicted spectra for radii less than 50 μm. This low-*R* fall-off may be due to the influence of the effective resolution of this photographic technique. Baldy and Bourguel[33] used a sophisticated laser bubble probe[34] to measure the bubble spectrum at a depth of 0.05m in a large wind-wave flume. This fresh water bubble spectrum, associated with a wind speed of 13ms^{-1}, is reproduced at Curve I on Figure (2).

CONCLUSIONS

The success to date of the aerosol flux and bubble flux models supports the contention that a modelling approach in which the individual whitecap, or bubble cloud, is taken at the link that makes it possible to combine laboratory simulation tank results with field observations of oceanic whitecap coverage to generate predictions relevant to the global ocean is a fundamentally valid one.

ACKNOWLEDGEMENT

Our laboratory and field studies of oceanic whitecaps have been supported throughout by the Office of Naval Research. Our current research is supported by ONR Contracts N00014-87-K-0185 and N00014-87-K-0169. This paper is Contribution No. 203 from the Marine Sciences Institute of the University of Connecticut.

REFERENCES

1. E. C. Monahan, 1968: "Sea Spray as a Function of Low Elevation Wind Speed", *J. Geophys. Res.*, **73**, 1127-1137.
2. R. S. Bortkovskii, 1987: **Air-Sea Exchange of Heat and Moisture During Storms**, D. Reidel Pub. Co., Dordrecht, The Netherlands, 194 pp.
3. G. DeLeeuw, 1986: "Size Distributions of Giant Aerosol Particles Close Above Sea Level", *J. Aerosol Sci.*, **17**, 293-296.
4. G. DeLeeuw, 1987: "Near-Surface Particle Size Distribution Profiles Over the North Sea, *J. Geophys. Res.*, **92**, 14631-14635.
5. C. F. Kientzler, A. B. Arons, D. C. Blanchard, & A. H. Woodcock, 1954: "Photographic Investigation of the Projection of Droplets by Bubbles Bursting at the Water Surface", *Tellus*, **6**, 1-7.
6. D. C. Blanchard, 1963: "The electrification of the atmosphere by particles from bubbles in the sea", *Prog. Oceanog.*, **1**, 71-202.
7. E. C. Monahan, D. E. Spiel, & K. L. Davidson, 1983: "Model of Marine Aerosol Generation via Whitecaps and Wave Disruption", 9th Conf. on Aerospace and Aeronautical Meteorology, 6-9 June, 1983, Omaha, Nebraska, Am. Meteorol. Soc., Preprint Vol., 147-158.
8. E. C. Monahan., K. L. Davidson, & D. E. Spiel, 1982: "Whitecap Aerosol Productivity Deduced from Simulation Tank Measurements", *J. Geophys. Res.*, **87**, 8898-8904.
9. E. C. Monahan, 1988: "From the Laboratory Tank to the Global Ocean, in Climate and Health Implications of Bubble-Mediated Sea-Air Exchange", in **Connecticut Sea Grant Program**, E. C. Monahan & M. A. Van Patten, eds. (in press).
10. E. C. Monahan, M. B. Wilson, & D. K. Woolf, 1988: "HEXMAX Whitecap Climatology: Foam Crest Coverage in the North Sea", in **Proc. of HEXMAX workshop** held 16 Oct.-23 Nov. 1986, Univ. of Washington Tech. Report (in press).
11. E. C. Monahan, 1969: "Fresh Water Whitecaps", *J. Atmos. Sci.*, **26**, 1026-1029.

12. E. C. Monahan, 1971: "Oceanic Whitecaps", *J. Phys. Oceanog.*, **1**, 139-144.
13. E. C. Monahan, I. G. O'Muircheartaigh, & M. P. FitzGerald, 1981: "Determination of Surface Wind Speed from Remotely Measured Whitecap Coverage, a Feasibility Assessment", in **Proceedings of an EARSeL-Sea Symposium, Application of Remote Sensing Data on the Continental Shelf**, Voss, Norway, 19-20 May 1981, European Space Agency, SP-167, 103-109.
14. E. C. Monahan, R. J. Cipriano, W. F. Fitzgerald, R. Marks, R. Mason, P. F. Nolan, T. Torgersen, M. B. Wilson, & D. K. Woolf, 1988: "Oceanic Whitecaps and the Fluxes of Droplets from, Bubbles to, and Gases Through, the Sea Surface", *Whitecap Report No. 4*, to ONR from MSI, Univ. of Conn. (in press).
15. E. C. Monahan 1986: "The Ocean as a Source for Atmospheric Particles", in **The Role of Air-Sea Exchange in Geochemical Cycling**, P. Buat-Menard, ed. D. Reidel Publ. Co., Dordrecht, 129-163.
16. E. C. Monahan & I. G. O'Muircheartaigh, 1980: "Optimal Power-law Description of Oceanic Whitecap Coverage Dependence on Wind Speed", *J. Phys. Oceanog.* **10**, 2094-2099.
17. E. C. Monahan & I. G. O'Muircheartaigh, 1986: "Whitecaps and the Passive Remote Sensing of the Ocean Surface", *Int. J. Remote Sensing*, **7**, 627-642.
18. Y. Toba & M. Chaen, 1973: "Quantitative Expression of the Breaking of Wind Waves on the Sea Surface", *Records, Oceanogr. Works Japan*, **12**, 1-11.
19. Burk, S.D., 1984: "The Generation, Turbulent Transfer, and Deposition of the Sea-Salt Aerosol", *J. Atmos. Sci.*, **41**, 3041-3051.
20. M. Stramska, 1987: "Vertical Profiles of Sea Salt Aerosol in the Atmospheric Surface Layer: Numerical Model", *Acta Geophys.Pol.*, 35, 87-100.
21. S. A. Slinn & W. G. N. Slinn, 1980: "Predictions for Particle Deposition on Natural Waters", *Atmos. Envir.*, **14**, 1013-1016.
22. S. A. Slinn & W. G. N. Slinn, 1981: "Modeling of Atmospheric Particulate Deposition to Natural Waters", in **Atmospheric Pollutants in Natural Waters**, S. J. Eisenreich, ed. Ann Arbor, MI: Ann Arbor Sci. Pub., 23-53.
23. E. C. Monahan, C. W. Fairall, K. L. Davidson, & P. Jones Boyle, 1983: "Observed Inter-Relations Between 10m Winds, Ocean Whitecaps and Marine Aerosols", *Qtrly. J. R. Meteor. Soc.*, **109**, 379-392.
24. A. H.Woodcock,1953: "Salt Nuclei in Marine Air as a Function of Altitude and Wind Force", *J. Meteor.*, 10, 362-371.
25. D. K. Woolf, E. C. Monahan, & D. E. Spiel, 1988: "Quantification of the Marine Aerosol Produced by Whitecaps", in **Preprint Volume: Seventh Conference on Ocean-Atmosphere Interaction**, 31 Jan.-5 Feb. 1988, Anaheim, CA, Am. Meteor. Soc., 182-185.
26. E. C. Monahan, 1988: "Whitecap Coverage as a Remotely Monitorable Indication of the Rate of Bubble Injection into the Oceanic Mixed Layer", in **Sea Surface Sound**, B. R. Kerman, ed., Kluwer Academic Pubs., Dordrecht, 85-96.
27. Woolf, D.K., P. A. Bowyer, & E. C. Monahan, 1987: "Discriminating Between the Film-Drops and Jet-Drops Produced by a Simulated Whitecap, *J. Geophys. Res.*, **92**, 5142-5150.
28. R. J. Cipriano & D. C. Blanchard, 1981: "Bubble and Aerosol Spectra Produced by a Laboratory 'Breaking Wave'." *J. Geophys. Res.*, **86**, 8085-8092.
29. S. A. Thorpe, 1982: "On the Clouds of Bubbles Formed by Breaking Wind-Saves in Deep Water, and Their Role in Air-Sea Gas Transfer", *Phil. Trans. R. Soc. London, A***304**, 155-210.
30. P. A. Kolovayev, 1976: "Investigation of the Concentration and Statistical Size Distribution of Wind-Produced Bubbles in the Near-Surface Ocean Layer. *Oceanology*, **15**, 659-661.
31. V. P. Glotov, P. A. Kolovayev, & G. G. Neuimin, 1962: "Investigation of the Scattering of Sound by Bubbles Generated by an Artificial Wind in Sea Water and the Statistical Distribution of Bubble Sizes", *Sov. Phys. Acoustics*, **7**, 341-345.
32. B. Johnson & R. C. Cooke, 1979: "Bubble Populations and Spectra in Coastal Waters: Photographic Approach", *J. Geophys. Res.*, **84**, 3761-3766.
33. S. Baldy & M. Bourguel, 1985: "Measurements of Bubbles in a Stationary Field

of Breaking Waves by a Laser-Based Single-Particle Scattering Technique", *J. Geophys. Res.*, **90**, 1037-1047.

34. F. Avellan & F. Resch, 1983: "Scattering Light Probe for the Measurement of Oceanic Air Bubble Sizes", *Int. J. Multiphase Flow*, **9**, 649-663.

FIGURE CAPTIONS

Figure 1. Model of whitecap decay and bubble cloud evolution

Figure 2. Near-surface resident bubble spectra for winds of 10-13 ms^{-1}. Region W x (B_D - C_C), range of spectra predicted by model. Curve G, oceanic spectrum, depth of 1.5m[30]. Curve H, oceanic spectrum, depth of 0.7m[32]. Curve I, flume spectrum, 0.05m depth[33]. See text for details.

8. Bubbles, Shells, & Encapsulations

Robert E. Apfel & Xavier J. R. Avula
Session Chairs

OPTICS OF BUBBLES IN WATER: SCATTERING PROPERTIES, COATINGS, AND LASER RADIATION PRESSURE

Philip L. Marston, W. Patrick Arnott[a], Stefan M. Bäumer[b], Cleon E. Dean, and Bruce T. Unger[c]
Department of Physics, Washington State University
Pullman, WA 99164-2814

ABSTRACT

Experiments and theory pertinent to light scattered by bubbles in water are examined. Topics include: the transition to total reflection at the critical angle, complex angular momentum theory of this transition, colors seen in sunlit bubble clouds, optical effects of surface films on bubbles, Brewster angle scattering, and the backscattering patterns and caustics of spherical and oblate bubbles. Responses of bubbles to optical radiation pressure include: levitation and optically stimulated acoustic emissions.

1. INTRODUCTION

The scattering of light from particles and liquid drops is a field of research which dates at least from work of Descartes and Newton on the understanding of rainbows.[1] That field of research has remained active.[2,3] By comparison, the optical properties of bubbles in liquids is a relatively new field of research. For a bubble or "bubble-like" scatterer, the refractive index of the scatterer is less than that of the surroundings so that phenomena are present which differ from those for drop-like scatterers. Attempts to quantitatively model the optical scattering properties of gas bubbles in water essentially began with the geometrical approach of Davis[4] in 1955. Since bubbles in water are almost always very much larger than the wavelength of light, it might be thought that a geometrical approach would give quantitatively useful results. It turns out, however, that the analysis of scattering based on the flux conservation laws of geometrical optics gives a rather incomplete description of several of the phenomena of interest. (Other shortcomings of Davis's analysis will be noted in Section 2.) The emphasis of the present paper is on phenomena which cannot be described quantitatively from elementary geometrical optics. Such phenomena have applications and are not just scientific curiosities.

One approach to improving geometrical optics is to use it only in modeling until the wave leaves the scatterer. Propagation to the distant observer is accounted for by evaluating appropriate diffraction integrals as in Airy's rainbow theory.[2] In scattering theory, such an approach is commonly known as a *physical optics approximation* (POA). It was first applied to the reflection properties of bubbles by Marston[5] in 1979. The present paper primarily considers scattering properties and results obtained subsequent to a brief review[6] published in 1982. While the understanding of optical phenomena is of primary concern, specific applications to problems of practical interest will be noted. As in typical discussions of scattering, we emphasize the dependence of the far-field irradiance on angle. The angular dependence may be used to understand the appearance of bubbles in photographic images.[6]

---

[a] Present address: National Center for Physical Acoustics, University, MS 38671.
[b] Present address: Optisches Institute der Technischen Universität Berlin, West Germany.
[c] Present address: Wenatchee Valley College, Wenatchee, WA 98801-1799.

The paper is organized as follows: Section 2 reviews phenomena associated with the total reflection of light from bubbles. Section 3 describes colors associated with reflection seen in clouds of sunlit bubbles. Section 4 summarizes an extension of recent analytical results of Ferrari and Nussenzveig[7,8] that give an asymptotic series for critical angle scattering. Section 5 is concerned with the reflection of light from bubbles in water coated by a thin insoluble film. Films are thought to stabilize microbubbles in the ocean[9,10] and knowledge of their optical properties may be useful for film and bubble characterization.[11] Section 6 describes novel observations of a minimum in the reflectivity of polarized light from bubbles in water associated with the Brewster angle.[12] Section 7 describes observations and models of laser light backscattered from freely rising spherical and spheroidal bubbles in water.[13] The scattering is enhanced by a focal phenomena[1] (glory scattering) previously observed for bubbles in oil.[14] Section 8 describes certain effects of optical radiation pressure on bubbles including levitation[15] and the emission of sound.[16,17] Forward glory and scattering properties of bubbles[6,18] will not be reviewed in detail.

Optical properties of bubbles are germane not only to the optical detection of bubbles but also to optical remote sensing and communication through bubbly water. Light scattering has been used in conjunction with acoustic levitation for size measurements[19] and should be useful for detecting small oscillations in the radius[20] or shape of bubbles by extending various experimental methods used to measure small oscillations of drops.[21] Scattering has been used to give precise measurements of the size of rising bubbles.[22,23]

2. TOTAL REFLECTION OF LIGHT FROM BUBBLES IN WATER AND SOME APPLICATIONS

A salient feature of the Davis model[4] is an enhancement of the scattered irradiance when the angle of deviation θ of the scattered rays is less than a critical value of $\theta_c \approx 82.8°$. The reason for this prediction is evident by inspecting the ray diagram shown in Figure 1. Let s denote the impact parameter of a ray relative to the center C of the bubble. The local angle of incidence at the surface of the bubble is $i = \arcsin(s/a)$ where a is the radius of the bubble. Since the angle of reflection is equal to the angle of incidence, the scattering angle for the reflected ray is $\theta = 180° - 2i$. (The reader is cautioned that here, and elsewhere in the present paper, the notation differs from that used in several of the references.) The refractive index n_w of the water which surrounds the bubble exceeds the index n_i of the gas within the bubble. If the effects of curvature and tunneling on the reflection coefficient are neglected, the usual flat surface Fresnel reflection coefficient can be used in the calculation of the amplitude of the reflected field. The reflection becomes total when i exceeds the critical value[5,23]

$$i_c = \arcsin(n_i/n_w), \qquad \text{critical scattering angle} = \theta_c = 180° - 2i_c = 82.8° . \qquad (1,2)$$

For the purposes of the present discussion, in Equation (1) we may take $n_w = 4/3$ and $n_i = 1$. The present level of approximation predicts that all rays reflected into the region $0 \le \theta \le \theta_c$ will be totally reflected. A more detailed analysis[24] suggests that tunneling will frustrate this total reflection from sufficiently small bubbles; however, the above considerations are sufficient to appreciate the cause of the observed enhancement[5,23] of the scattering into the region $\theta < \theta_c$. As reviewed below, to understand the detailed structure in the scattering pattern, the effects of interference of the reflected contributions with the contributions of other transmitted rays must be taken into account[23-25] as well as the effects of diffraction[5] in angular regions close to θ_c. Ordinary forward diffraction also enhances the scattering, but its effect is limited to a relatively narrow range of angles in near forward directions.[6,11]

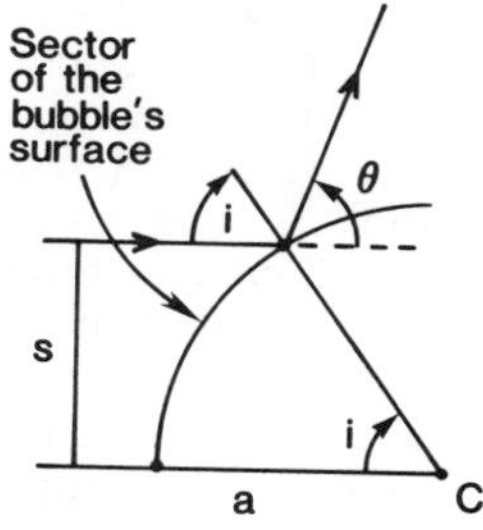

Fig. 1. Reflection of a ray from a spherical bubble. The ray lies in the scattering plane.

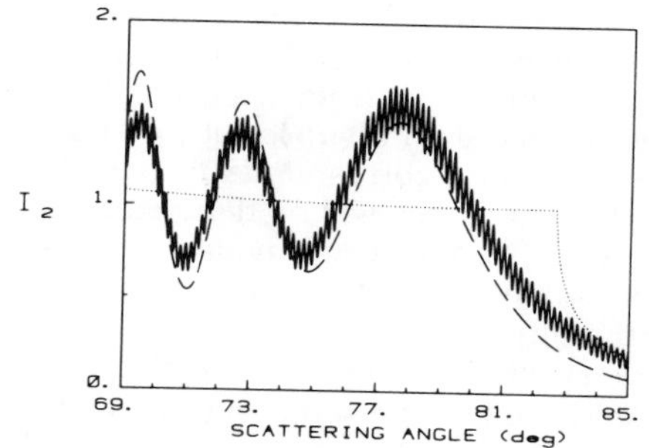

Fig. 2. Normalized irradiance (with j=2) from a bubble with ka = 1633 from Mie theory (solid curve), physical optics (dashed) and geometric optics (dotted).

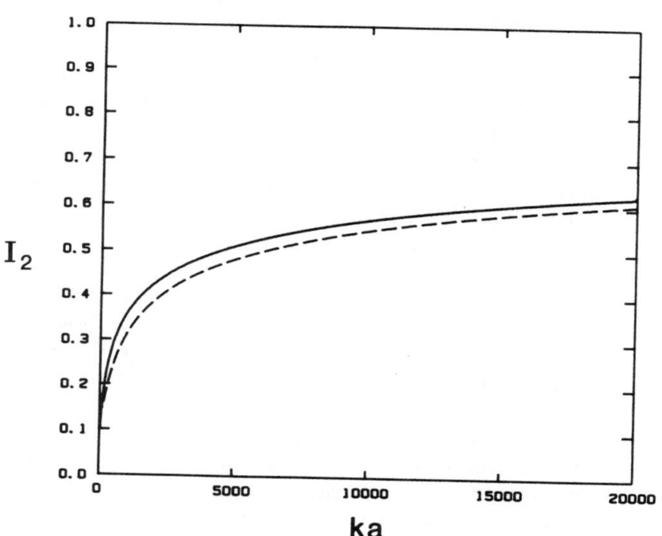

Fig. 3. Normalized irradiance (with j=2) at $\theta = \theta_c$ as given by Mie theory (see text), solid curve, and the leading order approximation from complex angular momentum theory, dashed curve. For red light, a size parameter ka of 10000 gives $a \approx 750\ \mu m$.

The scattering pattern of a spherical bubble may be computed exactly by numerical evaluation of the Mie series.[6,22-25] The relevant parameters are the relative refractive index $m = n_i/n_w = 0.75$ of the gas bubble and the *size parameter* $ka = 2\pi a/\lambda_w = 2\pi a/m\lambda_i$ where $k = 2\pi/\lambda_w$, λ_w is the wavelength of light in water and λ_i is the wavelength in gas. For example, for red light from a He Ne laser $\lambda_i = 632.8$ nm, $\lambda_w = 474.6$ nm, and $ka = 1000$ gives $a = 75$ μm. The incident light is taken to be linearly polarized with its electric field either perpendicular to the scattering plane (to be denoted with a subscript $j = 1$) or parallel to the scattering plane ($j = 2$). (the scattering plane corresponds to the plane of Figures 1 and 4). The normalized scattered irradiance will be denoted by I_j for the aforementioned choices of polarization; this irradiance normalization is such that the geometric reflection of light from a totally reflecting sphere gives a constant value of unity. Let i_j and i_{inc} denote the physical irradiances (having units of W/m^2) of the scattered and incident light, respectively. These are related to I_j and to the complex scattering amplitudes $S_j(\theta, ka)$ from the Mie series via the relations[23-25]

$$i_j = i_{inc} I_j a^2/4R^2, \qquad I_j = |S_j|^2 (2/ka)^2 \tag{3,4}$$

where R denotes the distance from the bubble to the far-zone observation point. It should be emphasized that the evaluation of the S_j usually requires that a large number of terms of the Mie series must be evaluated. The number of terms is somewhat[3,25] larger than ka.

Figure 2 (which is adapted from Reference 23) shows the transition to total reflection for light polarized with $j = 2$. The geometric-optics approximation is an extension of the Davis model[4] to polarized light. As θ decreases to θ_c, the derivative $dI_j/d\theta$ diverges unphysically as is evident by comparison with Mie theory. An important feature of the physical-optics approximation is that it removes that divergence[5] and describes the positions[24] (and general amplitude) of the coarse angular structure present in Mie theory[25] and observations.[23] The coarse structure is partially caused by the interference of rays refracted through the same (*or near*) side of the bubble as the reflected ray.[24] It has been photographed for laser light scattered from bubbles rising freely through water with radii ranging from 46 μm to 0.8 mm. The data show a general agreement with theory.[23]

What distinguishes the Mie theory from the POA in Figure 2 is a superposed fine structure. This fine structure, seen in the original observations with laser illumination,[5] has been modeled and may be used to measure the bubble size.[23] The fine structure corresponds to the interference fringes of the reflected ray with rays refracted by (and internally reflected from) the *far side* of the bubble.[23,25] (See subsequent discussion of Figure 4.) Near 82.8°, the angular period of this structure is $\approx 0.823\ \lambda_w/a$ radians. For polarization $j = 1$, the fine structure is more prominent.

Consider now the power P_j received by a photodetector which collects light scattered into a narrow range of angles about some angle θ. The incident light is assumed to be polarized with the meaning of the index j as noted above and P_j is proportional to a narrow angular average of i_j denoted by $\langle i_j \rangle$. If P_j increases monotonically in the bubble radius a, measured P_j can be used to infer the bubble radius provided the illumination is sufficiently uniform. Consider first the case where the detector is centered on $\theta_c \approx 82.8°$. The POA predicts that[5,24] $I_j(\theta = \theta_c) \approx 1/4$ for all ka so that P_j and $\langle i_j \rangle$ should increase monotonically as a^2 provided the angular width of the detector exceeds that of the fine structure. (A deviation from the a^2 proportionality described in Section 4 does not remove the needed monotonicity.) That $\langle i_j \rangle$ does increase monotonically with radius a for θ close to θ_c has been evident from the Mie computations of Hansen[19] at 80°. Consider now the case where θ is somewhat less than θ_c so that coarse angular structure is present. As the consequence of the interference of *near side* rays, $<i_j>$ will generally not increase monotonically with radius unless the angular aperture is quite large or unless white light is used.[26] This is evident from Hansen's computations for a detector centered on 55°. *A salient feature of critical angle scattering is that the scattered power is relatively large and it also increases monotonically in the bubble radius.*

Consider now the case in which the illumination is monochromatic but unpolarized and the detector is not sensitive to polarization. The scattered irradiance[3] is given by the average $(i_1+i_2)/2$. This is noteworthy since the method of averaging polarizations introduced by Davis[4] is incorrect. Davis performs a polarization average of the reflection or transmission coefficients at each interface instead of treating each polarization component separately until the light leaves the bubble. The results are erroneous.

Some comments on the effects of bubble shape are merited. This is because freely rising bubbles having radii in excess of 150 μm can be slightly oblate (see Section 7). A slice through the equatorial plane of such a bubble reveals a circular cross section. If the scattering plane contains the equatorial plane, treating the bubble as a sphere is an accurate approximation as is evident from data shown in Reference 23. Nevertheless, if the scattering plane is oriented in analogy with experiments on oscillating drops,[21] modulations of the coarse or fine structures could be useful for detecting shape oscillations of bubbles. Radial pulsations may also be detected in that way.[20]

3. COLORS OBSERVED WHEN SUNLIGHT IS SCATTERED FROM BUBBLE CLOUDS

The locations of the maxima and minima of the coarse structure in the scattering pattern in Figure 2 depend on the wavelength. Furthermore, dispersion causes n_w to depend slightly on the wavelength[5,26] such that θ_c for blue light is greater than θ_c for red light by ~0.7°. Consequently the scattering from a sunlit bubble should be colored in the angular region close to 82°. If, instead, a cloud of bubbles is illuminated by sunlight, a given region of the cloud corresponds to a given scattering angle θ and regions of the cloud will appear colored for θ close to 82°. Colors in the critical scattering region of individual bubbles in glass were photographed.[26] During the course of that work, a description of colors seen when viewing clouds of small bubbles rising in water was uncovered in an 1888 paper by Pulfrich.[27] With the unaided eye, Pulfrich observed sunlit water in a glass box. Bubbles were created by the agitation of filling the box. The observations were made with one eye to simulate observation from a distance. The following is a translation[26] of the reported sequence of events after the filling of the box: "...we see after a few moments a reddish hue, and after a few more moments, all the colors of the spectra. After 1 or 1.5 minutes, we can see the whole phenomena with all the supernumerary bows. With moving the head back and forth, the main bow, two of the supernumerary bows, and the red of the third supernumerary bow, are clearly visible." Pulfrich recognized that the cause of the colors was not like that for ordinary rainbows of drops but that it was instead associated with the transition to total reflection.

One of us (P.L.M.) recently photographed colors in clouds of bubbles in sea water for the angular region associated with the transition to total reflection. The bubbles, which had diameters of 1 mm and smaller, were produced by the periodic inflow of water into a large outdoor aquarium in Hawaii. Sunlight entered the water via the free surface and the clouds were observed and photographed by viewing through a vertical window which formed one side of the aquarium. Light was totally reflected to the observer for bubbles in the upper part of the cloud but only partially reflected for bubbles in the lower part. In the transition region the bubbles had yellow-reddish hues with the yellowish region at a slightly larger θ than the reddish region. Had a polarizer been available, the colored band of bubbles would have presumably been even more distinct. After several seconds, the bubbles rose to the surface and the band disappeared.

4. ASYMPTOTIC SERIES FOR SCATTERING AT THE CRITICAL ANGLE FROM CAM THEORY

While the aforementioned physical optics approximation (POA) gives a simple and quantitative understanding of the coarse structure, comparisons with Mie theory[6,25] show that the POA noticeably underestimates the irradiance at θ_c when ka is quite large (roughly $ka>2000$). The error is a consequence of the auxiliary assumptions made in the POA so that the relevant diffraction integral could be reduced to the form of a Fresnel integral.[5] To facilitate a more rigorous analysis Ferrari and Nussenzveig[7,8] extended the modified Watson transformation of the Mie series to the case of bubbles. The results is known as the complex angular momentum (CAM) theory. They obtained excellent agreement with angular scattering patterns computed from Mie theory.[7] (In the comparison, the fine structure was reduced by subtracting off the most insignificant contribution of far-side rays; S_j from Mie computations was replaced by $S_j - S_{j,2}^{f.s.}$ where $S_{j,2}^{f.s.}$ is the relatively weak far-side contribution.) The important result is an approximation for the amplitude of reflected light in terms of integrals denoted as P_F and F_F and given the names Pearcey-Fock and Fresnel-Fock integrals. The integrals depend on ka, θ, and θ_c. The integrals were evaluated numerically by Ferrari.[7]

To obtain a simple series approximation to the scattering at $\theta = \theta_c$, Cleon Dean found series expansions of P_F and F_F in the special case $\theta = \theta_c$. The contribution of the reflected light to the normalized irradiance becomes[28]

$$I_j(\theta_c, \beta) \approx \left| \sum_{q=0}^{M_A} A_{j,q}\, \beta^{-q/4} + \sum_{q=0}^{M_B} B_{j,q}\, \beta^{-q/4} \right|^2 \tag{5}$$

where $\beta = ka$, $A_{j,0} = B_{j,0} = 1/2$ and the coefficients $A_{j,q}$ and $B_{j,q}$ are complex for $q \geq 1$. Note that $\beta^{-q/4} = 1$ for $q = 0$. Because the analysis leading to the $A_{j,q}$ and $B_{j,q}$ is tedious, it was necessary to terminate the series at small M_A and M_B.

The basic results of this analysis were confirmed by comparison with Mie theory in which the contribution of a weak far-side ray has been subtracted off as described above. This comparison is shown in Figure 3 for two different orders of approximation. The Mie theory gives the solid curve which appears to vanish as $\beta \to 0$. (To remove some residual fine structure from the Mie computation, a sliding local average over ka was used.) The dashed curve is given by our approximation, Equation (5) with $M_A=1$ and $M_B=1$. Hence this curve contains corrections to the normalized amplitude through $O(\beta^{-1/4})$. Unless β is quite small, this approximation is more accurate than the POA which gives $I_j=1/4$. Even for the CAM result, the dependence of i_j on the bubble radius is dominated by the a^2 factor in Equation (3).

The CAM analysis leads to the following interpretation. Super-critical and sub-critical rays reflected from a *region* of the bubble's surface contribute to the scattering in the critical direction $\theta = \theta_c$. Super-critical rays are those which have angles of incidence $i \geq i_c$. They give rise to the integral P_F and the terms $B_{j,q}$ in Equation (5). Sub-critical rays have i_c and give rise to the integral F_F and the terms $A_{j,q}$. In the POA, the contribution from all sub-critical rays were neglected, which is equivalent to setting $A_{j,q} = 0$, while the approximation of the diffraction integral by a Fresnel integral is equivalent to taking $B_{j,q}=0$ for $q \geq 1$. With those simplifications, Equation (5) reduces to the POA result $I_j \approx 1/4$. As θ decreases below θ_c,

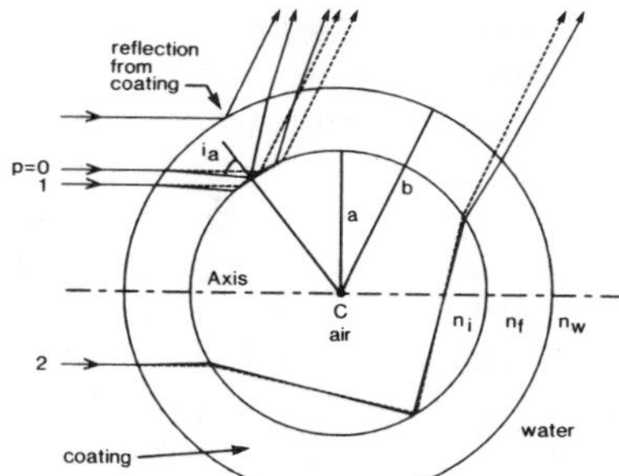

Fig. 4. Reflection and refraction of rays by a coated bubble. For comparison, the effects of the coating are omitted for the dashed rays.

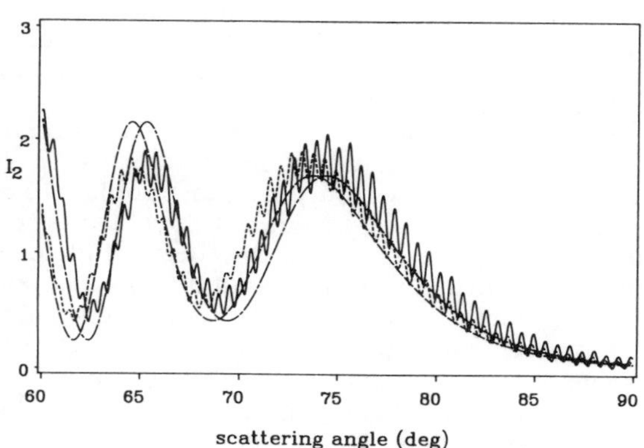

Fig. 5. The solid curve is the exact I_2 for a bubble with $a = 37.8\ \mu m$ and $h = 1.0\ \mu m$ in red light. Eq. (8) gives the alternating dashes while the two other curves are for an uncoated bubble (see text).

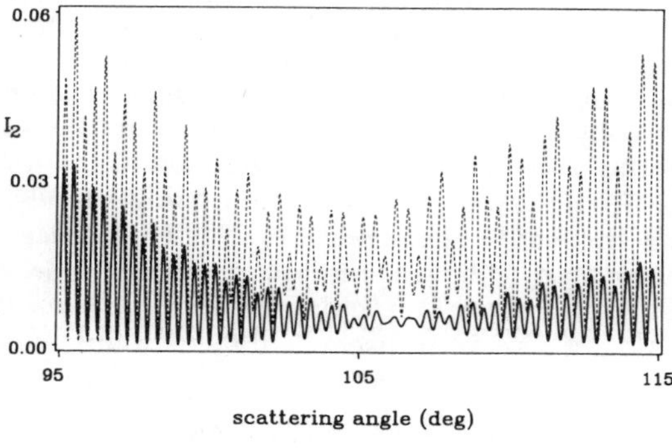

Fig. 6. Normalized scattering of polarized light from Mie theory with ka = 1000 (solid curve). With a coating of thickness $h \approx 0.007a$ (dashed curve) the fine structure is more visible.

the diffractive corrections are greatly diminished in their importance so that the POA gives a satisfactory description of the interference of the reflected ray with the (near-side) transmitted ray as is evident in Figure 2. Notice that as $\beta \to \infty$, Equation (5) predicts that $I_f(\theta_{c,\beta}) \to 1$ in agreement with geometrical optics; inspection of Figure 3 shows the limit is approached more slowly than in usual asymptotics where corrections are $O(\beta^{-1})$.

5. SCATTERING THEORY FOR COATED SPHERICAL GAS BUBBLES IN WATER

Some of the microbubbles which occur naturally in sea water may be stabilized by a thin insoluble coating on the bubble's surface.[9,10] The coatings can be sufficiently thick to inhibit the diffusion of gas from the bubble into the surrounding water. The thickness h of such coatings has been estimated to range from 0.01 μm to 1 μm. Theoretical scattering patterns of coated spherical gas bubbles in water were calculated by adapting the exact partial-wave series of Aden and Kerker[3,29] to the present problem.[11,30] The motivations are essentially three-fold: (i) It is desirable to design optical instruments to count and measure the size of bubbles in such a way that significant errors are not introduced due to the presence of coatings.[31] (ii) If coatings were predicted to have an orderly influence on specific optical scattering properties, it may be possible to invert scattering data to measure poorly understood properties of the coatings and/or study the time evolution of such coatings. (iii) It is worth understanding the gross features of coatings on scattering properties previously studied for clean bubbles.

To anticipate some of the effects of a coating, consider the ray diagram shown in Figure 4. The coating has a refractive index n_f and an outer radius $b = a+h$. It is anticipated that most substances in nature which coat the bubble will have $n_f > n_w$. For the purposes of the present discussion we take $n_f = 1.50$ which is typical of the refractive index of naturally occurring oils. To make the effects of the coating on the rays clearly seen in Figure 4, the thickness to radius ratio h/a was taken to be much larger than that for bubbles of interest. For the dashed outgoing set of rays, the changes in path due to the coating were omitted and the impact parameters were selected so that the scattering angle of each ray was $\theta = 65°$. The index p denotes the number of internal cords of the ray within the gas bubble where $p = 0$ is the reflected ray. For the same set of impact parameters, Figure 2 shows the rays deviated by a coating with $n_f = 1.5$ as solid outgoing lines. The coating also gives rise to a new class of rays which are due to reflections from the coating. For a coated bubble, the relevant interface at which there is an abrupt transition to total reflection becomes the *inner surface of the coating*. Denote the local angle of incidence for a ray at this surface by i_a where the subscript a denotes the radius of the relevant interface as in Figure 4. There will be an abrupt transition to total reflection here when $i_a \geq i_{ca}$ where the new critical angle is $i_{ca} = \arcsin(n_i/n_f)$. Now the ray which reflects from the inner surface will be refracted by the coating-water interface and leaves the bubble with a well-defined scattering angle relative to the direction of the incident beam. For that ray having $i_a i_{ca}$, denote the resulting scattering angle by θ_c'. Evidently θ_c' is the critical scattering angle of the coated bubble. The presence of the coating shifts the critical scattering angle by an amount $\Delta = \theta_c' - \theta_c$. Now Δ is a function of the coating-thickness to radius ratio h/a as well as n_w and n_f. Exact calculation of Δ requires the solution of a set of transcendental equations; however, the leading term[11,31] in a power series expansion of $\Delta(h/a)$ is

$$\Delta = \frac{2h}{a}\left[\left(n_w^2-1\right)^{-1/2} - \left(n_f^2-1\right)^{-1/2}\right] + O(h/a)^2, \quad \Delta \text{ in radians,} \tag{6}$$

where $n_i = 1$ and $O(h/a)^2$ is a function which vanishes as rapidly as $(h/a)^2$ as $h \to 0$. Notice that $\Delta \to 0$ if the coating index $n_f \to n_w$ or if the thickness $h \to 0$.

Numerical evaluation of the *Aden-Kerker Series* (AKS) gives scattering amplitudes S_j which, when inserted in Equations (3) and (4), give the exact unnormalized and normalized irradiances respectively.[11,30] It is appropriate here to reference these to the radius a of the gas pocket within the coating (see Figure 4) since it is often the volume of gas which is the quantity of interest instead of the outer radius b of the coating. The input parameters to the AKS are: the relative refractive indices, $m = n_i/n_w = 0.75$ and $m_f = n_f/n_w$; ka; and a/b. From these parameters and the wavelength λ_i in the gas interior of the bubble,

$$\text{the coating thickness} \quad h = b\text{-}a = \lambda_i\, mka(b/a)[1\text{-}a/b)]/2\pi \tag{7}$$

In the examples to be given, h and a for the specified ka are calculated by taking $\lambda_i = 632.8$ nm, corresponding to light from a He Ne laser; Equation (7) may be used to find the appropriate value of h for which each example applies for other λ_i. Figure 5 illustrates the effects of a coating on the near-critical-angle scattering from a bubble with ka= 500. The solid curve is the AKS result for h =1.0 μm while the curve with the short dashes is the Mie series result for an uncoated bubble. Two effects of the coating are to (i) *shift the coarse structure* in the direction of increased θ, and (ii) *increase the visibility of the fine structure fringes* superposed on the coarse structure. In the case of an uncoated bubble, the fine structure fringes are principally due to the interference of rays which leave the bubble widely spaced.[25] It may have been anticipated that even a thin coating increases the visibility of the fine fringes since the coating tends to increase the magnitude of the far-side p =2 scattering contribution because of an enhancement of the effective reflectance from within the bubbles. Effects (i) and (ii) are also evident in similar calculations which were carried out for ka of 100, 500, 1000, and 2500 for h as thin as 0.08 μm. We chose h of 1 μm in Figure 5 so that the shift Δ predicted by Equation (6) has the relatively large value of 0.72°. To better illustrate the shift, Figure 5 shows two additional curves. The smooth curve with dashes of intermediate length is the *physical-optics approximation* for an uncoated bubble,[23,24] to be designated here as $I_j(\theta,ka)$. (The comparison of the POA with Mie theory is much like the example in Figure 2.) To approximate the effect (i), the following *shifted* POA is plotted as the curve with alternate long and short dashes:

$$I_j^{CPOA}(\theta,ka) = I_j^{CPOA}(\theta - \Delta,ka) \tag{8}$$

for the present case of j = 2 and ka=500. The coating parameters n_c and h/a enter into this approximation only through the approximation of Δ by the first term in Equation (6). The main idea behind Equation (8) is that the principal effect of a *thin coating* (at least for polarization j = 2 as noted below) is to shift the scattering angles of the critical ray and the single-chord ray in Figure 4 by similar amounts. This shift results in a corresponding shift of the critical-angle diffraction and coarse interference pattern. It is evident that this CPOA describes the general rise in irradiance of the first coarse peak as θ decreases from 85° to 75°. The shift in the location of the peak originally near 65° is also well approximated by the model. This, and several other comparisons done for j = 2 with ka from 100 to 1000 show that the CPOA does about as well in approximating the coarse structure for a thinly coated bubble as the POA does for an uncoated bubble.

The effects of polarization, film refractive index n_f, and optical absorption by the coating material on the scattering pattern were also explored and recommendations were made concerning the design of optical devices to measure the size of bubbles in the ocean.[11,30,31] For polarization j = 1, which is the case of E-field perpendicular to the scattering plane, the effect of the coating on the coarse structure is more pronounced than the angle shift illustrated in Figure 5. This is evidently a consequence of the ray which reflects from the outside of the coating (see Figure 4). Its amplitude is much larger for j = 1 than it is for j = 2. Consequently, optical devices which use either the spacing or locations of the coarse features or the monotone dependence of irradiance on size for $\theta \approx \theta_c$, should use polarized incident light and select the scattering plane such that $j = 2$. When $j = 2$, the effects of changes in n_f and of small amounts of absorption can generally be understood with elementary arguments.[11,30,31]

6. BREWSTER ANGLE SCATTERING FROM BUBBLES: THEORY AND EXPERIMENT

When the electric field is polarized parallel to the plane of incidence, the Fresnel reflection coefficient of a flat clean interface vanishes when the angle of incidence is at Brewster's condition[6,11]

$$i_B = \arctan(n_i/n_w), \qquad \text{Brewster scattering angle} = \theta_B = 180° - i_B = 106.2° \qquad (9,10)$$

where Equation (10) follows from Figure 1 and the assumption that the Brewster condition is not significantly affected by curvature provided $a \gg \lambda_w$. This local reduction in the reflectivity manifests itself in the I_2 for an uncoated bubble as shown in Figure 6 which shows the calculated I_2 for uncoated (solid curve) and coated (dashed curve) bubbles for $ka = 1000$ and $a = 75.5$ μm; for the coated case $a/b = 0.993$, $h = 0.53$ μm, and $n_f = 1.5$. The coating is seen to increase the visibility of the fine structure fringes in the Brewster region relative to the uncoated case. This increase may be understood by noting the cause of decreased fringe visibility near θ_B for the uncoated case. Away from θ_B, these fringes arise from the interference of the $p = 0$ reflection with the reflection associated with the far side $p = 2$ ray (see Figure 4). The reduction in visibility near θ_B occurs because the reflection is suppressed at the Brewster condition, Equation (9). When the bubble is coated, the Brewster effect is quenched so that a reflection contribution is restored near 106.2°. Plots of I_2 for $ka = 1000$ with $h = 0.23$ μm and 0.99 μm also show significant increases in the fringe visibility in the Brewster region.[30] Therefore, measurements of the visibility of the fringes near 106° may be used to optically characterize coatings.[11] Calculations of I_1 for this angular region for uncoated bubbles show no local reduction in the fringe visibility since for polarization $j = 1$, there is no minimum of the Fresnel coefficient.[6]

The scattering patterns in the angular region near 106° were photographed for bubbles rising freely through distilled water.[12] This was done to verify for the first time, that a local reduction in fringe visibility is observable. The bubbles were illuminated by a horizontally propagating 300 mW beam from an Ar-ion laser for which $\lambda_i = 514.5$ nm and $k = 2\pi/(m\ 514\ \text{nm}) \approx 16.29\ \mu\text{m}^{-1}$. The already polarized output of the laser was made more perfectly polarized (with a horizontal E field) by passing the beam through an ellipsometer-grade polarizer. Light in the horizontal scattering plane passed through a window into a Polaroid filter and a camera focused on infinity. The apparatus was generally similar to the ones described in References 5 and 23, except for modifications needed to facilitate viewing the Brewster region and the reduction in the level of spurious and background scattering (note the low value for normalized irradiance predicted in Figure 6). After rising through the beam, the bubble floated against a glass slide and its radius was measured directly. The photographic negatives of the scattering were scanned with a microdensitometer from which an exposure E_N as a function of the scattering angle θ could be determined much like in Reference 23. The angle calibration of the photograph was achieved by use of a mirror-goniometer technique.[11,23] Even with no bubble in the beam, the photograph would be weakly exposed. The smoothed irradiance pattern inferred from a background photograph will be denoted by $f(\theta)$; it is a fourth order polynomial fitted to the photograph. It was desired to compare the scattering patterns from bubbles with predictions from Mie theory for clean (uncoated) bubbles. To facilitate this comparison, it was necessary to introduce three positive adjustable parameters α, β, and C, from which the normalized irradiance $I_2(\theta)$ could be inferred from the exposure $E_N(\theta)$ for a given bubble via

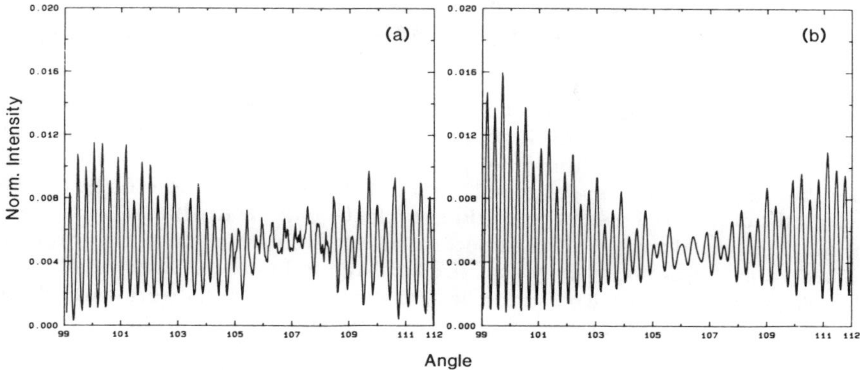

Fig. 7. Normalized irradiance for a bubble of radius 78.5 μm in green light: (a) is inferred from a photograph and confirms the local reduction of fringe visibility near the Brewster condition; (b) is a narrow angular average of I_2 from Mie theory for ka = 1278.

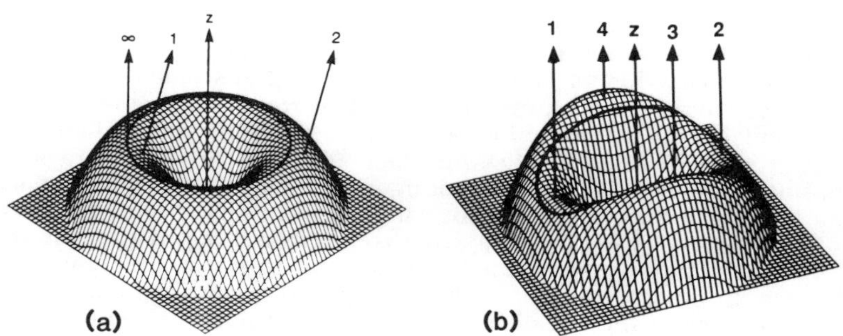

Fig. 8. (a) Shape of the outgoing wavefront associated with glory scattering from spheres. For bubbles, the focal radius b depends on the number (p-1) of internal reflections. (b) Perturbed wavefront for backscattering of horizontal light from an oblate rising bubble. Rays 3 and 4 correspond to glory rays in Fig. 9(b).

$$I_2(\theta) = \beta[E_N(\theta) - \alpha f(\theta) + C] \quad (11)$$

The parameter α was selected so as to remove a component of E_N attributable to background; C was selected to eliminate an apparent offset. The scaling constant β was adjusted to facilitate comparison with a locally averaged form of $I_2(\theta)$ from Mie theory. A similar local average was performed on the "raw" $E_N(\theta)$ from the negatives so as to reduce noise associated with the graininess of each photograph.

Figure 7 is a representative comparison of data with theory for a case where the local average was over ±0.04°. Outside of the region from about 105° to 108°, the quasiperiod of the irradiance oscillations is fairly uniform. Measurement of that quasiperiod shows that the oscillations are primarily associated with the interference of a $p = 0$ ray with a (far-side) $p = 2$ ray. For θ close to $\theta_B \approx 106°$, there is a noticeable reduction in the fringe visibility and the quasi-period changes. The $p = 0$ reflected ray is suppressed there and the fringes can be attributed to the interference of other (weak) rays.[11] The important result is that the same general features are observable. It is noteworthy that the local reduction of the fringe visibility near θ_B is evident in the "raw" $E_N(\theta)$. Similar detailed comparisons were carried out for bubbles having radii ranging from 60 to 80 μm, while reduction in fringe visibility was noted with radii from 50 to 100 μm. Since the experiments were carried out in distilled water, no optically significant coating was expected and no evidence for such a coating was found. This confirmation of Brewster effects for uncoated bubbles suggest that the effects of a film (Figure 6) may be observable.

Since geometric features are important near θ_c and θ_B, it may be argued that the ratio $\langle I_2(\theta_c)\rangle / \langle I_2(\theta_B)\rangle$ depends only weakly on ka for bubbles where $\langle\ \rangle$ denotes a local average about the indicated scattering angle and that measurement of that ratio may be useful for discrimination of bubbles from particles-in-water.[31]

7. BACKSCATTERING FROM FREELY RISING SPHERICAL AND SPHEROIDAL AIR BUBBLES IN WATER: GLORY SCATTERING AND THE ASTROID CAUSTIC

For a spherical bubble, some of the rays incident with a non-vanishing impact parameters leave the bubble scattered exactly backwards.[6,14,32] As a consequence of the azimuthal symmetry of spheres, the scattering associated with such rays is weakly focused along the backward axis.[2,22] This weak focusing is generally termed "axial focusing" or "glory scattering" and manifests itself in a local enhancement of backscattering from cloud droplets;[1,2] the detailed cause of glory rays is different for drops and bubbles. For backscattering from spheres, each type of glory ray produces a backward directed wavefront with the local shape of a torus; Figure 8(a) shows this shape where z denotes a horizontal axis directed towards the source. There are an infinity of rays from this wavefront in the z direction but only two parallel rays in other directions. The geometrical analysis of Davis[4] predicts an infinite irradiance in the backward direction. Real bubbles rising through water take on an oblate shape as a consequence of hydrodynamic forces on the bubble.[13,33] For a sufficiently oblate bubble, the outgoing wavefront becomes distorted as shown in Figure 8(b), and the focal properties of the scattering are changed. Previous experiments[14] had confirmed that bubbles rising through a viscous oil can produce backscattering patterns characteristic of glory scattering from spheres. It was not obvious, however, how small a bubble in *water* needed to be if the scattering pattern was to retain its symmetry. Furthermore, it was of general interest to see how the scattering pattern unfolded in response to a wavefront perturbation leading to the shape in Figure 8(b). Consequently, an experimental and theoretical study of backscattering from bubbles in water was undertaken.[13,22,33] The full analysis of the wavefields and caustics is lengthy and the results will be only briefly summarized here.

For the case of bubbles which are sufficiently small to be spherical, a three-way

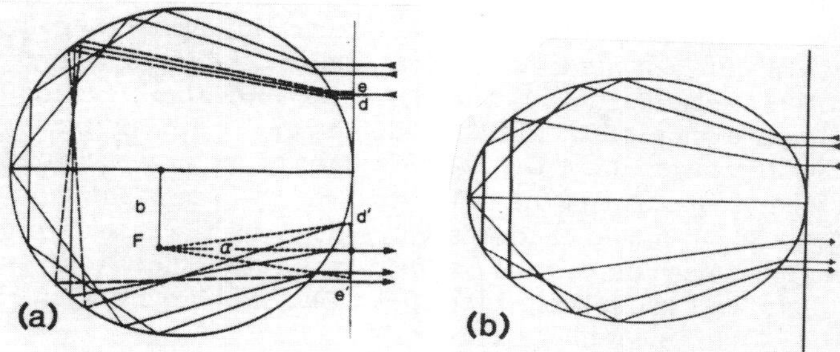

Fig. 9. For backscattering from spheres, (a) shows the glory rays with p = 3, 4, and 5. The line on the right is where the exit plane intersects the figure. For a rising bubble with an exaggerated oblateness, (a) is the horizontal equatorial plane while (b) is for a vertical plane containing the axis of rotational symmetry of the bubble.

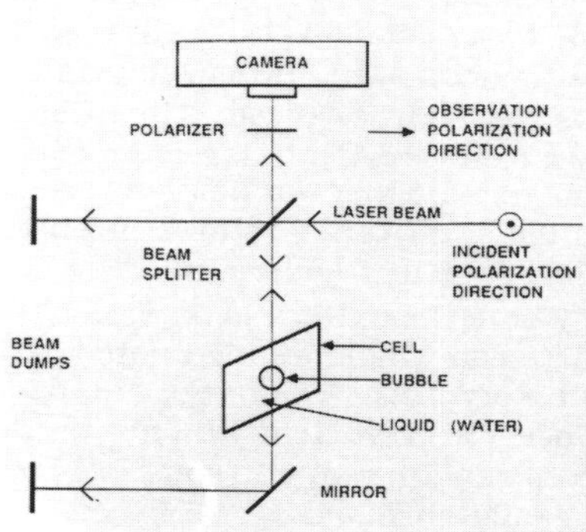

Fig. 10. Top view of the experimental apparatus for observing the cross-polarized backscattering. The bubble size relative to the other apparati and the tilt of the scattering cell have been exaggerated.

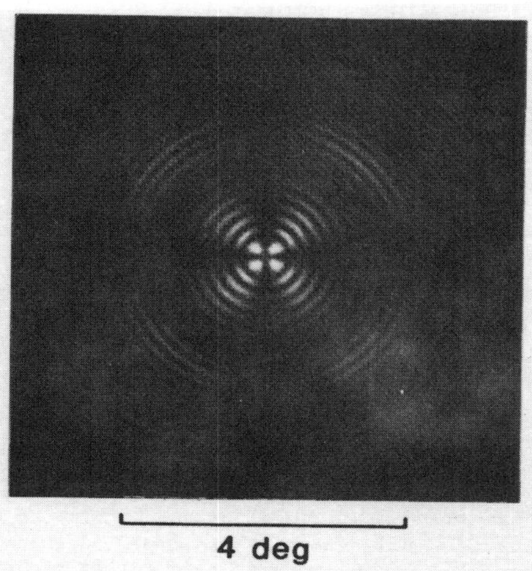

Fig. 11. Photograph of cross-polarized backscattering pattern for a bubble of radius 122.0 μm with ka = 1988, showing the fourfold symmetric azimuthal dependence characteristic of spherical scatterers.

agreement was found between observations of patterns for bubbles in distilled water, Mie computations, and a physical optics approximation (POA).[22] The first stage of the POA is the geometrical analysis of glory rays shown in Figure 9(a). The incoming wavelet *de*, which encompasses the three-chord glory ray becomes the outgoing wavelet *d'e'*; the outgoing wavefront is toroidal since the figure may be rotated about the horizontal symmetry axis. The field of the outgoing wave may be approximated in an *exit plane* which contacts the bubble by use of geometry and Fresnel reflection coefficients.[14] The POA propagates this wave to a distant observer via the approximation of a diffraction integral. For simplicity, we limit our discussion to the case where the incident light is linearly polarized and only the cross-polarized component of the scattered field $E_{2,p}{}^{g}$ is of interest; p denotes the number of chords of the associated glory ray ($p = 3$ for the case described above) and the superscript g designates that the field is due to a glory ray. At observers for which the distance r (from the center of exit plane) is large and the backscattering angle γ is <<1 radian, the analysis gives[13,32]

$$E^{g}_{2,p}(r, \gamma, \varphi) = \frac{a}{r}\,(ka)^{1/2}\, E_i K_p \exp[i\,(kr-\omega t) + i\,\tau_p(\gamma)] J_2(u_p)\sin 2\varphi \qquad (12)$$

where: φ denotes the azimuthal angle of the observer relative to the polarization direction of the incident wave, E_i is the amplitude of the incident wave, K_p is a complex constant, J_2 is a Bessel function, $u_p = kb \sin\gamma$, and b is the impact parameter of the pth glory ray. The angle-dependent phase factor is $\tau_p = -ik\alpha(1-\cos\gamma)$ where α is the distance to the virtual focus shown in Figure 9(a). The fields from significant glory rays ($3 \leq p \leq 16$) have been summed and used to compute the cross-polarized irradiance. This has been done for bubbles for various ka from 100 to 2000; the results agree with Mie theory.[13,32]

The $\sin 2\varphi$ factor in (12) shows the cross-polarized irradiance is proportional to $\sin^2 2\varphi$ so that backscattering pattern is four-fold symmetric. Observations of irradiance patterns for freely rising bubbles in water confirm the patterns are four-fold symmetric (provided $a \lesssim 150$ μm) and the model has the proper dependence on γ. Figure 10 shows a diagram of the apparatus viewed from above. The bubble rises into a horizontally propagating 200 mW beam from an Ar-ion laser for which $\lambda_i =$ 514.5 nm. The incident beam is vertically polarized while the polarizer in front of the camera is set to *block* vertically polarized light. Figure 11 shows a representative backscattering pattern which is manifestly four-fold symmetric. Exact backscattering ($\gamma = 0$) is at the center of symmetry where the pattern is dark since $J_2(0) = 0$ in Equation (12). Away from $\gamma = 0$, fringes are seen which are modulated as a consequence of the interference between different glory rays. The fringe angular spacing tends to decrease with increasing size since the coefficient kb in u_p increases. While the aforementioned predictions were confirmed for small rising bubbles, one aspect of the model which could not be as easily observed is the prediction that $E_{2,p}{}^{g}$ increases as $a(ka)^{1/2}$ where the factor $(ka)^{1/2}$ is a consequence of axial focusing.

Our observations show that if the bubble radius is increased above ≈150 μm, deviations from a $\sin^2 2\varphi$ dependence are evident and that eventually the symmetry and other features of the pattern differ radically from that of Figure 11. The bubble takes on the shape of an oblate spheroid in which the symmetry axis is vertical so that the outgoing wavefront no longer has the toroidal symmetry of Figure 8(a). The profile of the bubble remains circular in the horizontal equatorial plane so that the rays in that plane remain as shown in Figure 9(a). The profile of the bubble in a plane containing the vertical axis is an ellipse for which c denotes the semi-minor axis; c is the distance to the top of the bubble from the equatorial plane. Figure 9(b) shows rays in such a vertical plane which are backscattered. The resulting wavefront is shaped as shown in Figure 9(b), since the optical path length of the rays in

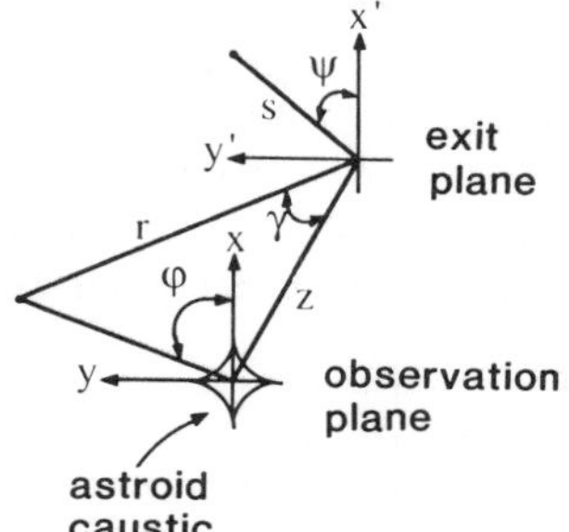

Fig. 12. The wavefront described by Eq. (13) gives an astroid caustic where the irradiance diverges in the geometric optics limit. The exit plane is at the bubble.

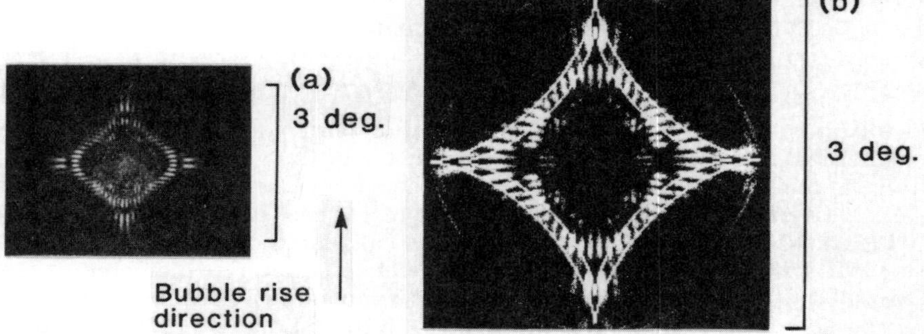

Fig. 13. Observed (a) and calculated (b) cross-polarized backscattering patterns for an oblate freely rising bubble in water. Astroids from Eq. (17) are superposed on (b).

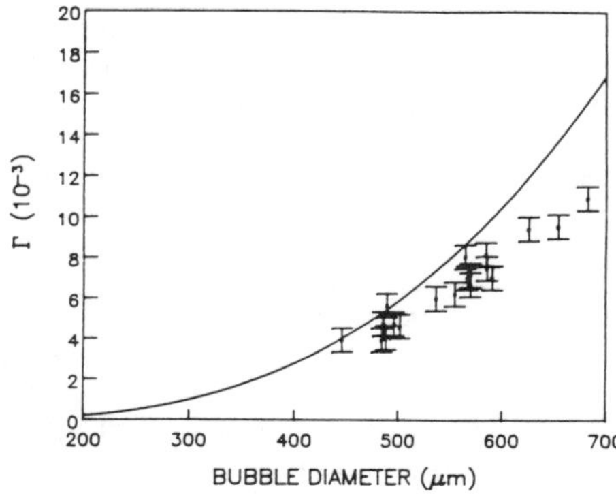

Fig. 14. The curve gives the oblateness $\Gamma \approx 9We/32$ where We is the Weber number as estimated from the drag law of Schiller and Nauman. The points are Γ inferred from observed caustics.

the vertical plane (rays 3 and 4) is less than the corresponding rays (1 and 2) in the equatorial plane. This wavefront shape may be well approximated by the function[13] where s and ψ are polar coordinates

$$W(s,\psi) = -\Lambda - \frac{(s-b)^2}{2\alpha} - \frac{\delta}{2}(1+\cos 2\psi) \tag{13}$$

and $\Lambda<0$, b, α, and δ are parameters which may be interpreted by inspection. It is convenient to define the oblateness Γ of the bubble which is related to δ via

$$\Gamma = (a^2/c^2)-1 \ , \qquad \delta \approx a A_p \Gamma, \qquad A_3 \approx -0.613 \tag{14, 15, 16}$$

where a denotes the equatorial radius of the bubble and the form of (15) and the values of the dimensionless coefficient A_p and ratios b/a have been calculated by ray tracing.[13]

Propagation of a wavefront described by (13) gives rise to an astroid caustic having the orientation shown in Figure 12 where the x and x' axes are vertical and the polar coordinates (s, ψ) are shown in the exit plane. This caustic (or bright region) of the scattering pattern is the salient feature of backscattering from oblate bubbles and is germane to other perturbed glory scattering problems. When the distance z to the observation plane is large, the directional coordinates $U = x/z$ and $V = y/z$ may be found by first locating points on the wavefront $W(s,\psi)$ where the *Gaussian curvature* vanishes. Rays from such points are in the direction of the caustic.[22] This yields the following approximation (where $b/a = 0.477$ when $p = 3$) for the caustic[13,33]

$$U^{2/3} + V^{2/3} \approx (2\delta/b)^{2/3} \tag{17}$$

which is an astroid. The wavefront perturbation parameter δ tends to be somewhat smaller than b, which is the impact parameter of the glory ray in the equatorial plane. In addition to the caustic, detailed cross-polarized backscattering patterns were calculated by extending the POA to the present case of slightly oblate bubbles.

Figure 13(a) and (b) show observed and calculated cross-polarized back-scattering patterns for a freely rising bubble whose equatorial diameter $d = 2a$ was measured with a microscope to be 654 μm. The apparatus was as shown in Figure 10. Superposed on Figure 13(b) are the caustics given by Equation (17) where the outer and inner astroids are for $p = 3$ and 4. The irradiance is greatest near these caustics. The oblateness parameter Γ was adjusted so that the angular width of the calculated pattern agreed with the observation; this gave $\Gamma = 0.00954$. The general features of the observed pattern are described by the model. The angular width $U_{max} \approx |2\delta/b|$ of the caustic decreases rapidly with decreasing d; U_{max} was measured for several bubbles with d from 440 to 680 μm and used to estimate Γ via Equation (15). The estimated Γ are compared in Figure 14 with a curve based on a hydrodynamics approximation.[13] The comparison generally supports Equation (15) since various assumptions in the hydrodynamics approximations become suspect when $d \gtrsim 600$ μm. In addition to Figure 13, other comparisons of observations with computed patterns were completed for d ranging from 304 to 660 μm and these generally support the POA. Patterns for a different incident polarization were also studied.[13,33]

8. EFFECTS OF LASER RADIATION PRESSURE: LEVITATION AND ACOUSTIC EMISSIONS

The natural upward buoyancy of bubbles in water was counteracted by the optical radiation pressure of a downward directed laser beam.[15] Bubbles having radii in the 10 to 30 μm range were "trapped" or "levitated" in beams whose optical power ranged from 1 to 3.5 W. The magnitude of the power required to levitate could be understood from Debye theory for the radiation force (which is exact when the incident wave is a plane wave[2,3]) and from a heuristic theory[15] (which includes only the force due to

Fig. 15. Photograph of a 15 μm radius bubble trapped at a depth of ~5 mm. The beam path in the water is visible due to fluorescing and scattering in the water. The bubble is located at the top of the bright spot caused by reflection off the bubble's surface. A microscope objective, used to focus the beam, is visible above the water tank.

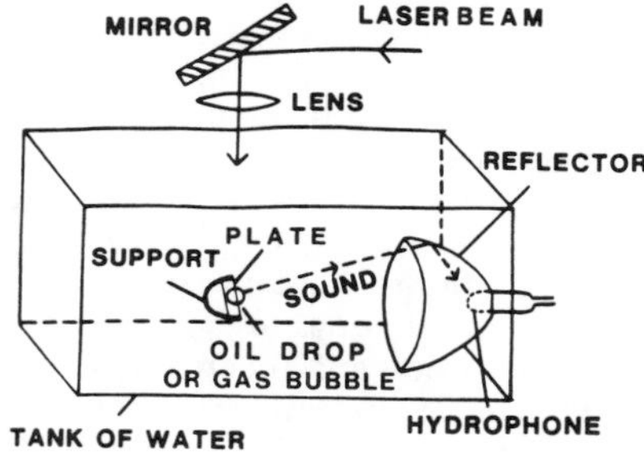

Fig. 16. Experiment for observing optoacoustic signals. The light was focused close to the bubble or drop. Sometimes it was directed up instead of down as shown here. Signal pressure amplitudes were referenced to what the pressure would be a distance of 103 mm from the bubble if no reflector were present.

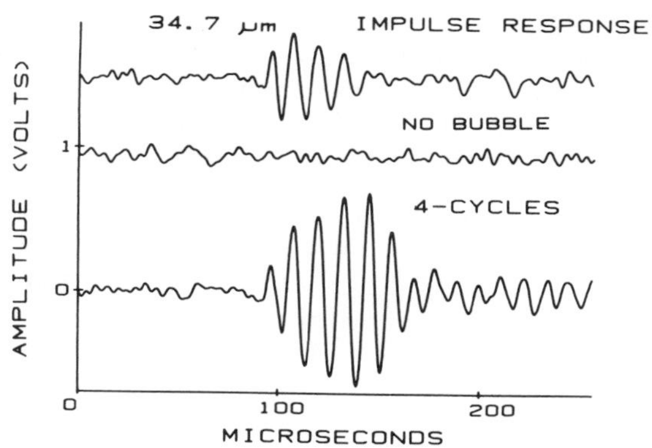

Fig. 17. The top and bottom traces are signals for a short light pulse and for a 4-cycle burst. The bubble radius was 34.7 μm while the peak beam power was 1.8W. The peak-to-peak pressure for the bottom trace was ≈ 8 mPa while the theoretical estimate was 0.9 mPa. The impulse response in the absence of a bubble is the central trace.

light totally reflected from the bubbles). The rough agreement of these theories suggests that momentum transfer is primarily due to rays incident on the bubble where the local angle of incidence (i in Figure 1) exceeds i_c of Equation (1). For horizontal stability the beam had an irradiance minimum at the center, while for vertical stability the bubble is levitated below the focus of a downward directed beam. Figure 15 is a photograph of a trapped bubble. The beam power is not modulated in time. This method of levitation is acoustically quiet.

When the power of a laser beam incident on a bubble is modulated, the bubble radiates sound.[16,17] The resonance described below indicates the acoustic emission is associated with volume pulsations of the bubble; the pulsations appear at least partially due to the compression of the bubble by optical radiation stresses. The apparatus, which was also used to detect sound from dyed drops, is shown in Figure 16. Bubble radii were in the 20-135 μm range. Bubbles were usually produced by electrolysis and float up to a thin transparent glass or Mylar support. The light was green, having a peak pulse power $\lesssim$ 2W. The acoustic signal was enhanced by placing the hydrophone at the focal point of an ellipsoidal reflector with the bubble at the conjugate focus. Figure 17 shows representative signals from a bubble of radius a = 34.7 μm. Here, and for other bubbles, the first record obtained was the impulse response: the sound radiated in response to a short pulse of light having a duration $\lesssim$5 μs. The arrival time for the signal is essentially the calculated propagation delay. The damped ringing of an oscillator is evident in the impulse response shown in Figure 17. Sounds from bubbles are strongly affected by a monopole oscillation in which the water provides the inertia and the gas in the bubble acts like a spring. The resulting natural frequency for bubble pulsations is calculated roughly to be $f_n \approx 3250/a$ (a in μm, f_n in kHz). This gives $f_n \approx$ 94 kHz while an FFT of the record gave a measured frequency $f_{mn} \approx$ 77 kHz. The bubble was subsequently driven by a burst of four square light pulses, each of duration $(2f)^{-1}$, where the fundamental frequency f was set equal to f_{mn}. The resulting record, the lower trace of Figure 17, shows an amplitude enhancement, which is the predicted resonance response of an oscillator. The excitation of monopole oscillations by the light was confirmed by measurements of f_{mn} with radii from 20 to 135 μm and by the reduced amplitude observed when f for the burst was detuned from f_{mn}.

The response of bubbles to modulated light gives a novel optoacoustic mechanism. Even in the absence of absorption, light energy is converted to acoustical energy. To understand that such a conversion of energy is plausible, note that the light scattered by the bubble is Doppler-shifted downward in frequency as a consequence of the reaction of the bubble's surface. A theory has been developed for bubble pulsations based on the momentum transferred to the bubble's surface in the total reflection region. The theory appears to describe how the amplitude is maximized as a function of a/ρ_o, where ρ_o is the beam radius. The peak-to-peak pressure is underestimated by an order of magnitude.[16] It is plausible that the effects of electrostriction are not fully accounted for, but (unlike the case for dyed oil drops) it appears implausible that direct absorption of light is significant.

ACKNOWLEDGEMENT

The research described in this paper was supported by the Office of Naval Research. Parts of Sections 4-6 were supported by the Naval Ocean Research and Development Activity. D. S. Langley and S. C. Billette assisted in computations.

REFERENCES

1. R. Greenler, **Rainbows, Halos, and Glories** (Cambridge U. Pr., Cambridge, 1980).
2. H. C. van de Hulst, **Light Scattering by Small Particles** (Wiley, New York, 1957).
3. C. F. Bohren & D. R. Huffman, **Absorption and Scattering of Light by Small Particles** (Wiley, New York, 1983).

4. G. E. Davis, *J. Opt. Soc. Am.* **45**, 572-581 (1955).
5. P. L. Marston, *J. Opt. Soc. Am.* **69**, 1205-1211 (1979); (E) **70**, 353 (1980).
6. P. L. Marston, D. S. Langley, & D. L. Kingsbury, *Appl. Sci. Res.* **38**, 373-383 (1982).
7. N. F. Ferrari, Jr., Ph.D. thesis, Univ. of São Paulo Brazil, 1983 (in Portuguese).
8. H. M. Nussenzveig, "Recent developments in high-frequency scattering," *Rev. Bras. Fiz.* (Brazilian Reviews of Physics, special issue) 302-320 (1984).
9. B. D. Johnson and R. C. Cooke, *Science* **213**, 209-211 (1981).
10. V. K. Goncharov, S. N. Kuznetsova, G. G. Nevimin, & N. A. Sorokina, *Sov. Phys. Acoust.* **30**, 273-275 (1984).
11. P. L. Marston, S. C. Billette, & C. E. Dean, "Scattering of light by a coated bubble in water near the critical and Brewster scattering angles," **Ocean Optics IX,** M. A. Blizard, ed., *Proc. SPIE* **925**, 308-316 (1988).
12. S. M. Bäumer, M.S. thesis, Washington State University, 1988.
13. W. P. Arnott, Ph.D. dissertation, Washington State University, 1988.
14. D. S. Langley & P. L. Marston, *Phys. Rev. Lett.* **47**, 913-916 (1981).
15. B. T. Unger & P. L. Marston, *J. Acoust. Soc. Am.* **83**, 970-975 (1988).
16. B. T. Unger, Ph.D. dissertation, Washington State University, 1987.
17. B. T. Unger & P. L. Marston, "Optically stimulated sound from oil drops and gas bubbles in water: thermal and radiation pressure optoacoustic mechanisms," **Ocean Optics IX,** M. A. Blizard, ed., *Proc. SPIE* **925**, 326-333 (1988).
18. P. L. Marston & D. S. Langley, "Forward optical glory from bubbles (and clouds of bubbles) in liquids and other novel directional caustics," in **Multiple Scattering of Waves in Random Media and Random Rough Surfaces,** V. V. Varadan & V. K. Varadan, eds. (Pennsylvania St. U. Pr., University Park, PA, 1987), pp. 419-429.
19. G. M. Hansen, Appl. Opt. **24**, 3214-3219 (1985).
20. P. L. Marston, *Annual Report*, 1984 (available from Defense Tech. Info. Ctr., Cameron Station, Alexandria, VA, Accession No. AD-A146703) pp. 51-53.
21. P. L. Marston, *Appl. Opt.* **19**, 680-685 (1980).
22. W. P. Arnott & P. L. Marston, *J. Opt. Soc. Am.* **A5**, 496-506 (1988).
23. D. S. Langley & P. L. Marston, *Appl. Opt.* **23**, 1044-1054 (1984).
24. P. L. Marston & D. L. Kingsbury, *J. Opt. Soc. Am.* **71**, 192-196, 917 (1981).
25. D. L. Kingsbury & P. L. Marston, *J. Opt. Soc. Am.* **71**, 358-361 (1981); Appl. Opt. **20**, 2348-2350 (1981).
26. P. L. Marston, J. L. Johnson, S. P. Love, & B. L. Brim, *J. Opt. Soc. Am.* **73**, 1658-1664+ Plate X (1983).
27. C. Pulfrich, Ann. Phys. Chem. (Leipzig) **33**, 209-212 (1888).
28. C. E. Dean & P. L. Marston: work in progress. The authors invite correspondence regarding it.
29. A. L. Aden & M. Kerker, J. Appl. Phys. **22**, 1242-1246 (1951).
30. S. C. Billette, M.S. thesis, Washington State University, 1986.
31. P. L. Marston, *Light Scattering Theory for Bubbles in Water: Inverse Scattering, Coated Bubbles, and Statistics*, 1986 (available from Defense Tech. Info. Ctr., Alexandria, VA, Accession No. AD-A174997).
32. P. L. Marston & D. S. Langley, "Strong backscattering and cross polarization from bubbles and glass spheres in water," **Ocean Optics VII,** M. A. Blizard, ed., *Proc. SPIE* **489**, 130-141 (1984).
33. W. P. Arnott & P. L. Marston, "Backscattering of laser light from freely rising spherical and spheroidal air bubbles in water," **Ocean Optics IX,** M. A. Blizard, ed., *Proc. SPIE* **925**, 296-307 (1988).

FIGURE CAPTIONS

1. Reflection of a ray from a spherical bubble. The ray lies in the scattering plane.
2. Normalized irradiance (with $j=2$) from a bubble with $ka = 1633$ from Mie theory (solid curve), physical optics (dashed) and geometric optics (dotted).
3. Normalized irradiance (with $j = 2$) at $\theta = \theta_c$ as given by Mie theory (see text), solid curve, and the leading order approximation from complex angular momentum theory, dashed curve. For red light, a size parameter ka of 10 000 gives $a \approx 750$ μm.

4. Reflection and refraction of rays by a coated bubble. For comparison, the effects of the coating are omitted for the dashed rays.
5. The solid curve is the exact I_2 for a bubble with a= 37.8 μm and h = 1.0 μm in red light. Equation (8) gives the alternating dashes while the two other curves are for an uncoated bubble.
6. Normalized scattering of polarized light from Mie theory with ka = 1000 (solid curve). With a coating of thickness $h \approx 0.007a$ (dashed curve) the fine structure is more visible.
7. Normalized irradiance for a bubble of radius 78.5 μm in green light: (a) is inferred from a photograph and confirms the local reduction of fringe visibility near the Brewster condition; (b) is a narrow angular average of I_2 from Mie theory for ka = 1278.
8. (a) Shape of the outgoing wavefront associated with glory scattering from spheres. For bubbles, the focal radius b depends on the number (p-1) of internal reflections. (b) Perturbed wavefront from backscattering of horizontal light from an oblate rising bubble. Rays 3 and 4 correspond to glory rays in Figure 9(b).
9. For backscattering from spheres, (a) shows the glory rays with p = 3, 4, and 5. The line on the right is where the exit plane intersects the figure. For a rising bubble with an exaggerated oblateness, (a) is the horizontal equatorial plane while (b) is for a vertical plane containing the axis of rotational symmetry of the bubble.
10. Top view of the experimental apparatus for observing the cross-polarized backscattering. The bubble size relative to the other apparati and the tilt of the scattering cell have been exaggerated.
11. Photograph of cross-polarized backscattering pattern for a bubble of radius 122.0 μm with ka=1988, showing the fourfold symmetric azimuthal dependence.
12. The wavefront described by Equation (13) gives an astroid caustic where the irradiance diverges in the geometric optics limit. The exit plane is at the bubble.
13. Observed (a) and calculated (b) cross-polarized backscattering patterns for an oblate freely rising bubble in water.
14. The curve gives the oblateness $\Gamma \approx 9\mathrm{We}/32$ where We is the Weber number as estimated from the drag law of Schiller and Nauman. The points are Γ inferred from observed caustics.
15. Photograph of a 15 μm radius bubble trapped at a depth of ~5 mm. The beam path in the water is visible due to fluorescing and scattering in the water. The bubble is located at the top of the bright spot caused by reflection off the bubble's surface. A microscope objective, used to focus the beam, is visible above the water tank.
16. Experiment for observing optoacoustic signals. The light was focused close to the bubble or drop. Sometimes it was directed up instead of down as shown here. Signal pressure amplitudes were referenced to what the pressure would be a distance of 103 mm from the bubble if no reflector were present.
17. The top and bottom traces are signals for a short light pulse and for a 4-cycle burst. The bubble radius was 34.7 μm while the peak beam power was 1.8W. The peak-to-peak trace was ≈8mPa while the theoretical estimate was 0.9 mPa. The impulse response in the absence of a bubble is the central trace.

UNSTEADY THERMOCAPILLARY MIGRATION OF BUBBLES

Loren H. Dill[a]
National Aeronautics and Space Administration
Lewis Research Center, Cleveland, Ohio 44135

R. Balasubramaniam[b]
Case Western Reserve, Cleveland, Ohio 44106

ABSTRACT

Upon the introduction of a gas bubble into a liquid possessing a uniform thermal gradient, an unsteady thermocapillary flow begins. Ultimately, the bubble attains a constant velocity. This theoretical analysis focuses upon the transient period for a bubble in a microgravity environment and is restricted to situations wherein the flow is sufficiently slow such that inertial terms in the Navier-Stokes equation and convective terms in the energy equation may be safely neglected (*i.e.*, both Reynolds and Marangoni numbers are small). The resulting linear equations were solved analytically in the Laplace domain with the Prandtl number of the liquid as a parameter; inversion was accomplished numerically using a standard IMSL routine. In the asymptotic long-time limit, our theory agrees with the steady-state theory of Young, Goldstein, and Block. The theory predicts that more than 90% of the terminal steady velocity is achieved when the smallest dimensionless time, *i.e.*, the one based upon the largest time scale—viscous or thermal—equals unity.

1. INTRODUCTION

Forces other than gravity are expected to often dominate the migration of bubbles in outer space. In particular, thermocapillary forces will cause bubbles to migrate in a thermal gradient. Consider the bubble depicted in Figure 1. The interface next to the hot liquid will be hotter than that next to the cold liquid. Assuming that surface tension varies inversely with temperature, as is often the case, the cold interface will have a higher surface tension than the hot interface. The thermocapillary stresses are such that near the interface liquid flows from the hot side to the cold side, as indicated by the arrows. Viscous stresses within the fluid oppose these thermocapillary stresses, causing the bubble to migrate toward the hot side. If the thermal gradient is constant throughout the region, the analysis of Young *et al.* (1959) revealed that the bubble will ultimately migrate with a steady velocity.

In the processing of materials in outer space, nonisothermal bubble-liquid systems will be common. Control of these systems will require a good understanding of thermocapillary migration. For example, bubbles form in glass melts during the manufacturing process; their removal is essential for the glass to be useful. On Earth, gravitational forces aid in their removal. During the containerless processing of glass in the microgravity environment of space, thermocapillary forces are expected to assist their removal (Mattox *et al.* 1982, Subramanian 1981). Thermocapillary migration may also be important in the design of two-phase heat exchangers for use in outer space. A poor design may unexpectedly permit bubbles to migrate *en masse* to the heating surface. A layer of gas next to the hot surface would act as a thermal insulator and prevent the efficient transfer of heat to the liquid phase.

In contrast to prior experimental and theoretical studies on thermocapillary

---

[a]National Research Council - NASA Research Associate.
[b]NASA Resident Research Associate.

migration (Balasubramaniam and Chai 1987, Hardy 1979, Mattox *et al.* 1982, Subramanian 1981, Szymczyk *et al.* 1987, Young *et al.* 1959), which focus upon *steady* systems, we here consider the *unsteady* development of thermocapillary migration. Analysis of unsteady thermocapillary migration is important for several reasons. First, to study the steady state, one must be able to estimate the time required for transients to die out. Also, some unexplained results of supposedly steady-state experiments (Neuhaus and Feuer-bacher 1986, Siekmann *et al.* 1986), *e.g.*, velocities less than predicted steady-state values, may possibly be explained by an unsteady analysis. Lastly, the transient period is expected to be long for relatively large bubbles; a theory for their transport will be useful.

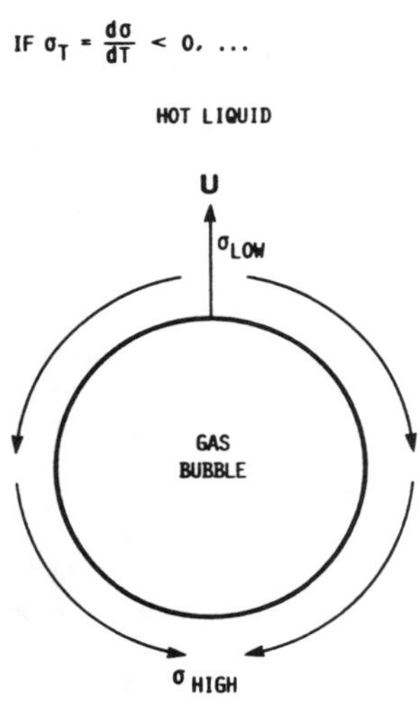

FIGURE 1. - ORIGINS OF THERMOCAPILLARY MIGRATION. VARIATIONS IN SURFACE TENSION ALONG A NONISOTHERMAL INTERFACE CAUSE THERMOCAPILLARY SHEARING STRESSES, WHICH VISCOUS SHEARING STRESSES TRANSMIT TO THE LIQUID. LIQUID IMPOSES A REACTION FORCE UPON THE BUBBLE, CAUSING IT TO SWIM TOWARDS THE HOT LIQUID WITH A VELOCITY **U**(t).

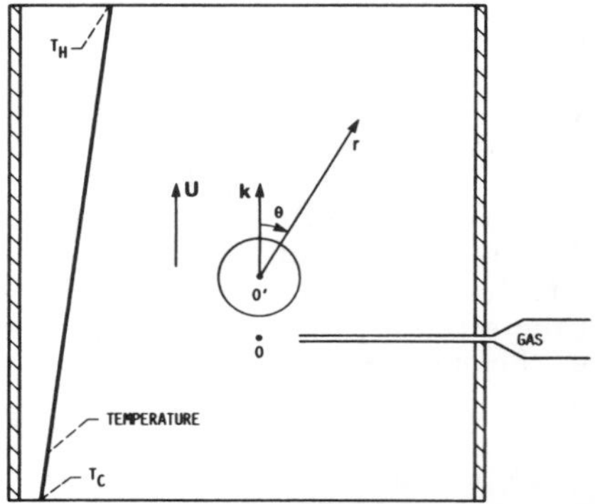

FIGURE 2. - SCHEMATIC OF A THERMOCAPILLARY EXPERIMENTAL CELL IN ZERO GRAVITY. IN TYPICAL EXPERIMENTS, A LINEAR TEMPERATURE FIELD IS ESTABLISHED PRIOR TO THE INTRODUCTION OF A GAS BUBBLE AT POINT 0. THERMOCAPILLARY ACTION CAUSES THE BUBBLE TO MIGRATE WITH THE UNSTEADY VELOCITY **U**(t) IN THE DIRECTION OF THE THERMAL GRADIENT. EVENTUALLY, THE MIGRATION VELOCITY BECOMES STEADY.

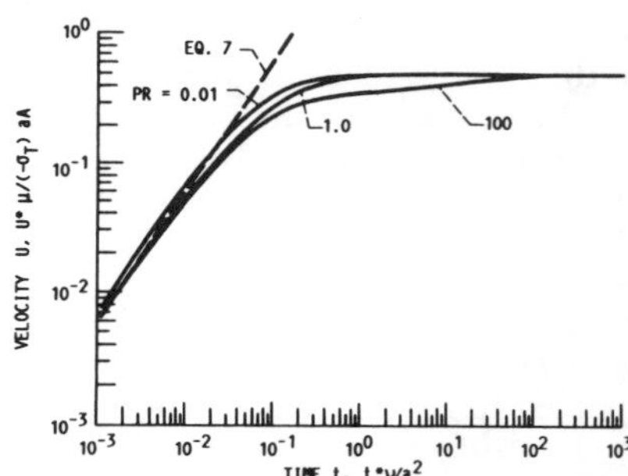

FIGURE 3. - DEVELOPMENT OF THERMOCAPILLARY MIGRATION. EACH SOLID CURVE REPRESENTS A PLOT OF THERMOCAPILLARY MIGRATION VELOCITY U (t) VERSUS TIME t FOR A SPECIFIED PRANDTL NUMBER PR = ν/α. THE DASHED LINE PROVIDES A GOOD ESTIMATE OF THE VELOCITY FOR SMALL VALUES OF t.

2. FORMULATION

Consider a thermocapillary experiment conducted in outer space as shown in Figure 2. An enclosure consisting of end plates maintained at different temperatures and of insulated side walls is filled with a liquid. In the absence of buoyancy effects, there is no fluid motion and, as depicted in the figure, the temperature field becomes linear when thermal equilibrium is obtained. The experiment then begins with the introduction of a small gas bubble. We neglect the small immediate effect of the bubble's introduction upon the velocity and temperature fields. Soon thereafter the conduction of heat to the interface causes thermocapillary stresses to develop and migration to begin. As is commonly done, we assume that both viscosity and density of the liquid phase are essentially constant throughout the volume, that transport processes within the bubble phase may be neglected relative to those in the liquid phase, and that surface tension is a linear function of interfacial temperature. To be definite, we assume that $\sigma T = d\sigma/dT$ is negative such that migration is toward the hotter fluid. We also assume that the bubble remains spherical during the acceleration; this is subject to *a posteriori* verification. Finally, we assume that both fluid inertia and the convective transport of heat may be neglected. These latter assumptions restrict the analysis to low Reynolds and Marangoni (thermal Péclet) numbers.

Given the above assumptions, the dimensionless temperature field T obeys the unsteady energy equation

$$Pr\,\frac{\partial T}{\partial t} = \nabla^2 T \qquad (1)$$

in the region outside of the bubble. Here, the liquid's Prandtl number $Pr = \nu/\alpha$, the ratio of kinematic viscosity to thermal diffusivity, appears multiplying the left-hand side. The variable t is time divided by the viscous time scale a^2/ν, with the characteristic length scale "a" being the radius of the bubble. The dependent variable T is the scaled difference between the actual temperature and that undisturbed temperature which prevails in the plane of the bubble's center. The characteristic temperature scale is the product aA, where A is the undisturbed temperature gradient in the system. Equation (1) is written with the bubble's center as the origin of the coordinate system.

Similarly, the velocity field obeys the unsteady Stokes equation

$$\frac{\partial \mathbf{v}}{\partial t} = -\left(\nabla p + \frac{d\,\mathbf{U}}{dt}\right) + \nabla^2 \mathbf{v} \qquad (2)$$

in a reference frame moving with the bubble. The vector $\mathbf{v}$ is the scaled difference between the fluid's and bubble's velocities. The velocity scale for both $\mathbf{v}$ and $\mathbf{U} = \mathbf{k}U(t)$, the scaled bubble velocity, is the positive quantity $-\sigma_T aA/\mu$. Because the bubble's reference frame is noninertial, there appears in Equation (2) the fictitious force $d\mathbf{U}(t)/dt$. This force adds to the gradient of the hydrodynamic pressure p and vanishes when a steady bubble velocity is obtained.

At the interface, viscous and thermocapillary stresses balance at each point. Now both T and $\mathbf{v}$ are axisymmetric with respect to an axis that passes through the bubble's origin and is parallel to the unit vector $\mathbf{k}$, which points in the direction of the temperature gradient. It follows that $T = T(r, \theta, t)$ and $\mathbf{v} = \mathbf{v}\,(r, \theta, t)$ and that the stress balance is given by

$$\left.\frac{\partial}{\partial r}\left(\frac{v_\theta}{r}\right)\right|_{r=1} = \left.\frac{\partial T}{\partial \theta}\right|_{r=1} \qquad (3)$$

Here, r is the dimensionless radial coordinate and θ is the polar angle; these are depicted in Figure 2. Since the flow is axisymmetric, the vector $\mathbf{v}$ has only two non-zero components: v_r and v_θ, the velocities in the radial and polar directions, respectively.

Because we neglect bubble inertia and gravitational forces, the total hydrodynamic force F^h on the bubble vanishes for all $t > 0$:

$$F^h(t) = \int_0^{2\pi} d\Phi \int_0^{\pi} \mathbf{n} \, l \, \mathbf{P}(l, \theta, t) \sin\theta \, d\theta = 0 \tag{4}$$

Here, $\mathbf{n}$ is a unit vector normal to the spherical bubble surface and $\mathbf{P}(r, \theta, t)$ is the hydrodynamic pressure dyadic (tensor).

We briefly mention the remaining conditions on the temperature and velocity fields. The *undisturbed* temperature and velocity fields prevail (1) initially for all space outside the bubble and (2) far from the bubble for all time. The neglect of thermal transport within the bubble requires a zero thermal flux normal to the bubble surface for $t > 0$. Because the bubble does not change in size, the radial velocity v_r vanishes at the interface. Lastly, the assumption of incompressibility requires $\nabla \cdot \mathbf{v} = 0$.

3. SOLUTION

To solve this system of equations, we defined a modified pressure field such that its gradient is given by the quantity within parentheses in Equation (2), introduced the streamfunction for the axisymmetric flow, and then applied the Laplace transform technique to replace the variable t with the parameter s. Analytic functions were found for the transformed temperature and streamfunction fields that satisfy the differential equations and all conditions. The solution yielded the following expression for the transformed, dimensionless bubble velocity $/U(s)$:

$$/U(s) = \frac{\frac{1}{s}\left(1 + \frac{1 + \sqrt{s\,Pr}}{2 + 2\sqrt{s\,Pr} + s\,Pr}\right)}{3 + \frac{s(3 + \sqrt{s})}{6(1 + \sqrt{s})}} \tag{5}$$

Multiplication of this expression by s and taking the limit as s approaches zero yields a dimensionless terminal velocity of 1/2, in agreement with the analysis of Young *et al.* (1959). We also applied the initial-value theorem twice to Equation (5) to determine dU/dt at $t = 0$. The first use of the theorem gave the expected result that $U(0) = 0$. The second use gave

$$\left.\frac{dU}{dt}\right|_{t=0} = 6 \, . \tag{6}$$

The step-function character of the acceleration precluded further application of the initial-value theorem.

To confirm that the bubble remains spherical throughout the acceleration period, we applied our analytic result for the transformed stream function to the transformed normal stress condition. The exact satisfaction of this condition indicates that there is no deviation from the spherical shape.

Numerical inversion via an IMSL routine gave the bubble velocity as functions of time with Prandtl number as a parameter. Figure 3 is a log-log plot of the dimensionless velocity vs. dimensionless time for Prandtl numbers of 0.01, 1.0, and 100.

4. DISCUSSION

The two most interesting features of Figure 3 are that (1) all three curves are nearly linear before a transition to the terminal velocity and (2) the dimensionless time to the terminal velocity ranges from order one to order 100, depending upon the Prandtl number. These features are discussed below.

For $t \ll 1$, the dimensionless velocity $U(t)$ is given approximately by

$$U(t) \sim 6t \tag{7}$$

The value of 6 (cf Equation (6)) derives from the acceleration calculated from application of the initial value theorem. Equation (7) is represented in Figure 3 as the dashed line; it fits the data reasonably well for small values of t.

Written in dimensional form, Equation (7) becomes

$$U^*(t^*) \sim 6\left(\frac{-\sigma_T A}{\rho a}\right)t^* \tag{8}$$

for $t^* \ \nu/a^2 << 1$. (The raised asterisks denote dimensional quantities.)

According to Equation (8), the leading term in the short-time expansion for the dimensional velocity is independent of viscosity. This result is analogous to that for a drop or solid sphere accelerating from rest due to buoyancy forces (Chisnell 1987). Though viscosity does not appear explicitly in Equation (8), it does appear in the time scale and thus affects the values of t^* for which the relation is valid. For highly viscous liquids or small bubbles, the relation is valid only briefly. On the other hand, for low-viscosity liquids or large bubbles, Equation (8) is likely to apply for a much longer period of time.

Another interesting feature of Equation (8) is that the bubble's acceleration is inversely proportional to its radius. Consider two bubbles of different radii released simultaneously into the same system. During the time regime for which Equation (8) is applicable for both bubbles, the smaller bubble may be expected to move faster than the larger one, even though the terminal velocity of the smaller bubble,

$$U^*(\infty) = \frac{-\sigma_T\, aA}{2\,\mu} \tag{9}$$

is less.

Our focus now shifts to estimating the duration of unsteady migration. According to Figure 3, the time t to a terminal velocity is of order unity if the $Pr \leq 1$, but of order 100 if $Pr = 100$. (Recall that the factor used to nondimensionalize time was the viscous scale a^2/ν.) It thus appears that for $Pr \leq 1$, the viscous scale is appropriate, but for $Pr \geq 1$, the product of the Prandtl number and the viscous scale is appropriate. This product equals a^2/α, which is the characteristic thermal time scale for the system. Observe that for any value of Pr, the larger of the viscous or thermal scales appears to be a good estimate of the transient period.

The above rule to predict the time to steady state was based upon graphical data for three curves. For a more quantitative picture, we examined numeric data for five liquids having Prandtl numbers in the range $0.01 \leq Pr \leq 100$. For each Prandtl number, we determined the dimensionless times $\tau 90$ and $\tau 95$ (based upon the appropriate scale) for a bubble to attain 90 and 95 percent of the terminal velocity. All results (Table I) are of order unity and thus confirm that the rule is correct.

TABLE I
Dimensionless times to terminal velocity

$Pr = \nu/\alpha$	τ_{90}	τ_{95}
0.01	0.28	0.47
0.10	0.30	0.52
1.00	0.58	1.10
10.00	0.29	0.68
100.00	0.26	0.65

To understand why the above scaling works, observe that if $Pr << 1$, the thermal field attains its steady-state value while the fluid motion is just beginning to develop. The time for the velocity field to reach its steady state is thus governed by the viscous time scale. If $Pr >> 1$, the fluid motion, which is driven by thermocapillary forces and thus by the temperature field, becomes quasistatic with a slowly-developing thermal field. In this case, it is the thermal time scale that determines when the bubble reaches its terminal velocity.

5. SUMMARY

The unsteady thermocapillary migration of a bubble within an otherwise quiescent liquid was studied via analytical methods. Within the same system, small bubbles are predicted to initially move more rapidly than larger ones, even though their terminal velocities are less. The larger of the viscous and thermal time scales is a good estimate of the duration of the unsteady migration period.

ACKNOWLEDGEMENT

This work was done while L. H. Dill was a National Research Council-NASA Research Associate at Lewis Research Center.

REFERENCES

1. R. Balasubramaniam & An-Ti Chai: Thermocapillary Migration of Droplets: An Exact Solution for Small Marangoni Numbers. *J. Colloid Interface Sci.*, **119**:2, Oct. 1987, 531-538.
2. R. F. Chisnell: The Unsteady Motion of a Drop Moving Vertically Under Gravity. *J. Fluid Mech.*, **176**, Mar. 1987, 443-464.
3. S. C. Hardy: The Motion of Bubbles in a Vertical Temperature Gradient. *J. Colloid Interface Sci.*, **69**:1, Mar. 15, 1979, 157-162.
4. D. M. Mattox *et alia*: Thermal-Gradient-Induced Migration of Bubbles in Molten Glass. *J. Am. Ceram. Soc.*, 65, 9, Sept. 1982, 437-442.
5. D. Neuhaus & B. Feuerbacher: Bubble Motions Induced by a Temperature Gradient. **Proc. of the 6th European Symposium on Material Sciences under Microgravity Conditions,** ESA SP-256, European Space Agency, Paris, France, 1986, 241-244.
6. J. Siekmann *et alia*: Experimental Investigation of Thermocapillary Bubble and Drop Motion Under Microgravity. **Proc. of the 6th European Symposium on Material Sciences under Micro-gravity Conditions,** ESA SP-256, European Space Agency, Paris, France, 1986, 179-182.
7. R. S. Subramanian: Slow Migration of a Gas Bubble in a Thermal Gradient. *AIChE J.*, **27**:4, July 1981, 646-654.
8. J. A. Szymczyk, G. Wozniak, & J. Siekmann: On Marangoni Bubble Motion at Higher Reynolds- and Marangoni-Numbers Under Microgravity. *Appl. Microgravity Tech.*, **1**:1, 1987, 27-29.
9. N. O. Young, J. S. Goldstein, & M. J. Block: The Motion of Bubbles in a Vertical Temperature Gradient. *J. Fluid Mech.*, **6**:3, Oct. 1959, 350-356.

FIGURE CAPTIONS

1. The Origins of Thermocapillary Bubble Migration. Variations in surface tension along a non-isothermal interface cause thermocapillary shearing stresses, which viscous shearing stresses transmit to the liquid. The liquid imposes a reaction force upon the bubble, causing it to swim towards the hot liquid with a velocity $\mathbf{U}(t)$.
2. Schematic of a Thermocapillary Experimental Cell in Zero Gravity. In typical experiments, a linear temperature field is established prior to the introduction of a gas bubble at point O. Thermocapillary action causes the bubble to migrate with the unsteady velocity $\mathbf{U}(t)$ in the direction of the thermal gradient. Eventually, the migration velocity becomes steady.
3. The Development of Thermocapillary Migration. Each solid curve represents a plot of thermocapillary migration velocity $\mathbf{U}(t)$ vs time t for a specified Prandtl number $Pr = \nu/\alpha$. The dashed line provides a good estimate of the velocity for small values of t.

FLUID- AND CHEMICAL-DYNAMICS RELATING TO ENCAPSULATION TECHNOLOGY

J. M. Kendall and M. Chang[a]
Jet Propulsion Laboratory, California Institute of Technology, Pasadena, CA 91109

T. G. Wang
Center for Materials Research and Applications, Vanderbilt University, Nashville, TN 37235

ABSTRACT

We have developed certain new technology for the encapsulation of various biological materials, including living cells. Submillimeter-size droplets bearing the encapsulant are given a biocompatible membrane covering by immersion into a bath containing a reactant. We present in this paper a brief description of the methods and materials of the process, and emphasize the results of a photographic study of the droplet deformation during submersion and of the concurrent membrane formation.

I. INTRODUCTION

The microencapsulation of living cells may have considerable utility in several medical and biological areas. An application of particular interest concerns the encapsulation of islets of Langerhans for diabetes control, as demonstrated by Lim and Sun[1]. For this, the membrane must be permeable to nutrients and to cell products, but not to antibody-size molecules. The immuno-isolation prevents the islets from being attacked by the host immune system. Encapsulated islets have been found to remain viable for two years in laboratory testing. Other applications involve the encapsulation of pituitary cells, of thyroid hormone adrenocortical cells, as described by Shen[2], and of other secretion cells for transplantation as described by Lim[3]. Another application concerns hybridoma encapsulation for monoclonal antibody production with higher concentration and reduced contamination, as described by Posillico[4]. Also, there are many applications being developed for the controlled release of pharmaceuticals, agricultural chemicals, etc.

While there exists an extensive literature on the encapsulation of various commodities, most materials utilized are unsuitable for biological applications. The technique developed by Lim and Sun[1], however, uses materials selected specifically for compatibility with living cells. All constituents are in aqueous solution and involve no monomers or organic solvents. The cells to be encapsulated are suspended in sodium alginate solution (polyanionic), droplets of which, formed by means of an air jet, are immersed into calcium chloride solution to harden. Polylysine solution (polycationic) is then added and a polymeric membrane is formed by ionic interaction between polycations and polyanions.

The technology developed by us and described below started from Lim's work as a basis, and while it resembles his in several ways it differs significantly. In the present scheme, a small-diameter jet flow of alginate solution carrying the encapsulant is caused to pinch into uniformly sized droplets at a very high rate due to instability, and "hardening" occurs only at the droplet surface, where a tough, compliant, yet permeable skin is formed as each droplet in turn passes beneath the surface of a liquid bath containing the reactant. The permeability of the membrane,

[a]*Present affiliation:* Advance Polymer Laboratory, 2038 East Foothill Blvd., Pasadena CA 91107.

which is of particular concern for most applications, is governed by the concentration of the solutions, the molecular weight of the polyanions and polycations, the distance of the adjacent ionic groups, and the number of membrane layers. In principle, as many layers as desired can be added, so long as the coating sequence alternates between cationic and anionic species.

A goal of the present work has been to examine the fluid- and chemical dynamics of the immersion and of the membrane formation in sufficient detail to provide guidance in optimizing the process.

II. MATERIALS AND METHODS

The principal materials used for capsule formation were sodium alginate (Sigma Chemical Co, St. Louis, MO.), used as the droplet material, and chitosan (Protan Laboratories, Redmond, WA.), used for forming the membrane. Chitosan is a product resulting from the deacetylation of chitin. It is structurally similar to cellulose except that some of the hydroxyl groups have been replaced by amine groups. The solution concentrations used in the experiments ranged up to three percent by weight. The viscosity was measured by the Cannon-Fenske method, which is applicable to Newtonian liquids. The chitosan solution, at less than 10 cP, was water-like in general behavior. The surface tension of the solutions was measured by the du Nouy ring method.

The general scheme for droplet production by jet instability and for forming the membrane is shown in Figure 1. As is well known, a cylindrical column of liquid will spontaneously break up to form droplets due to Rayleigh instability[5]. Although analysis indicates that there exists a most probable droplet diameter, approximately 1.8 times that of the column diameter for inviscid liquids, a rather wide distribution of droplet sizes usually results when a column disrupts. This tendency was circumvented by utilizing an electro-mechanical exciter to accelerate the nozzle at a single frequency near that to which the jet was most unstable. The droplet diameter was established jointly by the 175 μm nozzle diameter, the jet velocity, and the perturbation frequency. The procedure used here was to vary the jet velocity, the excitation frequency, and amplitude in an iterative manner during observation under stroboscopic illumination for most stable droplet formation. Formation rates of 2-6 kHz were typical. The stream was directed vertically downward into a shallow container of chitosan solution located below the nozzle for forming the membrane.

Droplets in the 2- to 5-mm size range were utilized for study of the membrane formation dynamics. A single pendant drop of desired size was generated at the tip of a fine capillary tube, and an electro-mechanical solenoid provided an upward-directed impact to the supporting structure for parting the drop. An optical-quality cuvette, 2.5 cm/s by 3.5 cm deep, was placed below such that the droplet fell 10 cm before impact.

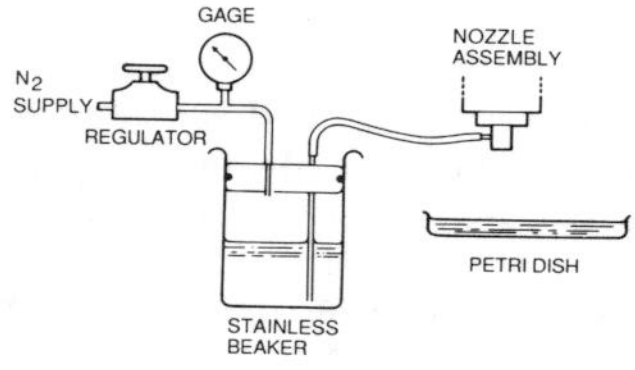

Figure 1

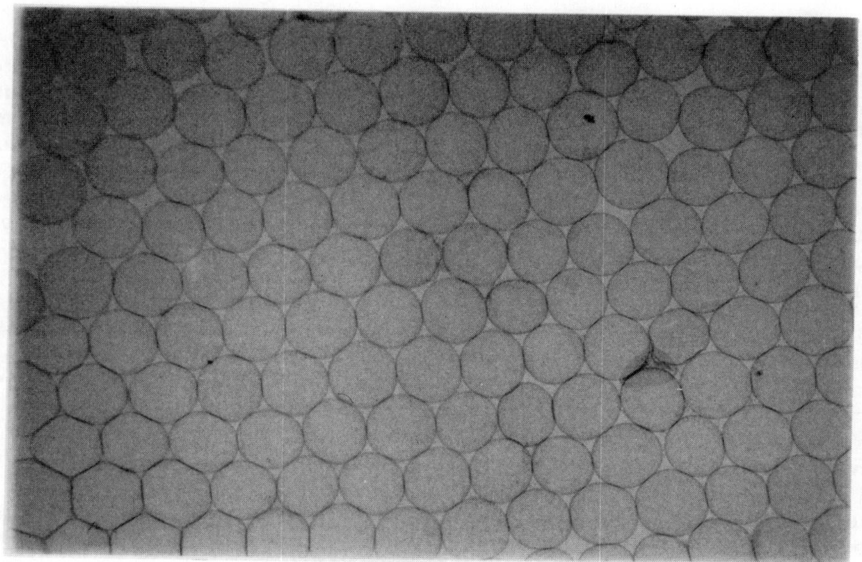

Figure 2

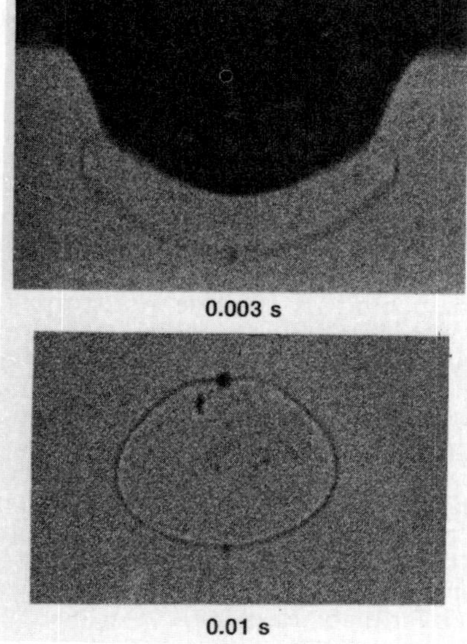

Figure 3

The reaction dynamics therein were studied by cinephotography. A 16-mm cine-camera (Teledyne Camera Systems, Arcadia, CA, Model DBM-55) was operated at its maximum rate of 500 frames-per-second, approximately, with a corresponding exposure duration of 0.33 ms. The image size on the film was set to either full or half actual size. Because the membrane was both thin and colorless, and because the indices of refraction of the chemical solutions differed little between themselves, the obtaining of satisfactory images was somewhat difficult. The method of illumination adopted was to provide strongly collimated light which passed through the cuvette and thence into the camera lens. This was differentially refracted near the interfaces as in shadowgraphy, and provided good images of the droplet profile, as long as the droplets remained within the depth-of-field.

The droplet deformation was analyzed for eight different film runs. Droplet outline profiles were magnified 16-fold and digitized frame-by-frame on a computer-interfaced device (Hewlett-Packard graphics plotter, Model 7225A.) Closely-spaced points around the entire periphery of the images were recorded for each frame, with points along the membrane portion of the boundary being distinguished from those along the air/alginate interface. These boundary data were analyzed to obtain the prevailing maximum diameter, the vertical on-axis dimension, and the arc-lengths of the membrane and air-interface portions of the profiles. Also, the droplet volume and constituent surface areas were computed by assuming circular symmetry about the vertical axis. The computed volume typically varied less than 3% throughout a given run, indicating good accuracy.

The permeability of 300 μm capsules was assessed for three molecules of particular interest: glucose (mol. wt. 180 Daltons), insulin (mol. wt. 6000 Daltons), hemoglobin (mol. wt. 64 000 Daltons), and immunoglobulin (mol. wt. 150 000 Daltons.) Glucose was detected by use of commercial reagent strips in which visual comparison of the color of a wetted strip with that of preprinted colors is made. Insulin diffusion was estimated by use of fluorescent-tagged insulin, the presence of which could be observed visually under ultraviolet light. Large-molecule diffusion was assessed by use of immunoglobulin antibodies, with detection again being by observation under ultraviolet light.

III. RESULTS

A. Encapsulation Results and Tests

Figure 3 shows an array of 300 μm capsules produced by the jet-instability method and which contain no encapsulant. The specimens had been placed between accurately-spaced microscope slide glasses and photographed using strongly collimated illumination from behind as for cinephotography. The only selection of sample was performed by passing the finished batch through a mesh in order to remove a few accumulations of amorphous material. It may be judged that the capsules are spherical to within two percent, approximately, and also that a high percentage fell within a relatively small departure from mean diameter.

On account of the near-invisibility of the membrane, a detailed evaluation of surface quality and uniformity cannot yet be made, but the membrane strength was such that it was quite difficult to puncture a specimen using a sharp-tipped probe. Qualitative assessment of the wall permeability of these capsules to glucose, insulin, and immunoglobulin was made according to methods described above. The walls were found to be porous to the first two of these molecules, but not to the last, all of which are as desired for diabetes control. Further, the capsules were found to be permeable to bovine hemoglobin, which represents another intermediate-size molecule. This result may have consequence for various applications involving the slow release of hormones, enzymes, or drugs.

Various model materials were encapsulated by the method in order to test its potential utility. These included 10 μm mono-disperse polystyrene microspheres, 40 μm magnetic iron microspheres, porcine pancreatic islets, and sea-urchin eggs of

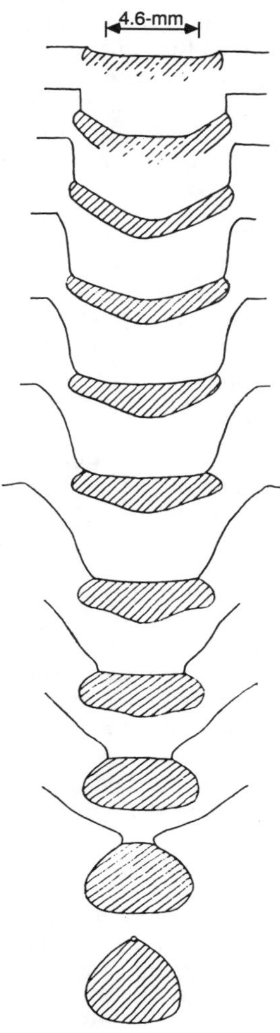

Figure 4

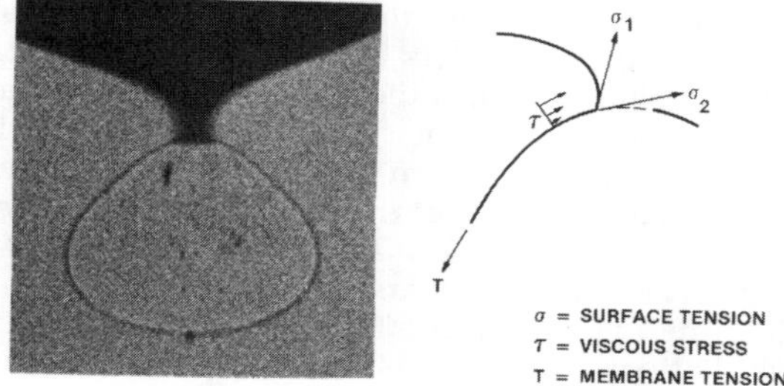

Figure 5

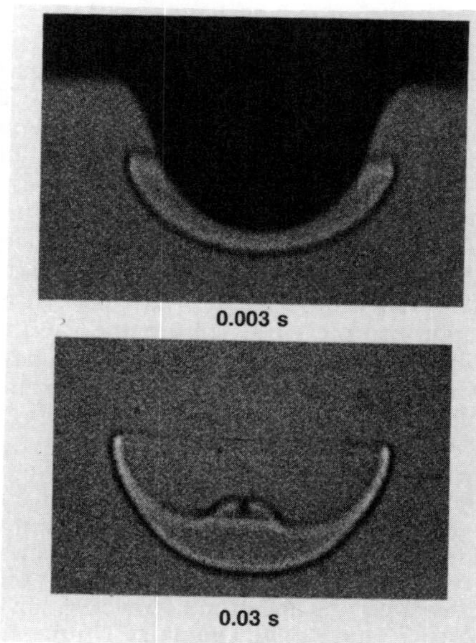

Figure 6

75 μm diam. The former two rigid materials presented no problem as long as the particle concentration was low enough to avoid clumping and blockage of the nozzle orifice, nor did the islets. The sea-urchin eggs, which were extremely fragile and easily damaged in handling, were encapsulated without apparent degradation. This is believed to represent a particularly significant test.

B. Study of the Membrane Formation

Studies were made to understand in detail how the membrane is formed. Photography was used to record the metamorphosis of a droplet into a capsule. Singly-produced droplets were used because their larger size reduced the time-scale of immersion to values more compatible with the framing rate and exposure duration of the camera. The type of results obtained are illustrated in Figure 3, where two frames from a film on the formation of a 2-mm capsule are shown. Because the camera shutter was not synchronized with the contact of the droplet with the free surface, the times shown have an uncertainty of approximately 1 msec. The boundary between the black and gray areas in the first frame represents the free surface, while the membrane profile appears as a black line. The membrane was resolved less distinctly where it is nearly horizontal than where it is vertical because the shutter speed was inadequate to freeze the vertical motion. The second of the two frames shows the membrane profile after submersion was complete. Also disclosed is the presence of a minute air bubble near each pole. The lower one is believed to result from the trapping of air between the rapidly falling droplet and the free surface, and the cause of the upper one will become apparent. The elongated black object is an artifact.

Figure 4 shows a series of tracings for a 4.6-mm droplet. The lower portion of the membrane has been omitted in the first two frames because of inadequate definition. The first frame indicates that the droplet had become substantially flattened within less than one camera period and that it had attained a diameter in the horizontal direction approximately fifty percent greater than that of the initial sphere. The flat shape, in its downward motion, generated a remarkably cylindrical cavity within the bath. It may be remarked that the energy associated with the surface tension and vertical area, and the energy due to displacement of liquid against the action of gravity, were at this moment each of the order of ten percent of the droplet impact energy. Subsequent frames show that the cavity widened at the top and narrowed at the bottom, with the overrunning liquid capturing the aforementioned air bubble and leaving it attached to the upper surface of the capsule. Upon completion of immersion the droplet briefly assumed a fairly spherical form but later approached a more oblate one as its terminal form, presumably because of built-in stresses resulting from membrane formation during a flattened state. Bubble trapping was typical for droplets of this size, but unusual for small ones.

A significant result of the membrane presence can be discerned in all frames prior to the conclusion of submersion. There, the curvature of the air-liquid interface was considerably greater near locations where the alginate-chitosan interface met the free surface than in adjacent ones. This indicates that the membrane had formed rapidly and that it was capable of sustaining tension from the earliest moments. Were the membrane not present or too weak to pull against the interface effectively, surface tension would have smoothed the curvature. The sketch of Figure 5 depicts the geometry shortly before the end of submersion, when the effect was most pronounced. Tangential components of force are applied to the membrane in the vicinity of the air interface by the surface tensions of alginate and of chitosan, and by viscous shear stress applied by the overrunning bath. These provide membrane tension which is balanced by internal pressure arising from inertial forces. The action of the tangential forces can be likened to the tightening of the drawstring of a fabric bag.

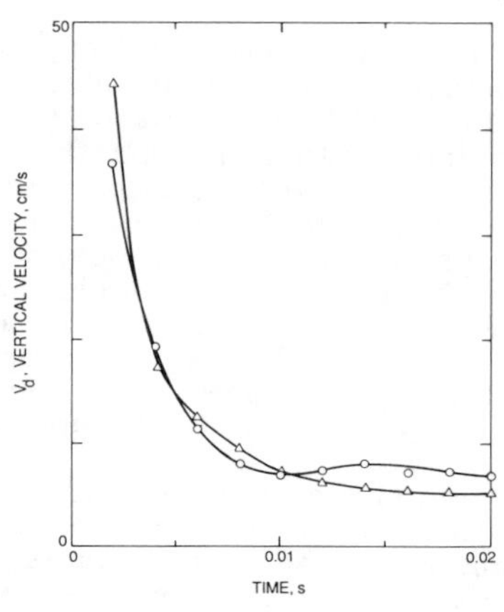

Figure 7

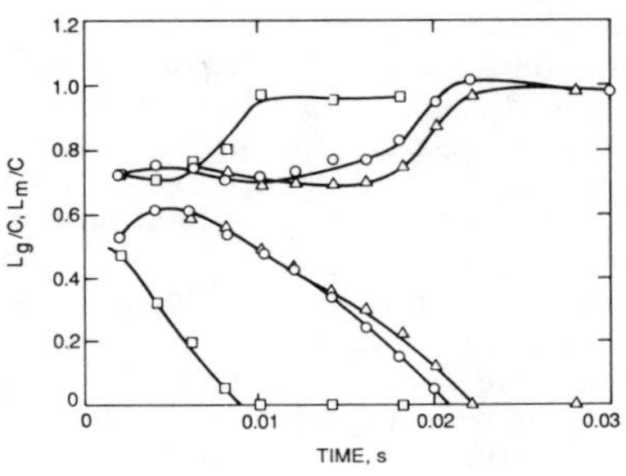

Figure 8

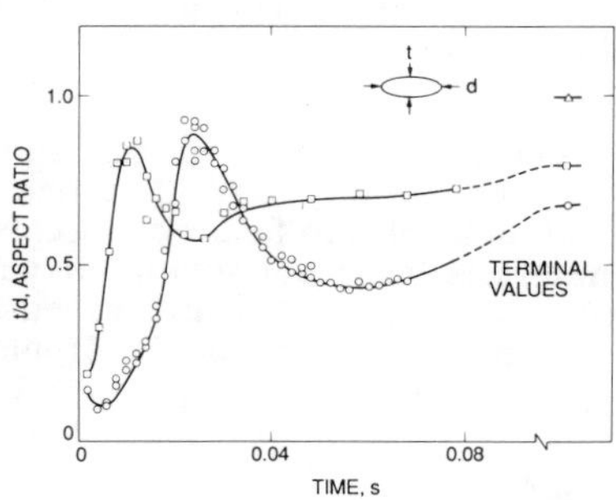

Figure 9

It is instructive to compare the aforementioned results with the interface configuration and the drop deformation in the absence of reaction, i.e., without membrane formation. Figure 6 shows two frames from a run in which a droplet, sized as in Figure 3, entered a chemically inert polysaccharide solution whose density and viscosity equalled that of chitosan. Apparent contrasts between the two cases are the inequalities in the liquid/liquid interface thicknesses, the evident lack of pull on the air-liquid interface by the membrane, and the greatly differing shape of the droplet after immersion was complete. Here, the shape evolved into an increasingly complex figure of revolution.

Frame analysis results are described here. The droplet vertical velocity during impact was characterized by subtracting the positions of the lower membrane surface of successive frames. The results, shown in Figure 7, do not warrant further differencing to obtain deceleration, but it can be inferred that that quantity was of the order of 10 g at the earliest observation. This is small in comparison with fluid acceleration in passing through the nozzle orifice. The flattening of the droplet during impact is indicated in Figure 8, where the aspect ratio, i.e., the ratio of vertical thickness to maximal diameter, is presented. The 2.2-mm droplet exhibited a much shorter time scale of deformation and recovery than did the 4.6-mm one, as expected. A recovery of roundness, followed by a quasi-oscillatory motion are observed for both sizes. Of particular interest is the terminal aspect ratio, shown at the right of the figure. This was taken to be unity for the 300 μm capsules in accordance with Figure 2.

The profile arc-lengths of the air-interface and of the membrane were examined for evidence of extension or contraction for three drops. As shown in Figure 9, one of the larger ones was undergoing flattening during the earliest frames, inasmuch as the air interface was increasing. Thereafter, and throughout the observation for the other drops, the interface length diminished as the droplet assumed a rounder shape and as the bath engulfed the surface. The membrane maintained nearly constant periphery until the overwash lengthened it, implying that it was inextensible. However, a general consequence of the droplets return from a flattened shape to a more spherical one is that the upper regions of the membrane necessarily contracted in the circumferential direction. If this were accomplished by wrinkling and foldover, as opposed to elastic contraction, there might be a resulting effect upon the localized surface smoothness and permeability. Methods for examining this matter have yet to be identified.

SUMMARY

The encapsulation method and materials described here have been shown to produce capsules in the 300 μm size range of high apparent quality. The membrane was strong and exhibited permeability to certain molecules in favorable accordance with anticipated requirements. Delicate materials were encapsulated without noticeable degradation. Photography revealed how a droplet is transformed into a capsule. It was found that the membrane formation proceeded while the droplet was in a flattened state, but that the capsule returned to a nearly spherical form, nevertheless.

ACKNOWLEDGMENT

The research described in this paper was carried out by the Jet Propulsion Laboratory, California Institute of Technology, under a contract with the National Aeronautics and Space Administration.

REFERENCES

1. F. Lim & A. M. Sun. *Science*, **210**, p. 908, 1980
2. J. C. Shen, *Diss. Abstr. Int. B*, **45 (4)**, p. 1072, 1984
3. F. Lim, **Biomedical Applications of Microencapsulation**, CRC Press, Cleveland, OH, 1984
4. E. C. Posillico, *Biotechnology*, **4**, p. 47, 1986
5. H. Lamb, **Hydrodynamics**, Dover Publications, New York, 1932

FIGURE CAPTIONS

1. Apparatus for capsule formation. The encapsulant is suspended in alginate within the beaker; chitosan is placed in the dish.
2. Photograph of 300 μm diameter unfilled capsules.
3. Two frames from a film on the submersion of a 2.2-mm alginate droplet in chitosan. The times are 0.003 and 0.01 s, approximately, after initial contact.
4. Tracings from a film on the submersion of a 4.6-mm droplet in chitosan. Time of the first frame is estimated to be 1 msec after contact and the increment between frames is 2 msec, approximately. Bubble appearing in last frame remains attached to the capsule.
5. One frame from film used to prepare Figure 4, and sketch showing forces applied near edge of membrane. σ, τ, and T are the surface tension, viscous stress and membrane tension, respectively.
6. Two frames from a film on the submersion of a 2.2-mm alginate droplet in an inert liquid of same viscosity as chitosan. The times are 0.003 and 0.03 s, approximately.
7. Vertical velocity of 2.2-mm alginate droplets during submersion process. Impact velocity was 140 cm/s, approximately. O, submersion in chitosan; Δ in inert liquid of same viscosity as chitosan.
8. The ratio of droplet axial thickness, t, to maximum diameter, d, for three alginate droplets entering chitosan. O, 4.6 mm initial droplet diameter; □, 2.2 mm; Δ, 0.3 mm.
9. Arc-lengths of membrane (L_m, upper curves) and of air/alginate interface (L_g, lower curves) in ratio to droplet initial circumference, C. □, 2.2 mm droplet; O, 4.1-mm droplet; Δ, 4.6 mm droplet.

DYNAMICS OF THIN LIQUID SHEETS

Chun P. Lee & Taylor G. Wang[a]
Jet Propulsion Laboratory, California Institute of Technology, Pasadena, California 91109

ABSTRACT

The motion of a thin liquid sheet subjected to the effects of surface tension and external gaseous pressure is studied. Equations of motion are constructed by essentially ignoring the internal flow of the liquid in the layer. For an axisymmetric case, for example, the system is reduced to a set of coupled nonlinear partial differential equations depending on time and only one spatial variable, which can be solved numerically as an initial-boundary-value problem. Here, a review of the application of this approach to two different problems is presented. The first is the instability of an annular jet, in which a cylindrical liquid sheet emanating from an annular nozzle and enclosing a gas stream breaks up into liquid shells downstream. The second is the centering mechanism of such a liquid shell under capillary oscillations. Reasonable results have been obtained from the numerical studies of these problems.

1. INTRODUCTION

It was observed by T. G. Wang, one of the authors, and D. D. Elleman, of the Jet Propulsion Laboratory, on board a NASA KC-135 aircraft flying parabolically to simulate zero gravity, that an oscillating liquid droplet containing an air bubble tends to become concentric. It was then thought that useful products might be made of these liquid shells (Lee *et al.* 1986). This led to the design and study of the annular jet (Kendall 1982, 1986), which is a cylindrical liquid jet enclosing a gas stream breaking up rapidly downstream through an instability to form liquid shells. Both the centering mechanism of the shell (Lee & Wang 1988) and the instability of the jet (Lee & Wang 1986, 1988) have been studied theoretically in the limit that liquid layer is reasonably thin.

In this paper, we present a review of these theoretical works, after making some generalization about how to formulate problems involving thin free liquid layers.

2. WAVES ON A THIN LIQUID SHEET

G. I. Taylor (1959) studied the wave formation on their fluid sheets. Let us consider a flat thin liquid layer of thickness D, lying between $y = D/2$ and $y = -D/2$, extending to infinity with the x- and z-directions. With two stress-free surfaces, the liquid can be considered as inviscid, such that its motion can be described by a potential flow. Let us consider the capillary waves on it in the x-direction. We are only interested in the case $kD \ll 1$, where k is the wavenumber of the disturbance. So the following are the results obtained from the potential theory expanding in kD after ignoring $O(kD)^2$.

There are two wave modes. One is the symmetric wave in which the two surfaces oscillate opposite to each other:

$$\eta_{\pm} = \pm\eta_0 \exp(-i\omega t) \sin kx, \tag{1}$$

where η_0 is the displacement amplitude of the wave on either surface; η_+ and η_- are the disturbance on the upper and lower surfaces, respectively; ω is its angular frequency; and t is time. The dispersion relation is

[a]*The authors' current address*: Center for Microgravity Research and Applications, Vanderbilt University, Nashville, TN 37235

$$\omega = \left(\frac{\sigma D}{2\rho}\right)^{1/2} k^2, \tag{2}$$

where σ and ρ are the surface tension and density of the liquid, respectively. The velocity in the x-direction is

$$u = -\frac{2i\,\omega\eta_0}{kD} \exp(-i\omega t) \cos kx, \tag{3}$$

and that in the y-direction is of the order of u multiplied by ky ($<kD/2$), which can be averaged to zero across the thickness. The pressure inside the layer is

$$p = \sigma k^2 \eta_0 \exp(-i\omega t) \sin kx, \tag{4}$$

Since Equations (3) and (4) are independent of y, it is a good approximation to consider that the symmetric mode of motion in the layer is only along the layer driven by a pressure gradient in the same direction.

The other mode is the antisymmetric wave in which the two surfaces oscillate in the same direction:

$$\eta_{\pm} = \eta_0 \exp(-i\omega t) \sin kx, \tag{5}$$

The dispersion relation is

$$\omega = \left(\frac{2\sigma}{\rho D}\right)^{1/2} k \ . \tag{6}$$

The velocity of the layer is mainly in the y-direction:

$$v = -i\omega\eta_0 \exp(-i\omega t) \sin kx, \tag{7}$$

while that in the x-direction is of the order of ky ($< kD$) times v, which can be reduced to zero when averaged over y. The pressure is also proportional to ky and can be similarly averaged to zero. But because of the factor ky, there is a pressure difference between the bottom and the top of the layer

$$\Delta q = -2\sigma\eta_0 k^2 \exp(-i\omega t) \sin kx, \tag{8}$$

which drives the motion in the y-direction. Since Equations (7) and (8) do not have a dependence across the layer thickness, we can consider that the layer moves as a structureless sheet in the normal direction subjected to surface tension which appears in the form of Δq.

Furthermore, we can see from Equations (2) and (6) that the symmetric mode has a time scale about kD times smaller than that of the antisymmetric mode.

3. THIN SHEET APPROXIMATION

Consider a free liquid layer which is thin in the sense that its thickness is much smaller than its characteristic dimension. Given its outer surface R_o and inner surface R_i, we can define a mid-surface R in a suitable way. Then we can define a unit vector $\mathbf{n}$, normal to R; and another unit vector $\mathbf{s}$, tangential to R; which forms the basis of a curvilinear coordinate system. The motion of the sheet, considered to be located at R, is governed by two pressure forces. Let us draw a line through R parallel to $\mathbf{n}$ intersecting R_o and R_i at points Q_o and Q_i, respectively. The tangential motion is caused by the tangential gradient of the mean pressure p which is the average of the pressures at Q_o and Q_i, on the liquid sides of the surfaces. The normal motion is similarly caused by the difference, Δq, of the pressures at the two points. We also need the continuity equation for the sheet, which is relatively straightforward. The dynamics of gas on the two sides is ignored. Other details of the formulation depends on the system we work with, and is best illustrated by the following examples.

4. ANNUAL JET INSTABILITY

An annular jet breaks down into shells at a frequency near that of an anti-symmetric wave. The symmetric wave is a much slower motion and can be ignored. Thus we can treat the liquid layer as a structureless sheet moving under the influence of surface tension. At any time, the cylindrical sheet emerging from the nozzle has to close somewhere downstream because of surface tension, at what we call the "first closure point". We refer to the sheet lying between this point and the nozzle as the "envelope". If the liquid velocity is faster than the wave speed c given by Equation (6), with D conveniently given by the annular nozzle gap distance, then the first closure point which leaves the nozzle ahead of the rest of the envelope is relatively well-adjusted to equilibrium, and can be considered as moving uni-formly. We choose this point as the origin, with the z-axis pointing upstream along the axis of symmetry. If the initial position of the nozzle is L_0, then at t, the nozzle is at $L = L_0 + U_J t$, where U_J is the liquid jet velocity. Let ΔU be the gas jet velocity relative to that of the liquid.

Let the position of the sheet be described by $R(z,t)$, its slope by $\theta(z,t)$ where $\tan\theta = \partial R/\partial z$, the normal and tangential velocity of the sheet by $v_n(z,t)$ and $v_s(z,t)$ respectively, and its surface density by $m(z,t)$. But because m is singular where R is zero, we replace μ by $m(z,t) = R\mu$.

Then, kinematic consideration leads to

$$\frac{\partial R}{\partial t} = SS_R\left\{\frac{v_n}{\cos\theta}\right\} \tag{9}$$

where $S_R = 1$ if $R > 0$; and 0 if $R = 0$; and $S = 1$, if $0<z<L$ and 0 if $L<z<L_z$. S_R and S suppress the motion of the sheet if it falls on the axis, and if it is still inside the nozzle, respectively. To prevent the sheet from moving after it hits the axis we require further that $v_n \rightarrow S_R v_n$ at every time step.

Consideration of the continuity of the sheet gives

$$\frac{\partial m}{\partial t} = SS_R\left[-K\frac{\partial m}{\partial z} - \cos\theta\left(\frac{\partial v_s}{\partial z} - v_n\frac{\partial^2 R}{\partial z^2}\cos^2\theta\right)m\right], \tag{10}$$

where

$$K = \cos\theta\left(v_s - v_n\frac{\partial R}{\partial z}\right) \tag{11}$$

represents the convective effect. The tangential momentum equation has no force and is given by the convective effect K and the curvilinear effect H:

$$\frac{\partial v_s}{\partial t} = SS_R\left(-K\frac{\partial v_s}{\partial z} + Hv_n\right), \tag{12}$$

where

$$H = \cos\theta\left(\frac{\partial v_n}{\partial z} + v_s\frac{\partial^2 R}{\partial z^2}\cos^2\theta\right). \tag{13}$$

The normal momentum equation is in addition governed by Δq, or more directly, the surface tensional pull of the sheet and the excess pressure P of the core:

$$\frac{\partial v_n}{\partial t} = SS_R\left\{-K\frac{\partial v_n}{\partial z} - Hv_s + \frac{R}{m}\left[P - 2\sigma\left(\frac{\cos\theta}{R} - \frac{\partial^2 R}{\partial z^2}\cos^3\theta\right)\right]\right\}. \tag{14}$$

P can be found by using the normal momentum equation requiring that the envelope grows in volume in consistence with the gas flow rate at the nozzle:

$$P = \left(\int_0^L F\,dz\right)\left(\int_0^L dz\frac{R^2}{m\cos\theta}\right)^{-1} \tag{15}$$

where

$$F = -\frac{v_n^2}{\cos^2\theta} + 2R\frac{\partial v_n}{\partial z}\left(v_s - v_n\frac{\partial R}{\partial z}\right) + R\frac{\partial^2 R}{\partial z^2}\cos^2\theta\left[v_s^2 - v_n^2\left(\frac{\partial R}{\partial z}\right)^2\right] + \frac{2\sigma R}{m}\left(1 - R\frac{\partial^2 R}{\partial z^2}\cos^2\theta\right). \tag{16}$$

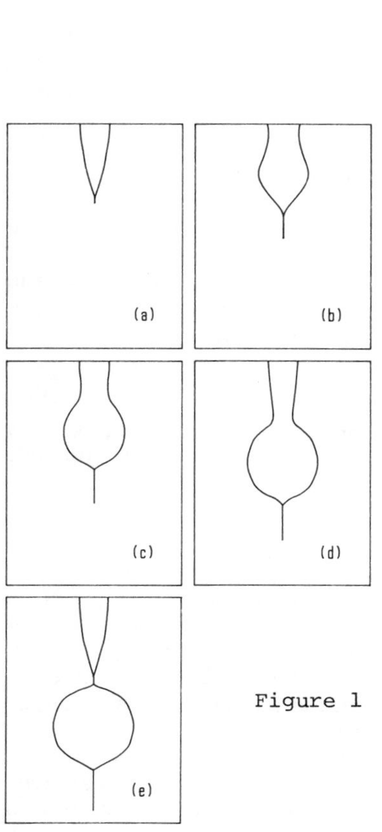

Figure 1

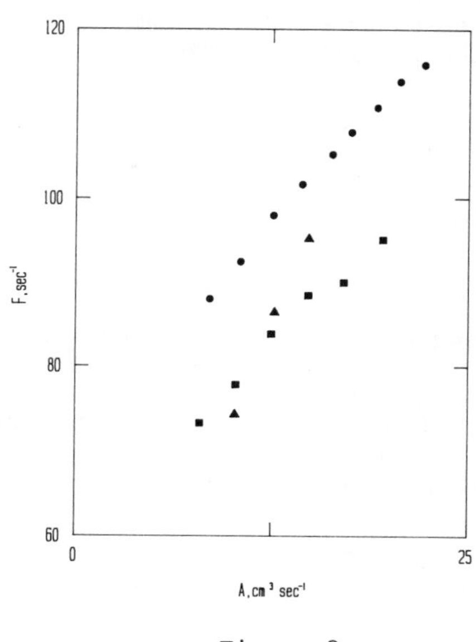

Figure 2

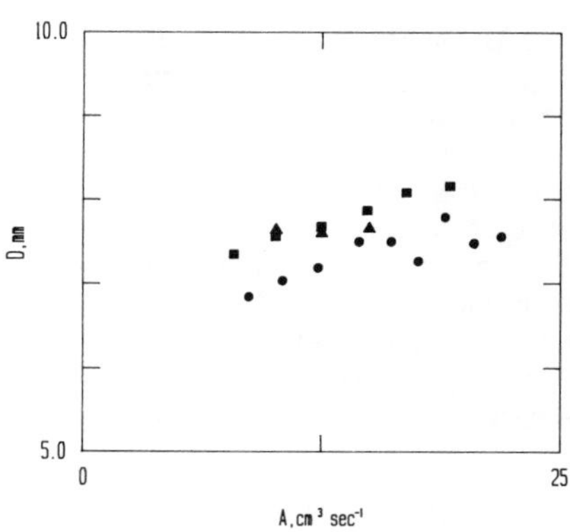

Figure 3

To suppress the motion of the sheet after it falls on the axis where the sheet model is no longer valid, we require further that $v_s \to S_R v_s$.

The boundary conditions are $(R, v_n, v_s, m) = (0,0,0,1)$ at $z = 0$ and $(R, v_n, v_s, m) = (1,0,0,1)$ at $z = L_z$, where L_z is the upper limit of the range of z in the program chosen such that the range covers all the envelope throughout a cycle. The initial conditions for the first cycle are, for $0<z<L_0$:

$$R = (z/L_0)(2 - z/L_0), \tag{17a}$$

$$v_n = B\Delta U (z/L_0)(1 - z/L_0), \tag{17b}$$

$$v_s = 0, \tag{17c}$$

and

$$m = 1, \tag{17d}$$

where

$$B = \left[2L_0 \int_0^1 \frac{dy}{\cos\theta} y^2 (1-y)(2-y) \right]^{-1}, \tag{18}$$

and for $L_0<z<L_z$: $R = m = 1$ and $v_s = v_n = 0$. But those for a latter cycle are given by the values of R, v_n, v_s and m at the collapse of the preceding cycle, with v_n reset as follows. When the envelope collapses, the closing of the bottleneck means that the gas flow at the nozzle becomes discontinuous because the sheet on the new envelope falls toward the axis everywhere. In order to maintain the gas flow, v_n must be reversed by a pressure pulse in the new envelope to become:

$$v_n = (v_n)_0 + RQ/m, \tag{19}$$

where $Q=\int Pdt$ is the impulse/area delivered by the pulse to the new envelope, and $(v_n)_0$ is the original v_n in the new envelope. Requiring that the gas flow is constant we have

$$Q = \left[\frac{\Delta U}{2} - \int_0^L dz \frac{R(v_n)_0}{\cos\theta} \right] \left[\int_0^L dz \frac{R^2}{m\cos\theta} \right]^{-1}. \tag{20}$$

By scaling (R, z, L, L_0, L_z) with the nozzle inner radius R_0, t with $\tau = R_0/c$, $(v_n, v_s, U_J, \Delta U)$ with c, m with $m_0 = R_0 D\rho$, and P with $2\sigma/R_0$ we have a dimensionless set of equations which can alternatively be obtained by letting $R_0 = c = \tau = m_0 = 2\sigma = 1$. The system depends only on U_J and ΔU, such that whatever the initial value of L_0 is, the solutions of the equations almost always converge onto some definite periodic natural patterns after a few cycles.

In Figure 1 we show the characteristic profile of the jet in quarter-cycles until it collapses. Figures 2 and 3 compare experimental results with the theoretical ones, showing good agreement. Figure 4 suggests that at high U_J, the dimensionless bubble-formation frequency f_b is independent on U_J but varies linearly with ΔU. Approximately, the relation is f_b=0.3 + 0.007ΔU.

5. SHELL CENTERING

Consider a thin liquid shell centered near the origin. Let us use spherical polar coordinates and assume azimuthal symmetry. Given the outer surface $R_o(\theta,t)$ and inner one $R_i(\theta,t)$, we can define the mid-surface as $R(\theta,t) = (R_o + R_i)/2$, and construct the unit vectors $\mathbf{n}(\theta,t)$ pointing outward and $\mathbf{s}(\theta,t)$ pointing in the θ-direction. The slope of the sheet in θ is given by $\tan\psi(\theta,t) = (1/R)\,\partial R/\partial\theta$. The velocity components in the normal and tangential directions are $v_n(\theta,t)$ and $v_s(\theta,t)$ as before, and the surface density $\mu(\theta,t)$ is used this time because the sheet is not expected to reach the origin.

Consideration of the continuity of the sheet gives

$$\frac{\partial \mu}{\partial t} = -\frac{K}{R}\frac{\partial \mu}{\partial \theta} - \frac{\mu}{R}\left[\cos\psi \frac{\partial v_s}{\partial \theta} + f_v \cos\psi + v_s \sin\psi\right] - \mu C v_n \ . \tag{21}$$

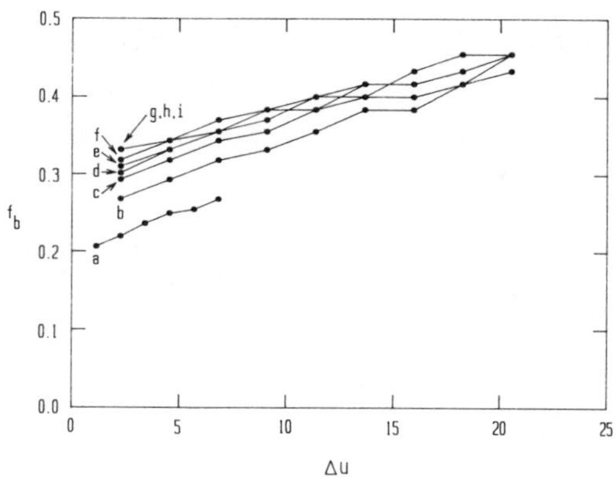

Figure 4

where
$$K = v_s \cos\psi - v_n \sin\psi \tag{22}$$
is the convective velocity, $f_v = v_s \cos\theta/\sin\theta$ for $0<\theta<\pi$ and $\partial v_s/\partial\theta$ for $\theta = 0$ or π,

and
$$C = \frac{1}{R}\left[2\cos\psi + \sin^2\psi\cos\psi - f - \frac{\cos^3\psi}{R}\frac{\partial^2 R}{\partial \theta^2}\right] , \tag{23}$$

in which $f = \sin\psi\cos\theta/\sin\theta$ for $0<\theta<\pi$, and $(\partial^2 R/\partial\theta^2)/R$ for $\theta = 0$ or π, is the curvature of the sheet. From kinematic consideration we have
$$\frac{\partial R}{\partial t} = \frac{v_n}{\cos\psi} \ . \tag{24}$$

Given R and μ we can find R_o and R_i by
$$R_{o,i} = R \pm \frac{\mu}{2\rho\cos\psi} \ . \tag{25}$$

Then, using **n** and **s**, we can find p and Δq. If we write $\Delta q = \Delta p + P$, where P is the excess pressure of the core which serves to keep the core incompressible, then Δp is the pressure difference due entirely to the wave. In a lengthy derivation involving Taylor series expansion, we found that
$$p = \frac{\sigma}{2}\left[C_o - C_i - \frac{\partial C}{\partial \theta}\frac{\mu\sin\psi}{\rho R}\right] + \frac{P}{2} , \tag{26}$$

$$\Delta p = \sigma\left[-(C_o + C_i) + \frac{\sin\psi}{\rho R}\left(\mu_o\frac{\partial C_o}{\partial \theta} - \mu_i\frac{\partial C_i}{\partial \theta}\right) - \frac{\partial C}{\partial \theta}\frac{\mu^2\sin\psi\cos\psi}{2\rho^2R^2} - \left(\frac{\mu\sin\psi}{2\rho R}\right)^2\frac{\partial^2 C}{\partial\theta^2}\right] . \tag{27}$$

The tangential and normal momentum equations are given by
$$\frac{\partial v_s}{\partial t} = -\frac{K}{R}\frac{\partial v_s}{\partial \theta} - v_n H - \lambda_1\frac{\cos\psi}{\rho R}\frac{\partial p}{\partial \theta} , \tag{28}$$

$$\frac{\partial v_n}{\partial t} = -\frac{K}{R}\frac{\partial v_n}{\partial \theta} + v_s H + \frac{\lambda_1}{\mu}\left[P + \Delta p\right] , \tag{29}$$

where

$$H = \frac{K}{R} - \frac{\cos\psi}{R}\frac{\partial v_n}{\partial \theta} + \frac{\cos\psi\sin\psi}{R}(v_s\sin\psi + v_n\cos\psi) - \frac{v_s\cos^3\psi}{R^2}\frac{\partial^2 R}{\partial\theta^2} + \frac{v_n\sin^3\psi}{R} \tag{30}$$

is the curvilinear effect, and λ_1 is equal to 1 now, but might take on other values later. From the normal momentum equation, and the requirement of incompressibility of the core, we have

$$P = \left[\int_0^\pi d\theta \sin\theta \, F \right] \left[\int_0^\pi d\theta \, \sin\theta \frac{R^2}{\mu \cos\psi} \right]^{-1} \tag{31}$$

where

$$F = \lambda_2 \left[\frac{2RK}{\cos\psi} \frac{\partial v_n}{\partial\theta} + \left(v_s^2 \cos^2\psi - v_n^2 \sin^2\psi \right) \frac{\partial^2 R}{\partial\theta^2} - R v_s^2 \left(1 + \sin^2\psi\right) - R v_n^2 (2 + \sin^2\psi) \right] - \frac{R^2 \Delta p}{\mu \cos\psi}\,, \tag{32}$$

in which λ_2 is unity now, but can be other values later.

The shell is not concentric in general. Its eccentricity is readily observed experimentally, but to describe it within the context of the thin sheet model we need to consider the center of mass

$$Z_{cm} = \frac{2\pi}{M} \int_0^\pi d\theta \sin\theta \frac{R^3 \mu \cos\theta}{\cos\psi}, \tag{33}$$

and its "geometrical center"

$$Z = \left[\int_0^\pi d\theta \sin\theta \frac{R^3 \cos\theta}{\cos\psi} \right] \left[\int_0^\pi d\theta \sin\theta \frac{R^2}{\cos\psi} \right]^{-1} . \tag{34}$$

The centering condition of the shell is described by $Z_s = Z - Z_{cm}$, with $Z_s = 0$ meaning concentricity.

The boundary conditions are $\delta R/\delta\theta = \delta v_n/\delta\theta = \delta\mu/\delta\theta = v_s = 0$ at $\theta = 0$ or π. There are two types of initial conditions. If the shell oscillates with a symmetric, or "sloshing", mode, the conditions are

$$R = R_S \,, \tag{35a}$$

$$\mu = \mu_S [\, 1 - \Delta \cos\theta + \varepsilon P_n(\cos\theta)] \,, \tag{35b}$$

and

$$v_S = v_n = 0 \,, \tag{35c}$$

where R_S is the equilibrium radius of the sheet, μ_s is the corresponding surface density when it is concentric; Δ lying between -1 and +1 represents the displacement of the core; and ε between -1 and +1 is the wave amplitude. For the antisymmetric, or "bubble", mode, the conditions are

$$R = R_S' + \varepsilon R_S P_n(\cos\theta), \tag{36a}$$

$$\mu = \mu_S'(1 - \Delta \cos\theta), \tag{36b}$$

and

$$v_S = v_n = 0, \tag{36c}$$

where R_s' and μ_s' are slightly modified from R_S and μ_S such that R and μ give volume and mass of $4\pi R_S^3/3$ and $4\pi R_S^2 \mu_S$, respectively, and Δ and ε are as defined before.

Let us define $\omega_0 = (\sigma/pR_S^3)^{1/2}$, and $\delta = D_s/R_S << 1$ where D_s is the equilibrium thickness of the shell when it is concentric. If $\Delta = 0$ and $|\varepsilon| << 1$, Equations (35) give an eigenfrequency $\omega = \omega_s[n(n+1))(n-1)(n+2)/2]^{1/2}$ where $\omega_s = \omega_0 \delta^{1/2}$, and Equations (36) give $\omega = \omega_b[2(n-1)(n+2)]^{1/2}$ where $\omega_b = \omega_0/\delta^{1/2}$. These agree with results from potential theory if $\delta << 1$ (Saffren *et al.* 1982). To make the equations dimensionless, we scale lengths with R_S, μ with μ_S, and time with $1/\omega_s$ or $1/\omega_b$ for the sloshing or bubble mode, respectively. Alternatively, we can simply let $\sigma = R_S = \mu_S = 1$, $1/\rho = \delta$, and $\lambda_1 = 1/\delta^2$ and $\lambda_2 = \delta^2$ or $\lambda_1 = \lambda_2 = 1$ for the sloshing or bubble mode, respectively, The problem then depends on three parameters: Δ, ε, and δ.

In Figures 5(a-e) and 6(a-e) we display how a nonconcentric shell oscillates in sloshing and bubble modes, respectively. In Figure 7 we show for the sloshing mode how Z_S oscillates in time, with a fast frequency due to the wave, and a slow one which comes from nonlinearity. The latter means that the core oscillates slowly relative

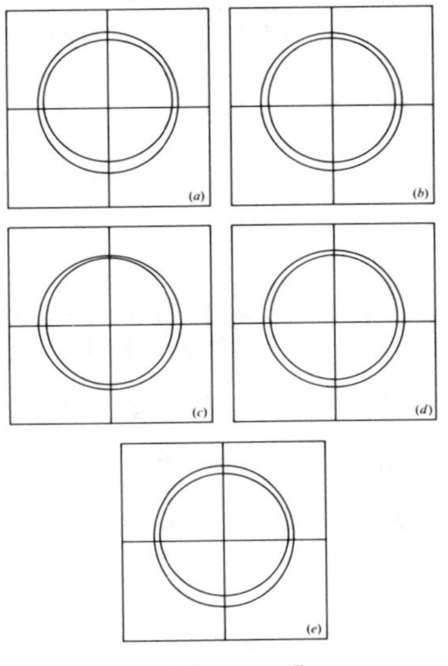

Figure 5

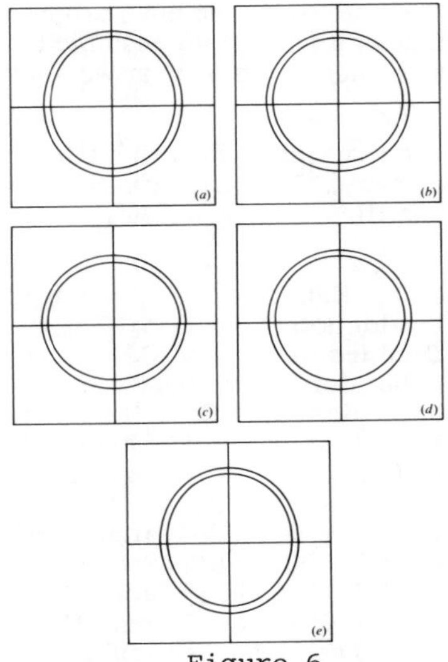

Figure 6

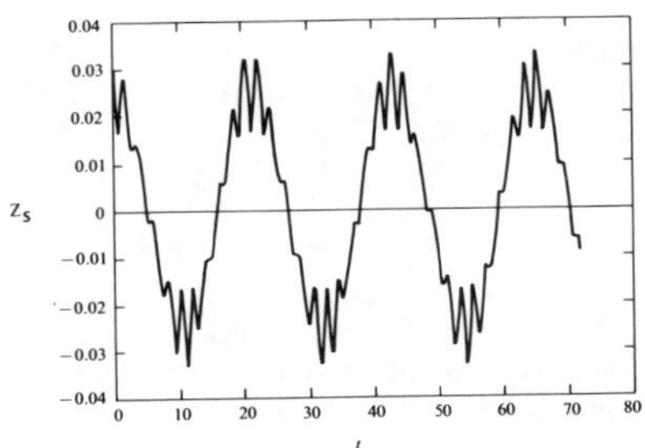

Figure 7

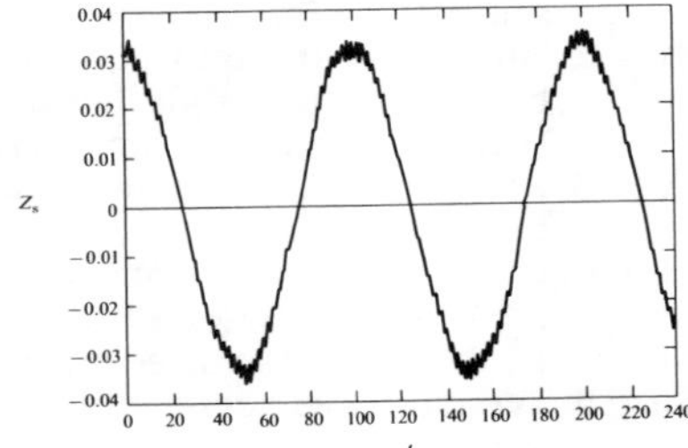

Figure 8

to the shell, and is the centering effect we want to look for. For the bubble mode we have a similar result as shown in Figure 8. From a parameter study we found that the slow frequency Ω is insensitive to Δ and δ, but is proportional to ε. We have, approximately, $\Omega = 0.6\varepsilon$ for the sloshing mode and $\Omega = 1.3\varepsilon$ for the bubble one.

6. CONCLUDING REMARKS

The approach we have proposed here is for a thin liquid layer in free space. But somehow, it should be possible to extend it to the case in which the layer rests on a solid surface. This is what we shall attempt to do next.

ACKNOWLEDGEMENT

The research described in this work was carried out at the Jet Propulsion Laboratory, California Institute of Technology, under contract with the National Aeronautics and Space Administration.

REFERENCES

1. J. M. Kendall, 1982, "Hydrodynamic Performance of an Annular Liquid Jet: Production of Spherical Shells," in **Proc. Second Int'l Colloquium on Drops and Bubbles,** Monterey, CA, November 1981 (ed. D. H. LeCroissette) (Jet Propul-sion Laboratory Pubn. 82-7), 79-87.
2. J. M. Kendall, 1986, "Experiments on Annular Liquid Jet Instability and on the Formation of Liquid Shells," *Phys. Fluids* **29**, 2086-2094.
3. C. P. Lee & T. G. Wang, 1988, "Centering of a Thin Liquid Shell in Capillary Oscillations," *J. Fluid Mech.* **188**, 411-435.
4. C. P. Lee & T. G. Wang, 1986, "A Theoretical Model for the Annular Jet Instability," *Phys. Fluids* **29**, 2076-2085.
5. C. P. Lee & T. G. Wang, 1988, "A Theoretical Model for the Annular Jet Instability -- Revisited," *Phys. Fluids* (in press).
6. M. C. Lee, J. M. Kendall, P. A. Bahrami, & T. G. Wang, 1986, "Sensational Spherical Shells," *Aerospace Am.* **24**, 72-76.
7. M. Saffren, D. D. Elleman & W. K. Rhim, 1982, "Normal Modes of a Compound Drop," in **Proc. Second Int'l Colloquium on Drops and Bubbles,** Monterey, CA, November 1981 (ed. D. H. LeCroissette) (Jet Propulsion Laboratory Pubn. 82-7), 7-14.
8. G. I. Taylor, 1959, "The Dynamics of Thin Sheets of Fluid: II. Waves on Fluid Sheets," *Proc. R. Soc. London* **A253**, 296-312.

FIGURE CAPTIONS

Figure 1. Evolution of the jet profile at quarter-cycles until collapse for $U_J = 2.27$ and $\Delta U = 4.54$.

Figure 2. Bubble-formation frequency versus the gas fill rate: circles for experimental results (Kendall 1982), ▲ for our previous theoretical results (Lee & Wang 1986), and ■ for our recent theoretical results (Lee & Wang 1988).

Figure 3. Bubble diameter versus the gas fill rate. See Figure 2 for the symbols used here.

Figure 4. Dimensionless f_b versus ΔU at various U_J. The values for U_J are: (a) 2.27, (b) 3.405, (c) 4.54, (d) 5.675, (e) 6.81, (f) 7.945, (g) 9.08, (h) 10.215, (i) 11.35. The last three curves essentially overlap one another.

Figure 5. Typical wave profiles for the $n = 2$ sloshing mode of a nonconcentric shell with the origin at the CM at quarter-cycles: $\epsilon = 0.5$, $\delta = 0.1$, $\Delta = 0.3$, dimensionless wave period $T = 1.8$.

Figure 6. Typical waves profiles for the n=2 bubble mode of a nonconcentric shell with the origin at the CM at quarter-cycles: $\epsilon = 0.05$, $\delta = 0.1$, $\Delta = 0.15$, dimensionless wave period $T = 2.2$.

Figure 7. Centering for the $n = 2$ sloshing mode, $\epsilon = 0.05$, $\delta = 0.1$, $\Delta = 0.1$. Plot of Z_s vs. t.

Figure 8. Centering for the $n = 2$ bubble mode, $\epsilon = 0.05$, $\delta = 0.1$, $\Delta = 0.1$. Plot of Z_s vs. t.

PRESENCE AND ABSENCE OF A WATER FILM BETWEEN MOVING AIR BUBBLES AND A PLATE

Carl J. Remenyik
Department of Engineering Science and Mechanics The University of Tennessee, Knoxville

ABSTRACT

The thickness of water films between an inclined Lucite plate submerged in water and air bubbles moving beneath it was measured with a small impedance probe. The instrument was calibrated with a laser interferometer built for this purpose. The bubbles released beneath the plate varied in size from 10 cc to 100 cc. At a plate inclination angle of 0.98°, and in tap water, an uninterrupted water film covered most of the bubbles. Some bubbles, however, dewetted the plate, and the water film covered only a forward part of the bubble. When the film was uninterrupted, its thickness was very uniform from front to rear. When the bubble dewetted the plate, a large forward section of the film had the same uniform thickness, but this was followed by a hump on the film the rear slope of which ended at the plate surface. For some of the experiments, the surface tension of the water was reduced by admixing a detergent. In these experiments, dewetting was not observed. In a second set of experiments, a hand held transparent container filled with water and a 1.3 cm3 air bubble was used to observe visually the behavior of the moving bubble and its associated water film.

INTRODUCTION

A bubble moving beneath a solid surface may, or may not, be separated from it by a thin liquid film. Presence or absence of a film makes a big difference in the speed and shape of the bubble, the drag acting on it, the pattern of the flow field around and inside it, the heat exchange with the solid surface, and the complexity of the mathematical analysis of the fluid motion. This paper presents quantitative results of measurements of these liquid film thicknesses, and qualitative results of visual observations of bubbles and their associated liquid films.

INSTRUMENTATION

Thicknesses of water films (mentioned in the introduction) were measured in a 254 cm long, 41 cm wide, and 43 cm deep water tank (Figure 1). A Plexiglas plate of 230 cm length and 30 cm width was suspended from adjustable supports to make the plate's angle of inclination variable. A movable mechanism released the bubbles. This mechanism pressed the flat end of a short Plexiglas pipe of 10 cm inside diameter against the underside of the plate (mounted near center of plate in Figure 1). Designed to accept and hold known volumes of air injected with a hypodermic needle, this pipe released bubbles when lowered.

The instrument measuring the film thicknesses is an impedance probe with small platinum electrodes (Figure 2). The cylindrical body is inserted through a hole along the plate's center line and fastened so that its flat end is flush with the plate's underside. The electrodes (Figure 3) are embedded at the center of this flat end. The line between the electrodes in Figure 3 is an edge-on view of an electrically "driven" shield embedded inside the probe and is designed to isolate capacitively one electrode from the other. The mean sensitivity of this instrument is 1.0 V/μm in the range of film thicknesses from zero to 10 μm decreasing gradually from its highest value at zero thickness. Beyond 10 μm, the sensitivity diminishes rapidly; and the mean sensitivity from 20 to 300 μm is only between 0.98 and 2.45 mV/μm depending

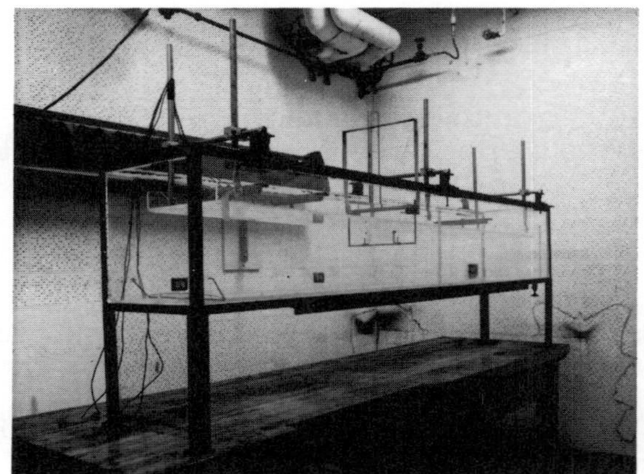

Figure 1

Figure 2

Figure 3

on water conductivity.

A second probe was also installed along the center line of the plate 10.1 cm from the first probe down the path of the bubbles. With this distance and with the arrival times at the two probes obtained from the recordings of the output signals, one can calculate the speed of the bubbles.

The impedance probe was calibrated employing light interference. The surface of the probe was polished to optical flatness and quality. For calibration, a probe was inserted from below through one of the holes in the transparent plate on the stand of the calibration instrument shown in Figure 4. A water film covered the electrodes. A light beam from a helium neon laser (at left in Figure 4; at right in Figure 5) was split, and one component (1 in Figure 5) was directed through the water film and reflected on the polished surface of one of the electrodes. Part of the returning beam (2) was transmitted (3) through the beam splitter. The other component (4) of the laser beam was transmitted through the beam splitter. It was then reflected (5) by a mirror and partially reflected (6) by the beam splitter. The two beam components (3 and 6) were aligned to produce interference. The intensity of the recombined beam depended on the phase relation of the two components. Alternating cancellations and enhancements were produced by the changing delay of the beam traversing the film as water was siphoned off and the film thickness decreased. Between two successive cancellations, or enhancements, the film thickness changed 0.950 μm. The recombined beam was directed into a light detector, the output of which, along with the output of the film probe's signal conditioner, were recorded on tape for later evaluation by computer.

The instruments described above permitted the measurement of water film thicknesses. In another part of this study, the behavior of bubbles and water films generated by them were observed visually.

A hand held 45 cm square and 2.5 cm deep Lucite container (Figure 6) served for that purpose. When it was filled with water and sealed, an air space of about 1.3 cc was left inside. This air space formed a moving bubble when the position of the container was changed. With proper illumination and viewing angle, the border of the liquid film, i.e., the common or contact line of all three phases, was clearly visible, and its behavior could be followed as the bubble moved at various speeds.

RESULTS AND CONCLUSIONS

The 1.3 cc bubble in the hand held container (Figures 7, 8, 9 and 10) is made to move by tilting the container. If the angle of tilt is sufficiently large, the bubble moves with a speed at which a continuous water film remains between the bubble and the container wall. When, subsequently, the container is leveled, the bubble comes to rest. Usually after the bubble has been at rest for a short time, but often when it is still moving, the film breaks at some point and from there it recedes rapidly radially outward with an irregular contact line. As this irregular line approaches the periphery of the bubble, the bubble briefly shakes randomly in all directions. This is probably caused by surface tension that is neither steady nor uniform along the periphery due to the irregularity of the approaching contact line. As the contact line settles to a smooth circular shape (Figure 7 and Figure 10a), the bubble diameter grows slightly. If now this container is tilted gradually, at some angle the bubble begins to move slowly, and a narrow crescent shaped film extends backward from the leading edge of the bubble (Figure 10b). The edge of this film, i.e., the contact line, is serrated and undulating. These wave forms are unsteady and irregular. At any constant bubble speed, the contact line has a fixed mean shape and a fixed mean position relative to the bubble. More tilt increases the bubble speed and moves the film's contact line further away from the front of the bubble. If the bubble moves with higher and higher speeds, the unsteady irregularities gradually

Figure 4

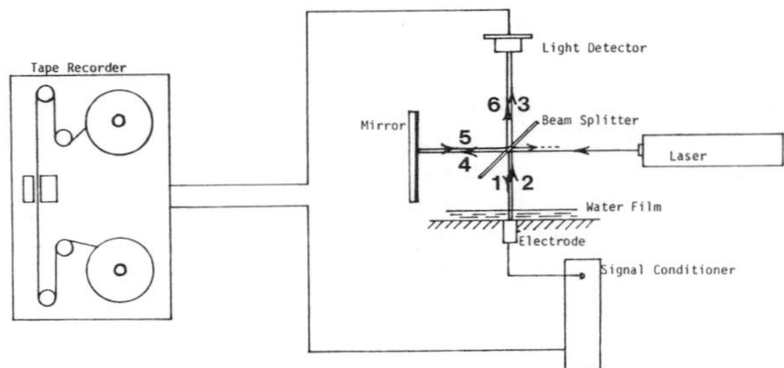

Figure 5

Figure 6

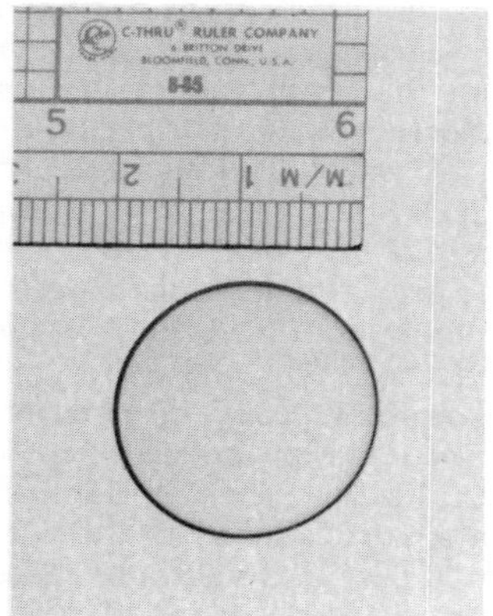

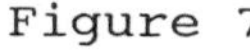
Figure 7

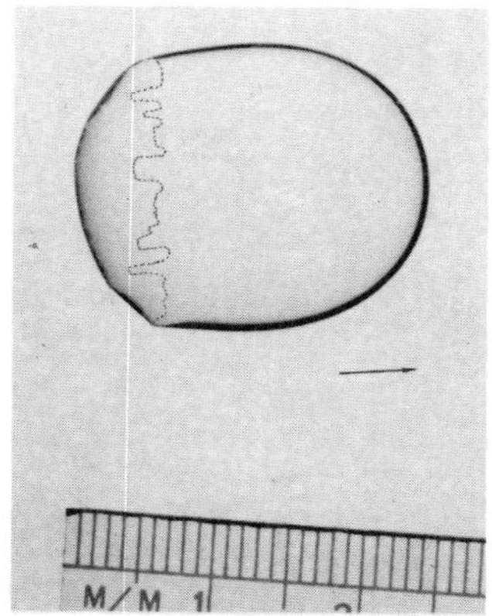

Figure 8

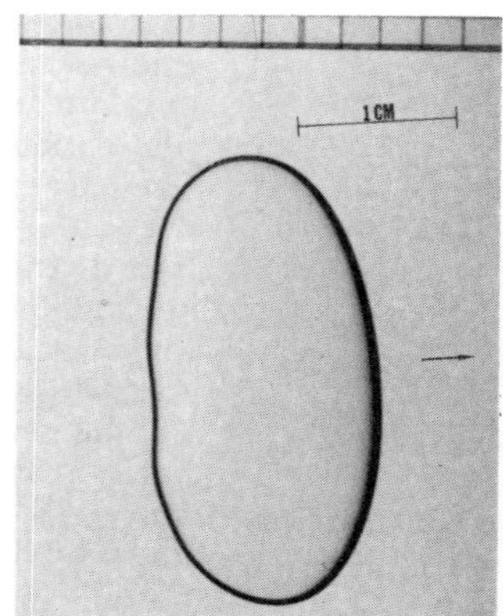

Figure 9

elongate to "fingers" (Figure 10c), the tips of which occasionally become unstable, probably under the effect of surface tension, and tiny droplets of various sizes are pinched off. With increasing bubble speed, these droplets become more numerous, with a concomitant increase in size variation.

The contact line of the film remains stationary in the mean relative to the bubble until it is displaced about halfway between the front and the rear of the bubble. If the speed increases further, the bubble "rolls over" the film, and the film becomes continuous over the entire bubble. Figures 8 and 10d show the bubble when it moved faster than the contact line of the film, and the contact line at the trailing edge of the bubble was about to reach the contact line of the film. An instant later, the film covered the entire bubble, and there was no contact line. The dotted line in Figure 8 indicates approximately the contact line of the film that did not show up photographically owing to the illumination needed.

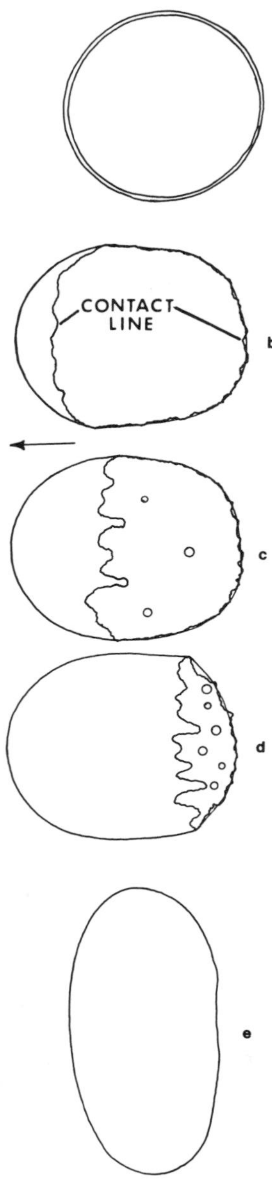

Figure 10

The instant the contact line ceases to exist, the drag on the bubble drops and it accelerates. During this acceleration, the curvature of the leading edge decreases, and the bubble deforms to an oval elongated perpendicular to the direction of motion (Figure 9, Figure 10e). Previously, the bubble was a wider and shorter oval with long axis in the direction of motion.

In another group of experiments, the impedance probe installed in the water tank described under "Instrumentation" measured the film thicknesses of several series of bubbles. To illustrate some features of the films, the measured film thicknesses of two 100 cc bubbles were plotted in Figure 11. These are two successive bubbles in a series that had the same volume and were released under identical experimental conditions beneath the plate inclined at an angle of 0.98°. They are much bigger then the one in the sealed container. The first bubble had already a continuous film. The second followed three minutes later and it dewetted the plate. Both were released 8.5 cm from the electrodes. The second 100 cc bubble was probably at the same stage of film development as the 1.3 cc bubble of the hand held container shown in Figure 8 and Figure 10d, namely just before the formation of the uninterrupted film was complete.

The forward portions of both films have very uniform and almost exactly the same 12 μm thicknesses. The uninterrupted film remains uniform all the way to the rear of the bubble. The other has a marked hump before it thins out and dewets the plate.

The following is a tentative interpretation of the observations. When dewetting occurs, surface forces drive the film's contact line forward. If the bubble moves slowly, the contact line moves as fast as the bubble and it maintains a position near the bubble's leading edge. The liquid swept

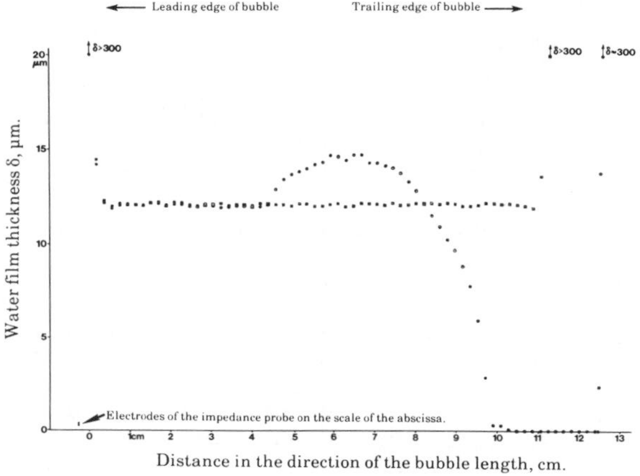

Figure 11

up from the plate first moves vertically away from the plate, then it turns forward, flows along the air-liquid interface of the film, and finally it rounds the front of the bubble. The flow in the film has also a lateral component, and part of the liquid clears the bubble to the sides.

If the bubble moves too fast, the film's contact line falls behind, the film becomes essentially stationary on the plate, and the liquid motion in the region of the contact line becomes decoupled from the flow field in front of the bubble. The liquid swept up from the plate accumulates and forms the hump on the film.

When the film is uninterrupted, it remains nearly stationary relative to the plate. What little motion there is, it is probably generated by shear stresses caused by motion of air inside the bubble.

At lower left in Figure 11, a small drawing on the scale of the abscissa shows the electrodes of the impedance probe and their separation. It permits comparison of their size and separation with the length of the bubble and of the film. In the plot, the film thickness is exaggerated nearly 5000 times relative to its length.

For a series of experiments, a detergent was dissolved in the water. It had the effect that dewetting was never observed. Due to the high conductivity of this solution, the instrument output at 15 µm film thickness approaches saturation, and the measured values become uncertain at greater thicknesses. Since the smallest film thickness measured with the detergent was more than 15 µm, these results were not plotted.

ACKNOWLEDGMENT

This work was supported by the National Science Foundation and by the Oak Ridge National Laboratory.

The author wishes to thank Dr. Stephen C. Traugott, who initiated this study, for many valuable suggestions made during numerous discussions. He also wishes to thank Kimberly Jones Holt who assisted in the experiments through many long hours.

LIST OF FIGURES

ULTRASONIC DETECTION OF BUBBLES USING TIME DELAY SPECTROMETRY

James A. Rooney,
Jet Propulsion Laboratory, California Institute of Technology, Pasadena, CA 91109

ABSTRACT

A capability to quantify the size of bubbles has been developed using time delay spectrometry. This signal processing technique uses swept frequencies and matched filter detection. It can provide either ranging or frequency response data based upon fundamental or second harmonic signal processing. This system has been tested using bubbles in the 1-50 micron range. The use of this signal processing provides a technique that is both quantitative and capable of uniquely characterizing bubbles and their size distributions.

The detection and quantification of microbubbles in soft tissues is a continuing problem. Ultrasound techniques based on the Doppler effect, the backscattering or absorption of sound by the bubbles, as well as their nonlinear oscillation, have been widely applied. The present paper reports recent results using new swept frequency ultrasonic techniques for bubble characterization.

Ultrasound has unique advantages for bubble detection over other potential detection modalities because it uses mechanical waves to detect a compressible mechanical target. The detection system can therefore interact with and couple directly to the target of interest. The detection schemes can either use continuous wave or pulse-echo technologies and can be classified by the particular aspect of the physics of interaction of sound with bubbles that they exploit.

Doppler detectors measure the change in frequency caused by the bubbles because of their motion within the blood vessel. These detectors have the advantages that they are relatively simple, inexpensive and operate in a continuous wave mode. The first such technique utilized an ultrasonic Doppler flowmeter which relied on the fact that the presence of bubbles in the blood stream changed the characteristics of the Doppler signal received from that of normal blood flow[1]. The technique has now been refined and quantified to some degree[2]. However, it has severe disadvantages since it can only detect bubbles in motion. Therefore, it can detect only bubbles that have already formed and entered the blood stream. Thus, they cannot locate nucleation sites or monitor tissues other than blood. Nor can they localize or image bubbles or determine their sizes. Johnson and Postles[3] have reported on the difficulties in accurate determination of Doppler signals in decompression studies in divers. These result from the lack of operator familiarity with the signal, an inability to hear the signal due to a frequency related hearing loss in some operators, failure to receive the signal due to improper probe positioning, noise and motion artifacts or any combination of these. A more recent study of Doppler techniques has been completed by Olsen *et al*[4].

A second type of bubble detector uses the pulse-echo instrumentation and detects the amplitude of the backscattered signal alone. These systems have been used to find the range to the bubbles in blood and for imaging of bubbles in soft tissues in decompression applications. The best bubble detection system of this type described to date is an 8 MHz pulse-echo imaging system developed by S. Daniels *et al.*[5]. They found that the minimum bubble size that they could detect was 10 microns in diameter. Using this system, they were able to detect and monitor the presence of both moving and stationary bubbles in a variety of tissue types. In spite of these improvements their system has severe limitations for application to decompression studies. While their system is capable of detecting bubbles of 10 microns in diameter, they analyzed their data by classifying bubbles into bubble diameter ranges of: (1) less than 100 microns; (2) between 100 and 500 microns; and (3) greater that 500 microns. The reason for this classification scheme is that, since they were operating at only a single frequency, precise measurements of bubble size could not be made[6,7]. Because of this problem, only estimates of bubble sizes based upon comparative measurements with model systems were possible with the resulting imprecise classification scheme being used in analysis. In addition, their pulse-echo type of system is subject to motion artifacts. Thus, while better than the Doppler system, the single-frequency pulse-echo system still lacks the capability to precisely size bubbles, is not free from motion artifacts and cannot quantitatively study bubble growth and population dynamics.

In an attempt to find a more unique acoustic characterization of bubbles, Miller[8] developed a system based upon the detection of the second harmonic oscillation of bubbles. He used a source transducer driven at levels sufficient to cause nonlinear oscillation of the bubbles if they were of the appropriate size. The system included a detecting receiving transducer that was tuned to a frequency that was twice that of the driving transducer. The system was able to detect bubbles only of a narrow size distribution but was capable of more easily differentiating between those bubbles and other objects within the insonified region.

The present research effort has as its objective the development of swept-frequency ultrasonic techniques for quantification and sizing of bubbles that are associated with decompression sickness. Swept frequency ultrasound systems for medical applications have been under development at JPL for many years and were first reported in 1974[9]. Since then the systems have been used for both materials characterization and imaging applications[10-13]. The system may be used in either transmission or reflection modes. The use of the swept-frequency methodologies has advantages since they can exploit properties associated with resonances of bubbles. At their resonance frequencies the scattering and absorption cross-sections of bubbles are many times that of their geometric cross-section.

The swept frequency technique being used is called Time Delay Spectrometry (TDS) and is capable of providing either spectral information of improved time (range) resolution. The details of the mathematical development of TDS are given in Reference 10. TDS consists of a swept source frequency and a matched filter. The received signals arrive at the receiver with a time delay that depends on the path length and propagation velocity. Since the transmitter frequency is swept, the time delay is equivalent to a frequency offset. Thus, by selecting and appropriate offset frequency and bandwidth for the tracking filter, any equivalent time interval can be selected.

In the current implementation of the TDS, the system uses linearly swept frequencies from 0.3 to 5 MHz. The signal is transmitted into the test medium and the echo from the region of interest is received and coherently mixed with the input signal. This signal is low-pass filtered and a fast Fourier transform is used to convert the frequency differences into an A-scan single scan line. A two dimensional cross-sectional B-scan can be obtained by moving the acoustic beam across the region of interest.

The signal processing associated with this system has the additional advantage that it is designed to utilize the information obtained in the analytic signal[14]. the analytic signal $h(t)$ is a complex function defined by the equation.

$$h(t) = f(t) + ig(t) \tag{1}$$

In this expression $f(t)$ is the in-phase acoustic response of the system to a Dirac delta stimulus and $g(t)$ is the quadrature response. The total energy $E(t)$ is composed of two components $V(t)$ and $T(t)$ such that $E(t) = V(t) + T(t)$. It has been shown by Heyser that the energy-density is proportional to the square of the magnitude of the analytic signal so that

$$E(t) = K/h(t)/^2 \tag{2}$$

The physical significance of the energy time curve is that it is a strictly positive quantity that identifies the time course of the energy density response of the medium. It is not modulated by the transducer resonance frequency and is a smoother, sharper and more readable signature than that obtained using conventional signal processing.

In order to test the feasibility of applying the JPL swept frequency system for bubble detection and quantification, we have performed experiments using bubble targets in various media. Specifically, the backscattered and attenuated signals were obtained from air entrapped in micropores of hydrophobic membranes similar to those described by Miller[15]. These membranes have a pore radius of 2.0 ± 0.14 microns and thickness of 12 microns. The "bubbles" used in this experiment were formed by mounting the membrane in a test tank perpendicular to the direction of the propagation of sound. The transducer of the TDS system was aligned so that the sound was incident normal to both the membranes and the wall of the tank behind the membrane. The backscattered measurements were made using the energy time curve and the attenuation measurements were made by tuning the TDS system for the range of the tank wall.

Results of the experiment are shown in Figure 1. The energy-time curve for a scan line through the region where the bubbles were present is shown in Figure 1a. The echo associated with a plastic membrane on the front surface of the chamber is located near 3.75 and the walls of the test tank are located near 5.25 and beyond. The strong echo located near 4.5 is related to the backscattered energy of the resonant bubbles. We note that the amplitude of that signal is significantly greater than that associated with the first plastic membrane.

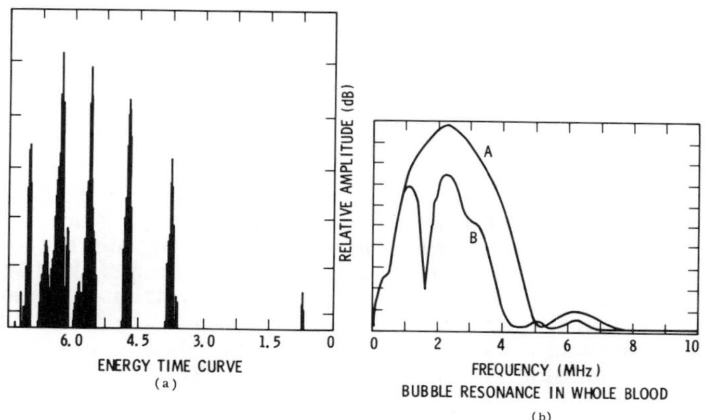

Figure 1
(a) & (b)

The resonant nature of the scattering can be more clearly seen in Figure 1b. The two curves shown represent the amplitude of the received signal (in arbitrary units) as a function of frequency for two cases. Curve A corresponds to the signal received from the far wall when no bubbles were present and therefore represents the transducer and the tank wall. The sharp reduction in signal near 1.5 MHz is the result of the resonant attenuation of the ultrasound signal. The results shown in Figure 1b are those obtained when the membrane was placed in a sample of human whole blood. Similar results have been obtained in water and synovial fluid.

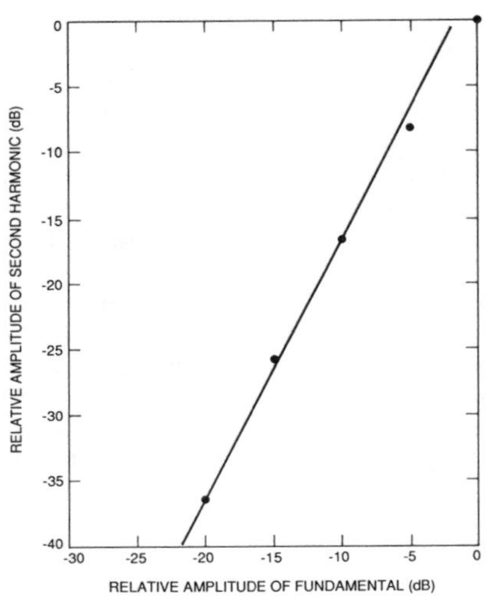

Figure 2

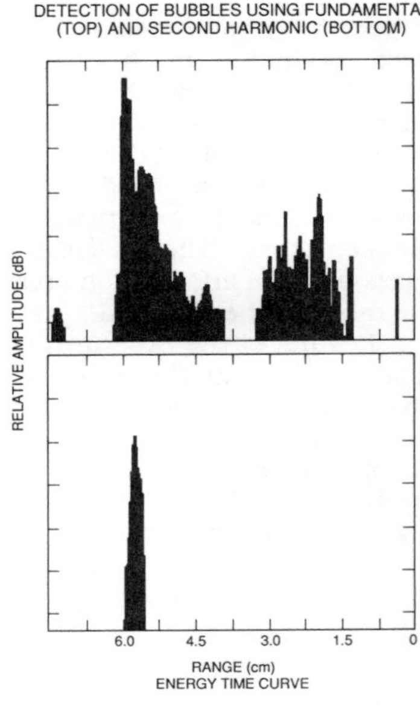

Figure 3

In addition, measurements have been made of the second harmonic emissions from the target bubbles. For these measurements, the sweep program of the tracking receiver was driven synchronously with the transmitting program, but at a rate of 1 GHz/sec and a delay offset corresponding to the arrival time of the backscattered sound from the membrane. The measured frequency spectrum of the second harmonic backscatter from the bubbles corresponds closely to the predicted spectrum shape showing a dominant peak at twice the frequency at which a strong absorption dip occurs for sound passing through the membrane. Results for the measurement of the amplitude of the second harmonic as a function of amplitude of the fundamental are shown in Figure 2. The slope of the fitted line is 2, as expected from theory.

Energy time curves have also been obtained using both fundamental and second harmonic detection as shown in Figure 3. Using the fundamental as the detection signal the backscattering from several targets including bubbles can be seen. When the TDS processing uses the second harmonic signal only the bubbles are detected. Thus the system can be used to located the range to the bubbles, and then the frequency response for the targets at that range will give the size distribution of the bubbles. This capability provides a significant improvement over existing bubble detection systems.

ACKNOWLEDGMENT

This work was carried out at the Jet Propulsion Laboratory, California Institute of Technology, under contract to the National Aeronautics and Space Association. It was supported in part by the Naval Medical Research and Development Command.

REFERENCES

1. M. P. Spencer & S. D. Campbell (1968), "Development of Bubbles in Venous and Arterial Blood During Hyperbaric Exposures." *Bull. Mason Clinic,* **22**, 26
2. M. P. Spencer (1977), **Proc. Twelfth Undersea Medical Society Workshop, U.M.S. Meeting 12 May 1977**, 1.
3. D. C. Johanson & W. F. Postles (1977), **Proc. Twelfth Undersea Medical Society Workshop, U.M.S. Meeting 12 May 1977**, 98.
4. R. M. Olsen, R. W. Kutz, G. A. Dixon, & K. W. Smead (1988), "An Evaluation of Precordial Ultrasonic Monitoring to Avoid Bends at Altitude," *Aviat. Space Environ. Med.*, **59**, 635.
5. S. Daniels, J. M. Davies, W. D. M. Paton, & E. B. Smith (1980), "The Detection of Gas Bubbles in Guinea-Pigs After Decompression From Air Saturation Dives Using Ultrasonic Imaging." *J. Physiol.*, **308**, 369.
6. T. W. Beck, S. Daniels, W. D. Paton, & E. B. Smith (1978), "Detection of Bubbles in Decompression Sickness," *Nature (London)* **276**, 173.
7. S. Daniels, W. D. Paton, & E. B. Smith (1979), "An Ultrasonic Imaging System for the Study of Decompression Induced Gas Bubbles," *Undersea Biomed. Res.*, **6**, 197.
8. D. L. Miller (1981), "Ultrasonics Detection of Resonant Cavitation Bubbles in a Flow Tube by Their Second Harmonic Emissions," *Ultrasonics* **19**, 217.
9. D. H. Le Croissette & R. C. Heyser (1974), "A New Ultrasonic Imaging System Using Time Delay Spectrometry," *Ultrasound Med. Biol.* **1**, 119.
10. D. H. Le Croissette & R. C. Heyser (1976), "Attenuation and Velocity Measurements Using Time Delay Spectroscopy," NBS. Special Pubn. 453, 81.
11. P. M. Gammell, D. H. Le Croissette, & R. C. Heyser (1979), "The Temperature and Frequency Dependence of Ultrasonic Attenuation in Selected Tissues," *Ultrasound Med. Biol.* **5**, 269.
12. J. A. Rooney, P. M. Gammell, J. D. Hestenes, H. P. Chin, & D. H. Blankenhorn (1982), "Velocity of Sound in Arterial Tissues," *J. Acoust. Soc. Am.* **71**, 462.
13. J. A. Rooney, P. M. Gammell, J. D. Hestenes, H. P. Chin, & D. H. Blankenhorn (1981), "The Use of Ultrasonic Spectroscopy to Characterize Calcified Lesions," *IEEE Trans. Sonics Ultrason.*, **SU-28**, 291.
14. R. C. Heyser (1971), "Determination of Loudspeaker Signal Arrival Times: Part 1 J. Aud. Eng. Soc. 19, 829 and Determination of Loudspeaker Signal Arrival Times: Part III," *J. Aud. Eng. Soc.* **19**, 902.
15. D. H. Miller (1982), "Experimental Investigation of the Response of Gas-filled Micropores to Ultrasound." *J. Acoust. Soc. Am.* **71**, 471.

FIGURE LEGENDS:

Figure 1: *a)* Energy-time curve for ranging bubbles and other targets using detection of the fundamental. *b)* Frequency response for transmission mode when bubbles are (B) and are not (A) in insonified region.

Figure 2: The response of the second harmonic from bubbles as a function of amplitude of fundamental excitation. The slope of the fitted line is 2.

Figure 3: Energy-time curves shown for detection of bubbles among other targets using the fundamental (top) and second harmonic (bottom).

SELECTION AND STABILITY OF BUBBLES IN A HELE-SHAW CELL

Saleh Tanveer
Mathematics Department, Virginia Polytechnic Institute & State University, Blacksburg VA 24061

ABSTRACT

The effect of surface tension on the determination of shape and velocity of a steadily translating bubble in a Hele-Shaw cell as well as its linear stability are presented in this paper. For zero surface tension, there is a continuum of steady bubble solutions for which the bubble velocity and the distance of the bubble centroid from the channel centerline are arbitrary. However, for arbitrarily small surface tension, only a discrete set of bubble solutions are possible with the bubble being symmetric about the channel centerline. Of this discrete set, only the branch of the solution for which the bubble velocity is the largest is linearly stable in accordance with similar results for the finger. Transcendentally small terms in surface tension play a crucial role in the determination of bubble velocity, symmetry as well as linear stability. The discrepancy of experimental observations with these theoretical results based on a simplified set of boundary conditions are pointed out. Recent theoretical developments that may account for a part of the discrepancy is discussed and open problems pointed out.

1. INTRODUCTION

The general area of two phase flow in a Hele-Shaw cell has received considerable attention over the last few years because of its connection to the porous media flow, dendritic crystal growth and the general area of pattern formation. Most of the attention has been devoted to the phenomena of fingering and the resulting steady finger shapes, its time evolution and the stability of the steady finger. Reviews by Saffman[1], Homsy[2] and Bensimon *et al.*[3] summarize the state of affairs as of 1986. Since then, a fair amount of progress has been made in the understanding of some of the outstanding questions pertaining the determination of the steady state finger and its linear stability.

Here, we will concentrate on concurrent developments in the analogous problem of the motion of a finite bubble in a Hele-Shaw cell. Features affecting the front of a long bubble are expected to be in common with those of the finger. Indeed, historically, Taylor & Saffman[4] first studied the bubble in an indirect attempt to understand why there is no restriction on the theoretical finger width (Saffman & Taylor[5]) when surface tension is entirely neglected. From a theoretical standpoint, some features common to both a finger and a bubble are easier studied in the context of a bubble because the finger has a geometrical singularity at the tail. An example of such a feature that is yet to be explained theoretically may be the dendritic instability observed in the experiment of Couder[6] and Maxworthy[7]. Aside from the connection to finger motion, the study of bubble is interesting in its own right because of the experimentally observed diversity both in the shapes and dynamics of motion (Maxworthy[7] and Kopf-Sill & Homsy[8]). Understanding of all these features is far from complete.

In this paper, some recent developments in the study of the motion of a single bubble will be summarized. We will be concerned only with the prediction of the shape and the corresponding velocity and the linear stability of a steadily moving bubble. Theoretically, very little is known about the more complex features such as the nonlinear evolution and formation of fractal structures of Hele-Shaw cell interfaces or the dynamics of the motion of multiple bubbles (Maxworthy[7]). There appears to be a wide range of open questions and puzzles that need to be addressed.

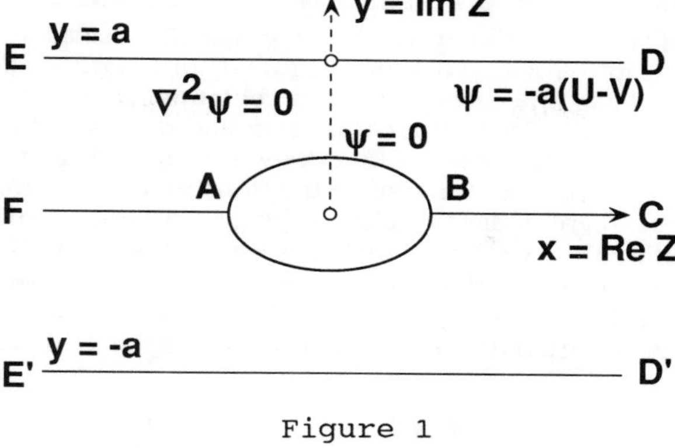

Figure 1

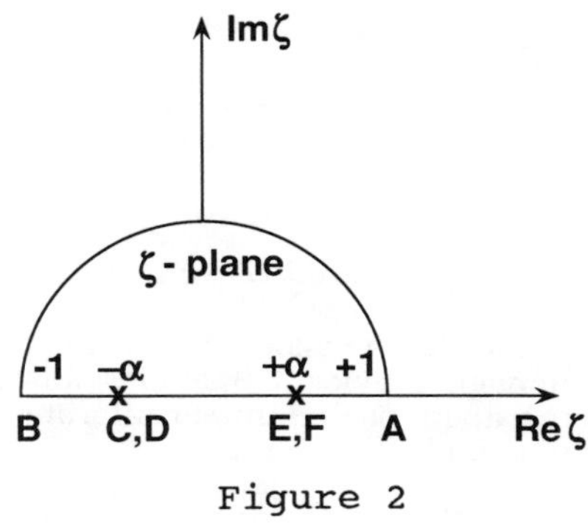

Figure 2

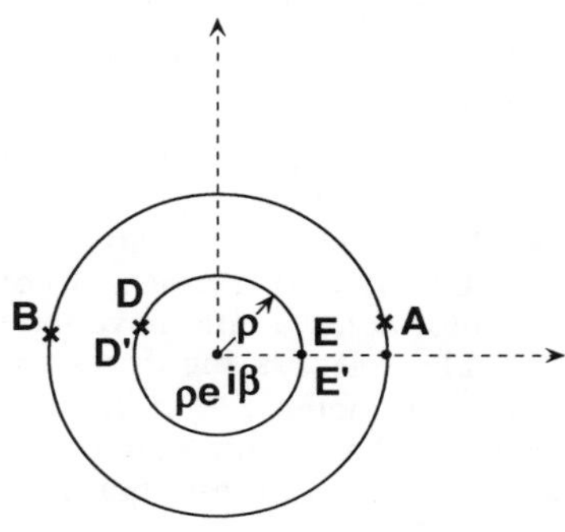

Figure 3

2. MATHEMATICAL FORMULATION AND DEGENERACY FOR $T = 0$

We consider a finite sized bubble of negligible viscosity steadily translating with velocity U in the x-direction (see Figure 1) in a parallel sided Hele-Shaw cell of width $2a$ and gap b between the plates containing a viscous fluid with viscosity μ which moves with velocity V far upstream and downstream. We ignore gravity and any thin film of the viscous fluid between the bubble and the plates in the narrow gap direction. We also assume the transverse curvature is a constant. The theory presented here under these simplifying assumptions will be referred to later as the simplified theory. The simplifying assumptions will be discussed later in Section 7 in connection to discrepancies with experimental observation.

The averaged fluid flow across the gap in a frame moving with the steady bubble is described by a harmonic potential $\phi(x,y)$ (see Taylor & Saffman[4] and Tanveer[9] for details) where

$$\phi\,(x,y) = -\frac{b^2}{12\mu}\nabla p\,(x,y) - Ux \tag{1}$$

in which p is the mean pressure across the stratum. The averaged velocity across the gap is $\nabla\phi$. We will set $a = 1$ and $V = 1$ without any loss of generality, since this is equivalent to nondimensionalizing all length and time scales. We introduce the stream function $\psi(x,y)$, which is the harmonic conjugate of ϕ. The boundary condition at $x \pm \infty$ is

$$\phi = -(U-1)x + O(1) \tag{2}$$

and on the walls at $y = \pm 1$

$$\psi = \mp (U-1)\,. \tag{3}$$

On the bubble boundary, we have the following kinematic and dynamic boundary conditions:

$$\psi = 0 \tag{4}$$

$$\phi + Ux = \frac{b^2 T}{12\mu V a^2}\,\kappa + constant \tag{5}$$

where T is the interfacial surface tension and κ is the curvature in the xy-plane.

For zero surface tension, Taylor & Saffman[4] found an exact solution for bubbles that are symmetric about the channel centerline. Introducing $z = x + iy$, the Taylor-Saffman solution can be written as

$$z = z_0(\zeta,\alpha) \equiv \frac{1}{\pi}\ \ln\left(\frac{\zeta-\alpha}{\zeta+\alpha}\right) + \frac{1}{\pi}\left(\frac{2}{U} - 1\right)\ \ln\left(\frac{1+\alpha\zeta}{1-\alpha\zeta}\right) \tag{6}$$

$z\,(\zeta)$ is the conformal map that maps the interior of the unit semi-circle (Figure 2) into the physical flow region exterior of the bubble on one side of the channel centerline ($y \geq 0$) such that $z = \mp\infty$ corresponds to $\zeta \pm \alpha$; while the two stagnation points of the bubble B and A in Figure 1 correspond to $\zeta = \pm 1$. The semi-circle in the ζ-plane corresponds to half the bubble boundary for which $y \geq 0$. The complex velocity potential $W \equiv \phi + i\psi$ is given by

$$W = \frac{U-1}{\pi}\ \ell n \frac{(\zeta-\alpha)(1-\alpha\zeta)}{(\zeta+\alpha)(1+\alpha\zeta)} \tag{7}$$

In this solution both U, the bubble velocity, and α, a parameter characterizing the bubble size, remain arbitrary over a range. In order that the mapping function in Equation (6) remain conformal, we only have to require $U > 1$. The limits ∞ and 1 for U correspond to bubble degenerating into straight lines along the x and y directions respectively. The parameter α is in the interval (0,1) with the limits 0 and 1 corresponding to infinitesimal and infinitely long bubbles, respectively. Thus the specification of bubble area leads to a one parameter set of degenerate solutions. Each value of the parameter U for a given α corresponds to a specific bubble shape. The degeneracy of solutions at $T = 0$ in this case is identical to that of the finger

(Saffman & Taylor[5]), where the finger velocity U (or equivalently the finger width λ since $\lambda = 1/U$) is arbitrary. This is not surprising since the finger can be viewed as the front of the $\alpha \rightarrow 1$ bubble. Taylor & Saffman[4] proposed an *ad hoc* hypothesis which singled out the $U = 2$ bubble. However, no justification could be given as to the validity of the hypothesis.

The two parameter Taylor-Saffman solutions are only a special case of a more general three parameter family of solutions (Tanveer[10], Kadanoff (private communication[11]). The extra parameter results from relaxing the assumption of bubble symmetry about the channel centerline. Originally (Tanveer[10]), it was stated that Kadanoff found a four-parameter family of solutions is disagreement with ours. Since then, a relation between Kadanoff's four-parameter that was overlooked previously has been found and the conclusions of the two concurrent works are identical on the number of free parameters. Kadanoff also proves rigorously that there are no other solutions given reasonable conditions on flow singularities at $\pm\infty$. Following the Tanveer[10] notation, we can write the three-parameter family of exact solutions as

$$z = z_0(\rho, \beta, \xi) \equiv -i\frac{(\pi-\beta)}{\pi} + \frac{2}{\pi}\ \ell n \left\{ e^{-i\beta/2} \frac{\theta_1\left(-i/2\ \ln \xi + i/2\ \ln \rho,\ \rho^2\right)}{\theta_1\left(-i/2\ \ln \xi - \beta/2 + i/2\ \ln \rho,\ \rho^2\right)} \right\}$$
$$+ \frac{2(U-2)}{\pi U}\ \ell n \left\{ e^{i\beta/2} \frac{\theta_1\left(-i/2\ \ln \xi - i/2\ \ln \rho,\ \rho^2\right)}{\theta_1\left(-i/2\ \ln \xi - \beta/2 - i/2\ \ln \rho,\ \rho^2\right)} \right\}. \tag{8}$$

where $z(\xi)$ is the conformal map that maps an annular region in the ξ plane (Figure 3) between radius $\rho (\rho < 1)$ and 1 into the entire flow region exterior of the bubble in the physical domain such that $\xi = \rho,\ \rho e^{i\beta}$ corresponds to $z = \mp \infty$ respectively, and $\theta_1\ (t, q)$ denotes the Jacobi θ function of the first kind with argument t and nome q. For purposes of numerical evaluation, we found it convenient to use the following rapidly convergent power series representation

$$\theta_1\ (t, q) = 2q^{1/4} \sum_{n=0}^{\infty} (-1)^n\ q^{n(n+1)} \sin\left(2\ (n+1)\ t\right) \tag{9}$$

In Equation (8), the parameter ρ characterizes the size, while $\beta/2\pi$ has the physical meaning of the ratio of the fluid that moves over the bubble to the total fluid moving past the bubble in the frame of the steady bubble, and this ratio can take any value between 0 and 1. For a symmetric bubble $\beta = \pi$, and we recover the Taylor-Saffman solution in a different representation. Generally, for a given bubble area, both the bubble velocity U and the asymmetry parameter β are arbitrary over some range.

3. APPARENT CONTRADICTION BETWEEN PERTURBATION AND NUMERICS AND ITS RESOLUTION

When surface tension $T \neq 0$, no exact solution is known. However, one can try a perturbation expansion in powers of T. One finds that a consistent perturbation expansion in powers of T is possible without any restriction on the parameters (see Tanveer[9] for details). This is true for bubbles that are non-symmetric as well, though Tanveer[9] only considers bubbles that are assumed to be symmetric. Thus it may seem that the bubble velocity and symmetry remain arbitrary even with the inclusion of surface tension. However, this contradicts the numerical results (Tanveer[9] and Tanveer[10]) which suggests that for nonzero surface tension the bubble symmetry and velocity are determined for specified bubble area. Only symmetric bubbles with a discrete set of possible velocities appeared to be possible. We will discuss the numerical method and more detailed results in Section 4. Thus, there is an apparent contradiction between the perturbation series results and numerics similar to the case of a finger (Mclean & Saffman[12]).

We now discuss the resolution of this conflict. We first point out that the ability to construct a consistent regular perturbation expansion need not imply that a solution exists. This is illustrated in the following example (Tanveer[9]).

We want to find an analytic function f within the unit semi-circle whose first two derivatives are continuous on the boundary and satisfies the following boundary condition on semi-circular arc $\zeta = e^{iv}$:

$$\varepsilon \, Re\left[\frac{d}{d\zeta}(\zeta^3 f')\right] + 4\, Re f = Re\, \zeta^2 \tag{10a}$$

and on the boundary of the semi-circle coinciding with the real axis

$$Im f = 0 \tag{10b}$$

This problem as posed above will be referred to as problem O. Note that the Schwarz reflection principle is applicable in view of Equation (10b), and such a solution f will be analytic within the entire unit circle.

If a solution f indeed exists to problem O, then such an f must satisfy the differential equation

$$\varepsilon \frac{d}{d\zeta}(\zeta^3 f') + 4f - \zeta^2 = 0 \tag{11}$$

This follows from rewriting Equation (10a) as the real part of some analytic function which is equal to zero on the circular boundary and noticing that the imaginary part of the same quantity is zero on the real diameter in view of (10b). Now, it is clear that every solution to Equation (11) cannot be a solution to (10a,b) though it is true the other way. We must look for solutions to Equation (11) that are analytic inside the unit circle with its first two derivatives continuous at the boundary. Note that the higher derivative terms in Equation (11) which vanish when $\varepsilon = 0$ does not lead to any nonuniformity of Equation (12) in the domain $\zeta \leq 1$ since the boundary conditions are automatically satisfied by each f_n. For small ε, we can construct a formal asymptotic solution of Equation (11) in powers of ε:

$$f = \sum_{n=0}^{\infty} \varepsilon^n f_n \tag{12}$$

It is easily seen that $f_0 = \zeta^2/4$, $f_1 = -\zeta^3/2$, and so on. At each stage we are able to construct f_n which is analytic in $\zeta \leq 1$ and the ratio of f_n/f_m for fixed n and m stays bounded for all $|\zeta| \leq 1$. Thus the perturbation series is consistent and appears to solve problem O asymptotically since Equation (11) is equivalent to O if f is analytic with continuous second derivatives on the boundary.

We now show that problem O has no solution. Indeed if there was a solution f solving problem O, then it would have a convergent power series expansion. Substitution of this expansion into the left hand side of Equation (11) must result in a convergent series as well, since the left hand side of Equation (11) is clearly analytic when f is. This resulting series must have every coefficient equal to zero. We obtain a recurrence relation for the different coefficients of the expansion of f and solving the recurrence relation, we find that

$$f = \sum_{n=2}^{\infty} \frac{(-1)^n}{6} (n+1)!\,(n-1)!\, \frac{\varepsilon^{n-2}}{4^{n-1}} \zeta^n \tag{13}$$

which is a divergent series for all $\zeta \neq 0$. Thus we have a contradiction in the assumption that f is analytic within the unit circle and its derivatives continuous at the boundary. Thus there is no solution to problem O for nonzero ε.

To gain more insight why the formal perturbation series (12) failed to indicate that the problem O had no solution, we consider the exact solution to the differential Equation (11) whose asymptotic behavior for small ε is given by Equation (12). Tanveer[9] showed that such a solution was

$$f = \frac{16\pi i}{\varepsilon^2}\left(u^2 H_2^2(u) \int_\infty^u dt\, H_2^1(t) t^3 - u^2 H_2^1(u) \int_\infty^u dt\, H_2^2(t) t^3\right) \tag{14}$$

where $\zeta = 16/(\varepsilon u^2)\alpha$ and the ∞ in each of the integral limits in Equation (14) is on the positive real axis. Here H_2^1 and H_2^2 denote Hankel functions of the 1st and 2nd kind. For small ε, the asymptotic behavior of (14) is given by the series (12) as can be seen on integration by parts. It is clear that solution (14) has a singularity at $\zeta = 0$. One finds that $f(\zeta\, e^{i\,2\pi}) - f(\zeta)$ is transcendentally small in ϵ in the region of interest: $|\zeta| \leq 1$. Thus, transcendentally small terms neglected in the regular perturbation expansion in ϵ is actually non-analytic at $\zeta = 0$. Thus, if we are able to supplement the asymptotic information in Equation (12) by inclusion of transcendentally small terms, then the non-analyticity of the transcendental correction within the flow domain suggests the nonexistence of solution to problem O.

Another point of interest in the above problem that will be seen to apply to other problems is that the transcendentally small correction does not appear directly as a solution to problem O. Indeed, because of the non-analyticity of such a correction, such terms cannot be allowed for in the solution to problem O. Here, we are finding specific solutions to Equation (11) which is a generalization of problem O, but not equivalent to it, since (11) allows for non-analytic f in $|\zeta| \leq 1$. The two arbitrary constants in the general solution to (11) are determined in such a way that the asymptotic series for small ε in the region $|\zeta| \leq 1$ to the leading order is given by Equation (12). This requirement gives us a unique solution to (11) as noted by the very specific choice of limits in the integrals in Equation (14).

Using the results from the analysis of problem O, it was possible to show analytically for small bubbles (Tanveer[9]) that there was indeed a constraint on the bubble velocity U. Without this restriction, we found that we could have a solution to a more generalized equation which allowed for flow singularities at ∞, which is inconsistent to the original fluid flow problem. Thus the apparent contradiction between numerics and perturbation for the case of a small bubble was resolved. For nonzero surface tension, the constraint on the set of parameters that were arbitrary for the zero surface tension solution is generally referred to as "selection".

The resolution of the apparent contradiction between perturbation and numerics for the case of a finger took place at about the same time as this work on small bubbles. Combescot *et al.*[13], Shraiman[14], and Hong & Langer[15] found the selection mechanism for the case of a symmetric finger. Tanveer[16] considered the more general case of an asymmetric finger and showed that surface tension selected both the symmetry and finger width. For a generally nonsymmetric bubble of arbitrary size, the analytical determination of bubble velocity and symmetry through the extraction of transcendentally small terms has been carried out recently by Combescot & Dombre[17]. Concurrently, for the restricted case of symmetric bubbles of arbitrary size, similar analytical results were obtained by us (Tanveer[18]). We will outline our method (Tanveer[18]) in Section 5.

In each of these works, the leading order transcendentally small term in the asymptotic expansion of the solution to some generalized equation was determined and found to be an unacceptable part of the solution to the original problem, because they are responsible for non-smooth behavior of the interface such as a corner unless very specific relations were satisfied by the parameters. On imposing conditions on the parameters, one ensures that the leading order transcendental correction is smooth on the finger or bubble boundary. This is taken to be the evidence that the solution to the originally posed problem exists under certain constraints on the parameters. Theoretically, we should check that with appropriate choice of the parameters, the total transcendental correction is smooth; but this is never done in view of the complexity of the problems. Thus, the selection criteria that all the investigators have found based on the leading order transcendental correction is not totally fool proof. Whenever possible, the results of such analysis should be confirmed by numerics in order to have a firmer evidence on the existence of a solution. There is, of course, no alternative to rigorous mathematical proof; but given the complexity of the nonlinear free boundary problem, that does not seem to be an easy task.

4. NUMERICAL METHOD FOR STEADY STATE DETERMINATION AND RESULTS

For a general value of surface tension, numerical calculations seem to be the only way of obtaining solutions to the nonlinear free-boundary problem. Two different methods have been used (Tanveer[9,10]) in the calculations. The first one (Tanveer[9]) is not suitable for large bubbles and is restricted to symmetric bubbles. However, it is very simple and efficient for small bubbles and easy to code. The other method (Tanveer[10]) is more general, though a little more complicated, and is suitable for large bubbles or for bubbles not necessarily symmetric about the channel centerline.

For the sake of simplicity of exposition, only the first method is outlined here. Assuming the bubble boundary to be smooth, the determination of the boundary shape and the flow field is equivalent to determining the analytic function f for $|\zeta| \leq 1$ (Tanveer[9] for details) such that

$$Im f = 0 \tag{15}$$

for ζ on the real axis between [-1, 1] and for $\zeta = e^{i\nu}$ on the arc of the unit semi-circle

$$Re f = -\frac{\gamma}{|f' + h|} \left\{1 + Re\, \zeta \frac{d}{d\zeta}\, \ell n\, (f' + h)\right\} \tag{16}$$

where γ is a surface tension parameter defined by

$$\gamma = \frac{\pi^2 U}{4\alpha^2} \frac{b^2 T}{12\mu} \frac{1}{\left[U\left(1+\alpha^2\right) - 2\alpha^2\right]^2} \tag{17}$$

and

$$h(\zeta) = \frac{\left(1 - p^2 \zeta^2\right)}{\left(\zeta^2 - \alpha^2\right)\left(1 - \zeta^2 \alpha^2\right)} \tag{18}$$

in which

$$p^2 = \frac{U\left(1+\alpha^2\right) - \alpha^2}{U\left(1+\alpha^2\right) - 2\alpha^2} \tag{19}$$

Note that the definition of the functions f and h here is different from the original paper (Tanveer[9]) just by multiplicative constants. The surface tension parameter γ here and ϵ in the Tanveer[9] paper also differ. The notation here is consistent to the Tanveer[18] paper.

Since f is analytic in and on the unit circle, and Equation (15) holds, it follows from Schwarz reflection principle that f is analytic for $|\zeta| \leq 1$, and therefore the power series representation

$$f = \sum_{n=0}^{\infty} a_n \zeta^n \tag{20}$$

with a_i's all real in view of Equation (15). The series (20) has a radius of convergence larger than unity. For numerical calculation, we truncate Equation (20) to N-1 terms and satisfy Equation (16) at N uniformly spaced out point on the upper half unit circle including $\zeta = \pm 1$. We get a nonlinear system of equations for the unknowns $a_0, a_1, \ldots a_{N-1}$ and U which is solved by Newton iterative procedure for specified α and γ.

As mentioned before the numerical investigations (Tanveer[9,10]) suggested that for for specified bubble area (equivalently α) and surface tension (equivalently γ), only symmetric bubbles with a discrete set of velocities were possible. Subsequent analytical work (Combescot & Dombre[17] and Tanveer[18]) suggested that there was actually a discrete infinity of bubble solutions, though the numerical calculations were done only for the first few branches in order of decreasing U. The possibility of isolated non-symmetric bubbles is not totally ruled out since the convergence of the Newton iterative procedure in the numerical scheme depends on the initial guess. However, Combescot & Dombre's[17] recent analysis appears to rule that out. In the limit of zero surface tension, all the bubble velocities tend to 2. If we introduce B =

$\frac{b^2T}{12\mu Va^2}$ as a surface tension parameter, it is found that

$$2 - U = kB^{2/3}$$

where k is different for different branches of solutions, but curiously enough, was found to be independent of the bubble size for the first few branches that were calculated numerically. This has been verified analytically[18] for all branches of solution provided T is small enough so that $(2 - U)$ is small. Because of the independence of the relation on the bubble size, the same relation is expected to hold for a finger and this has been confirmed analytically (Tanveer[16]). In the large bubble area, when the bubble was long, comparison of the bubble shapes for different branches were made with the finger branches calculated by Mclean & Saffman[12] as well as by Romero[19] and Vanden-Broeck[20]. We found that the front of each of the bubble branch coincided with the finger shapes on some previously calculated finger branch. If integer n going from 0 to ∞ denotes the bubble branches in order of decreasing U, the $n = 0$ bubble branch was found to be equivalent to the Mclean-Saffman finger branch, $n = 1$ equivalent to the first Romero-Vanden-Broeck branch, $n = 2$ to the 2nd Romero-Vanden-Broeck branch, and so on. We thought it appropriate to call these bubble branches by the names of their finger counterparts.

However, as far as identifying the branches of bubble solution by their finger counterpart, we have an omission on our part that we wish to rectify here. Originally (Tanveer[9]), we had called the Mclean-Saffman branch the main branch, and another branch the extraordinary branch in view of the rather unusual bubble shapes we obtained over some range of parameters. Subsequently, we found other branches (Tanveer[10]) and made connection with the finger solutions of Romero-Vanden-Broeck. However, we had not made the connection of the extraordinary branch with the Romero-Vanden-Broeck branches since we could not obtain convergence of the numerical solution when the bubble size was large. However, using the different numerical procedure (Tanveer[10]), we have continued the extraordinary branch of bubble solutions to larger bubble area and found that it is in our nomenclature the $n = 3$, or the 3rd Romero-Vanden-Broeck branch.

For larger B, solutions did not seem to exist. This is in agreement with the Dombre & Hakim[21] analysis for the finger, which showed that solutions do not exist beyond some critical B. Similar analysis is apparently possible for large bubbles which cannot be fitted into the cell as a circle, though this has not been done yet. However, under some conditions, analytical arguments by Tanveer[9] showed that solution could not exist for large B.

5. DETERMINATION OF TRANSCENDENTAL TERM FOR STEADY STATE DETERMINATION

Now we come to the determination of transcendentally small terms in γ which determine the bubble velocity U. We will only outline the method without going into details. The reader is referred to Tanveer[18] for details. Combescot & Dombre[17] have considered the general case of non-symmetric bubble and showed how transcendentally small terms in surface tension determine the symmetry and bubble velocity. Their method is in some sense more general than ours because they address both symmetry and bubble velocity selection mechanism. However, the formalism here is more powerful because it has also been used to address the associated problem of linear stability for the bubble (Tanveer[18]). It is quite similar to our earlier work on finger (Tanveer[22]).

As noted in section 3, for given α, a perturbation expansion in powers of surface tension of the form

$$f = \sum_{n=1}^{\infty} \gamma^n f_n \qquad (21)$$

satisfying Equations (15) and (16) is possible (Tanveer[18]) without any other constraints on the parameter p^2 (equivalent to U). Each of the f_n can be shown (Tanveer[18]) to have a singularity at $\zeta = \pm 1/p$ outside the unit circle in the neighborhood of which the expansion in Equation (21) is not uniformly valid. We note that $|\zeta| > 1$ does not correspond to the physical flow region and so Equation (21) would be consistent in the physical flow region. The formal perturbation solution described by Equation (21) is therefore analytic at least on $|\zeta| = 1$. However as shown before, this need not imply that there is a solution f analytic for $|\zeta| \leq 1$ to Equations (15) and (16) for arbitrary p and α.

In the procedure outlined below and discussed in details in (Tanveer[18]), we calculate the leading order transcendental correction to the regular perturbation expansion to a more general Equation than (15) and (16). It is the equivalent to Equation (11) of problem O. The solution to the generalized equation only solves (15) and (16) when it is analytic.

We find that for general value of p, the best we can do is to construct a solution to the generalized equation that is non-analytic at just one point, $\zeta = -1$, corresponding to the tip of the bubble. The non-analyticity is in the form of a branch point corresponding to finite angle at the tip. It is clear that a bubble with a cusp cannot actually solve the physical equations since the curvature at the cusp in infinite. However, solving the generalized equations is equivalent to solving for the bubble shape by relaxing the pressure equalling curvature condition right at the tip, as Vanden-Broeck[20] did for the finger. Only for specific values of U do we recover the analyticity of the leading order transcendental correction. This gives us the selection rule to the leading order in surface tension.

In order to calculate transcendental terms in surface tension, we use the idea that transcendental terms in the physical region arise due to source of nonuniformity in the unphysical region on analytic continuation. This idea in the context of nonlinear problems was first used by Kruskal & Segur[23] for extracting transcendental terms in a third order non-linear O.D.E. In the our context, the unphysical region is $|\zeta| > 1$. Note that, in problem O, the source of non-uniformity of perturbation expansion was at $\zeta = \infty$; so the transcendentally small terms for $|\zeta| < 1$ can be viewed as the effect of the source of nonuniformity at ∞.

Therefore, the first step in the determination of transcendental terms will be to analytically continue the Equations (15) and (16) outside the unit circle.

This is done by using the Poisson's integral formulas relating a harmonic function and its conjugate in the interior of the unit circle to its boundary value. These two real formulas can be combined into one complex expression for an analytic function g within the unit circle:

$$g(\zeta) = \frac{1}{2\pi} \int_0^{2\pi} dv' \left(\frac{\zeta + \zeta'}{\zeta' - \zeta} \right) Re\, g(\zeta') \tag{22}$$

where $\zeta' = e^{iv'}$ and $|\zeta| < 1$. We then define analytic function g by requiring that on the unit circle

$$Re\, g = |f' + h|\, Re\, f \tag{23}$$

Equation (23) defines g up to an arbitrary additive imaginary constant. We choose that constant by using the formula (22). We note that for $\zeta = e^{iv}$,

$$|f' + h| = (f'(\zeta) + h(\zeta))^{1/2}\, (f'(1/\zeta) + h(1/\zeta))^{1/2} \tag{24}$$

Using (22), (23) and (24) we obtain for $|\zeta| < 1$,

$$g(\zeta) = \frac{1}{4\pi i} \int_{C_o} d\zeta' \left\{ \frac{(l_1(\zeta') l_2(\zeta'))^{1/2}}{\zeta'(\zeta'^2 - \alpha^2)(1 - \zeta'^2 \alpha^2)} \right\} \cdot \{(f(\zeta') + f(1/\zeta')\} \left\{ \frac{\zeta + \zeta'}{\zeta' - \zeta} \right\} \equiv I(f, \zeta) \tag{25}$$

where C_o is the anticlockwise closed contour coinciding with the unit circle and

$$l_1 = [\zeta^2 (\zeta^2 - p^2) + (1 - \alpha^2 \zeta^2)(\zeta^2 - \alpha^2) f'(1/\zeta)] \tag{26}$$

$$l_2 = [(1 - p^2 \zeta^2) + (\zeta^2 - \alpha^2)(1 - \alpha^2 \zeta^2) f'(\zeta)] \tag{27}$$

The analytic continuation of g for $|\zeta| > 1$ is given by

$$g(\zeta) = I(f, \zeta) + \frac{l_1^{1/2} l_2^{1/2}}{(\zeta^2 - \alpha^2)(1 - \alpha^2 \zeta^2)} ((f(\zeta) + f(1/\zeta)) \tag{28}$$

Now Equation (16) can then be written as

$$Re [1 + \zeta (f'' + h')/(f' + h) + \frac{g}{\gamma}] = 0 \tag{29}$$

on the unit circle. By adding suitable terms which are purely imaginary on $\zeta = e^{i\nu}$, one can subtract of the simple pole singularities within the unit circle of the function appearing within the square parenthesis in Equation (29). This results in

$$Re \left[1 + \zeta (f'' + h')/(f' + h) + g/\gamma + \frac{2\zeta^2}{\zeta^2 - \alpha^2} - \frac{2}{1 - \alpha^2 \zeta^2}\right] = 0 \tag{30}$$

from which it follows that

$$1 + \zeta (f'' + h')/(f' + h) + g/\gamma + \frac{2\zeta^2}{\zeta^2 - \alpha^2} - \frac{2}{1 - \alpha^2 \zeta^2} = 0 \tag{31}$$

For $|\zeta| > 1$, Equation (31) is valid as well when g is related to f through Equation (28).

Equation (31) is then a nonlinear integro-differential equation for f for some region in $\zeta > 1$. Because of the process of analytic continuation that is used, we emphasize that the solutions to Equation (31) are actually solutions to the originally posed problem in Equations (15) and (16), only if such solution is analytic in $|\zeta| < 1 + \delta$ for some δ as assumed in the process of analytic continuation. Otherwise a solution to Equation (28) will not necessarily be a solution to Equations (15) and (16). It has been found (Tanveer[18]) that Equation (28) has an asymptotic solution that is analytic is the upper half plane for $|\zeta| < 1/p$. However, the imaginary part of this solution is not zero on the real axis for $\zeta < -1$. We immediately conclude that such asymptotic solution to (28) is not an acceptable solution to Equations (15) and (16), since the function f could not be analytic at $\zeta = -1$, or else (16) would hold in some interval $[-1-\delta, -1]$for some positive δ.

The integro differential equation simplifies by noting that the transcendental small terms in f in the physical domain that we are seeking are not small near the singular points $\zeta = \pm \frac{1}{p}$ and all the terms in Equation (30) containing f and its derivatives at ζ outside the unit circle will have to be retained. But for the integrand in the term I involve values of f and f' on the unit semi-circle where any correction to the perturbation expansion of Equation (21) is assumed to be transcendentally small and therefore it is legitimate to substitute (21) for f in the integral I. Again, for the same reason, the term $f(1/\zeta)$ can be replaced by the perturbation expansion (21) with transcendental error since $1/\zeta$ is inside the circle when ζ is outside. Thus Equation (31) reduces to a non-linear second order differential equation, for which we construct a solution which matches with the regular perturbation expansion as $|\zeta| = 1$ is approached from outside. We were able to construct such a solution whose imaginary part vanished on the real axis to the immediate right of $\zeta = 1$. However to the immediate left of -1, we found that the the leading order transcendental correction to (21) is

$$C_3 H^{1/2} L^{-1/4} e^{\gamma^{-1/2} p} \tag{32}$$

where C_3 is a constant involving α and p and

$$H = \frac{(1 - p^2 \zeta^2)^{3/2}}{(\zeta^2 - p^2)^{1/2} (\zeta^2 - \alpha^2)(1 - \alpha^2 \zeta^2)} \tag{33}$$

$$L = \frac{\left(1-p^2\zeta^2\right)^{3/2}\left(\zeta^2-p^2\right)^{3/2}}{\left(\zeta^2-\alpha^2\right)^2\left(1-\alpha^2\zeta^2\right)^2} \tag{34}$$

$$P(\zeta) = \int_{1/p}^{\zeta} i\,L^{1/2}(\zeta')\,d\zeta' \tag{35}$$

The requirement that (32) be real for ζ in $[-1-\delta, -1]$ for some δ is equivalent to a condition on the phase of C. For the branches of bubble solutions characterized by large n compared to 1 but small enough so that 2- U is small compared to unity, we find that

$$\frac{(2-U)^{3/2}}{B} = (n - \text{constant})\,\pi \tag{36}$$

where the constant is independent of any parameters and can be determined numerically from a second order ordinary differential equation as shown by Dorsey & Martin[24] in the context of a finger. Generally, for small n, the right hand side is some positive constant dependent only on the constant n.

6. LINEAR STABILITY OF BUBBLES

Tanveer & Saffman[25] studied the linear stability of the steady bubbles (Tanveer[9,10]) found earlier. Numerical computations for arbitrary sized bubbles suggest that only the Mclean-Saffman branch of bubble solutions is stable, while all the Romero-Vanden-Broeck branches are unstable for any surface tension. For the finger, analogous numerical (Kessler & Levine[26,27]) and analytical results (Tanveer[22]) were found. The numerical results for arbitrary sized bubble has now been confirmed analytically (Tanveer[18]) in the limit of small surface tension. Earlier, Tanveer & Saffman[25] showed that for small sized bubbles of the Mclean-Saffman branch, which are circular, closed form expressions are possible for arbitrary surface tension. For any nonzero surface tension, the small bubble is stable; whereas at exactly zero surface tension the bubble was unstable with a continuous spectrum. For an arbitrary sized bubble, the zero surface tension modes could still be described in a closed form (Tanveer & Saffman[25]) and it is found that the spectrum of the stability operator is the entire complex plane as for the small bubble. We can immediately conclude that the linearized time-dependent problem must be not be well posed. As soon as any amount of surface tension is included in the analysis, a drastic change in the linearized time evolution operator takes place. Generally, the associated stability operator does not have eigenmodes which are small perturbation to an arbitrary mode of the zero surface tension stability operator. We say that the zero surface tension eigenmodes are nonperturbative in the same way that the zero surface tension steady state solutions are. Only very specific zero surface tension eigenmodes have neighboring eigenmodes at small but nonzero surface tension. Indeed, this can be viewed as a problem of selection of eigenvalue and corresponding eigenmode rather than the selection of steady bubble velocity U. It has been shown (Tanveer[18]) that a regular perturbation expansion in powers of surface tension is possible for each of the eigenmodes without any restrictions on the eigenvalues. However, once again the existence of a consistent perturbation expansion for eigenmodes does not imply that a proper stability mode exists for the stability operator. Indeed, we find that if we take a generalized stability equation then the transcendentally small terms in surface tension in general are non analytic at the tip of the bubble corresponding to a corner. Therefore, they could not possibly be proper eigenmodes for arbitrary eigenvalues. Requirement that the transcendental correction to the regular perturbation expansion of the eigenmodes be also analytic at the bubble tip determines the eigenvalues. The reader interested in the details is referred to the paper by Tanveer[18].

7. COMPARISON WITH EXPERIMENTAL RESULTS AND MODIFICATION OF BOUNDARY CONDITIONS

The ultimate test of a good theory is the agreement with experiment. On this account, the simplified bubble theory gets a mixed review. The two experiments of which we are aware are by Maxworthy[7] and Kopf-Sill & Homsy[8]. We will first discuss the Maxworthy experiment and then the Kopf-Sill experiment since the nature of discrepancy with the simplified boundary condition may be different in the two cases.

In the Maxworthy experiment, relatively small sized air bubbles (diameter typically 10 - 20% of the channel width) were driven by buoyancy through viscous silicone oil bounded by lucite sheets. The simplified theory presented in the last few sections can take into consideration the buoyancy effect by a transformation of parameters, originally due to Saffman & Taylor[5], that makes the flow with gravity equivalent to a flow without gravity. The equivalent bubble velocity U in Maxworthy's experiment was found to be significantly larger than 2 (up to 40% in some cases) where as the theory predicts U close to but less than 2. Maxworthy noted that if the experimentally observed velocity is used as a given parameter in the Taylor-Saffman exact solution, then reasonably good agreement in the shape resulted. The situation is reminiscent of the Saffman-Taylor finger experiments, where if the experimentally determined finger width is used as a given parameter in the Saffman-Taylor zero surface tension solution, good agreement resulted between theoretical and experimental shapes at relatively large capillary number. Thus, even though the simplified boundary conditions may appear unrealistic in that it neglects the thin film in the transverse direction between the bubble and the plates, the agreement of Taylor-Saffman theoretical shape with experiment for some U cannot be dismissed entirely. When Mclean & Saffman[12] include surface tension in their calculations the predicted width as a function of capillary number did not agree with experiment. However, recent numerical work of Reinelt[28] shows that inclusion of the thin film effect in the boundary conditions as studied earlier by Park & Homsy[29] and Reinelt[30] makes the theoretical prediction on finger width quite close to the experiment. Thus it may be expected that inclusion of these thin film effects for the bubble problem will cure the deficiencies of the theory as far as explaining Maxworthy bubble observations.

In the limit of small $Ca = \frac{\mu U}{T}$, Bretherton[31] studied the thin film effect in the context of the the motion of a long bubble in a tube. It is clear from the work of Reinelt[30] and Park & Homsy[29] that in the limit $Ca \to 0$, when the Bretherton analysis is valid, the boundary condition (5) has to be modified to include an additive term on the right hand side of (5) which is proportional to $Ca^{2/3}$. The effect on the thin film on the kinematic boundary condition (4) involves a higher order correction in the capillary number. The same is expected to be true for the bubble though the difference between the advancing and receding interface will make a difference in the functional form of the thin film effect on the 2-d boundary condition, which is yet to be worked out. It is clear that in the limit of small capillary number, where the Bretherton analysis is valid and in the limit of $\frac{b^2}{a^2}$ so small that $\frac{b^2 T}{12\mu V a^2}$ is also small, the Taylor-Saffman zero surface tension with appropriate selection theory U is relevant to experiments. However, our understanding is that such an experiment is difficult to design, owing to the extremely large pressure that would be necessary to drive the fluid through the cell. The pressure difference between the different parts of the cell may also cause the plates to bend, making the Hele-Shaw equations invalid.

As far as stability, Maxworthy's bubbles were unstable for large bubble size, which is not in agreement with the stability analysis of the simplified theory. The form of the two dimensional interface condition that takes into consideration the thin film effect for the dynamic situation is yet to be studied for the finger as well as the bubble. Thus it is unclear how the thin film effect is going to effect the stability features of a bubble or a finger. Further study in this area is warranted.

Kopf-Sill & Homsy[8] experimented with air bubble that are driven by a pressure gradient through a glycerine-water mixture between glass plates. They found the possibility of a wide variety of bubbles, some of them having rather unusual shapes that have very little resemblance to Taylor-Saffman zero surface tension solutions. They categorize the bubble according to their shapes into six different classes: near circle, flattened, elongated, long-tail, short-tail and the so-called Tanveer bubbles. The shapes of the long tail or the short tail variety of bubbles are not similar to any of the predicted shapes based on the simplified theory. However, the shapes of the near circle, elongated, flattened and "Tanveer" bubbles have some resemblance to non-zero surface tension calculations of Tanveer[10,11].

Though the theoretical shapes of the simplified theory are not very different from the experimentally observed shapes for some classes of bubbles, the value of U in almost every case is their experiment is significantly smaller than 1, sometimes by a factor of 3 or 4. This observation is not consistent with the simplified theory exact solutions at zero surface tension as discussed earlier. Although, we cannot rigorously rule out the possibility of $U < 1$ for non-zero surface tension, we suspect that this is not possible for the solutions of the simplified boundary conditions in the absence of drastic deformations since in order for the bubble to move much slower than the fluid flow velocity at infinity, the fluid around the bubble should be moving faster since the fluid flux in the x direction is the same at any location. Thus there must exist a large pressure difference between the front and the back sides of the bubble in order for the bubble to move much slower than the fluid at ∞. We know that the bubble shapes in the limit of zero surface tension do not generate such a pressure differential and so for small values of the surface tension parameter that are in the range of experimental values, it is plausible that the pressure differential cannot be drastically different. Thus, the simplified theory cannot produce the large pressure differences without large changes in shape.

In order to see if the thin film effect could be strong enough to produce these small bubble velocities, Saffman & Tanveer[33] solved a simple model equation that accounted for in a qualitative manner the pressure differences due to difference in film thickness between the advancing and receding parts of the bubble interface; call it the Bretherton effect. We ignored the effect of the thin film on the 2-d kinematic boundary condition since this is a higher order effect for small capillary number. We found that the Bretherton effect was not nearly strong enough to account for the large discrepancy in U between theory and experiment. Instead, we hypothesize that the thin film for the slowly moving experimental bubbles is actually punctured so that the air is in direct contact with glass. We also hypothesize that there is a significant variation of contact angle between the front and back of the bubble. We modelled this effect by including such a term on the right hand side of Equation (5) to account for variation of transverse curvature at different points on the bubble. The numerical solutions to this model equation showed that it was possible for the bubble to move with a velocity U of around 0.3, as observed in one experimental case. These results should not be taken too seriously, since the model equation accounts for contact angle variation in a rather *ad hoc* manner. This is the best we could do in view of the uncertainty of the contact angle in a dynamic situation. However, from a phenomenological view point, we believe we have pointed out a possibility for the experimentalist to look for.

As far as stability, the observation of several different types of bubbles at for the same values of the parameter suggests an obvious disagreement with simplified theory which predicts that only bubbles on one branch of solutions are stable. It is quite unclear what the origin of the discrepancy is. We can guess that the contact angle and the thin film effect have something to do with it, but your guess is as good as mine on this matter.

CONCLUSION

We have presented recent results on the theory of selection and stability of bubbles in a Hele-Shaw cell. From a mathematical standpoint, we have pointed out how transcendental terms can be so crucial in controlling the steady state and stability features of the bubble. Discrepancy with experiment has been pointed out and some open questions presented. It is clear that a lot has to be learnt before we can say we understand the Hele-Shaw cell bubble.

ACKNOWLEDGEMENT

This work has been supported by the National Science Foundation grant DMS-8713246. Partial support was also provided by NASA Langley research center (NAS1-18605) while the author was in residence at the Institute of Computer Applications in Science and Engineering.

REFERENCES

1. P. G. Saffman, 1986, Viscous fingering in a Hele-Shaw cell, *J. Fluid Mech.*, **173**, 73-94
2. G. M. Homsy, 1987, *Ann. Rev. Fl. Mech.* **19**, 271
3. D. Bensimon, L. P. Kadanoff, S. Liang, B. I. Shraiman, & C. Tang, 1986, *Rev. Mod. Phys.* **58**, 977
4. G. I. Taylor & P. G. Saffman, 1959, A note on the motion of bubbles in a Hele-Shaw cell and porous medium, *Q. J. Mech. Appl. Math.*, **12**, 265
5. P. G. Saffman & G. I. Taylor, 1958, The penetration of a fluid into a porous medium of Hele-Shaw cell containing a more viscous fluid, *Proc. R. Soc. London A* **245**, 312
6. Y. Couder, A. Cardoso, D. Dupuy, P. Tavernier, & W. Thom, 1986, Dendritic growth in a Saffman Taylor experiment, *Euro Phys Lett.*
7. T. Maxworthy, 1986, Bubble formation, motion and interaction in a Hele-Shaw cell, *J. Fluid Mech.*, **173**, 95
8. A. Kopf-Sill & G. M. Homsy, 1988, *Phys.Fluids* **31**.
9. S. Tanveer, 1986, The effect of surface tension on the shape of a Hele-Shaw cell bubble, *Phys.Fluids* **29**, 3537
10. S. Tanveer, 1987, New solutions for steady bubbles in a Hele-Shaw cell bubble, *Phys.Fluids* **30**, 651
11. L.P.Kadanoff, 1987, private communication. Please contact author for details.
12. J. W. McLean & P. G. Saffman, 1981, The effect of surface tension on the shape of fingers in a Hele-Shaw cell, *J. Fluid Mech.* **102**, 455-69
13. R. Combescot, T. Dombre, V. Hakim, Y. Pomeau, & A. Pumir, 1986, Shape selection for Saffman-Taylor fingers, *Phys.Rev Lett.* **56**, 2036
14. B. I. Shraiman, 1986, *Phys.Rev.Lett.*,**56**,2028
15. D. C. Hong & J. S. Langer, 1986, Analytic theory for the selection of Saffman-Taylor finger, *Phys.Rev.Lett.*,**56**, 2032
16. S. Tanveer, 1987b, Analytic theory for the selection of symmetric Saffman-Taylor finger, *Phys.Fluids* **30**, 1589
17. R. Combescot & T. Dombre, 1988, Selection in the Saffman-Taylor bubble and asymmetrical finger problem, submitted. Please contact author for details.
18. S. Tanveer, 1988, Analytic theory for selection and linear stability of bubbles in a Hele-Shaw cell, Submitted to the *J. Theor. Comput. Fluid Dyn.*.
19. L. A. Romero, 1982, Ph.d thesis, California Institute of Technology.
20. J.-M. Vanden-Broeck, 1983, *Phys. Fluids*, **26**, 2033.
21. T. Dombre & V. Hakim, 1986, Saffman Taylor finger at large surface tension, preprint. Please contact author for details.

22. S. Tanveer, 1987c, Analytic theory for the linear stability of Saffman-Taylor finger, *Phys. Fluids* **30**, 2318
23. M. Kruskal & H. Segur, 1986, Private communication.. Please contact author for details.
24. A. T. Dorsey & O. Martin, 1987, University of Illinois preprint.. Please contact author for details.
25. S. Tanveer & P. G. Saffman, 1987, Stability of bubbles in a Hele-Shaw cell, Phys. Fluids 30, 2624.
26. D. Kessler & H. Levine, 1985, *Phys.Rev. A* **32**, 1930
27. D. Kessler & H. Levine, 1986, *Phys.Rev. A* **33**, 2632
28. D. A. Reinelt,, 1987b, *Phys.Fluids* **30**.
29. Park, C.W. & G. M. Homsy, 1984, Two phase displacement in a Hele-Shaw cell, *J. Fluid Mech.* **139**, 291-308
30. D. A. Reinelt, 1987a, *J. Fluid Mech.* **180**
31. F. P. Bretherton, 1961, The motion of long bubbles in a tube, *J. Fluid Mech.*, **10**, 166-188
32. P. G. Saffman & S. Tanveer, 1989, *Phys. Fluids A* **32**, 219

FIGURE CAPTIONS

Figure 1: The physical flow domain in the bubble frame

Figure 2: The ζ plane

Figure 3: ξ plane. $|\xi| = 1$ corresponds to the bubble boundary and $|\xi| = \rho$ to the walls

BUBBLES RISING IN A FLUID

Jean-Marc Vanden-Broeck
Department of Mathematics and Center for the Mathematical Sciences, University of Wisconsin-Madison, WI 53706

ABSTRACT

This paper reviews the effect of surface tension on three different free surface flow problems. Each flow is characterized by a continuum of solutions when surface tension is neglected. When surface tension is taken into account there is a discrete set of solutions. As the surface tension approaches zero all these solutions reduce to a unique solution. The corresponding free surface profiles are found to be in good agreement with experimental data.

1. INTRODUCTION

Recent numerical computations have uncovered an unexpected effect of surface tension. It was discovered numerically that some nonlinear two-dimensional free surface flow problems, which are characterized by a continuum of solutions when surface tension is neglected, possess a discrete set of solutions when surface tension is taken into account. More importantly it was found that this discrete set of solutions reduces to a unique solution as the surface tension tends to zero. Therefore an arbitrary small amount of surface tension can be used to remove the degeneracy of some free surface flow problems.

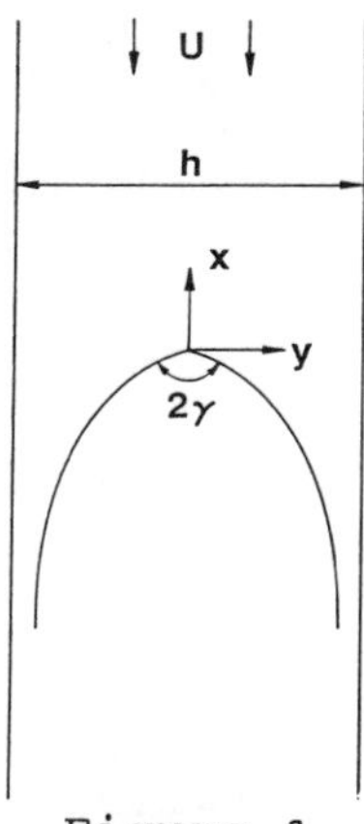

Figure 1

These problems include the well known Saffman-Taylor model for fingering in a Hele-Shaw cell. Saffman & Taylor (1958) derived an exact solution for the problem without surface tension. This solution leaves the ratio λ of the width of the finger to the width of the channel undetermined. Vanden-Broeck (1983) considered the problem with surface tension and provided numerical evidence that solutions exist only for a countably infinite number of values of the parameter λ. This discrete set of solutions include as special cases those of McLean & Saffman (1981) and Romero (1982). Similar results were obtained by Tanveer (1986) for a bubble in a Hele-Shaw cell. For a review on fingering in Hele-Shaw cells see Saffman (1986).

In order to compute the discrete set of solutions, Vanden-Broeck (1983) used the following simple numerical technique. In the first stage a modified problem is defined by allowing the slope of the surface of the finger to be discontinuous at the apex. Solutions of this modified problem are computed for all values of λ. The discrete set of solutions is then obtained by selecting among the solutions of the modified problem those for which the slope is continuous at the apex.

In the present paper we review three other free surface flow problems for which this numerical technique was used successfully. For each of them there is continuum of solutions when surface tension is neglected and a discrete set of solutions when surface tension is taken into account.

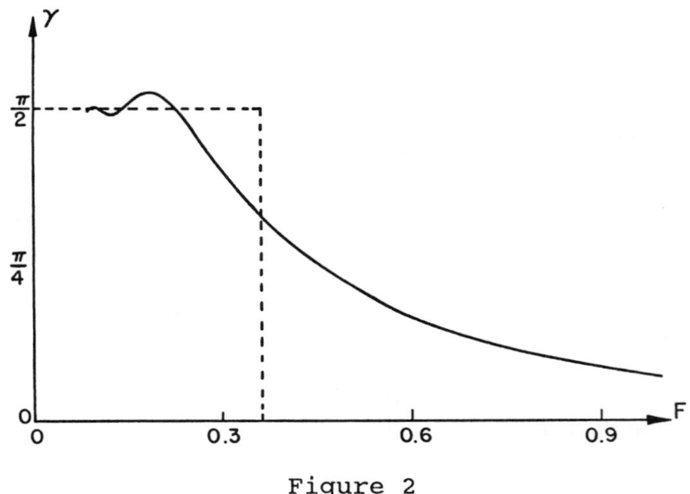

Figure 2

2. BUBBLES RISING IN A TWO-DIMENSIONAL TUBE

Let us consider the steady two-dimensional potential flow of an inviscid incompressible fluid past a bubble in a tube of width h (see Figure 1). The pressure in the bubble is assumed to be constant. We introduce Cartesian coordinates with the origin at the top of the bubble and we assume that the bubble is symmetric about the x-axis. Gravity acts in the negative x-direction. As $x \to +\infty$, the velocity approaches the constant U. We define dimensionless variables by taking U as the unit velocity and h as the unit length. The problem is characterized by the Froude number

$$F = U/(gh)^{1/2} \qquad (2.1)$$

and the Weber number

$$\alpha = \rho U^2 h/T \ . \qquad (2.2)$$

Here g is the acceleration of gravity, T the surface tension and ρ the density of the fluid. Vanden-Broeck (1984a, 1984b, 1986) solved the problem numerically by series truncation. The solutions were selected by using the technique derived by Vanden-Broeck (1983) in his investigation of the effect of surface tension on the shape of fingers in a Hele-Shaw cell. A modified problem is defined by allowing the slope to be discontinuous at the apex of the bubble. The angle at the apex is denoted by 2γ (see Figure 1). Here, γ is to be found as part of the solution. The modified problem is solved for all values of F. The solutions of the original problem are obtained by selecting among the solutions of the modified problem those for which $\gamma = \pi/2$.

When surface tension is neglected ($\alpha = \infty$), the numerical computations show that

$$\gamma = \pi/2 \qquad F < F_c \sim 0.36 \qquad (2.3)$$

$$\gamma = \pi/3 \qquad F = F_c \sim 0.36 \qquad (2.4)$$

$$\gamma = 0 \qquad F > F_c \sim 0.36. \qquad (2.5)$$

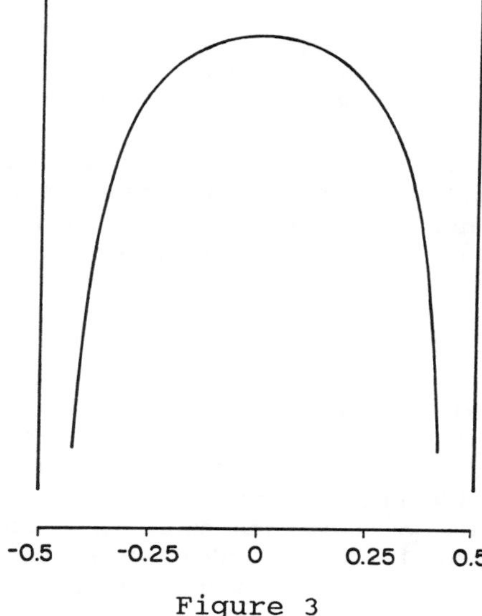

Figure 3

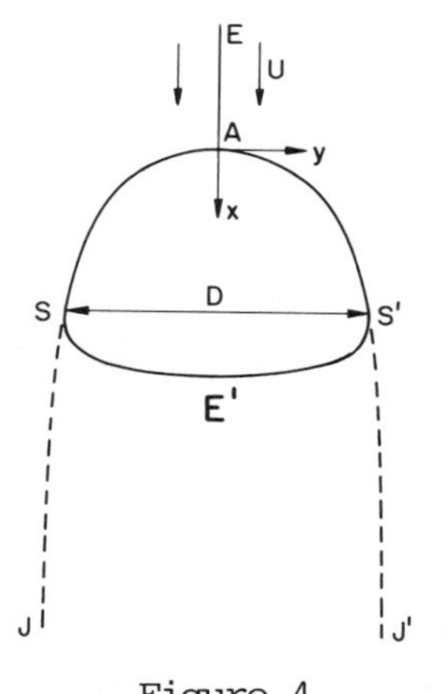

Figure 4

Relations (2.3)-(2.5) show that all solutions corresponding to $F<F_c$ are solutions of the original problem. This finding agrees with the analytical work of Garabedian (1957) who proved that a solution exists for all values of F smaller than a critical value. The solutions for $F \geq F_c$ are only solutions of the modified problem. They are characterized by a discontinuity in slope at the apex of the bubble.

Vanden-Broeck (1984b) solved the problem with surface tension ($\alpha \neq \infty$). In Figure 2 we present values of γ versus F for $\alpha = 10$. As F tends to infinity, γ tends to zero. As F approaches zero, γ oscillates often around $\pi/2$. Figure 2 suggests that there exists a countably infinite number of values of F for which $\gamma = \pi/2$. The solutions corresponding to these values of F are the solutions of the original problem. Similar results were found for other values of α. As α increases, the amplitudes and wavelengths of the oscillations in Figure 2 decrease. As $\alpha \to \infty$, the discrete set of solutions of the original problem reduces to a unique solution characterized by $F = F^* \sim 0.23$. Therefore a solution in the interval $0<F<F_c$ is selected by introducing surface tension and then taking the limit as $T \to 0$. The profile corresponding to $F = F^*$ and $T = 0$ is shown in Figure 3.

Collins (1965) performed some experiments and obtained the experimental value $F = 0.25$. In addition, he measured the ratio of the radius of curvature at the top of the bubble to the width h of the tube and obtained the value 0.305. The corresponding ratio for the theoretical profile of Figure 3 is 0.32. Finally, let us mention that similar results were recently obtained by Couet and Strumolo (1987) for a bubble rising in an inclined tube.

3. BUBBLES RISING IN AN UNBOUNDED FLUID

3.1 A free streamline model

We consider the steady two-dimensional potential flow past a bubble in an unbounded region. We assume that there is a wake of stagnant liquid extending to infinity below the bubble (see Figure 4). The problem is characterized by the Froude number

$$F = U/(gD)^{1/2} \tag{3.1}$$

and the Weber number

$$\alpha = \rho U^2 D/T \ . \qquad (3.2)$$

Here g is the acceleration of gravity, T the surface tension, ρ the density of the liquid, and D the width of the bubble (i.e.; the distance between the separation points S and S').

Vanden-Broeck (1986) solved the problem numerically by series truncation. The modified problem was defined by allowing the slope of the bubble profile to be discontinuous[5] at the apex. The angle at the apex is denoted by 2γ. The numerical results were found to be qualitatively similar to those of the previous section.

When surface tension is neglected it is found that

$$\gamma = \pi/2 \qquad F < F_c \sim 0.9 \qquad (3.3)$$

$$\gamma = \pi/3 \qquad F = F_c \sim 0.9 \qquad (3.4)$$

$$\gamma = 0 \qquad F > F_c \sim 0.9. \qquad (3.5)$$

For a given value of $\alpha \neq \infty$, there is a discrete set of solutions for which $\gamma = \pi/2$. As $\alpha \to \infty$ this discrete set reduces to a unique solution characterized by $F = F^* \sim 0.51$. Therefore a unique solution is selected by introducing surface tension, and taking the limit as $T \to 0$. The free surface profile corresponding to $F = F^*$ is shown in Figure 5. The broken line corresponds to a two dimensional equivalent of the spherical cap approximation of Davies and Taylor (1950) (see Vanden-Broeck (1986) for details). The agreement between the solid curve and the broken line in Figure 5 is quite satisfactory.

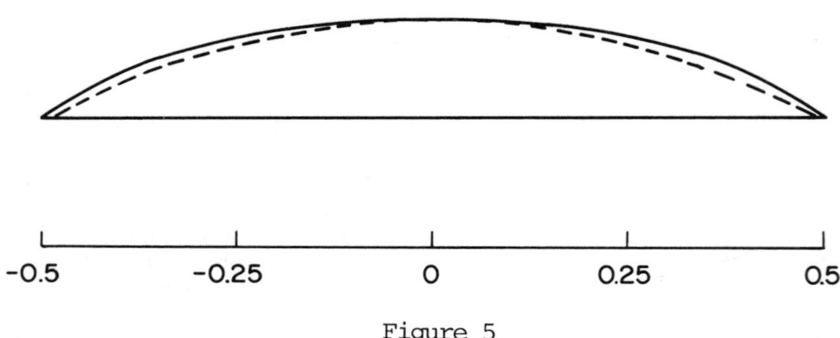

Figure 5

3.2 Joukovskii's Model

We now consider the flow problem of Section 3.1 with a different model for the wake. We assume that the wake is approximated by a region bounded by two vertical walls (see Figure 6). The problem is characterized by the Froude number.

$$F = U/(gL)^{1/2} \qquad (3.6)$$

and the Weber number

$$\alpha = \rho U^2 L/T \ . \qquad (3.7)$$

Here g is the acceleration of gravity, L the distance between the vertical walls, ρ the density of the fluid, and T is the surface tension.

This problem was considered before by Joukovskii (1891) (see also Garabedian (1961) and Gurevich (1966)). By using an inverse method, Joukovskii obtained an exact solution. This solution corresponds to $\alpha = \infty$ and $F = (2\pi)^{-1/2}$.

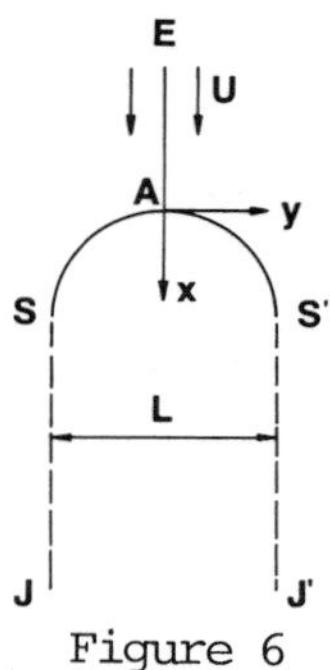

Figure 6

The results presented in Section 3.1 suggest that there are additional solutions of the Joukovskii model for $F \neq (2\pi)^{-1/2}$. Vanden-Broeck (1988) solved the Joukovskii model numerically. As in the previous sections the modified problem was defined by allowing the slope to be discontinuous at the apex of the bubble. The angle at the apex is denoted by 2γ.

For $\alpha = \infty$, Vanden-Broeck (1988) found

$$\gamma = \pi/2 \qquad F < F_c \sim 0.66 \tag{3.8}$$

$$\gamma = \pi/3 \qquad F = F_c \sim 0.66 \tag{3.9}$$

$$\gamma = 0 \qquad F > F_c \sim 0.66\ . \tag{3.10}$$

Relations (3.8)-(3.10) show that all the solutions corresponding to $F<F_c$ are solutions of the original problem. This set of solutions includes Joukovskii's exact solution.

When surface tension is taken into account there is a discrete set of solutions. Each of these solutions corresponds to a different value of F. As the surface tension tends to zero, all these solutions approach a unique solution. Interestingly the numerical results indicate that this limiting[6] solution is Joukovskii's exact solution.

REFERENCES

1. R.Collins, 1965, *J. Fluid Mech.* **22**, 763.
2. B. Couet & G.S.Strumolo, 1987, *J. Fluid Mech* **184**, 1.
3. R.M.Davies & G.I.Taylor, 1950, *Proc. R. Soc. London* **A 200**, 375.
4. R.R. Garabedian, 1957, *Proc. R. Soc. London* **A 241**, 423.
5. R.R. Garabedian, **Modern Mathematics for the Engineer**, (ed. E. F. Beckenbach) (McGraw-Hall, New York, 1961) p.365.
6. M.I.Gurevich, **The Theory of Jets on an Ideal Fluid** (Pergamon, New York, 1966).
7. N.W. Joukovskii, 1891, *J. Russian Physico. Chem. Soc.* 22, 19.
8. J.W. McLean &P.G. Saffman, 1981, *J. Fluid Mech*,102, 455.
9. L. Romero, Ph.D. thesis, California Institute of Technology, 1982.
10. P.G. Saffman, 1986,*J. Fluid Mech* 173, 73.
11. P.G. Saffman & G.I. Taylor, 1958, *Proc. R. Soc. London* **A 245**, 312.
12. S. Tanveer, 1986, *Phys. Fluids* 29, 3537.
13. J.-M. Vanden-Broeck, 1983, *Phys. Fluids* 26, 2033.
14. J.-M. Vanden-Broeck, 1984a, *Phys. Fluids* 27, 1090.
15. J.-M. Vanden-Broeck, 1984b,*Phys. Fluids* 27, 2604.
16. J.-M. Vanden-Broeck, 1986, *Phys. Fluids* 29, 1343.
17. J.-M. Vanden-Broeck, 1986, *Phys. Fluids* 29, 2798.
18. J.-M. Vanden-Broeck, 1988, *Phys. Fluids* 31, 974.

CAPTIONS FOR FIGURES

Figure 1: Sketch of a bubble rising in a tube.

Figure 2: Values of γ versus F for $\alpha=10$. The broken line corresponds to the solution defined by (2.3)-(2.5).

Figure 3: Bubble profile for $F = F^{*} \sim 0.23$ and $T = 0$.

Figure 4: Sketch of the flow past a bubble in an unbounded region. The solid line corresponds to the bubble profile and the broken line to the wake profile.

Figure 5: Bubble profile for $F = F^{*} \sim 0.51$. The broken line corresponds to an arc of circle of radius $4(F^{*})^2$.

Figure 6: Sketch of the flow past a bubble in an unbounded region. The solid line corresponds to the bubble profile and the broken lines to the wake profile.

Third International Colloquium on Drops and Bubbles
Monterey, California

Members*

Andreas Acrivos, Director
Benjamin Levich Institute for Physico-Chemical Hydrodynamics, Steinman Hall 202
City College, City University of New York
New York, NY 10031
(Keynote speaker, Hydrodynamics & Physics of Droplets)

J. Iwan D. Alexander
Ctr. for Microgravity & Materials Research
University of Alabama in Huntsville
R. I. Bldg M-65
Huntsville AL 35899
(Keynote speaker, Microgravity Science & Space Experiments)

Carol T. Alonso
Lawrence Livermore National Laboratory
L-13, A Division
P. O. Box 808
Livermore CA 94550

Robert E. Apfel
Dept. of Engineering & Applied Science
Yale University
New Haven CT 06520
(Bubbles, Shells & Encapsulations session co-chair, author)

Nasser Ashgriz
Dept. of Mechanical & Aerospace Engineering
SUNY Buffalo
Buffalo NY 14260

C. Thomas Avedisian
Sibley School of Mechanical & Aerospace Engineering
Dept. of Mechanical Engineering
Cornell University
Ithaca NY 14850

Xavier J. R. Avula
Dept. of Mechanical & Aerospace Engineering & Engineering Mechanics
University of Missouri, Rolla
Rolla MO 65401
(Bubbles, Shells & Encapsulations session co-chair)

R. Balasubramaniam
NASA Lewis Research Center M/S 500-217
21000 Brookpark Road
Cleveland OH 44135

Osman A. Basaran
Oak Ridge National Labs M/S 4501/224
P. O. Box 2008
Oak Ridge TN 37831

James C. Baygents
Biophysics Research, Space Science Lab
ES76 Bldg. 4481
Marshall Space Flight Center AL 35812

Robert Bayuzick
Ctr. for the Space Processing of Engineering. Mat'ls
Vanderbilt University
Box 1680-B
Nashville TN 37235
(Undercooling & Solidification session chair, author)

Kenneth V. Beard
Dept. of Atmospheric Sciences
University of Illinois
105 South Gregory Avenue
Urbana IL 61801

Thomas P. Bernat
Lawrence Livermore National Lab. L-482
P. O. Box 5508
Livermore CA 94550

Robert A. Brown
Massachusetts Institute of Technology
77 Massachusetts Ave., Room 66-352
Cambridge MA 02139
(Computational Fluid Dynamics session chair, author)

Bradley M. Carpenter
Bionetics Corp./NASA HQ
Microgravity Science & Applications Div.
Code EN Room 233
Washington DC 20546
(Microgravity Science & Space Experiments chair)

Georges L. Chahine
Dynaflow, Inc.
7210 Pindell School Road
Laurel MD 20707

C. K. Chan
TRW, M/S 01-2050
1 Space Park
Redondo Beach, CA 90278

I-Dee Chang
Aeronautics & Astronautics Dept.
Room 365, Durand Building
Stanford University
Stanford CA 94305
(Hydrodynamics & Physics of Droplets session chair)

Catherine Chuang
Dept. of Atmospheric Sciences
University of Illinois
105 South Gregory Avenue
Urbana IL 61801

Sang Chung
Jet Propulsion Laboratory MS 183-401
California Institute of Technology
4800 Oak Grove Drive
Pasadena CA 91109

*address at press or last contact

Third International Colloquium on Drops and Bubbles
Monterey, California

Members*

Robert Cole
Dept. of Chemical Engineering
Clarkson College of Technology
Potsdam, NY 13676

Joseph Crowley
Electrostatic Applications
16525 Jackson Oaks Drive
Morgan Hill CA 95037

Eric W. Curtis
Dept. of Materials Engineering
University of Wisconsin
1509 University Avenue
Madison WI 53706

David S. Dandy
Sandia National Labs, Div. 8363
P. O. Box 969
Livermore CA 94551-0969

Robert H. Davis
Dept. of Chemical Engineering
University of Colorado
Boulder CO 80309-0424

Loren H. Dill
NASA Lewis Research Center MS 500-217
21000 Brookpark Road
Cleveland OH 44135

Fred L. Dryer
Dept. of Mechanical & Aerospace Engineering
Princeton University
Princeton, NJ 08544

Jacques L. Duranceau
Massachusetts Institute of Technology
77 Massachusetts Ave., Room 66-301
Cambridge MA 02139

Hany Elghazaly
Faculty of Engineering
Cairo University
Cairo, EGYPT

Daniel D. Elleman
Jet Propulsion Laboratory MS 183-401
California Institute of Technology
4800 Oak Grove Drive
Pasadena CA 91109
(Steering Committee, Poster Session chair, author)

Scott A. Elrod
Xerox PARC
3333 Coyote Hill Rd.
Palo Alto CA 94304

Michele Emmer
Dipartimenta di Matematica
Universita di Roma "La Sapienza"
Piazzale A. Moro
00185 Roma ITALIA

Patrick V. Farrell
Dept. of Materials Engineering
University of Wisconsin
1509 University Avenue
Madison WI 53706

Robert Finn
Dept. of Mathematics
Stanford University
Stanford CA 94305

William F. Flanagan
Box 17 Station B
Vanderbilt University
Box 1680-B
Nashville TN 37215

Pascale Gillon
Dept. of Chemical Engineering
Stanford University
Stanford CA 94305

Martin E. Glicksman
Materials Engineering Dept.
Rensselaer Polytechnic Inst., School of Engineering
Troy NY 12180-3590

Iskender Gökalp
CNRS--Centre de Recherche sur la Chimie de la Combustion & des Hautes Temperatures
1C Avenue de la Recherche Scientifique
45071 Orléans Cedex 2 FRANCE

Babur Hadimioglu
Xerox PARC
3333 Coyote Hill Rd.
Palo Alto CA 94304

John Hallett
Desert Research Institute
University of Nevada at Reno
Reno NV 89506
(Keynote speaker, Meteorology)

Juan Heinrich
Dept. of Aerospace & Mechanical Engineering
University of Arizona
Tucson, AZ 85721

W. H. Hofmeister
Ctr. for the Space Processing of Engineering. Mat'ls
Vanderbilt University
Box 6309-B
Nashville TN 37235
(Keynote speaker, Undercooling & Solidification)

R. J. Hung
Ctr. for Microgravity & Materials Research
University of Alabama in Huntsville
Huntsville AL 35899

In Sook Kang
Dept. of Chemical Engineering 206-41
California Institute of Technology
Pasadena CA 91125

*address at press or last contact

Third International Colloquium on Drops and Bubbles
❦ *Monterey, California*
Members*

James M. Kendall, Jr.
Jet Propulsion Laboratory MS 183-601
California Institute of Technology
4800 Oak Grove Drive
Pasadena CA 91109

Kyekyoon Kim
Dept. of Electrical & Computer Engineering
University of Illinois
1406 W. Green St. EEB/155
Urbana IL 61801

Dean S. Langley
Dept. of Physics
St. John's University
Collegeville MN 56321

Chung K. Law
Dept. of Mechanical Engineering
Princeton University
Princeton NJ 08544
(Keynote speaker, Combustion)

L. Gary Leal
Dept. of Chemical and Nuclear Engineering
University of California, Santa Barbara
Santa Barbara CA 93106
(Keynote speaker, Computational Fluid Dynamics)

Norman Lebovitz
University of Chicago
Ryerson 359a
5734 So. University Avenue
Chicago IL 60637
(Steering committee, Astrophysics session chair, author)

Chun P. Lee
Ctr. for Microgravity Research & Applications
Vanderbilt University
Box 6079 Station B
Nashville TN 37238

Francis C. Lee
IBM Almaden Research Center
650 Harry Road, K41/803
San Jose, CA 95120-6099

Mark C. Lee
NASA HQ, Code EN
Microgravity Science & Applications Div.
Washington DC 20546
(Steering committee)

Emily W. Leung
Jet Propulsion Laboratory MS 183-401
California Institute of Technology
4800 Oak Grove Drive
Pasadena CA 91109

Ernst G. Lierke
Battelle Institute
Am Romerhof-Institut e. V.
D-6000 Frankfurt am Main 90
Frankfurt, West Germany

Thomas A. Lundgren
Dept. of Aerospace Engineering & Mechanics
University of Minnesota
Minneapolis MN 55455

Philip L. Marston
Dept. of Physics
Washington State University
Pullman WA 99164-2814
(Keynote speaker, Bubbles, Shells & Encapsulations session)

Mario J. Martinez
Dept. of Thermal and Fluid Sciences
Sandia National Laboratory Dept. 1511
P. O. Box 5800
Albuquerque NM 87185

Edward C. Monahan
Marine Sciences Institute, Avery Point
University of Connecticut
Groton CT 06340

Harry T. Ochs
Illinois State Water Survey
2204 Griffith Dr.
Champaign IL 61820

Tadeusz W. Patzek
Shell Development Corporation
P. O. Box 481
Houston TX 77001

Ned J. Penley
c/o Physics Department
Utah State University
Logan UT 84322-4415

John H. Perepezko
Dept. of Materials Engineering
University of Wisconsin
1509 University Avenue
Madison WI 53706
(Keynote speaker, Undercooling & Solidification)

Cal F. Quate
Xerox PARC
3333 Coyote Hill Rd.
Palo Alto CA 94304

Eric G. Rawson, Mgr. AIP Area
Xerox PARC
3333 Coyote Hill Rd.
Palo Alto CA 94304

*address at press or last contact

Third International Colloquium on Drops and Bubbles
Monterey, California
Members*

Carl J. Remenyik
Dept. of Engineering Science & Mechanics
Perkins Hall Room 310
University of Tennessee
Knoxville TN 37996-2030

Ivan Rezanka
Xerox Corp. 0114-44D
Webster NY 14580

Won-Kyu Rhim
Jet Propulsion Laboratory MS 183-401
California Institute of Technology
4800 Oak Grove Drive
Pasadena CA 91109

James A. Rooney
Jet Propulsion Laboratory MS 183-401
California Institute of Technology
4800 Oak Grove Drive
Pasadena CA 91109

Carl E. Rosenkilde
Lawrence Livermore National LaboratoryL-84
P. O. Box 808
Livermore CA 94550

James Ross
Dept. of Mathematics
San Diego State University
San Diego CA 92182

Jack Salzman, Chief,
Microgravity Science Branch MS 500-217
Lewis Research Center
21000 Brookpark Road
Cleveland OH 44135
(Combustion session chair)

Jun'ichi Sato
IHI Research Institute
1-15, 3-chome, Toyosu, Koto-ku
Tokyo 135 JAPAN

Clive Saunders
University of Manchester Inst. of Sci. & Technology
POB 88 (Sackville St.)
Manchester M60 1QD ENGLAND
(Steering committee, Meteorology session chair)

Christian Schön
Dept. of Mathematical Sciences P-236
San Diego State University
San Diego CA 92182

Timothy C. Scott
Oak Ridge National Labs
Box 2008 MS44
Oak Ridge TN 37831-6224

Mark Seaver
Naval Research Lab Code 6540
Washington DC 20375

Nicholas K. Sheridon
Xerox PARC
3333 Coyote Hill Rd.
Palo Alto CA 94304

Jan J. Sojka
Physics Department
Utah State University
Logan UT 84322-4415

Howard A. Stone
c/o Dept. of Applied Math. & Theoretical Physics
Silver Street
University of Cambridge
Cambridge CB3 9EW ENGLAND

Ming-Yang Su
NORDA Code 331
NSTL, MS 39529

Shankar Subramanian
Dept. of Chemical Engineering
Clarkson University
Potsdam, NY 13676

Saleh Tanveer
Mathematics Department
Virginia Polytechnic Institute
Blacksburg VA 24061-4097

James Tegart
Martin Marietta Corporation
381 W. Davies Ave.
Littleton CO 80120

Bob Thompson
NASA Lewis Research Center MS 49-8
21000 Brookpark Road
Cleveland OH 44135

Joel Tohline
Dept. of Physics and Astronomy
Louisiana State University
Baton Rouge LA 70803-4001
(Keynote speaker, Astrophysics)

Eugene H. Trinh
Jet Propulsion Laboratory MS 183-401
California Institute of Technology
4800 Oak Grove Drive
Pasadena CA 91109

John A. Tsamopoulos
Dept. of Chemical Engineering
SUNY Buffalo
Buffalo NY 14260
(Keynote speaker, Computational Fluid Dynamics)

Jean-Marc Vanden-Broeck
Center for the Mathematical Sciences
University of Wisconsin
610 Walnut St.
Madison WI 53705

*address at press or last contact

Third International Colloquium on Drops and Bubbles
Monterey, California
Members*

Peter W. Voorhees
Dept. of Material Science
Northwestern University
Evanston IL 60203
(Keynote speaker, Microgravity Science & Space Experiments)

Taylor G. Wang, Director
Ctr. for Microgravity Research & Applications
Vanderbilt University
Box 6079 Station B
Nashville TN 37238
(Colloquium Chair, author)

G. Wozniak
Un.-GH-ESSEN
FB 12 Mechan D-4300/Essen 1
Schutzenbahm 70-P 1037-64
West Germany

K. Wozniak
Un.-GH-ESSEN
FB 12 Mechan D-4300/Essen 1
Schutzenbahm 70-P 1037-64
West Germany

*address at press or last contact